Programming the
PIC Microcontroller with MBasic

Please note that the previous printing included a CD-ROM;
for all references to this in the text please read 'companion website.'

The CD-ROM material is now available on the companion website:
http://www.elsevierdirect.com/companion.jsp?ISBN=9780750679466

Programming the
PIC Microcontroller with MBasic

by Jack R. Smith

AMSTERDAM • BOSTON • HEIDELBERG • LONDON
NEW YORK • OXFORD • PARIS • SAN DIEGO
SAN FRANCISCO • SINGAPORE • SYDNEY • TOKYO

Newnes is an imprint of Elsevier

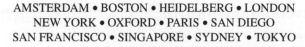

Newnes is an imprint of Elsevier
30 Corporate Drive, Suite 400, Burlington, MA 01803, USA
Linacre House, Jordan Hill, Oxford OX2 8DP, UK

 Recognizing the importance of preserving what has been written,
Elsevier prints its books on acid-free paper whenever possible.

Library of Congress Cataloging-in-Publication Data

(Application submitted.)

British Library Cataloguing-in-Publication Data
A catalogue record for this book is available from the British Library.

ISBN: 978-0-7506-7946-6

For information on all Newnes publications
visit our website at www.books.elsevier.com

Transferred to Digital Printing in 2012

Contents

Preface x

Acknowledgments xii

What's on the companion website? xiii

CHAPTER 1: What is a PIC®? 1

PICs "101" 1

How Do I Tell Them Apart? 2

Which One Should I Use? 4

How Do I Pick One? 7

So, Which One Do I Really Want to Use? 8

Basic Micro's MBasic876 Compiler 8

References 9

CHAPTER 2: MBasic Compiler and Development Boards 10

The Compiler Package 10

BASIC and Its Essentials 11

Development Boards 13

Programming Style 15

Building the Circuits and Standard Assumptions 16

Pins, Ports and Input/Output 17

Pseudo-Code and Planning the Program 23

Inside the Compiler 25

References 27

CHAPTER 3: The Basics – Output 28

Pin Architectures 28

LED Indicators 31

Switching Inductive Loads 34

Low Side Switching 36

Isolated Switching 45

Special Purpose Switching 50

Fast Switching—Sound from a PIC 50

References 51

CHAPTER 4: The Basics – Digital Input 53

Introduction 53

Switch Bounce and Sealing Current 58

Isolated Switching 62

Contents

Reading a Keypad 63
References.. 66

CHAPTER 5: LCD Modules.. **67**
Selecting a Display... 67
VFD Displays.. 69
Connection to PIC 69
Hello World 72
LCD Module Memory, Shifts and Lines... 74
Font Selection 79
Custom Characters... 80
References.. 85

CHAPTER 6: Reading Complex Input Switches **86**
Pin Saving Techniques.. 86
Rotary Encoders 91
Reading a Relative Encoder... 95
Dual Encoders and LCD 100
References.. 106

CHAPTER 7: Seven-Segment LED Displays...................................... **107**
LED Display Selection.. 107
Circuit Design 108
References.. 119

CHAPTER 8: Introductory Stepper Motors **120**
Stepper Motor Basics.. 120
Programs 133
References.. 150

CHAPTER 9: RS-232 Serial Interface **151**
How to Connect to Your PC 151
Voltage Levels in RS-232 and Level Conversion 152
Standard Pin Connections... 154
Asynchronous Transmission, Start Bits, Stop Bits and Bit Order........ 154
MBasic's Procedures for Serial Communications 156
Programs 159
References.. 186

CHAPTER 10: Interrupts and Timers in MBasic................................ **187**
Interrupts and Timers—Overview.. 187
Interrupts 188
Timers 194
Capture and Compare 203
References.. 210

CHAPTER 11: Analog-to-Digital Conversion **211**
Introduction to Analog-to-Digital Conversion.................................. 211
Resolution and Accuracy 212
Self-Contained DVM 218
References.. 230

CHAPTER 12: Digital Temperature Sensors and Real-Time Clocks............................**231**
DS18B20 Temperature Sensor... 231
DS1302 Real-Time Clock 243
Combination Date, Time and Temperature 252
References.. 259

CHAPTER 13: Assembler 101 ....**260**
The Basics 260
OpCodes.. 267
References.. 280

CHAPTER 14: In-Line Assembler..**281**
Adding Assembler to MBasic Programs.. 281
Bolt-In Assembler Functions... 295
References.. 316

CHAPTER 15: Interrupt Handlers and Timers in Assembler **317**
ISRASM – MBasic's Gateway to Assembler Interrupt Service Routines............... 317
Program Examples 323
References.. 334

CHAPTER 16: Digital-to-Analog Conversion **335**
Introduction to Digital-to-Analog Conversion... 335
Resolution – Accuracy and Signal-to-Noise Ratio... 336
Henry Nyquist and his Sampling Theorem 337
DAC Circuit Design.. 339
Alternative Analog Output Solutions 352
References.. 358

CHAPTER 17: DTMF Tone Decoding and Telephone Interface **360**
What is Touch-Tone Signaling? 360
Generating Touch-Tone Signals.. 361
Decoding a Touch-Tone Signal 361
References.. 388

CHAPTER 18: External Memory...**389**
I²C-Bus Devices.. 389
Practical Use of External EEPROM 403
Parallel Access Memory 408
References.. 416

CHAPTER 19: Advanced Stepper Motors..**418**
Microstepping 418
Programs.. 420
References.. 452

CHAPTER 20: X-10 Home Automation..**453**
How X-10 Works.. 453
Programs.. 459
References.. 486

Contents

CHAPTER 21: Digital Potentiometers and Controllable Filter **487**

Getting Started with an MCP41010 489

RS-232 Control of an MCP41010 493

Daisy Chaining Multiple MCP42010 Devices 498

RS-232 Command of Multiple Daisy Chained MCP42010 Devices 501

Logarithmic Response for Audio Volume Control 506

Electronically Tunable Low-Pass Filter Using MCP42010 511

References 515

CHAPTER 22: Infrared Remote Controls **517**

Common Encoding Standards 518

IR Receiver 520

Characterizing Wide/Narrow Pulse Intervals 522

Decoding a REC-80 Controller 532

References 541

CHAPTER 23: AC Power Control **542**

Introduction to Triacs 543

Snubberless versus Standard; dV/dt and dl/dt Issues 545

Triggering a Triac 548

Phase and Cycle Control 549

Power Control Board 551

Programs 555

References 565

CHAPTER 24: DC Motor Control **567**

Introduction to Control Theory 567

Measure Motor Speed (Tachometer Output Pulse Width) 568

Error = Target Width – Measured Width 572

The Control Algorithm 572

Motor Control Programs 573

References 594

CHAPTER 25: Bar Code Reader **595**

Bar Codes "101" 595

Bar Code Wand 599

Programs 602

References 631

CHAPTER 26: Sending Morse Code **633**

Morse Code 101 633

Programs 635

References 660

CHAPTER 27: Morse Code Reader **661**

Sending and Receiving Morse 661

Tone Detector Circuit 663

Programs 668

References 689

CHAPTER 28: Weather Station and Data Logger *691*
Sensor Selection 691
Connecting the Sensors and Memory 698
Initial Tests 700
References 728

CHAPTER 29: Migrating from v5.2.1.x to 5.3.0.0 and the Undocumented MBasic *729*
Migrating from v5.2.1.x to 5.3.0.0 729
Undocumented MBasic 733

APPENDIX A: Parts List and Suppliers *745*
Suppliers 745
Generic Components Required 746
Specific Components 748

APPENDIX B: Function Index *755*

About the Author *760*
Index *761*

Preface

My introduction to computers was in the days of IBM's Model 29 card punch. You first carefully printed your FORTRAN code on a coding sheet, then punched a card deck and finally walked your cards over to the campus computer center. There, one of the high acolytes of the IBM 360—in reality a grad student—accepted the deck with a faint look of disdain. You might even catch a glimpse of the computer itself through the glass wall of the computer center. The following day, if you were fortunate, your card deck was ready for pick-up, wrapped in the green bar paper output your job elicited. If you were really lucky, the output made sense and you could go on to your next task. If you were less fortunate, the printout identified your errors. And, if you were really having a bad day, your card deck was hidden inside an inch-thick core dump printout, dense with hexadecimal register and memory values.

Today, we have as much computing power on our desktops as was behind the glass wall when I was punching card decks. Computers are now embedded in almost every imaginable electronic device. One of the pioneers in embedded computers was General Instruments, which in 1976 released the 1650 "programmable intelligent computer," the grandfather of today's PICs. (There is a raging debate among the cognoscenti over the "true" name behind the PIC acronym, with "peripheral interface controller" often being cited. GI's 1977 data sheet for the PIC1650, though, confirms the term "programmable intelligent computer." Microchip Technology Incorporated, who acquired GI's PIC business in the mid 1980s, wisely stays out of the debate and just calls its products "PICmicro® microcontrollers.")

This book focuses on programming Microchip's mid-range PIC line with MBasic, a powerful, but easy to learn programming language, developed by Basic Micro of Murrieta, California. Since a PIC by itself is not all that useful, I will illustrate MBasic's abilities through a series of construction projects, some simple and some more advanced. I will also dip into assembler language, as there are some applications that require us to become more intimate with the PIC's internals than possible in MBasic.

The projects assume the user has MBasic Professional version 5.3.0.0 compiler, the associated ISP-PRO programmer and a 2840 development board, all available from Basic Micro. Almost all examples use a 16F877A PIC and a 20 MHz resonator. However, both the code and supporting circuitry are easily portable to many other PICs supported by MBasic. More importantly, almost every project in this book can be built with the free MBasic876 compiler included in the companion website. In a few cases, the 16F876 doesn't have enough I/O pins to support the project.

Reading Basic Micro's message board, and questions from beginners posted to the PIC microcontroller discussion list, reveals a need for information showing how the smorgasbord of functions, procedures and code snippets found in the MBasic User's Guide might be put together to actually do something useful. And, since doing "something useful" with a PIC inevitably requires some associated circuitry, electronics questions are sprinkled liberally throughout these fora as well.

I've tried to address both the software and hardware aspects of working with PICs, with my imagined reader having an interest in both programming and electronics, but without specialized training. Although I've tried to err on the side of inclusion over brevity, this book can't replace a basic understanding of electronics, nor an elementary grasp of how one goes about writing BASIC programs. I trust that readers experienced in electronics will forgive the simplifications necessitated in this endeavor, and that experienced programmers understand that they will not necessarily find elegant algorithms or code in every case. But, this work is not intended to replicate Knuth's *The Art of Computer Programming*, nor Horowitz and Hill's *The Art of Electronics*. And, I couldn't duplicate either if I tried my best for the next decade.

Finally, I've never found the impersonal passive technical writing style conducive to learning a new subject. After all, "the code was transferred to the PIC" isn't what actually happened was it? Someone—probably you, but certainly not some disembodied entity—programmed the PIC using the MBasic software. Why not say so? Likewise, although we know that a PIC's output pin doesn't "see" a load resistance through physical eyes, these anthropomorphic analogies are easier to understand than reading "the equivalent resistance that would be measured by an appropriate impedance measuring instrument connected in place of the pin and applying a +5 volt dc stimulus signal to the load resistance." Hence, I make no apologies for the chatty style.

This book is not an "official" publication of Basic Micro, and its contents reflect my views, not those of Basic Micro, or its employees or owners.

In a book of this length and detail, there will inevitably be errors and omissions, despite the best efforts of the author and editors. Some are the unavoidable byproduct of simplifying complex subjects for an introductory level presentation and others are just plain dumb mistakes. Regardless of the category, I accept full responsibility. I may be contacted by e-mail at Jack.Smith@cox.net to report errors or omissions.

Jack Smith
June 2005, Clifton, VA

Acknowledgments

Acknowledgments are due to my wife Janet, who has tolerated with good grace the innumerable hours I've spent in the basement workshop or in front of the computer screen. I also wish to thank the people at Basic Micro, including Nathan Scherdin, and Dale Kubin, for their assistance in writing this book. I also wish to thank Larry and Janet Phipps for the hospitality shown to me.

What's on the companion website?

The content of the companion website was developed with Windows® XP, and has been verified as readable with Windows® 2000. It has not been tested with other operating systems.

Its contents are organized in a series of directories:

MBasic876

Basic Micro Inc. has provided a free MBasic compiler, MBasic876, with all features of MBasic Professional, but restricted to program only 16F876 and 16F876A devices. An installation program, MBasic876_Setup, to add MBasic876 to your computer is contained in this directory. Launching MBasic876_Setup will start the installation process.

The directory *MBasic\Documents* contains the MBasic User's Guide and data sheets on Basic Micro's programming board and development and prototype boards.

Basic Micro's website, *http://www.basicmicro.com/*, has an active MBasic user's forum that I highly recommend.

Linear Technology Circuit Simulation Software

Linear Technology Corporation has provided two programs for circuit simulation. Both programs were used in developing and illustrating the circuits in this book.

The directory *Linear Technology Circuit Simulation Software\FilterCAD* contains the installation program, FilterCADv300.exe, which installs FilterCAD version 3.00 on your computer. FilterCAD is a powerful tool for designing and simulation active filters.

The directory *Linear Technology Circuit Simulation Software\SWCADIII* contains the installation program swcadiii.exe, which installs a program called *LTspice/SwitcherCAD III* on your computer. This is a full-featured general-purpose electronic circuit simulation program. After installing LTspice, I recommend you use the automatic update feature to download the most recent version.

You may also wish to join the LTspice user's group, via the home page at http://groups.yahoo.com/group/LTspice/.

MBasic Programs

All of the programs in this book are contained in this directory. The programs are organized with a separate directory for each chapter. Within each chapter directory are separate directories with program versions compatible with MBasic (and MBasic876) version 5.3.0.0 and with earlier versions (5.2.1.1.) I developed the programs with version 5.2.1.1 originally, but have revised and tested each for compatibility with version 5.3.0.0.

Data Sheets and Application Notes

Data sheets for many of the transistors, diodes and integrated circuits used in this book's circuits are provided in the directory *data sheets* and *application notes*. In addition, I have included a selection of relevant application notes and other material from key semiconductor manufacturers.

Each manufacturer's data sheets are contained in a directory named after the manufacturer.

PICs "101"

What is a PIC®?

PICs are inexpensive one-chip computers designed and manufactured by Microchip Technology, Inc. The acronym originally stood for Programmable Intelligent Computer, but Microchip's official name for these devices is now PICmicro® microcontrollers. We will call them PICs. In 1977, General Instruments, Microchip's predecessor, developed the original PIC, the PIC1650. The PIC1650 can be though of as the grandfather of today's PICs, and its architecture, programming approach and other features directly correspond to those found in modern PICs. Its instruction set and register arrangement mirror current PICs with only minor differences.

General Instruments sold its microcontroller business in the mid-1980s to the entity that later became Microchip. Microchip's current product line includes nearly 200 PIC models with MBasic supporting more than half. Microchip has sold more than 2 billion PICs since the mid-1980s, and in 2002 was number one worldwide in 8-bit microcontroller sales, based on number of units shipped.

PICs are microprocessors, akin to the ones inside personal computers, but significantly simpler, smaller and cheaper, optimized to deal with the real world—operating relays, turning lamps off and on, measuring sensors and responding to changed readings with specific actions—instead of running word processing or spread sheet programs. To emphasize the outside world connection, the term "microcontroller" was coined to distinguish it from a "microprocessors." GI envisioned its PIC1650 as a means to replace dozens of discrete logic chips in computers using its CP1600 microprocessor, but immediately recognized the power of its flexible, programmable design serving as a stand-alone microcontroller. Figure 1-1 illustrates the main elements inside a PIC:

- *A processing engine*: The central processing unit, or CPU, is the microcontroller's intelligence. It performs the logical and arithmetic functions of the PIC following instructions it reads from the program memory. It reads from and writes to data memory and the input/output module.
- *Program memory*: Holds instructions for the CPU. The CPU reads program memory but is physically prevented in most model PICs from writing to program memory.
- *Data memory*: Holds memory that the programmer may use for variables. The CPU reads from and writes to data memory.
- *Input/output*: How the PIC communicates with the world outside the chip; for example, pins that go between logical 0 and logical 1.

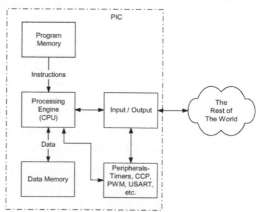

Figure 1-1: Main elements of a PIC.

- *Peripherals*: Special purpose functions built into the PIC, such as timers, analog-to-digital converters and pulse width modulators.

If you are familiar with the Intel microprocessors used in IBM-compatible personal computers, you may notice one striking difference in Figure 1-1; the program memory and data memory are separate. In computer techno-speak, PICs follow the Harvard architecture model, while Intel's microprocessors (and those of most other manufacturers as well) implement von Neumann's architecture, sharing common memory between program and data as necessary. Fortunately, MBasic hides the details of this difference from us and we seldom need to delve into it. One place this difference is critical, though, is since data memory and program memory capacities are separately specified in PICs, both must be sized to accommodate the job at hand.

How Do I Tell Them Apart?

Microchip identifies PICs with a multipart identifier such as a 16F877A-E/P:

Microchip groups its PIC line in three performance and three memory type categories:

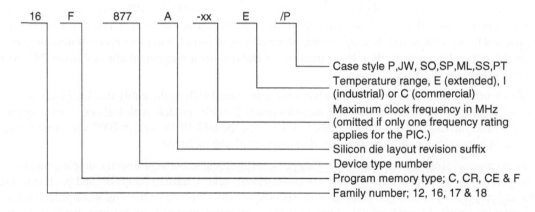

Microchip's General Purpose PIC Line Name (Instruction Word Length) and MBasic Support			
Program Memory Type	*Base-Line (12-bit)*	*Mid-Range (14-bit)*	*High-End (16-bit)*
EPROM/EEPROM		*12C, 12CE, 16C and 16CE-series EPROM and EEPROM*	*17C-series and 18C-series. Not supported by MBasic*
Read-only Memory (ROM)	*None supported by MBasic*	*16CR-series. Not supported by MBasic*	*Not produced by Microchip*
Flash (electronically erasable)		*Some 12F-series and 16F-series*	*18F-series. Not presently supported by MBasic (See Note 1)*
1. 18F-series support is under development by Basic Micro and will be included in a future MBasic release.			

It would have been logical for Microchip to use the series identifier to point to the instruction word length, but it missed that opportunity. Thus, we have the 12C508A, a 12-bit device and the 12F629, a 14-bit device. And, we have the 16C54C, a 12-bit device and the 16C554, a 14-bit device. In almost all—but unfortunately not entirely all—instances a 16-series device is a mid-range PIC with a 14-bit instruction word, but to be sure we must consult Microchip's reference documents.

The *instruction word length* is not related to the program memory size, but rather defines how many unique machine code instructions may be implemented. It isn't necessary to go into details as MBasic takes care of this for us, but many machine level operands include an 8-bit literal value, such as moving a defined byte value (the "literal") into the CPU. Since the 8-bit literal is part of the instruction, a 12-bit instruction word leaves only four bits for instructions containing a literal, resulting in only 16 possible unique instructions. Moving to a 14-bit word increases the potential instruction set with a literal to 64. (The problem of limited program instruction width also shows up in assembler jump or "go to" instructions.) In any event, since MBasic currently supports only midlength (14-bit) PICs, we can file this information in the "interesting but not immediately useful" category in the back of our minds, at least until we start to mix assembler language routines with MBasic.

Program memory in a PIC may consist of three types:

- **Read-only**: Read-only memory means exactly that; the memory is configured at the time of manufacturing to contain the program code and may not be subsequently altered, something economically feasible only in high volume products. MBasic does not support PICs with read-only memory.
- **EPROM and EEPROM**: EPROM (erasable programmable read-only memory) and EEPROM (electrically erasable programmable read-only memory) memory may be written to electronically through the application of a programming voltage to the PIC. Once written, EPROM memory may not be re-written, and is thus becomes read-only afterwards. Microchip refers to these devices as "one time programmable" or OTP products. EEPROM devices, however, may be erased through several minutes' exposure to ultraviolet light. Electrically, Microchip's EPROM and EEPROM chips use the same technology, with EPROM chips being encapsulated in opaque epoxy. EEPROM chips have a quartz window through which UV light may reach the chip surface. (After programming, you cover the window with an opaque label to prevent erasure through ambient sunlight or fluorescent light exposure.) EPROM PICs may be useful in small to medium volume production, but both EPROM and EEPROM devices are rapidly being supplanted by flash memory PICs.
- **Flash**: Flash memory may be written to and erased electronically through the application of a programming voltage to the PIC. Flash memory may be written to hundreds of thousands of times without error and, at room temperature, based on extrapolated life testing, will retain data for 100 years. Flash is ideal for developing programs and learning MBasic, as revising code and writing the revised program to flash requires well under a minute for all but the longest programs.

Looking at the price of chips of similar performance and capacity with EPROM, EEPROM and flash memory types, it's easy to see why flash devices are taking an increasing share of the market.

Memory Type	Part Number	Packaging	Unit Cost
EEPROM (UV erasable)	16CE625/JW	Ceramic windowed 18-pin DIP (CDIP) Type "JW"	$10.64
EPROM (one-time programming)	16CE625/P	Plastic 18-pin DIP (PDIP) Type "P"	$4.38
Flash	16F628A	Plastic 18-pin DIP (PDIP) Type "P"	$3.05

Finally, within each category, Microchip offers standard voltage (5 volt nominal) and extended voltage (minimum voltage dependent upon memory type; compatible with 5 volt supply; some with built-in regulator for operation from higher voltages.) PIC's also have a wide variety of memory size, internal peripheral options, temperature ranges, maximum operating frequency and packaging. These variants are identified through associated alphanumeric designators.

Memory and Voltage Designators	
Memory/Voltage Letter	Memory/Voltage Type
C	EPROM
CR	ROM
CE	One-time programmable (EPROM) and EEPROM (erasable)
F	Flash
HV	High Voltage (15V)
LF	Low Voltage Flash
LC	Low Voltage One-time programmable
LCR	Low Voltage ROM

Temperature Range Designators	
Temperature Letter	Temperature Range
C	Commercial 0°C to +85°C
I	Industrial -40°C to +85°C
E	Extended -40°C to +125°C

Partial List of Package Designators	
Package Option Letter	Package
JW	Ceramic window (EEPROM only)
P	Plastic DIP SP/PJ for 28 pin x 0.3 ("skinny-dip")
SN, OA, SM, SL, OD, SO, SI	SOIC-plastic small outline; surface mount
PQ	QFP-Plastic quad flatpack surface mount
SS	SSOP-plastic shrink small outline surface mount
ML	Chip scale package
ST	TSSOP-Plastic thin shrink small outline surface mount
PT	TQFP-plastic thin quad flatpack

Which One Should I Use?

Let's look at the PICs supported by MBasic.

PICS Supported by MBasic							
Device	Data RAM	ADC	Program Memory	Serial I/O	Speed	Timers	Low Voltage Device
PIC12CE673	128	4	1024	–	10	1+WDT	PIC12LCE673
PIC12CE674	128	4	2048	–	10	1+WDT	PIC12LCE674
PIC12F629	64	-	1024	–	20	2+WDT	PIC12F629
PIC12F675	64	4	1024	–	20	2+WDT	PIC12F675
PIC16C554	80	-	512	–	20	1+WDT	PIC16LC554

(continued)

PICS Supported by MBasic							
Device	*Data RAM*	*ADC*	*Program Memory*	*Serial I/O*	*Speed*	*Timers*	*Low Voltage Device*
PIC16C558	*128*	*–*	*2048*	*–*	*20*	*1+WDT*	*PIC16LC558*
PIC16C620	*80*	*–*	*512*	*–*	*20*	*1+WDT*	*PIC16LC620*
PIC16C620A	*96*	*–*	*512*	*–*	*40*	*1+WDT*	*PIC16LC620A*
PIC16C621	*80*	*–*	*1024*	*–*	*20*	*1+WDT*	*PIC16LC621*
PIC16C621A	*96*	*–*	*1024*	*–*	*40*	*1+WDT*	*PIC16LC621A*
PIC16C622	*128*	*–*	*2048*	*–*	*20*	*1+WDT*	*PIC16C622*
PIC16C622A	*128*	*–*	*2048*	*–*	*40*	*1+WDT*	*PIC16LC622A*
PIC16C62A	*128*	*–*	*2048*	*I²C, SPI*	*20*	*3+WDT*	*PIC16LC62A*
PIC16C62B	*128*	*–*	*2048*	*I²C, SPI*	*20*	*3+WDT*	*PIC16LC62B*
PIC16C63	*192*	*–*	*4096*	*USART, I²C, SPI*	*20*	*3+WDT*	*PIC16LC63*
PIC16C63A	*192*	*–*	*4096*	*USART, I²C, SPI*	*20*	*3+WDT*	*PIC16LC63A*
PIC16C642	*176*	*–*	*4096*	*–*	*20*	*1+WDT*	*PIC16LC642*
PIC16C64A	*128*	*–*	*2048*	*I²C, SPI*	*20*	*3+WDT*	*PIC16LC64A*
PIC16C65A	*192*	*–*	*4096*	*USART, I²C, SPI*	*20*	*3+WDT*	*PIC16LC65A*
PIC16C65B	*192*	*–*	*4096*	*USART, I²C, SPI*	*20*	*3+WDT*	*PIC16LC65B*
PIC16C66	*368*	*–*	*8192*	*USART, I²C, SPI*	*20*	*3+WDT*	*PIC16LC66*
PIC16C662	*176*	*–*	*4096*	*–*	*20*	*1+WDT*	*PIC16LC662*
PIC16C67	*368*	*–*	*8192*	*USART, I²C, SPI*	*20*	*3+WDT*	*PIC16LC67*
PIC16C71	*36*	*4*	*1024*	*–*	*20*	*1+WDT*	*PIC16LC71*
PIC16C710	*36*	*4*	*512*	*–*	*20*	*1+WDT*	*PIC16LC710*
PIC16C711	*68*	*4*	*1024*	*–*	*20*	*1+WDT*	*PIC16LC711*
PIC16C712	*128*	*4*	*1024*	*–*	*20*	*3+WDT*	*PIC16LC712*
PIC16C715	*128*	*4*	*2048*	*–*	*20*	*1+WDT*	*PIC16LC715*
PIC16C716	*128*	*4*	*2048*	*–*	*20*	*3+WDT*	*PIC16LC716*
PIC16C717	*256*	*6*	*2048*	*I²C, SPI*	*20*	*3+WDT*	*PIC16LC717*
PIC16C72	*128*	*5*	*2048*	*I²C™, SPI™*	*20*	*3+WDT*	*PIC16LC72*
PIC16C72A	*128*	*5*	*2048*	*I²C, SPI*	*20*	*3+WDT*	*PIC16LC72A*
PIC16C73A	*192*	*5*	*4096*	*USART, I²C, SPI*	*20*	*3+WDT*	*PIC16LC73A*
PIC16C73B	*192*	*5*	*4096*	*USART, I²C, SPI*	*20*	*3+WDT*	*PIC16LC73B*
PIC16C745	*256*	*5*	*8192*	*USB, USART*	*24*	*3+WDT*	*–*
PIC16C74A	*192*	*8*	*4096*	*USART, I²C, SPI*	*20*	*3+WDT*	*PIC16LC74A*
PIC16C74B	*192*	*8*	*4096*	*USART, I²C, SPI*	*20*	*3+WDT*	*PIC16LC74B*
PIC16C76	*368*	*5*	*8192*	*USART, I²C, SPI*	*20*	*3+WDT*	*PIC16LC76*
PIC16C765	*256*	*8*	*8192*	*USB, USART*	*24*	*3+WDT*	*–*
PIC16C77	*368*	*8*	*8192*	*USART, I²C, SPI*	*20*	*3+WDT*	*PIC16LC77*
PIC16C770	*256*	*6*	*2048*	*I²C, SPI*	*20*	*3+WDT*	*PIC16LC770*
PIC16C771	*256*	*6*	*4096*	*I²C, SPI*	*20*	*3+WDT*	*PIC16LC771*
PIC16C773	*256*	*6*	*4096*	*USART, I²C, SPI*	*20*	*3+WDT*	*PIC16LC773*
PIC16C774	*256*	*10*	*4096*	*USART, I²C, SPI*	*20*	*3+WDT*	*PIC16LC774*
PIC16C923	*176*	*–*	*4096*	*I²C, SPI*	*8*	*3+WDT*	*PIC16LC923*
PIC16C924	*176*	*5*	*4096*	*I²C, SPI*	*8*	*3+WDT*	*PIC16LC924*
PIC16CE623	*96*	*–*	*512*	*–*	*30*	*1+WDT*	*PIC16LCE623*
PIC16CE624	*96*	*–*	*1024*	*–*	*30*	*1+WDT*	*PIC16LCE624*
PIC16CE625	*128*	*–*	*2048*	*–*	*30*	*1+WDT*	*PIC16LCE625*

(continued)

| PICS Supported by MBasic | | | | | | | |
Device	Data RAM	ADC	Program Memory	Serial I/O	Speed	Timers	Low Voltage Device
PIC16F627	224	–	1024	USART	20	3 + WDT	PIC16LF627
PIC16F628	224	–	2048	USART	20	3 + WDT	PIC16LF628
PIC16F73	192	5	4096	I²C, SPI, USART	20	3+WDT	PIC16LF73
PIC16F74	192	8	4096	I²C, SPI, USART	20	3+WDT	PIC16LF74
PIC16F76	368	5	8192	I²C, SPI, USART	20	3+WDT	PIC16LF76
PIC16F83	36	–	512	–	10	1+WDT	PIC16LF83
PIC16F84	68	–	1024	–	10	1+WDT	PIC16LF84
PIC16F84A	68	–	1024	–	20	1+WDT	PIC16LF84A
PIC16F870	128	5	2048	USART	20	3+WDT	PIC16LF870
PIC16F871	128	8	2048	USART	20	3+WDT	PIC16LF871
PIC16F872	128	5	2048	I²C, SPI	20	3+WDT	PIC16LF872
PIC16F873	192	5	4096	USART, I²C, SPI	20	3+WDT	PIC16LF873
PIC16F873A	192	5	4096	USART, I²C, SPI	20	3+WDT	PIC16LF873A
PIC16F874	192	8	4096	USART, I²C, SPI	20	3+WDT	PIC16LF874
PIC16F874A	192	8	4096	USART, I²C, SPI	20	3+WDT	PIC16LF874A
PIC16F876	368	8	8192	USART, I²C, SPI	20	3+WDT	PIC16LF876
PIC16F876A	368	5	8192	USART, I²C, SPI	20	3+WDT	PIC16LF876A
PIC16F877	368	8	8192	USART, I²C, SPI	20	3+WDT	PIC16LF877
PIC16F877A	368	8	8192	USART, I²C, SPI	20	3+WDT	PIC16LF877A

This list may seem bewildering at first, so let's go through the table's parameters:

Device—This is simply a short form of the device part number.

Data RAM—Data RAM specifies the amount (in bytes) of random access memory available to hold variables in your MBasic program. Since MBasic requires some RAM for its internal use not all the Data RAM will be available for your programs. RAM contents are lost whenever the power is removed from the PIC. (Many devices include nonvolatile EEPROM memory as well. We'll use EEPROM memory in several sample programs in later chapters.)

Program Memory—Since PICs are Harvard architecture devices, the program and data memory are separate. The Program Memory column, following Microchip's documentation, identifies the program memory size in *program words*. In the case of the PICs supported by MBasic, the word length is 14 bits. The MBasic compiler, however, reports program memory use in 8-bit *bytes*, as shown in Figure 1-2. Should you wish to convert between the two, the compiler reports one 14-bit word as 1.75 bytes, and conversely, 1 byte represents 0.57143 14-bit words.

ADC—An analog-to-digital converter (ADC) allows the PIC to read the value of an analog voltage and convert it to a numerical value. Depending upon the model, the ADC may have 8-bit, 10-bit or 12-bit resolution. Chapter 11 shows how to use the ADC to build a digital voltmeter.

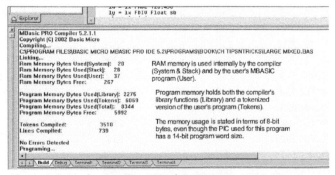

Figure 1-2: Understanding MBasic's memory usage report.

Serial I/O—Certain PICs have specialized hardware support of up to three serial standard protocols: The *USART* (universal synchronous asynchronous receiver transmitter) supports the common RS-232-type asynchronous protocol, as well as others; I^2C (inter-integrated circuit) and *SPI* (serial peripheral interface) are both primarily used to communicate between the PIC and other integrated circuits, such as add-on memory, temperature sensors, serial number generators and the like. MBasic however, implements RS-232-type serial communications (the **SerIn** and **SerOut** procedures), as well as I^2C (the **I2Cin** and **I2Cout** procedures) and SPI (via the **ShiftIn** and **ShiftOut** procedures) in software, so all three protocols are available whether or not the PIC has the associated specialized hardware. Indeed, in some respects MBasic's software solution is superior, as it permits user-defined pin assignments, while Microchip's hardware implementation is tied to specific pins. However, MBasic supports, through the **HserOut** and **HSerIn** procedures, certain aspects of the USART hardware for those PICs so equipped. We'll see how this works in later chapters.

Speed—The maximum clock speed in MHz that the PIC device type supports. Microchip produces lower than maximum speed versions of some devices, however, so when purchasing a PIC check the speed suffix. Don't buy a PIC16F876-04/SP (4 MHz maximum) if you want a PIC16F876-10/SP (10 MHz) or a PIC16F876-20/SP (20 MHz) product! The price difference between the slower speed version of a device type and the maximum speed version is usually modest.

Timers—Timers are programmable internal counters. Among their many uses is to set up a periodic interrupt signal that causes the PIC to perform the code at the interrupt handler. The watch dog timer (WDT) is a specialized timer that may be used to detect and take action upon the main software freezing. Timers and interrupts are the subject of Chapter 10.

Low Voltage Device—Historically, PICs have required a 5-volt power supply, or V_{DD} voltage. With the trend towards lower voltage logic, Microchip has responded with low voltage alternatives of its standard PIC lineup. The low voltage chips are identified with an "L" in the suffix, and operate with as little as 2.0 V V_{DD}, although slower speed may be necessary at the lower end of the operating voltage range. Fortunately, Microchip's low voltage PICs also function with the traditional 5 V supply so they may be used with Basic Micro's ICP and development boards.

How Do I Pick One?

The first step is to identify your requirements and then find the matching devices.

- How many I/O pins do you need?
- How much RAM is required? Each byte variable declared in MBasic consumes one byte of RAM, each word variable two bytes and each long, four bytes.
- How much program memory is required? As a crude estimate of the MBasic program size that fits into a particular program memory capacity, you may assume 400–1300 words for library functions and 8–20 words per line of executable code, depending upon the compiler optimization choice (minimum size or maximum speed), the mix of instructions used and the length of the program. The more different procedures and functions used, the larger the library requirement.
- How fast a device meets your speed requirements?
- Do you need special purpose functions, such as an A/D converter, a USART, specific timers or an internal clock oscillator?
- How does the cost fit into the project budget?
- Do you want a one-time-programmable or a flash memory device?
- Are there physical package preferences?
- Is low-voltage operation necessary?

If your requirements are uncertain, start with the largest, most fully equipped PIC available, and refine your device selection later as you are further along the process. MBasic makes this process especially easy, as the same MBasic code runs on any supported PIC, except, of course, for a handful of instructions dependent upon particular hardware features.

So, Which One Do I Really Want to Use?

While your choice of PIC may be critical if you are planning a production run of 100,000 products, for general experimentation and education, I prefer the 16F876 and 16F877 devices, either the original or the '876A/'877A versions. (For most purposes, there isn't a significant difference between the original and A suffixed '876 and '877 PICs.) Of the chips currently usable with MBasic, these four devices offer the maximum available program memory (8192 words), the maximum RAM (368 bytes) and have the other "bells and whistles" offered by Microchip, such as a USART and an A/D converter. For a particular project, the choice between the two is driven by the number of input/output pins required, with the '876 chips having a maximum of 22 possible I/O pins, while the '877 chips increase to 33 possible I/O pins. And, of course, these are flash memory devices so we need not worry about UV erasers.

For smaller projects, the 16F628 is worthy of consideration. It is available in an 18-pin package, so it must be used with Basic Micro's 0818 development board. The '628 has a maximum I/O capacity of 16 pins, and has generous 224 bytes RAM and 2048 words of program memory. It does not have an ADC.

Finally, for jobs that require a tiny PIC, the 12F629 and 12F675 devices are useful. Both have a small footprint (8-pin DIP package), 1024 words of program memory and 64 bytes of RAM. The 12F629 does not have an A/D converter, while the 12F675 does. Both permit up to six of their eight pins to be used for I/O purposes. Either chip may be used with Basic Micro's 0818 development board.

Basic Micro's MBasic876 Compiler

The CD-ROM accompanying this book includes a free MBasic876 compiler from Basic Micro. MBasic876 is a complete, 100% functional version of MBasic, limited in that it works only with the 16F876 and 16F876A devices. To use MBasic876 as intended, with integrated debugging and interactive programming, you will need to purchase Basic Micro's in-circuit programmer (ISP-PRO) and its 2840 Development Board. Or, if you are willing to sacrifice integrated debugging and interactive programming—both features of great benefit—you may use MBasic876's output HEX code with third-party PIC programmers. We'll look at the ISP-PRO and 2840 Development Board in Chapter 2.

You should not regard MBasic876's restriction to the '876 and '876A devices as a serious limit, as these chips are feature and performance rich and, in fact, are the most advanced mid-range PICs available in a 28-pin package. With only a handful of exceptions—where more I/O pins are required than are available in a 28-pin package—every circuit in this book can be constructed with an '876/876A, and the associated programs compiled by MBasic876.

References

[1-1] Microchip provides a wealth of PIC information, available for free downloading, on its Internet website, http://www.microchip.com. All are worth reading, but of particular interest to beginners are the introductory tutorials found at http://www.microchip.com/1010/suppdoc/design/toots/index.htm including:

- Analog-to-digital conversion: http://www.microchip.com/download/lit/suppdoc/toots/adc.pdf
- Device configuration: http://www.microchip.com/download/lit/suppdoc/toots/config.pdf
- Power considerations: http://www.microchip.com/download/lit/suppdoc/toots/power.pdf
- On-chip Memory: http://www.microchip.com/download/lit/suppdoc/toots/ramrom.pdf

Product line card: http://www.microchip.com/1010/pline/picmicro/index.htm contains a detailed table identifying the capabilities of the Microchip product line.

[1-2] A complete data sheet for most PICs comprises two elements; (a) a detailed "family" reference manual and (b) the particular device data sheet. MBasic supports only PICs from Microchip's "midrange" family and the associated PICmicro™ Mid-Range MCU Family Reference Manual may be downloaded at http://www.microchip.com/download/lit/suppdoc/refernce/midrange/33023a.pdf. This is a 688-page document, in almost mind numbing detail, but nonetheless is an essential reference to a complete understanding of PICs. For individual PIC family member datasheets, the easiest source is to go to http://www.microchip.com/1010/pline/picmicro/index.htm and select either the PIC12 or PIC16 group and from that link then select the individual PIC device.

General note on web addresses: Manufacturers periodically reorganize their websites, so the URLs in this book may change from those given as references. The documents, however, may be easily found through the manufacturer's home page search function, or through a general search engine such as Google.

MBasic Compiler and Development Boards

The Compiler Package

A Note on Compiler Versions

By the time this book is published, Basic Micro will have released an updated MBasic compiler (version 5.3.0.0) and rationalized its compiler family, dropping its "standard" version compiler, making the former "professional" version its flagship PIC compiler. (If you are still using version 5.2., check with Basic Micro for upgrade information. Owners of MBasic Professional version 5.2 qualify for a free upgrade, while MBasic Standard owners qualify for a reduced price upgrade to MBasic-Professional.) In addition, Basic Micro has a made available a free version of its MBasic Professional compiler, MBasic876 on the CD-ROM associated with this book. MBasic876 is a complete, 100% functional version of MBasic Professional, limited to working only with the 16F876 and 16F876A devices.

All programs in this book were originally developed and tested with MBasic Professional, version 5.2.1.1 and have been verified with a pre-release version of 5.3.0.0. However, bug-fixes and other "tweaking" to the official release version 5.3.0.0 may occur that introduce minor incompatibilities between the code in this book and Basic Micro's ultimately released compiler. The CD-ROM associated with this book provides both 5.2.1.1-compliant and 5.3.0.0-compilant source code. Chapter 29 summarizes the differences between version 5.3.0.0 and 5.2.1.1.

Unless specifically noted, this book assumes you are using MBasic or MBasic876, version 5.3.0.0. The printed program listings are for version 5.3.0.0.

MBasic Compiler

As used in this book, Basic Micro's MBasic compiler comprises three main elements:

1. *MBasic Compiler Software*—From version 5.3.0.0 onward, Basic Micro offers one version of its MBasic compiler, the "Professional" version. MBasic runs under Microsoft's Windows operating system in any version from Windows 95 to Windows XP. The computer requires an RS-232 port for connection to the ISP-PRO programmer board. A second RS-232 port, although not essential, is useful to capture any serial information from the program you are developing. If your computer does not have a second serial port, but does have a USB port, you may wish to add one using an inexpensive USB-to-serial converter.

2. *ISP-PRO Programmer*—MBasic, after the assembly stage completes, generates Microchip-compatible standard HEX code file that must be loaded into the PIC. Basic Micro offers a programmer, the ISP-PRO, well integrated with the MBasic compiler that automatically loads HEX code file. A major plus of Basic Micro's ISP-PRO is real-time debugging through its "in-circuit debugging" or ICD capability. Although it would be possible to substitute a third-party programmer for the ISP-PRO, losing both seamless integration with the compiler and ICD ability more than offsets any cost savings. The ISP-

PRO communicates with the computer running MBasic via an RS-232 cable, and with the PIC to be programmed through a 6-wire RJ11 telephone-type cable for Basic Micro's development and prototype boards, or a 10-pin standardized header for other boards.

3. *Development Board*—Basic Micro offers plug board style development boards and solder-in prototype boards for 8- and 18-pin and 28- and 40-pin PICs. The experiments in this book assume the user has Basic Micro's development boards. These boards have an RJ11 connector for the ISP-PRO connection and an uncommitted RS-232 port that may be used by the PIC for communications to the outside world.

Note on Serial Ports: The single largest source of trouble reported in calls to Basic Micro's help line concerns unreliable serial port connections with laptop computers. The built-in serial port on many laptop computers cannot reliably operate at 115.2 kb/s, the default speed at which the PC-to-ISP-PRO communications link operates. In those cases, Basic Micro suggests using an inexpensive add-on USB-to-serial adapter to substitute for the built-in serial port and recommends Bafo Technologies' BF-180 USB-to-serial adapter. A slightly more expensive alternative that I have had reliable results with is Belkin's F5U109, sold as a "USB PDA Adapter," but which is, in fact, a straight USB-to-serial adapter. Many other USB-to-serial adapters likely will provide reliable results.

In addition to the development and prototype boards, the ISP-PRO is compatible with Basic Micro's Universal Adapter. The Universal Adapter, however, does not contain an oscillator or the other circuitry needed to actually run a PIC program, and is intended for programming only.

Figure 2-1: ISP-PRO and RJ-11 jumper cable.

BASIC and Its Essentials

This book is not intended to teach BASIC programming from the ground up. There are many good "BASIC programming for the beginner" books and we assume the reader has at least passing familiarity with program control statements, mathematic procedures and variable assignment and structure. It also assumes the reader has installed the MBasic compiler (either the full version or MBasic876, version 5.3.0.0 as of the date of writing) and has familiarized himself with the first 80 pages or so in the MBasic User's Guide. Incidentally, because MBasic is, in some respects, a return to the early days of micro computer language implementation, I've found 20-year old reference documents for IBM's Personal Computer BASIC beneficial in refreshing my memory on some of the finer points of BASIC syntax or procedure and of considerably more help than modern books detailing, for example, Visual Basic. A visit to your local used bookstore may turn up useful reference material. I've provided the names of a few of my favorite long-out-of-print BASIC references in this chapter's reference section.

As a guide to finding the appropriate procedure, Table 2-1 groups MBasic's commands into a logical classification.

Table 2-1: Taxonomy of MBasic functions and procedures.

Group	Procedure	Group	Procedure

Program Flow	`Repeat/Until`
	`While/WEND`
	`Do/While`
	`For/Next`
	`If/Then/Else/EndIf`
	`GoTo`
	`GoSub/Return`
	`Branch`
		Hardware Related	`ADIN`
			`ADIN16`
			`Count`
			`HPWM`
			`SetCapture`
			`GetCapture`
			`SetCompare`

Table 2-1: Taxonomy of MBasic functions and procedures.

Group	Procedure	Group	Procedure
Pin Related	Button Low PulsIn PulsOut RCTime Reverse Toggle SetPullups INxx Outxx Dirxx	Miscellaneous	DeBug End Let Nap Sleep Stop
EEPROM	Data Read ReadDM Write WriteDM	Variables	Clear Swap
I/O	I2Cin I2Cout Owin Owout SerDetect SerIn SerOut ShiftIn ShiftOut HSerIn HSerOut	Sound and Sound Related	DTMFOut DTMFOut2 FreqOut PWM Sound Sound2
LCD	LCDWrite LCDRead LCDInit	Data Table	LookDown LookUp
Timing	Pause PauseUs PauseClk	Memory Related	Peek Poke
Random Generator	Random	Program Memory	ReadPM WritePM
On Reset	OnPOR OnBOR OnMOR	Explicit External Device Support	Servo SPMotor Xin Xout
Interrupts	Enable Disable OnInterrupt SetExtInt SetTmr0 SetTmr1 SetTmr2 IsrASM GetTimer1	Assembler	ASM {} ISRASM

Table 2-1: Taxonomy of MBasic functions and procedures.

Group	Procedure	Group	Procedure
Command Modifiers	Dec Hex Bin Str Sdec Shex Sbin Ihex Ibin ISHex ISBin REP Real WaitStr Wait Skip	Math Operators and Functions	+ - * LowMult HighMult FractionalMult / // ABS SIN COS DCD SQR BIN2BCD BCD2BIN Max Min Dig Rev
Bitwise Operators	! & | ^ >> <<	Comparison Operators	= <> < > <= >=
Logical Operators	And Or Xor Not And Not Or Not Xor	Floating Point Conversion	ToInt ToFloat FloatTable

Development Boards

Basic Micro offers two breadboard style development boards; models 0818 for 8- and 18-pin DIP PICs, (Figure 2-2), and the 2840 for 28 and 40-pin DIP PICs, (Figure 2-3). Both boards have a small solderless plug-in area for additional components and are full assembled with surface mount components. Sockets are installed for the PICs. An expanded development board, is under development and may be available by the time this book is published.

Figure 2-2: Basic Micro's 0818 development board.

Figure 2-3: Basic Micro's 2840 development board.

Additionally, Basic Micro offers corresponding semi-permanent prototype boards, models 08/18, Figure 2-4 and 28/40, Figure 2-5 differing from the development boards in that additional components are to be soldered in rather than plugged into a solderless breadboard. These are sold as bare boards, but Basic Micro also offers an inexpensive complete parts kit. The prototype boards use through-hole components.

Figure 2-4: Basic Micro's 08/18 prototype board.

Figure 2-5: Basic Micro's 28/40 prototype board.

All four boards permit in-circuit programming—that is, the PIC may be programmed without removing it from your board, or disconnecting its pins from whatever you may have connected to them. Figure 2-6, a simplified block diagram of the 28/40 prototype board, shows how this is possible. Three of the pins required for programming, RB4, RB6 and RB7, are switched through a 74HC4053 analog multiplexer/de-multiplexer between their normal connection to the PIC pin header and the RJ11 socket that connects to the ISP-PRO programmer. For our purpose, the 74HC4053 can be regarded as an electronic three-pole double throw switch, controlled by the ISP-PRO. The MCLR (master clear) pin is the fourth connection required for programming and is directly connected to the RJ11 programming socket.

The 0818 and 08/18 boards follow a similar design, but with extra configuration jumpers necessitated by the multiple functions Microchip assigned to certain pins of PICs produced in 8 and 18-pin packages. The 0818 and 08/18 data sheets should be consulted before programming these small PICs.

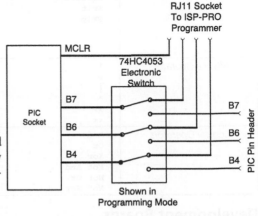

Figure 2-6: Simplified block diagram of 28/40 prototype board.

All four boards bring the various PIC pins to logically labeled headers; for example, A0, A1, so you don't have to continually cross-reference physical pin numbers with their logical assignments.

In working with Basic Micro's development boards and ISP-PRO programmer watch out for the following:

• These are sold as bare boards, with unprotected traces on the bottom. Don't put them down on conductive surfaces or the board may be damaged and watch for stray wires or component leads as well. (I watched my ISP-PRO board be dragged by its serial cable across the metal edge on the table and looked on help-lessly as sparks flew. Needless to say, the ISP-PRO didn't work after that.) It helps to add small stick-on rubber feet to the bottom of all boards.

• It is possible to damage the 74HC4053 electronic switch, as is rated at a maximum switched current of 25 mA. The most likely damage scenario comes from forcing the PIC to sink excessive current. Additionally, unlike a mechanical relay, the 74HC4053 introduces approximately 80 to 100 ohms of series resistance.

Another difficulty beginners often have is confusing V_{DD} and V_{SS} when wiring circuits. V_{SS} is ground in Basic Micro's development boards. V_{DD} is the supply voltage and is +5 volts in the development boards. Thus a schematic reference to +5 V is the same as V_{DD} and a reference to ground corresponds to V_{SS}. (This terminology comes from V_{DD} as the "drain" voltage and V_{SS} as the "source" voltage for a field effect transistor, the basic building block of PICs.)

Programming Style

Every program printed in the text is also provided as a file in the accompanying CD-ROM with two versions supplied—the originally developed programs compatible with MBasic version 5.2.1.1. and a revised version compatible with 5.3.0.0. There may be differences between the printed program and the CD-ROM for several reasons:

- The CD-ROM is quicker to update and may have a later or corrected version of the text program.
- Page width and overall length restrictions make it necessary to limit the comments and blank spaces used on the printed page. The data files have no similar restriction and hence may have additional comments and may be formatted for greater readability. (Although not documented in the User's Guide, MBasic uses the vertical bar " | " as a continuation symbol, thus allowing one logical line of BASIC code to be split over two or more physical lines.)

Standard Program Layout

As an aid in readability and maintainability, I like to follow a standard layout when programming, as exemplified in the following template:

```
;Program Sample.Bas - File name
;Version 1.00
;14 September 2003 - original version
;
;Constants
;-----------------
Define constants here

;Variables
;-----------------
Declare variables here

;Initialization
;---------------
Initialization code here - this code is executed only once

Main
;---------
Main code segment here
If something GoSub Sub1
If something else GoSub Sub2
If something totally different happens GoSub Sub3
GoTo Main ;if appropriate to have a continuous loop

Sub1
;-----
        Subroutine code here
Return

Sub2
;-----
        Subroutine code here
Return

Sub3
;-----
        Subroutine code here
        Includes GoSub SubSub1
Return

SubSub1
;------
        Subroutine code goes here
Return

End
```

The structure is logical; first define constants and variables and conduct any necessary initialization, then write the main program segment. I use subroutines as necessary, with the goal of keeping the main segment short and to the point. MBasic has no restriction on the number of subroutines that may be nested so one subroutine may call another.

With version 5.3.0.0, MBasic introduced user-defined functions, albeit with global, not local, variables. This feature was added too late to be used in the programming examples, except for Chapter 29.

Constants, Variable and Subroutine Names

Intelligently selecting constant, variable and names aids program readability. MBasic permits variables and constants to have names up to 1024 characters long, should you so wish. (MBasic has several hundred reserved words and you shouldn't use these names. In most cases you will receive a warning or error message if you try to redefine a reserved word.) All names are case insensitive, so we can use capitalization to improve readability without worrying about consistency. The naming conventions I've developed include:

- Index or counting variables, for example, control variables used in For/Next loops, start with letters in the range i…n. (Yes, this is a holdover from the early days of FORTRAN when integer variable names had to start with one of these letters.) Keep these names short, particularly if they will be used frequently. In many cases, the single letters i, j…n are perfectly suitable.
- Names should reflect contents or activities, without being overly long. Suppose we use the A/D converter to read a voltage followed by a subroutine call to average this reading with the last 15 readings—that is, a moving average of 16 readings and that the individual reading isn't elsewhere used. Since the voltage value being read will be discarded after the averaging process, we may use **TempVolt**, so the A/D read statement would be: **ADIN AN0, CLK, ADSETUP, TempVolt**. I like to name variables that have limited scope with a two-part name, starting with Temp. To make the name more readable, use upper and lower case, thus **TempVolt** is easier to read than **tempvolt** or **TEMPVOLT**. Or, insert an underscore as a separator; **Temp_Volt**. (Permissible separator characters are _,@,$,%,? and `.) Subroutines should be named according to the activity performed, in this example, **TakeAverage** is an appropriate name, or even **TakeAverageVolt**. I would reserve the name **AvgVolt** for the variable holding the average of the last 16 readings, thus keeping the suffix name **Volt** and changing the prefix to describe the type of voltage parameter in the variable. Short, concise and informative variable names can often be constructed of the form "adjective–noun" while subroutine names are often of the form "verb-noun" where the verb describes the action and the noun describes the subject of the action.
- Although very long names, such as **AverageTheLast16VoltageReadings** may seem descriptive at first, they actually hinder, rather than assist comprehension if they are used with abandon.

These are guidelines, not hard and fast rules, and even my observance isn't 100%, but following a logical consistent naming approach will pay dividends in the long run in terms of fewer errors and improved readability.

Building the Circuits and Standard Assumptions

In addition to the MBasic compiler (either the full or MBasic876 free versions), an ISP-PRO programmer and a 2840 development board, and any associated parts required for specific projects, you should have access to:

- A second larger plug board to hold overflow circuitry.
- Assorted jumper wires. You can purchase a kit of jumper wires, or make your own from a short length of scrap 25-pair or 50-pair telephone cable.
- A second adjustable regulated power supply, preferably one that has two independent outputs so that positive and negative voltages may be provided.
- A digital multimeter.
- Your ability to troubleshoot, experiment and verify operation of circuits will be greatly enhanced if you have a triggered oscilloscope.

Instead of buying resistors and capacitors one or two at a time, consider buying assortment kits. Possible suppliers are identified in Appendix A. Almost all resistors used with a PIC will be 10K or less in value and are ¼ watt dissipation rating, so even a limited assortment of values will be quite beneficial.

Since we use prebuild development boards, certain things are omitted in the schematics provided in this book.

* Basic Micro's prototype and development boards include bypass capacitors on the V_{DD} supply headers. Hence, they will not normally be illustrated in schematic diagrams. However, if parts of the test circuitry is build on an second plug board, or if you are designing a printed circuit board to hold your design, good design practice says you should liberally bypass V_{DD}.
* The development boards are designed for ceramic resonator frequency determining elements and are shipped by Basic Micro with a 10 MHz resonator. The circuits and software in this book use a 20 MHz resonator. In many cases, there will be no difference in performance, but for maximum compatibility with the programs in this book, use a 20 MHz resonator. Resonators cost under $1 and are available from suppliers identified at Appendix A.

Choice of PIC

I've used a 16F877A to develop the circuit in this book, but none of the programs depend upon the "A" version's added features, so a 16F877 will work equally well. Except for those few circuits that are input/output pin constrained, you may substitute a 16F876 or '876A in any design and use Basic Micro's free MBasic876 compiler for almost every program we develop.

Pins, Ports and Input/Output

Since every useful program must read from or write to a PIC's input/output pins, let's summarize how MBasic handles pins and ports. It can be confusing because some pins have triple or quadruple or even more duties and because MBasic provides several ways to address any given pin. And, the word pin itself has dual usage, as it refers to the physical packaging (an 8-pin DIP package for example) and to those physical pins that may be used for various purposes. To simplify our discussion we will limit ourselves to PICs that are supported by MBasic and plug into either the 0818 or 2840 development boards.

PICs communicate with external circuitry through intermediary "ports." Ports are treated internally by the PIC's CPU, and by MBasic, as byte (8-bit) variables with each bit corresponding to a particular pin. For example, the most significant bit in PortB's byte value corresponds to pin RB7, while the least significant bit corresponds to RB0. (In some PICs, not all bits of each port variable have physical pin assignments.)

Letters from A…E identify ports, except in DIP8 packaged PICs which have only one port, called GPIO (general purpose input/output). Thus, we have GPIO, PortA, PortB…PortE as predeclared variables in MBasic. (Port identifiers are written without a space, for example, PortA, not Port A.) Of course, not all PICs physically support all of these ports, and in some cases not all eight bits of a port have associated pins. For example, the PIC16F876 has PortA (but only bits 0…5 are mapped to pins), PortB and PortC. MBasic's configuration files, fortunately, ensure that only legitimate port variables are predeclared for the particular PIC being programmed.

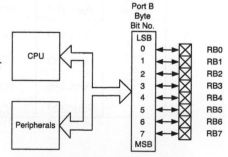

Figure 2-7 illustrates, as an example, Port B and its pin assignments. Each general-purpose I/O pin is identified with a consistent naming convention. For example, RB0 is PortB, bit 0. The "R" in RB stands for "register," which is synonymous with "file" or "variable." MBasic also pre-defines constants associated with each

Figure 2-7: PortB to pin assignments.

of these pins, so we have constants B0, B1…B7 available to us. MBasic gives us a second way to reference pins through a sequential numbering system, for example, A0 = 0, A1 = 1 up through E7 = 39. Finally, to provide backwards compatibility with the Basic Stamp™, MBasic includes the dedicated functions **INxx**, **OUTxx** and **DIRxx** where xx is the bit, nibble, byte or word identifier as reflected in Table 2-2.

Table 2-2: Port and bit I/O variables, constants and dedicated functions.

Variables			Constants		Dedicated Functions			
Port at a Time	**Nibble at a Time**	**Bit at a Time**						
					INS/OUTS/DIRS			
Port Variable	**Nibble**	**Port.Bit**	**Pin Constant**	**Pin Value**	**Bit**	**Nibble**	**Byte**	**Word**
PortA	PortA. Nib0 or PortA. LowNib	PortA.Bit0	RA0	0	IN0/OUT0/DIR0	INA/OUTA/ DIRA	INL/OUTL/ DIRL	INS/ OUTS/ DIRS
		PortA.Bit1	RA1	1	IN1/OUT1/DIR1			
		PortA.Bit2	RA2	2	IN2/OUT2/DIR2			
		PortA.Bit3	RA3	3	IN3/OUT3/DIR3			
	PortA. Nib1 or PortA. HighNib	PortA.Bit4	RA4	4	IN4/OUT4/DIR4	INB/OUTB/ DIRB		
		PortA.Bit5	RA5	5	IN5/OUT5/DIR5			
		PortA.Bit6	RA6	6	IN6/OUT6/DIR6			
		PortA.Bit7	RA7	7	IN7/OUT7/DIR7			
PortB	PortB.Nib0 or PortB. LowNib	PortB.Bit0	RB0	8	IN8/OUT8/DIR8	INC/OUTC/ DIRC	INH/OUTH/ DIRH	
		PortB.Bit1	RB1	9	IN9/OUT9/DIR9			
		PortB.Bit2	RB2	10	IN10/OUT10/ DIR10			
		PortB.Bit3	RB3	11	IN11/OUT11/ DIR11			
	PortB.Nib1 or PortB. HighNib	PortB.Bit4	RB4	12	IN12/OUT12/ DIR12	IND/OUTD/ DIRD		
		PortB.Bit5	RB5	13	IN13/OUT13/ DIR13			
		PortB.Bit6	RB6	14	IN14/OUT14/ DIR14			
		PortB.Bit7	RB7	15	IN15/OUT15/ DIR15			
PortC	PortC. Nib0 or PortC. LowNib	PortC.Bit0	RC0	16				
		PortC.Bit1	RC1	17				
		PortC.Bit2	RC2	18				
		PortC.Bit3	RC3	19				
	PortC. Nib1 or PortC. HighNib	PortC.Bit4	RC4	20	INxx: Read status, whether in input or output mode			
		PortC.Bit5	RC5	21	OUTxx: Write value			
		PortC.Bit6	RC6	22	DIRxx: Set direction 1=input, 0=output			
		PortC.Bit7	RC7	23				

(continued)

Table 2-2: Port and bit I/O variables, constants and dedicated functions.

Variables			Constants		Dedicated Functions			
Port at a Time	**Nibble at a Time**	**Bit at a Time**						
					INS/OUTS/DIRS			
Port Variable	**Nibble**	**Port.Bit**	**Pin Constant**	**Pin Value**	**Bit**	**Nibble**	**Byte**	**Word**
PortD	PortD. Nib0 or PortD. LowNib	PortD.Bit0	RD0	24	*xx as appropriate for bit, nibble, byte or word* *Note: DIRxx command is reversed from Basic Stamp*			
		PortD.Bit1	RD1	25				
		PortD.Bit2	RD2	26				
		PortD.Bit3	RD3	27				
	PortD. Nib1 or PortD. HighNib	PortD.Bit4	RD4	28				
		PortD.Bit5	RD5	29				
		PortD.Bit6	RD6	30				
		PortD.Bit7	RD7	31				
PortE	PortE.Nib0 or PortE. LowNib	PortE.Bit0	RE0	32				
		PortE.Bit1	RE1	33				
		PortE.Bit2	RE2	34				
		PortE.Bit3	RE3	35				
	PortE.Nib1 or PortE. HighNib	PortE.Bit4	RE4	36				
		PortE.Bit5	RE5	37				
		PortE.Bit6	RE6	38				
		PortE.Bit7	RE7	39				

MBasic permits us to reference a port or a pin as an *address or as a variable*. As an address, the port or pin is an argument to certain functions. As a variable, the value of the port (either in reading or writing) can be used like any other variable. There are also the dedicated functions identified in Table 2-2 that operate on specific ports or pins without an explicit port or pin reference, such as IN0. We must remember that MBasic automatically initializes all I/O pins as inputs and that before reading from or writing to a port or a pin we must follow some simple rules:

> **First, set the direction of the port or pin to be either an input or output;**
> **Second, read the port or pin if an input, or write to the port or pin if an output;**
>
> *or,*
>
> **Read from or write to a port or pin with a procedure that automatically sets the direction.**

Output Mode

Let's see how many different ways we can assign a pin to an output and to make its value 0. We'll use pin RB0 as our example.

```
;Direct Pin Addressing
;--------------------
Output B0              ; First make it an output. B0 is a constant
PortB.Bit0 = 0         ; PortB.Bit0 is a variable

Dir8 = 0               ; Special purpose function, DIR8 is for pin B0
Out8 = 0               ; Likewise for Out8
```

```
Low B0                  ; Low function automatically makes it an output
                        ; no need to separately make it into an output

Output 8                ; B0 is an alias for 8 so can use 8 directly
PortB.Bit0 = 0          ; Make the variable assignment

Low 8                   ; B0 is an alias for 8 so can use 8 directly
                        ; LOW switches to output mode and outputs 0

TRISB.Bit0 = 0          ; TRISB variable controls PortB I/O direction,
                        ; 0=output & 1=input.
PortB.Bit0 = 0          ; PortB.Bit0 is a variable

;Byte at a time addressing to deal with multiple pins
;in one instruction
;-------------------------------------------------------
TRISB = %00000000       ; Sets all 8 pins to 0, i.e., output
PortB = %00000000       ; Assign all 8 bits (pins) to 0.

DIRH = %00000000        ; Make all 8 Pins in PortB output
OUTH = %00000000        ; Set all 8 bits (pins) to 0
```

Input Mode

To make RB0 an input and read its value, we have the choice of a similar set of options:

```
;Direct Pin Addressing
; Assume we have already declared:
;       BitVar  Var     Bit
;       ByteVar Var     Byte
; to hold the value being read
;---------------------
Input B0                ; First make it an input. B0 is a constant
BitVar = PortB.Bit0     ; PortB.Bit0 is a variable

Dir8 = 1                ; Special purpose function, DIR8 is for pin B0
BitVar = In8            ; Likewise for IN8

Input 8                 ; B0 is an alias for 8 so can use 8 directly
BitVar = PortB.Bit0     ; Make the variable assignment


TRISB.Bit0 = 1          ; TRISB variable controls PortB I/O
                        ; direction,0=output & 1=input.
BitVar = PortB.Bit0     ; PortB.Bit0 is a variable

;Byte at a time addressing to deal with multiple pins
;in one instruction
;-------------------------------------------------------
TRISB = %11111111       ; Sets all 8 pins to 1, i.e., input
ByteVar = PortB         ; Read all 8 bits (pins) into ByteVar.

DIRH = %11111111        ; Make all 8 Pins in PortB input
ByteVar = INH           ; Read all 8 bits (pins) into ByteVar
```

Pin Variables vs. Addresses

One common error by beginners is confusing pin *variables* with pin *addresses*. The functions **Output**, **Low** and **Input** require a pin *address* as their argument. The pin address may be one of MBasic's pre-defined constants, for example, **B0**, or its equivalent numerical value, **8**. The pin address may also be the *value* of a variable, such as:

```
For I = B0 to B7        ; I goes from 8 (B0) to 15 (B7)
        Low I           ; Makes B0 low, then B1 through B7 sequentially
Next
```

In the **Low I** statement, **Low** operates on the value of **I**, which it interprets as the *address* of a pin. When **I** is **8**, for example, **Low** operates on pin RB0. Thus, the above code fragment is identical with:

```
For I = 8 to 15
    Low I
Next
```

Pin *variables* are used to read the value of a pin or of a port and to write to a pin or port via an *assignment* (the "=" sign). Thus we have `ByteVar = PortB`, or `PortB = $FF`. We also may use `PortB` like any other byte variable, such as `x=2*PortB`.

If we try to read the *value* of pin B0 as an input with the statement `BitVar = B0`, the compiler will produce no error, but `BitVar` will not hold the desired result. Rather, this statement is identical with `BitVar = 8`. If testing for a pin value condition in a loop statement, it's important that the variable construct be used.

```
;To test Pin B0
;--------------
If PortB.Bit0 = 1 Then
; execute code goes here
EndIf
;The following code will compile but won't work
;since it's the same as writing If (8 = 1), which is always false
If B0 = 1 Then
;execute code goes here
EndIf
```

Finally it's possible to *read* from a pin or port that is set for *output* in whole or in part, and to *write* to a pin or port that is set for *input* in whole or in part. No error message will be generated. If you are experiencing strange or unstable results reading or writing to pins or ports, check to ensure the correct direction is set and that you are correctly using pin variables and pin constants.

PICs equipped with analog-to-digital converters apply the designators AN0, AN1... to pins that also have an analog function. Thus, a 16F876 pin name RA0/AN0 indicates that the pin has three possible uses: digital output, digital input (RA0) and analog (AN0) input. The process of assigning a pin to be an analog input is discussed in detail in Chapter 11.

Run Time vs. Program Time Pin Assignments

All the pin assignments we have discussed to this point are *run time* alterable, i.e., their status may be altered by the program on the fly. In one part of your program a pin may be an input and later in the program the same pin may be an output. However, in some PICs—most often those in the 8 and 18-pin packages—certain pin configurations may only be established at *program time,* a task usually accomplished via an option dialog box in MBasic before compiling your code. (This permits Microchip to make their smaller package devices more flexible, but at the cost of confusion to beginning programmers.) Then, depending upon the program time configuration, further run time changes may be possible. Program time pin setup is highly device specific and reference to the data sheet for your specific PIC will be beneficial.

We'll explore the difference between run time and program time alterable pins in the context of the 12F629, which has 3 pins that must be configured at program time:

12F629 Example of Pins Configured at Program Time		
Pin Name	*Program Time Configuration*	*Run Time Configuration*
GP3/MCLR/Vpp	GP3 (general-purpose I/O)	GP3: Input GP3: Output
	MCLR/Vpp (master clear / Vprogram)	None
GP4/T1G/OSC2/CLKOUT	GP4 (general-purpose I/O)	GP4: Input GP4: Output T1G (timer 1 gate)
	OSC2 (second resonator connection)	None
	CLKOUT (clock out)	None

12F629 Example of Pins Configured at Program Time		
Pin Name	*Program Time Configuration*	*Run Time Configuration*
GP5/T1CKI/OSC1/CLKIN	GP5 (general-purpose I/O)	GP5: Input GP5: Output T1CKI (Timer 1 clock in)
	CLKIN (external clock input)	None
	OSC1 (first resonator connection)	None

If the GP3/MCLR/Vpp pin is defined at *program time* to be a general-purpose I/O pin, it may be used for input or output exactly as we have earlier discussed, and changed from input to output under program control. However, if at *program time* it is defined as MCLR, it is unavailable for any other purpose. This selection is accomplished with the **MCLR** check box found in MBasic's Configuration dialog box, as shown in Figure 2-8.

The two oscillator pins also must be defined at program time, but are linked. If you plan to use an external resonator or crystal, the OSC1 and OSC2 pin configuration must be active. If you plan to use an external clock source, then the CLKIN option must be active. If you wish to use the internal RC oscillator, then the CLKOUT pin may either be GP5 or OSCOUT. If an external RC oscillator is used, the RC network must connect to the OSC1 pin. Table 2-3 shows how these options interact and how they are selected in the MBasic configuration box of Figure 2-8. (The MBasic configuration options correspond to the first column in Table 2-3.)

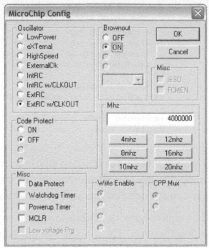

Figure 2-8: Program time pin options for 12F629.

Table 2-3: Configuration Dialog for 12F629 – Oscillator Configuration.

Configuration Dialog Box	Oscillator Configuration	GP4/T1G/OSC2/CLKOUT Function	GP5/T1CKI/OSC1/CLKIN Function
LowPower	LP	OSC2—crystal connection	OSC1—crystal connection
External	XT	OSC2—crystal connection	OSC1—crystal connection
HighSpeed	HS	OSC2—crystal connection	OSC2—crystal connection
ExternalClk	EC	GP4 (general-purpose I/O)	CLKIN (external clock input)
IntRC on GP4	INTOSC	GP4 (general-purpose I/O)	GP5 (general-purpose I/O)
IntRC on ClkOut	INTOSC	CLKOUT (clock waveform output)	GP5 (general-purpose I/O)
ExtRC on GP4	RC	GP4 (general-purpose I/O)	CLKIN-RC circuit on pin
ExtRC on ClkOut	RC	CLKOUT (clock waveform output)	CLKIN-RC circuit on pin

Further complicating an already complex matter, an external clock source may be used in the LP, XT and HS modes by feeding it into OSC1, in which case, OSC2 is unused. The LP, XT and HS modes set internal parameters in the oscillator section of the 12F629 and establish the maximum resonator or crystal frequency and associated capacitor values. Section 9 of Microchip's PIC12F629/675 Data Sheet should be consulted for specifics.

LVP Programming Pin Selection

One compile time feature shared by 16F876/877 chips (including the "A" versions), along with many other flash program memory PICs, is the low voltage programming (LVP) option. Historically, flash memory has required application of a programming voltage two or three times that of the normal operating voltage, typically 12V for a PIC operating with V_{DD} of 5V, known as high voltage programming (HVP). Newer PICs, such as the 16F876/877/876A/877A may be optionally programmed in LVP mode, using only +5V. Whether a PIC that supports LVP actually has LVP enabled is determined by a configuration bit, the value of which is keyed to the LVP checkbox in MBasic's configuration setup dialog box seen in Figure 2-8. If the chip does not support LVP, as is true in Figure 2-8, the LVP check box is grayed out.

To ensure maximum flexibility when programming both older and newer model PICs, Basic Micro's ISP-PRO and its 0818 and 2840 Development Boards use high voltage programming and *for the programs in this book you should not select LVP mode in MBasic's programming options. Microchip enables LVP as part of the manufacturing process, so when programming a new PIC for the first time you will find it necessary to clear the LVP selection box in MBasic's configuration menu, if that model PIC has LVP functionality.*

Basic Micro's ISP-PRO does not support LVP and programs only using HVP mode. But, since a PIC with LVP enabled is still programmable via HVP, you can, nonetheless, select LVP and program 16F876/877/ 876A/877A chips with MBasic and the ISP-PRO. However, if you do so, pin RB3 becomes the LVP control pin and is no longer available as a general-purpose I/O pin. (The specific pin used for LVP control varies; for example, a 16F628 uses RB4.) Not every LVP-capable model PIC behaves so nicely and you may find some model devices refuse to program if you inadvertently leave the LVP option selected. I've even seen different samples of the same model PIC behave differently with the LVP option selected. In this case, clearing the LVP check box, followed by several cycles of MBasic's "erase" function usually restores programmability.

Weak Pull-Up

One last remark and we may leave this overly long discussion of pins. Many PICs have built-in "weak" pull-up resistors for Port B, usable when set to be an input. We'll deal with floating input gates and the need for pull-up resistors in Chapter 4, but MBasic's procedure for controlling Port B's internal pull ups is **SetPullUps** **<mode>** where mode is one of two pre-defined constants, **Pu_Off** or **Pu_On** for de-activating or activating, respectively, internal pull-up resistors. Pull-up resistors for all eight pins of Port B are activated or deactivated by this command, not individual pins. For Port B pins that are set to be outputs, **SetPullUps** has no effect.

Pseudo-Code and Planning the Program

In describing an algorithm or even a complete program, we will often use a mixture of English and MBasic statements called "pseudo-code." Pseudo-code is a useful tool when developing an idea before writing a line of true code or when explaining how a particular procedure or function or even an entire program works. To distinguish it from MBasic or assembler, pseudo-code will appear in bold, italic Courier typeface.

Let's illustrate the benefit of pseudo-code with a simple example. Suppose we wish to illuminate an LED for 1 second whenever a button is pushed. We'll assume that the button is connected to the RB0 pin on a PIC, that pressing the button takes RB0 low and that the LED is illuminated by taking pin RB1 high. After any button push, the program repeats and waits for the next push. (We'll ignore button debounce, multiple button presses and a few other real-world concerns.) A pseudo-code version of our program is:

```
        Initialization Routine
        Initialize button pin to be an input
        Initialize LED pin to be an output
        Set LED pin to not illuminate LED

        Main Program Loop Start
                Read button pin and determine state
                        If button not pressed do nothing
                        If button pin is pressed, set LED pin to illuminate
                        If button has been pressed wait 1 second
                        Turn LED Off after 1 second
        Go back to Main Program Loop and test for next press
```

This pseudo-code example offers a easy to understand statement of the program structure and once we are satisfied that its logic matches our desired operation, we can easily transform it into an MBasic program, with each line of pseudo-code expanding into one or two lines of real code:

```
;Initialization
;--------------
Input B0          ; Initialize button pin to be an input
Output B1         ; Initialize LED pin to be an output
Low B1            ; Set LED pin to not illuminate LED

Main
;-----
        If PortB.Bit0 = 0 Then ; Read pin and determine state
            High B1                 ; If pressed, LED illuminated
            Pause 1000              ; If button pressed wait 1 second
            Low B1                  ; Turn LED Off after 1 second
        EndIf                       ; If button not pressed do nothing
                                    ; Go back to Main Program Loop and
GoTo Main                           ; test for next press

    End
```

For all but the simplest programs, I start with a high level pseudo-code definition of overall program flow—perhaps aided by a simple flow chart—but at a high level of abstraction. The first pseudo-code draft concentrates on the high level program flow and logic, with subroutines, initialization, input/output concerns being a line or two of pseudo-code. It shows the desired input and output data, even if it is something as simple as "read user's keypad press." I may even write and debug MBasic code implementing the high level design, substituting dummy sub-routines for the detailed ones, and hard-coding user inputs.

Once the high-level program flow is functioning, I write pseudo-code for each main subroutine, followed by an MBasic realization. Each subroutine is written and debugged before starting on the next, to the extent permitted by the program logic. Where one subroutine depends on the next, follow a top down approach.

If you have to boil down writing good code into one rule, it would be:

☞ *Think first, code later.*

If we are permitted a few more rules, they would be:

- Define the problem, including the "goes intos" and "comes out ofs," that is, the information to go into the program, such as switch readings, sensor readings and the conditions that cause those readings to be generated, and the desired output, such as logic level pin changes, analog outputs through a digital-to-analog converter, and the actions these outputs cause, such as turning on a motor or activating a solenoid.
- Document the problem and the solution and keep the documentation up to date as you develop answers, even if you are programming for personal satisfaction, or education and not with the though of developing a commercial product.
- Think first, code later—the single most important thing for a programmer to do is to resist the siren call of writing code and instead study, understand and plan what to do. Coding is often the simplest task and can be almost mechanical, if the problem is first properly understood and defined.
- Program modularly, proceeding from the top down, an essential philosophy if you are to efficiently produce readable, stable code. It's possible to program from the bottom up, starting with details and working towards a general structure that fits together the details, but it's never the right way to proceed. And if you do manage to make it work, a change in details can upset the general structure you've cobbled together. Top-down programming is much more tolerant to the inevitable changes that occur as a project progresses. And, write code in modular subroutines, not as one large omnibus program. Code development, maintainability and debugging are all immensely aided by coding in subroutines. As seen

in many examples in this book, our modular programming takes the form of simple programs to test concepts, hardware and code, but which are then combined into more complex constructions.

Inside the Compiler

The process of compiling an MBasic program and having the resultant code programmed into the PIC is illustrated in Figures 2-9 and 2-10. The process is transparent to users, as Basic Micro's integrated development environment automatically invokes the assembler, the linker and the loader program and deletes any unnecessary intermediate files. We will find reason, however, to examine the intermediate assembler (.ASM) file when we examine how assembler programming meshes with MBasic.

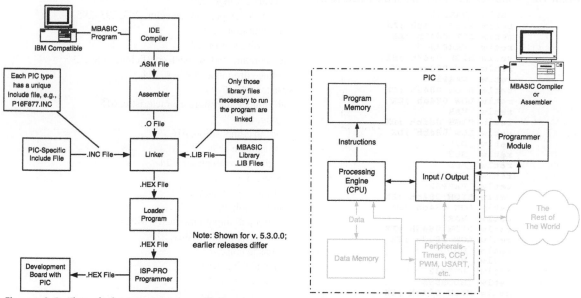

Figures 2-9: Flow during program compilation.

Figure 2-10: PIC to compiler flow.

Compiler vs. Interpreter

We've referred to the MBasic software as a "compiler." Since the software runs under Windows on an Intel-compatible PC, and the target is a PIC it should be called a "cross-compiler," as the term "compiler" is usually understood to mean that the target processor is the same as the processor upon which the compiler executes.

A more fundamental question, though, is MBasic a compiled language or an interpreted language? In the purest case, a compiler takes the high level language source code and translates all of it at once to machine code. A pure interpreter, in contrast, translates the high level language statements one at a time. All else being equal, a pure interpreter runs the same source code significantly slower than a pure compiler. However, the interpreter yields a much more compact result, occupying a fraction of the code space in the target machine required for its compiled equivalent. There are, in the real world, few pure compilers and few pure interpreters, and MBasic falls on the continuum between the two extremes, albeit closer to the interpreter end than the compiler end, a decision necessitated by the small code and variable size in PICs. If you examine the intermediate assembler file output from a simple program, you will find the MBasic statements are converted to tokens and ordered according to Reverse Polish Notation, which will be familiar to those having used an older Hewlett-Packard handheld calculator.

Let's look at a very simple example to illustrate how MBasic converts a statement into RPN tokens. Here's a simple mathematical expression in MBasic:

```
Ax      var     byte
Bx      var     byte
Cx      var     byte
Dx      var     byte
Ex      var     byte
Fx      var     byte
Gx      var     byte

Gx = Ax+(Bx-Cx/Dx)-Ex*Fx
```

The resulting assembler output, extracted from the ASM file, after compiling under MBasic version 5.2.1.1 (removing nonessential lines) from the MBasic is:

```
        retlw _VARP32
        retlw HIGH 0848h ;AX
        retlw LOW 0848h ;AX
        retlw _VARP32
        retlw HIGH 0849h ;BX
        retlw LOW 0849h ;BX
        retlw _VARP32
        retlw HIGH 084ah ;CX
        retlw LOW 084ah ;CX
        retlw _VAR
        retlw HIGH 084bh ;DX
        retlw LOW 084bh ;DX
        retlw _DIV
        retlw _SUB
        retlw _ADD
        retlw _PUSH32
        retlw _VARP32
        retlw HIGH 084ch ;EX
        retlw LOW 084ch ;EX
        retlw _VAR
        retlw HIGH 084dh ;FX
        retlw LOW 084dh ;FX
        retlw _MULL
        retlw _SUB
        retlw _PUSH32
        retlw _ADDRESS
        retlw HIGH 084eh ;GX
        retlw LOW 084eh ;Gx
        retlw _LET
```

We won't spend a lot of time on this topic and will save for later discussion several important concerns, but let's go through it quickly. **Retlw** statements (return literal in the w register) return a literal value to the calling function; for example, **retlw** 1 returns the value 1 when it is called. (If you compile this statement under version 5.3.0.0, the **retlw** statements are replaced by chip-specific macro calls, but the code structure remains the same.)

The statement group:

```
        retlw _VARP32
        retlw HIGH 0848h ;AX
        retlw LOW 0848h ;AX
```

can be understood to accomplish placing the value of the MBasic variable **Ax** into a form usable by MBasic's mathematical library.

The statement **retlw _DIV** instructs MBasic's mathematical library to execute the division operator on the two values immediately preceding it.

Based on this understanding, we can simplify the meaning of the assembler code:

```
        AX
        BX
```

```
CX
DX
DIVIDE
SUBTRACT
ADD
EX
FX
MULTIPLY
SUBTRACT
Gx
LET (=)
```

Parentheses may help show how this is evaluated, and it will be immediately clear if you have used an HP calculator with RPN:

```
AX ADD (BX Subtract(CX DX Divide)) Subtract ((EX FX Multiply)) (GX=)
Or, Gx = Ax+(Bx-Cx/Dx)-Ex*Fx
```

References

The following two references are long out of print, but may turn up from time to time in used bookstores.

[2-1] *BASIC Reference, Personal Computer Hardware Reference Library, Third Ed.*, (May 1984). Any of IBM's BASIC handbooks from the late 1970's to mid '80s are worth acquiring as they document, in IBM's exceedingly thorough fashion, the nuts and bolts of BASIC programming. Obviously some of the information is highly specific to PCs of the era, but a surprisingly high amount of information is directly applicable to MBasic.

[2-2] Nagin, Paul and Ledgard, Henry, *Basic With Style—Programming Proverbs*, Hayden Book Company, Rochelle Park, NJ (1978). This thin volume is packed with more good advice on programming than most books three times its size. It isn't as much a "how to code" book as it is a "how to think about coding" book, and is thus that much more valuable, as "how to code" books are plentiful.

[2-3] Brown, P.J., *Writing Interactive Compilers and Interpreters*, John Wiley & Sons, New York, NY (1979).

[2-4] Aho, A.V., et al, *Compilers Principles, Techniques and Tools,* Addison-Wesley Publishing Co., Reading, MA (1986).

[2-5] *PIC12629/675 Data Sheet*, Microchip Technology, Inc., Document No. DS41190C (2003).

[2-6] *PIC16F87X Data Sheet*, Microchip Technology, Inc., Document No. DS30292C (2001).

CHAPTER **3**

The Basics – Output

A great deal of your work with PICs will involve turning things on and off. The action may be as simple as illuminating an LED to show program status, or as complex as sequencing multiple motors. You may accomplish these actions with the PIC's input/output pins, unaided, or with external electronic or electromechanical devices. The action may require "sourcing" or "sinking" current or voltage. A high state pin *sources* current into an external load, while a low state output pin receives or *sinks* current from an external load. In this chapter, we will review a few elementary electronics principles and learn how to use them to allow PICs to control external devices.

This chapter deals with the electronic characteristics of PIC pins as *output* devices.

Diagrams and discussions in this book assume positive or classical current flow, in which current flow from positive to negative, as shown in Figure 3-1. Traditional circuit equations, as well as the arrow symbol for diodes and transistors follow this convention as well.

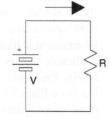

Figure 3-1: Conventional Current Flow

Before building any of the sample circuits please download and read the relevant data sheets from the device manufacturer's Internet website. (The CD-ROM supplied with this book also contains some datasheets.) To keep this chapter at a manageable length, I've had to gloss over many subtleties in the specifications and application of these devices, hitting only the highlights. Careful advance study of data sheets and any associated application notes will reduce the time spent designing and debugging your designs.

Pin Architectures

At first glance, Microchip's simplified schematic of the I/O pins may seem confusing. Chapter 3 of the 16F87x Reference, for example, requires ten figures to illustrate the internals of I/O pin construction. At the beginning and intermediate stages of programming with MBasic and concentrating only on the output mode, though, we can simplify things further, reducing the essentials to those of Figure 3-2. In the 16F87x series, and in other mid-range PICs, when in output mode, pins are connected to a classical complementary metal oxide semiconductor (CMOS) configuration. In some cases, such as for RA0...RA3, Microchip's documents show the CMOS transistors directly; in others, such as RB0...RB3, they are not shown but are imbedded in a logic gate symbol.

For many purposes, we can regard the PMOS and NMOS transistors of Figure 3-2 as simply switched resistors; they are either very high resistances, amounting to almost open circuits, or a low value resistor, as illustrated in Figure 3-3. When the output is low, the pin appears to be a low value resistor, approximately 25 ohms. When the output is high, the pin appears to be the V_{DD} source connected through resistor of about 85 ohms, as long as the sourced current doesn't exceed 15 mA or so.

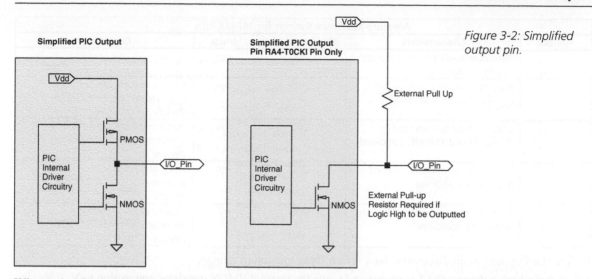

Figure 3-2: Simplified output pin.

When using Basic Micro's development or prototype boards, the 74HC4053 multiplexer needed to permit in-circuit programming adds approximately 50–100 ohms series resistance to pins RB4, RB6 and RB7. In many cases this additional resistance can be ignored.

One pin, RA4, is different; it is configured as an open drain MOSFET. When set to low, it performs identically with the other pin architectures. However, when set to high, there is no internal connection with V_{DD} and hence it will not directly source voltage. If it's necessary to use RA4 as a sourcing output pin, you can add an external "pull-up" resistor, typically in the range of 470 ohms–4.7K ohms. The sourced current then comes from the pull-up resistor. Unlike all other pins that cannot exceed V_{DD}, RA4's open drain is rated to 12 volts.

When either sourcing or sinking current, the safe operating limits of the PIC must be observed. The following maximum safe parameters apply to the 16F87x series, and the Electrical Characteristics section of Microchip's data sheet for your target PIC should be consulted. Exceeding these limits may cause damage to the device, or reduce its reliability.

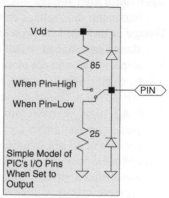

Figure 3-3: For many analyses, the output pins appear to be simple resistors.

Absolute Maximum Ratings for 16F87X PICs				
Symbol	**Characteristic**	**MaximumValue**	**Units**	**Conditions**
V_{OD}	Open drain high voltage	14	V	Applies to pin RA4 only
	Voltage on any pin with respect to Vss	-0.3V to V_{DD}+0.3V	V	
	Total chip supply current into V_{DD} supply pin	250	mA	
	Total chip current out of V_{SS} pin	300	mA	
I_{OK}	Output clamp current (Vf <0 or Vf>V_{DD})	±20	mA	
	Maximum output current sunk by any I/O pin	25	mA	

Absolute Maximum Ratings for 16F87X PICs				
Symbol	*Characteristic*	*MaximumValue*	*Units*	*Conditions*
	Maximum output current sourced by any I/O pin	25	mA	
	Maximum current sunk by PortA, PortB and PortE, combined	200	mA	PortD and PortE are not implemented on 16F873/876 devices
	Maximum current sourced by PortA, PortB and PortE, combined	200	mA	PortD and PortE are not implemented on 16F873/876 devices
	Maximum current sunk by PortC and PortD, combined	200	mA	PortD and PortE are not implemented on 16F873/876 devices
	Maximum current sourced by PortC and PortD, combined	200	mA	PortD and PortE are not implemented on 16F873/876 devices

Before starting our circuit discussion, let's review these maximum ratings.

Open drain high voltage—RA4 is unique and omits the internal PMOS transistor connection to V_{DD}. V_{OD} is maximum safe voltage that may be applied to RA4.

Voltage on any pin with respect to Vss—In the normal circuit, V_{ss} will be at ground potential. Your circuit should be designed so that when in output mode, pins will not be taken more than 0.3 V negative with respect to ground nor more than 0.3 V above the PIC's positive supply voltage, V_{DD}. Should these voltages be significantly exceeded, the protective diodes shown in Figure 3-2 will start to conduct, potentially causing the pin or chip maximum current limit to be exceeded, unless otherwise current limited.

Total chip supply current into V_{DD} supply pin—In addition to sourcing current limits on individual pins, this parameter establishes a global maximum available current for the entire PIC. It is, with negligible error, the sum of all pin sourcing currents.

Total chip current out of V_{SS} pin—In addition to sinking current limits on individual pins, this parameter establishes a global maximum for the entire PIC. It is, with negligible error, the sum of all pin sinking currents.

Output clamp current (Vf <0 or Vf>V_{DD})—If a pin is taken above V_{DD} or below ground, it must be current limited, commonly with a series resistor, so that the output clamp current is not exceeded.

Maximum output current sunk by any I/O pin—The maximum safe sinking current when a pin is low. Sinking current is not internally limited and is governed by the external circuit parameters.

Maximum output current sourced by any I/O pin—The maximum current that may be safely sourced by a high pin. Internal circuitry limits sourcing current to approximately 25-30 mA so it is safe, but not good design practice, to operate an output pin into a short circuit.

Maximum current sunk by PortA, PortB and PortE, combined—Another composite limit, applying to sinking current by all Ports A, B and E pins combined.

Maximum current sourced by PortA, PortB and PortE, combined— Another composite limit, applying to sourcing current by all Ports A,B and E pins combined.

Maximum current sunk by PortC and PortD, combined— Another composite limit, applying to sinking current by all Ports C and D pins combined.

Maximum current sourced by PortC and PortD, combined— Another composite limit, applying to sourcing current by all Ports C and D pins combined.

A high output will source between 25 and 30 mA into a short circuit indefinitely, but when sinking current, the maximum safe current rating must be observed. Figures 3-4 and 3-5 illustrate the typical voltage

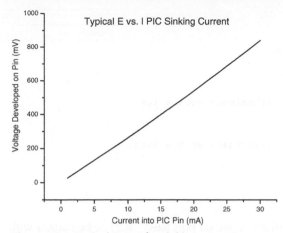

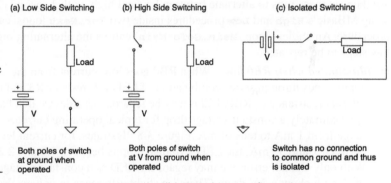

Figures 3-4: Typical E vs. I for sinking current.   *Figures 3-5: Typical E vs. I for sourcing current.*

versus current relationship for both sourcing and sinking current. Also remember that when using Basic Micro's 2840 Development Board, pins RB4, RB6 and RB7 are switched through the 74HC4053 multiplexer which has a 25 mA maximum current limit.

(a) Low Side Switching   **(b) High Side Switching**   **(c) Isolated Switching**

Load   Load   Load

V   V

Both poles of switch at ground when operated   Both poles of switch at V from ground when operated   Switch has no connection to common ground and thus is isolated

One final bit of terminology and we'll be onto circuitry. Figure 3-6 shows three possible switching configurations. For clarity, the drawing shows a mechanical switch. We, of course, will use a variety of electronic substitutes.

Figure 3-6: Possible switching configurations.

Low side switching—The switch is between the load and ground. When closed, both sides of the switch are at ground potential.

High side switching—The switch is between the voltage being switched and the load. When closed, both sides of the switch are at the switching voltage.

Isolated switching—There is no common connection between the circuit being switched and the controlling PIC. Many devices suitable for isolated switching also work for low side or high side switching.

LED Indicators

In learning how to programming a PC in a high level language, the traditional first program writes "Hello World" to the screen. Since PICs don't have a screen, the first MBasic program traditionally blinks an LED. We'll do that idea one better, building up to four states with one LED and one PIC pin. But, first we'll start with two LEDs and two pins as shown in Figure 3-7.

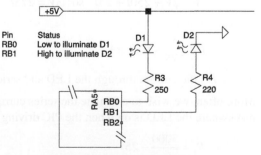

Pin	Status
RB0	Low to illuminate D1
RB1	High to illuminate D2

Figure 3-7: LED connections.

Program 3-1

```
i         var      byte

For i = B0 to B1        ;LEDs are on B0...B1
    Output I    ;so we make them outputs
Next

Main
        For i = B0 to B1        ;some will illuminate with a low
            Low i
        Next
        Pause 1000
        For i = B0 to B1        ;some will illuminate with a high
            High i
        Next
        Pause 1000
GoTo Main

    End
```

The code is straightforward; After declaring our index variable **i**, we set pins RB0...RB1 to be outputs with the **Output** procedure inside a **For...Next** loop. The **Output(i)** procedure takes the pin address as its argument, with **i** ranging from **B0** to **B1**, pre-defined in MBasic as the numerical addresses of pins RB0...RB1. We then set these pins to alternate between low and high, with 1 second (1,000 milliseconds) in each state using MBasic's **High** and **Low** procedures inside two **For...Next** loops, each followed by a **Pause(1000)** procedure. An endless loop (**Main...GoTo Main**) causes the alternating high/low steps to be repeated.

D1 illuminated when RB0 low—When RB0 goes low, current from the +5 V supply goes through series combination of LED D1, resistor R3 and the internal resistance of RB0. LEDs may be regarded as a device that have approximately a constant voltage drop for typical operating currents in the range from 1 mA to tens of mA. Figure 3-8 illustrates, for current levels between 1 and 50 mA, the LED's voltage drop is between 1.7 and 2.2 V. With only a small error, we may regard the LED as a constant voltage device, with about a 2 V drop. (There's a slight difference in voltage drop for different output colors, but for almost all red, green and yellow LEDs, we may calculate the current limiting resistors assuming a 2 volts drop.)

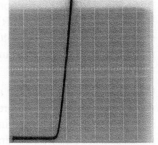

Figure 3-8: E/I curve trace of red LED. Horiz: 0.5V/div Vert: 5mA/div.

We may now solve the current loop equation for the circuit involving D1, remembering that a low pin is functionally equivalent to a 25 ohm resistor:

$$5V = 2V + 250I + 25I$$

rearranging

$$5V - 2V + 250I + 25I \quad \text{so} \quad 3V = 275I$$

or

$$I = \frac{3}{275} = 10.9 \ mA$$

Where *I* is the current through the LED and series resistor.

More often, we wish to calculate the series current limiting resistor needed for a particular LED current *I* (in mA) where the LED is on when the PIC driving pin is low:

$$R_3 = \frac{3000}{I_{mA}} - 25$$

We fudged a bit by assuming the voltage drop across D1 is constant regardless of current, but these simple equations will be within 10% of a more detailed calculation, more than accurate enough for determining the current through an LED indicator.

D2 Illuminated when RB1 high—When RB1 goes high, current from the V_{DD} (the +5 V supply in Basic Micro's development boards) goes through series combination of LED D2, resistor R4 and the internal resistance of RB1. This is only a slight rearrangement of our earlier analysis of D1, with the internal equivalent resistance of the high pin being 85 ohms. Hence,

$$5V = 2V + 220I + 85I$$

rearranging

$$5V - 2V + 220I + 85I \quad so \quad 3V = 305I$$

or

$$I = \frac{3}{305} = 9.8 \ mA$$

Where I is the current through the LED and series resistor.

Or, to calculate the series current limiting resistor where the LED is on when the PIC driving pin is high (in mA):

$$R_4 = \frac{3000}{I_{mA}} - 85$$

In addition to the constant voltage drop fudge, this analysis assumes a high pin is modeled accurately by as an 85 ohm resistor in series with V_{DD}. As Figure 3-5 shows, this assumption starts to fail as the sourced current exceeds 15 mA and the plot of I versus E diverges from a straight line.

Two LED's on one pin—We can connect two LEDs to one pin using the circuits we just developed as shown in Figure 3-9. The current for each LED is calculated using the same equations for individual pin connections.

Four states from one pin—Using the connection of Figure 3-10, it's possible for one pin to produce four states in a 2-pin dual LED. (Most dual LEDs have two pins, but some dual LEDs have three pins permitting the circuit of Figure 3-9 to be used.) Fairchild's MV5491A two-pin dual LED is configured as a red and green LED in anti-parallel whereby current flow in one direction provides red light while the opposite direction provides green light.

In the circuit of Figure 3-10 when RB2 is high, current flows from RB2 through D1 and R2. When RB2 is low, current flows from the +5 V supply through R1, D2 and is sunk at RB2. The suggested resistors yield 6.9 mA current for the green LED (D1) and 8.6 mA for the red LED (D2).

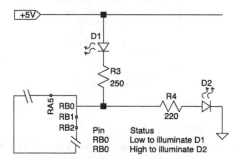

Figure 3-9: Two LEDs on one pin.

Pin	Status
RB0	Low to illuminate D1
RB0	High to illuminate D2

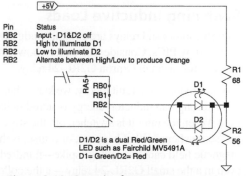

Pin	Status
RB2	Input - D1&D2 off
RB2	High to illuminate D1
RB2	Low to illuminate D2
RB2	Alternate between High/Low to produce Orange

D1/D2 is a dual Red/Green LED such as Fairchild MV5491A
D1= Green/D2= Red

Figure 3-10: One pin, four states.

It's possible to get a third color out of this design as well. By rapidly switching between the red and green LEDs, the eye perceives orange. The following code fragment will accomplish this, switching at approximately 100 Hz.

```
Main
        High B2
        Pause 5
        Low B2
        Pause 5
GoTo Main
```

Finally if a fourth condition, LED off, is desired, switch RB2 to input. As an input, RB2 is essentially an open circuit, and neither D1 nor D2 will be illuminated. This trick will not work with the configuration of Figure 3-9, as both diodes will illuminate in that state.

Program 3-2 exercises all four states of Figure 3-10's dual LED.

Program 3-2

```
;Four states from one dual color LED and one PIC Pin
;Assumes bi-color LED on RB2
;With voltage divider circuit

i       Var             Byte


Main
        High B2                     ;Green
        Pause 1000
        Low B2                      ;Red
        Pause 1000

        For i = 0 to 255            ;Orange
            High B2
            Pause 5
            Low B2
            Pause 5
        Next

        Input B2                    ;no illumination
        Pause 1000

    GoTo Main

    End
```

Program 3-2 first illuminates the green LED for 1 second followed by red for 1 second, followed by 2.5 seconds of orange when both the red and green diodes are sequentially active for 5 ms. Finally, the diode is dark for 1 second.

Switching Inductive Loads

Stepper motors and relays are common inductive loads switched by PICs. Consider the circuit shown in Figure 3-11 that controls a small Omron G2RL-24 relay.

From introductory circuit theory, we know that when current flows through an inductor, energy is stored in its magnetic field. When the circuit is switched off, the stored energy must go "somewhere." What happens, of course, is the collapsing magnetic field causes a voltage spike—hundreds of volts even in a the small G2RL-24 relay—at the collector of Q1. In the absence of protective circuitry, Q1 will temporarily break down when the spike exceeds its V_{CEO} rating (40 V

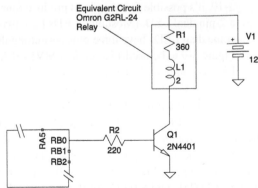

Figure 3-11: Switching an inductive load.

for a 2N4401) and the stored energy is dissipated in Q1. Even if Q1 isn't damaged by the repeated over-voltage breakdown, good design practice says that we should limit the over voltage to safe limits. Fortunately, as shown in Figure 3-12 it's easy to add a protective diode. D1 is called a "clamping diode" because it clamps the voltage spike. You may also see it referred to as a "snubbing diode" or "snubber."

When the magnetic field collapses as Q1 is switched off, the induced voltage causes diode D1 to conduct, and the stored energy is dissipated in the internal resistance of the inductor, R1 in Figure 3-12, and an optional external resistor, R3.

As with many things in electronics, there is a trade off here. The current resulting from the magnetic field decay doesn't drop to zero instantaneously, but rather as a function of the total resistance in the D1-R3-R1 circuit. The faster we make the current drop to zero (smaller the series resistance) the higher the voltage spike. Conversely, if we limit the voltage spike to its minimum level by setting R3 at zero ohms, we find the longest time for the induced current to decay. Figure 3-13 illustrates how the peak voltage spike and current decay times interact for the circuit of Figure 3-12. If you are familiar with elementary calculus, this relationship is obvious since the voltage E across an inductor of value L Henries is proportional to the time rate of change of the instantaneous current i through the inductor:

$$E = L\frac{di}{dt}$$

A faster decay (greater di/dt) means more induced voltage and vice versa.

If we are concerned with the relay release time, we want the current to decay below the release current as quickly as possible. This suggests a higher series resistor, perhaps with a Q1 possessing a higher V_{CEO} to accept the resulting higher voltage spike. Perhaps more of a concern exists with when driving a stepper motor. We wish the magnetic field to collapse as quickly as possible when current is removed from a winding, particularly if we are interested in running the motor near its maximum steps per second rating.

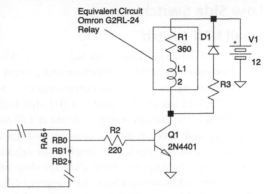

Figure 3-12: D1 and R3 are added protective components.

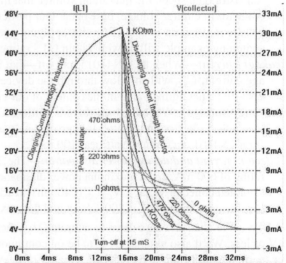

Figure 3-13: "Shutdown at 15 ms, current and voltage vs. clamping circuit resistance.

For critical applications, and particularly for stepper motors, the clamping diode should be a fast switching device, such as a Schottky diode. Alternatively, it is possible to use a Zener diode, set to avalanche upon turn off. By delaying the onset of current flow until the Zener diode avalanches, significantly faster decay is possible. (The Zener is in series with D1, polarized so that D1 prevents forward current flow through the Zener.) We'll look at stepper motor driving circuits in detail in a later chapter.

It is possible to calculate the inductive spike level and decay time analytically, but it's much easier to use a SPICE circuit simulation program such as Linear Technology's LTSpice.[Ref 3-4]

The remainder of this chapter won't mention inductive spike protection, unless it is appropriate because of device characteristics.

Low Side Switching

Small NPN Switch

Figure 3-14 depicts a simple low side switch. When RB0 is high, Q1 is forward biased into conduction and current flows through the load. Let's work through a few design concerns with this simple circuit. We will assume the load is a 100 ohm resistor and V is 12 volts. Hence, the current being switched is 120 mA. We'll treat the current through this circuit as a constant and ignore the voltage drop across the switching device to simplify our calculations. Since our switch circuits will operate with a voltage drop of well under 0.5V, our simplifications will not introduce appreciable error.

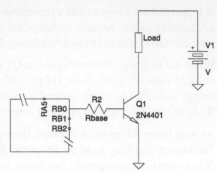

Figure 3-14: 2N4401 NPN low side switch.

Voltage rating—When RB0 is low, Q1 appears as an open circuit and thus has the full load supply voltage V across it. In a transistor data sheet, the maximum voltage that may safely be applied in this mode is V_{CEO} or maximum collector to emitter voltage, base open. Our particular device, a 2N4401 is rated for 40 V V_{CEO}. It should be safe to use it up to about 25 volts, applying a reasonable safety margin to the rated value. Our 12 V switching example will be well within Q1's ratings.

Leakage current—When the 2N4401 is cut off—that is, the base voltage is less than about 0.4 V, some leakage current, I_{CEX}, will still flow through the device's collector. I_{CEX} is rated not to exceed 0.1 µA in the 2N4401, a negligible value in the context of our circuits.

Saturation voltage—When the base drive is sufficient to saturate a bipolar transistor, the voltage drop between the collector and emitter is approximately a constant, referred to in data sheets as $V_{CE(SAT)}$, 0.4 V for a 2N4401 at current levels near 100 mA.

Collector current and device power dissipation—The 2N4401 has a maximum continuous collector current rating of 600 mA, and a maximum power dissipation rating of 625 mW at room temperature (25°C). The saturated collector voltage, $V_{CE(SAT)}$ is 400 mV at 150 mA and 750 mV at 500 mA. The thermal resistance junction to ambient $R_{\theta JA}$ is 200°C/watt and the maximum junction operating temperature is +150°C.

We'll assume adequate base drive to saturate Q1, hence we expect the collector voltage at 120 mA to be 400 mV or less. We'll also assume Q1 is to be on continuously—continuously in this context means long enough for thermal equilibrium to be reached, a matter of a few seconds for a 2N4401 size device. Hence:

The device dissipation will be 120 mA × 400 mV, or 48 mW.

The junction temperature rise over ambient will thus be 200°C/watt × 0.048 watts, or 9.6°C. Assuming the ambient air temperature is 120°F (49°C), the maximum junction temperature will thus be 48° + 9.6° = 57.6°C.

To determine case temperature, we use the thermal resistance junction to case $R_{\theta JC}$ specification, 83.3°C/watt. The case will thus be at 83.3°C/watt × 0.048 watts, or 4°C above ambient temperature. Our design thus is well within the safe operating parameters of the 2N4401.

If the 2N4401 is being cycled off and on at a rapid rate, the duty cycle will enter into certain of these ratings. For example, suppose the 2N4401 is driving a multiplexed LED display, on for 2 ms and off for 8 ms, for a duty cycle of 0.20. The average power dissipation of Q1 will thus be 20% of the peak power and the permissible peak power dissipation limit may be as much as five times the continuous value. Of course, for this averaging effect to work, the on time must be short compared with the time it takes for the device to reach thermal equilibrium.

Base current drive—As a rough approximation, we may regard Q1 as a current operated switch—that is, for every milliampere of current we wish to be sunk by Q1's collector, we must inject into the base a certain current level. (This is a highly simplified approximation of semiconductor operation, but adequate for our purpose.) The ratio of collector current to base current is known as h_{FE} or "DC current gain." The DC current gain varies from device type to device type, is not well controlled from example to example of the same device type and, finally, varies with current even for a particular transistor. 2N4401 devices, for example, have an h_{FE} that varies from 20 to 300, depending on the collector current.

If we are not concerned with switching time, or power minimization, the simplest design approach is to assume the worst case h_{FE} and design accordingly. To sink 120 mA, for example, since the minimum specified h_{FE} at 100 mA is 100, the target base current should be 1.2 mA. However, we note that at both 500 mA and 10 mA collector currents, the minimum h_{FE} drops to 40. Hence, as a matter of perhaps excessive caution, and to ensure Q1 is driven well into saturation, we will design for h_{FE} of 40, representing 3 mA base current.

The base to emitter junction voltage, V_{BE}, for a 2N4401 is specified at 750 mV for a base current of 15 mA and collector current of 150 mA, so we will use this value in calculating the base resistor, R2. (Since the base to emitter junction is modeled as a forward biased silicon diode, 700 mV is a commonly used rough estimate for the base to emitter voltage over a wide range of base currents for all silicon bipolar junction transistors.) R2's value (neglecting the PIC's approximately 85 ohm series resistance when sourcing current) is thus:

$$R2 = \frac{5V - 0.750V}{0.003A} = 1.4\ kohm$$

Switching Speed—We've alluded to Q1's switching speed concerns several times in our design. If we are switching an LED, or relay or stepper motor, these problems are unlikely to concern us. However, there are times where it is critical to switch a load as fast as possible. Figure 3-15 shows what happens when very short switching intervals are used in the circuit of Figure 3-14. RB0 emits a fast rise and fall 200 ns wide pulse and Q1 turns on with less than 20 ns delay. However, when RB0 goes low, Q1 exhibits nearly 500 ns turn off delay. The turn-off delay results from the "stored charge" effect, where excess minority carrier charge is stored in the base region of the transistor junction structure and must be removed before the transistor turns off and collector current ceases. We assume that anyone desirous of switching speeds in the sub-microsecond range knows about stored charges and the mitigating techniques to deal with the problem.

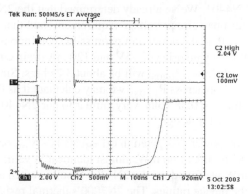

Figure 3-15: 2N4401 switching time Ibase = 6 mA, Ic = 40 mA; Ch1: PIC pin to base drive; Ch2: 2N4401 collector.

One final point should be noted with the bipolar transistor design of Figure 3-14—the voltage V being switched is immaterial. Of course, Q1 must be rated to withstand the voltage, but with a suitable transistor, the circuit of Figure 3-14 could switch 500 volts as easily as it switches 5 volts. The current required to saturate or cut off Q1 is not affected by the voltage it switches.

Small N-Channel MOSFET Switch

At the risk of considerable oversimplification, Q1 in Figure 3-14 may be thought of as a current controlled switch; current injected into the base causes the collector to be pulled close to ground potential. There is a similar voltage controlled switch, the MOSFET, whereby voltage applied to the device's gate causes the drain voltage to be pulled close to the source, or ground potential in a low side switch.

Figure 3-16 is the MOSFET counterpart of Figure 3-14. A 2N7000 MOSFET compares favorably with the 2N4401 in the maximum permissible voltage, with a 60 V rating. However, the 2N7000 is rated at 200 mA maximum continuous current and 500 mA maximum pulsed current with a total device maximum dissipation of 400 mW. Let's look at the areas of difference between the MOSFET and NPN bipolar transistor.

When saturated, a MOSFET acts like a low value resistor between the drain and source, referred to $R_{DS(ON)}$, with the corresponding voltage between the drain and source determined by the product of the drain current I_D and $R_{DS(ON)}$. Recall that in the 2N4401, the corresponding voltage $V_{CE(SAT)}$ is approximately a constant value over a wide current range.

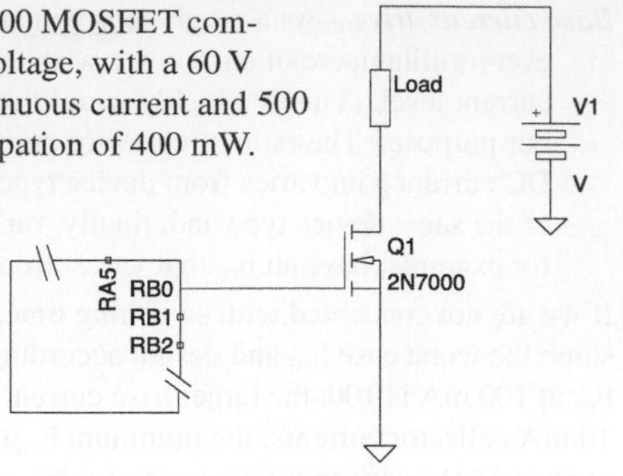

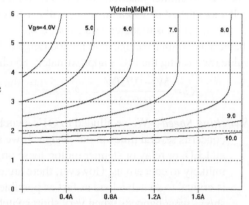

Figure 3-16: 2N7000 NPN Low Side Switch.

The relationship between $R_{DS(ON)}$ and the gate voltage is, as illustrated in Figure 3-17, complex. The point to be taken away from Figure 3-17 is that since we can drive Q1's gate only to +5 volts with a high on an output pin, we exit the saturation region with only modest drain current.

Let's run through the same 120 mA sink design we did for the 2N4401. We've already determined that the 2N7000 meets our open circuit voltage requirements and that 120 mA is less than the maximum permissible continuous drain current. With a gate drive of +5V and 120 mA drain current, Figure 3-17 shows $R_{DS(ON)}$ will be about 3.2 ohms. Since V_{DS} is the IR drop across $R_{DS(ON)}$, we may calculate it as 0.120 A × 3.2 ohms, or 0.38 volts, very similar to our 2N4401 NPN bipolar transistor design.

Figure 3-17: 2N7000 predicted on-resistance variation with gate voltage and drain current.

The power dissipated in Q1 equals $I_D \times V_{DS}$, or 0.38 volts × 0.120 mA or 46 mW, almost identical in value with the 48 mW we found for the 2N4401 bipolar switch and well within the device ratings. The 2N7000's thermal resistance junction to ambient $R_{\theta JA}$ is 312.5°C/watt and the maximum junction operating temperature is +150°C. The temperature rise at the case will thus be 312.5°C/watt × 0.046 watt, or 14.4°C. Assuming the ambient air temperature is 120°F (49°C), the maximum junction temperature will thus be 48°C + 14.4°C = 62.4°C, all quite acceptable values.

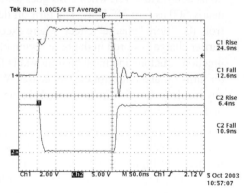

If we were to repeat this series of calculations for, say a 400 mA load, we will find $R_{DS(ON)}$ is 3.6 ohms, V_{DS} is 1.44 V and Q1's dissipation is 576 mW, well over the maximum permissible value for a 2N7000. The problem is that 5 V is inadequate gate voltage to fully turn the MOSFET at 400 mA.

Figure 3-18: 2N7000 driven by PIC turn-on/turn-off speed Ch1: PIC output; Ch2: 2N7000 drain.

When looking at nanosecond switching with a 2N4401, we found significant turn-off problems due to stored charge. As Figure 3-18 shows, both the turn on and turn off times for a 2N7000 are quite respectable. But, a close examination of the leading edge of the PIC output foreshadows the main difficulty of driving MOSFETs, gate charge. The small plateau or kink in the rise time of the PIC output reflects the fact that the gate of a MOSFET behaves like a nonlinear capacitive load to its driving circuitry. Accordingly, the switching time performance is defined by the ability of the driving PIC to change the gate voltage. The 2N7000's gate, assuming a 5 V gate drive and 120 mA load, looks approximately like a 200 pF capacitor. We will examine MOSFET gate driving problems when we look at a higher power device, the IRF510. For now, we simply note that a PIC is capable of directly driving a 2N7000 to quite respectable switching speeds.

High Power Bipolar Low Side Switching

Both the 2N4401 and 2N7000 are low power devices, good for continuous currents of a few tenths of an ampere at most. Suppose we wish to switch an ampere or two with a bipolar transistor.

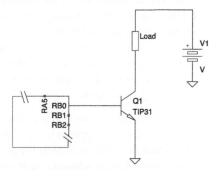

In most instances, we will not be able to build a high power switch by simply replacing the 2N4401 with a high power device, as the 25 mA or so maximum current output of a high PIC pin isn't enough base drive to reliably saturate a simple bipolar power transistor. Consider Figure 3-19. The TIP31's maximum current rating is 3.0 A, but at this level the minimum guaranteed h_{FE} is only 10, so 300 mA base drive would be required for saturation. Even at 1.5 amperes collector current the PIC is short of achieving full saturation with the maximum possible PIC output current as its base drive (This, of course, would require h_{FE} to be at least 60.). Figure 3-20 shows the V_{CE} is 920 mV, instead of the expected 300 mV shown in the TIP31's data sheet for 1.5A collector current.

Figure 3-19: High current switching with TIP31.

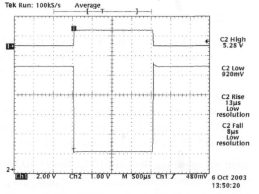

A Darlington transistor solves our base drive problem. As reflected in Figure 3-21, a Darlington transistor uses one transistor as a emitter follower current booster to drive the base of the second transistor. The Darlington configuration may use two separate devices or, as in the case of the TIP120, the driver transistor may be integrated onto the same die as the power transistor. The composite h_{FE} of the Darlington pair is approximately the product of the h_{FE} of the two transistors. Since the driver transistor doesn't handle high currents, its h_{FE} may be very large, thus ensuring a composite h_{FE} in the thousands. For a collector current of 1A, the TIP120's h_{FE} is approximately 3300, permitting collector saturation with a base current of as little as 300 µA.

Figure 3-20: TIP31 at 1.5A IC Ch1: Base Ch2: collector.

So far, so good. The TIP120 is rated at 60 V V_{CE}, 5A maximum collector current I_C and 65 watts dissipation, with an appropriate heat sink. The price to be paid for the extra h_{FE}, though, is found in the saturation voltage, $V_{CE(SAT)}$. For an I_C of 3A, $V_{CE(SAT)}$ is 2.0 V while at 5A, $V_{CE(SAT)}$ is 4.0 V. A quick glance in Figure 3-21 reveals why $V_{CE(SAT)}$ is so poor. The driver transistor relies upon $V_{CE(SAT)}$ as its source of current to supply base drive to the output transistor.

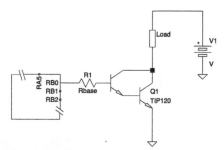

Figure 3-21: Switching with TIP120 Darlington.

Recall that the base of the output transistor's base must be at about 0.7V above its emitter before base current flows. And, the driver transistor itself introduces another 0.4V or so minimum voltage drop even if it is saturated. Hence, if $V_{CE(SAT)}$ drops below 1.1V, the driver transistor no longer can supply base current to the output stage.

Another area we expect to see a performance problem with the Darlington is turn off time as power transistors are slower than the small switching devices we have looked at so far. Figure 3-22 confirms our pessimism, as the TIP120's turn off time is nearly two microseconds. However, we should remember that for many applications, 2 µs turn off time is more than adequate.

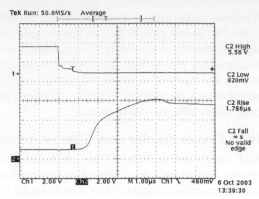

Figure 3-22: TIP120 turn off time Ch1: PIC output; Ch2: TIP120 collector.

Calculating the base series resistor Rbase, and the other parameters follow the approach used for the 2N4401 and won't be repeated. However, attention must be paid to proper heat sink selection for power devices. Although the TIP120 is rated at 65 watts dissipation, that value presumes an adequate heat sink. Absent a heatsink, its rating is only 2 watts. Operating without a heatsink, in continuous operation, a TIP120 should not sink more than 1A or so—and even that is pushing it.

High Power MOSFET Low Side Switching

IRF510 Switch

Just as there are high power relatives of the 2N4401, the 2N7000 has many big brothers as well. We'll quickly examine one of the original power MOSFETs but then turn our attention to a new power MOSFET device that solves many of the shortfalls of earlier devices.

The IRF510 is one of the earliest inexpensive power MOSFETs and is still in production. It's rated at 100V V_{DS}, maximum I_D of 5.6A and a power dissipation of 43 watts, assuming proper heat sinking. It is supplied in a TO220 package. $R_{DS(ON)}$ is under 1 ohm at room temperature. (Modern devices are in the tens of milliohm range, but we'll stick with the IRF510 to illustrate the point on gate voltage concerns.)

When we substitute the IRF510 for the 2N7000, as shown in Figure 3-23 again we are limited to 5 V gate drive. If we consult the $R_{DS(ON)}$ versus gate voltage and current plot of Figure 3-24, we see that for 5 V V_{GS} and

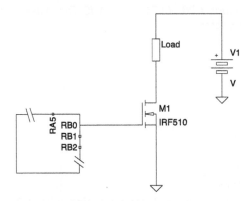

Figure 3-23: Low side switching with IRF510.

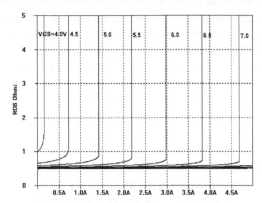

Figure 3-24: IRF510 predicted on-resistance variation with gate voltage and drain current.

1A I_D, we expect $R_{DS(ON)}$ to be about 0.5 ohm. But, suppose we wish to sink 3.0A of current representing, say a 12 volt supply and a 4 ohm load, well within the IRF510's ratings. As Figure 3-24 shows, the minimum V_{GS} to obtain saturation at 3A exceeds 5 V. If all we have to drive the gate is the +5 V high from our PIC, $R_{DS(ON)}$ will be many ohms indicating the IRF510 is out of saturation and operating in the linear region. In fact, under these conditions with V_{GS} = 5 V we will find V_{DS} approximately 6.2 V and I_D 1.4A. The IRF510 will dissipate 8.6 watts and its R_{DS} is 4.4 ohms. If we expected the IRF510 to act as a saturated switch, with an $R_{DS(ON)}$ of a few tenths of an ohm, we will be in for a surprise. And, if our design didn't use a heatsink because it wasn't necessary with a low $R_{DS(ON)}$, we'll have an even greater surprise as the IRF510 destroys itself from overheating.

Finally, rounding out the difficulties in driving an IRF510 directly from a PIC, we see the expected turn on problem. Figure 3-25 shows nearly 750 µs turn on delay.

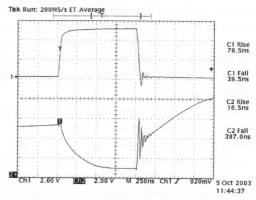

Figure 3-25: PIC Output; Ch2: IRF510 Drain.

IPS021 Switch

Many suppliers have addressed the shortcomings seen when we examined the IRF510. We'll look at one particular device, International Rectifier's IPS021 intelligent power MOSFET switch which is rated at 50 V V_{DS} and has several interesting features:

- Logic level input with built-in voltage multiplier and level converter to ensure saturation of the switching MOSFET;
- $R_{DS(ON)}$ 0.125 ohms or less at room temperature and 5 V input;
- Over-temperature protected with automatic shutdown;
- Over-current protected, at 5.5 A nominal;
- Built-in snubbing protection for inductive loads.

The IPS021 is supplied in a TO-220 package and has the same pin connection as the IRF510. In almost all cases it can be directly substituted for an IRF510 with little or no change in circuit design or physical layout. As shown in Figure 3-26, is connected in the same fashion. International Rectifier recommends a series resistor of 500 ohms to 5K ohms between the driving pin and the input pin of the IPS021, although it may not be necessary in all cases.

What's not to like about the IPS021? The main issue is switching speed, with turn-on and turn-off speeds in the several microsecond range. For many purposes, where a few microseconds of turn-on or turn-off time is immaterial, the IPS021 is a good choice. Figure 3-27 shows excellent performance in switching 1.5A in the circuit of Figure 3-26. The series resistance of 240 milliohms computed from Figure 3-27 includes wiring and plugboard components as well as the IPS021's $R_{DS(ON)}$.

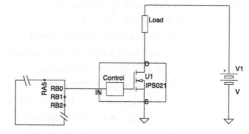

Figure 3-26: Low side switching with IPS021 intelligent power switch.

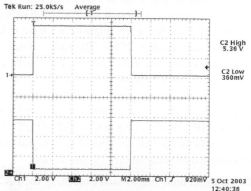

Figure 3-27: IPS021 Switching 1.5A Ch1: PIC Output; Ch2: IPS021 Drain.

High Side Switching

Small PNP Switch

Figure 3-28 depicts a simple high side switch. When RB0 is low, Q1 is forward biased into conduction and current flows through the load. Let's work through a few design concerns with this simple circuit. We will assume the load is a 47 ohm resistor and V is 2 volts. Hence, the current being switched is approximately 45 mA.

We calculate the base resistor and power dissipation exactly as for a low side switch, but we must be careful to select the correct voltage sources. The 2N4403 h_{FE} at $-2V$ V_{CE} and -150 mA I_C is specified at a minimum of 100. We will wish the bias current to be at least 450 µA. (In a PNP transistor, the collector is negative with respect to the emitter; hence the sign of the voltages and currents are reversed from the NPN case.) When conducting, the base bias source (RB0) is at 0 volts and the emitter is at V volts (2.0 V in our design example) positive. Hence in our example the voltage to drive the base current is 2.0 V. We will use the standard 700 mV assumption for Q1's base-emitter junction voltage drop and we'll ignore RB0's 26 ohm equivalent output resistance when low. Hence, the net voltage across Rbase is -2.0 V $+ 0.7$ V or -1.3 V. To obtain 450 µA current flow, Rbase should be 1.3 V/450 $\times 10^{-6}$ or 2.9 Kohm. If we wish to ensure saturation under varying temperatures and component tolerances, we will increase the base current to, say 750 µA (Rbase calculated as 1.7 Kohm) and use the nearest standard 5% resistor value, 1.8 Kohm. Figure 3-29 shows our design in operation.

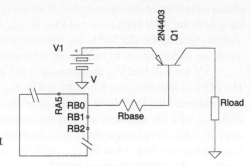

Figure 3-28: 2N4403PNP high side switch.

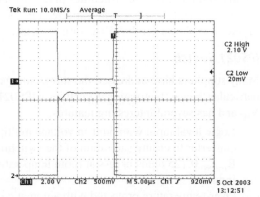

Figure 3-29: 2N4403 PNP high side switch Ch1: PIC Output; Ch2: 2N4403 collector.

In low-side NPN transistor switching design, the voltage being switched was immaterial, as long as the transistor was within its ratings. When RB0 is at ground, the NPN transistor is reversed biased and hence cut off; both the emitter and collector are at the same potential, ground so there is no voltage difference between the base and emitter. Looking at the PNP high side circuit when RB0 is high and we wish Q1 to be cut off shows a different story, however. Suppose we wish to high side switch 12 V and RB0 is high, at 5 V. The end of Rbase connected to Q1's base is one base-emitter diode drop lower at +11.3 V while the end connected to RB0 is at +5 V. Hence current will flow out Q1's base-emitter junction; Q1 becomes forward biased and will conduct. This is the central problem in high side switching with a PNP transistor or a positive MOSFET (PMOS); we must have available control voltages at least equal to the voltage to be switched in order to turn off the device.

To control a high side switch exceeding the PIC's V_{DD}, we may modify the circuit of Figure 3-28 by adding an 2N4401 NPN transistor to control the base of Q1, as shown in Figure 3-30.

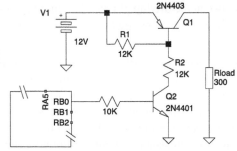

Figure 3-30: Use both PNP and NPN transistors to implement high side switch.

When RB0 is high, Q2 is forward biased and its collector is essentially at ground ($V_{CE(SAT)}$ is around 400 mV). Q1 is then biased into conduction by current flowing through R2 and Q2. When RB0 is low, Q2 is cut off and may be regarded as an open circuit. Hence, Q1's base is pulled to 12 V by R1, making V_{CE} zero, cutting Q1 off. The values given in Figure 3-30 provide –0.9 mA base drive, more than adequate to saturate Q1 for the load current. Note that adding the 2N4401 driver inverts the control sense compared with the circuit of Figure 3-28.

High Power High Side Switching

Rather than duplicate the low side switching discussion but with a high power PNP transistors, or PMOS devices, we will immediately jump to an integrated high side MOSFET switch, International Rectifier's IPS511. The IPS511 is a member of IR's intelligent power switch family, but with some features not included in the IPS021 we earlier examined. (If you wish to experiment with a high side version of the earlier low side switching designs, a TIP125 is the PNP complement of the TIP120 Darlington and a TIP32 is the PNP complement of the TIP31. Finally, a BS250P is a good p-channel analog of the 2N7000 and an IRF9510 is the p-channel complement of the IRF510.)

If you construct high side switches with p-channel devices, you will soon discover they exhibit limited operating range when the gate voltage swing is limited to the 0…5 V available from a direct connection to a PIC's output pin. To reliably turn off an IRF9510 with a 5 V gate swing, for example, requires it to be switching less than approximately 8 volts, while to ensure the device is saturated requires it to be switching at least 4.5 volts. In the high side switching configuration, V_{GS} equals the voltage being switched minus either 0 V (PIC at low) or 5 V (PIC at high). Since V_{GS} must be at least –4.5 V to to bring $R_{DS(ON)}$ to reasonable levels (turn-on; PIC output is low so V_G is 0 V) the minimum voltage to be switched (V_S) must be 4.5 V. Conversely, to ensure turn off, V_{GS} must be at least –3 V when the PIC output is high. When high, V_G is +5 V, so V_S can't exceed +8 V. Hence, an IRF9510 high side switch with its gate driven directly by a PIC's output pin is of limited practical use. An auxiliary n-channel gate control device will be necessary in most cases, such as the 2N7000 shown in Figure 3-31. To avoid exceeding the IRF9510's V_{GS} limit, the maximum voltage being switched must not exceed 20 V.

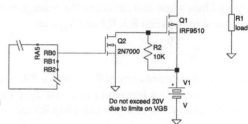

Figure 3-31: Suggested 2N7000/IRF9510 high side switch.

The IPS511 is a much cleaner alternative and offers many useful features, including:

- Logic level input with built-in voltage multiplier and level converter to ensure saturation of the switching MOSFET;
- $R_{DS(ON)}$ 0.135 ohms or less at room temperature and 5 V input;
- Over-temperature protected with automatic shutdown;
- Over-current protected, at 5.0A nominal;
- Built-in snubbing protection for inductive loads.
- Status feedback reporting of normal operation, open load, over current and over temperature.

The IPS511 is available in a 5-pin TO220 package and is rated at 50 V V_{DS}. Figure 3-32 shows a typical connection. As with IR's IPS021 low side intelligent switch, the IPS511 provides crisp, low drop switching as seen in Figure 3-33. Rise and fall times for the IPS511 are well under 100 µs.

Unlike the IPS021, the IPS511 has a status pin which, when read in conjunction with the In and Out pins on the IPS511, provides useful information. In Figure 3-32, the output voltage is sensed through a voltage

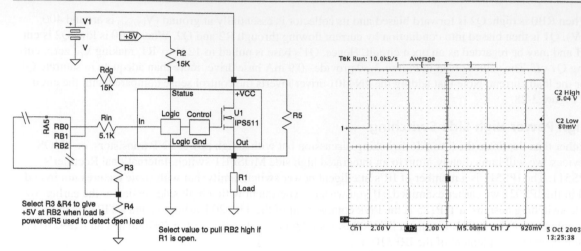

Figure 3-32: High side switching with IPS511 intelligent power switch.

Figure 3-33: IPS511 switching 50mA Ch1: PIC output; Ch2: IPS511 drain.

divider consisting of R3 and R4, assuming, as is likely the case, the IPS511 is switching greater than 5V. Applying Ohm's law and summing the voltage drops, if we wish the voltage at RB2 to be 5V, we can state the relationship between R3, R4 and V_{OUT} as:

$$R_3 = \frac{V_{OUT}R_4}{5} - R_4$$

Since no significant current flows into a PIC input pin, we may select R4 as some convenient value, say 10K. V_{OUT} will be only a few tenths of a volt different from V1, so we may use V1 instead of V_{OUT} in our calculation. Assume V1 is 12V. We calculate R3 as:

$$R_3 = \frac{12 \times 10,000}{5} - 10,000 = 14,000$$

The nearest standard 5% value is 15K.

If we wish to detect an open load condition, we must also install the pull-up resistor R5. R5's value depends, in part, upon the load resistance and the acceptability of the resulting parasitic current through the load when the IPS511 is off as well as the values of R3 and R4. Let's assume Rload is 12 ohms. If we set R5 to 1200 ohms, 100 times Rload, the parasitic current will be not more than 1% of the normal on current and the voltage across the load will be less than 1% of V1. For many purposes, this will be more than acceptable. If the load should become an open circuit, then when the IPS511 is off the voltage at RB2 will then be derived through a voltage divider where the top resistor consists of R3 in series with R5. If we can keep R5 to approximately 10% of R3, we will still maintain a good high condition on RB2. R3 is 15K, so as long as R5 is at least 1500 ohms, we have met this objective. Hence, we may pick R5 as 1300 ohms, a standard 5% value between our minimum desired value of 1200 ohms and our upper objective of 1500 ohms. We can quickly check the voltage at RB2 under the four possible conditions:

Voltage at RB2 Under Conditions of Load Status

In State	Load	RB2 Voltage	RB2 Status
H	Normal	4.8	H
H	Open	4.8	H
L	Normal	0.1	L
L	Open	4.6	H

As we will learn later, when an input, the maximum voltage level a PIC is guaranteed to be read as a low is 0.8 V, while the minimum voltage that will be read as a high is 2.0 V on a PortB pin, assuming V_{DD} is in the range 4.5 V...V_{DD}...5.5 V. Hence, RB2 will be read as low only under the condition at the third line in the table.

To determine the status of the IPS511 we would implement the following pseudo-code:

```
;Possible routine to read status of IPS511
;Assumes connection to PIC as at Fig 3-32
;
Status   var      PortB.Bit1
Load     var      PortB.Bit2

Input B1
Input B2
;
Main
        High RB0        ;turn on the IPS511 and apply power to load
        Pause 2         ;Delay to ensure IPS511 is fully on

        ;now check the status when the IPS511 should be on
        If (Status=1) AND (Load=1) Then GoSub NormalOn
        If (Status=0) AND (Load=0)Then GoSub Overload

        Low RB0         ;turn the IPS511 off
        Pause 2         ;now check the status when off
        If (Status=0) AND (Load=0) Then GoSub NormalOff
        If (Status=1) AND (Load=1) Then GoSub OpenLoad


NormalOn ;Subroutine if all is OK at the load when on
;-------
        Code to be executed if result is OK
Return

NormalOff ;Subroutine if all is OK at load when off
;--------
        Code to be executed if result is OK
Return

OpenLoad ;Subroutine to be executed if load is open
;-------
        Code to be executed if the load is open
Return

OverLoad ;Subroutine to be executed if excessive current
;----------     or over temperature
        Code to be executed for over current condition
        Or over temperature condition. Determine the difference
        Between the two by checking for cycling or steady state
        Low on Load
        If Load is cycling—problem is over temperature
Return
```

Separating certain faults, such as over temperature and over current may require repeated polling of the status and output pins to determine whether the fault periodically clears itself (as the device cools down and the thermal trip resets) or remains static. Additionally, if the onset of current limiting must be detected it may be necessary to alter R3, R4 and R5 to cause abnormally low, but not zero, voltage across the load to read as a low on RB2.

Isolated Switching

Although we will discuss a variety of devices under the isolated switching category, of course these may also be used for low side or high side switching.

Relay Switching

Relays pre-date electronics, as they were developed to extend the range of manual Morse telegraph systems in the mid 1800s. Nonetheless, we should not quickly discard the relay solution to switching. Relays are available in a rich variety of contact configurations, contact material, power rating, voltage rating and coil rating.

Good things about relays	Not so good things about relays
• *Resistant to damage from overloads and polarity reversal*	• *Limited life, although many relays are rated for tens of millions of operations*
• *Excellent isolation between switched load and controlling circuitry*	• *Noise*
• *Wide range of ratings*	• *Speed of operation and release, usually in the millisecond range.*
• *Can switch AC, DC, video, RF, low level audio, etc. by proper selection of device.*	• *Size may be an issue*
	• *Contact bounce*
	• *Requires power to hold relay operated, unless it is a latching relay.*

We'll look at three relays:

Make/Model	Coil Rating	Contact Configuration	Contact Rating	Comments
Standex JG102-12-1	*12V / 24mA*	*SPST (1A)*	*48V/1A*	*Very high speed reed relay*
Omron G5V-2-H1	*12V/12.5mA*	*DPDT (2C)*	*125VAC/0.5A* *30VDC/2A*	*Low signal relay with bifurcated contacts; ultra sensitive*
Omron G2RL-24	*12V/33.3mA*	*DPDT (2C)*	*250VAC/8A* *30VDC/8A*	*General-purpose power relay*

To test the contact closure and operate/release time, we'll use the circuit of Figure 3-34. Figure 3-34 should be familiar; a 2N7000 low side switch drives the relay coil. As we earlier determined, the current and voltage required to operate the relays under test is well within the limits for a 2N7000. To sense contact closure and bounce, we use a 5 V source in series with a 56 ohm resistor to pass approximately 90 mA through the contacts. In order to make measurements, V2's negative terminal is connected to ground; but in order to emphasize the ability of relays to switch isolated circuits, Figure 3-34 shows the load in its most general "floating" form.

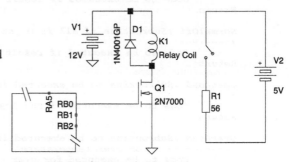

Figure 3-34: Relay test circuit.

Since a relay coil is an inductive load, we must use a snubbing diode to avoid damaging the 2N7000. To illustrate the voltage spike even a small inductive load generates, compare Figures 3-35 and 3-36. Without a snubbing diode, the inductive spike exceeds 76 V, at which point the 2N7000 breaks down. Adding a 1N4001 snubbing diode reduces the spike to 12.7 V—that is, 0.7 V above the 12 V relay supply voltage.

Finally, although the operating and release times for these relays is long compared with pure electronic switches, for many applications a few milliseconds delay between PIC output and relay pull-in is of little consequence.

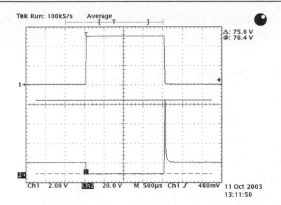

Figure 3-35: JG102-12-1 Relay Without Snubber Ch1: PIC Output; Ch2: 2N7000 Drain.

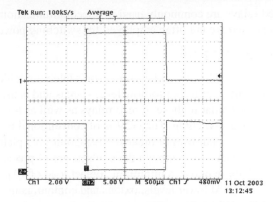

Figure 3-36: JG102-12-1 Relay with Snubber Ch1: PIC Output; Ch2: 2N7000 Drain.

Standex JG102-12-1

Reed relays are well known for high speed operation, and the JG102-12-1 (since replaced by model JG100-12-1) meets our expectations. Figures 3-37, 3-38 and 3-39 show operating times for this device. The JG102-12-1 relay turns on in 250 µs, but an additional 150 µs is necessary for contact bounce to cease. Turn off time is also approximately 250 µs. Since the turn-on and turn-off times are approximately equal, the overall relay high time is very close to the PIC control pin high time, although delayed by approximately 400 µs. (The terms "operate time" or "pull-in time" and "release time" are usually used when discussing relay speeds, instead of turn-on and turn-off. However, to illustrate the commonality with transistor switching, we'll use the turn-on and turn-off terminology as well.)

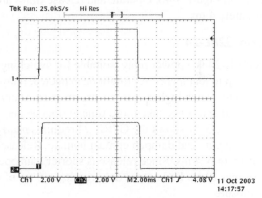

Figure 3-37: JG102-12-1 Reed relay Ch1: PIC output; Ch2: relay contact.

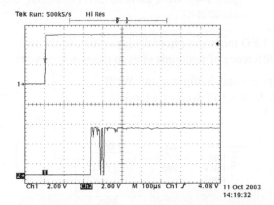

Figure 3-38: JG102-12-1 Reed relay turn-on delay and contact bounce Ch1: PIC output; Ch2: relay contact.

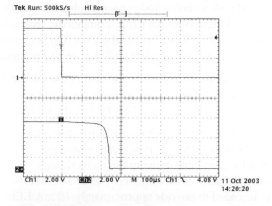

Figure 3-39: JG102-12-1 Reed relay turn-off delay Ch1: PIC output; Ch2: relay contact.

Reed relays are commonly used in telecommunications equipment to switch voice, data and high frequency signals and are not often employed for switching power circuits.

Omron G5V-2-H1

The G5V-2-H1 is a member of Omron's telecommunications family, optimized for low level signals and is used for purposes similar to those of reed relays as well as for low level power switching. The G2V is of conventional relay construction, but with bifurcated cross-point gold plated silver contacts. Even though the contacts are of precious metal, Omron quotes a minimum contact load of 10 μA and 10 mV DC. At lower levels, oxides and contaminants may prevent reliable operation.

Although constructed with bifurcated cross-point contacts, contact bounce is still apparent with the G5V-2-H1, as reflected in Figure 3-40. Operate and release times are approximately 4 ms.

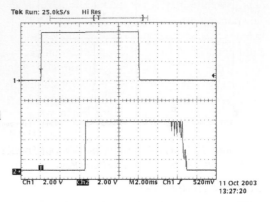

Figure 3-40: G5V-2-H1 relay Ch1: PIC output; Ch2: relay contact.

Omron G2RL-24

The G2RL-24 is a power relay, suitable for switching AC and DC up to 8 amperes. As might be expected, relays designed to switch higher currents are more substantially constructed and hence take more time to operate and release. Figure 3-41 shows the G2RL-24 requires nearly 8 mS to operate and dampen contact bounce. Release time is shorter, only 4 ms. Since release is shorter than operate, the original 10 ms PIC output only results in 6 ms of useful relay closure.

4N25 Optical Isolated NPN Switch

Optical couplers or "optoisolators" consist of an LED packaged with a photo-diode or a phototransistor. When illuminated, light from the LED saturates the receptor and it conducts. There is no electrical connection between the LED input and photo device output, so the two circuit halves are independent and may be at a potential difference of hundreds or even thousands of volts.

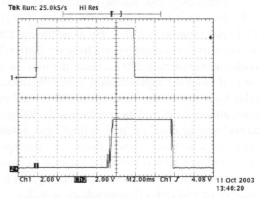

Figure 3-41: G2RL-24 relay Ch1: PIC output; Ch2: relay contact.

Optoisolators are available in a wide range of operating speeds and configurations. We'll first look at a low power optoisolator, the venerable 4N25 device, followed by a modern high power optically coupled MOSFET, the PS710A-1A-1.

Figure 3-42 shows how simple it is to connect an optoisolator to a PIC. R1 is selected to achieve the desired LED on-current using the methodology developed earlier in this chapter. The 220 ohm resistor is intended to provide approximately 10 mA LED on-current. The 4N25's output transistor is configured as a low side switch in the example.

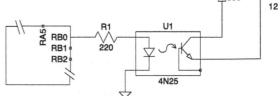

Figure 3-42: 4N25 Optoisolator connection.

With a maximum current rating of only 150 mA, the 4N25 is not intended to switch large currents so it will often be used as the first stage in a multistage switching arrangement, such as that shown in Figure 3-43. In this design, the 4N25 operates as an emitter follower. When the LED is illuminated via RB0 going high, current flows through the 4N25's output transistor and R3, taking the gate of Q1 into conduction. Current then flows through the load and Q1. When the LED is

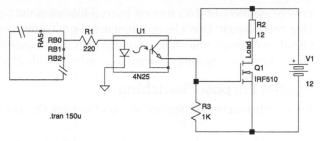

Figure 3-43: 4N25 with IRF510.

dark, the 4N25's output transistor is cut off and Q1's V_{GS} is close to zero and Q1 is cut off. Of course, any of the more modern IPS products may be substituted for the IRF510.

Low power optoisolators are also often used to isolate data or signal circuits from the PIC, and an optical coupled RS-232 circuit permits isolating a PIC from the associated computer or controlled devices.

PS710A-1A AC/DC Optically Isolated MOSFET

NEC's PS710A-1A is a high power MOSFET optoisolator. Unlike the 4N25, the PS710A is a power device, capable of switching loads up to 1.8 A at 60 V, and its series MOSFET design will switch both AC and DC. Each MOSFET has an $R_{DS(ON)}$ of 0.1 ohm and, for DC switching, may be paralleled handle 3.6A.

The PS701A-1A is connected as illustrated in Figure 3-44 for AC or DC switching. Other configurations are possible for DC switching and you should consult the data sheet for additional information.

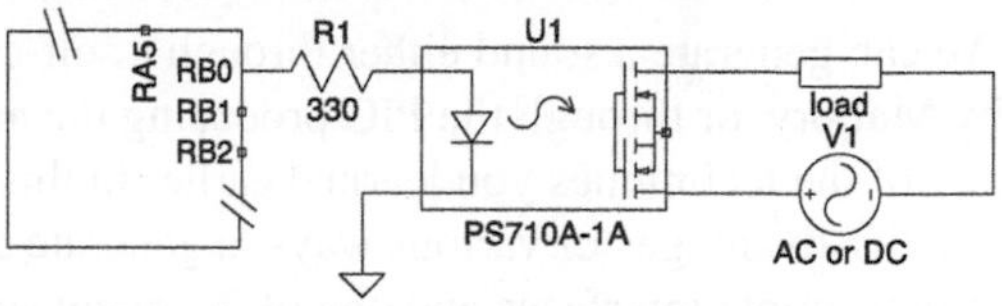

Figure 3-45 shows the results for the circuit of Figure 3-44, switching 1 A at 5 V. Figure 3-45 confirms the data sheet's 1 ms typical turn-on time and 50 µs typical turn-off time.

Figure 3-44: AC/DC isolated switching with NEC's PS710A-1A.

In switching high currents in microsecond times, undesired transients and oscillations are often seen, particularly when using a plug-in board and lab power supplies with long leads. When the design is transferred to a printed circuit board with wide power traces and integrated power distribution filtering the problem is often solved. In some cases additional filtering and bypassing will be necessary. Figure 3-46 shows an expanded

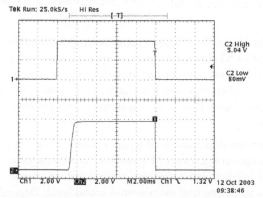

Figure 3-45: Switching with a PS107A-1A Ch1: PIC output; Ch2: load.

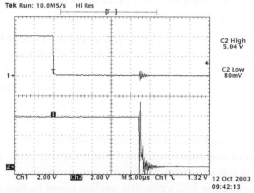

Figure 3-46: Undesired oscillation in plug board layout of PS107A-1A Ch1: PIC output; Ch2: load.

view of the PS710A-1A's turn-off interval illustrates the type of extraneous transients sometimes seen in a plug board layout. Don't be surprised to see similar unwanted oscillations in some of your layouts. Even in a breadboard layout, you can usually stop or at least reduce these extraneous signals by better attention to lead dress and grounding jumper locations, combined with additional power lead bypass capacitors

Special Purpose Switching

We've only touched the surface of ways to switch DC and AC power with a PIC. In later chapters, we will address:

- H-bridge motor drivers
- Integrated Darlington transistor arrays
- Triac switching and controlling AC power loads

And, there are special purpose integrated circuits suitable for low level signal, audio and video switching, such as the 74HC4053, what we'll save for a future book.

Fast Switching—Sound from a PIC

So far in this chapter, our emphasis has been on relatively slow switching. But, if we switch a loudspeaker off and on at an audio rate, we can produce sound, perhaps to be used as an alert tone, or a beep to confirm an action or status. (We will need fast switching to control a DC motor's speed through pulse width modulation, and to control stepper motors, both topics dealt with in later chapters.)

We can generate a sound either through a self-contained sounder, such as the Sonalert® products introduced by Mallory, or through the PIC producing the audio signal itself. A Sonalert may be driven by a PIC using any of the techniques you learned earlier in this chapter. Later chapters explore in some detail the advantages and disadvantages of various ways to generate audio signals using MBasic. Here, however, we will just look at two simple interfaces and one of the many audio output procedures available in MBasic.

We'll assume you don't intend to produce ear splitting, high fidelity output from a PIC. Rather, you are interested in beeps and other alerting tones. In some cases, it may be possible to obtain adequate volume levels by driving the speaker directly from a PIC, as illustrated in Figure 3-47. When thinking of a speaker, low impedance designs most often come to mind, with 3.2, 4 and 8 ohm devices being common. I've generally been disappointed with the volume levels when a low impedance speaker is directly connected to a PIC. Indeed, a series resistor, R1 in Figure 3-47, of 50 ohms or so is necessary to produce useful sound output.

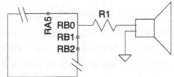

Figure 3-47: Driving a speaker directly from a PIC.

Remember, however, that a high output pin may be thought of as a 5V source in series with approximately 85 ohms. Working into, for example, a 3.2 ohm speaker with a 50 ohm series resistor, approximately 99.8% of the theoretical maximum output power of the pin will be lost and not transferred to the speaker. This still may produce an acceptable volume level. If the speaker has higher impedance, say at least 32 ohms, a much greater proportion of the available power will produce useful sound. If necessary, a simple series resistor, shown as R1 in Figure 3-47, can serve as a volume control.

If you've examined a loudspeaker you know the typical construction consists of a paper cone that moves in or out in response to current through the voice coil. Our simple connection of Figure 3-47 moves the cone only in one direction, either in or out, depending on which speaker connection you ground and which you connect to the PIC's pin. The unidirectional motion throws away one half the potential sound level. Depending on your desired sound level and speaker rating, this may or may not be important. Figure 3-48 shows we are able to develop a peak current of 27 mA through a 3.2 ohm speaker. This particular speaker yielded a weak sound with 27 mA current. We may calculate the power delivered to the speaker by recalling that the RMS power a square wave is equal to the one-half the peak power. (The RMS of the on period is equal to the peak; but since half the cycle is off, the RMS reduces by one half.) Hence, the RMS power delivered to the speaker is approximately 1.2 mW. (This is based upon the speaker's nominal 3.2 ohm impedance. Measurements of the particular speaker I tested showed its true impedance at 1000 Hz is 3.09 ohms, representing 2.95 ohms resistance in series with 149 μH inductance.)

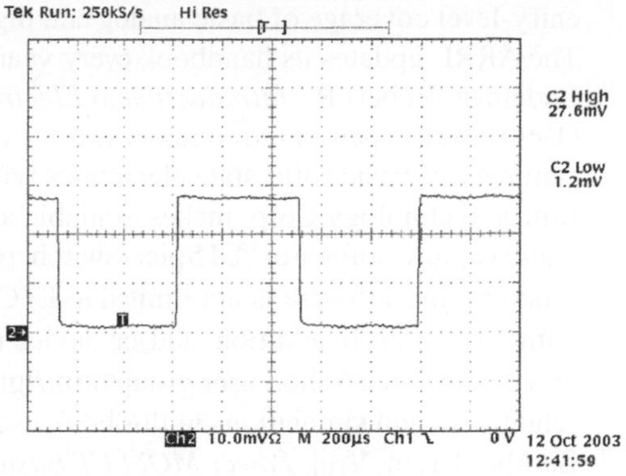

Let's look at a higher power driver for a low impedance speaker. Since we are not overly concerned with the sound quality—the PIC sound procedure we use outputs a square wave, after all—we will use a 2N4401 emitter follower to drive the 3.2 ohm speaker, using the circuit shown in Figure 3-49. And, to permit the speaker's voice coil to have both in and out excursions, we use C1 to block the DC component.

Figure 3-50 shows the resulting current thorough the speaker. The RMS power delivered to the speaker is now approximately 45 mW, yielding nearly 16 dB more sound output, a very noticeable improvement over the direct drive connection.

Program 3-3 uses MBasic's **sound** procedure to output a 1000 Hz square wave for 1,000 milliseconds on RB0. The tone output is repeated endlessly through the **GoTo Main** loop.

Program 3-3

```
;Program 3-03
Main
        ;burst of 1000 Hz for 1 second
            w/ endless loop
        Sound B0,[1000000\1000]
GoTo Main

    End
```

References

[3-1] Horowitz, Paul and Hill, Winfield, *The Art of Electronics, 2nd. Ed.*, (1989). If you have only one book on electronics in your library, this should be it. A long-awaited 3rd edition is rumored to be in the works, but that shouldn't discourage you from purchasing the 2nd edition.

[3-2] American Radio Relay League, *The ARRL Handbook for Radio Communications 2003 ed.*, American Radio Relay League (2003). Although aimed at radio amateurs, the ARRL Handbook provides good

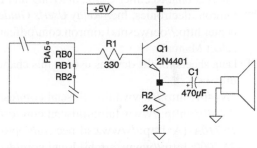

Figure 3-48: Direct drive of 3.2 ohm speaker with PIC and 56 ohm series resistor Ch2: Speaker current (mA).

Figure 3-49: 2N4401 emitter follower speaker driver.

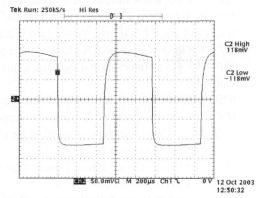

Figure 3-50 : 2N4401 Follower Drive of 3.2 Ohm Speaker with PIC and 56 Ohm Series Resistor Ch2: Speaker Current (mA).

entry-level coverage of basic analog and digital electronics, test equipment and construction practices. The ARRL updates its handbook every year, so purchase the most recent version available.

[3-3] Ludeman, Robert R., *Introduction to Electronic Devices and Circuits*, Saunders College Publishing (1990). Written as an introductory text for community college electronic technician students, it's a good summary of basic solid-state electronics without requiring advanced mathematics.

[3-4] Linear Technology Corp. makes available a free Microsoft Windows-based SPICE simulator and schematic capture software, "LTSpice/SwitcherCAD III." Although aimed as a design tool supporting LTC's products, the software is not limited to LTC devices. It may be downloaded at http://www.linear-tech.com/software/. In addition, add-on device libraries and explanatory material for LTSpice are available in the associated Yahoo user group http://groups.yahoo.com/group/LTspice/ in the files folder. The schematics and simulations in this book use LTSpice.

[3-5] Barkhordarian, Vrej, *Power MOSFET Basics,* International Rectifier Corp. Technical Note (undated).

[3-6] International Rectifier Corp., *The Do's and Don'ts of Using MOS-Gated Transistors*, AN-936 (v. Int). (Undated)

[3-7] International Rectifier Corp., *Current Ratings of Power Semiconductors*, AN-949 (v. Int), (Undated).

[3-8] International Rectifier Corp., *Selecting and Designing in The Right Schottky*, AN-968, (Undated)

[3-9] Omron Electronics, Inc., *Relay User's Guide* (1990). Available for free download at Omron's reference center http://oeiwcsnts1.omron.com/pdfcatal.nsf. From this page, select Relays. From the relays page select Manual.

[3-10] Data sheets for the devices used in this chapter are available for downloading at the following URL addresses:

2N4401: http://www.fairchildsemi.com/ds/2N/2N4401.pdf

2N4403: http://www.fairchildsemi.com/ds/2N/2N4403.pdf

PS710A-1A: http://www.csd-nec.com/opto/english/pdf/PN10268EJ01V0DS.pdf

2N7000: http://www.fairchildsemi.com/ds/2N/2N7000.pdf

4N25: http://www.fairchildsemi.com/ds/4N/4N25.pdf

TIP31: http://www.fairchildsemi.com/ds/TI/TIP31.pdf

TIP120: http://www.fairchildsemi.com/ds/TI/TIP120.pdf

MV5491A Dual LED: http://www.fairchildsemi.com/ds/MV/MV5094A.pdf

IRF510: Go to International Rectifier's home page http://www.irf.com/ and enter IRF510 in the search box.

IRF9510: Go to International Rectifier's home page http://www.irf.com/ and enter IRF9510 in the search box.

IPS021: Go to International Rectifier's home page http://www.irf.com/ and enter IPS021 in the search box.

IPS511: Go to International Rectifier's home page http://www.irf.com/ and enter IPS511 in the search box.

G5V Relay: Go to Omron's home page for US relay products http://oeiweb.omron.com/ and enter G5V-2-H1 in the search box.

G2RL-24 Relay: Go to Omron's home page for US relay products http://oeiweb.omron.com/ and enter G2RL-24 in the search box.

Standex JG100 Relay: http://www.standexelectronics.com/serjg.htm

The Basics – Digital Input

After Chapter 3's examination of the output mode, we'll now turn to PIC pins used as *digital input* devices. Many PICs include analog-to-digital converters and we'll cover analog inputs in Chapter 11.

Digital signal levels are either a logical low (0) or a logical high (1)—what could be simpler? As with the rest of this book, we assume V_{DD} is 5 volts and V_{SS} is 0 volts. In PIC logic, a 0 volt input corresponds to a logical low. Likewise, a 5V input is a logical high. But, suppose our input is 1.7V. Is it a logical low or is it a logical high? Does the answer to this question depend on our choice of an input pin? And, does it depend on the voltage at the input pin earlier in time? We'll find out in this chapter.

Introduction

First, let's define a few terms, as illustrated in Figure 4-1:

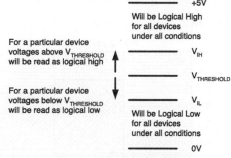

V_{IL}—The maximum voltage on an input pin that will be read as a logical low.

V_{IH}—The minimum voltage on an input pin that will be read as a logical high.

Undefined region—The undefined region the voltage level between V_{IL} and V_{IH}. Input voltages in the undefined region may be read as a low or as a high, as the PIC's input circuitry may produce one result or the other. (Obviously the voltage will be read as either a low or a high; it's just that we have no assurance which one it will be.)

Figure 4-1: Input level relationships.

Threshold voltage—The input voltage that separates a low from a high; the threshold voltage, V_T, minus a small increment is read as a low while the threshold voltage plus a small increment is read as a high. The threshold voltage differs from logic family to logic family and somewhat between different chips of the same type. The differences between V_T and V_{IL} and V_{IH} are design margins accounting for device-to-device process tolerances, temperature effects and the like.

In an ideal logic device, V_{IL} equals V_{IH} and there is no undefined region.

Microchip has chosen to build PICs with varying input designs and associated varying V_{IL} and V_{IH} values. We will not consider a few special purpose pins, such as those associated with the oscillator, in this chapter. Even so, the 16F87x series PICs have three input pin variations:

TTL level— Of the many logic families introduced in the early days of digital integrated circuits, transistor-transistor logic (TTL) was by far the most successful with TTL and its descendants still used today. Port A and Port B input pins mimic TTL logic input levels, except for RA4, which has a Schmitt trigger input. (Of course, TTL style Schmitt trigger inputs exist; but since they are not found in the PICs we consider, we won't further consider them.)

Schmitt trigger inputs—Almost all other input pins are of Schmitt trigger design. A Schmitt trigger has different transition voltages, depending on whether the input signal is changing from high to low or low to high, as illustrated in Figure 4-2.

Special Schmitt trigger inputs—Pins RC3 and RC4 are software selectable Schmitt trigger or SMBus configuration. (SMBus is a protocol developed by Intel for data exchange between integrated circuits. We will not further discuss SMBus communications in this book.) When RC3 and RC4 are used as normal, general purpose, input pins, their parameters differ slightly from those of the other Schmitt trigger input pins.

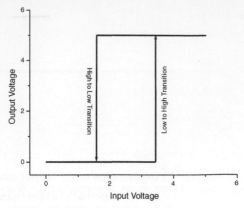

Figure 4-2: Schmitt trigger input.

Input Logic Level Specifications: 16F87x with 4.5V < V_{DD} < 5.5V		
Input Type	V_{IL} (max)	V_{IH} (min)
TTL	0.8V	2.0V
Schmitt	0.2 V_{DD}	0.8V_{DD}
RC3/RC4 Schmitt	0.3V_{DD}	0.7V_{DD}

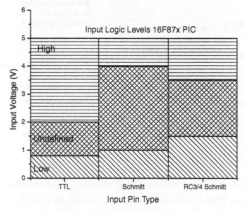

Figure 4-3: Input logic level comparison.

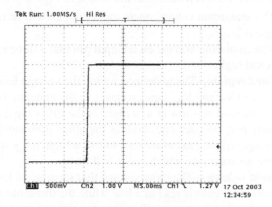

Figure 4-4: SN7400 TTL logic levels Ch1: X-axis (gate input); Ch2: Y-axis (gate output).

Figure 4-3 illustrates the differences between the three input types. To see how these input designs differ in practice, we can apply a time varying 0…5 V signal to an input pin and plot the input voltage against the output value using an oscilloscope. If we have a stand-alone logic gate, this is a straightforward exercise. Figure 4-4, for example, shows the input versus output result for a SN7400 TTL quad NAND gate. (To obtain a noninverting output, the test configuration places two NAND gates in series, and to yield the full 5 V logic high, a 2.2K ohm pull-up resistor was added to the output gate.) We see a clear, narrow transition, with $V_{TRANSITION}$ around 1.6 V, fully consistent with Microchip's V_{IL} and V_{IH} parameters for TTL mimicking input pins.

If we wish to obtain a similar plot with a 16F877, however, we must somehow determine the logical state—high or low—of the input pin corresponding to the input voltage to. The easiest way to do this is simply echo the input pin's value to an output pin. In pseudo-code the algorithm is:

```
ReadPin:
Read Input Pin - If Input Pin <> 1 Make Output Pin = 0
Read Input Pin - If Input Pin <> 0 Make Output Pin = 1
Goto ReadPin
```

It turns out that to be useful, the process of reading the input pin and making the output pin must be done more quickly than possible using MBasic, we'll add some high speed assembler into the mix. The reason for the unusual pseudo-code structure (the not equal operator) will become clear when we look at the real code.

Program 4-1

```
;Program 4-01

Input B0
Output B1

Main

ASM
{
ReadIn
        btfss    PortB,0        ;If RB0=1 then skip setting it to 0
        bcf      PortB,1        ;make RB1=0
        btfsc    PortB,0        ;If RB0=0 then skip setting it to 1
        bsf      PortB,1        ;make Rb1=1
        GoTo     ReadIn         ;Repeat the loop
}

GoTo Main
End
```

Don't worry if you don't understand the assembler portions of Program 4-1, as we will learn more about mixing assembler and MBasic later. However, the actual program tracks the pseudo-code. We first make RB0 an input and RB1 an output using MBasic. Then, the main program is an endless loop.

The **btfss PortB,0** statement reads the 0'th bit of Port B (RB0) and if it is "set," i.e., if it reads high, the immediately following statement is skipped. If it is low, the statement immediately following is executed. (The mnemonic is Bit Test File, Skip if Set, or **btfss**).

```
btfss    PortB,0        ;If RB0=1 then skip setting it to 0
bcf      PortB,1        ;make RB1=0
```

The above code reads RB0 and branches, depending on its value. If RB0 is set—that is, RB0 = 1, then the next statement (**bcf PortB,1**) is skipped. Conversely, if RB0=0, then **bcf PortB,1** is executed. The **bcf**—or bit clear file—operator clears or sets to zero a bit, in this case, RB1. These two statements, therefore, read RB0 and set RB1 to zero if RB1 is zero.

```
btfsc    PortB,0        ;If RB0=0 then skip setting it to 1
bsf      PortB,1        ;make Rb1=1
```

The next two statements perform the inverse operation. RB0 is read a second time and tested—but this time for the clear state—using the **btfsc** operation. (The **btfsc** operator works just like **btfss**, except the next instruction is skipped if the tested bit is clear.) If RB0 is high, then the operation **bsf PortB,1** is performed, thereby setting RB1.

Execution continues with the **GoTo ReadIn**, which loops back to reading RB0.

To terminate this program, power must be removed from the PIC, or another program written into its memory.

Program 4-1 isn't the fastest way to transfer an input pin level to an output pin, but we'll look at more efficient techniques later. It also uses multiple read actions, thereby creating the possibility of mishandling if the input changes value between the two read operations.

With a 20 MHz clock, this program has a 1.2 μs operating cycle. (The SN7400 gate, performing these tasks in hardware, requires less than 10 ns, 120 times faster than the PIC's software.) Hence, when we examine the input/output relationship with the same set up we used for the SN7400 gate, we will expect to see horizontal smearing, where the output lags the input by this delay. See Figure 4-5. We see $V_{THRESHOLD}$ is approximately 1.5 V, quite close to the value measured for the SN7400 true TTL gate.

Modifying Program 4-1 to accept RC0 as the input and running the same input/output sweep, we see in Figure 4-6 the hysteresis of the Schmitt trigger. The low-to-high transition occurs at approximately 3.1 V, while the high-to-low transition occurs at approximately 1.8 V. The beauty of separate high-low and low-high transition levels is noise rejection. Suppose the input signal has noise riding on it, perhaps induced from high-speed logic chips on the same board. Once a transition from one state to the other has occurred, it takes a 1.3 V noise excursion to cause a reverse transition. As we shall see later, this hysteresis adds greatly to noise rejection.

We understand Microchip's decision to use TTL mimicking inputs as the product of backward compatibility with earlier PICs and access to the huge base of TTL devices. But, why has Microchip designed the rest of its PIC inputs with Schmitt trigger inputs instead of normal CMOS inputs, such as that seen in Figure 4-7 for a CD4001BE quad NOR gate? After all, PICs are CMOS devices so it makes sense to give them standard CMOS characteristics, right?

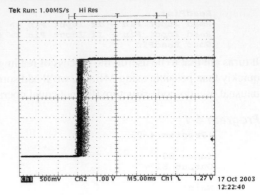

Figure 4-5: 16F876 TTL logic levels Ch1: X-axis (RB0 input); Ch2: Y-axis (RB1 output).

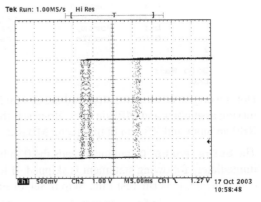

Figure 4-6: 16F876 Schmitt trigger logic levels Ch1: X-axis (RC0 input); Ch2: Y-axis (RB1 output)

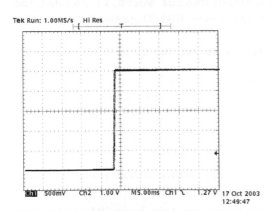

Figure 4-7: CD4001BE quad 2-input NOR CMOS logic levels Ch1: X-axis (input); Ch2: Y-axis (output).

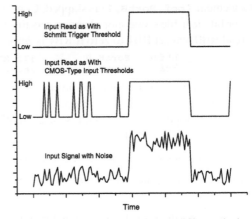

Figure 4-8: Noisy input read differently by standard CMOS and Schmitt inputs.

If Microchip though its PICs would communicate only with other integrated circuits installed on the same board, it likely would have adopted standard CMOS input levels. But, PICs often must communicate with the real world, through sensors and switches, operating in a much less benign environment where the noise rejecting properties of the Schmitt trigger come to the fore. Figure 4-8 illustrates the ability of a Schmitt trigger input to correctly read an input signal in the presence of large noise voltages.

In order to ensure that levels are correctly read, regardless of the input type, we should aim for logic 0 input levels not exceeding 0.8 V and logic high levels of at least 4.0 V. If we meet these objectives, all input port types will correctly read the input levels. If these levels cannot be achieved, it will be necessary to further investigate the design to ensure reliable data transfer.

Regardless of their type, PIC input pins represent a high input impedance to the outside world. As reflected in Figure 4-9, the only current that flows in or out of an input pin is due to leakage and does not exceed 1 μA. (In the 16F876 family, pin RA4's leakage current may be up to 5 μA.) For low impedance circuits we may safely consider an input pin as an open circuit, with zero current flow in or out of the pin. A high imped-ance input pin carries with it the

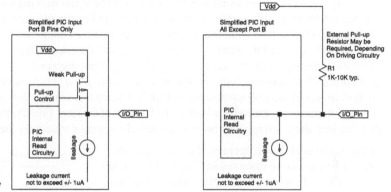

Figure 4-9: Simplified model of input pins.

possibility of static damage, as even a small static charge produces high voltages into a high impedance pin. Although the clamping diodes to V_{DD} and V_{SS} help prevent damage, it's still a good idea to follow anti-static precautions when handling PICs. It also means that input pins should be protected when exposed to outside world voltage and currents. We will briefly cover some of these real world issues in this and later chapters.

The exception to this assumption relates to the software enabled internal pull-up resistors on pins RB0… RB7. If the weak pull-up feature on Port B is enabled via MBasic's procedure **SetPullUps**, the input pins are connected to V_{DD} through an internal 20 Kohm resistor. If enabled through **the SetPullUps PU_On** or disabled through **SetPullUps PU_Off** procedure, the action applies to all PortB pins. In this case, the input pin sources 250 μA.

We've referred to pull-up resistors without explaining why and where they are used. Suppose we wish to read a switch and determine if it is open or closed. If we simply connect the switch to an input pin, as in Figure 4-10a, we cannot be assured of RB0's status when the switch is open. Certainly, if the switch is closed, RB0 will be at ground potential and will be read as a logical low. When open, however, the voltage at RB0 results from the PIC's internal random leak-age current, plus whatever stray signal that outside circuitry may induce in the connections between the switch and

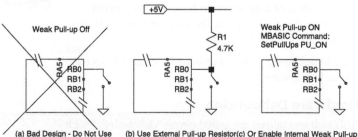

Figures 4-10a,b,c: Pull-up required to read switch.

RB0. The result may be a logical high, or low, and we are without any assurance that the switch's position will be correctly or consistently read.

If instead, we connect RB0 to V_{DD}, either through the internal pull-up source (Figure 4-10c) or through an external pull-up resistor (Figure 4-10b), when the switch is open RB0 will be "pulled up" to V_{DD} and the switch position will correctly be read as a logical high.

Switch Bounce and Sealing Current

Beyond assuring that an open switch is correctly read, a pull-up resistor provides the switch with enough current to reliably operate. In addition to the mechanical wiping action of switch contacts, a small DC current greatly assists in maintaining reliable conduction between moving switch contacts. And, if the switch is connected through wiring with mechanical splices and crimp or screw pressure terminals, the current helps clean oxidizing film from the mating conductors. From telephone terminology, we often refer to this as a "sealing" current or a "wetting" current. Hence switch circuits are often referred to as wet—carrying a current well in excess of that necessary to pull an input pin high or low—or dry—carrying only a minimal signal current. For most mechanical switches, unless we are trying to save power such as in a battery powered system, select the pull-up resistor to supply 5 to 20 mA current flow through the switch when closed. For 5 V systems, use a pull-up resistor of 1K ohm to 250 ohms. Finally, the smaller the pull-up resistor value, the less stray voltage will be induced into the wiring connecting the PIC with the switch.

When a mechanical switch operates, the same "contact bounce" phenomena we observed in Chapter 3 with relay contacts is seen. In most switches, the contact separation operation ("break") is relatively clean, but the contact closure ("make") exhibits multiple bounce events. Figure 4-11 shows nearly 1.5 ms is required to reach a steady state closed condition in the microswitch SPDT switch being tested.

Why should we be worried about switch bounce? The answer is that in many cases, we don't care. Bounce isn't a concern, for example, where a baud rate switch is read only at start-up, or where a limit switch senses the position of a part and activates a software stop sequence. In the first case, the switch isn't operated while the program executes and in the second multiple activations of stop sequence isn't a concern. Suppose, however, the switch counts the number of times an action is performed. Clearly we want one switch operation to increment the count once, not once for each of a dozen or so bounces. To prevent switch bounce from repeatedly triggering an operation, we must debounce the switch.

We can debounce a switch either with external electronics, or in software.

Hardware Debouncing

Although specialized integrated circuit "debouncers," such as Fairchild's FM809 microprocessor supervisor devices, exist we'll look at a simple circuit described in a recent issue of *EDN Magazine*, as shown in Figure 4-12.[4-1]

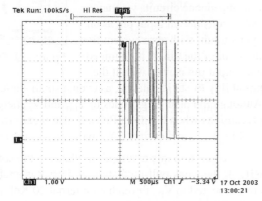

Figure 4-11: Contact bounce upon closure of microswitch.

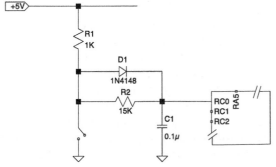

Figure 4-12: Hardware debounce.

When the switch is open, C1 charges to V_{DD} and RC0 is read as high. When the switch makes, C1 is discharged through R2 (D1 is reversed biased and may be neglected) and will be read as low when the voltage across C1 drops below high-to-low transition voltage, approximately 1.8 V. If the time constant of R2-C1 is long compared with the individual bounce intervals, the decay will be smooth and only one transition through $V_{THRESHOLD}$ will occur. But, even if spikes of several hundred millivolts occur at RC0 the output will stay low, as in order to change its read state, RC0 must see the low-to-high transition voltage of approximately 3.1 V.

When the switch is closed, the input pin is connected to ground through R2. The leakage current from the PIC input pin is rated not to exceed 1μA, so in the worst case with R2 at 15K ohm, the input pin will be at 0.015 V, well within the logical low range.

When the switch is opened after closure, C1 charges through R1 and R2 in parallel with D1. Until the voltage across C1 reaches 0.7 V from V_{DD}, C1 charges mostly through R1 and D1. Hence, the charge cycle is significantly shorter than the discharge cycle. This design assumes that the switch has bounce problems only on make and therefore little or no anti-bounce effect is required on break.

The relationship between C1, R2 and the desired debounce time T_B is given by:

$$R2C1 = -\frac{T_B}{Ln\dfrac{V_{THRESHOLD}}{V_{DD}}}$$

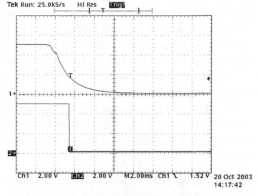

If we design for a Schmitt input where $V_{THRESHOLD}$ for a high to low transition is approximately 1.8 V, and if we make C1 0.1 μF, a convenient value, we may simplify this equation and solve for R2 in terms of T_B:

$$R2 \approx 10T_B$$

R2 is in Kohm, and T_B is in milliseconds. If the desired debounce time is 1.5 ms, R2 should be 15K ohm. Figure 4-13 shows how well this simple circuit removes the bounce from the same switch shown in Figure 4-11.

Figure 4-13: External debounce circuit operation Ch1: RC0 PIC input; Ch2: RB0 PIC output.

To study the effect of the debounce circuit, we simply make RB0 equal to RC0 and repeat the test in an endless loop.

Program 4-2

```
;Program 04-02
;Program to echo read of C0 to B0
;Variables
Temp     Var            Byte

Input   C0
Output  B0

Main
        PortB.Bit0 = PortC.Bit0
GoTo Main
End
```

The key statement in the program is **PortB.Bit0 = PortC.Bit0** where the assignment operator forces a read of RC0 and a subsequent write of the resulting value to RB0. This read/write sequence is repeated every time the loop executes. Program 4-2 runs quite a bit slower than Program 4-1, with the switch pin RB0 being read once every 68 μs, compared with the once every 1.2 μs in the assembler code.

Software Debouncing

We may also debounce the switch in software. In the simplest case, we simply pause program execution to allow the switch contacts to stabilize.

Program 4-3

```
;Program 04-03
;Variables
Temp    Var             Byte

Input   B0
Output  B1

Main
    PortB.Bit1 = PortB.Bit0
    Pause 2
GoTo Main

End
```

Figure 4-14 shows the effect of pausing 2 m after reading the input pin. The actual reading of RB0 in the statement **PortB.Bit1 = PortB.Bit0** still occurs very quickly; in 0.2 μs for a 20 MHz clocked PIC, even though the entire statement requires approximately 14 μs to execute. However, these quick reads are now spaced approximately every 2 ms apart. Hence, if the read happens to detect a high during the bounce period, the output stays high until at least the next read period. By that time, the switch will have had time to stabilize and the next read will correctly be a low. If, on the other hand, RB0 reads a low during a bounce, the output drops to low immediately. The next read, 2 ms later, will also read a low, since the switch will have by then stabilized. In either case the switch is successfully debounced.

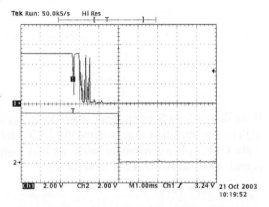

Figure 4-14: Debounce via Pause2 Ch1: RB0 PIC input; Ch2: RB1 PIC output.

MBasic also provides a built-in switch debounce procedure **Button**, invoked as:

Button pin, downstate, delay, rate, bytevariable, targetstate, address

Pin—is the address of the pin to be read and may be either a pre-defined constant, such as **B0**, or a variable set to the pin's address.

Downstate—**Button** tests for a specific input, either high or low. **Downstate** is a constant or variable (either 0 or 1) that defines which condition represents the switch being operated. If the switch operation causes a low to be applied to **pin, Downstate** should be 0; if operation causes a high, **Downstate** should be 1.

Delay—a constant or variable (0…255) that controls how long **Button** waits before starting auto-repeat. Auto-repeat means that **Button** acts as if the switch is repeatedly cycled, similar to the repeated letters obtained when you hold a key down on your computer keyboard. If **Delay** is set to 0, debounce and auto-repeat are disabled; if **Delay** is set to 255, debounce is enabled, auto-repeat is disabled. Delay is in units of the time **button** is called. For example, if **button** is used in a loop that requires 2 ms to execute, and if **Delay** is set to 10, the net delay is 20 ms.

Rate—is a constant or variable (0…255) that determines the time between auto-repeats. **Rate** is also in units of time **button** is called.

ByteVariable—is a byte variable used by **Button** for workspace. You must declare this variable.

TargetState—is a constant or variable (0 or 1) that specifies which state **pin** must be in to cause program execution to branch to **Address**.

Address—A label to which execution will jump upon **pin = TargetState**.

Program 4-4 illustrates how **Button** may be used.

Program 4-4

```
;Program 04-04
;Variables
Temp    Var             Byte

Input  B0
Output B1


Main
       High B1
       Button B0, 0, 255, 100, Temp, 1, UpButton
       GoTo Main

UpButton:
       Low B1
       Pause 10
GoTo Main

       End
```

Program 4-4 reads the status of a switch connected as shown in Figure 4-15. Program 4-4 makes an output pin, B1, track the debounced input pin B0. We first set B1 high and test to see if B0 is low via the **Button** debounce procedure. If the switch is open, B0 remains high, and program execution resumes after **Button** with the **GoTo Main** statement which loops back to setting B1 high and reading B0. If, however, the switch is closed and **Button** reads a low on B0, program execution branches to **UpButton. UpButton** is a dummy procedure that simply makes B1 low for 10 ms, after which execution resumes with setting B1 high and reading B0 through the **GoTo Main** statement. If you wish the button press to do something useful, the appropriate code would be at the procedure **UpButton**.

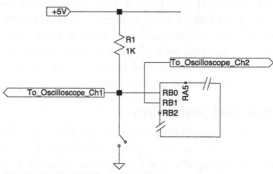

Figure 4-15: Switch connection.

Figure 4-16 shows the output of Program 4-4 where **Button** is set for automatic repeat operation with **Delay 250** and **Rate 250**. Since the program loop reading **Button** takes 140 μs to execute, it will start repeating after 250 × 140 μs, or 35 ms. It will continue to repeat every 250 × 140 μs or 35 ms. Upon activation (either initially or following each auto-repeat) program execution branches to a simple routine that drops the output low for 10 ms.

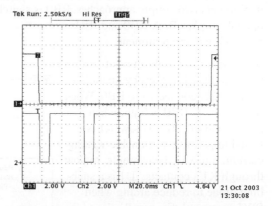

Figure 4-16: Button with Delay=250,Rate=250
Ch1: RB0 PIC Input; Ch2: RB1 PIC Output.

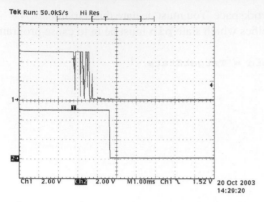

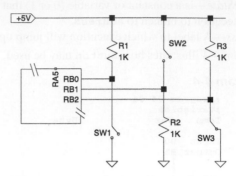

Figure 4-17: Button with delay=255, rate=100 Ch1: RB0 PIC input; Ch2: RB1 PIC output

Figure 4-18: Alternative Switching Connections.

In many cases, we need debounce only, with one operation for every button push. Figure 4-17 shows the effectiveness of **Button** at debouncing.

Figure 4-18 shows three possible switch connections. SW1 and SW2 correspond to State 0 and State 1 designs in MBasic's User's Guide button discussion. When SW1 is closed, RB0 goes low while when SW2 is closed, RB1 goes high. If SW2 is installed on the same printed circuit board as the PIC, the connection shown in Figure 4-18 is acceptable. However, if the switch is remote from the PIC, perhaps even connected by a lengthy cable, the design associated with SW2 requires unprotected +5 V to be run to SW2, thus exposing it to possible problems. If a normally closed switch is available—perhaps the NC contacts of a SPDT switch—the alternate configuration of SW3 is better. In the worst case should either side of SW3 be inadvertently grounded no damage will occur.

Isolated Switching

Just as with switching isolated loads, there are occasions when you may wish to read a switch closure that cannot have a common ground with your PIC circuit. Or, you may have a long run of cable between the switch and the PIC and wish to ensure against stray induced voltages or ground currents.

We can apply several techniques developed in Chapter 3 isolating PIC inputs from switches. Figure 4-19 implements two simple approaches. RB0 is switched through a relay, while RB1 is switched through an optical coupler. Both RB0 and RB1 need debouncing treatment.

SW1 makes or breaks current through relay K1, thereby operating the contacts connected to RB0. When SW1 is open, RB0 is high; when SW1 is closed, RB0 is low. R1 is selected to permit a reasonable wetting current, 5 mA in the example, to flow through K1's contacts. D1 is a snubbing diode to reduce arcing when SW1 is opened. For small relays operated from low voltage sources, D1 may be safely omitted.

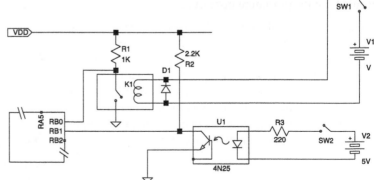

Figure 4-19: Isolating remote switches from a PIC.

SW2 makes or breaks current through the LED half of optoisolator U1. When illuminated, U1's LED saturates the phototransistor and pulls its collector—and RB1—low. When not illuminated, U1's phototransistor is off and RB1 is taken high R2. R3 should be selected to ensure the phototransistor is fully saturated when the LED is illuminated, with 15 to 20 mA being a typical current value for a 4N25 optoisolator. R2 should be selected to ensure the phototransistor's collector voltage is under 0.5 V when the LED is illuminated. For a 4N25, we can accomplish this by a collector current of 1 mA or so. The component values in Figure 4-19 result in an LED-on voltage of 0.15 V at the phototransistor's collector.

Reading a Keypad

So far, we've looked at isolated switches, with one pin for each switch. We'll look at one type of multiple switch assembly—called a rotary encoder—in Chapter 7, and a different type—the keypad—in this chapter.

Keypads, such as those found on telephones and calculators, are almost always matrix switches, as illustrated in Figure 4-20. Pressing a button establishes a connection between its column and row terminals. For example, pressing the "5" button in Figure 4-20 connects column 2 with row 2. Many inexpensive keypads, including the Velleman 16KEY model we'll use in our experiments, are constructed with a conductive elastomer design rather than physical make/break switches. These switches typically have an on resistance of 100–200 ohms, compared with the milliohm range resistance found in mechanical switches.

Let's see how we might go about reading the keyboard. Suppose we connect it as shown in Figure 4-21. Then, we sequentially make pins B4…B7 output a high and after each high, we read pins B0…B3 to see if any are high. If so, we can determine the row number and column number and from that identify which button has been pressed. This process is illustrated in Figure 4-22.

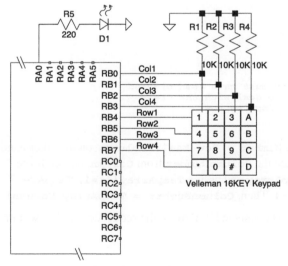

Figure 4-21: Keyboard connection to PIC.

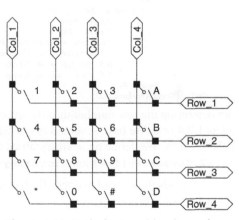

Figure 4-20: Typical 16-position keypad switch.

R1…R4 are "pull-down" resistors and ensure that input pins RB0…RB3 do not drift upwards towards a logical high from internal leakage currents or possible leakage across the keypad's open contacts. R5 limits the current through LED D1 to approximately 10 mA when RA0 is high.

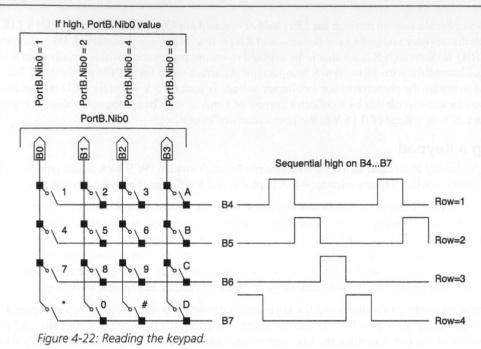

Figure 4-22: Reading the keypad.

In pseudo-code, we would:

```
ScanKeyboard
        For Row = B4 to B7
                High Row
                Pause 10
                Column = PortB.Nib0
                If Column <> 0 then key must have been pressed
                So go to a subroutine to deal with the press
                If not, then keep scanning the row
        Next Row
GoTo ScanKeyboard
```

We can determine the row number directly since `RowNumber = Row-B4+1`. `Column,` on the other hand, is read as 1,2,4 or 8. We have several ways to calculate the `ColumnNumber` from `Column`, but we'll use a simple one now. We define an auxiliary array, `TempArray` and assign `TempArray(1)=1`, `TempArray(2)=2`, `TempArray(4)=3` and `TempArray(8)=4`. Then, `ColumnNumber = TempArray(Column)`.

Program 4-5 reads the keypad and then flashes an LED connected to RA0 in the sequence <flash row number> pause <flash column number>.

Program 4-5

```
;Program to test keypad reading
;
;the keypad output pulses a LED with the row/column value. LED is
;powered by the port and goes to ground. For example pressing the ;"8" key pulses
the LED in a 3 - pause - 2 pattern.

Column          Var             Byte    ;Counter for For/Next loops
Row             Var             Byte    ;Counter for For/Next loops
```

```
Temp            Var             Byte        ;holds the column binary value
LED             Con             A0          ;Have an LED hanging off A0
RIndex          Var             Byte        ;Holds the Row Value 1..4
CIndex          Var             Byte        ;Holds the Column value 1..4
i               Var             Byte        ;Counter for various For/Next loops
TempArray       Var             Nib(9)      ;Use for conversion

;
;
;Initialize
;=========
Column = 0
Row = 0
Temp = 0

Output LED      ;set up the LED pin
Low LED         ;we want the LED off


For Column = B0 to B3
    Input Column                ;Set these for input
Next ;Next Column


For Row = B4 to B7
    Output Row              ;Set these for output
    Low Row                ;we will pulse a high, start them out as lows
Next ; Next Row


;holds the actual column number. 1,2,4 & 8 are only legal values
;so other elements of the array can have random values
TempArray(1) = 1
TempArray(2) = 2
TempArray(4) = 3
TempArray(8) = 4
;
;
;Main Here we read the keypad. Put a 1 sequentially on the rows
;====    and see which column has the high.
;Since the called subroutine takes a long time to run and is only
;triggered once, no additional debounce is required. This may not
;be the case for other called subroutines.

Loop:
    For Row = B4 to B7          ;Scan the rows
        High Row                ;pulse a 1 across each row
        Pause 10                ;wait a bit
        Temp = PortB.Nib0       ;read all 4 columns at once
        Low Row                 ;restore the low
        If Temp > 0  Then GoSub LED_On        ;button pushed
    Next ;Row
GoTo Loop ;check for more keypresses


;Execute upon keypress --at the moment it just flashes an LED
;-------------------------------------------------------------
LED_On:
   ;pulse LED number of row times
   Rindex = Row - B4 + 1 ;the row number, from 1..4

   For i = 1 to RIndex ;now flash the LED
       High LED
       Pause 150
       Low LED
       Pause 150
   Next          ;Next i
```

```
               ;now we convert column value (held in Temp) to column number
               ;Column value is 1,2,4,8 corresponding to Column 1,2,3 or 4.

               CIndex = TempArray(Temp) ;conversion via array indexing

               Pause 500              ;pause to permit the user to distinguish rows
                                      ;from columns when watching the LED

           ;same approach to flash the column number using the LED
           For i = 1 to CIndex
               High LED
               Pause 150
               Low LED
               Pause 150
           Next ;Next i
        Return   ;LED_On


           End
```

The central portion of the program is the key scan loop, which implements the pseudo-code almost directly:

```
        Loop:
           For Row = B4 to B7          ;Scan the rows
               High Row                ;pulse a 1 across each row
               Pause 10                ;wait a bit
               Temp = PortB.Nib0       ;read all 4 columns at once
               Low Row                 ;restore the low
               If Temp > 0  Then GoSub LED_On ;button pushed
           Next ;Row
        GoTo Loop ;check for more keypresses
```

The subroutine **LED_On** is a dummy routine that simply flashes an LED to show the row and column numbers of the button being pressed. You may wish to use this keypad routine as one building block in a more complex and useful program.

References

[4-1] Mancini, Ron, "Examining Switch-Debounce Circuits," *EDN*, p. 22, Feb. 21, 2002.

LCD Modules

Blinking an LED is fine, as far as it goes, but it limits our communications opportunity with the outside world. If users are not to learn Morse code, a text-based message display expands our communications horizon greatly. Liquid Crystal Displays have become the display of choice for PIC output for good reasons; they are relatively inexpensive, have built-in support in MBasic and are available in a variety of sizes.

We'll deal exclusively with LCD *modules* supported by MBasic. An LCD *module* includes both the display and a controller board, as shown in Figure 5-1. Don't buy inexpensive LCD displays sold by some surplus stores without the controller board—they are not compatible with MBasic's LCD functions. Almost universally, LCD modules use an Hitachi HD44780 controller/driver chip, or a derivative chip compatible with the HD44780 command set, such as Samsung's KS0066 or Epson/ Seiko's SED1278. (An HD44780 only supports a 16-character display, but with auxiliary chips will control up to an 80-character display.)

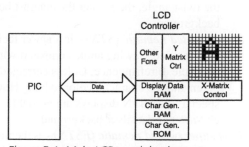

Figure 5-1: Main LCD module elements.

The particular display we'll use in our experiments is a Tianma TM162YBC6, 2-line, 16 characters per line display, super twisted nematic technology with LED backlighting, available from Basic Micro, as shown in Figure 5-2. However, almost any LCD module may be substituted for this display with little or no modifications to the hardware or software developed in this chapter.

The TM162YBC6's controller chip is a KS0066, U1 in Figure 5-3. In this product, the KS0066 is supplied as an un-mounted die and the "black blob" is the chip's encapsulated housing. (A second similarly packaged chip is underneath the quality control sticker.)

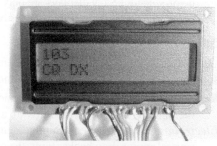

Figure 5-2: TM162YBC6 LCD display.

Selecting a Display

We'll look at three aspects of selecting an LCD module; size, technology and backlighting.

Display Sizes

There are three elements to display sizing:

1. *Numbers of characters*—Displays are standardized in 1, 2 and 4 line configurations, with 8, 16, 20 and 40 characters per line, with 40 characters per line available 1, 2 or 4 line sizes. A few

Figure 5-3: KS0066 controller chip on LCD board.

10 and 12 characters per line displays are also available. (The 4 × 40 display is configured as two 2 × 40 displays with common data and control lines, but independent E, or clock, lines.)

2. *Character size*—Sizes of individual characters range from the almost invisible (0.072") up to approximately 0.25" wide.

3. *Dot matrix size*—Most displays use a 5 × 7 pixel matrix, usually with an 8th line for the underscore, but a few premium price displays offer a 5 × 8 matrix, with a 9th line reserved for underscore. Figure 5-4 shows how characters are formed from the TM162YBC6's 5 × 7 pixel display, with an 8th line, not active in Figure 5-4, reserved for the underscore.

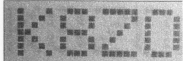

Figure 5-4: Characters are formed from small square pixels.

Crystal Technology

LCD manufacturers have, over time, improved the quality of displays through, among other things, new liquid crystal chemistry. When selecting a display, you may find

Twisted nematic (TN)—is an older technology, still used today, with a twist angle of 90° or less. The greater the twist angle, the greater the contrast between "on" and "off" pixels. Characters are black with a gray background.

Super twisted nematic (STN)—is a newer technology with a twist angle between 180° and 270°, offering an improved viewing angle compared with TN devices. STN devices function using a different physical principle, birefringence. Unless corrected, however, birefringence shifts the background color to yellow-green and the character color to blue. However, the background color can be changed to a gray with a special filter. STN displays are available in two colors, black characters on a green background or black characters on a silver background.

Film super twisted nematic (FSTN)—is the most recent advance, adding a retardation film to the STN display to compensate for the color added by the birefringence effect. This allows a black and white display to be produced. FSTN displays are only available in black characters on a white background. It is the most expensive technology, but improves both viewing angles and contrast compared with STN devices.

Backlighting

LCDs are passive devices; they do not emit light but rather are only visible through light from another source. Three lighting options are commonly available for small LCD modules:

Reflective—Reflective LCDs have no backlighting and are viewed only through ambient light reflected from the display. Reflective modules are a bit cheaper and for low power applications may be desirable.

Electroluminescent—Electroluminescent illuminated displays have a very pleasing appearance, but have several major drawbacks. First, the electroluminescent panel requires several hundred volts AC operating power, usually supplied through an inverter power supply operating at 400Hz. Many inexpensive surplus electroluminescent illuminated LCDs are sold without the associated inverter. Second, electroluminescent panels have a limited life, typically a few thousand hours, while the LCD display itself may last 20 years in continuous operation. It may be necessary, therefore, to replace the electroluminescent source many times during the display's lifetime. Electroluminescent lifetimes continue to improve, but older devices likely to be found in the surplus market are generally not good bets for longevity.

LED—The most popular backlighting illumination source is an LED array. An array of LEDs edge-illuminates a thin diffuser sheet underlying the LCD glass and provides uniform background lighting. LED backlighting is available in red, green, yellow, blue and other colors and may be dimmed or turned off by varying the LED drive current. LED backlighting has two major disadvantages—high power consumption and reduced LCD life at high temperatures due to the extra heat added by the LEDs. LED backlighting is, however, much longer lived than electroluminescent devices, with typical expected lifetimes well over 50,000 hours.

It isn't generally possible to install your own backlighting in a reflective display, as their back surfaces are opaque to light coming from the rear. Transmissive display intended for backlighted operation have a rear neutral density polarizer permitting light to pass through.

Large LCD displays, such as found in laptop computers, are back illuminated with cold cathode fluorescent lamps, but this technology is seldom, if ever, found in small alphanumeric displays.

LCD Environmental Considerations

We sometimes forget that the "L" in LCD stands for liquid. When it gets cold enough, the liquid crystals are frozen in place and the display stops working. Fortunately freezing seldom damages the display. If you're planning on outdoor operation, a standard LCD, rated to only 0°C, may present problems. Extended temperature LCDs are available permitting operation down to −20°C, but often require a separate −7 V display bias power supply. Check the data sheet to determine your display's requirements.

VFD Displays

We've concentrated on LCD displays, but another technology, the vacuum fluorescent display, or VFD, should be mentioned. VFDs are active emitting displays, generating a soft blue glow when a pixel is active. Each pixel is a miniature cathode ray vacuum tube, active when electrons strike a phosphor. VFDs offer high brightness and a wide viewing angle—typically 140°—and operate over a wide temperature range. Their primary drawbacks are expense—between two and four times the price of a similar backlighted LCD module—and relatively high power consumption. A typical lifetime specification for a VFD is 50,000 hours to half brightness.

VFD modules are available in the same configurations as LCD modules and their display controllers are HD44780 compatible. Indeed, many VFD modules are pin-for-pin drop-in replacements for LCD modules. Accordingly, we won't further mention VFDs, as our LCD discussion directly applies.

Connection to PIC

Figure 5-5 shows the connection we'll use for our test programs. Let's run through the function and connection of each pin. The PIC pin assignments are for our test circuit and may be adjusted as needed for your projects. The LCD pin number assignments apply to the vast majority of LCD modules, but, of course, you should check your particular display to verify the assignments.

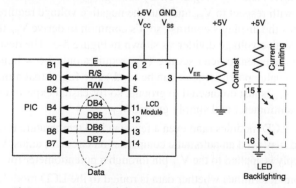

Figure 5-5: LCD module connection to PIC.

LCD Pin Name	LCD Pin Number	PIC Pin	Comments
V_{SS}	1	None	Ground.
V_{CC}	2	None	To regulated +5V source.
V_{EE}	3	None	For contrast adjustment; connect to 20K potentiometer or in many cases can be directly grounded. Note that extended temperature range displays may require V_{EE} to be connected to a negative supply.
RS	4	B0	Register (control/display) select bit: R/S=0—Data is written/read to/from control register. R/S=1—Data is written/read to/from display RAM.
R/W	5	B2	Read/Write select bit: R/W=0—Write data to RAM. R/W=1—Read data from RAM. If you do not intend to read from the LED memory (very often the case) you may omit this connection from the PIC and directly ground the LCD's R/S pin.
E	6	B1	Enable—the clock pin for reading and writing data.
DB0	7	None	Data bit 0 (LSB). 8-bit data transfer to LCD is not supported by MBasic's LCDRead and LCDWrite procedures so DB0...DB3 are not used.
DB1	8	None	Data bit 1—not used in 4-bit transfer.
DB2	9	None	Data bit 2—not used in 4-bit transfer.
DB3	10	None	Data bit 3—not used in 4-bit transfer.
DB4	11	B4	
DB5	12	B5	DB4...DB7 are grouped together into a nibble for LCDRead and LCDWrite procedures. We'll use PortB.HighNib (B4...B7).
DB6	13	B6	
DB7	14	B7	
LED+	15	None	Connect to LED supply; need not be regulated, but does require series current limiting resistor.
LED–	16	None	Connect to ground.

A few of these connections deserve further discussion.

Power, ground and bias—V_{SS} and V_{CC} (labeled V_{DD} in some LCD modules) are straightforward; V_{SS} is ground, while V_{CC} is connected to +5 V. (Some lower voltage display modules are now available; of course check your module's data sheet to verify that it requires a +5 V supply.) The LCD's logic chips are powered from this voltage. However, the voltage applied to the crystals to cause them to rotate is obtained from V_{CC} (positive) and V_{EE} (negative). By setting V_{EE} a few tenths of a volt above ground, V_{DD} becomes negative with respect to V_{EE} and thus the negative voltage required for crystal rotation is obtained. V_{EE} determines the display's contrast, so it's common to derive V_{EE} through a 10 Kohm to 20 kohm potentiometer resistive voltage divider, as shown in Figure 5-5. The desired display contrast can be obtained by varying the potentiometer setting. (V_{EE} requires only a few hundred microamperes current, so a relatively high value potentiometer can be used.) I've found that almost all displays work well at room temperature if V_{EE} is just connected to ground, so for casual experimenting I'll often omit the contrast adjustment potentiometer and simply ground V_{EE}.

Extended temperature range LCD modules (and even a few standard temperature modules) require a separate –7 V supply to drive V_{EE} and may need an automatic compensation circuit to adjust V_{EE} as the ambient temperature changes. The –7 V supply is applied to the V_{EE} pin through a potentiometer for contrast adjustment.

RS—The Register Select pin determines whether data is routed to the LCD module's control register or display RAM. For example, if the RS pin is high, sending the data value (hex) $28 to the LCD module can cause the left parenthesis character "(" to be displayed; if the RS pin is low, the same $28 character sets the display to two line mode. Since we must both write normal display data and command instructions, the RS pin must be connected to a PIC output pin. **LCDWrite** automatically sets or clears the RS pin, based on the value being written to the display.

R/W—Data to be displayed is held in read/write random access memory in the LCD module. MBasic allows us to read from (**LCDRead**) is well as write to (**LCDWrite**) the LCD module's RAM. The read/write (R/W) pin determines whether we are writing data to the module (R/W is low) or reading data from the module (R/W is high). Quite often we need not read data from the LCD module and we may simply connect the LCD module's R/W pin to ground. For generality, this chapter's sample programs connect the R/W pin to a PIC output pin so both **LCDRead** and **LCDWrite** operate.

E—Data is exchanged between the PIC and the LCD module is parallel format—that is, either eight or four bits of a byte are sent simultaneously. The enable (E) pin, by changing state from high to low, informs the receiving device (the LCD module if data is being sent to for display) that the data pins should be read. We can consider the E pin as a data clock.

Data—Normal LCD modules are parallel transfer devices; they can exchange data as either one 8-bit byte or two 4-bit nibbles. MBasic's **LCDRead** and **LCDWrite** functions support only 4-bit (nibble) transfer mode, so we'll concentrate on that. In 4-bit mode, only the data lines DB4…DB7 are active. To transfer a byte the high nibble is transferred first followed by the low nibble. Fortunately, **LCDRead** and **LCD-Write** take care of these details for us. If you've been keeping track of how many pins we've used to communicate with the LCD module, you understand why Basic Micro chose to implement the 4-bit mode. Even so, seven pins must be devoted to LCD communications in the general case, dropping to six if we wish to only write to the module and accordingly ground the LCD's R/W pin.

The only thing left is to calculate the LED current limiting resistor's value. Figure 5-6 shows the standard LED configuration. The backlight consists of a dozen or more LED pairs, series connected, with a forward voltage of typically 4.1 V when operating at rated current.

Our TM162YBC6 backlight array is specified at 4.7 V at the rated 93 mA current, a higher voltage than typically seen. Assuming the LED supply is connected to a +5 V regulated output, we may quickly calculate the required current limiting resistor:

$$R1 = \frac{5V - 4.7V}{0.093A} = 3.2\ \Omega$$

We will use the closest standard value, 3.3 ohms.

The power dissipated by the limiting resistor is:

$$P = I^2 R1 = (0.093)^2 \times 3.2 = 0.027\ watts$$

A one-quarter watt resistor is adequate.

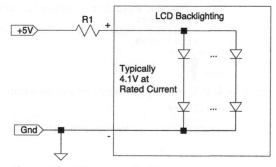

Figure 5-6: LED current limiting resistor.

If the backlight LED array is rated at 4.2 or 4.3 V, we may use a 1A silicon power diode in series with the LED array instead of a resistor to limit the current.

We should be cautious in connecting devices to the +5 V regulated supply in Basic Micro's development boards. The 5 V regulator has no heat sink and its measured input voltage is 13 V (nominal 12 V "wall wart" power supply). The 7805's thermal resistance (junction to air) $R_{\theta JA}$ is 65°C/W and the maximum operating junction temperature is +125°C. If the ambient air temperature is 25°C, we can permit no more than a 100°C junction temperature rise over ambient. A 100°C rise in junction temperature is caused by 100°C/65°C/W, or

1.5 W dissipation. Since the voltage drop across the regulator is 13 V–5 V, or 8 V, 1.5 W device dissipation is reached with a total 192 mA current draw. Well before this current level the 7805 will become hot enough to burn your finger, so it's not desirable to push it to the limit. But, if you do, 7805's are well protected by internal circuitry, including over-temperature shutdown, so the chances of damaging the regulator are minimal.

Hello World

Now that we've wired the LCD module according to Figure 5-5, let's look at some code. The canonical first program displays "Hello World," so here's our LCD version as seen in Figure 5-7. Program 5-1 writes "Hello World" on the LCD module, waits 1.5 seconds, clears the screen and writes "Hello World" again in an endless loop.

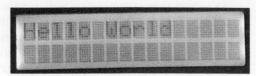

Figure 5-7: Hello World.

Program 5-1

```
;Program 5-01 -- Hello World

;Varibles
;--------
LCDNib          Var PortB.Nib1

;Constants
;---------
RegSel Con      B0
Clk             Con     B1
RdWrPin         Con     B2

;Initialization
;--------------
Pause 500       ;Allows the LCD to initialize

LCDINIT RegSel\Clk\RdWrPin,LCDNib

Main
    LCDWrite RegSel\Clk\RdWrPin, LCDNib,["Hello World"]
    Pause 1500
    LCDWrite RegSel\Clk\RdWrPin, LCDNib,[CLEAR]
    Pause 1500
Goto Main

End
```

Let's look at some key elements of the program.

```
;Varibles
;--------
LCDNib          Var PortB.Nib1

;Constants
;---------
RegSel  Con     B0
Clk             Con     B1
RdWrPin         Con     B2
;Initialization
;--------------

Pause 500       ;Allows the LCD to initialize
```

The **Pause 500** statement gives the LCD module 500 ms to go through any internal power-on startup sequence it has before receiving commands and data from our MBasic program.

```
LCDINIT RegSel\Clk\RdWrPin,LCDNib
```

The **LCDINIT** command, new in version 5.3.0.0, initializes the LCD module with the most common setup parameters, clears the display memory, and positions the cursor at the top line, leftmost position. If your initialization requirements are not met by **LCDINIT**, you can custom initialize the module using the **LCDWrite** function.

Rather than separately discuss **LCDINIT**'s arguments, we will consider them in conjunction with **LCDWrite**.

All three LCD function, **LCDInit**, **LCDWrite** and **LCDRead** use the first four arguments to map the physical connections between the PIC and the LCD module. To understand the physical connection to software mapping, we'll look at **LCDWrite** in detail. (The **RdWrPin** is optional; if omitted, only three arguments are required for these three LCD functions.)

LCDWrite RegSel\Clk\RdWrPin,LCDNib,[{mods} Exp]

The first three arguments for **LCDWrite**—**RegSel**, **Clk** and **RdWrPin**—identify the pins to which the LCD module's RS (**RegSel**), E (**Clk**) and R/W (**RdWrPin**) connections are attached. Following Figure 5-5, we've connected RS to RB0, E to RB1 and R/W to RB2. Hence, we define three new constants, **RegSel**, **Clk** and **RdWrPin** and set them equal to **B0**, **B1** and **B2**, respectively. (If you hard wire the LCD module's R/W pin to ground, you may omit the **RdWrPin** from the **LCDWrite** argument list.)

Of course, we could have simply written **LCDWrite B0\B1\B2**... every time we invoke **LCDWrite**. Suppose, though, we wish to copy our code to another project, but one where the LCD is connected to **C0**, **C1** and **C2**, instead of **B0**, **B1** and **B2**. Instead of trying to find perhaps dozens of places in our code where **LCDWrite** is used and change every one, instead all we need to is change the three constant declaration statements, a much easier task.

LCDWrite's fourth argument, **LCDNib**, defines the Port nibble to which the LCD module's DB4...DB7 pins are connected. Since we've used three pins from PortB for control, we'll stick with PortB for the data nibble. We've connected RB4 to DB4, RB5 to DB5, etc. so the data nibble is **PortB.HighNib** (equivalent to **PortB.Nib1**). Although the MBasic User's Guide suggests that **LCDNib** may be either a variable or a constant, I found that anything other than a variable generates a compiler error.

Now, let's look at the final part of **LCDWrite**; the "stuff" within the square brackets. The first four arguments inform **LCDWrite** where to send the data; the information within the square brackets is the data to be sent.

To understand what **LCDInit** does, we'll detour by examining how to initialize an LCD using the older method, sending an initialization message with **LCDWrite**.

To duplicate the functionality of **LCDInit**, we send the initialization sequence **[INITLCD1, INITLCD2, TWOLINE, CLEAR, HOME, SCR]**.

Referring to the MBasic User's Guide, we find these initialization commands cause the following action in the LCD module:

Hex Value	Name	Action Caused in LCD Display
$103 (note)	*INITLCD1*	*First initialization command.*
$102 (note)	*INITLCD2*	*Second initialization command.*
$128	*TWOLINE*	*Set display for two-line mode.*
$101	*CLEAR*	*Writes a space ($20) character into all display RAM memory. It returns the display address counter to 0. The increment/decrement mode is set to increment. The result is to clear the display of all text and position the cursor at the left edge of the display in the first line and cause the next text character to be displayed to the right of the last character.*
$102	*HOME*	*Sets the display address counter to 0 and removes any display shift that might have been in place. Display RAM contents are not altered.*
$10C	*SCR*	*Display on, underscore and block cursors disabled.*

The MBasic User's Guide says the value associated with **InitLCD1** is $133, while that of **InitLCD2** is $132. However, examination of the resulting assembler code shows the User's Guide is in error. (The pre-defined LCD initialization constant **Home** is also $102.) In any event, the sequence established by **InitLCD1, InitLCD2** must be used to set the LCD module to 4-bit operation and properly initialize its operation.

The next four initialization constants, **TWOLINE, CLEAR, HOME, SCR**, set the display to operate in two-line mode; erases the display memory; sets the start of text to the top line, left position; and turns the display on, with no cursor. If we wish to use this older initialization methodology, the corresponding function is:

```
LCDWRITE RegSel\Clk\RdWrPin, LCDNib, [INITLCD1,INITLCD2, TWOLINE, CLEAR,
HOME, SCR]
```

This initialization sequence provides the same functionality as the **LCDInit** command added in version 5.3.0.0.

```
Main
      LCDWrite RegSel\Clk\RdWrPin, LCDNib,["Hello World"]
      Pause 1500
      LCDWrite RegSel\Clk\RdWrPin, LCDNib,[CLEAR]
      Pause 1500
Goto Main
```

Having initialized the LCD module—either with the **LCDInit** command or **LCDWrite** and an initialization string—we now write our "Hello World" string to the display. (Regardless of how we initialize the LCD, we need to initialize it but once.) Note that we use the same **LCDWrite** procedure as might we do for sending command sequences; if the values inside the brackets are 255 or less, **LCDWrite** recognizes them as characters, not commands, and appropriately clears the RS line. Remember that MBasic treats a string sequence as a series of individual characters, each with a byte value between 0...255. After writing "Hello World," the program waits 1.5 seconds and clears the display with the **LCDWrite[CLEAR]** command. After another 1.5 second pause, "Hello World" is written again to the display.

LCD Module Memory, Shifts and Lines

Before we become more adventuresome with **LCDWrite**, we must understand how display memory is organized in LCD modules.

The HD44780 and compatible controller chips have 80 bytes of display data RAM (DDRAM)—the memory that holds the message you want displayed—regardless of whether they are connected to a 16 × 1, a 16 × 2, a 20 × 2 or some other size LCD display. It's possible to write to all 80 bytes of display memory, but only that part of the memory that fits into the display window will be seen. As we will see, however, it's possible to offset the display window to show different segments of the total memory.

Some LCD modules may add external DDRAM memory; so check the data sheet for your particular display. Others, such as a 40 × 4 display, fit two HD44780 control chips, with separate E (E1 and E2) connections but with all other connections paralleled. These displays can be written to with **LCDWrite** by treating them as two independent 40 × 2 displays, with common pins, except for different E1 and E2 connections.

Memory layout for 16 × 2 and 20 × 2 displays seems reasonably standardized. Here's the 16 × 2 display layout for the default screen window position:

Characters	0	1	2	3	4	5	6	7	8	9	10	11	12	13	14	15
Line 1	$00	$01	$02	$03	$04	$05	$06	$07	$08	$09	$0A	$0B	$0C	$0D	$0E	$0F
Line 2	$40	$41	$42	$43	$44	$45	$46	$47	$48	$49	$4A	$4B	$4C	$4D	$4E	$4F

The values in the table are the LCD module's memory address that correspond to the default screen window position—that is, a character written into memory position $41 will appear as the second character from the left in the second line of the display. Memory for line 1 and line 2 continues; line 1 to memory location $3F and line 2 to memory position $7F, but these are outside the display window width.

A 20 × 2 display has the same starting points for Lines 1 and 2, but the visible character display, of course, extends for four additional characters. Even the first two lines of a 20 × 4 display map to $00 and $40. Nonetheless, you should verify the memory arrangement for your display, as some variations exist. Incidentally, although not documented in the User's Guide, **LCDWrite** works for 20 × 4 displays, as we shall see in Chapter 6.

ScrLeft and ScrRight

To understand how the scrolling the display window works, suppose we have an 8 × 1 display, with 8 characters available on one line. After initialization, we send a 16-character string to the display (for clarity omitting the pin and nibble address details from **LCDWrite**; we'll indicate this omission by adding a ellipse "…"):

LCDWrite…["HELLO WORLD!!!!!]

The result is to load the first 16 DDRAM positions in the display with this message.

Characters	0	1	2	3	4	5	6	7	8	9	10	11	12	13	14	15
Memory	$00	$01	$02	$03	$04	$05	$06	$07	$08	$09	$0A	$0B	$0C	$0D	$0E	$0F
Message	H	E	L	L	O		W	O	R	L	D	!	!	!	!	!

Since we have an 8-character window, we'll indicate which characters are displayed by boxing them with a heavy outline. We'll also assume the display has been initialized with the same constants in Program 5-1, except that it is a one-line display. Hence the display window shows:

Message	H	E	L	L	O		W	O	R	L	D	!	!	!	!	!

We will see HELLO WO on the display.

However, we can scroll the memory left or right underneath the display window by sending a **ScrLeft** or a **ScrRight** command to the display with **LCDWrite**. Suppose we execute **LCDWrite…[ScrLeft]** four times in immediate sequence. The result is:

Message	H	E	L	L	O		W	O	R	L	D	!	!	!	!	!

Our display now shows O WORLD!, as the text has scrolled four positions to the left.

We'll demonstrate scrolling with Program 5-2. Program 5-2 assumes we use a 16 × 2 display and implements two successive full screen width scrolls to display all 40 characters in line 1's DDRAM.

Program 5-2

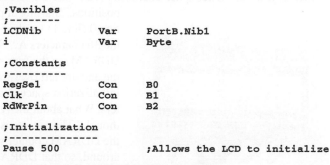

```
        ;Varibles
        ;--------
        LCDNib          Var         PortB.Nib1
        i               Var         Byte

        ;Constants
        ;---------
        RegSel          Con         B0
        Clk             Con         B1
        RdWrPin         Con         B2

        ;Initialization
        ;--------------
        Pause 500                   ;Allows the LCD to initialize

        LCDInit RegSel\Clk\RdWrPin, LCDNib
```

```
Main
    LCDWrite RegSel\Clk\RdWrPin, LCDNib,[Clear,ScrRAM]
    LCDWrite RegSel\Clk\RdWrPin, |
        LCDNib,["ABCDEFGHIJKLMNOPQRSTUVWXYZ0123456789"]
    Pause 15000

    For i = 0 to 15
        LCDWrite RegSel\Clk\RdWrPin, LCDNib,[ScrLeft]
    Next

    Pause 15000

    For i = 0 to 15
        LCDWrite RegSel\Clk\RdWrPin, LCDNib,[ScrLeft]
    Next

    Pause 15000
Goto Main
```

After the same initialization routine developed in Program 5-1, we write 36 characters A...9 into the LCD module's DDRAM, at positions $00...$23 with **LCDWrite..."ABCDEFGHIJKLMNOPQRSTUVWXYZ0123456789"]**. (The initialization commands set the DDRAM write position to $00 and enable automatic increment for each character added to DDRAM.) Since we're using a 16 × 2 display, we expect to see A...P on the top line of the screen, and Figure 5-8 confirms our expectation.

After pausing 15 seconds, we scroll left 16 times:

```
For i = 0 to 15
        LCDWrite RegSel\Clk\RdWrPin, LCDNib,[ScrLeft]
    Next
```

As Figure 5-9 confirms, the next 16 characters—Q...5—are now visible. After 15 more seconds, we scroll left 16 more times.

```
For i = 0 to 15
        LCDWrite RegSel\Clk\RdWrPin, LCDNib,[ScrLeft]
    Next
```

Figure 5-8: First 16 characters.

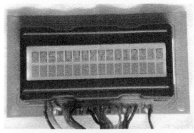

Figure 5-9: Scroll left 16 positions.

Figure 5-10: Scroll left 16 positions a second time.

So, what do we see now? We expect the display to start with 6, and at run through 9. Figure 5-10 shows, however, an additional four spaces and A...H. To understand why we see what we do, remember that the HD44780 has 80 bytes of DDRAM, split equally between line 1 (positions 0...39) and line 2 (positions 64...104). (We're bouncing back and forth between decimal and hexadecimal representations of memory addresses, so don't get lost or confused. Hex values are preceded with a $ symbol.) We first cleared the memory, which writes space characters ($20) into all 80 DDRAM positions. Starting at position 0 (line 1) we then wrote the 36 characters A...9 into DDRAM, so the four spaces at the end are left over from the initialization space characters. What about the A...H though? The answer is that the DDRAM counter wraps around, so that DDRAM address 40 is treated by the HD44780 as position 00, 41

76

as 01, etc. Hence, the A…H are the characters we wrote at positions 0…7, wrapped around when we tried to display positions 40…47.

We've illustrated **ScrLeft**, but **ScrRight** works the same way, but with opposite movement.

Why would you want to scroll? If your message is too long to fully display, scroll it across the display, one step every half-second. Or, perhaps you want your message to receive attention. Scroll it left and then right several times as an eye catcher. And, suppose you wish to display a stream of continuous characters, perhaps from monitoring an RS-232 line, or maybe from an amateur radio Morse code decoder. It's a lot easer to read continuous text if the old text scrolls one position left with each new characters added at the right.

Writing at a Defined Location

We've looked at scrolling the display, but suppose we wish to keep the display text fixed and write new information at a specific location. For example, suppose our display is reporting the outdoor temperature. We might like the display to read Outdoor 72 F when it's 72 degrees outside and Outdoor 73 F when it's 73 degrees.

We have a couple of ways to display this message. One would be to write it directly using the **LCDWrite** procedure, for example., **LCDWrite…["Outdoor ",DEC OTemp," F"]** every time we read our temperature sensor (with the resulting value held in **OTemp**) and update the display. Before writing this string, we might clear the display. In pseudo-code this approach is:

```
Initialize things that need initialization
Main
        Read the temperature sensor and put value into OTemp
        Clear LCD
        Write "Outdoor" + String(OTemp) + " F" to LCD
GoTo Main
```

However, there are some advantages in writing the fixed text parts "Outdoor" and " F" just once and writing only the temperature value when read. If nothing else, we can save a bit of execution time, and our code may well be easier to read as we can separate fixed initialization type activities (writing the "Outdoor" and " F") from repetitive actions, such as reading the temperature sensor and writing its value.

Using our 16 × 2 display and looking at line 1, the fixed part of our message is thus:

Characters	0	1	2	3	4	5	6	7	8	9	10	11	12	13	14	15
Memory	$00	$01	$02	$03	$04	$05	$06	$07	$08	$09	$0A	$0B	$0C	$0D	$0E	$0F
Message	O	u	t	d	o	o	r					F				

What we need, therefore, is a way to write the temperature starting at DDRAM address $08. Fortunately, we may accomplish that easily through **LCDWrite…[ScrRAM+Offset]** command. The command **LCDWrite… [ScrRAM+Offset]** moves the DDRAM address counter to the memory position **Offset**. Hence, to write the temperature beginning at DDRAM position $8, we first execute **LCDWrite…[ScrRAM+$8]**. Program 5-3 shows the complete code to execute this task. (If we wanted to position the DDRAM address counter to line 1's start (0), we would simply use **LCDWrite…[ScrRAM]** with no offset.)

Program 5-3

```
;Variables
;--------
LCDNib          Var PortB.Nib1
OTemp           Var     Byte

;Constants
;--------
RegSel Con      B0
Clk             Con     B1
RdWrPin Con     B2
```

```
;Initialization
;--------------
Pause 500                ;Allows the LCD to initialize

LCDInit RegSel\Clk\RdWrPin, LCDNib
OTemp = Random OTemp

LCDWrite RegSel\Clk\RdWrPin, LCDNib,[Clear,ScrRAM]
LCDWrite RegSel\Clk\RdWrPin, LCDNib,["Outdoor     F"]
Pause 1000

Main

    LCDWrite RegSel\Clk\RdWrPin, LCDNib,[ScrRAM +$7]
    LCDWrite RegSel\Clk\RdWrPin, LCDNib,["    "]
    LCDWrite RegSel\Clk\RdWrPin, LCDNib,[ScrRAM +$8]
    LCDWrite RegSel\Clk\RdWrPin, LCDNib,[DEC OTemp]
    Pause 10000
Goto Main
End
```

We don't have a real temperature reading routine, so we substitute a random number. We also don't worry about minus signs (we should use a signed variable for **OTemp**) and we won't concern ourselves with some other presentation niceties, as our focus is upon how to write at a specific DDRAM address.

```
LCDWrite RegSel\Clk\RdWrPin, LCDNib,[Clear,ScrRAM]
LCDWrite RegSel\Clk\RdWrPin, LCDNib,["Outdoor     F"]
```

Here we write the fixed part of the display. We allow four spaces for the temperature reading; three for the digits and one blank to separate the word "Outdoor" from the first temperature digit.

```
Main

    LCDWrite RegSel\Clk\RdWrPin, LCDNib,[ScrRAM +$7]
    LCDWrite RegSel\Clk\RdWrPin, LCDNib,["    "]
    LCDWrite RegSel\Clk\RdWrPin, LCDNib,[ScrRAM +$8]
    LCDWrite RegSel\Clk\RdWrPin, LCDNib,[DEC OTemp]
    Pause 10000
Goto Main
```

Now we write the temperature reading. First, because we don't know how many digits the preceding reading might have occupied, we clear the old reading by writing four spaces starting at $7. First, we position the DDRAM memory pointer at $7 and then write four blanks. Now we are ready to write the actual temperature reading. (If this were a real thermometer program we would first get a temperature reading, of course.) We position the DDRAM memory pointer where we wish the first digit of the temperature reading to begin ($8) with the **LCDWrite...** **[ScrRAM +$8]** procedure. Finally, we write the actual temperature value with **LCDWrite...[DEC OTemp]**. Figure 5-11 shows the result.

Figure 5-11: Simulated temperature display.

Blinking and Cursor Selection

Another attention getter is to flashing the text off and on. While we could do this by clearing the display and re-writing the information, it's easier to send off/on commands to the LCD module. The following code fragment endlessly flashes the current text on for one second and off for one second.

```
FlashText

LCDWrite RegSel\Clk\RdWrPin,LCDNib,[Off]
Pause 1000
LCDWrite RegSel\Clk\RdWrPin,LCDNib,[Scr]
Pause 1000

GoTo FlashText
```

The **Off** command does exactly what its name suggests; it turns the display image off. We then wait 1,000 milliseconds and turn the display image back on. There is no "on" command; rather you must choose from four commands that turn the display image on and simultaneously control the cursor.

Hex Value	Name	Action Caused in LCD Display
$10C	SCR	Display on, no cursor displayed,
$10D	SCRBLK	Display on, no underscore cursor, blinking solid block shows next character position
$10E	SCRCUR	Display on, steady underscore cursor shows next character position
$10F	SCRCURBLK	Display on, steady underscore cursor and blinking solid block show next character position.

These commands, of course, can also be used at initialization. If so, it will be necessary to use an appropriate initialization string, sent via **LCDWrite**, instead of the **LCDInit** function.

Figures 5-12 and 5-13 show the results of **SCR** (line 1 in both figures), **ScrCur** (line 2, Figure 5-12) and **SCRBLK** (line 2, Figure 5-13). The fourth option, **ScrCurBlk** has both a steady underscore and a flashing block at the next character position.

Figure 5-12: Underscore cursor.

Figure 5-13: Block cursor.

If your application is display-only—users do not enter information—a cursor isn't necessary and the **SCR** initialization constant is appropriate. If the user interacts with your program, however, a cursor is a valuable cue to show where the data is to be entered or changed. We'll see examples of user interaction in later chapters.

Font Selection

In addition to the normal upper and lower case English alphabet, numbers and punctuation symbols, HD44780-compatible controllers have an extended font set. Usually the extension is Katakana (Japanese) or, less often, generic Western European accented characters. Other font extensions, such as Cyrillic, Greek and Hangul (Korean) are also available from some manufacturers. (If you are willing to purchase a large enough quantity, semiconductor houses will manufacture an LCD controller chip with your own custom designed characters.)

The extended character set isn't a true second font set, where, for example, the English character "A" appears as a different symbol following a font selection command. Rather, these are characters with a value of $80 to $FF, above the standard character byte range of space ($20) to delete ($7F). To write the extended character corresponding with a value of $9B, you simply use **LCDWrite...[$9B]**. Multiple characters are separated with commas, e.g., **LCDWrite... [$9B,$9C,$8F]**. Program 5-4 displays extended characters, together with their value, with the results in Figure 5-14.

Figure 5-14: Extended character set display.

Program 5-4

```
;Variables
;---------
LCDNib          Var PortB.Nib1
i                       Var     Byte

;Constants
;---------
RegSel          Con     B0
Clk             Con     B1
RdWrPin Con     B2

;Initialization
;--------------
Pause 500               ;Allows the LCD to initialize

LCDInit RegSel\Clk\RdWrPin, LCDNib

Main
    For i = $80 TO $FF
        LCDWrite RegSel\Clk\RdWrPin, LCDNib,[Clear,Home,"Value: ",DEC i,"   ",ISHEX i]
        LCDWrite RegSel\Clk\RdWrPin, LCDNib,[ScrRAM+$40,"Char ",i]
        Pause 2000
    Next
Goto Main

End
```

By now, this program should require little explanation, as we've seen all of it earlier in this chapter. The key lines write the numerical value (in decimal and hex) on the top line, while the corresponding character appears on bottom line. The normal English character set stops at $7F, so we start at $80 and go to the maximum possible byte value, $FF. If you change the starting point to $20, instead of $7F, you can see the normal English character set as well.

Custom Characters

If the normal and extended characters aren't enough, you can define up to eight custom characters. Once defined, these characters are fully accessible just as any pre-defined character. Custom characters are contained in character generator memory, or CGRAM. Since CGRAM is volatile, your program must rewrite the custom characters to CGRAM if power is removed from the LCD module. We will load our custom characters at initialization.

We'll demonstrate custom characters in the context of a bar graph display. If we're measuring a parameter, such as an analog voltage with a PIC's built-in analog-to-digital converter, can show the resulting value as a digit string on our display—12.56 volts, for example, updated several times a second. But, if you are adjusting a variable voltage control, you will find digital displays aren't well suited to showing trends or short-term changes. Indeed, seeing a trend or spotting a momentary fluctuation are areas where an old-fashioned moving needle meter is often better than a digital display. Many digital meter manufacturers have added a quick responding bar graph mimicking an analog meter scale. Our bar graph example duplicates this feature.

Let's start with a close up of a character, "T" as if it were added to CGRAM:

Row	Value
0	31
1	4
2	4
3	4
4	4
5	4
6	4
7	0

The character "T" is formed from a 5x8 matrix, with the bottom row reserved for the underscore cursor. We've numbered the rows 0...7, top to bottom and columns 0...4, right to left. To store this character information, the 44780 controller maps each row into a memory byte, with the bottom five bits mapped into the display cells—1 for an opaque pixel and 0 for a transparent pixel. We can calculate the value of the corresponding byte by summing the opaque pixels. For example, the cap of the "T" has the binary value 00011111, which we may convert to decimal as $16 + 8 + 4 + 2 + 1$, or 31. (The three most significant bits may be anything, as they are not used in character mapping. We'll make them 000 for convenience.)

The eight custom characters are assigned values 0...7. To write all eight custom characters to the display, therefore, the syntax is `LCDWrite…[0,1,2,3,4,5,6,7]`.

To create custom characters, we must write their row byte values to CGRAM. We'll accomplish this through MBasic's all-purpose LCDWrite procedure, `LCDWrite… [CGRAM+Offset,Value]`. `CGRAM` is the address of the first byte of CGRAM memory, and **value** is the byte value associated with the particular address. Each byte follows in row order, and each character follows in character order. The following table provides the offset for each row, for each character. Fortunately, we rarely must keep track of individual CGRAM values, as the HD44780 controller chip does it for us.

If we wish character number 0 to contain our new "T," we would simply execute `LCDWrite…`
`[CGRAM,31,4,4,4,4,4,4,0]`.
This procedure first sets the writing to occur at CGRAM position 0. (We could have written this `CGRAM+0`.) The next eight arguments are the byte values of the character rows, in order row 0...row 7. If we have used the normal initialization string for the

Row	Character Number							
	0	1	2	3	4	5	6	7
0	0	8	16	24	32	40	48	56
1	1	9	17	25	33	41	49	57
2	2	10	18	26	34	42	50	58
3	3	11	19	27	35	43	51	59
4	4	12	20	28	36	44	52	60
5	5	13	21	29	37	45	53	61
6	6	14	22	30	38	46	54	62
7	7	15	23	31	39	47	55	63

LCD module, the LCD module is in auto-increment mode and each byte written is automatically placed in the next memory location by the HD44780. If we have a second custom character to define, we would start it at `CGRAM+8`, and the third at `CGRAM+16`. (If it immediately followed the first definition, we could skip the `CGRAM+8` and allow the internal auto-increment feature to define the address.)

Our bar display is based on work by Larry Phipps, a friend of the author, who developed a quick acting display reference used in adjusting his amateur radio equipment. We start by defining four custom characters:

Character 0

Row	Value
0	16
1	16
2	16
3	21
4	16
5	16
6	16
7	0

Character 1

0	20
1	20
2	20
3	21
4	20
5	20
6	20
7	0

Character 2

0	21
1	21
2	21
3	21
4	21
5	21
6	21
7	0

Character 3

0	0
1	0
2	0
3	21
4	0
5	0
6	0
7	0

Custom character 3 is a visual placeholder; it fills the display with dots. The other three characters show activity, with one, two or three vertical bars. Here's how we might write these custom characters into CGRAM:

```
LCDWRITE RegSel\Clk\RdWrPin, LCDNib,[CGRAM,16,16,16,21,16,16,16,0]
LCDWRITE RegSel\Clk\RdWrPin, LCDNib,[CGRAM+8 20,20,20,21,20,20,20,0]
LCDWRITE RegSel\Clk\RdWrPin, LCDNib,[CGRAM+16 21,21,21,21,21,21,21,0]
LCDWRITE RegSel\Clk\RdWrPin, LCDNib,[CGRAM+24 0,0,0,21,0,0,0,0]
```

A 16 character display, used totally for bars, permits displaying 0…48. Larry's adjacent bars have a space since the LCD module has a one-pixel gap between adjacent characters. The result is uniformly spaced vertical bars across the full display width.

We'll assume that the bar graph is to display a value held in a byte variable **Temp** and that **Temp** is scaled 0…48. Our strategy for displaying bars is thus:

```
Get new value of Temp
Determine which characters will be all bars (Char(2))
        Write those characters as all bars
Determine how many bars will be displayed to the right of the last all bars
character
            Display that as either 0 (Char(3), 1 (Char(0) or 2 (Char (1) bars
Fill any remaining space with all dots (Char(3))
```

Our demonstration program cycles through 0…48 bars, with one second between successive bar displays. If we remove the intentional one second pause, we can update the bar approximately 100 times/second assuming our PIC has a 20 MHz clock and there are no lengthy calculations involved to determine **Temp**.

Program 5-5

```
;Program Bar Graph Sample

;Varibles
```

```
;--------
LCDNib          Var     PortB.Nib1
i               Var     Byte
j               Var     Byte
x               Var     Byte
y               Var     Byte(3)
Temp            Var     Byte


;Constants
;---------
RegSel  Con     B0
Clk     Con     B1
RdWrPin Con     B2


;Initialization
;--------------

Pause 500               ;Allows the LCD to initialize
y(0) = 3
y(1) = 0
y(2) = 1


LCDInit RegSel\Clk\RdWrPin,LCDNib

GoSub LoadBar

Main
        For i = 0 to 48         ;execution time 10.08mSec
                Temp = i
                x = Temp/3      ;div


                LCDWrite RegSel\Clk\RdWrPin,LCDNib,[SCRRAM]

                j = 0
                While j < x
                LCDWrite RegSel\Clk\RdWrPin,LCDNib,[2]
                j = j+1
                WEND

                LCDWrite RegSel\Clk\RdWrPin,LCDNib,[y(Temp//3)] ;mod

                For j = x+1 to 15
                LCDWrite RegSel\Clk\RdWrPin,LCDNib,[3]
                Next

                LCDWrite RegSel\Clk\RdWrPin,LCDNib,[SCRRAM+$40," "]
                LCDWrite RegSel\Clk\RdWrPin,LCDNib,[SCRRAM+$40,Dec Temp]

                Pause 1000
        Next
        GoTo Main
End


LoadBar         ;load the bar characters into character RAM
;--------------
        LCDWRITE RegSel\Clk\RdWrPin, LCDNib,[CGRAM,16,16,16,21,16,16,16,0]
        LCDWRITE RegSel\Clk\RdWrPin, LCDNib,[20,20,20,21,20,20,20,0]
        LCDWRITE RegSel\Clk\RdWrPin, LCDNib,[21,21,21,21,21,21,21,0]
        LCDWRITE RegSel\Clk\RdWrPin, LCDNib,[0,0,0,21,0,0,0,0]
Return
```

As before, we've seen much of Program 5-5, so we'll concentrate on the new elements.

```
LoadBar ;load the bar characters into character RAM
;--------------
        LCDWRITE RegSel\Clk\RdWrPin, LCDNib, [CGRAM,16,16,16,21,16,16,16,0]
        LCDWRITE RegSel\Clk\RdWrPin, LCDNib, [20,20,20,21,20,20,20,0]
        LCDWRITE RegSel\Clk\RdWrPin, LCDNib, [21,21,21,21,21,21,21,0]
        LCDWRITE RegSel\Clk\RdWrPin, LCDNib, [0,0,0,21,0,0,0,0]
    Return
```

Subroutine **LoadBar** is called during the initialization process and loads our four custom characters into CGRAM. Since we load these characters in their character number order, without any other intervening data being sent to the LCD module, we need only set the CGRAM address once, and permit the HD44780's auto-increment feature to take care of the rest.

```
Main
        For i = 0 to 48          ;execution time 10.08mSec
            Temp = i
            x = Temp/3      ;div
```

The **For i** loop cycles us through all possible values of bar length. For this demonstration program, we let the bar length variable, **Temp**, equal **i**. The variable **x** determines the break point between all three bars and all three dots. Suppose **Temp=11**. The correspond bar graph will have three "|||" characters followed by one "||." character. The remaining characters will be the blank identifiers "...". Since each "|||" corresponds to a value of 3 bars, we can determine how many 3-bars are contained in **Temp** by simple integer division. In this case **11/3=3**. Rather than repeat this division, we'll do it once and use the variable x to hold the result via **x = Temp/3**. So, we can now write **x** "|||" characters to the display, after first resetting the display address to the top line, character 0:

```
LCDWrite RegSel\Clk\RdWrPin,LCDNib,[SCRRAM]

j = 0
While j < x
LCDWrite RegSel\Clk\RdWrPin,LCDNib,[2]
j = j+1
WEND
```

We use the **While/WEND** construction here, because it's possible that there should no "|||" characters displayed, as is the case where **Temp** has the value 0, 1 or 2. Since the conditional test is made at the outset in a **While/WEND** loop, we have properly dealt with the case where no "|||" is to be displayed. We simply execute **LCDWrite…[2]** **x** times to put up the correct leading "|||" characters.

```
LCDWrite RegSel\Clk\RdWrPin,LCDNib,[y(Temp//3)] ;mod
```

Now we deal with the variable number of bars in the next character. There's a lot packed into this one line of code, so pay attention. To determine how many bars are in the next character, we look at the remainder after integer division by 3, the number of bars per character. (In MBasic, the remainder or "modulus" operator uses the // symbol.) The result is 0, 1 or 2. If the remainder is 0, we display character 3. If the remainder is 1, we display character 0. If the remainder is 2, we display character 2. We could index into a series of **LCD-Write** statements, based on the value of the remainder, but that's slow. Instead, we use a three-element byte array we defined and initialized earlier:

```
y(0) = 3
y(1) = 0
y(2) = 1
```

Thus, if we use the remainder as the index, the **y** array gives us the correct character to write to the screen. Hence, we have **LCDWrite…[y(Temp//3)]**.

We may have avoided array index mapping by a better selection of character order, but this example is a fragment of a larger program and display that called for the order we used. Besides, array mapping is a useful trick that will be handy in other instances.

```
For j = x+1 to 15
LCDWrite RegSel\Clk\RdWrPin,LCDNib,[3]
Next
```

Now, we finish up by filling whatever is left of the display with our blank placeholder, character 3. We do this with a **For/Next** loop, starting one position past the variable bar character.

```
LCDWrite RegSel\Clk\RdWrPin,LCDNib,[SCRRAM+$40," "]
LCDWrite RegSel\Clk\RdWrPin,LCDNib,[SCRRAM+$40,Dec Temp]

        Pause 1000
    Next
    GoTo Main
```

Finally, we write the decimal value of **Temp** on the bottom line, pause 1 second to allow you to see the bar versus digital count and then go to the next test loop. Figure 5-15 shows the results of Program 5-5. In a program doing useful work, we would delete the Pause 1000 statement. Figure 5-16 shows the application resulting from Larry's work.

Figure 5-15: Bar graph display.

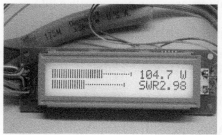

Figure 5-16: Using a bar graph display.

References

[5-1] Hitachi, *HD44780 (Dot Matrix Liquid Crystal Display Controller/Driver)*, (undated). If you are interested in the nuts and bolts of LCD work, you will need the HD44780 data sheet. Hitachi recently spun its semiconductor operations into Renesas Technology and the HD44780 data sheet is available for download at http://america.renesas.com/products/supportdocs/hd44780.pdf. Although Renesas does not recommend the HD44780 for new designs, newer LCD controllers are downward compatible with the HD44780 command set.

[5-2] Tianma Microelectronics Co., Ltd., *Specification for LCD Module, Module No. TM162YBC6*, (undated). This data sheet is available for download at http://www.tianma.com/spec_sheets/TM162YBC6%20SPEC.pdf.

[5-3] Epson Electronics, *SED1278 LCD Controller/Drivers Technical Manual*, (undated).

[5-4] Samsung Electronics, *KS0066U 16COM/40SEG Driver & Controller for Dot Matrix LCD*, (undated).

Reading Complex Input Switches

Chapter 4 looked at reading individual switches and a matrix keypad. We'll extend our switch reading to look at additional pin saving techniques. We'll also see how rotary encoders—both internally decoded and incremental versions—can be read with a PIC.

Pin Saving Techniques

It's a common requirement to read a configuration switch during start-up. Perhaps the switch defines the baud rate for a serial output, or it may set some other parameter. Regardless, to dedicate several pins to sense a start-up configuration is a waste of scarce pin resources and may force us to use a larger, more expensive PIC than otherwise desired.

Simple Approach

How might we read a 16-state switch? We'll immediately discard hooking 16 individual switches to 16 input pins. We recall that 4-bits give us 16 unique combinations, so we can use four switch contacts to generate 16 states. This might be either a 4-section DIP switch with individual switch selections, or it might be a rotary DIP switch with positions 0...F internally coded into one common connection and four switched poles, such as the simple connection is shown in Figure 6-1. A closed switch places a high on the corresponding input pin, and an open switch places a low on the corresponding pin. Reading RB7...RB4 as `PortB.`
`HighNib`, the switch sections correspond to the 4 bits in the nibble.

To determine the switch closure pattern corresponding to any particular value, we look at the binary combinations. For example, if `PortB.HighNib` is to return the value 13, we simply note that 13 in binary is 1101. Hence, B7, B6 and B4 must be high and B5 must be low. So, the switches associated with B7, B6 and B4 will be closed and B5's switch will be open. Another way to look at this is to see that B7 has the value of either 8 (closed) or 0 (open), B6 is either 4 or 0, B5 is either 2 or 0 and B4 is either 1 or 0. To sum to 13 requires 8 + 4 + 1, again corresponding to switch closures on B7, B6 and B4. Figure 6-1's value labels correspond to this weighting.

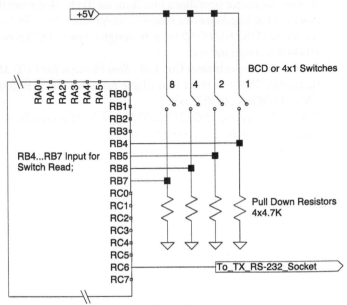

Figure 6-1: Reading a 4-section switch.

Let's see how we might read the arrangement of Figure 6-1 and use the configuration switch reading to set the serial output baud rate.

Program 6-1

```
;Show reading baud rate
;from switches on B4...B7
;8       4       2       1
;B7      B6      B5      B4

TRISB = $F0
EnableHSerial

;Initialize Baud Rate
Branch PortB.HighNib, [S0, S1, S2, S3, S4, S5, S6, S7, |
S8, S9, S10, S11, S12, S13, S14, S15]

        S0  :   SetHSerial H1200    : GoTo SendMessage
        S1  :   SetHSerial H2400    : GoTo SendMessage
        S2  :   SetHSerial H4800    : GoTo SendMessage
        S3  :   SetHSerial H7200    : GoTo SendMessage

        S4  :   SetHSerial H9600    : GoTo SendMessage
        S5  :   SetHSerial H14400   : GoTo SendMessage
        S6  :   SetHSerial H19200   : GoTo SendMessage
        S7  :   SetHSerial H24000   : GoTo SendMessage

        S8  :   SetHSerial H26400   : GoTo SendMessage
        S9  :   SetHSerial H28800   : GoTo SendMessage
        S10 :   SetHSerial H31200   : GoTo SendMessage
        S11 :   SetHSerial H33600   : GoTo SendMessage

        S12 :   SetHSerial H36000   : GoTo SendMessage
        S13 :   SetHSerial H38400   : GoTo SendMessage
        S14 :   SetHSerial H57600   : GoTo SendMessage
        S15 :   SetHSerial H115200  : GoTo SendMessage

Main
        SendMessage
        HSerOut ["Switch Value: ",DEC PortB.HighNib ,13]
        Pause 100
GoTo Main

    End
```

We're saving serial output for a later Chapter, so we'll simply note that certain PICs, including the 16F876/877 series we use in our experiments, have built-in hardware support for serial output, and that these devices the transmit output appears on RC6. So, it's necessary to connect a jumper from C6 to the "TX" pin of the serial socket in Basic Micro's 2840 development board. The DB9 serial output port connector should then be hooked to a second computer running a serial terminal program, such as HyperTerminal included with Windows, or, even better, the excellent freeware program Terminal [6-9].

Before using the hardware serial output, it's necessary to initialize it with the **EnableHSerial** command, which instructs the compiler to load the hardware serial input/output code module. Don't forget that before actually sending or receiving serial data, you must specify the data speed with the **SetHSerial** function, which takes as its argument a pre-defined parameter that defines the desired baud rate. These constants are of the form Hnnnn where nnnn is the data speed. After initialization and speed setting, text may be sent over the serial port through the **HserOut** procedure. (We'll cover serial data in much more detail in Chapter 9, so if you may wish to skip ahead for more information.)

```
TRISB = $F0

;Initialize Baud Rate
Branch PortB.HighNib, [S0, S1,S2, S3,S4, S5, S6, S7, |
        S8, S9, S10, S11, S12, S13, S14, S15]
```

We initialize pins B4...B7 to be inputs through the **TRISB=$F0** assignment. (See Chapter 2 for a review of the TRIS statement.)

We then read the four switch poles simultaneously by reading **PortB.HighNibble**. **PortB.HighNibble** has a value of 0...15 corresponding to the binary weighted values of the switch positions. The **Branch PortB.HighNib [...]** procedure transfers program execution to **S0** if **PortB.HighNib = 0**, to **S1** if **PortB.HighNib = 1** through **S15 if PortB.HighNib = 15**.

Each jump label is similar in structure, so we'll look at **S0**:

```
    S0  :  SetHSerial H1200     : GoTo SendMessage
```

First, we set the hardware serial port baud rate and then execution branches to **SendMessage**.

```
    Main
            SendMessage
            HSerOut ["Switch Value: ",DEC PortB.HighNib ,13]
            Pause 100
    GoTo Main
```

The code following **SendMessage** sends the switch value on the serial output port. If the switch is set for 13, corresponding to a baud rate of 38400, the following message will be displayed on the connected terminal: "Switch Value 13." This message repeats every 100 ms.

Of course, your terminal program must be set to match the baud rate set by the switches. The character "13" at the end of the message string is a carriage return. If you want these messages to each start on a new line, make sure your terminal program is configured for this action. (Some terminal programs require a line feed character to start a new line and this cannot be changed. If you receive all the messages on one line, change the output data to **[Switch Value: ",DEC PortB.HighNib ,13,10]**, as 10 is the character value for line feed.)

If you see only gibberish on the terminal screen, try resetting the PIC with the reset button on the development board to force the switches to be freshly read. Or, briefly remove power from the development board to force a power-on reset.

Our discussion and program structure assumes the switch has binary coded contacts, as is by far the most frequent arrangement for inexpensive DIP-packaged configuration switches.

Doubled-Up LCD/Configuration Switch

Program 6-1 reads the configuration switch once, at start-up only, but the four pins connected to the switch poles are unavailable for other uses. In some cases, we can reuse those pins for other purposes.

Figure 6-2 shows how we may use the same pins to read four switch contacts and output data to an LCD module. (The diodes are ordinary silicon signal diodes, such as 1N4148s.) This circuit operates in two modes as shown in the following table.

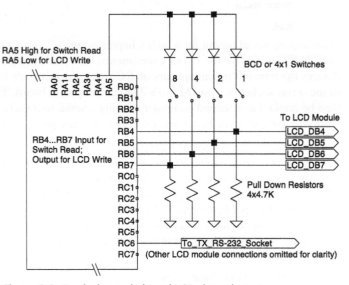

Figure 6-2: Duplexing switch and LCD data pins.

Mode	Pins RB4...RB7	Pin RA5
Read Switches	*Input*	*Output – High*
Output to LCD	*Output*	*Output - Low*

To read the four switch contacts, make RB4...RB7 inputs and take RA5 high. Any switch that is closed causes a high on its connected pin just as in the circuit of Figure 6-1. Pins connected to open switches are pulled low by the pull-down resistors. The LCD module initializes in input mode, so its data input pins DB4...DB7 are effectively open circuits and may be disregarded.

After reading the switch contacts, we take RA5 low and make RB4...RB7 outputs. (The pull-down resistors do not significantly burden the PIC's output pins and may be disregarded.) Each of the output pins RB4... RB7 may be either high (approximately at V_{DD}) or low (approximately at V_{SS}) and the associated switch may be either open or closed. Thus, we have four possible combinations:

Pins RB4...RB7	Associated Switch Open	Associated Switch Closed
High	Switch isolates extraneous circuitry from the connection.	Diode is reversed biased so it presents a high impedance circuit and isolates the extraneous circuitry.
Low	Switch isolates extraneous circuitry from the connection.	Both the anode and cathode of the diode are at approximately same voltage, so no significant current flows and the extraneous circuitry may be disregarded.

Under all possible combinations of RB4...RB7 outputs and switch open/closed, the connection between the PIC and LCD module acts as if the switch were not there.

Program 6-2 demonstrates duplexing the switch and LCD. The LCD connections are the same as we used in Chapter 5.

Program 6-2

```
;Show reading baud rate
;from switches on B4...B7
;8      4      2      1
;B7     B6     B5     B4

;Varibles
;--------
LCDNib  Var PortB.Nib1
SWValue Var     Byte

;Constants
;---------
RegSel Con     B0
Clk    Con     B1
RdWrPin Con    B2

Initialize
;----------
TRISB = $F0
EnableHSerial
Output A5

;Initialize Baud Rate
        High A5
        Pause 5
        SWValue = PortB.HighNib
        Branch SWValue,[S0, S1, S2, S3, S4, S5, S6, S7, |
            S8, S9, S10, S11, S12, S13, S14, S15]
```

89

```
S0  :   SetHSerial H1200       : GoTo InitLCD
S1  :   SetHSerial H2400       : GoTo InitLCD
S2  :   SetHSerial H4800       : GoTo InitLCD
S3  :   SetHSerial H7200       : GoTo InitLCD

S4  :   SetHSerial H9600       : GoTo InitLCD
S5  :   SetHSerial H14400      : GoTo InitLCD
S6  :   SetHSerial H19200      : GoTo InitLCD
S7  :   SetHSerial H24000      : GoTo InitLCD

S8  :   SetHSerial H26400      : GoTo InitLCD
S9  :   SetHSerial H28800      : GoTo InitLCD
S10 :   SetHSerial H31200      : GoTo InitLCD
S11 :   SetHSerial H33600      : GoTo InitLCD

S12 :   SetHSerial H36000      : GoTo InitLCD
S13 :   SetHSerial H38400      : GoTo InitLCD
S14 :   SetHSerial H57600      : GoTo InitLCD
S15 :   SetHSerial H115200     : GoTo InitLCD


InitLCD
        Low A5
        LCDInit RegSel\Clk\RdWrPin,LCDNib
        Pause 500               ;Allows the LCD to initialize
SendMessage
        HSerOut ["Switch Value: ",DEC SWValue,13]
        LCDWRITE RegSel\Clk\RdWrPin,LCDNib,["Switch Value: ",DEC SWValue]
        Pause 5000
        LCDWRITE RegSel\Clk\RdWrPin,LCDNib,[Clear,Home]
        Pause 500
GoTo SendMessage

    End
```

We've seen large parts of this program earlier in this chapter and in Chapter 5, so we'll concentrate on the differences.

Because we can only read the switch under specific conditions, that is, RA5 being high and RB4…RB7 being inputs, if we wish to use the switch value elsewhere in the program it's necessary to store it after reading. Hence we declare a new variable **SWValue** to hold the value. We also make RA5 an output pin with the **Output A5** function.

```
High A5
Pause 5
SWValue = PortB.HighNib
Branch SWValue, [S0, S1, S2, S3, S4, S5, S6, S7, S8, |
S9, S10, S11, S12, S13, S14, S15]
```

The actual read and branch procedure is straightforward; take A5 high, pause for 5 ms to stabilization (the pause statement can be removed if time is an issue; it's added as insurance and isn't strictly necessary) and then read **PortB.HighNib** through the assignment operator **SWValue = PortB.HighNib**. We then branch to the appropriate baud rate selection just as in Program 6-1, but using **SWValue** as the index. (Remember, the | is a continuation character allowing a single logical line of MBasic code to be broken over two or more physical lines.)

```
S0  :   SetHSerial H1200       : GoTo InitLCD
```

Note, however, that after setting the baud rate we jump to **InitLCD**.

```
InitLCD
        Low A5
        LCDWRITE RegSel\Clk\RdWrPin,LCDNib,[InitLCD1,InitLCD2,TwoLine,CLEAR,HOME,SCR]
        Pause 500               ;Allows the LCD to initialize
```

The routine **InitLCD** duplicates the code we explored in Chapter 5. After initializing the LCD, execution then passes to **SendMessage**.

```
SendMessage
        HSerOut ["Switch Value: ",DEC SWValue,13]
        LCDWRITE RegSel\Clk\RdWrPin,LCDNib,["Switch Value: ",DEC SWValue]
        Pause 5000
        LCDWRITE RegSel\Clk\RdWrPin,LCDNib,[Clear,Home]
        Pause 500
GoTo SendMessage
```

We have expanded **SendMessage** to also write the switch value to the
LCD. After displaying the switch value for 500 ms, we erase the LCD,
pause for another 500 ms and repeat the **SendMessage** routine. Figure
6-3 shows the result.

To make this circuit work the LCD module pins must be inputs. If your
program requires reading data from the LCD module, you must read the
switch only during times when the LCD module is in input mode.

Although we have cast duplexing the LCD and switch in the context of
a one-time initialization read, the circuitry and program logic is not so
limited. There is nothing stopping us from duplexing the LCD and switch

Figure 6-3: LCD module displays the switch value.

with multiple reads, such as where the switches control user-adjustable functions that are polled periodically
as long as the program is operating, perhaps along the following lines:

```
Initialization Routines
Initialize LCD
Main Loop
        Read switches
                Make B4...B7 inputs
                Take RA5 high
                Read B4...B7 and store value in SWValue
                Take RA5 low
                Make B4...B7 Outputs
        Do stuff based on SWValue
        Write messages to LCD
GoTo Main Loop
```

Rotary Encoders

The term "rotary encoder" is a fancy name for a specialized switch. A switch, after all, converts mechanical
motion into something—contact operations—that may be sensed electronically. A rotary encoder converts
motion—shaft rotation—into contact operation. We'll look at rotary encoders, but linear encoders also exist where a straight-line motion, such as a rod moving back and forth, is converted into contact operations.
We'll use the word "encoder" as shorthand for "rotary encoder," but remember linear encoders also exist.

Rotary encoders, also known as "shaft encoders" and "position encoders," have many uses. For example,
the position of a machine tool spindle may be located through a rotary encoder and leadscrew mechanism.
(More often, this is done with linear encoders, however.) A rotary encoder on its shaft may measure the
RPM speed of a motor. Or, it may measure the angle of a robotic arm joint, or the azimuth of a rotating radar
antenna. We'll look at a more mundane rotary encoder application, allowing a user to scroll through an LCD
menu and select options. These are low-end applications, but the concepts we learn carry over into precision
and high speed encoders.

Encoders may be grouped into two main categories, *absolute* and *relative*.

Absolute—Absolute encoders provide an output that uniquely identifies the shaft angle. An absolute encoder
can be thought of as a multiposition rotary switch; as you rotate the switch one set of contacts breaks
and a new set of contacts make. Hence, there is a direct relationship between angle and contacts; if we
know which contacts are closed we know the shaft position. The prior position of the shaft is immaterial; to uniquely determine the current shaft angle all we need to know is which contact is closed. (We'll
see how real encoders are made later.)

Relative—Relative encoders (sometimes called "incremental encoders") provide information on the direction of shaft rotation, for example, did the shaft move clockwise or counterclockwise and, with many devices, how many degrees did the shaft rotate? In order to know the angle the shaft is at for this instant, we must know its position at some starting time and remember all the changes reported by the relative encoder since the starting time.

An analogy to absolute and relative encoders is found in house addresses. Suppose you are trying to find a particular house. The absolute address of the house is 123 Main Street. The relative address of the house is: "from where you are now, go two blocks north, three blocks west and it's the fourth house on the right." Both addresses will get you to the same place, but the relative address is useful only from a defined starting point and requires you to keep track of all turns and the distance traveled. Worse yet, should you make a mistake and take even a single wrong turn the absolute address doesn't change; but the relative addressing information becomes worthless.

The same is true with rotary encoders; if we make a mistake and miss one of the relative reports, all future position information will be in error. And, during power-up and initialization, we must index the relative shaft rotation to a known reference. The absolute encoder, in contrast, acts like a nonvolatile memory; it reports its position at start-up just as it did when last running.

For our simple programs where the user rotates an encoder shaft to select items from a menu, we can often assume a relative encoder starts at a default menu. If we instead wish track the azimuth of a rotating radar antenna or if we wish our device to have a front panel with labeled switch positions, we will almost certainly decide to use an absolute encoder. If we use an incremental encoder, we will find it necessary to provide a synchronization mechanism, such as a contact closure every time the device passes a certain azimuth position.

Encoder Construction

Inexpensive encoders are almost always mechanical, similar to rotary switches. Figure 6-4 shows typical relative and absolute mechanical encoders. The CTS Series 288 devices shown cost less than $5.00 in single lot purchases and offer a typical operating lifetime of a 15,000 actuations, quite adequate for manual menu selection and data entry in most applications. Absolute mechanical encoders are rarely found with more than 32 steps/revolution and are limited to moderate rotation speed. They are available with and without detents. (Detents force the shaft to preferred positions while encoders without detents are stable at any angle.)

Figure 6-4: CTS Series 288 mechanical rotary encoders.

Higher end encoders are predominantly optical, with prices ranging from $25 up into the thousands of dollar range. Their lifetime is measured in the tens of millions of rotations and some are rated for high RPM operation. Resolution is available in hundreds or even thousands of positions per revolution. Multiturn rotary encoders are also available, similar to multiturn potentiometers. Since optical encoders have no mechanically moving switch contacts, they have a smooth, almost silky, feel when rotated by hand. Consequently they are used for controls that must have a smooth feel or seek to mimic an analog control, such as the tuning knob on a short wave receiver. Optical encoders are, of course, the product of choice for motor RPM sensors and when used for this purpose are often called "optical tachometers."

Absolute optical encoders use multiple LED/photo diode pairs to read a glass or plastic disc with a printed scale. Relative optical encoders use the same technology, but with a much simpler internal structure. (We'll skip over the details of absolute optical encoders at our level of discussion.)

Figures 6-5, 6-6 and 6-7 show the construction of a special purpose optical relative encoder developed by Sherline Products, Inc. as part of its digital readout option for its line of tabletop milling machines and lathes. This particular device reads the rotation of a calibrated handwheel on a small inexpensive milling machine manufactured by Sherline to determine X, Y and Z movement. (The leadscrews to which the handwheels are attached are 20 threads/inch, and the encoder outputs 100 pulses per revolution, so the position *resolution* is 0.0005." To determine the resulting linear *accuracy* requires analysis of backlash and errors in leadscrew rolling and similar mechanical issues.) The designer of this particular readout system opted for relative encoders, since an integral part of the machining process is assigning a 0, 0, 0 reference location to the part being machined and there would be little gained with a vastly more expensive rotary absolute encoder. Professional digital readouts for milling machines use linear encoders, often with glass scales, so that position information is independent from the leadscrews. These readouts often cost several times as much as the price of a complete Sherline milling machine, including the digital readout.

Let's go through how Sherline's design works. Figure 6-5 shows the handwheel with the interrupter disk attached, as well as an un-mounted interrupter disk. The interrupter disk fits into the sensor holder of Figure 6-6. The sensor holder has two LED/photo receptor pairs, spaced a carefully controlled distance apart. Figure 6-7 shows the interrupter disk loosely inserted into the sensor holder. The leadscrew and shaft assembly, of course, are not shown in these figures.

Figure 6-5: Handwheel and interrupter wheel.

Figure 6-6: LED and photo sensor housing.

Figure 6-7: Interrupter wheel shown in housing.

Suppose we have one LED/photo-sensor pair, as shown in Figure 6-8. Since the interrupter wheel has 25 spokes and 25 slots, we have 50 unique positions for each revolution. But, we can't determine if the rotation direction is clockwise or counter-clockwise. To both improve resolution and provide direction information, we can add a second LED/photo-sensor pair, spaced an integer number of interrupter teeth apart, plus one-quarter tooth. This arrangement is often called a quadrature sensor or quadrature encoder, since the extra one-quarter spacing represents a 90-degree phase shift of one complete opaque/clear cycle. Figure 6-9 shows how this technique both doubles angular resolution and provides rotation direction information. One

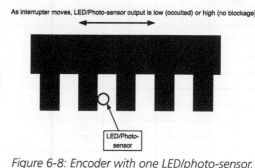

As interrupter moves, LED/Photo-sensor output is low (occulted) or high (no blockage)

LED/Photo-sensor

Figure 6-8: Encoder with one LED/photo-sensor.

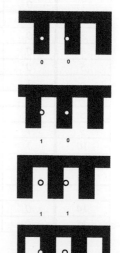

Figure 6-9: Quadrature sensing of rotation.

cycle, i.e., rotation through one solid spoke, gives four outputs, and rotation direction can be derived from the order in which the outputs occur. Since we have 25 spokes, each revolution yields 100 unique outputs.

Direction	Output	Direction
↑	00	↓
	10	
	11	
	01	

Although we've looked at Sherline's relative encoder, the same principles apply to off-the-shelf rotary encoders you purchase from an electronics parts house, including the ones in Figure 6-4.

Code Output

Absolute encoders are available with a selection of output coding, with the two most commonly available being Gray code and straight hexadecimal binary code. Gray coding (named after Frank Gray, its inventor) changes by only one bit for every state change. Hexadecimal code can change with more than one bit (think about the rollover in a 4-bit code from $F (1111) to 0 (0000); all four bits simultaneously change state). Gray-coded encoders permit no improper intermediate output states. For example, if our algorithm reads a 4-bit hexadecimal-coded encoder starting at the instant any bit changes value, we may get an erroneous result unless all bits change simultaneously, or we adequately debounce the output. Since only one bit changes for each state transition in a Gray-coded device, readings triggered by a single bit change automatically produce the correct result. If our encoder is turned by hand, there may be little advantage to a Gray-coded output, but if we are reading a motor shaft spinning at 3600 RPM, a Gray-code output is preferable.

The difference between hexadecimal and Gray code can be seen for a CTS Series 288 mechanical 4-bit (16 positions per revolution) encoder.

Position	4-Bit Gray Code Contact Closure to Common					4-Bit Hexadecimal Code Contact Closure to Common				
	A (LSB)	B	E	F (MSB)	Value	A (LSB)	B	E	F (MSB)	Value
0	0	0	0	0	0	0	0	0	0	0
1	0	0	1	0	4	1	0	0	0	1
2	0	0	1	1	12	0	1	0	0	2
3	0	0	0	1	8	1	1	0	0	3
4	0	1	0	1	10	0	0	1	0	4
5	0	1	1	1	14	1	0	1	0	5
6	1	1	1	1	15	0	1	1	0	6
7	1	1	0	1	11	1	1	1	0	7
8	1	0	0	1	9	0	0	0	1	8
9	1	0	1	1	13	1	0	0	1	9
10	1	0	1	0	5	0	1	0	1	10
11	1	0	0	0	1	1	1	0	1	11
12	1	1	0	0	3	0	0	1	1	12
13	1	1	1	0	7	1	0	1	1	13
14	0	1	1	0	6	0	1	1	1	14
15	0	1	0	0	2	1	1	1	1	15

Reading a Relative Encoder

After all this theory, let's look at some code. Program 6-3 will read a 16 position, two-bit relative shaft encoder (CTS model 288T232R161A2) and illuminate an LED indicating whether the shaft is rotated clockwise or counter-clockwise. We'll also increment a counter for clockwise rotation and decrement it for counter-clockwise rotation and output the counter value over the RS-232 port. The circuit associated with Program 6-3 is shown in Figure 6-10.

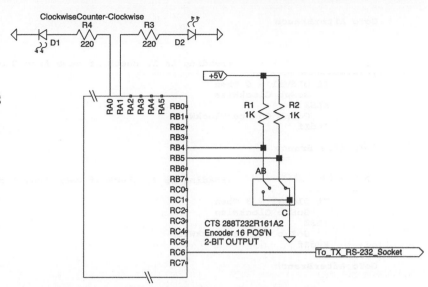

Figure 6-10: Schematic for incremental encoder read program.

Program 6-3

```
;Variables
;-----------
CurVal          Var            Byte    ;current encoder
OldVal          Var            Byte    ;old encoder
Counter         Var            Sbyte   ;counts state changes

;Initialize
;-----------
Clear
Output A0                ;clockwise LED
Output A1                ;counter-clockwise LED

Low A0                  ;initialize both LEDs off
Low A1
EnableHSerial
SetHSerial H115200      ;change if lower speed needed


CurVal = (PortB & %00110000) >> 4      ;get current reading
OldVal = CurVal                        ;initialize for future change
HSerOut ["Ctr ",SDEC Counter,13]       ;output the counter

Main
;--------------
    CurVal = (PortB & %00110000) >> 4  ;returns 0...3
    If CurVal <> OldVal Then
        Branch CurVal, [S0,S1,S2,S3]   ;if changed, look at old
    EndIf
    AfterBranch
GoTo Main


S0                      ;reading is 0, check if came from 2 or 1
;----------
        If OldVal = 2 Then
            GoSub ClockWise
        ELSE
            GoSub CounterClockWise
        EndIf
```

```
        GoTo AfterBranch


S1                        ;reading is 1, check if came from 0 or 3
;-----------
        If OldVal = 0 Then
            GoSub ClockWise
        ELSE
            GoSub CounterClockWise
        EndIf

GoTo AfterBranch


S2                        ;reading is 2, check if came from 3 or 0
;-----------
        If OldVal = 3 Then
            GoSub ClockWise
        ELSE
            GoSub CounterClockWise
        EndIf

GoTo AfterBranch


S3                        ;reading is 3, check if came from 1 or 2
;-----------
        If OldVal = 1 Then
            GoSub ClockWise
        ELSE
            GoSub CounterClockWise
        EndIf

GoTo AfterBranch


ClockWise
;--------------
        High A0                ;turn CW LED on
        Low A1                 ;turn CCW LED off
        OldVal = CurVal        ;ready to check for next input
        Counter = Counter + 1  ;bump up counter
        HSerOut ["Ctr ",SDEC Counter,13]        ;output the counter
Return


CounterClockWise
;--------------
        Low A0                 ;turn CW LED off
        High A1                ;turn CCW LED on
        OldVal = CurVal        ;ready for next input
        Counter = Counter -1   ;decrement counter
HSerOut ["Ctr ",SDEC Counter,13]        ;output the counter
Return

        End
```

Let's go through the interesting portions of the code.

```
;Variables
;-----------
CurVal          Var          Byte
OldVal          Var          Byte
Counter         Var          SByte
```

We'll read and hold the encoder value in **CurVal**. **OldVal** holds the previous encoder value, and we will increment or decrement **Counter** depending on the direction of shaft rotation. **Counter** is a signed byte, with permitted values from –128 to +127, thus allowing us to increment or decrement `Counter` from its initialized value of 0.

```
CurVal = (PortB & %00110000) >> 4
OldVal = CurVal
```

We initialize by reading the current encoder value and set the old value equal to the current value. This permits us to detect the first rotation by checking for `CurVal <> OldVal`. First, we perform a bitwise AND ("`&`") of the read value with %00110000. The AND function masks off all bits read on PortB except for those corresponding to 1's in the mask operator `%00110000`, B5 and B4. Masking all except B5 and B4 permits us to ignore the values returned by the remainder of PortB's pins. Since the encoder is connected to B4 and B5, the read value is shifted to the left by four bits, which we remove by shifting the returned value four bits to the right. (The reason we connected the encoder to B4 and B5 instead of B0 and B1 will be seen when we look at the interrupt version of Program 6-3.) To force the correct order of operation, we enclose the logical AND operation with parentheses. In the absence of the parentheses, MBasic 5.3.0.0 would first divide %00110000 by 16 and then logically AND the result with the value read on PortB. This represents a change from earlier versions of MBasic, where the order of precedence was different.

CTS's data sheet shows the output sequence (assuming clockwise rotation) for the first eight positions as:

Contact Closure	B	0	0	1	1	0	0	1	1
To Common	A	0	1	1	0	0	1	1	0
Position No.		0	1	2	3	4	5	6	7

In our circuit a contact closure (data sheet 1) results in a logical low at the PIC's input pin, and an open (data sheet 0) is a logical high. Hence, our circuit inverts the 0's and 1's from the CTS's data sheet's prospective. As shown in Figure 6-10, pin A is connected to B4, while pin B is connected to B5. Consequently, our PIC input pin sequence looks like:

PIC Pin	B5	1	1	0	0	1	1	0	0
	B4	1	0	0	1	1	0	0	1
Position No.		0	1	2	3	4	5	6	7
PortB & b00110000		48	32	0	16	48	32	0	16
PortB & b00110000 / 16		3	2	0	1	3	2	0	1

We now rewrite (`PortB & b00110000`) `>> 4` in two rows, representing clockwise and counter-clockwise rotation, starting with 0 for convenience:

| Clockwise | 0 | 1 | 3 | 2 | 0 | 1 | 3 | 2 |
| Counter-Clockwise | 0 | 2 | 3 | 1 | 0 | 2 | 3 | 1 |

To understand how we differentiate clockwise and counter-clockwise rotation, suppose the encoder's current value is read as 0. If the previous value was 0, the encoder shaft has not moved since its last read. If the previous value was 2, the rotation is clockwise; if the previous value was 1, the rotation is counter-clockwise.

```
Main
;---------------
    CurVal = (PortB & %00110000) >> 4
    If CurVal <> OldVal Then
        Branch CurVal, [S0,S1,S2,S3]
    EndIf
    AfterBranch
GoTo Main
```

With this understanding, the main program should be clearer. We have two sequential comparisons to make. First, has the encoder moved, which we test for with the `If CurVal <> OldVal Then`… test. If the encoder shaft has rotated since its last read, the conditional is true and we branch to one of four program labels based on the current encoder value, `S0` if the value is `0`, and so on.

```
        S0                          ;reading is 0, check if came from 2 or 1
        ;----------
                If OldVal = 2 Then
                    GoSub ClockWise
                ELSE
                    GoSub CounterClockWise
                EndIf

        GoTo AfterBranch
```

Program execution reaches **S0** only if **CurVal = 0**. Hence, to determine the rotation direction, we only need test the prior value, which we do with the **If OldVal = 2** test. If true, the direction is clockwise and we execute the **ClockWise** subroutine; if false, the direction must be counter-clockwise, so we execute the **CounterClockWise** subroutine.

```
        ClockWise
        ;--------------
                High A0                      ;turn CW LED on
                Low A1                       ;turn CCW LED off
                OldVal = CurVal                 ;ready to check for next input
                Counter = Counter + 1   ;bump up counter
        HSerOut ["Ctr ",SDEC Counter,13]        ;output the counter

        Return


        CounterClockWise
        ;--------------
                Low A0                       ;turn CW LED off
                High A1                      ;turn CCW LED on
                OldVal = CurVal                 ;ready for next input
                Counter = Counter -1    ;decrement counter
        HSerOut ["Ctr ",SDEC Counter,13]        ;output the counter

        Return
```

The **ClockWise** and **CounterClockWise** subroutines are mirror images of each other; they set and clear the direction LEDs and bump up or decrement the counter and write the updated value of the position counter to the hardware serial port.

After reading Chapter 4, you may wonder where the debounce routine is. After all, we are sensing a mechanical switch whose data sheet quotes a 5 ms maximum bounce period. The answer is that the execution time of the program, including the serial output statement, provides sufficient delay to provide trouble free reading. In most cases, a real program would branch into enough code that consumed enough time to provide debounce as an incidental benefit. If not, we could add a **Pause 5** statement after each switch read. Or, we might add debounce circuitry to our design.

As it stands now, with a 20 MHz clock, Program 6-3's main program loop executes in 400 µs where the encoder has not changed value, and in 750 µs where it has changed value. Thus, it accurately reads well over 1,000 steps/second, or 3700 RPM, far faster than someone twisting a knob attached to the shaft will operate the encoder.

Figures 6-11 and 6-12 show the input waveforms as I turned the shaft quickly by hand. Incidentally, you may read references to quadrature encoder outputs being read by comparing the phase of the two outputs for clockwise and counter-clockwise rotation. Figure 6-11 shows Channel 2 (B5) leading Channel 1 (B4) during clockwise rotation, while Figure 6-12 shows Channel 1 (B4) leads Channel 2 (B5) during counter-clockwise rotation. And, in fact, the waveforms at B4 and B5 are 90° out of phase. I view this phase comparison analysis far more confusing than beneficial, particularly in a beginning level discussion. Hence we have instead focused on reading the encoder value and comparing it with the prior value.

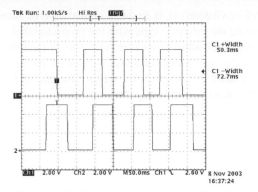

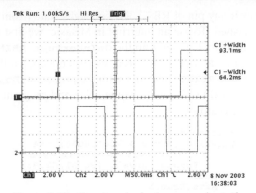

Figure 6-11: Clockwise manual rotation Ch1: B4 Ch2: B5.

Figure 6-12: Counter-clockwise manual rotation Ch1: B4 Ch2: B5.

Program 6-3 spends a great deal of time potentially doing nothing; in a typical program that uses a rotary encoder to select menu items, operator interaction is infrequent. How many times do you change the thermostat setting in your house? Certainly not hundreds of times a second. Wouldn't it be nice if we our main program could spend its time making important computations and branch to deal with a changed user input only when the encoder is changed, without checking the encoder's value every time we go through our main loop? The answer, of course, is that not only would this be nice, the folks at Microchip and Basic Micro have come up with a way to do exactly that through the mechanism of interrupts. We're going to save interrupts until Chapter 10, but here's a sneak preview.

Whenever the state of input pins RB4…RB7 change, we can cause the PIC to stop whatever code it is running and instead run new code. This process is called an "interrupt" in computer-speak and it does exactly that; it interrupts the current code and switches execution to an "interrupt handler."

Delete the Code Below from Program 6-3	*Replace it with the Following Interrupt Version*

<pre>CurVal = (PortB & %00110000) >> 4
OldVal = CurVal
 	
Main
;--------------
   CurVal = (PortB & %00110000)>>4
      ;returns 0…3
      If CurVal <> OldVal Then
      Branch CurVal, [S0,S1,S2,S3]
      ;if changed, look at old
   EndIf
   AfterBranch
GoTo Main</pre>
	<pre>CurVal = (PortB & %00110000) >> 4
	OldVal = CurVal
	
	OnInterrupt RBINT, ReadEncoder
	Enable RBINT
	
	Main
	;--------------
	GoTo Main
	
	Disable
	
	ReadEncoder
	CurVal = (PortB & %00110000)>>4
	
	    If CurVal <> OldVal Then
	    Branch CurVal, [S0,S1,S2,S3]
	  EndIf
	  AfterBranch
	Resume</pre>

Whenever any pin of RB4…RB7 changes state (from 0 to 1 or vice versa), if enabled, an **RBINT** type interrupt is issued. To use an interrupt in MBasic, we first associate the interrupt type with the program label of the interrupt handler with the **OnInterrupt RBINT, ReadEncoder** procedure. This simply says that whenever an **RBINT** is triggered, we execute code following the label **ReadEncoder**. Next, we turn on or "enable" the **RBINT** interrupt with the procedure **Enable RBINT**. Now, whenever we turn the encoder shaft, either B4 or B5 will change value and program execution will jump to **ReadEncoder**.

```
Disable

ReadEncoder
CurVal = (PortB & %00110000) >>4
        If CurVal <> OldVal Then
        Branch CurVal, [S0,S1,S2,S3]
    EndIf
    AfterBranch
Resume
```

In order to avoid interrupts of interrupts, we must turn off additional interrupts before the interrupt handler via the **Disable** function. The interrupt procedure itself duplicates the code we developed in Program 6-3. At the end of the interrupt procedure, we return to normal program flow through a **Resume** statement.

The code executed in the **Main … GoTo Main** loop in our interrupt program is simply a serial output. But, it could have been anything, such as lengthy mathematical calculations, LCD writes, or anything else you can do in MBasic. When the interrupt occurs, MBasic will seamlessly execute the interrupt handler and return flow to your main program. Hence, it is up to your code to recognize and accommodate changes to **Counter**—as a result of user input commands—between program statements.

There's another subtlety here as well. MBasic will not break execution of a statement to service an interrupt. Suppose, for example, your main code has a one second delay in it through a **Pause 1000** procedure. If you twist the shaft on the encoder just as the **Pause 1000** procedure starts execution, the interrupt handler will *not* be invoked until after the **Pause 1000** completes. So, if this happens you may twist the shaft encoder knob through a dozen positions, but the interrupt procedure won't recognize them. (In a later chapter we'll later see that an assembler language interrupt doesn't share this problem.)

Dual Encoders and LCD

Let's see how we might meld our LCD and rotary encoder work. Suppose we wish to make a custom thermostat with a different setting for each room in our house. We'll see how you might make a display that shows the room name on one line of an LCD and the temperature set point on the second line. One rotary encoder selects the room and a second rotary encoder sets the de-

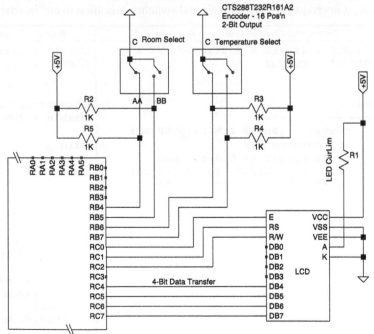

Figure 6-13: Two encoders and LCD.

sired temperature. We'll build only a skeleton program, omitting many useful elements, such as storing set points in EEPROM to preserve settings after power outages.

Our wiring diagram is shown in Figure 6-13. The LCD we will use is a 20 × 4 model, but since the program uses only the top two lines a smaller two-line display may be substituted without change. The resulting display can be seen in Figure 6-14. To operate, one encoder selects the room. Turning it clockwise advances through the room names and counter-clockwise reverses through the name list. A second encoder varies the room's temperature set point; clockwise to increase and counter-clockwise to decrease the set point temperature.

Figure 6-14: Sample output of Program 6-4.

Program 6-4

To save space, we have omitted a printed listing of Program 6-4. Its full content is available on the CD-ROM.

Let's look at the parts of Program 6-4 we haven't seen before:

```
CurVal          Var     Byte    ;current room encoder value
OldVal          Var     Byte    ;old room encoder value
Counter         Var     SByte   ;up/down room counter

TempCtr         Var     SByte   ;up/down temperature counter
CurTempEnc      Var     Byte    ;current temperature encoder reading
OldTempEnc      Var     Byte    ;old temperture encoder reading

EncFlag         Var     Byte    ;is interrupt from temperture or room
TempOld         Var     Byte    ;temporary S0...S3 index
```

The room selection variables have the same name and same function developed in Program 6-3. We also set up a parallel variable structure to hold the temperature encoder values. These three variables have the word **Temp** in them. We also declare a sentinel or flag variable, **EncFlag**, to identify whether the encoder being read is the room selection or temperature adjust encoder. Finally, since we wish to reuse the branch and the prior encoder test routines, we declare a new variable **TempOld** (temporary old) that replaces **OldVal** in these routines.

```
WriteDM 0, [11,"Living Room"]
StrIndex(0) = 0
```

Since we wish to select and display room names using an up/down counter, a simple data structure would be an array of strings, accessed through the counter as the index variable. Indeed, in Pascal terminology, we might prefer to define an array of records to hold not just the room name, but also the set point temperature and other related information. Alas, MBasic doesn't permit arrays of strings, or, for that matter, other two-dimensional arrays, let alone arrays of records. And, if it did, we would quickly consume RAM for static messages, unnecessarily wasting precious system resources. So, we need an alternative way to display String(0), String(2)…

We've briefly mentioned EEPROM memory in Chapter 2. To recap, EEPROM is nonvolatile memory available in many, but not all, PICs. Rather than consume RAM with a series of byte arrays containing room names, we'll instead write the strings to EEPROM and retrieve them as necessary. MBasic provides several mechanisms for writing to EEPROM, and we'll use the **WriteDM** procedure.

```
WriteDM Address [{mod} Expression,{mod}Expression...]
```

We are responsible for keeping track of how much memory when we write to EEPROM, and we must define the starting memory address, **Address**. The 16F876/877 PIC family we're using has 256 bytes of EEPROM and **Address** accordingly ranges from 0…255. **WriteDM** automatically writes the contents within the brackets to consecutive EEPROM addresses, starting with **Address**. We write a package to EEPROM containing the length of the string (excluding this length byte) followed by the string contents, so that we may simplify the process of writing each string to the LCD. (We're borrowing this idea from early versions of Borland's Pascal compiler. I view it as a better approach for short strings and limited storage space than the "terminate the string with a 0" approach used in C compilers and later versions of Pascal.) We store the starting address of each packaged string in an array **StrIndex** so that we may access each string with an index number.

All of this work gives us the functional equivalent of RoomName(i), where RoomName(0) = "Living Room," RoomName(1) = "Dining Room," and so on. We use the room selection encoder to vary i from 0 to 7 thereby permitting us to select and write the selected room name to the LCD. We've also declared the array **CurValue()** to hold the desired set point temperature for each room.

```
;Initial read of both encoders
CurVal = (PortB & %00110000) >> 4
OldVal = CurVal

CurTempEnc = (PortB & 11000000) >> 6
OldTempEnc = CurTempEnc
```

We follow the concept introduced in Program 6-3 to read the temperature encoder. Since the temperature encoder is connected to pins B7 and B6, the mask and divisor change, as seen in the previous code fragment.

```
Main
        ;Keep Counter to legal room numbers
        If Counter > MaxRoomNo Then
                Counter = MaxRoomNo
        EndIf

        If Counter < 0 Then
                Counter = 0
        EndIf

    MsgID = Counter
```

The main program loop first ensures that the room encoder counter does not return a value outside the permitted index range. We have eight rooms, indexed from 0…7, so we correspondingly limit **Counter** to 0…7. We set **MsgID = Counter**, since we **MsgID** is a dummy index variable used to display strings in the subroutine **WritelnMSG**.

```
        ;Changed Room Encoder?
        If MsgID <> OldMsgID Then
            LineNo = 1
            GoSub WritelnMsg              ;Write new room
            OldMsgID = MsgID
            TempCtr = CurSetting(MsgID)   ;Display temp
            LineNo = 2
            GoSub WritelnTemp
        EndIF
```

As in Program 6-3, if the room selection encoder has changed, then **MsgID**, has changed and we execute the statements following the **IF…Then** statement. Let's take a quick look at the subroutine **WritelnMsg**.

```
WritelnMsg  ;Writes EEPROM string to LCD. Starting adr is Adr
;---------- ;Line is LineNo
    ReadDM StrIndex(MsgID),[StrLen]
    ReadDM StrIndex(MsgID)+1,[Str TempStr\StrLen]
    LCDWrite RegSel\Clk\RdWrPin,LCDNib,[ScrRAM + LineOff(LineNo)]
    LCDWrite RegSel\Clk\RdWrPin,LCDNib,["                        "]
    LCDWrite RegSel\Clk\RdWrPin,LCDNib,[ScrRAM + LineOff(LineNo)]
    LCDWrite RegSel\Clk\RdWrPin, LCDNib,[Str TempStr\StrLen]
Return
```

Since we can't write custom procedures and functions with local variables in MBasic, we'll use global variables and subroutines. **WritelnMsg** displays the message pointed to in **StrIndex(MsgID)** on the LCD line identified in **LineNo**. First, we read the length of the string to be displayed and store it in **StrLen** via the statement **ReadDM StrIndex(MsgID), [StrLen]**. Next, we retrieve the string itself (starting one byte past the starting address) and store it in **TempStr** via the statement **ReadDM StrIndex(MsgID)+1, [Str TempStr\StrLen]**. Note that we use the string length to read the correct number of bytes from EEPROM.

The remainder of this subroutine should be familiar after thrashing through LCD workings in Chapter 5. To keep the subroutine flexible enough to write on any of the four lines in our display, we index the starting memory to the line number via the array **LineOff(LineNo)**. (The values of **LineOff(1)…LineOff(4)** are taken from the display's data sheet and are established in the initialization section of Program 6-4.) We then clear the line by writing 20 space characters, followed by resetting the LCD write address to the start of the line and write the desired room name.

Going back to the main routine, after writing the room name on line 1, we set the temperature encoder counter to the current value of the set point for the displayed room with the assignment statement **TempCtr = CurSetting(MsgID)**.

Finally, we write the set point temperature on line 2 with a call to **GoSub WritelnTemp**. Subroutine **WritelnTemp** follows the format of **WritelnMsg**, but instead of writing a string, it writes the set point temperature for the selected room with **LCDWrite RegSel\Clk\RdWrPin,LCDNib, ["Setting ",DEC CurSetting(Counter),"F"]**. This statement write the set point temperature held in **CurSetting()** to the line defined by the value of **LineNo**.

```
        ;Changed temp setting?
        If TempCtr <> CurSetting(MsgID) Then
            CurSetting(MsgID) = TempCtr
            LineNo = 2
                GoSub WritelnTemp
        EndIf

        Goto Main
```

The main routine now checks to see if the temperature encoder has been changed. (Remember we set the temperature counter to the current set point whenever we display a new room) The temperature counter is compared with the set point of the selected room, and if it differs, we update the set point through **CurSetting(MsgID) = TempCtr** and display the new value with a call to **WritelnTemp**.

The remainder of the work is done through the interrupt handler, **ReadEncoder**.

```
    ReadEncoder
    ;Invoked on change in B4...B7
        CurVal = PortB & %00110000 / 16    ;Room encoder
            If CurVal <> OldVal Then        ;If changed, check it out
                    EncFlag = RoomEnc        ;ID source as Room Encoder
                    TempOld = OldVal
            Branch CurVal, [S0,S1,S2,S3]
        EndIf

        CurTempEnc = PortB & %11000000 / 64    ;Temperature encoder
            If CurTempEnc <> OldTempEnc Then     ;If changed
                    EncFlag = TempEnc            ;ID as Temperature Encoder
                    TempOld = OldTempEnc
                    High A2
            Branch CurTempEnc, [S0,S1,S2,S3]
        EndIf
        AfterBranch
    Resume
```

ReadEncoder is invoked whenever B4…B7 change state, caused by rotating either the room selection or temperature selection encoders. We distinguish which encoder has changed by comparing their respective current values with their last read values. The interrupt handler then sets the encoder flag to either `RoomEnc` or `TempEnc` and sets the variable `TempOld` to either `OldVal` or `OldTempEnc`. Execution then jumps to S0...S3 via the `Branch` instruction, just as in Program 6-3.

```
S0
;----------
        If TempOld = 2 Then
            GoSub ClockWise
        ELSE
            GoSub CounterClockWise
        EndIf

    GoTo AfterBranch
```

S0...S3 are identical with their corresponding components in Program 6-3, except the old value being tested is held in `TempOld`. Using `TempOld` permits us to use the same code to handle either encoder. Otherwise, we would have to construct parallel and virtually identical code, to analyze each encoder's state.

```
ClockWise
;--------------
        ;Check which encoder is changed
        If EncFlag = RoomEnc Then       ;Room Encoder
                OldVal = CurVal
                Counter = Counter + 1
        EndIf

        If EncFlag = TempEnc Then        ;Temperature encoder
                TempCtr = TempCtr + 1
                If TempCtr > MaxTemp Then
                        TempCtr = MaxTemp
                EndIf
                If TempCtr < MinTemp Then
                        TempCtr = MinTemp
                EndIf
                OldTempEnc = CurTempEnc
        EndIf
    Return
```

The subroutines `ClockWise` and `CounterClockWise` are based upon the subroutines in Program 6-3, but expanded to differentiate between the two encoder sources. Based upon the encoder source flag variable `EncFlag`, these two subroutines increment (clockwise) or decrement (counterclockwise) the correct counter (`Counter` for the room selection encoder; `TempCtr` for the temperature set encoder). These subroutines also limit the possible values of `TempCtr` to the pre-established maximum (`MaxTemp`) and minimum (`MinTemp`) temperature values.

Ideas for Modifications to Programs and Circuits

- Modify Program 6-4 to write the temperature set points to EEPROM and to read the set points from EEPROM upon start-up.
- Expand `WritelnMsg` and `WritelnTemp` to add a third parameter, starting character number. Make an appropriate change to the erase functions of these subroutines.
- Rewrite Program 6-4 to use up/down left/right push button switches instead of rotary encoders. After a button is held in for a certain time, the action should repeat. Add a fifth button "Enter" to accept a changed temperature set point value.
- Expand Program 6-4 to include other typical thermostat functions, such as Mode: Heat/Cool/Off/Manual Fan. Add a current room temperature reading to the display.
- It's occasionally necessary to read several BCD switches, such as where the PIC is the interface between user settable switches and a serial-programmed frequency synthesizer chip. Figures 6-15 and 6-16 show how many switches may be multiplexed to reduce the number of pins that must be dedicated to switch reading. A pseudo-code algorithm to read the configuration of Figure 6-15 is:

```
Make RB0...RB3 Outputs
Make RB4...RB7 Inputs
For switch = 0 to 3
        Make RB0+switch high
        All other of RB0...RB3 make low
        Read switch value on RB4...RB7
Display switch value on LCD
Next switch
```

Breadboard up the circuits and develop code to read the switches and write the results on an LCD display driven from PortC as in Program 6-4.

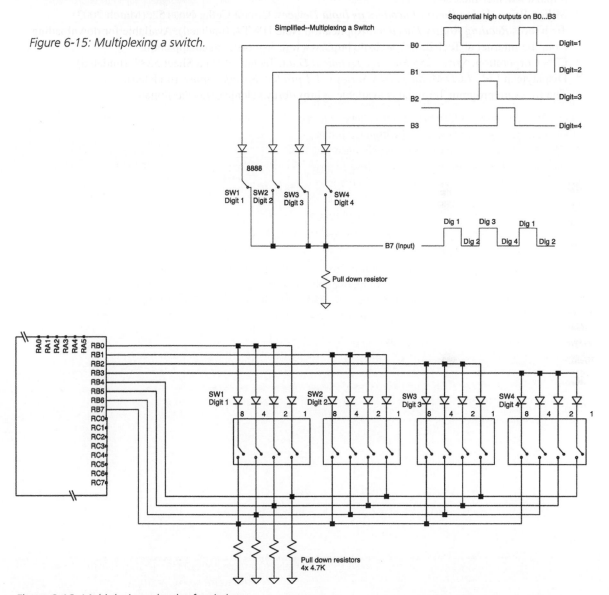

Figure 6-15: Multiplexing a switch.

Figure 6-16: Multiplexing a bank of switches.

References

[6-1] Julicher, Joseph, *Hardware Techniques for PICmicro Microcontrollers*, Microchip Technologies, Inc. Application Note AN234 (Document no. DS00234A) (2003).

[6-2] D'Souza, Stan, *Four Channel Digital Voltmeter with Display and Keyboard*, Microchip Technologies, Inc. Application Note AN557 (Document no. DS00557C) (1997).

[6-3] Microchip Technologies, Inc., *Tips 'n Tricks*, (Document no. DSS40040B) (2003).

[6-4] More information on Sherline's tabletop machinery products can be found at its website, http://www.sherline.com.

[6-5] Miller, Brian, *Using Rotary Encoders as Input Devices*, Circuit Cellar No. 152, (March 2003).

[6-6] TechTools *Reading Rotary Encoders*, Application Note HWT4, (undated). Available for downloading from the Internet at TechTools's website http://www.tech-tools.com/catapps.htm.

[6-7] CTS Corporation, *Series 288 Encoder Technical Data*, Technical Data Sheet 5288, (undated).

[6-8] Displaytech Ltd., *LCD Module 204A Series v. 1.1 product Specifications*, (undated).

[6-9] The freeware program Terminal is available at http://bray.velenje.cx/avr/terminal/.

Seven-Segment LED Displays

In many respects, an LCD module is an ideal display; easy to use, available in a wide choice of sizes and capacity and supported in MBasic with one easy-to-use function, **LCDWrite**. But, as good as LCD modules are, there are times when they just won't do. Chapter 7 looks at an earlier—but still common—technology, seven-segment LED displays. (Indeed, much of what we learn in this chapter may be directly applied to an even older technology, the Nixie tube.) Seven-segment displays have been built with other technologies, including incandescent filament and vacuum fluorescent, but these are rarely used today.

LED displays are available in several formats, including dot matrix (5x7, 8x8 and other pixel arrangements) and alphanumeric multisegment (16-segment, for example). We're going to work with a simpler device, though, the humble 7-segment numeric display, intended to display the digits 0…9. (It's possible, after a fashion, to display the letters "a","b","c","d","E" and "F" on a seven-segment display, should you desire.) The particular device used in this chapter is an Agilent Technologies HDSP-5721, two digit, common anode yellow LED display, with a 0.56" character height. Any common anode display may be substituted for the HDSP-5721 with, at most, minor changes to the circuitry. (The main difference will be the value of current limiting resistors.) We'll also use a variant seven-segment display in Chapter 11.

LED Display Selection

DigiKey's catalog shows nearly 500 different seven-segment LED displays, with a wide range of size, color, packages and configuration. We'll assume that you can sort out the size and color display necessary for your application. Several selection parameters are worthy of further consideration, however.

Mounting style—The overwhelming majority of seven-segment displays are made with connection pins exiting at the back of the housing. In most cases, however, your main PCB will be horizontal, thus requiring either a separate vertically mounted PCB to hold the displays, or expensive right angle display sockets. A few displays, however, are available with bottom pins for right angle mounting. (We'll use these in Chapter 11's digital voltmeter.)

Seven-segment displays are also manufactured as individual digits, paired digits, triple digits and quad digits. For experimentation and education purposes, there's little to be gained from anything other than individual digits. I used HDSP-5721 dual digit displays because I happened to have a quantity on hand.

Special characters—Sometimes the display must include a special character, such as "+" and "–" signs. The "–" can be simply the center segment of a normal seven-segment display, but if a "+" sign is necessary a special display configuration is required, or "three-quarters" of a "+" can be made from a standard display. You can also find partial digit displays, with a "+", "–" and the digit "1". Other common display arrangements include clock style displays, with colons separating the hour and minute digits. Decimal points are another option, with their placement being either left or right of the digit (or no decimal point at all).

Common anode or common cathode—Figure 7-1 shows a typical seven-segment display, with the segments identified A…G and the decimal point. Each segment contains one or more LEDs.

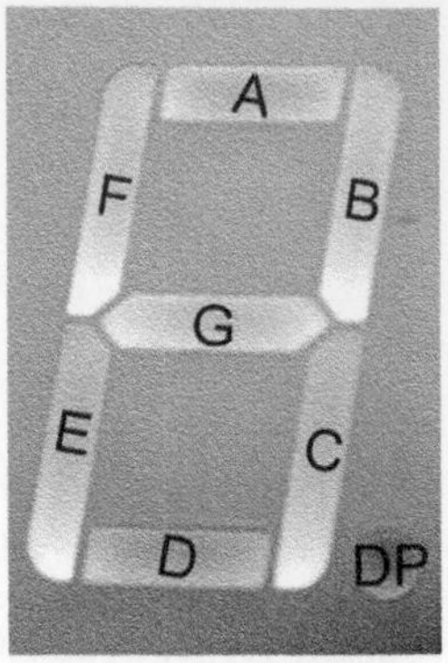

It isn't necessary to connect to both the cathode and anode of each individual diode, so to save pins, individual segment LEDs are constructed with either their anodes or cathodes in common, as shown in Figure 7-2. Common cathode displays require individually controlled high side segment drivers and a common low side digit selector. Common anode displays use a high side digit selection switch and individual low side digit selectors. Either configuration, of course, requires a current limiting resistor for each segment. We'll use a common anode display, but the program logic and hardware can be easily revised to work with a common cathode display.

There's one reason to prefer common anode displays. Our design uses separate digit and segment driver transistors permitting high segment drive current for improved brightness. If your display provides adequate brightness with 25 mA or less segment current, you may directly connect the segments of a common anode display to the PIC's output pins, as they will safely sink 25 mA. In a common cathode display, however, the PIC would be required to source current to each segment and, as we found earlier, PICs are better at sinking current than sourcing current.

Figure 7-1: Seven-segment display segment identification.

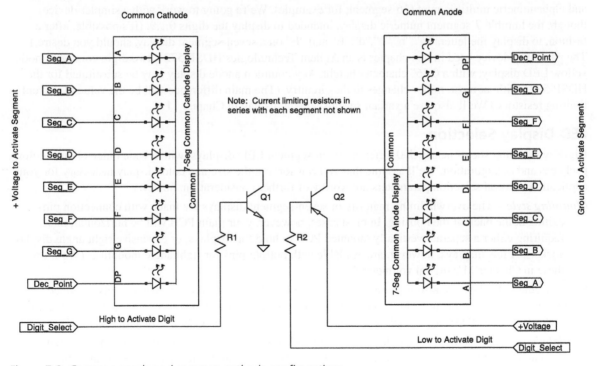

Figure 7-2: Common anode and common cathode configurations.

Circuit Design

Figure 7-3 shows our test circuit. It's a familiar mix of high-side and low-side drivers as discussed in Chapter 3. Let's look at it in detail.

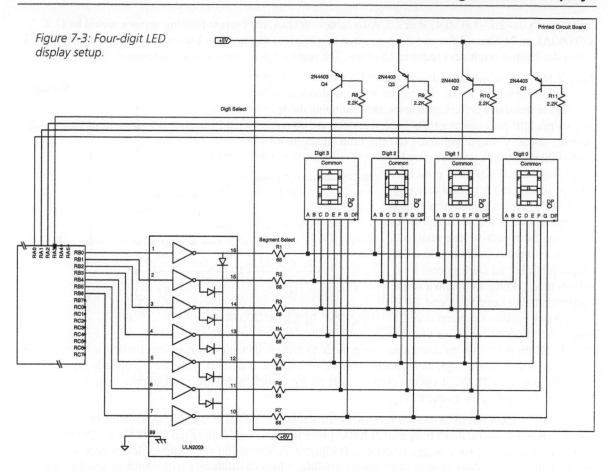

Figure 7-3: Four-digit LED display setup.

Pins A0…A3 are connected to 2N4403 PNP transistors set up as high-side digit switches. When A0…A3 go low, Q1…Q4 are forward biased, and +5V, less V_{SAT}, appears on the common anodes of Digits 3…0. We've selected R8…R11 (2.2K ohm) to limit the base current drive to around 2.5mA.

Each digit has its cathode segments A…G connected in parallel, with each segment going to an output pin of a ULN2003A NPN Darlington array. The corresponding inputs of the ULN2003A are connected to pins B0…B6. (The decimal point isn't connected, as it isn't needed in our application. But, don't worry; we'll see a seven-segment display with automatic decimal point setting in Chapter 11.)

When any pin B0…B6 goes high, the attached Darlington is saturated and thus completes the circuit.

Thus, to turn on Segment(i) of Digit(j), the associated B0…B6 pin must be high and the associated digit selection pin A0…A3 must be low. This permits us to individually control each segment of each digit. To give the appearance of simultaneous operation, we cycle through all four digits quicker than our persistence of vision recognizes flicker.

Each segment of a HDSP-5721 is rated for 60 mA peak current, and 20 mA continuous current. Since we are multiplexing four digits, we'll be limited by the peak current rating. The HDSP-5721's forward voltage is specified as 2.1V typical. As we'll see below, we may expect around 1.0 V drop across the ULN2003A driver. From Chapter 3, we know the 2N4403's V_{SAT} is around 0.4V. Hence, the voltage across the limiting resistor is (5.0 – 2.1 –1.0 –0.4) V or 1.5 V. To achieve 60 mA, the current limiting resistor would be (1.5 V/0.060A), or 25 ohms. I found 60 mA resulted in a brighter display than I wanted, so by experimentation I found the desired brightness required 68 ohms. The resultant peak current is approximately 25 mA.

The ULN2003A is an interesting and useful device; so let's take a quick look at what's inside. It consists of seven identical NPN Darlington transistor drivers, and associated snubbing diodes in a 16-pin DIP package. Each Darlington is rated at 500 mA current, and 50 V output voltage. (You can't run all seven drivers each sinking 500 mA without grossly exceeding the ULN2003A's power ratings, of course.) A typical driver appears in Figure 7-4. Again, this should be familiar from Chapter 3. The driver transistor has a built-in base resistor, so we may directly connect it to a PIC output pin. Like all Darlingtons, it suffers in V_{SAT} compared with a single transistor, being rated at around 1.0 V for 100 mA collector current, com-pared with the 0.4 V or so we find for a

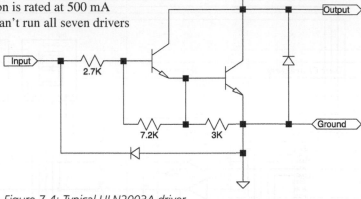

Figure 7-4: Typical ULN2003A driver.

2N4401. Each driver has an associated clamp diode to prevent damage when driving inductive loads. We've connected it to V_{DD}, although this connection could be omitted, if desired, since there's no inductive spike when driving an LED. We'll use this chip for later projects, such as driving small stepper motors, where it's important to connect the clamp to V_{DD}.

We're using a ULN2003A instead of seven separate 2N4401s or 2N7000s purely as a matter of wiring convenience. So, if you don't have a ULN2003A, wire up the circuit with seven 2N7000s, or 2N4401s with 2.2K base resistors. It isn't necessary to use a Darlington configuration at the current levels necessary to drive small LED displays. If your LED display provides adequate brightness with 25mA or less current, omit any external segment driver and connect the current limiting resistor directly to pins B0…B6. In this case, you will need to recalculate the dropping resistor, and invert the sense of the B0…B6 pins. (With an external driver, a high on B0…B6 activates a segment, but with a direct connection a low on B0…B6 is required. The current limiting resistor should be around 100ohms.)

I developed the circuit of Figure 7-3 as part of a larger project, and Figure 7-5 shows the resulting prototype PCB. The device that looks like an integrated circuit above the displays are 68-ohm current limiting resistors in the form of a Bournes 4116R-1-680 resistor network. Separate resistors would work just as well. Of course, you don't need to design and build a PCB to experiment with this circuit; a solderless plugboard works perfectly well. You can't squeeze all the parts onto the small 2840 development board's plugboard, however, so a second, larger, plugboard is required.

Figure 7-5: Completed 4-digit LED display.

Test Program

Now that you've wired up the circuit of Figure 7-3, let's test it.

Program 7-1

```
;Program to demonstrate 4 digit 7-segment LED display
;Sequentially show each segment and each digit


i        Var              Byte
j        Var              Byte

For i = B0 to B6
    Output i
    Low i
Next

For i = A0 to A3
    Output i
    High i
Next


Main
        For i = A0 to A3
                Low i
                For j = B0 to B6
                        High j
                        Pause 250
                        Low j
                Next
                High i
        Next
GoTo Main
```

Program 7-1 exercises each segment and each digit. The segments illuminate, one at a time in order A…G for 250 ms each, starting with Digit 0 (rightmost) and moving left.

The code itself requires little analysis; after setting **B0…B6** and **A0…A3** for output, we sequentially scan the digits and segments. The outer loop, **i**, steps through the digits. For each digit, the inner **j** loop scans the segments. Recall that in order to be illuminated, a segment requires its digit line (**A0…A3**) to be low (applies +5 V to the display common) and its segment pin (**B0…B6**) to be high (saturates the ULN2003 Darlington). Hence, the **i** loop sets one of **A0…A3** low and the **j** loop sets one of **B0…B6** high. To turn off unwanted digits, all other **A0…A3** pins are set high and to turn off unwanted segments, all other **B0…B6** pins are set low.

Now, let's look at displaying actual digits, 0…9. Program 7-2 continuously cycles the rightmost digit from 0…9, displaying each digit for one second.

Program 7-2

```
;Shows digits 0...9 in last place
;LED display is common anode
;cathodes through ULN2003A array

;Variables
;--------------
i               Var             Byte
Decode          Var             Byte(11)        ;holds segment pattern
;Digit values [D3][D2][D1][D0]


;Initialization
;--------------
        Clear

        DeCode(0) = %00111111    holds segments
        DeCode(1) = %00000110        +---A---+
        DeCode(2) = %01011011        |       |
        DeCode(3) = %01001111        F       B
        DeCode(4) = %01100110        |       |
```

```
                DeCode(5)  = %01101101           +---G---+
                DeCode(6)  = %01111100           |       |
                DeCode(7)  = %00000111         E |       | C
                DeCode(8)  = %01111111           |       |
                DeCode(9)  = %01100111           +---D---+
                DeCode(10) = %00000000  ;all blank    XGFEDCBA is bit order

                TRISB = $00
                TRISA = $F0
                PortA.LowNib = %1111

        Main
                Low A0
                For i = 0 to 9
                        PortB = DeCode(i)
                        Pause 1000
                Next
        GoTo Main


        End
```

The major addition to this program is the array **DeCode** to map the desired digits to the segments that must be illuminated to display that particular digit.

We've connected the segments in the order of segment A to B0...segment G to B6. Hence, to illuminate segment A, we must write **%00000001** to Port B. To illuminate segment B, we write **%00000010** to Port B, and so forth. If we wish to display the character "0", for example, we must write **%00111111** to Port B, thereby illuminating all segments except G. (If we directly connect the display cathodes to pins B0...B6, we must invert this mapping, for example, to display the character "0" we would write **%11000000** to Port B.)

Another change we've made is to initialize Port A and Port B through **TRISA** and **TRISB** statements, as outlined in Chapter 2. This change illustrates that MBasic often gives us several ways of accomplishing the same task. (We could omit the **TRISA** statement, as the **Low** and **High** statements automatically change the pin to output, but it's cleaner to initialize our input and output pins directly.) We also set all digits to off through **PortA.LowNib = %1111**.

```
        Main
                Low A0
                For i = 0 to 9
                        PortB = DeCode(i)
                        Pause 1000
                Next
        GoTo Main
```

The main loop first activates Digit 0 by dropping **A0** low. Then, we loop through the digits 0...9. Each digit is displayed through the assignment **PortB = DeCode(i)**.

Now that we've verified that all segments illuminate and that the digits 0...9 can be displayed, let's do something more useful with our display. Program 7-3 reads three momentary contact switches; one increments the count, one decrements the count and one will clear the count to zero. Making it a bit fancier, we'll blank leading zeros. We'll debounce the switches and set up the read loop so that the counter only increments once for each closure.

First, as shown in Figure 7-6 add three momentary contact, normally open switches and pull-up resistors to the circuit of Figure 7-3.

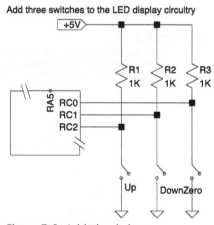

Figure 7-6: Added switches.

Program 7-3

```
;Program to demonstrate 4 digit 7-segment LED display
;Counts up/down on 4-digit LED Display
;LED display is commmon anode
;cathodes through ULN2003A array

;Constants
DelayCount      Con             3

;Variables
;--------------
i               Var             Byte
j               Var             Byte
Decode          Var             Byte(11)        ;holds segment pattern
Digit           Var             Nib(4)          ;holds digit values
Counter         Var             Word            ;value to disply
TempWord        Var             Word
ArmUp           Var             Nib
ArmDown         Var             Nib
;Digit values [D3][D2][D1][D0]


;Initialization
;--------------
        Clear

        DeCode(0) = %00111111       holds segments
        DeCode(1) = %00000110           +---A---+
        DeCode(2) = %01011011           |       |
        DeCode(3) = %01001111           F       B
        DeCode(4) = %01100110           |       |
        DeCode(5) = %01101101           +---G---+
        DeCode(6) = %01111100           |       |
        DeCode(7) = %00000111           E       C
        DeCode(8) = %01111111           |       |
        DeCode(9) = %01100111           +---D---+
        DeCode(10) = %00000000          ;all blank     XGFEDCBA is bit order

        TRISB = $00                 ;segments on PortB
        TRISA = $F0                 ;Digit drive on PortA
        TRISC = %00000111           ;Switches on PortC

        PortA.LowNib = 0
        PortB = DeCode(8)
        Pause 1000
        PortB = DeCode(10)
        Pause 250

        ;blank leading 0s
        Digit(3) = 10
        Digit(2) = 10
        Digit(1) = 10

Main
        ;multiplex right to left
        For i = A0 to A3
                ;order is important to void ghosting
                PortB = DeCode(Digit(i-A0))
                Low i
                ;GetDigits is part of the delay
                GoSub GetDigits
                Pause 1
                High i
        Next
GoTo Main


GetDigits
```

```
                    ;Check the up count switch
            If (PortC.Bit0 = 0) AND (ArmUp = 0) Then
                Counter = Counter +1
                ArmUp = 1
            EndIf

            ;Is it off? If so delay
            If (ArmUp > 0) AND (PortC.Bit0 = 1) Then
                ArmUp = ArmUp + 1
                If ArmUp = DelayCount Then
                        ArmUp = 0
                EndIf
            EndIf

            ;Check the down count switch
            If (PortC.Bit1 = 0) AND (ArmDown = 0) Then
                Counter = Counter -1
                ArmDown = 1
            EndIf

            ;If switch off? If so delay
            If (ArmDown > 0) AND (PortC.Bit1 = 1) Then
                        ArmDown = ArmDown + 1
                        If ArmDown = DelayCount Then
                            ArmDown = 0
                        EndIf
            EndIf

            If PortC.Bit2 = 0 Then
                Counter = 0
            EndIf

            ;Avoid rollover
                If Counter > 9999 Then
                    Counter = 0
                EndIf

                ;Digits are individual BCD
                TempWord = Bin2BCD Counter
                Digit(0) = TempWord.Nib0
                Digit(1) = TempWord.Nib1
                Digit(2) = TempWord.Nib2
                Digit(3) = TempWord.Nib3

                ;Following is for blanking leading 0s
                If Counter < 1000 Then
                    Digit(3) = 10
                EndIf

                If Counter < 100 Then
                    Digit(2) = 10
                EndIf

                If Counter < 10 Then
                    Digit(1) = 10
                EndIf
        Return

        End
```

We've declared several new variables, including:

```
    Digit          Var          Nib(4)
    Counter        Var          Word
```

Counter is the value that we increment, decrement or zero with the added switches. Since we have four display digits, **Counter** may run from 0...9999. We accordingly store it in a word length variable. The array **Digit** holds the value of the individual digits to be displayed. If **Counter** is 1234, for example, **Digit(0)** = 4, **Digit(1)** = 3, **Digit(2)** = 2 and **Digit(3)** = 1. (The digits are numbered with 0 at the rightmost position.)

```
PortA.LowNib = 0
PortB = DeCode(8)
Pause 1000
PortB = DeCode(10)
Pause 250

;blank leading 0s
Digit(3) = 10
Digit(2) = 10
Digit(1) = 10
```

As part of initialization, we illuminate all segments by displaying "8888" for 250 ms, followed by blanking the leftmost three digits. (**DeCode(10)** is a blank.) Briefly illuminating all segments is a common initialization feature, as it permits the user to detect failed segments. We don't have to worry about operating all segments simultaneously as we use robust drivers. If your segments are directly connected to pins B0…B6, simultaneously operating all segments and all digits may overstress the PIC's ability to sink current, in which case modify your code to sequence the digits.

```
Main
        ;multiplex right to left
        For i = A0 to A3
            ;order is important to void ghosting
            PortB = DeCode(Digit(i-A0))
            Low i
            ;GetDigits is part of the delay
            GoSub GetDigits
            Pause 1
            High i
        Next
GoTo Main
```

The main loop sequentially steps through the digits, from right to left, with the **For i…Next** loop and illuminates the segments required to display the associated value held in **Digit(i)**. We'll assume for the moment that **Digit(i)** holds the correct value (0…9) for each of the four displayed digits. After Port B is set to the proper bit sequence to display the correct digit, execution branches to subroutine **GetDigits**, following which an extra 1ms pause is executed. (As we'll see shortly, subroutine **GetDigits** fills **Digit()** with the correct values.)

Several points are worth mentioning if you modify this code for other programs. First, in order to avoid flicker, the loop must execute relatively fast. As written, each digit is illuminated for about 4 ms, so all four digits are updated about 60 times per second. If this update rate significantly slows, due to added code, the display will start to flicker. Different people perceive flicker differently, but I find flicker objectionable if the update rate falls below 40 per second. Second, it's important to switch between digits as fast as possible to avoid "ghosting" where false segments are briefly illuminated. To avoid ghosting we first write the segment information to Port B and then take the associated A pin low to activate the digit. Reversing this order causes objectionable ghosting. Finally, there's a trade-off between the digit-on time and perceived brightness, of course, so some tinkering with the current limiting resistor may be necessary after your code is finished.

Let's look at **GetDigits**. This subroutine is called every 4 ms or so.

```
If (PortC.Bit0 = 0) AND (ArmUp = 0) Then
        Counter = Counter +1
        ArmUp = 1
EndIf

;Is it off? If so delay
 If (ArmUp > 0) AND (PortC.Bit0 = 1) Then
        ArmUp = ArmUp + 1
        If ArmUp = DelayCount Then
           ArmUp = 0
        EndIf
 EndIf
```

All switches are wired with the normal (unpressed) position open, so that a 0 represents switch pressed and 1 is switch un-activated. We have two almost identical routines to handle the up and down switch inputs. We'll only look at the "up" switch code.

Assume for the moment that **ArmUp = 0** (it's initialized that way) and that the "up" switch is pressed. The conditional **If (PortC.Bit0 = 0) AND (ArmUp = 0)** thus evaluates as true, **Counter** is incremented and **ArmUp** is assigned as **1**.

As long as the "up" switch continues to be pressed the second conditional **If (ArmUp > 0) AND (PortC.Bit0 = 1)** fails, as **PortC.Bit0 = 0**. **ArmUp** thus continues to equal **1** and the earlier conditional will fail, thus preventing **Counter** from being updated due to continued switch activation.

When the "up" switch is released, the second conditional **If (ArmUp > 0) AND (PortC.Bit0 = 1)** evaluates true, so **ArmUp** is incremented. If the switch bounces, it's possible that the first conditional may evaluate **PortC.Bit0 = 0** as true. However, **ArmUp** continues to be greater than **0**, so the **ArmUp = 0** part of the first conditional fails and **Counter** is not incremented. Eventually, however, assuming the "up" switch is not pressed, **ArmUp** reaches **DelayCount** at which time it is reset to **0** and we are ready for another press of the "up" switch. **DelayCount** is a constant that I've set at 3, thus representing two passes of **GetDigits**, or about 8 ms. You may need to vary **DelayCount**, depending on the bounce performance of your switches, but 3 is a good starting point.

```
If PortC.Bit2 = 0 Then
        Counter = 0
EndIf
```

We need not complicate reading the reset switch connected to pin C2; it doesn't matter if bounce causes two or three consecutive resets.

```
;Avoid rollover
If Counter > 9999 Then
    Counter = 0
EndIf
```

Since our four-digit display only shows 0…9999, we restrict **Counter** to those values that are capable of being displayed.

```
TempWord = Bin2BCD Counter
Digit(0) = TempWord.Nib0
Digit(1) = TempWord.Nib1
Digit(2) = TempWord.Nib2
Digit(3) = TempWord.Nib3
```

Let's look at how we get the individual digit values out of **Counter** and into the **Digit** array. It turns out that this is a rather common activity and MBasic includes a function, **Bin2BCD** to perform it for us. (There's an error in some editions of the User's Guide where this function is incorrectly identified as **Dec2BCD**.) Let's look at "binary coded decimal" a bit more closely.

Suppose **Counter** holds 1234. The value 1234 (decimal) is held in MBasic as hex $4D2. If we wanted our display to show hexadecimal (assuming we defined **DeCode($A)…DeCode($F)**), we could very easily find our digits. We've declared **Counter** as a word length variable, so it has four nibbles:

Word Value	Nibble 3	Nibble 2	Nibble 1	Nibble 0
$4D2	$0	$4	$D	$2

Hence, **Digit(0) = Counter.Nib0**, etc. But, you say, we want a decimal display, not a hexadecimal one. Fair enough. First, however, what is 1234? It's $1 \times 1000 + 2 \times 100 + 3 \times 10 + 4$. With this understanding, we can devise an algorithm to fill **Digit()**:

```
Digit(3) = Counter / 1000
Remainder = Counter // 1000
```

```
Digit(2) = Remainder / 100
Remainder = Counter // 100

Digit(1) = Remainder / 10

Digit(0) = Remainder // 10
```

In MBasic "/" is integer division and "//" is the remainder, or modulus, function. Let's see how this algorithm works if **Counter** is 1234.

```
1234 / 1000 = 1              -> Digit(3)
1234 // 1000 = 234

234/100 = 2                  -> Digit(2)
234//100 = 34

34/10 = 3                    -> Digit(1)

34//10 = 4                   -> Digit(0)
```

But, there's a faster way to accomplish this conversion than invoking six integer division and remainder operations in MBasic. First, a definition: binary coded decimal, or BCD, uses four bits (one nibble) to represent each decimal digit. Since a word is 16 bits long, we can "pack" four BCD digits, each four bits long, into each word. This arrangement, often called "packed BCD," is shown below, supposing, somehow, we are able to get each nibble to hold the desired digits.

Word Value	Nibble 3	Nibble 2	Nibble 1	Nibble 0
4660 $1234	1	2	3	4

The decimal value of $1234 is 4660, but looking at each nibble individually we immediately see what we are looking for, [1][2][3][4].

MBasic's **Bin2BCD** function converts a binary variable to packed BCD. Hence, filling **Digit()** is simply a matter of assigning each nibble to its **Digit()** counterpart the following the BCD conversion. We accomplish this with the following code snippet:

```
TempWord = Bin2BCD Counter
Digit(0) = TempWord.Nib0
Digit(1) = TempWord.Nib1
Digit(2) = TempWord.Nib2
Digit(3) = TempWord.Nib3
```

Finally, for aesthetic reasons we blank leading zeros.

```
;Following is for blanking leading 0s
If Counter < 1000 Then
    Digit(3) = 10
EndIf

If Counter < 100 Then
    Digit(2) = 10
EndIf

If Counter < 10 Then
    Digit(1) = 10
EndIf
```

We've defined **DeCode(10)** as a blank digit (all segments off), so to blank a digit we set **Digit()** as 10. This task is easily accomplished by checking the value of **Counter**. If **Counter** is under 1000, we know that at least the leftmost digit must be 0, and according assign **Digit(3)** as a blank. We repeat this test and assignment process with 100 and 10 for **Digit(2)** and **Digit(1)**.

After blanking leading zeros, the subroutine is finished and program execution returns to **Main**.

We can now test the program. Every time the "up" button is operated, the displayed count will increase by one. Every time the "down" button is operated, the displayed count will decrease by one, with a minimum of zero. Pressing the "clear" button resets the display to zero.

Program 7-3 calls the subroutine once every 3.8 ms as written. The anti-bounce delay limits the speed with which the counter will increment and decrement to about 20 to 30 operations per second, more than fast enough for hand-operated switches.

Ideas for Modifications to Program and Circuits

- Add segment definitions for "a", "b", "c", "d", "E" and "F" to **DeCode()**. Remove the **TempWord = Bin2BCD Counter** statement and display the count in hexadecimal.
- How does flicker change as you modify the **Pause 1** statement? (I found that if I changed the pause time to 5, flicker is quite objectionable, and with 4, flicker is barely noticeable. I could not detect any flicker for a pause time of 3 or below. If you wish to disable the pause, comment it out or delete the statement. **Pause 0** actually executes **Pause 255**.)
- How does perceived brightness change as you modify the **Pause 1** statement? What is the ratio of digit-on to digit-off time? If the overhead is zero, each digit should be on 25% of the time. (As Figure 7-7 shows, with the **Pause 1** statement operational, each digit is illuminated 22.1% of the time. The overhead is thus 11.6%.)

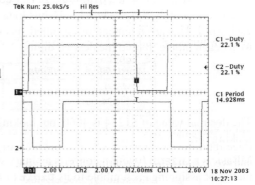

- Instead of triggering up/down counts with mechanical switches, substitute a function generator's square wave output. How fast can you update the counter before it misses input pulses? It is possible to increase the response speed? (You don't need to debounce a function generator's input, so the input software can be simplified.) Is it possible to replace the **Pause 1** statement with repeated subroutine calls to **GetDigits**?

Figure 7-7: Program 7-3 digit on time, Pause 1; Ch1: Pin A0; Ch2: Pin A1.

- An alternative to using **Bin2BCD** to fill the array **Digit** with the four individual digits of **Counter** is to use MBasic's **DEC** modifier:

```
Digit = DEC4 Counter\4
```

The **DEC** operator fills **Digit** with the ASCII values of **Counter**. If **Counter = 1234, Digit** will hold the following values:

```
Digit(0) = "1"
Digit(1) = "2"
Digit(2) = "3"
Digit(3) = "4"
```

To convert the ASCII values held in **Digit** to numerical values, simply subtract the value of ASCII "0" using a **For...Next** loop to cycle through all four elements of **Digit**. Since we are working with four digit values, we modify the **DEC** operator to specify a maximum of four characters by appending the digit 4 to the operator. (See Chapter 9 for more on using the **DEC** operator.)

Revise Program 7-3 to use this approach. How does the program speed compare with the original **Bin2BCD** approach?

References

[7-1] Agilent Technologies, *14.2 mm (0.56 inch) Seven Segment Displays*, Document no. 5968-9410E (2000).

[7-2] Allegro MicroSystems, Inc., *2003 thru 2024 High-Voltage, High-Current Darlington Arrays*, (undated).

Introductory
Stepper Motors

Stepper motors, as the name implies, rotate in discrete steps. Most conventional motors are continuous; if we make a mark on the shaft of a conventional motor and if we were able to precisely control the motor's excitation, we could make the shaft move to any angle. 123.456 degrees from the starting mark is just as achievable as 321.765 degrees. Of course, practical considerations make this degree of precision unlikely in a real motor. A stepper motor's shaft, in contrast, is moveable to only certain, pre-defined angles. A 48-step stepper motor, for example, may be positioned only in increments of 7.5 degrees (360°/48). Hence, we may command the shaft to go to 7.5° (one step) or 15° (two steps) but not to 8.432°. (Later in this chapter, and in Chapter 19, we'll see ways to step the shaft rotation one-half or a smaller fraction of the motor's normal step size, so the difference between conventional and stepper motors blurs.)

Why would we want a motor that only moves in steps? Suppose we wish to move an ink jet printer's print head across the paper, and that we must position the print head with an accuracy of 0.001". We'll assume the print head is attached to a nonslip, no stretch toothed belt and that the belt is driven through a toothed pulley system attached to a positioning motor. Let's attach a 200-step stepper motor to the pulley, through 5:1 step down gears, and pick the pulley size so that 1,000 steps of the motor moves the printing head 1.000 inch. Each motor step therefore corresponds to 0.001" and we may position the print head to 4.567" by initializing the print head at the start position and then advancing the motor by 4,567 steps. This "open loop" solution is much cheaper than a "closed loop" design that continuously monitors the position of the print head and stops the advance when the target position is reached.

We'll concentrate on stepper motor fundamentals in this chapter and advanced applications in Chapter 19.

Stepper Motor Basics

Introduction

Let's start by considering the pluses and minuses of stepper motors:

Stepper Advantages	Stepper Disadvantages
• *Precise control of position-one pulse advances one step, permitting open loop control.* • *Full torque from zero RPM.* • *Step accuracy typically 5% of step size, but errors are non-cumulative.* • *No brushes or other current carrying moving parts; lifetime is therefore limited only by the bearing life.* • *Easy to interface with microcontrollers.* • *The motor is self-locking if the windings are powered while not rotating; even if unpowered, most designs have appreciable residual torque.*	• *Limited speed.* • *Certain step rates may mechanically resonate with the motor causing loss of torque and undesired vibration.* • *Large steppers are not readily available. Most steppers are in the 0.0001 HP to 0.05 HP range.* • *Torque decreases as rotational speed increases; if the motor stalls, position location is lost.*

Operation

Let's see how a stepper motor works. We'll consider a highly simplified motor that doesn't match real motor designs but provides a useful mental model of how a stepper functions. Figure 8-1 illustrates our simple motor model. At the center is a bar magnet, free to rotate, surrounded by four electromagnets, spaced 90 degrees to each other. The electromagnets are wound over soft iron poles. In motor terminology, the bar magnet is the *rotor* and the surrounding electromagnets form the *stator*. In Figure 8-1, the motor is un-powered and the rotor has automatically rotated to the position of minimum magnetic energy; that is, the permanent magnet rotor positions itself so that its flux has the shortest air path and the longest iron path. If you try to rotate the rotor by twisting the shaft, you will note resistance; you have to supply the energy required to break the magnetic attraction between the stator and the nearest pole. You should feel the same resistance, followed by the motor snapping into a new stable position, if you twist the shaft of a real stepper motor. (One uncommon type of stepper, the variable reluctance motor, doesn't exhibit this behavior, as we'll note later.) In an unpowered motor the torque required to break the rotor free from its rest position is called the *detent torque*.

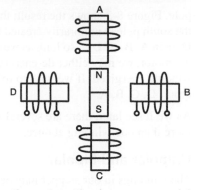

Figure 8-1: Simple stepper model: unpowered.

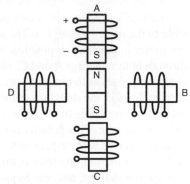

Figure 8-2: Winding A energized.

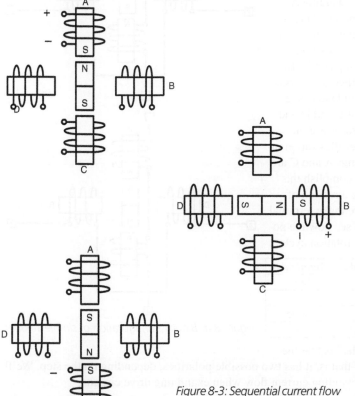

Figure 8-3: Sequential current flow in A, B and C causes rotation.

Now, let's energize winding A, as shown in Figure 8-2. We'll set the polarity of the current through A so that the inward facing pole is a magnetic south pole, which attracts the rotor's north pole. As long as winding A is energized, the rotor is held in place. The external torque necessary to override the magnetic attraction and move the rotor is known as the *holding torque* or *static torque*. For most motors the holding torque when operated at rated current is about 10 times the detent torque.

What happens if we energize the windings in sequence? Suppose we apply current to windings A, B and C, in that order, with the polarity so that the inside of each winding is a magnetic south pole. Figure 8-3 shows the result; the rotor rotates clockwise, as its north pole is sequentially attracted by the south poles temporarily created by energizing A, B and C. Of course, we may keep this up and energize D, then A, B, C, D… so long as we desire the rotor to continue stepping clockwise. Should we wish to stop the motor, we may either de-energize all windings or, if we need additional rest torque, we may keep one winding energized. If we wish to rotate the rotor counter-clockwise, we energize the windings in the reverse order: D, C, B, A.

As we'll see later, there are several variations on the order of energizing windings, including energizing more than one winding at once.

Unipolar and Bipolar

The windings in our stepper may be internally connected in several configurations. Two, however, two are of interest, the *unipolar* and *bipolar* connection. We'll start with the bipolar motor, sometimes called a *two-phase* stepper motor.

Figure 8-4 shows our simple model motor connected in the bipolar configuration. The motor has four terminals accessible to the user, 1 through 4. The upper diagram shows our motor in the starting position, with current flowing through both windings A and C. Note that windings A and C are wound so as to produce opposite field polarity; when current flows in the direction of the arrows, winding A presents a south pole to the rotor while winding C presents a north pole. This polarity is represented by terminal 1 being positive with respect to terminal 3. Unlike our earlier examination, the rotor is thus held in place by two energized windings, not one. Suppose we then de-energize windings A and C, energize windings B and D to rotate the rotor 90 degrees clockwise, de-energize B and D and then re-energize windings A and C. We now desire the magnetic polarity to match that of the lower illustration in Figure 8-4; the magnetic polarity of windings A and C are reversed from the upper illustration. We accomplish this magnetic polarity reversal by reversing the direction of current flow through windings A and C; we make terminal 3 positive with respect to terminal 1. Let's see how the polarity changes for one complete clockwise rotation cycle.

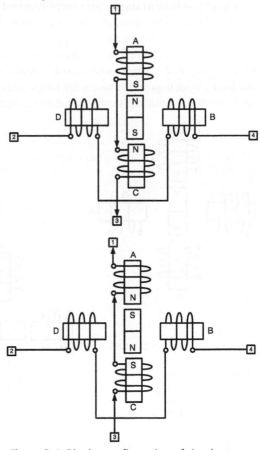

Figure 8-4: Bipolar configuration of simple stepper.

Step	Terminal 1	Terminal 3	Terminal 2	Terminal 4
1	+	–	None	None
2	None	None	+	–
3	–	+	None	None
4	None	None	–	+

The reason we term this connection "bipolar" is that the current polarity in the windings reverses—that is, it has two possible polarities, depending on the step. We'll consider how to accomplish reversing the winding current flow when examining drive circuits.

Figure 8-5 shows our simple stepper connected as a unipolar motor. (Unipolar motors are also known as *four phase* stepper motors.) The windings A-C and B-D remain in series, but the center taps, X and Y, respectively are also available, thus giving us six connections to the windings. (In some unipolar motors, both center taps are connected together and only five wires are brought out, as shown by the dashed connection line between terminals X and Y in Figure 8-5.) In the normal mode of operation, the center tap is always connected to the positive supply and we cause current to flow in the windings by connecting their free ends to the negative return, ground in most designs.

In the upper illustration in Figure 8-5, winding A is energized by placing positive voltage on terminal X and grounding terminal 1. From this is the starting point we'll assume that winding A is then been de-energized, winding B energized to pull the rotor 90 degrees clockwise and then winding B is de-energized and now winding C is energized by connecting terminal 3 to ground, as shown in the lower illustration in Figure 8-5. Current flow through the windings is always in the same direction; hence the name "unipolar" for this connection. Note that since windings A-C and B-D are in series just as in our bipolar configuration, we may take this unipolar motor and connect it to a bipolar drive circuit (making no connections to the center taps X and Y) and it will work. (This is true for real unipolar motors, not just for our simple model.)

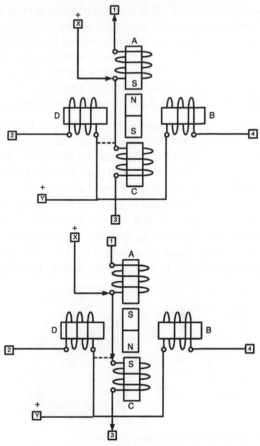

Figure 8-5: Unipolar configuration of simple stepper.

We can summarize the benefits and drawbacks of unipolar and bipolar connections as:

Configuration	Advantages	Disadvantages
Unipolar	• Simplest drive circuit.	• Less efficient use of motor windings.
Bipolar	• Efficient use of motor windings. • Greater torque than for same size unipolar motor.	• Requires special drive circuitry; most commonly an H-bridge arrangement.

Types of Stepper Motors

Figure 8-6 shows four typical stepper motors. The smaller two motors at the left of the figure are known as *tin can* or *can stack* or *permanent magnet* motors, while the two larger motors are *hybrid* constructed.

Tin can motors are inexpensive, constructed from a pressed or stamped case and with a smooth permanent magnet rotor magnetized with alternating north and south poles. Usually tin can motors have relatively coarse step sizes, with 24 and 48

Figure 8-6: Typical stepper motors.

steps/rev (15 and 7.5 degrees/step) being typical values. Tin can motors use sleeve bearings and are typically found in inexpensive electronic products, such as ink jet printers and fax machines. Most manufacturers use the case diameter (in millimeters) and number of steps as part of the model number. For example, the smallest motor in Figure 8-6 is a Nippon Pulse Motor model PF35-48L4 stepper. The case diameter is 35 mm (about 1-3/8") and it has 48 steps per revolution. The L4 suffix indicates the coil voltage (nonstandard) and rotor magnet type (Neodymium). Other manufacturers use different identifiers, of course, but case diameter and number of steps/revolution are commonly incorporated into the model identification.

The two larger motors in Figure 8-6 are of hybrid construction. Figures 8-7 through 8-9 show a partially disassembled hybrid motor. (I don't recommend disassembling a stepper motor unless absolutely essential, as some high performance rotors will be partially demagnetized if the rotor is removed from the motor case.) Figure 8-8 shows the toothed permanent magnet rotor. The rotor is constructed of two toothed segments, with one segment offset by one-half tooth width from the other, thereby effectively halving the step size. (The objects at the end of the rotor shaft are a ball bearing and a Belleville washer.) Figure 8-9 shows the stator, which has several noteworthy features. First, the four windings are clearly visible, just like our mental motor model. However, the poles are segmented, with each pole having four projecting pieces. (In motor terminology, these are *salient* poles.) It isn't necessary—or even desirable—for the poles to be continuous around the inner periphery of the motor; the rotor is continuous, which is sufficient.

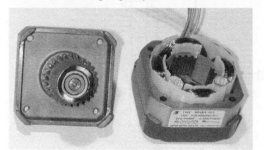  

Figure 8-7: Hybrid motor disassembly.   *Figure 8-8: Hybrid motor toothed rotor.*   *Figure 8-9: Hybrid motor interrupted toothed stator.*

Hybrid motors are often manufactured in industry-standard case sizes, as defined by the National Electrical Manufacturers Association (NEMA). A motor manufactured by Company X in NEMA style 34 is mechanically interchangeable with another NEMA 34 case size motor manufactured by Company Y. The largest motor in Figure 8-6 is a NEMA 34 motor, while the one next to it is a NEMA 23 motor. Of course, the electrical and performance specifications of two motors with identical NEMA case sizes are not necessarily (or even usually) the same.

Hybrid motors are more expensive than tin can motors and feature higher quality construction, such as ball bearings instead of sleeve bearings, and cast or machined cases instead of pressed case. Additionally, the toothed construction permits much finer steps, with 180 and 200 step/rev being common values.

A third type of motor is the variable reluctance, resembling the hybrid in construction, but with a nonpermanent magnet toothed rotor. Variable reluctance motors are relatively uncommon and will not be further discussed.

There are other much less common stepper motor types, such as three-phase unipolar. These require specialized driving circuits and are beyond the scope of this text.

Identifying Stepper Motors

To identify a stepper motor that has no nameplate or for which a data sheet is not available, we may use the following steps.

1. Turn the motor shaft by hand. You should feel the detents; if you feel no detents the motor is not a stepper of the type considered in this chapter. Turn the motor shaft one complete revolution and count the number of detents you encounter. This gives the steps/rev value for the motor. Common step/rev values are 24, 48, 72, 100, 180, 200, 400 and 800, although the later two values are relatively unusual. Your count should be close to a multiple of 10 or 12. If you count 197, it almost certainly means you missed a count here or there and have a 200 step/rev motor.

2. How many wires or a connection does the motor have?
    a. Four—you likely have a bipolar motor.
    b. Five or six—you likely have a unipolar motor.

3. With an ohmmeter, identify the wire colors or terminal numbers corresponding to your windings and label them as in Figure 8-10 (bipolar) or Figure 8-11 (unipolar). Make a note of your resistance measurements. The resistance of windings A and B should measure within 5% or so of each other. Likewise, in a unipolar motor, the resistance from the common center tap to each winding end should be approximately equal and the resistance across each complete winding (A1 to A2 and B1 to B2) should be approximately equal and twice the value from the common to each end.

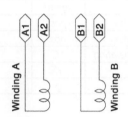

Figure 8-10: Winding labels for bipolar stepper motor.

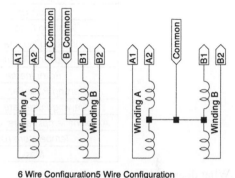

6 Wire Configuration 5 Wire Configuration

Figure 8-11: Winding labels for unipolar stepper motor.

4. If you have access to an inductance bridge, measure the inductance of the windings. If you don't have a bridge, you may safely skip this step.

5. Now we will attempt to "guestimate" the motor's voltage and current ratings. This step is necessary only if your motor doesn't have a nameplate or part number providing this information or if you can't find a data sheet for the motor. There is no magic way to accomplish reverse engineering the motor's rating with complete accuracy, but we can come close enough for experimentation purposes. Measure the physical size of the unknown motor and determine the construction type. Is it a tin can or a hybrid motor? Next, search the manufacturer's catalogs, either paper copies or on the internet, until you find a motor with the same physical size, construction type connection type (bipolar or unipolar) and number of steps. See if you can find a motor with similar coil resistance and (if you have measured it, coil inductance). If you can't find a match, then calculate the power dissipation (in watts) for several motors of the same case size as your motor using the formula $P = I^2R$, where P is the power in watts, I is the motor's current rating in amperes and R is the motor's winding resistance in ohms, with both I and R from the catalog values. Your calculated P will likely differ among the matching motors, so calculate an average value. Then, using the average power dissipation for the physically similar motors and your measured resistance value, calculate the resulting I for your motor, using the formula $I = \sqrt{P/R}$.

6. Now that we have determined the motor's rated current, I, and the measured R, calculate the motor's nominal operating voltage V from Ohm's law, $V = IR$.

7. Note any other important parameters from the closest matching data sheet, such as the maximum speed in steps/second or the maximum torque.

Reading a Stepper Specification Sheet

Let's look at a typical specification sheet for an inexpensive tin can motor, a Nippon Pulse Motor model PF35-48. I've reproduced the data sheet parameters below.

Parameters	Units	PF35-48			
Drive Mode		Unipolar		Bipolar	
Excitation Mode		Full-step (2-2 ex)			
Step Angle	°	7.5			
Step Angle Tolerance	%	± 5			
Steps per Revolution		48			
Voltage	V	12	5	12	5
Winding Resistance	ohm/Ø	90	16	100	17
Winding Inductance	mH/Ø	48	8.9	124	19
Holding Torque	mN•m	20	20	25	25
Rotor Inertia	kg•m²	4.5 x10⁻⁷			
Starting Pulse Rate, Max	pps	500			
Slewing Pulse Rate, Max	pps	530			
Ambient Temp. Range, Operating	°C	–10 ~ +50			
Temperature rise	K	55			
Mass	g	80			

What does each line mean?

Drive mode—The PF35-48 is available in either a bipolar or a unipolar configuration.

Excitation mode—As we will see later, a stepper may be operated in several modes, and certain parameters, such as torque and step angle, are different for different modes. The data sheet's statement "full-step (2-2 ex)" means that the performance data is based upon full step operation, with both coils energized. If this sentence doesn't mean much to you right now, put a star in the margin and come back to it after reading the rest of this chapter.

Step angle—the angle in degrees through which the shaft rotates when it advances one step while in full step mode. The value 7.5° corresponds to 48 steps/rev.

Step angle tolerance—the tolerance, as applied to the step angle, that is, the angle the motor shaft advances in one full step is 7.5° ±5%. It's important to remember this tolerance applies on a step-by-step basis and is not cumulative. After 48 steps, the motor will return to its original starting point with an accuracy of ±5% × 7.5° or ±0.375°. After 48000 steps (1000 complete revolutions), the motor will be at its original starting angle ±0.375 degrees. The noncumulative error performance of a stepper is the key to its ability to perform precision operations. If the error were cumulative, after being commanded to perform 48000 steps, or 1,000 revolutions, the shaft angle would be unknown within ±50 revolutions, quite a difference from the actual ±0.375 degrees!

Steps per revolution—the number of full steps required to return the motor shaft to its starting angle. Since there are 360° in one revolution, the step angle and steps per revolution are related by the formula:

$\theta = \dfrac{360}{N}$ where θ is the step angle and N is the number of steps per revolution.

Voltage and winding resistance—We'll consider these two parameters at the same time. You may recall from high school physics that the magnetic field of an electromagnet is proportional to the current in the windings multiplied by the number of turns (ampere-turns) and that the attractive force between two magnets is proportional to their magnetic fields. Hence, for a fixed number of turns, the shaft torque in the stepper motor is proportional to the current through the windings. If we double the current, we double the torque. And, we know from elementary circuit theory that resistive power dissipation is proportional to the square of the current; $P = I^2R$. If we double the current, the power dissipated in the motor goes up fourfold.

The motor designer must balance these two effects against each other; to make the motor more powerful for its size, the designer wishes to maximize the current. However, more current causes more internal heating and if the motor temperature exceeds a certain level the winding insulation may break down and the motor will fail. Alternatively, to increase the stator's magnetic field, the designer may decide to use smaller diameter wire, which allows more turns in a given space (increasing ampere turns), but the smaller diameter wire has greater resistance which means we must apply higher voltage to the stator coil to obtain the desired current. The trend is to lower voltage power supplies, and the motor manufacturers try to meet their customers' needs with lower voltage motor designs.

The motor's rated voltage and resistance allow us to calculate the nominal winding current using Ohm's law: $I = V/R$. As we will see when we look at driver circuits, usually we drive the motor through a quasi-constant current arrangement.

In this case, the PF35-48 motor has two winding options; a 12 V winding with 100 ohms resistance and a 5 V winding with 17 ohms resistance. We can calculate the corresponding currents as 120 mA and 294 mA, respectively. Since the torque specifications are identical for both winding options, we may safely assume that the 12 V winding has close to 2.45 more turns than the 5 V version, thereby keeping the ampere turns—and torque—identical. A quick check confirms that the PF35-48 is designed for identical power dissipation for both 12 V and 5 V windings. The 12 V coil dissipates 1.44 watts at the rated voltage, while the 5 V coil dissipates 1.47 watts.

The PF35-48 motor I used in this chapter has an "L4" suffix, meaning it is a "special" voltage rating. Since we know the motor dissipation is 1.4 watts, and since I measured the coil resistance as 20 ohms, we may determine the motor's rated voltage is:

$$P = \frac{E^2}{R}; E = \sqrt{PR}$$
$$E = \sqrt{1.4 \times 20} = 5.3V$$

Likewise, we calculated the rated current:

$$P = EI; I = \frac{P}{E}$$
$$I = \frac{1.4}{5.3} = 265\ mA$$

Where:

P is the power dissipation in watts;
I is the current in amperes;
V is the voltage in volts;
R is the resistance in ohms.

Finally, there is a "nondissipative" element of input power to the motor; the power that goes to perform mechanical work at the output shaft. We've neglected this, as for many stepper motor applications the mechanical work output is small compared with the resistive winding loss.

Winding inductance—As a consequence of producing the desired magnetic field, the motor windings are inductors. Their inductance limits the rate of rise of current through the winding when a winding is energized. From a performance prospective, we would like the winding current to immediately assume its final value when the voltage step applied. We'll look at this in more detail when we consider drive circuits, but let's examine the difference between the 5 V and 12 V versions of the PF35-48 motor.

In introductory electrical circuits class, we learn that if a voltage is applied across a series circuit of an inductor and a resistor, such as the windings of a stepper motor, the current starts at zero and increases according to an exponential function, with a limiting value determined by the series resistance. The current versus time relationship is:

$$i = I_f\left(1 - e^{\frac{Rt}{L}}\right)$$

Where:

I_f is the steady-state current, in this case the applied voltage V divided by the winding's resistance R, or

$$I_f = \frac{V}{R}$$

R is the winding's resistance (plus any other resistance in the circuit) in ohms;
L is the winding's inductance in Henries;
t is the time in seconds after the voltage is applied to the inductor;
e is the base for natural logarithms, 2.71828182…

We often can use a simpler calculation—the inductive time constant τ for circuit analysis:

$$\tau = \frac{L}{R}$$

Where:

τ is the time constant, in seconds.

After the time τ, the current will have reached approximately 63% of its steady state value, after time 2τ it will have reached 86.4%, etc.

Figure 8-12 shows the result of a SPICE circuit simulation of the current through the 5 V and 12 V winding of a bipolar PF35-48 motor. It follows the negative exponential form of the equation shown above. Although the magnitudes of the two currents are different, both reach equal percentages of the maximum at almost the same time. This can be seen either by examining Figure 8-12, or by comparing the time constants of the two windings:

$$\tau_{5V} = \frac{19 \times 10^{-3}}{17} = 1.12 \ ms$$

$$\tau_{12V} = \frac{124 \times 10^{-3}}{100} = 1.24 \ ms$$

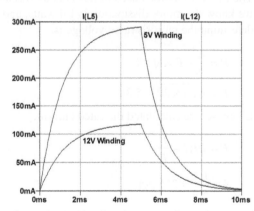

Figure 8-12: Simulated current through 5V and 12V stepper motor windings, PF35-48 motor.

Since the two time constants are almost identical, the current through the two coils will follow an almost identical time relationship, expressed in terms of the percentage of final current versus time. Since the motor's torque is proportional to ampere turns, and since the 12 V motor has proportionally more turns than the 5 V motor, we conclude that regardless of whether the 5 V or 12 V windings are used, the PF35-48's torque versus time performance will be almost identical. That the 5 V and 12 V windings have almost identical time constants is, of course, not an accident, but rather a product of the designer's intention to produce two motors that have essentially identical performance, regardless of whether the 5 V or 12 V version is used.

Holding torque—is the torque that must be applied externally to rotate the shaft from its position if power is applied to the windings. It's stated in the metric unit milli-Newton-meters (mN-m), and may be converted to the corresponding Imperial unit ounce-inches by the relationship: 1 oz-in = 7.06 mN-m. If you work in Imperial units, don't forget the difference between force and weight; torque is expressed in force units, not weight units.

Rotor inertia—is the moment of rotational inertia of the motor's rotor. For our purpose, we'll just note that this value is a measure of the effort required to get the rotor moving and to stop it, once it is moving. If you know how to use the rotor inertia value in your dynamic performance calculations, you don't need me to summarize it here and if you don't know how to make those performance calculations, then a short summary won't do you any good.

Starting pulse rate, max and slewing pulse rate max—Both values are measures of how fast the motor may be stepped—from either a standing start, or once rotating. Let's see how fast we may run the motor, in terms of revolutions per minute, a term perhaps more common in everyday usage. We'll assume the motor starts from rest, so the maximum rate at which we may pulse it is 500 pps.

In one second, therefore, the number of revolutions is equal to the number of steps divided by the number of steps/revolution:

$$R_{sec} = \frac{Steps/sec}{Steps/rev} = \frac{500}{48} = 10.4 \; rev/sec$$

To convert to revolutions per minute, multiply by 60:

$$RPM = \frac{60P}{N} = \frac{60 \times 500}{48} = 625$$

Where

RPM is the speed in rev/min;
P is the number of pulses/sec;
N is the number of steps per revolution for the mode being used.

By small motor standards, this is a modest speed indeed. But the advantage of a stepper is not its high speed, but rather its precision and its ability to provide torque at slow speeds.

If you have to wring the last possible bit of speed from the motor, you can ramp the speed up from a standing start.

Ambient temperature range – operating and temperature rise—The motor is rated to operate in an ambient temperature range of –10°C (+14°F) to +50°C (122°F). Its temperature rise is 55°K. A Kelvin degree is numerically equal to a Celsius degree, with the difference being the zero point—Celsius's zero is approximately the freezing point of water while 0°K represents absolute zero—approximately –273°C or –459°F. Hence, if operated at the maximum permitted ambient temperature, the motor's temperature will not exceed 105°C (222°F).

Mass—The motor's weight.

Operation Modes

We've alluded to various drive methods earlier, so let's see what's involved in controlling a stepper.

We'll start with a unipolar motor, as it's the easiest to understand. We'll assume it's a 6-wire motor and that it's connected to a power supply set to the motor's rated voltage. Figure 8-13 shows four SPST switches that en-

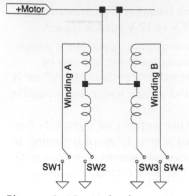

Figure 8-13: Unipolar motor connection.

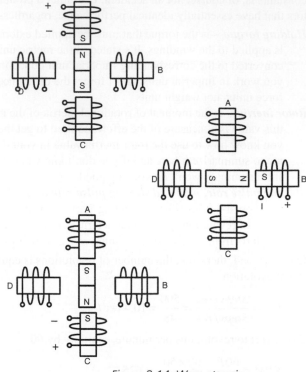

Figure 8-14: Wave stepping.

able us to connect any of the motor's four windings to ground; in a real circuit, we would use, for example, one of the low side driver circuits we saw in Chapter 3. But to understand the principles, we'll just think of the low side driver circuit we finally settle upon a nothing more than a way to either isolate the motor winding or connect it to ground.

We'll return to our simplified motor to examine how we step the switches. We'll see three patterns produce useful results; *wave, full step* and *half-step*.

Wave—Figure 8-14 shows how we may advance the rotor by sequentially energizing coils A, B, C and D (note that I've shown the switches in winding order, not numerical switch order):

Step	S1 (Winding A)	S3 (Winding B)	S2 (Winding C)	S4 (Winding D)
1	Closed	Open	Open	Open
2	Open	Closed	Open	Open
3	Open	Open	Closed	Open
4	Open	Open	Open	Closed

This mode is called "wave" because you can see a sequence of switch closures stepping down and across the table, causing a rotating magnetic wave within the motor. The stator poles are aligned with the rotor poles for each step.

Full step—In wave mode, only one winding is energized at any time. If we could simultaneously energize two windings, we could double the magnetic strength within the motor, thereby doubling the torque. Can we do this? Figure 8-15 shows how we might accomplish this. By simultaneously energizing windings A and B, for example, we create the equivalent of (almost) a double-strength coil, halfway between A and B. Thus, our stepping pattern is AB, BC, CD, DA:

Step	S1 (Winding A)	S3 (Winding B)	S2 (Winding C)	S4 (Winding D)
1	Closed	Closed	Open	Open
2	Open	Closed	Closed	Open
3	Open	Open	Closed	Closed
4	Closed	Open	Open	Closed

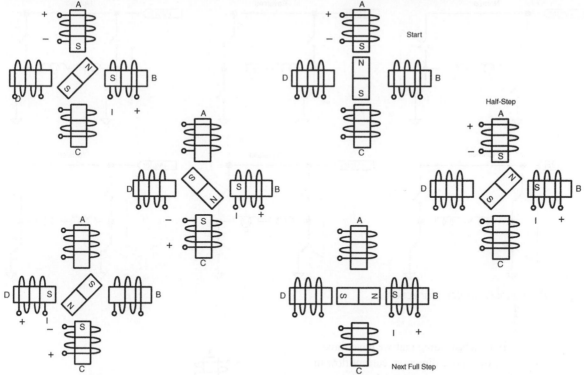

Figure 8-15: Full-step mode.   *Figure 8-16: Half-step mode.*

In fact, energizing two windings at 90 degrees to each other doesn't quite double the torque compared with a single winding, but it's close. The rotor poles align half-way between adjacent energized stator poles.

Half-step—We've seen that wave mode makes the rotor step in alignment with the poles, and full wave mode causes the rotor to move to half-way between the poles. Suppose we combined and interleafed wave and full wave mode. We could then cause the rotor to move in increments of one-half the normal step size. Figure 8-16 shows how this works. Our excitation mode is A, AB, B, BC, and so on:

Step	S1 (Winding A)	S3 (Winding B)	S2 (Winding C)	S4 (Winding D)
1	Closed	Open	Open	Open
2	Closed	Closed	Open	Open
3	Open	Closed	Open	Open
4	Open	Closed	Closed	Open
5	Open	Open	Closed	Open
6	Open	Open	Closed	Closed
7	Open	Open	Open	Closed
8	Closed	Open	Open	Closed

If you carefully look at the half step excitation pattern, you see that the wave and full step patterns are interleafed. You should also see that half stepping produces nonuniform torque—for half the steps only one winding is energized while for the remaining half of the steps two windings are energized. This effectively limits half stepping operation to torque requirements that may be met by only one winding.

Although we've looked at the patterns for a unipolar motor, in fact the identical patterns work for a bipolar motor. Let's look at how we might switch a bipolar motor. Figure 8-17 shows a conceptual view of controlling a

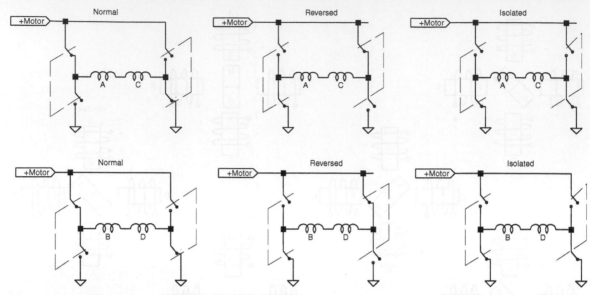

Figure 8-17: Bipolar motor connection.

bipolar motor. For each winding pair we have three options: current flow in one direction, current flow in the opposite direction and unenergized. For convenient reference, we'll identify the current flow direction by a plus sign. A+/C for example, means the free end of winding A is positive and the free end of winding C is negative, Figure 8-18 shows wave drive operations for the first three steps. The winding excitation pattern for one complete revolution is: A+/C, B+/D, C+/A, D+/B.

We'll use an electronic version of the switches shown in Figure 18-17, in particular a SN754410 H-bridge. The H-bridge, as we'll see later in this chapter, has four inputs, corresponding to each end of the two windings. When the corresponding input is set at logic high (1), that winding end is connected to the positive motor supply; when an input is logic low, the associated winding end is connected to ground. Hence, A+/C corresponds to a logic pattern of 10, while A/C+ is 01. If AC is de-energized, the logic pattern is 00. Using this terminology, the wave drive pattern for our bipolar stepper is:

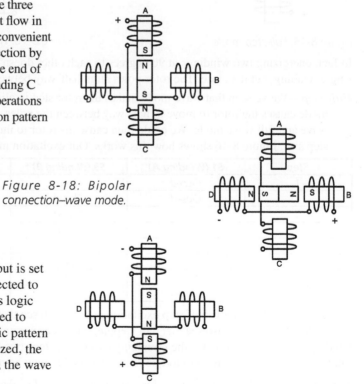

Figure 8-18: Bipolar connection–wave mode.

Step	S1 (Winding A)	S3 (Winding B)	S2 (Winding C)	S4 (Winding D)
1	1	0	0	0
2	0	1	0	0
3	0	0	1	0
4	0	0	0	1

Compare this with the drive pattern for wave drive of a unipolar motor. Substitute 1 for a closed switch and 0 for an open switch and you will see that the two patterns are identical. Since the full step and half-step patterns are also identical with their unipolar counterparts, we won't repeat the drawings and patterns charts for bipolar motors.

Comparing Figures 8-14 and 8-18 should reveal an advantage of bipolar motors; in wave drive, a unipolar motor has only one winding energized, while the bipolar motor has two windings energized, thus approximately doubling the available torque. If we make the same comparison for half-step operation, we find that the unipolar motor has either one or two windings energized while the bipolar motor has two or four windings energized, again about doubling the available torque. With respect to full step drive, the torque difference is less pronounced, as both motor have two windings energized for each step. However, due to rotor and stator alignment issues, a bipolar motor typically provides 20% to 30% more torque in full step mode than an otherwise identical unipolar counterpart. Of course, the motor's power and temperature ratings must also be considered and may be a limiting factor.

Programs

Let's start with a simple program demonstrating MBasic's built-in support for stepper motors, via the **SpMotor** function. Figure 8-19 shows how to connect the stepper to the 16F877A. We use no special functions, so almost any PIC may be substituted for the '877A. The motor you use must be a unipolar stepper and it must draw less than 500 mA. I used the PF35-48 unipolar stepper we earlier analyzed. (The PF35-48 I used has 5V windings.) Note carefully the winding connections in figure 8-19. Each pair of windings connects to alternating outputs,

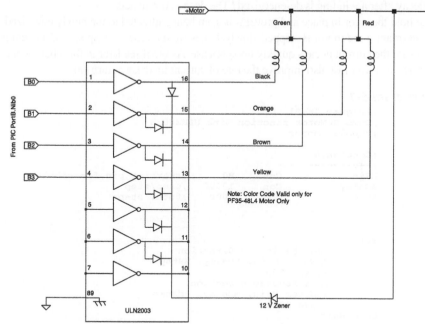

Figure 8-19: Unipolar stepper motor connection for Program 8-1.

not to adjacent outputs. Set the motor supply voltage to the stepper's rated voltage. Don't use the built-in +5V supply in the 2840 Development Board to run a stepper—it's not safely up to the task.

As we learned in Chapter 3, the PIC's output circuitry is not suitable to directly connect to a stepper. Hence we use a ULN2003A Darlington transistor driver. Do not connect a stepper directly to the PIC; it may well destroy the PIC. If you don't recall the ULN2003A, please review Chapter 7 where we went through its application. The ULN2003A contains seven Darlington transistors, with built-in base current limiting resistors. A logical 1 on the ULN2003A's input saturates the output, thereby providing the functional equivalent of closing one of the four mechanical switches in Figure 8-13.

As we learned in Chapter 3, when we switch an inductive load, the collapsing magnetic field causes a voltage spike that may damage the switching transistor. In addition, the coupled windings of a stepper motor act like transformer windings and can have induced in them a substantial voltage spike when other windings are turned on or off. To control the spike, we use the built-in diodes of the ULN2003A and an external Zener diode. The circuit will work without the Zener, but with slower current decay. Slower current decay may not be a problem if your stepper runs at slow step speeds, but to maximize speed add the Zener. Figure 8-20 is a SPICE simulation analysis of the current release improvement a Zener diode makes over using the ULN2003A's internal clamping diode. With a Zener, the motor current decays to zero in about 400 µs, versus 2.3 ms for a simple diode clamp. Why do we want the current to quickly decay after a winding is de-energized? The decay current acts

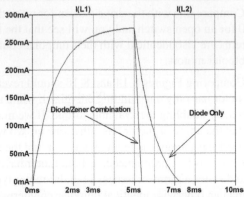

Figure 8-20: A Zener diode improves current release time.

to hold the rotor in place and prevents it from being attracted to the newly energized winding. This may not be so important if the motor is stepped slowly, but as we increase the step speed, it becomes increasingly important to ensure the current decays quickly upon release. As we'll see later in this chapter and in Chapter 19, for fast stepping rates we must also improve the rate of current increase at turn-on.

Program 8-1

```
;Program 08-01
;Demo SpMotor function with ULN2003A
;Bipolar motor

;Constants
;----------
MPin            Con      B0      ;starting pin for nibble
MDelay          Con      10000   ;usec/step
MStep           Con      1000    ;total steps



Main
        ;Rotate in one direction
        SpMotor MPin, MDelay, MStep
        Pause 1000
        ;Now reverse direction
        SpMotor MPin, MDelay, -MStep
        Pause 1000
GoTo Main
```

Program 8-1 exercises MBasic's built-in stepper motor function, **SPMotor**, invoked with three arguments:

    SPMotor Pin, Delay, Step

Pin—is a constant or variable that defines the first pin of four consecutive output pins used by **SPMotor**.

Delay—is a constant or variable defining the time between steps. Delay is in microseconds with a maximum 32-bit integer value.

Step—is a variable or constant and defines how many steps are to be taken, and the direction. Positive numbers step in one direction and negative numbers step in the reverse direction.

```
MPin        Con     B0      ;starting pin for nibble
MDelay      Con     10000   ;microseconds/step
MStep       Con     1000    ;total steps
```

We start by defining all three arguments to **SPMotor**; for our test, we'll use pins B0…B3, step at 10 ms/step and take 1000 steps. (10 ms is 10000 µs.)

```
Main
        ;Rotate in one direction
        SpMotor MPin, MDelay, MStep
        Pause 1000
        ;Now reverse direction
        SpMotor MPin, MDelay, -MStep
        Pause 1000
GoTo Main
```

Our program steps the motor 1000 steps in one direction, pauses for one second and steps 1000 steps in the reverse direction. After another one second pause, the cycle repeats endlessly.

Now is as good a time as any to discuss rotation direction. Although MBasic's User's Guide refers to "clockwise" and "counter-clockwise" rotation as being associated with positive and negative values of the step argument, in fact, the direction of rotation is also governed by the order in which the windings are connected to the driver. If the motor rotates in a direction opposite to the way you wish it to rotate, reverse any winding pair.

For example, referring to Figure 8-19, to reverse the motor's rotational direction, interchange the motor leads connected to pins 16 and 14 of the ULN2003A. Or, interchange the motor leads connected to pins 15 and 13. However, if you interchange both pairs of leads, you will restore the original rotational direction! Accordingly, should your motor rotate counterclockwise even though **SPMotor** has a positive value step argument, reverse any pair of winding leads to synchronize your motor's windings with **SPMotor**'s output pulse sequence.

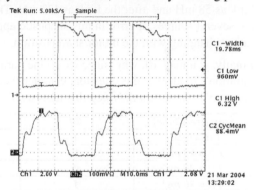

Figure 8-21: Waveforms of Program 8-1; Ch1: Pin 16 ULN2003A; Ch2: Current through motor winding connected to pin 16.

Let's look at the waveforms associated with Program 8-1 and the circuit of Figure 8-19. Figure 8-21 shows the voltage at pin 16 of the ULN2003A (Chan. 1) and the associated current (Chan. 2) through the motor winding. Since Channel 1 of the oscilloscope is connected to the collector of the drive transistor, current flow is associated with taking the collector low. Let's see what we may learn from Figure 8-21.

First, we note that when saturated, the ULN2003A's collector does not drop much below 1 V, with the measured value being 960 mV. We learned the reason that a Darlington transistor has a relatively high V_{SAT} in Chapter 3, so this value is expected. However, since we are applying only 5 V to the motor windings and ULN2003A, the motor has only 4 V applied across its windings, which reduces the current and hence its available torque.

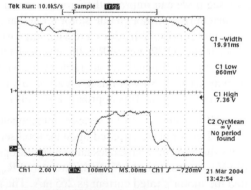

The current through the motor winding (Channel 2) shows a typical inductive current rise, but with two interesting aspects. Note the kink; current goes up rapidly, then reverses momentarily before resuming the rise. A similar effect is seen at the release. (Figures 8-21 and 22 were taken without

Figure 8-22: Expanded view of current waveform; Ch1: Pin 16 ULN2003A; Ch2: Current through motor winding connected to pin 16.

the Zener diode, so they show a relatively slow current release.) Also, the slope of the current rise is faster before the kink than afterward. Figure 8-22 provides an expanded view of the current waveform.

What causes the kink? There are two factors as work here. First, when the rotor moves, its magnetic field induces a current back into the windings, with the direction opposing the original current. Second, as the rotor moves, the internal geometry of the motor is different, giving rise to a change in inductance of the winding. From circuit theory, we know the relationship between a constant inductance L and the rate of change of current for a constant applied voltage V is:

$$V = L\frac{di}{dt}$$

Rearranging we find:

$$\frac{di}{dt} = \frac{V}{L}$$

This relationship may be interpreted as V/L being the slope of the curve of current versus time plot. If we substitute finite changes, i.e., $\frac{\Delta I}{\Delta T}$ for the infinitesimal $\frac{di}{dt}$, we determine the winding inductance before the kink as 58 mH and after the kink as 133 mH. These values should only be regarded as approximate, as there are several sources of error in this simplistic approach, not the least of which being L is not a constant but is rather changing with the motor shaft position.

Finally, you may ask why, since we set the parameter **Mdelay** at 10 ms, the current pulse measures 20 ms duration? To answer this question, consider the output pattern of a full step drive (to clarify the repetitive nature, I've include a fifth step that takes us back to the starting point):

Step (start time)	B0	B2	B1	B3
1 (t = 0 ms)	1	1	0	0
2 (t = 10 ms)	0	1	1	0
3 (t = 20 ms)	0	0	1	1
4 (t = 30 ms)	1	0	0	1
5 (t = 40 ms)	1	1	0	0

A high on B0...B3 corresponds to a low at the Darlington output, and current flow. As we see, each winding is held low for the duration of two consecutive steps, or 20 ms in this case.

Let's see if we can improve the current rise time. Figure 8-23 shows one simple approach to speeding up the current rise. Although our motor is rated at 5 V, we'll use a series resistor and higher voltage to speed up the current rise. Why does a resistor improve rise time? As we learned earlier, the current rise time is proportional to L/R, where R is the total resistance in the series circuit. Since L is governed by the motor's construction, and hence not alterable by us, we may reduce the quotient L/R by increasing R. Alternatively, you may wish to think of this approach as using a higher voltage and a series resistor to approximate a constant current source. This approach is sometimes called an "L/R drive" design.

We can't of course, select a series resistance and voltage randomly. The resistance must be chosen to provide the rated motor current, based upon the total series resistance, and the desired drive voltage. Figure 8-24 shows the relevant resistances and voltage drops we must consider. We've previously measured the voltage drop V_{SAT} across Q1 at 1 V. And, I measured $R_{internal}$ at 20 ohms for my PF35-48 motor. We previously calculated the motor's rated current as 265 mA. The relationship between R1, V1 and I_{motor} is:

$$R_1 = \frac{V_1 - V_{SAT}}{I_{motor}} - R_{Internal}$$

136

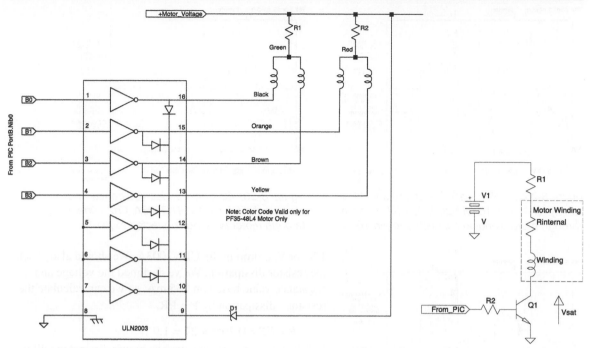

Figure 8-23: L/R drive for unipolar motor.

Figure 8-24: Calculating motor series resistor value.

Suppose we wish to feed the motor from a 12 V supply. What is R1's value?

$$R_1 = \frac{12-1}{0.265} - 20 = 21.5\,\Omega$$

To illustrate the difference in rise time versus series resistance, I conducted tests with 20 ohms, 40 ohms and 50 ohms series external resistance. The test voltages I used differ slightly from theoretically exact values, but are within reasonable tolerances. I've also measured the inductance of a winding as 8.5 mH at 1000 Hz with a General Radio 1650A RLC bridge and we'll use this value for calculating L/R. (Yes, I know this value doesn't come close to the values we estimated earlier from the rise time data, but it closely matches the data sheet value.)

Parameter	R1=0	R1=20	R1=40	R1=50
Supply Voltage	5	11	16	18
L/R (μs)	425	212	142	121
Figure	Fig 8-21	Fig. 8-25	Fig. 8-26	Fig. 8-27

Comparing the following four figures shows significant improvement in the initial current rise time. We also see diminishing returns setting in. The maximum step rate of the PF35-48 is 550 pps if we ramp up the speed from a standing start. If we use wave excitation, each motor current pulse would be 1.8 ms, as unlike full step excitation, the pulse length is not doubled. If we desire the rise time to be roughly 10% of the pulse length, we see that the series resistance should be between 20 and 40 ohms. If we calculate it exactly, for a rise time of 180 μs, we determine that the total resistance must be at least 47 ohms, which requires 27 ohms external resistance plus the windings internal 20 ohms. The required supply voltage is then 13.5V, allowing

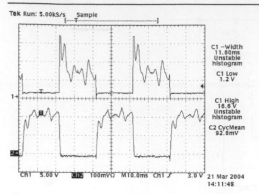

Figure 8-25: 20 ohms series resistance and 11 V supply Ch1: pin 16 ULN2003A; Ch2: current through motor winding connected to pin 16.

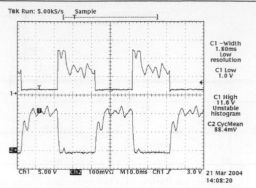

Figure 8-26: 40 ohms series resistance and 16 V supply Ch1: pin 16 ULN2003A; Ch2: current through motor winding connected to pin 16.

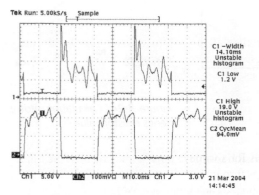

Figure 8-27: 50 ohms series resistance and 18 V supply Ch1: pin 16 ULN2003A; Ch2: current through motor winding connected to pin 16.

1 V for V_{SAT} drop in the ULN2003A. We should also check the resistor dissipation. We've designed the voltage and resistance value based on 265 mA, so we can calculate the resistor's dissipation by $P = I^2R$.

$$P = I^2R = 0.265^2 \times 27 = 1.9\ W$$

In full step operation, the average power dissipated will be one half this value. However, if we operate the motor in the locked position—that is, a winding powered up to lock the rotor in place with the motor's holding torque, R1 will dissipate the full 1.9 W. R1 should accordingly be a 27-ohm, 5 W resistor. When stopped, **SPMotor** retains the output value of the last step position, so two windings will be energized unless you explicitly drop all output pins low.

We will make one final measurement—how high a voltage spike might we expect without a clamping diode? If you wish to make this measurement, reconfigure your circuit to the arrangement shown in Figure 8-28. We are intentionally operating the ULN2003A in an unsafe environment, so there is a risk that your ULN2003A will be destroyed. If you don't want to risk your ULN2003A, skip this test. Figure 8-29 shows that without a clamping diode, spikes reach close to 30 V. The clamping diodes I used are 1 A, 50 V Schottky devices.

Program 8-2

From version 5.3.0.0 onward, MBasic's **SPMotor** can be commanded to operate in either full step or half step modes. By adding the line **SPMOTOR_Half CON 0** to the program, we may force **SPMotor** to operate in half step mode. The configuration line **SPMOTOR_Half CON 0** is a compile-time option, i.e., you may not switch between half and full step operation while your program is running. Rather, **SPMotor** will either cause full step operation (the default) or half step operation if you include the command **SPMOTOR_Half CON**. (The associated CD-ROM contains an example of half-step operation of **SPMotor** in a second version of Program 8-1. Of course, this program requires MBasic 5.3.0.0 or later to correctly operate.)

To provide more flexibility, let's look at writing our own routines for both wave and half-step modes. While we're at it, we'll include the full step mode. Program 8-2 is compatible with any of the circuits we used with Program 8-1.

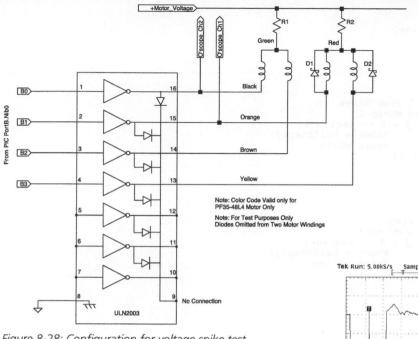

Figure 8-28: Configuration for voltage spike test.

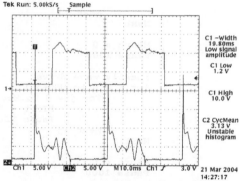

Figure 8-29: Spikes from collapsing field; Ch1: clamped with Schottky diode; Ch2: unclamped.

```
;Program 08-02
;Demo Full, Half and Wave with ULN2003A

;Constants
;----------
SeqLen          Con     8         ;how many patterns for repeat
MPin            Con     B0        ;starting pin for motor driver
MDelay          Con     5         ;milliseconds per step
                                  ;each MStep causes 8 motor cycles
MStep           Con     1000      ;number of cycle repeats to make

;Following hold the output patterns for each stated mode.
;Full and Wave are 4 steps long, so we repeat each to make
;all patterns 8 steps long.
HalfStep   ByteTable 8,12,4,6,2,3,1,9
FullStep   ByteTable 12,6,3,9,12,6,3,9
WaveStep   ByteTable 8,4,2,1,8,4,2,1


;Variables
;----------
i               Var     Word
j               Var     Word
MPort           Var     PortB.Nib0
```

139

```
;Initialization
;--------------
For j = MPin to Mpin+3
        Output j
Next
MPort = 0


Main
        ;First, full step drive
        For i = 0 to MStep-1
                For j = 0 to SeqLen-1
                        MPort = FullStep(j)
                        Pause MDelay
                Next ;j
        Next

        MPort = 0
        Pause 1000

        ;Next half step
        For i = 0 to MStep-1
                For j = 0 to SeqLen-1
                        MPort = HalfStep(j)
                        Pause MDelay
                Next ;j
        Next

        MPort = 0
        Pause 1000

        ;Lastly wave drive
        For i = 0 to MStep-1
                For j = 0 to SeqLen-1
                        MPort = WaveStep(j)
                        Pause MDelay
                Next ;j
        Next

        MPort = 0
        Pause 1000
    GoTo Main
```

The concept behind Program 8-2 is simple; we write the binary pattern corresponding to the desired motor step sequence to four consecutive pins on a port. Let's recap the stepping sequence for all three modes:

Mode	Step	Bit 3	Bit 2	Bit 1	Bit 0	Decimal
Full	1	1	1	0	0	12
	2	0	1	1	0	6
	3	0	0	1	1	3
	4	1	0	0	1	9
Half	1	1	0	0	0	8
	2	1	1	0	0	12
	3	0	1	0	0	4
	4	0	1	1	0	6
	5	0	0	1	0	2
	6	0	0	1	1	3
	7	0	0	0	1	1
	8	1	0	0	1	9

(continued)

	1	1	0	0	0	8
Wave	2	0	1	0	0	4
	3	0	0	1	0	2
	4	0	0	0	1	1

I've added a column "decimal" to the table to show the numerical value of the four pattern bits. Perhaps the most convenient form to hold these pattern bits is a byte table. We can construct three byte tables, one for half-step, one for full step and one for wave drive.

```
HalfStep    ByteTable 8,12,4,6,2,3,1,9
FullStep    ByteTable 12,6,3,9,12,6,3,9
WaveStep    ByteTable 8,4,2,1,8,4,2,1
```

The half-step table entries follow the sequence directly. The full and wave byte tables, however, double up the values, so that we use eight entries to hold two complete four-step patterns. We do this to make the three byte tables identical length, as it lets us use the same code to output any of the three modes. (We'll see the benefit of this approach more in Program 8-3.)

```
;Constants
;----------
SeqLen          Con     8       ;how many patterns for repeat
MPin            Con     B0      ;starting pin for motor driver
MDelay          Con     5       ;milliseconds per step
                                ;each MStep causes 8 motor cycles
MStep           Con     1000    ;number of cycle repeats to make
```

We've used the same names as in Program 8-1 for the key constants. However, we've deviated from how **Mstep** is used. In MBasic's **SPMotor** function, **MStep** defines how many full steps the motor makes. For simplicity, we use **MStep** to define how many 8-step cycles Program 8-2 outputs. (We'll get back to a step-based approach in Program 8-3.) Thus, if we set **MStep** to 1000, Program 8-2 outputs 1000*8, or 8000 motor steps.

```
MPort           Var             PortB.Nib0
```

We also define one variable, **Mport**, to use when addressing the output port. Since we only need four pins, we use a nibble. For consistency with Program 8-1, we use Port B's low nibble.

```
For j = MPin to Mpin+3
        Output j
Next
MPort = 0
```

We initialize the four output pins to be outputs.

```
Main
        ;First, full step drive
        For i = 0 to MStep-1
                For j = 0 to SeqLen-1
                        MPort = FullStep(j)
                        Pause MDelay
                Next ;j
        Next
```

To output full step, we loop through the byte array **FullStep**, sending each pattern in sequence. We repeat this **MStep** times.

```
        MPort = 0
        Pause 1000
```

At the completion of outputting the full steps, we de-energize all windings and pause for one second.

```
        ;Next half step
        For i = 0 to MStep-1
                For j = 0 to SeqLen-1
                        MPort = HalfStep(j)
                        Pause MDelay
                Next ;j
```

```
                Next

                MPort = 0
                Pause 1000
```

We repeat the same sequence, except outputting the half step pattern.

```
                ;Lastly wave drive
                For i = 0 to MStep-1
                        For j = 0 to SeqLen-1
                                MPort = WaveStep(j)
                                Pause MDelay
                        Next ;j
                Next

                MPort = 0
                Pause 1000
        GoTo Main
```

We conclude with wave drive, and when completed go back to **Main** and start the sequence over again.

After verifying Program 8-2 with one of the ULN2003A circuits we used with Program 8-1, let's try an alternative driver. In Chapter 3, we saw how efficient a small MOSFET, the 2N7000, is as a low side driver, so let's replace the ULN2003A with four 2N7000s. Figure 8-30 shows our circuit. I've shown four Schottky diodes as clamps, because I didn't have four suitable Zener diodes at hand. But, you can significantly improve the current release time by adding Zener diodes in reverse series with the four Schottky diodes. Incidentally, if your motor only has five windings, the connection arrangement shown in Figure 8-31 may be used. If we restrict ourselves to full step or wave drive, we can use a simpler arrangement; use one series resistor for all windings (but only one mode for any resistor arrangement). This is possible because the same number of windings will always be on; one in the case of wave drive or two in the case of full step drive. Thus, the circuit has a constant load and one current limiting resistor is adequate for each mode. (You can't use the same resistor for wave and for full step operation, of course.) However, if we operate half-step mode, the current alternates between one and two

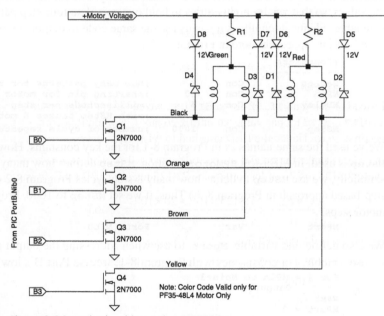

Figure 8-30: Unipolar driver using 2N7000s.

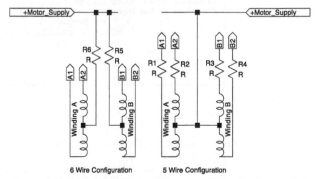

Figure 8-31: Connection arrangement differs for five and six wire unipolar steppers.

windings energized, so a single current limiting resistor is not acceptable. Figure 8-32 shows the overlap between two motor windings during full step operation.

Program 8-3

We'll wrap up our excursion into the world of stepper motors with a program that demonstrates an improved approach to the technique introduced in Program 8-2. We'll encapsulate the nuts and bolts of outputting the stepper pattern in a subroutine. We'll also look at bipolar drivers and show how to connect a unipolar motor as a bipolar device. To verify the operation of Program 8-3, you may start with any of the circuits we've developed for Programs 8-1 and 8-2.

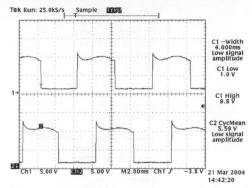

Figure 8-32: 2N7000 driver output during full step operation; Ch1: drain voltage Q1; Ch2: drain voltage Q2.

```
;Program 08-03
;Demo Full, Half and Wave with
;SN754410 unipolar and ULN2003A bipolar operation
;Quasi-Function subroutine

;Constants
;----------
MPin          Con     B0        ;starting pin for motor driver
MDelay        Con     5         ;milliseconds per step
                                ;unlike program 8-2, MStep is number of steps
MStep         Con     1000      ;number of steps to make
                                ;can't be more than 65535.

SeqLen        Con     8         ;how many patterns for repeat
ModeFull      Con     0         ;mode constants
ModeWave      Con     8
ModeHalf      Con     16

DFwd          Con     1         ;direction constants
DRev          Con     -1        ;fwd and rev are arbitrary

;0..7 posn = full, 8...15=wave 16-23 = half
;for full & wave repeat for cycle length = 8
StepTable  ByteTable |
12,6,3,9,12,6,3,9,8,4,2,1,8,4,2,1,8,12,4,6,2,3,1,9

;Variables
;----------
i             Var     Word      ;counts steps
j             Var     SByte     ;indexes patterns
MPort         Var     PortB.Nib0 ;output port pins
Direct        Var     SByte     ;direction
Mode          Var     SByte     ;mode type

;Initialization
;--------------
;set motor port to output
For j = MPin to Mpin+3
        Output j
Next
;all windings off
MPort = 0


Main
```

```
                ;Full step forward
                Mode = ModeFull
                Direct = DFwd
                GoSub BiPolar
                MPort = 0
                Pause 1000
                ;Full step reverse
                Direct = DRev
                GoSub BiPolar
                MPort = 0
                Pause 1000

                ;Half step forward
                Mode = ModeHalf
                Direct = DFwd
                GoSub BiPolar
                MPort = 0
                Pause 1000
                ;Half step reverse
                Direct = DRev
                GoSub BiPolar
                MPort = 0
                Pause 1000

                ;Wave step forward
                Mode = ModeWave
                Direct = DFwd
                GoSub BiPolar
                MPort = 0
                Pause 1000
                ;Wave step reverse
                Direct = DRev
                GoSub BiPolar
                MPort = 0
                Pause 1000

GoTo Main


;Called subroutine needs MPin, MStep and MDelay set
;Direct and Mode also must be set. Motor pattern in
;Byte Array StepTable()
;and needs i & j free variables
BiPolar
;-------
        j = Mode        ;starting point into table
        ;If reverse, then start at end of table entry
        If Direct = DRev Then
                j = j + SeqLen-1
        EndIf

        ;Step through the required number of steps
        For i = 0 to MStep-1
                MPort = StepTable(j)   ;output pattern
                Pause MDelay                  ;wait
                j = j + Direct              ;next pattern no.
                ;only have j < Mode if reverse stepping
                ;check and wrap back to starting point
                If (j < Mode) AND (Direct = DRev) Then
                                j = Mode + SeqLen -1
                EndIf
                ;Likewise j>= Mode+SeqLen if forward. Check for
                ;overrun and wrap back to starting point
                If (j >= (Mode+SeqLen)) AND (Direct = DFwd) Then
                                j = Mode
                EndIf
        Next
Return
```

Program 8-3 uses the same constant names as in Programs 8-1 and 8-2 to define the starting pin, the delay (in milliseconds) per step and the number of steps to take. In order to more closely mimic the functionality of the function **SPMotor**, **MStep** in Program 8-3 defines the number of steps, not the number of 8-step cycles as in Program 8-2.

```
MPin        Con     B0      ;starting pin for motor driver
MDelay      Con     5       ;milliseconds per step
                            ;unlike program 8-2, MStep is number of steps
MStep       Con     1000    ;number of steps to make
```

Although I've made **MDelay** and **MStep** constants, you may wish to make them variables and compute or assign their values based on user inputs or calculated parameters. Since we use a word value counter to keep track of the number of steps, **MStep** can't exceed 65535. If this is inadequate, change the counter variable **i** to be a type long.

It's also necessary to define a variable to address the output port nibble. I've named this variable **MPort**.

```
MPort           Var     PortB.Nib0      ;output port pins
```

If you use different pin connections, don't forget to change both **MPin** and **MPort**.

```
ModeFull    Con     0       ;mode constants
ModeWave    Con     8
ModeHalf    Con     16

DFwd        Con     1       ;direction constants
DRev        Con     -1      ;fwd and rev are arbitrary

Direct  Var     SByte       ;direction
Mode        Var     SByte   ;mode type
```

We set the direction and operational mode using two variables, **Direct** and **Mode**. In order to make our program more intelligible, we define three mode constants, **ModeFull**, **ModeWave** and **ModeHalf**, representing full step, wave and half-step operation, respectively. We'll see the reason for choosing values of 0, 8 and 16 for these constants a bit later. Likewise, we define two directional constants, **DFwd** and **DRev** for forward and reverse. Your choice of motor winding connection, as we covered in connection with Program 8-2, determines whether forward is clockwise or counterclockwise.

StepTable ByteTable 12,6,3,9,12,6,3,9,8,4,2,1,8,4,2,1,8,12,4,6,2,3,1,9

Rather than keep three separate byte tables to hold the pattern for full step, wave and half step, I've combined them into one 24-element byte table. Elements 0…7 are the full step patterns, 8…15 are the wave patterns and 16…23 are the half-step patterns. As with Program 8-2, **StepTable** holds two complete cycles of full step and wave patterns, so that full, wave and half-step all have eight entries. We also now see why the constants **ModeFull**, **ModeWave** and **ModeHalf** are defined with the values 0, 8 and 16—these numbers are the starting point for indexing into **StepTable** for their respective modes.

```
;Initialization
;--------------
;set motor port to output
For j = MPin to Mpin+3
        Output j
Next
;all windings off
MPort = 0
```

We initialize by setting output state the pins we have selected to control the motor and we de-energize all the windings. In the de-energized mode, the stepper has only its residual torque to hold the rotor against external torque, so you may instead wish to initialize the motor with one winding energized.

Let's now examine the subroutine **Bipolar**. I named this subroutine **Bipolar** because I originally intended to demonstrate Program 8-3 only with a bipolar motor, but later decided this is too limiting. *The step pattern, if you use the correct driver and motor connections, is identical for a unipolar and bipolar motor. Hence, if properly connected Program 8-3 functions identically with either motor type.*

Subroutine `BiPolar` assumes we have defined and set values for `MPin`, `MStep`, `MDelay`, `Direct`, `Mode`, and that `MPort` is aliased with the port nibble to be used. In addition, it uses two variables `i`, type word, and `j`, type sbyte.

```
BiPolar
;-------
        j = Mode            ;starting point into table
        ;If reverse, then start at end of table entry
        If Direct = DRev Then
                j = j + SeqLen-1
        EndIf
```

To reverse direction, we step through `StepTable` in reverse order, that is, emitting the values of `StepTable` in the sequence `StepTable(0)`, `StepTable(1)`...`StepTable(7)` is forward, while the sequence `StepTable(7)`, `StepTable(6)`...`StepTable(0)` is reverse. Hence, based upon the value of `Direct`, we start indexing into `StepTable` at 0,8 or 16 for forward, or 7, 15 or 23 for reverse. The constant `SeqLen` holds the number of elements in `StepTable`—8—for each mode. We enter the program control loop, therefore, with `j` properly set at the first element in the pattern we wish to send to the motor.

```
        ;Step through the required number of steps
        For i = 0 to MStep-1
                MPort = StepTable(j)   ;output pattern
                Pause MDelay                   ;wait
                j = j + Direct                 ;next pattern no.
```

We now emit `MStep` number of steps, using a loop construction to control program execution. (Since we start with 0, to send `MSteps` steps, we must stop at `MStep-1`.) After the pattern for each step is emitted we update `j` with the statement `j = j + Direct`. `Direct` is either +1 (forward) or –1 (reverse) so `j` increments or decrements based upon the chosen direction of rotation,

```
                ;only have j < Mode if reverse stepping
                ;check and wrap back to starting point
                If (j < Mode) AND (Direct = DRev) Then
                        j = Mode + SeqLen -1
                EndIf

                ;Likewise j>= Mode+SeqLen if forward. Check for
                ;overrun and wrap back to starting point
                If (j >= (Mode+SeqLen)) AND (Direct = DFwd) Then
                        j = Mode
                EndIf
        Next
Return
```

We must limit `j`'s value so that it wraps around when it attempts to go outside the range of values permitted for the selected mode. We do this with two `IF...Then` statements. If the direction is forward (`Direct=DFwd`) we check to see if `j` exceeds the maximum index, which is `Mode + SeqLen`. If this is the case, we reset the index by the statement `j=Mode`.

If the direction is reverse (`Direct=DRev`) we check for an under-run—that is, `j` is less than `Mode`. If this is the case, we reset the index by the statement `j=Mode+SeqLen-1`.

To show how subroutine `BiPolar` is called, we return to the main program loop.

```
        ;Full step forward
        Mode = ModeFull
        Direct = DFwd
        GoSub BiPolar
        MPort = 0
        Pause 1000
        ;Full step reverse
        Direct = DRev
        GoSub BiPolar
        MPort = 0
        Pause 1000
```

We set the variables that define the mode, **Mode**, and direction, **Direct**, and then call **BiPolar**. Of course, if we made **MStep** or **MDelay** variables instead of constants, they also would require setting to appropriate values.

The remainder of Program 8-3 demonstrates the remaining modes and directions and requires no further analysis.

After you have verified that Program 8-3 is functioning with a unipolar motor and any of the circuits developed for Programs 8-1 and 8-2, we may now turn to a bipolar driver. Figure 8-33 shows a suitable circuit. Note carefully the input and output connections to the SN754410. I've shown connections for either a pure bipolar motor, or for connecting a 6-wire unipolar motor as a bipolar device. And, the circuit of Figure 8-33 may be used with the Program 8-1 or 8-2, should you so desire. The H-bridge circuit permits MBasic's **SP-Motor** function to drive a bipolar motor, not just a unipolar motor as the User's Guide suggests.

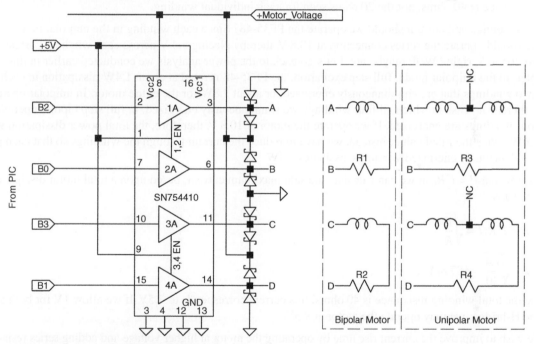

Figure 8-33: Bipolar motor driver.

Our circuit is based upon an SN754410 quad half-H bridge device. (By combining two half-H bridges, we get a full H-bridge.) Earlier, in Figure 8-17, we saw a conceptual overview of an "H" bridge. By setting one control pin high and the other low, the SN754410 permits current flow through the winding in one direction. If the two control pins reverse state, current flows through the winding in the reverse direction. Pins 1 and 9 are "enable" pins that must be held at +5 V for their associated bridge drivers to function. The SN754410 has two V_{CC} connections; one for +5 V and the second to be connected to the motor supply. Don't reverse these connections or the "magic smoke" inside the SN754410 will escape and you then can throw what's left into the trash!

The SN754410's superior ratings make it preferable to the more popular L293. And, making it even better, the SN754410 is few cents cheaper than the L293. The SN754410 H-bridge is constructed with Darlington transistors in both the high side and low side switches. Hence, we anticipate approximately a 1 V drop in both the high and low side motor connections. In computing the current limiting resistors, don't forget to include both voltage drops. In our application, the increased voltage drop is not of major concern, but it might be for larger motors. Fortunately, a wide variety of H-bridge circuits are available, including those with less voltage drop and higher current ratings than the SN754410.

Figure 8-33 also shows eight Schottky voltage clamp diodes, based on the recommended design in the SN754410's data sheet. The SN754410 contains internal clamping diodes and for low power stepper motors, the internal diodes may be adequate. However, adding external clamping diodes at the motor itself is an excellent safety precaution, and should be followed. Schottky diodes have the dual advantage of being much faster than conventional silicon power diodes of the 1N400X family and also provide a lower forward voltage drop.

To demonstrate how Program 8-3 operates in the bipolar mode, I used the same unipolar PF35-48 motor that appears throughout this chapter. In this case, however, we operate with two windings in series, so the winding resistance is 40 ohms, not the 20 ohms seen for each individual winding.

At what voltage and current should we operate the PF35-48? Since each winding in the unipolar mode is 5.4 V, we should operate the series connection at 10.8 V, thereby placing 5.4 V across each winding, corresponding to 265 mA, right? Well, maybe not. Let's go back to the power analysis we conducted earlier in this chapter. In the unipolar mode, full step excitation, the PF35-48 is rated at about 1.4W dissipation in each of the two windings that are simultaneously energized, or about 2.8W for the entire motor. In unipolar operation, there is no case in which all four windings are simultaneously energized. In full step bipolar operation, all four windings are energized. If we operate the motor at 10.8 V, therefore, the total power dissipation will be 5.6W, twice the rated value. Instead, we have to reduce the current through the windings so that each pair of simultaneously energized windings dissipates 1.4W.

We know that $P = I^2R$, so we can calculate the safe maximum current, based upon a total motor dissipation of 2.8 watts.

$$P = I^2R; I = \sqrt{\frac{P}{R}}$$

$$I = \sqrt{\frac{1.4}{40}} = 187\ mA$$

Since the total winding resistance is 40 ohms, this current corresponds to 7.5V. If we allow 1V for both sides of the H-bridge, we may operate the motor at 9.5V.

If we wish to improve the current rise time by operating the motor at higher voltage and adding series resistance, we should use 187mA as the safe operating current for full step operation. This calculation is based upon full step operation; if we were to operate the motor in bipolar wave drive, only two windings are energized at any time, so we may use the full unipolar current rating and operate the motor at 10.8V, plus the H-bridge drops. For half-step operation, the safe current will be between these two values, as the average number of windings energized is 3 (4 steps at 4 windings; 4 steps at 2 windings, for an average of 3 energized windings.)

When driving a unipolar motor in a bipolar circuit, you should be aware that the winding inductance also increases. The bipolar connected winding inductance is approximately four times the inductance of a single winding. The inductance doesn't simply double because the two windings are closely magnetically coupled and their mutual inductance must be taken into account. Since the series resistance only doubles, the L/R time constant of the bipolar connected winding is twice that of the unipolar connection.

Of course, you need not go through these calculations if the motor is of bipolar design, as its ratings will be based upon bipolar operation.

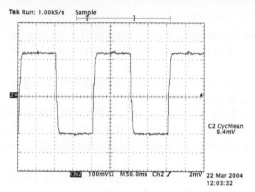

Figure 8-34: Bipolar operation—full step mode; Ch1: off; Ch2: motor current.

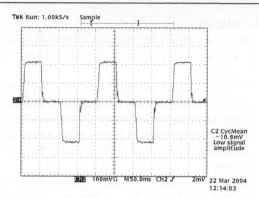

Figure 8-35: Bipolar operation—wave mode; Ch1: off; Ch2: motor current.

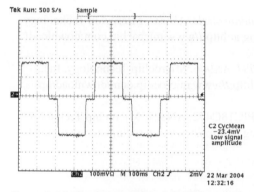

Figure 8-36: Bipolar operation—half-step mode; Ch1: off; Ch2: motor current.

Let's see how the current flow looks when running Program 8-3 with a bipolar driver. Figure 8-34 shows the current flow through one winding in full step operation. The current pulses are approximately equal and show the expected alternating direction, +/–/+/–… Figure 8-35 illustrates wave operation. The winding current pattern is +/0/–/0/…, again as expected. Finally, Figure 8-36 shows half-step operation. The current waveform looks unusual, doesn't it? But, it's exactly what we should expect. Let's look at a piece of the half-wave pattern table. The winding that is shown in Figures 8-34 through 8-36 is connected to the SN754410's pin 3. Hence, its current flow is controlled by Bits 0 and 2 of the output nibble, corresponding to the PIC's output pins B0 and B2 for the particular connection shown in Figure 8-33. (I've omitted Bits 3 and 1 from the table extract.) The SN754410 interprets input commands as **10** = current flow in one direction; **01** = current flow in opposite direction and **00** = no current flow. Based on this, I've added a current flow column, arbitrarily assigning the **10** direction as a plus.

Mode	Step	Bit 2	Bit 0	Current Flow
	1	0	0	None
	2	1	0	+
	3	1	0	+
Half	4	1	0	+
	5	0	0	None
	6	0	1	-
	7	0	1	-
	8	0	1	-

Thus, the current pattern we see should be 0/+/+/+/0/-/-/-, which is an alternating current flow, with a period of no current flow between the reversals. The quiescent period should be one-third the duration of the active current flow period. And, that's exactly what Figure 8-36 shows.

References

[8-1] Thompson Airpax Mechatronics, *Product Selection and Engineering Guide,* (undated), available for downloading at http://www.thomsonindustries.com/PDFs/Catalogs_and_Brochures/Airpax_Catalog_Idx.pdf.

[8-2] Thompson Airpax Mechatronics, *Stepper Motor Handbook*, (undated), available for downloading at http://www.allegromicro.com/techpub2/compumot/a04a08.pdf.

[8-3] Texas Instruments, *SN754410 Quadruple Half-H Driver Data Sheet*, Doc. No. SLRS007B, (Nov. 1995).

[8-4] Telefonaktiebolaget LM Ericsson, *Industrial Circuits Data Book and Stepper Motor Control Handbook; Stepper Motor Basics*, (1995) available for download at http://library.solarbotics.net/pdflib/pdf/motor-bas.pdf.

[8-5] Sax, H., *Application Note 235: Stepper Motor Driving*, SGS-Thomson Microelectronics (July 1988), available for downloading at http://library.solarbotics.net/pdflib/pdf/motorbas.pdf.

[8-6] Hopkins, Thomas L., *Application Note 460: Stepper Motor Driver Considerations Common Problems and Solutions*, SGS-Thomson Microelectronics (2003), available for downloading at http://www.st.com/stonline/books/pdf/docs/1675.pdf.

[8-7] Condit, Reston, & Jones, Douglas W., *Stepping Motors Fundamentals, AN907*, Microchip Technology Inc., Doc. No. DS00907A (2004), available for downloading at http://www.microchip.com/download/appnote/pic16/00907a.pdf.

[8-8] Condit, Reston, *Stepper Motor Control Using the PIC16F684, AN906*, Microchip Technology, Inc., Doc. No. DS00906A (2004), available for downloading at http://www.microchip.com/download/ap-pnote/pic16/00906a.pdf.

[8-9] Yedamale, Padmaraja, Stepper Motor Microstepping with PIC18C452, AN822, Microchip Technology, Inc., Doc. No. DS00822A (2002), available for downloading at http://www.microchip.com/download/appnote/pic16/00822a.pdf.

RS-232 Serial Interface

The computer science dictionary defines "serial" as "of or relating to the sequential transmission of all the bits of a byte over one wire." As we'll see, there are many serial protocols, such as I²C (Chapter 18), SPI (Chapter 21), one-wire (Chapter 12) and the infrared remote control protocol REC-80 (Chapter 22). In this chapter, we'll look at an older serial protocol, RS-232, and learn to use MBasic's **SerIn**, **SerOut**, **HSerIn** and **HserOut** procedures. Along the way, we'll document some undocumented features of MBasic and write an interactive menu program.

It's not correct to refer to the entire bundle of interaction and data exchange we'll explore in this chapter as "RS-232." The details are well beyond the scope of this book, but the open systems interconnection (OSI) reference model defines a seven-layer model of network and computer communications. In the OSI model, RS-232 applies to the lowest, or "physical," layer, where it defines voltage and signal levels, pin wiring, connector standardization and flow control. But nothing in the RS-232 specification says that we must transmit and receive 8-bit bytes over an RS-232 connection, or that we should use the ASCII code to represent text, or that we must one of the widely recognized standard data speeds. We could, for example, use a 1930's Model 15 Teletype machine to send five-bit Baudot code at 45.45 bits/sec over an RS-232 connection and, as long as we follow the pin and connector wiring and voltage standards, we would be in complete compliance with RS-232. If you connect the Model 15 Teletype to your PC, you will, of course, receive nothing but gibberish without additional protocol and speed conversion. These protocol and speed conversions are part of higher layers in the OSI stack, and if you correctly implement these higher layers, you will be able to communicate with the Model 15. Nonetheless, we're going to "go with the flow" and not distinguish those things defined in the RS-232 standard and those things defined in other standards relating to higher layers in the OSI model, except where absolutely necessary. Further, we'll often call this communications protocol simply "serial."

RS, by the way, means "recommended standard" and the original version, RS-232 with no letter suffix, was adopted by the Electronics Industry Association in 1962. The current version of the standard is known as EIA/TIA RS-232E. (In 1988, the EIA merged with a telecommunications industry group, and the merged entity added the Telecommunications Industries Association name to the standard.)

How to Connect to Your PC

Since you're programming in MBasic, you already have a PC running a recent version of the Windows operating system. Your PC has at least one RS-232 serial port, necessary to connect the ICP to the PC.

We're going to use a second RS-232 connection, the DB9 connector mounted on the 2840 Development Board, as well as the other development and prototype boards produced by Basic Micro. You'll find it far more convenient if your computer has a second RS-232 port to connect to the Development Board. If your PC has only one serial port, you can move the connecting cable between the ICP and the 2840 Board, or, you can purchase an inexpensive USB-to-serial adapter and use it to connect to the 2840 Board. You'll also need a second DB9-to-DB9 cable. This cable should be a "modem" cable, i.e., one where each pin on one end connects with the same numbered pin on the other end. (The cable sold by Basic Micro for computer-to-ICP connections is constructed this way.)

You'll need to run a program to communicate with the PIC's software. This program is usually called "terminal" software, as it allows the PC to emulate a dumb serial terminal. Windows comes packaged with a terminal program, Hyperterminal, which you will find under Program | Accessories | Communications. And, for some purposes, the terminal utility that's part of MBasic's IDE is quite adequate. (I observed problems with its implementation of hardware flow control, however.) My personal favorite for development work is Terminal[9-2] (freeware from Bray++) and I also find the diagnostic terminal program ComTest[9-3] (also freeware) from B&B useful for debugging serial problems. Terminal provides much more flexible control over the technical parameters of serial communications than either Hyperterminal or MBasic's IDE terminal utility, important considerations when developing and testing serial communications programs.

We'll also assume you understand the basics of serial communications, that the terminal program must match the PIC's settings for speed, number of bits, parity, stop pulse length and handshaking. (Not all terminal programs have all these parameters accessible to the user.) If not, we'll recap these settings:

Parameter	Set to Value	Comment
Com Port	See Comment	The Com Port setting must match the physical communications port you have connected to the PIC's serial connector.
Data Bits	8	Our programs will not use parity, so all 8 bits are data.
Parity	None	Parity is a rudimentary from of error checking. We will not use it (MBasic only supports parity for **SerOut** and **SerIn**).
Stop Bits	1	MBasic's **HserOut** sends 1 stop bit; **SerOut** is configurable for either 1 or 2 stop bits. Don't worry if this is not an option on your terminal program.
Handshaking	None	We'll change this when we examine flow control. Supported for **SerOut** and **SerIn**, not **HSerOut** or **HSerIn**.

One additional option deserves a bit more discussion than can be squeezed into the table—the end of line character(s).

Since the earliest days of mechanical Teletype machines in the 1920s, separate keys (and codes) existed for carriage return and line feed functions. Because it took longer for the heavy type basket assembly to return to the index position than for the platen to ratchet the paper up one line, the normal line ending sequence became <CR> <LF>, or sometimes even <CR> <CR> <LF> where the second carriage return ensured enough time for the carriage to fully return before printing resumed. And, if the Teletype machine was connected to a poor radio circuit, the double <CR> helped avoid the dreaded pile-up where a full line of text turned into a black blob at the right hand margin due to a missed <CR> command. (The angle brackets indicate a character command.)

Now that we use electronic, not mechanical, terminals, it isn't necessary to issue separate <CR> and <LF> commands, as the two functions may be combined into one through code in the terminal program. In ASCII, the carriage return code has a value of 13 (**$D**) and the line feed code value is 10 (**$A**). A traditional mechanical printer end-of-line sequence is thus <13>,<10>. We'll almost always omit the line feed character and use only the carriage return symbol <13> to indicate the start of a new line. In many terminal programs, you must select a button or check box control such as "CR = LF" or "CR = CR + LF" to enable this option.

Voltage Levels in RS-232 and Level Conversion

PICs work in the world of digital logic, with + 5 V and 0 V representing logic levels 1 and 0, respectively. The RS-232 standard predates widespread use of +5 V logic and, for a variety of reasons, it adopted signal

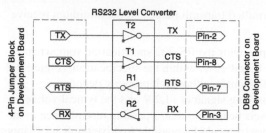

Figure 9-1: RS-232 level conversion in Basic Micro development board.

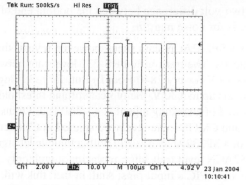

Figure 9-2: RS-232 and logic level Ch1: logic level from PIC; Ch2: RS-232 bipolar signal out of SP232ACN.

levels of +15 V and –15 V, a form of "bipolar" signaling. (Bipolar signaling had long been used in landline Teletype circuits, but with voltage levels of ±100 V, needed for transmission over long local loops and inter-city circuits.) Hence, it's necessary to convert the RS-232 levels to logic levels compatible with our PIC. Basic Micro's development boards include a conversion chip with two logic-to-RS-232 level and two RS-232-to-logic level converters, configured as shown in Figure 9-1. The RS-232 side of the level converter is hardwired to the TX (transmit), RX (receive), CTS (clear-to-send) and RTS (request-to-send) pins of the DB9 connector, while the logic side appears at four positions of a header, with corresponding labels.

Basic Micro used Sipex SP232ACN level converters in the boards I have, but pin-compatible devices are available from several manufacturers. The following table summarizes the relationship between the ±15 V RS-232 levels and the 0/+5 V logic employed by PICs, while Figure 9-2 shows the inversion in a typical bit stream.

RS-232		PIC Logic Levels	
Nominal Voltage	*Terminology*	*Voltage*	*Terminology*
+15V	Spacing, Unasserted/Negate	0	Logic 0, Logical Low
–15V	Marking, Asserted/Assert	+5	Logic 1, Logical High

The terms *Asserted/Unasserted* or *Assert/Negate* apply to control signals, and the terms *Marking and Spacing* describe transmit and receive data signals. One more bit of history; the terms marking and spacing derive from the earliest days of telegraphy, where an inking mechanism printed high-speed dots and dashes onto a moving paper tape, for later reading by a human telegrapher. Some of Thomas Edison's earliest inventions were for recording telegraph equipment. Marking meant that the paper tape had a mark; spacing meant there was no mark. These terms carried over into the punched paper tape era as well as today's nonmechanical communications.

Finally, note the inversion between RS-232 and logic; a positive voltage on the RS-232 line corresponds to 0 V in logic and a +5 V in logic corresponds to a negative voltage on the RS-232 line.

If you look closely at the RS-232 specification, you will find a wide range of permitted voltages:

Transmitted Levels (Maximum/Minimum)		Received Levels (Maximum/Minimum)	
+15V +5V	Space (Logic 0)	+15V +3V	Space (Logic 0)
–5V –15V	Mark (Logic 1)	–3V –15V	Mark (Logic 1)

The two-volt difference between the bottom end of the permitted transmit range and the minimum receive range is for noise margin.

Sipex's SP232ACN data sheet, however, reveals that it implements the RS-232 specification somewhat diffidently. The transmitted RS-232 levels are between ±9 V and ±5 V, within the specification. (Figure 9-2 shows ±8 V for my 2840 Board.) But, for receiving RS-232 signals, although it properly decodes signal levels up to ±30 V, the typical voltage threshold of a marking signal is +1.2 V and the typical voltage threshold of a spacing signal is +1.7 V. These transition levels bear more than a passing resemblance to the TTL and CMOS logic levels we studied in Chapter 4. Hence, it would be possible to drive the RS-232 input side of a SP232ACN directly with a logic level signal, implying that we might dispense completely with the SP232ACN level translator, replacing it with only some protective diodes and current limiting resistors to protect our PIC's input pins. And, in fact, this will, under certain circumstances, work. But, it's bad practice to say the least. Suppose your PIC circuit is communicating with a PC that implements the RS-232 standards more rigorously than does the SP232ACN, and that it requires at least –3 V to decode a valid mark. If your PIC outputs 0 V, but the correctly implemented RS-232 terminal requires between –3 V and –15 V to be recognized as a mark, your communications will fail.

Standard Pin Connections

The 2840 Board implements five lines of the full 9-pin RS-232 standard:

Computer (DTE)				PIC Development Board (DCE)		
Pin No.	Signal Name	RS-232 Name	Direction	Pin No.	Signal Name	RS-232 Name
1	Carrier Detect (DCD)	CD		1		
2	Receive Data (RX)	RD	←	2	Transmit Data (TX)	TD
3	Transmit Data (TX)	TD	→	3	Receive Data (RX)	RD
4	Data Terminal Ready	DTR		4		
5	Signal Ground	GND	←→	5	Signal Ground	GND
6	Data Set Ready	DSR		6		
7	Request to Send	RTS	→	7	Request to Send	RTS
8	Clear to Send	CTS	←	8	Clear to Send	CTS
9	Ring Indicator	RI		9		
Shield				Shield		

Asynchronous Transmission, Start Bits, Stop Bits and Bit Order

Serial transmission comes in two flavors: synchronous and asynchronous. These are fancy words that describe sending the data bits, one after another without any pause and without anything to distinguish one byte from the next (synchronous) and sending the data bits framed by a clear start-of-byte and end-of-byte marker with the possibility of pauses between successive bytes (asynchronous).

We're not going to further look at synchronous data transmission in this chapter, so let's see how an asynchronous signal employs start-of-byte and end-of-byte markers. (And yes, asynchronous data doesn't have to be sent in byte-size lengths, but that's how it's done for PC-to-PIC communications.)

Let's assume that no data has been sent and our data line is in the mark state. We indicate the start of a byte by sending a start (space) bit, followed by the eight data pulses (LSB first) and finish with a stop (mark) bit. In some cases, we might send two stop bits. Then, we do the same for the next byte, if there is one to be sent. Let's see how we would send the letter "A" (65, or %01000001).

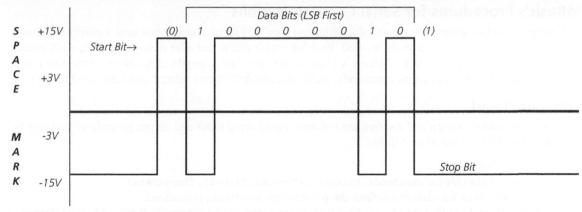

I've indicated the start and stop bit values in parenthesis, since we don't usually consider these to have a useful data value. The decoding algorithm (either embedded within **SerIn**, or within the PIC's hardware for **HSerIn**) can be thought of as:

```
Wait for a Start Bit (0)

Characterize the next 8 bits are 0's or 1's and store them in LSB order in a
storage byte

Stop characterizing after the 8th bit and verify that the next bit received is a
stop bit. If it is, then regard the byte as valid; otherwise it is invalid and set
a bad data flag.
```

OK, you say, that makes sense. But what happens if I start receiving data in the middle of a transmission? How does my algorithm decide if a 0 is a start bit or simply one of the eight data bits? It turns out that for normal text steams, this simple algorithm will correctly lock onto the start bit after not more than a half-dozen or so characters. There are some pathological cases, whereby certain characters repeated without a pause can cause a false lock, but these rarely happen in practical text or data transmission. And, more importantly, in most circumstances there will be a pause between characters or words from time to time and an idle ensures synchronization.

Although we've mentioned "parity," we've not defined it. For our purpose, even parity involves taking the eighth data bit (the MSB bit of the byte being sent) and using it to hold not data, but rather the binary sum of the previous seven data bits. Since we only have one bit to hold the sum, we can't carry, so **%1 + %1 = %0**, not **%10**. Another way of looking at it is the parity bit is **%0** when the 7 data bits have an even number of **%1**s but it is **%1** when there are an odd number of **%1**'s in the 7 data bits. Thus, the eight transmitted bits always will have an even number of **%1**s. (Odd parity is just the reverse of this.) If parity is enabled, the receiving device can detect a single bit error and flag the received byte as corrupt. (Note that 2 bit errors will not be detected, but 3 bit errors will, and so forth. Fortunately, single bit errors are far more common than multiple bit errors.) Using parity means we only can send seven information bits. If we limit our communications to standard alphanumeric text, this is not a problem, as all numbers, letters and punctuation marks in the ASCII English alphabet require only 7 bits.

One final point and then we'll look at MBasic's support for serial communications. The maximum permitted error between the sending and receiving data rates (also called *clock skew*) is only about 5.5%. If the PIC's clock is either crystal or ceramic resonator controlled, this objective is easy to meet. However, if the PIC uses its internal RC clock, achieving the needed tolerance over a wide temperature range is problematic, at best. And, remember, it's not good engineering practice to consume the entire permitted error rate at one end of the link, thereby forcing the other end to operate perfectly or else experience data errors.

MBasic's Procedures for Serial Communications

MBasic gives us two alternative means of sending and receiving serial communications; a purely software-implemented method and a hardware method. Both have their place, and after describing each, we'll recap their strong and weak points. Since MBasic's User's Guide provides a lengthy description of the nuts and bolts of the serial communications commands, we'll concentrate more on related undocumented features.

Software-Based

MBasic's procedures **SerIn** and **SerOut** are software-based serial input and output procedures and may be used with any PIC supported by MBasic.

The syntax for **SerIn** is:

```
SerIn Rpin\{Fpin},Baudmode,{Plabel},{Timeout,Tlabel},[InputData]
```

Rpin—is a constant or variable that defines the pin through which data is received.

Fpin—is an optional pin that may be used for flow control. **FPin** is an *output* pin that is at logical 0 when **SerIn** is ready to accept data and at logical 1 when it is unable to accept data. (We'll look at flow control in conjunction with Programs 9-1A and 9-1B).

Baudmode—is a constant or variable (16-bit) that defines the bit speed, output pin status, the bit sense polarity (normal or inverted) the number of bits, parity and the number of stop bits. Since we use the RS-232 level converter chip, we must use only inverted modes. And, to keep things simple, we will only use the 8-bit/no-parity option.

MBasic provides predefined baudmode constants that you will use to specify these parameters. From version 5.3.0.0. onwards, MBasic's **SerIn** and **SerOut** functions have a new parameter structure and revised baudmode constants, constructed as summarized below.

Sense		Mode		Data Bits		Parity		Stop Bits		Speed	
Ltr	*Fcn*	*Ltr*	*Fcn*	*Ltr*	*Fcn*	*Ltr*	*Fcn*	*Ltr*	*Fcn*	_(with leading)_	
N	Normal	Blank	Normal	7	7 data bits	N	None	1	1 stop bit	_300	_19200
I	Inverted	O	Open (Bus mode)	8	8 data bits	E	Even	2	2 stop bits	_600	_21600
										_1200	_24000
										_2400	_26400
										_4800	_28800
						O	Odd			_7200	_31200
										_9600	_33600
										_12000	_36000
										_14400	_38400
										_16800	_57600

For example, the baudmode constant for inverted, 8 data bits, no parity, one stop bit, standard mode is **I8N1_9600**.

The baudmode constants we use in this chapter are inverted, normal mode, eight data bits, no parity, one stop bit. Hence, they will be of the form I8N1_nnnn where nnnn is one of the standard data speeds.

We've covered all these elements earlier in this chapter, save for bus mode. In some circumstances, multiple serial sending pins (most likely in different PICs) may be bussed—that is, they share a common connection. In this case, except when actually transmitting data, we wish the transmitting pins to be set to their "disconnect" or high impedance mode. Otherwise, we could easily have one transmitting pin fighting a second transmitting pin. While we could accomplish the switch from high impedance mode to transmit mode with some extra code, we can also let **SerOut** handle this for us via an appropriate baudmode constant. To add bus mode operation to **I8N1_9600**, we simply add the letter "O" thus changing it to **IO8N1_9600**.

Plabel—is an optional program label to which execution will be transferred should a parity error occur.

Timeout & Tlabel—defines a period (32-bit value, in microseconds) in which to wait for an input and if no input is received within this time, program execution transfers to program label **TLabel**. If **Timeout** and **Tlabel** are not specified, **SerIn** will wait indefinitely for an input. If **Timeout** is set to zero, and if there is no data in the buffer, the timeout will immediately happen.

InputData—is a list of variables and modifiers. The modifiers instruct **SerIn** how the incoming data is formatted. The data is stored in the associated variable(s). (We'll look at this aspect of reading input data in more detail in several programs in this chapter.)

The syntax for **SerOut** (two alternative syntax forms) is:

```
SerOut Tpin,Baudmode,{Pace},[OutputData]
SerOut Tpin\{Fpin},Baudmode,{Timeout,Tlabel},[OutputData]
```

Tpin—is a constant or variable that defines the pin through which data is transmitted.

Fpin—is an optional pin that may be used for flow control. **FPin** is an *input* pin that controls **SerOut**'s data transmission; when **FPin** is at logic 1, **SerOut** is disabled from sending data; when **FPin** is at logic 0, data transmission is enabled. (We'll look at flow control in conjunction with Programs 9-1A and 9-1B.)

Baudmode—Same as for **SerIn**.

Pace—An optional 32-bit variable or constant that defines how many microseconds should be paused between sending sequential bytes.

Timeout &Tlabel—defines a period (32-bit value, in microseconds) in which to wait for a handshake from the receiving program, if flow control is activated. If no input is received within this time, program execution transfers to program label **TLabel**.

OutputData—is a list of variables and modifiers. The modifiers instruct **SerOut** how the outgoing data contained in the variables is to be formatted. (We'll look at this aspect of sending data in more detail in several programs in this chapter.)

A third procedure, **SerDetect**, is associated with software serial communications. It automatically detects the incoming baud rate. We will not further discuss **SerDetect**.

Hardware (USART)-Based

MBasic's procedures **HSerIn** and **HSerOut** are hardware-based serial input and output procedures and may be used only with PICs constructed with a universal synchronous/asynchronous receiver/transmitter (USART). Fortunately, most new PICs have USART hardware, and thus support **HSerIn/HserOut**. The PICs we use in this book, the 16F876/876A and the 16F877/877A, are all equipped with a USART. The USART is hard-wired to certain pins, and that pin assignment cannot be changed. For example, in the 16F876/877 PICs, the USART transmit output is on C6 and the receive input is on C7. Other PICs may have different USART pin assignments.

Before using either **HSerOut** or **HSerIn**, two initialization steps must be taken:
1. The compiler instruction **EnableHSerial** must appear; and
2. The function **SetHSerial** must be executed. The syntax for **SetHSerial** is:

SetHSerial mode

Mode is a predefined constant as listed in the User's Guide, of the form **H1200** (1200 bits/sec), **H2400** (2400 bits/sec), etc. through **H1250000** (1250000 bits/sec) where the numeric portion is the bit rate. **SetHserial** sets both transmit and receive bit rates and there are no options for inversion or parity. Unless there is a reason to alter the bit rate during program execution, **SetHSerial** need be executed only once, usually as part of an initialization sequence.

You should be aware of the interaction between `Clear` and `SetHSerial`. The following construction fails:

```
EnableHSerial
SetHSerial H38400
Clear
HSerOut ["P 28-50",13,13]
```

The desired string is not emitted over the serial port. Rather, `Clear` must precede `SetHSerial`:

```
EnableHSerial
Clear
SetHSerial H38400
HSerOut ["P 28-50",13,13]
```

The `Clear` command erases the PIC's RAM, including certain buffer information needed by `HSerOut` and `HSerIn`. Hence, if you use the `Clear` function, it must precede `SetHSerial`.

The syntax for `HSerIn` is simpler than `SerIn`, as it has no flow control options:

```
HSerIn {Tlabel,Timeout}, [InputData]
```

Timeout & Tlabel—defines a period (32-bit value, in microseconds) in which to wait for an input and if no input is received within this time, program execution transfers to program label `Tlabel`. If `Timeout` and `Tlabel` are not specified, `HSerIn` will wait indefinitely for an input.

InputData—is a list of variables and modifiers. The modifiers instruct `HSerIn` how the incoming data is formatted. The data is stored in the associated variable(s). (We'll look at this aspect of reading input data in more detail in several programs in this chapter.)

The syntax for `HSerOut` simple; no flow control, no timeout options:

```
HSerOut [OutputData]
```

OutputData—is a list of variables and modifiers. The modifiers instruct `HSerOut` how the outgoing data contained in the variables is to be formatted. (We'll look at this aspect of sending data in more detail in several programs in this chapter.)

`HSerStat` is an undocumented command. Its syntax is:

```
HSerStat cmd{,label}
```

Cmd	Function
0	Clear Input buffer (label not used)
1	Clear Output buffer (label not used)
2	Clear both buffers (label not used)
3	If Data available in input buffer GoTo label
4	If no Data available in input buffer GoTo label
5	If Data is waiting to be sent GoTo label
6	If no Data is waiting to be sent GoTo label

Programs 9-2A, 9-2B and 9-3, will show us that `HSerStat` is a very powerful addition to MBasic's suite of serial tools and allows us to write programs that execute other code and periodically check to see if any serial data has been received. This gives us a quasi-multitasking mode, where the PIC's serial data input functions run seemingly independently of the rest of the code.

Now that we've take a quick look at MBasic's software and hardware support of serial output, let's summarize the good and bad points of these two approaches.

	Good Things	Not So Good Things
Software Serial (SerIn/SerOut)	• Assignable to any pin • Can be used with any PIC • Can implement multiple serial in/out streams (but not active simultaneously) • Parity option • Inversion option • Bus option • Implements more speeds than hardware serial • Does not consume RAM for a buffer	• Maximum speed is 57,600 bits/sec • No input or output buffer • Can't execute other code while waiting for input
Hardware Serial (HSerIn/HserOut)	• Maximum speed is 1250000 bits/sec • Has a buffer so data can be received while the program is performing other functions	• Not all PICs have hardware serial support • Hardware serial in/out is hard-connected to specific pins • Only one HSerIn/HserOut stream per PIC • No parity option • No inversion option (always inverted) • No way to alter the buffer size; cuts into RAM memory otherwise available to the program. • No flow control options • No pacing option

Although the list of "not so good" things is much longer for hardware-based serial communications, in fact I prefer it to MBasic's software-based procedures. As we'll see later in this chapter, the **HSerStat** function and **HSerIn**'s automatic receive buffer gives us enormous flexibility in program structure compared with the software-based functions **SerIn** and **SerOut**.

Programs

We'll study serial communications with a series of small, single-purpose test programs. Then, we'll assemble a user-interactive serial menu program.

Program	Subject
9-1A	Simple **SerIn** and **SerOut**; RTS flow control
9-1B	CTS flow control when receiving STR format array
9-1C	Reading string input using **SerIn**; $0 string termination
9-1D	Using the parameter **WaitStr** with **SerIn**
9-1E	Formatting numerical and string outputs with **SerOut**
9-1F	Reading a numerical input value with **SerIn**
9-2A	Reading an input string with hardware serial functions and **HSerStat**
9-2B	Converting a string to numeric value using hardware serial functions
9-3	Interactive menu program using **HSerStat, HSerIn** and **HserOut**

Figure 9-3 shows the wiring diagram we'll use for all the programs in this chapter.

Figure 9-4 shows how the PC is to be connected to the 2840 Development Board. As mentioned earlier in this chapter, the DB9-to-DB9 cable is a "modem" or straight-through cable. Note that the 2840 Board and the PC have complementary connections; the transmit pin of the PC connects to a receive pin on the 2840 Board and vice versa. Using RS-232 terminology, the PC's connections are DTE (data terminal equipment) while the 2840 Board's connections are DCE (data communications equipment). Regardless of the terminology, the important thing is that transmit pins on one end connect to receive pins on the other, and vice versa. After all, your communications link isn't going to get very far with transmit pins tied only to other transmit pins and receive pins tied only to other receive pins.

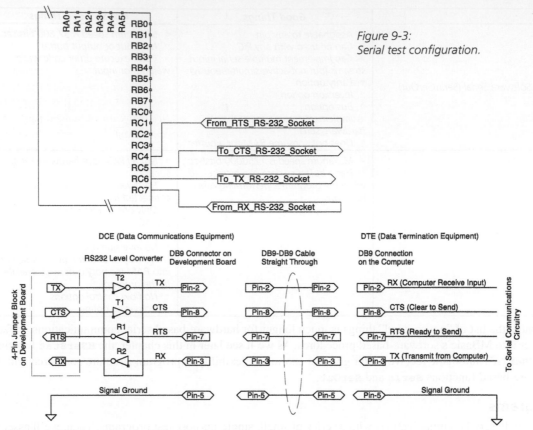

Figure 9-3:
Serial test configuration.

Figure 9-4: Connecting the PC to the development board.

Program 9-1A

We've briefly mentioned flow control; let's see how it works, or doesn't work, in practice. Flow control is necessary because the sending and receiving devices may have different storage and processing capabilities. For example, suppose the sending device is a PIC and that it has a batch of 10 characters to send using **SerOut**, and that the link is operating at 38400 bits/sec. At this speed, 10 characters can be sent in about 2.6 ms, or one character every 260 µs. Let's say the receiving device is also a PIC using **SerIn** to receive the data and that after receiving a byte, our program examines it, executes a program branch based on the byte value and then executes some additional code for each character. We'll also say that the program instructions consume 2 ms for each received character. And, since we're using **SerIn**, there is no general automatic buffer in which incoming characters are stored.

The problem is obvious; characters are arriving at a rate of one every 260 µs but it takes 2 ms to deal with that character and be ready for the next. And, since **SerIn** has no buffer store, the characters that arrive while the program is processing one character are lost. What we need, therefore, is a way for the receiving device to communicate to the sending device "I've received as many characters as I can handle—don't send more until I authorize it." This type of flow control is supported by MBasic through "hardware handshaking," that is, dedicated connections between the sending and receiving device are asserted or negated to control the flow of information. (Many other approaches to flow control exist; one of the most common alternatives is "X-on/X-off," where the receiving and transmitting devices send defined characters to each other over the TX and RX lines. This chapter only looks at hardware handshaking.)

MBasic implements hardware flow control with both **SerIn** and **SerOut**:

Function	Flow Control Pin on RS-232	Direction	RS-232 Status	PIC Logic Level	Function
SerOut	*RTS*	*Input to* **SerOut**	*Asserted*	*1*	**SerOut** *holds characters and stops sending.*
			Negated	*0*	**SerOut** *Sending is enabled.*
SerIn	*CTS*	*Output from* **SerIn**	*Asserted*	*1*	**SerIn** *not ready to accept data.*
			Negated	*0*	**SerIn** *ready to accept data.*

RTS and CTS are acronyms for "request to send" and "clear to send."

Now, let's see how flow control works in practice.

```
;Program 09-1A
;Software Serial I/O

;Constants
;--------------
TPin            Con     C6              ;transmit pin
RPin            Con     C7              ;RX Pin
CTSPin          Con     C5              ;CTS Pin
RTSPin          Con     C4              ;RTS Pin

TXBaudMode      Con     I8N1_38400
;inverted/normal pin/no parity/1 stop/38400 bit/sec
RXBaudMode      Con     I8N1_38400      ;same for receive

;Variables
;--------------
i               Var     Byte            ;loop counter
InByte          Var     Byte            ;Received character

Main
        For i = 0 to 9
                SerOut TPin,TXBaudMode,["i: ",DEC i,13]
        Next

        Pause 100

        For i = 10 to 19
                SerOut TPin\RTSPin,TXBaudMode,["i: ",DEC i,13]
        Next

        SerOut TPin\RTSPin,TXBaudMode,["Ready to Receive: ",13]

        i = 0

RXData
        SerIn RPin\CTSPin,RXBaudMode,[InByte]
        SerOut TPin\RTSPin,TXBaudMode, |
                ["InByte Value: ",DEC InByte," Char: ",InByte,13]
        i = i+1
        If i < 100 Then
                GoTo RXData
        EndIf

        End
```

Let's see how Program 9-1A works.

```
TPin        Con     C6          ;transmit pin
RPin        Con     C7          ;RX Pin
CTSPin      Con     C5          ;CTS Pin
RTSPin      Con     C4          ;RTS Pin
```

We've defined the transmit and receive pins as C6 and C7 respectively, to avoid rewiring the 2840 Board in later programs using the **HSerIn** and **HSerOut** functions employing the USART that is hardwired to C6/C7.

```
TXBaudMode  Con     I8N1_38400
RXBaudMode  Con     I8N1_38400
```

We wish to set the speed to 38400 bits/sec and operate with eight data bits, one stop bit, no parity. And, as we learned, the output must be inverted where we use a level converter to transform logic levels to RS-232 levels. Hence, our mode constant is **I8N1_38400** for both transmit and receive.

```
For i = 0 to 9
        SerOut TPin,TXBaudMode,["i: ",DEC i,13]
Next
```

First, we simply send the digits 0...9 to the serial port, with no flow control at all.

```
Pause 100

For i = 10 to 19
        SerOut TPin\RTSPin,TXBaudMode,["i: ",DEC i,13]
Next

SerOut TPin\RTSPin,TXBaudMode,["Ready to Receive: ",13]

i = 0
```

After a 100 ms pause, we send the digits 10...19, and the message "Ready to receive" but with flow control enabled.

I'll assume you are using Terminal v1.9b software and that it's configured to the communications port connected to the DB9 connector on the 2840 Board. Set Terminal to 38400 bits/sec, 8 data bits, no parity and handshaking (flow control) to *none*. The RTS button in the transmit window should be set to clear and the RTS indicator should not be illuminated.

Run Program 9-1A and Terminal's receive screen should display the following:

```
i: 0
i: 1
i: 2
i: 3
i: 4
i: 5
i: 6
i: 7
i: 8
i: 9
```

Now click on the RTS Set button. Two things should happen. First, the RTS indicator will illuminate green and second, the remainder of the output should be seen in the receive screen:

```
i: 10
i: 11
i: 12
i: 13
i: 14
i: 15
i: 16
i: 17
i: 18
i: 19
Ready to Receive:
```

If your display does not match, check the voltages at C4 and see if it toggles between 0 and +5 V as you set and clear the RTS button in Terminal. Disconnect C4 from the RTS socket and connect it to V_{DD} and rerun Program 9-1A. Then, connect it to V_{SS} and rerun Program 9-1A. This will illustrate the difference between asserting and negating the RTS line.

The reason that we manually set or clear the RTS line, instead of showing it via a sequence of automatic starts and stops in the displayed data results from the fact that PCs implement large receive buffers and are so much faster than a PIC that it's almost impossible for our PIC to output data so fast that a modern PC won't be able to keep up with it. Hence, we simulate transmit flow control manually.

```
            i = 0

RXData
            SerIn RPin\CTSPin,RXBaudMode,[InByte]
            SerOut TPin\RTSPin,TXBaudMode,   |
                  ["InByte Value: ",DEC InByte," Char: ",InByte,13]
            i = i+1
            If i < 100 Then
                  GoTo RXData
            EndIf
```

Next, we test the CTS and received data. This part of Program 9-1A receives a character with the **Se-rIn** function and then echoes the character back with a **SerOut** function. **SerIn** enables CTS. After 100 received characters, execution stops.

Since we have enabled CTS, turn select RTS/CTS handshaking in Terminal. And, enable RTS in the transmit window by clicking on the green RTS icon button, which will illuminate. Now try typing or automatically repeating a letter by holding it down. If you send a 0, you should see the following echoed back:

InByte Value: 48 Char: 0

Now, prefill the send control of Terminal with 0123456789 and press the "← SEND" button. We should see ten responses, 0 through 9, right? Each character is sent, and then the next is held until it is echoed back and **SerIn** is ready for the next character. In fact, we see something quite different:

InByte Value: 48 Char: 0
InByte Value: 50 Char: 2
InByte Value: 52 Char: 4
InByte Value: 54 Char: 6
InByte Value: 56 Char: 8

How can this be? Figure 9-5 shows the incoming data and CTS line for manual sending. As expected, after each character, the CTS line goes high for a few milliseconds while the received character is echoed back. It then goes low, indicating **SerIn** is ready for the next character. This is exactly as it should be. Why then are we missing every second byte sent automatically? The mystery is solved when we more closely examine the CTS line and the automatically sent data, as seen in Figure 9-6. Although the CTS line reacts properly, it reacts too slowly; by the time the CTS line is asserted, Terminal has already started to send the next character. It's too late to stop it from being sent. Hence, Terminal outputs data in spurts of two characters at a time. I tried slowing the data rate down to 600 bits/sec and found that CTS is not asserted quickly enough to prevent Terminal from sending a second character.

The point to be made from Program 9-1A is that you must test flow control with the terminal program and data rates you expect; don't assume that just because **SerIn** and **SerOut** support flow control, it works the way you think it should.

By the way, the terminal utility built into MBasic's IDE keeps RTS asserted, regardless of whether the flow control option is set at "no flow control" or "flow control." Since RTS is always asserted, **SerOut** will send no data.

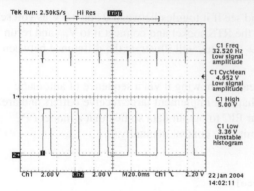

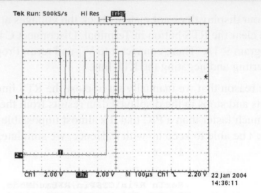

Figure 9-5: Received data and CTS line; Ch1: RX data out of SP232; Ch2: CTS into SP232.

Figure 9-6: Received data and CTS line; Ch1: RX data out of SP232; Ch2: CTS into SP232.

Program 9-1B

Program 9-1A accepted data one byte at a time. What happens when we use **SerIn** to read a string of multiple characters? Does flow control work better under these conditions?

```
;Program 09-1B
;Software Serial I/O

;Constants
;-------------
TPin            Con     C6              ;transmit pin
RPin            Con     C7              ;RX Pin
CTSPin          Con     C5              ;CTS Pin
RTSPin          Con     C4              ;RTS Pin

TXBaudMode      Con     I8N1_38400
;inverted/normal pin/no parity/1 stop/38400 bit/sec
RXBaudMode      Con     I8N1_38400      ;same for receive

;Variables
;-------------
i                       Var     Byte    ;loop counter
InByte                  Var     Byte    ;Received character
InStr                   Var     Byte(11);Received string

Main

        SerOut TPin,TXBaudMode,["Ready to Receive: ",13]

        i = 0

RXData
        SerIn RPin\CTSPin,RXBaudMode,[Str InStr\10]
        SerOut TPin,TXBaudMode,["InStr: ",Str InStr\10,13]
        i = i+1
        If i < 100 Then
                GoTo RXData
        EndIf

End
```

We've kept the skeleton of Program 9-1A, but modified the **SerIn** statement to read an input string of 10 characters.

```
        SerIn RPin\CTSPin,RXBaudMode,[Str InStr\10]
```

After reading the string, we echo it back with a **SerOut** procedure.

```
        SerOut TPin,TXBaudMode,["InStr: ",Str InStr\10,13]
```

164

The idea behind Program 9-1B is that since we found problems being able to quickly enough assert CTS when reading one character at a time, suppose we read ten characters at a time. We do this by using the I/O modifier **[Str InStr\10]**. This instructs **SerIn** to read an input stream of 10 characters in length, and to store the received characters in the array **InStr**.

This should solve our immediate problem of being able to read the sequence "0123456789" sent without a break between characters, right? Keeping the same Terminal configuration we used for Program 9-1A, let's run Program 9-1B. Load Terminal's send buffer with 0123456789 and hit the "← SEND" button. Here's the result:

```
Ready to Receive:
InStr: 01™&S"éÂÊÿ
```

Not what we expected is it? It looks like the internals of **SerIn** can't keep up with the received data. If this theory is correct, we should be able to see a difference as the data rate slows. Let's try it and see what happens.

Speed (bits/sec)	Mode Constant	Result
38400	I8N1_38400	InStr: 01™&S"éÂÊÿ
19200	I8N1_19200	InStr: 0123456789

Sure enough, if we cut the speed in half, the sent string is received correctly.

The fact, therefore, that **SerIn** supports data rates up to 57600 bits/sec doesn't mean that we can send a string of characters at 57600 bits/sec and expect them to be correctly received. We also see that **SerIn**'s flow control did not correct the problem. This again cautions us when using **SerIn** and consecutive characters sent without pause. If you send the 10 characters 0123456789 by typing as fast as you can, you'll not see a problem, as there is more than ample time between keystrokes for **SerIn** to receive each character and store it in **InStr**.

Program 9-1C

Let's look at a more practical string input routine.

```
;Program 09-1C
;Software Serial I/O

;Constants
;--------------
TPin            Con     C6              ;transmit pin
RPin            Con     C7              ;RX Pin
CTSPin          Con     C5              ;CTS Pin
RTSPin          Con     C4              ;RTS Pin

TXBaudMode      Con     I8N1_38400
;inverted/normal pin/no parity/1 stop/38400 bit/sec
RXBaudMode      Con     I8N1_38400      ;same for receive

;Variables
;--------------
i               Var     Byte    ;loop counter
j               Var     Byte
InStr           Var     Byte(11);Received string

Main

        SerOut TPin,TXBaudMode,["Ready to Receive: ",13]

        i = 0

RXData
        ;wait until 10 chars are received, or the carriage return
        For j = 0 to 10
```

```
            InStr(j) = $0
      Next ;j

      SerIn RPin\CTSPin,RXBaudMode,[Str InStr\10\13]
      SerOut TPin,TXBaudMode,["InStr: ",Str InStr,13]

      SerOut TPin,TXBaudMode,["InStr: ",Str InStr\10\0,13]

      j = 0
      While InStr(j) > $0
            j = j+1
      WEND

      SerOut TPin,TXBaudMode,["Length: ",Dec j,   |
      "  InStr: ",Str InStr\j,13]

      i = i+1
      If i < 100 Then
            GoTo RXData
      EndIf

    End
```

Our goal is to allow the user to enter up to 10 characters, and to terminate the input with the enter key if fewer characters are to be entered. We'll then echo the received characters back to the screen.

```
      ;wait until 10 chars are received, or the carriage return
      For j = 0 to 10
            InStr(j) = $0
      Next ;j
```

MBasic's User's Guide states that if we output a string with the **Str** modifier, output will stop when the first $0 byte is found. We'll test that statement in Program 9-1C and see if it's correct.

Therefore, first we'll preload **InStr** with $0 values. Since **InStr** has a length of 11 and since we will only allow 10 characters, we should always have a $0 as the last byte, even if we input the full 10 characters.

```
      SerIn RPin\CTSPin,RXBaudMode,[Str InStr\10\13]
      SerOut TPin,TXBaudMode,["InStr: ",Str InStr,13]
```

First, we read the input into the array **InStr** and then, following the suggestion in the User's Guide, output the array **InStr** as a string, without any length or explicit termination information.

```
      SerOut TPin,TXBaudMode,["InStr: ",Str InStr\10\0,13]
```

Next, we output **InStr** but this time explicitly defining the end-of-line character. The **Str** modifier parameters are:

```
      Str Variable\Length\EOL
```

Where:

Variable—is an array of byte holding the string information, one character per array element.

Length—is an optional parameter that defines the maximum number of characters to be outputted.

EOL—is an optional parameter that defines the "end-of-line" character. The string ends when either (a) **Length** characters have been outputted, or (b) the EOL character is found, whichever occurs first.

```
      j = 0
      While InStr(j) > $0
            j = j+1
      WEND

      SerOut TPin,TXBaudMode,["Length: ",Dec j,   |
      "  InStr: ",Str InStr\j,13]
```

Finally, we look at another way of defining the output length; we count the characters up to the point where we find the end-of-line terminator, $0, storing the length in the variable **j**. We then stop counting and output the string, using **j** to define the number of characters to be outputted.

Let's run Program 9-1C. In response to the "Ready to Receive" prompt, enter 123 <Enter>, i.e., 123 followed by the enter key.

We see the following in the receive window of Terminal:
Received:

InStr:
123<00><00><00><00><00><00><00><00>¶} I@mƒ&Ix̂ôÀ„S<00><00><00><00>0ƒ33I...@ŠII•;•o°I<00>I¾ISI•
<00><00>IX<00><00>

I<00><00>I"IIA<00>IžO<00><00><00> <00>O<00><00><00><00>O
<00><00><00><00>X
<00><00><00><00><00><00><00>@I<00>I
I<00><00><00><00><00><00>I<00><00><00>II<00><00><00><00><00>I<00><00><00><00><00><00><00><00><00>
<00><00<00><00>I<00><00>I<00><00><00><00><00><00>@<00>IF<00>I<00><00><00><00><00><00><00><00<00>
<00><00>"<00>@<00>„<00>I<00><00><00><00><00><00><00><00><00><00> <00> <00><00>@<00><00><00><00>
<00>I<00><00><00><00><00>I<00><00><00>@<00>"o

o

o

o

u

o

x

<00>I<00>I<00><00><00><00>HI<00><00><00><00><00>I<00><00>
InStr: 123
Length: 3 InStr: 123

The result shows that the **str** modifier without an explicit end-of-line definition does not automatically terminate output upon the first $0 byte and the User's Guide is in error in this regard. Rather, we get 255 bytes of output, regardless of the number of $0 bytes.

The last two lines show that we may properly limit the output by adding either an explicit end-of-line definition to the **str** modifier; or by determining the length of valid characters and adding the length to the **str** modifier.

Program 9-1D

We'll look at a specialized input modifier next, **WaitStr**.

```
;Program 09-1D
;Software Serial I/O

;Constants
;-------------
TPin        Con     C6      ;transmit pin
RPin        Con     C7      ;RX Pin
CTSPin      Con     C5      ;CTS Pin
RTSPin      Con     C4      ;RTS Pin
```

167

```
TXBaudMode        Con      I8N1_38400       ;inverted/normal pin/no parity/1 stop/38400
bit/sec
RXBaudMode        Con      I8N1_38400       ;same for receive

;Variables
;--------------
i                 Var              Byte     ;loop counter
j                 Var              Byte
InStr             Var              Byte(11);Received string

Main

        SerOut TPin,TXBaudMode,["Ready to Receive: ",13]

        i = 0

RXData
        For j = 0 to 2
                InStr(j) = "A" + j
        Next ;j
        For j = 3 to 10
                InStr(j) = $0
        Next ;j

        SerIn RPin\CTSPin,RXBaudMode,[WaitStr InStr\3\13]

        SerOut TPin,TXBaudMode,["Str Received - Value: ", |
        Str InStr\10\0,13]

        i = i+1
        If i < 100 Then
                GoTo RXData
        EndIf

        End
```

The **WaitStr** modifier instructs the receiving procedure to wait until the defined string has been received.

```
        For j = 0 to 2
                InStr(j) = "A" + j
        Next ;j
```

First, we must define the string we are to await. We'll use the string "ABC" and we put these letters into the first three elements of **InStr** with the above statements. Since "B" immediately follows "A" in the ASCII collating sequence, we just add 1 to "A" to get "B," and add 2 to "A" to get "C."

```
        For j = 3 to 10
                InStr(j) = $0
        Next ;j
```

We then fill the remainder of **InStr** with $0 characters.

```
        SerIn RPin\CTSPin,RXBaudMode,[WaitStr InStr\3\13]
```

We then wait for the received characters to match ABC. When they match, execution resumes and the received string is echoed via the serial port.

```
        SerOut TPin,TXBaudMode,["Str Received - Value: ", |
        Str InStr\10\0,13]
```

Let's challenge the **WaitStr** with some nonmatching characters. After each, we'll use a carriage return. The output we see matches expectation.

Sent Characters	Received	Comments
	Ready to Receive	*Prompt message*
A		*No response*
B		*No response*
C		*No response*
AB		*No response*
ABC	*Str Received - Value: ABC*	*Received characters match the expected; execution jumps to next statement and echoes the received information*
abc		*No response to lower case*

We can use **WaitStr** to pluck certain desired data items from a sequence with both wanted and unwanted items. For example, consider the following sample of the NMEA standard output from a Global Positioning Satellite receiver.

```
$GPGLL,3846.4266,N,07722.7589,W,195138.033,A*2D
$GPGGA,195138.03,3846.4266,N,07722.7589,W,1,05,4.1,00104,M,,,,*3E
$GPRMC,195138.03,A,3846.4266,N,07722.7589,W,00.0,000.0,250104,10,W*4D
$GPGSA,A,3,24,20,07,04,09,,,,,,,,05.1,04.1,03.1*0F
$PMGNST,04.00,3,T,716,09.7,+03846,00*40
```

Suppose we wish to read the $GPGLL string, which contains the geographic position latitude and longitude and time information, but ignore other strings. We might use **WaitStr** to monitor the input data stream until the message identifier $GPGLL has been received. We then would capture the rest of the line and decode the latitude and longitude information.

Program 9-1E

Let's now explore a poorly documented part of MBasic—the input and output modifiers. Here's my summary of the undocumented portions of input/output modifiers.

Modifier Family	Modifier	Base Value	Output	Comments
Decimal (Dec) (and others)	Dec	123	123	*The modifier* **Decn** *where n is a number determines the number of digits to display, starting from the right. Note that specifying more digits than required to display the number does not cause blanks or leading zeros.* **Dec** *without an "n" displays all digits of the number.*
	Dec1		3	
	Dec2		23	
	Dec3		123	
	Dec4		123	**This same "add n" to define displayed digits applies to:** `Dec, sDec` `Hex, iHex, sHex, isHex` `Bin, iBin, sBin, isBin`
Decimal (Dec) (and others)	DEC \1	123	123	*The output is forced to the length given by the value following the \ by adding leading zeros.*
	DEC \2		123	
	DEC \3		123	
	DEC \4		0123	*This applies to* **Dec, sDec** `Hex, iHex, sHex, isHex` `Bin, iBin, sBin, isBin`

(Continued)

Modifier Family	Modifier	Base Value	Output	Comments
Real (floating point numbers)	`Real`		123.4559936523	*Extra digits are resolution error. Without any modifier, we get the full 14 characters.*
	`Real1`		3.4559936523	*Used without anything else, `Realn` works the same way as for Dec. However, `n` applies only to digits to the left of the decimal point.*
	`Real2`		23.4559936523	
	`Real3`		123.4559936523	
	`Real4`	123.456	123.4559936523	
	`Real4 Var\1`		123.4	*To set the digits displayed on the right of the decimal point, use the /n syntax. (`Var` is a type Long)*
	`Real4 Var\2`		123.45	
	`Real4 Var\3`		123.455	
	`Real4 Var\4`		123.4559	
Signed Decimal	`sDec`	−123	−123	*If value is <0, − sign appended; no + sign if >0.*
		+123	123	
Signed Hex	`sHex`	7B	7B	*If value is <0, − sign appended; no + sign if >0.*
		−7B	−7B	
Indicated Signed Hex	`isHex`	7B	$7B	*If value is <0, − sign appended; no + sign if >0. Minus sign appears after the $ hex indicator.*
		−7B	$-7B	
Signed Binary	`sBin`	1111011	1111011	*If value is <0, − sign appended; no + sign if >0.*
		−1111011	−1111011	
Indicated Signed Binary	`isBin`	1111011	%1111011	*If value is <0, − sign appended; no + sign if >0. Minus sign appears after the % binary indicator.*
		−1111011	%−1111011	

Another undocumented reminder—when assigning a length value to a string output, MBasic does not check the length. For example, consider:

```
TestStr Var    Byte(10)
SerOut TPin,TXBaudMode,[Str TestStr\20,13]
```

Although the string array being written to the serial pin only has 10 valid elements, the \20 modifier writes 20 characters. No error message will be generated, and 20 characters will be written. The extra 10 characters are the contents of the next 10 RAM memory locations immediately following the last element of **TestStr**.

Program 9-1E exercises these undocumented modifiers.

```
CPU = 16F877A
MHZ = 20
CONFIG 16250


;Program 09-1B
;Software Serial I/O

;Constants
;-------------
TPin        Con     C6              ;transmit pin
RPin        Con     C7              ;RX Pin
CTSPin      Con     C5              ;CTS Pin
RTSPin      Con     C4              ;RTS Pin

TXBaudMode  Con     i38400 ;inverted driven no parity
RXBaudMode  Con     i38400 ;inverted driven no parity

;Variables
;-------------
i               Var     Byte    ;loop counter
```

```
j               Var     SByte
InStr           Var     Byte(11);Received string
RealVar Var             Long

Main
        ;String Output
        InStr = "ABCDEFGHIJK"
        ;Limit length with explicit length parameter
        SerOut TPin,TXBaudMode,["InStr: ",Str InStr\10,13]
        InStr = "ABCDEFGHIJK"
        ;Limit with embedded EOL
        InStr(3) = $0
        SerOut TPin,TXBaudMode,["InStr: ",Str InStr\10\0,13]
        ;Follow with an underline
        SerOut TPin,TXBaudMode,[REP "-"\10,13]

        SerOut TPin,TXBaudMode,[13,13]
        ;Decimal Integer Output
        i = 123
        SerOut TPin,TXBaudMode,["i:",Dec i,"F",13]
        SerOut TPin,TXBaudMode,["i:",Dec1 i,"F",13]
        SerOut TPin,TXBaudMode,["i:",Dec2 i,"F",13]
        SerOut TPin,TXBaudMode,["i:",Dec3 i,"F",13]
        SerOut TPin,TXBaudMode,["i:",Dec4 i,"F",13]
        ;Note the following produces no error code
        ;it outputs 30 variables as string, starting
        ;at address of i.
        i = 88
        SerOut TPin,TXBaudMode,[Str i\10,13]

        SerOut TPin,TXBaudMode,[13,13]
        ;Signed byte options
        j = -123
        SerOut TPin,TXBaudMode,["j: ",Dec j,13]
        SerOut TPin,TXBaudMode,["j: ",SDec j,13]
        SerOut TPin,TXBaudMode,["j: ",SDec1 j,13]
        SerOut TPin,TXBaudMode,["j: ",SDec2 j,13]
        SerOut TPin,TXBaudMode,["j: ",SDec3 j,13]
        SerOut TPin,TXBaudMode,["j: ",SDec4 j,13]

        SerOut TPin,TXBaudMode,[13,13]

        ;Signed byte options
        j = 123
        SerOut TPin,TXBaudMode,["j: ",Dec j,13]
        SerOut TPin,TXBaudMode,["j: ",SDec j,13]
        SerOut TPin,TXBaudMode,["j: ",SDec1 j,13]
        SerOut TPin,TXBaudMode,["j: ",SDec2 j,13]
        SerOut TPin,TXBaudMode,["j: ",SDec3 j,13]
        SerOut TPin,TXBaudMode,["j: ",SDec4 j,13]

        SerOut TPin,TXBaudMode,[13,13]
        ;hex options
        i = 123
        SerOut TPin,TXBaudMode,["i: ",Hex i,13]
        SerOut TPin,TXBaudMode,["i: ",Hex1 i,13]
        SerOut TPin,TXBaudMode,["i: ",Hex2 i,13]
        SerOut TPin,TXBaudMode,["i: ",Hex3 i,13]
        SerOut TPin,TXBaudMode,["i: ",Hex4 i,13]

                SerOut TPin,TXBaudMode,[13,13]
        ;indicated hex options
        i = 123
        SerOut TPin,TXBaudMode,["i: ",iHex i,13]
        SerOut TPin,TXBaudMode,["i: ",iHex1 i,13]
        SerOut TPin,TXBaudMode,["i: ",iHex2 i,13]
        SerOut TPin,TXBaudMode,["i: ",iHex3 i,13]
        SerOut TPin,TXBaudMode,["i: ",iHex4 i,13]
```

```
j = -123
SerOut TPin,TXBaudMode,["j: ",sHex j,13]
SerOut TPin,TXBaudMode,["j: ",isHex j,13]

j = 123
SerOut TPin,TXBaudMode,["j: ",sHex j,13]
SerOut TPin,TXBaudMode,["j: ",isHex j,13]

SerOut TPin,TXBaudMode,[13,13]

i = 123
;binary options
SerOut TPin,TXBaudMode,["i: ",Bin i,13]
SerOut TPin,TXBaudMode,["i: ",iBin i,13]
j = -123
SerOut TPin,TXBaudMode,["j: ",sBin j,13]
SerOut TPin,TXBaudMode,["j: ",isBin j,13]

SerOut TPin,TXBaudMode,[13,13]
RealVar = 123.456
SerOut TPin,TXBaudMode,["RealVar: ",Real RealVar,13]
SerOut TPin,TXBaudMode,["RealVar: ",iHex RealVar,13]

SerOut TPin,TXBaudMode,["RealVar: ",Real1 RealVar,13]
SerOut TPin,TXBaudMode,["RealVar: ",Real2 RealVar,13]
SerOut TPin,TXBaudMode,["RealVar: ",Real3 RealVar,13]
SerOut TPin,TXBaudMode,["RealVar: ",Real4 RealVar,13]

SerOut TPin,TXBaudMode,["RealVar: ",Real4 RealVar\1,13]
SerOut TPin,TXBaudMode,["RealVar: ",Real4 RealVar\2,13]
SerOut TPin,TXBaudMode,["RealVar: ",Real4 RealVar\3,13]
SerOut TPin,TXBaudMode,["RealVar: ",Real4 RealVar\4,13]

    End
```

Program 9-1E needs no analysis, as it simply tests many of the undocumented output modifiers summarized earlier. Here's the output from Program 9-1E: (To save space, I've reformatted the output into four columns):

InStr: ABCDEFGHIJ	j: 123	i: $7B	RealVar: 123.4559936523
InStr: ABC	j: 123	i: $B	RealVar: $8576E978
----------	j: 3	i: $7B	RealVar: 3.4559936523
	j: 23	i: $7B	RealVar: 23.4559936523
i:123F	j: 123	i: $7B	RealVar: 123.4559936523
i:3F	j: 123	j: -7B	RealVar: 123.4559936523
i:23F		j: $-7B	RealVar: 123.4
i:123F	i: 7B	j: 7B	RealVar: 123.45
i:123F	i: B	j: $7B	RealVar: 123.455
X...ABC<00>EFGH	i: 7B		RealVar: 123.4559
	i: 7B	i: 1111011	
j: 4294967173	i: 7B	i: %1111011	
j: -123		j: -1111011	
j: -3		j: %-1111011	
j: -23			
j: -123			
j: -123			

Another undocumented feature of MBasic is that modifiers such as **Dec**, **Hex**, and **Bin** (including their variants, such as **Dec2**, and their indicated and signed forms such as **iHex**, **sHex** and **isHex** and even their variant indicated and signed form such as **isHex3**) may be used as functions to convert a numerical value to its string representation.

172

Program 9-1F

We've looked at MBasic's undocumented numerical modifiers when formatting output values. These modifiers, such as **Dec2**, **isHex3**, etc., may be used when receiving data. How do they work when applied to numerical input variables?

What I've been able to distill from the results of Program 9-1G are:

Automatic buffer—MBasic assigns a 10-digit input buffer to values read into a formatted variable unless the modifier has an explicit length term. For example, **[Dec x]** will cause an automatic 10 digit buffer to be created; **[Dec2 x]** creates a *de facto* 2 digit buffer. The buffer is packed BCD format, one digit per nibble.

Termination of input—Input terminates upon the earlier of (a) filling the buffer (10 digits if not defined by an explicit length modifier) or (b) receipt of the first nonnumeric character.

Nonnumerical characters—Nonnumerical characters are ignored when preceding numbers, and terminate the input when they follow a number. Thus, if the input modifier is **[Dec x]**, entering xyz123 causes **x** = **123**. Entering 1x23 results in **x = 1**.

Indicated input—Where the indicated modifiers **iHex** or **iBin** are used, any characters other than the leading indicator, "$" or "%", respectively, are ignored. Legitimate input characters are 0...F for **iHex**, and 0...1 for **iBin**. No case conversion is performed, so **$ab** is rejected and **$AB** is accepted. Nonacceptable characters preceding the indicator, for example, **x$AB** will be ignored. Nonacceptable characters after the indicator, for example, **$xAB** cause the entry to be rejected.

Signed values—Signed modifiers, such as **sDec**, accept both + and – as valid leading indicators. (The + is ignored as a nonnumeric character.)

Real—As indicated in the User's Guide, the modifier **Real** is restricted to only output statements. However, it's possible to read the integer and fractional parts of a real number in a single statement, such as **[Dec IntegerIn, Dec2 FractionIn]**. If a number such as 123.45 is entered, the result will be to set **IntegerIn = 123** and **FractionIn = 45**. These may be recombined into a floating point number after suitable manipulation. (The tricky part is the fractional value.)

Program 9-1G demonstrates many of these possible input modifiers.

```
;Program 09-1F
;Software Serial I/O

;Constants
;-------------
TPin            Con     C6        ;transmit pin
RPin            Con     C7        ;RX Pin
CTSPin          Con     C5        ;CTS Pin
RTSPin          Con     C4        ;RTS Pin


TXBaudMode      Con     I8N1_38400        ;inverted/normal pin/no parity/1 stop/38400
bit/sec
RXBaudMode      Con     I8N1_38400        ;same for receive
;Variables
;--------------
InByte  Var     Byte    ;input
InWord  Var     Word
InLong  Var     Long
InSByte Var     SByte


Main
        SerOut TPin,TXBaudMode,[13,"Enter Byte Decimal Value: "]
        SerIn RPin\CTSPin,RXBaudMode,[Dec InByte]
        SerOut TPin,TXBaudMode,[13,"Received: ",Dec InByte,13]
```

```
            SerOut TPin,TXBaudMode,[13,"Enter Byte Dec2 Value: "]
            SerIn RPin\CTSPin,RXBaudMode,[Dec2 InByte]
            SerOut TPin,TXBaudMode,[13,"Received: ",Dec InByte,13]

            SerOut TPin,TXBaudMode,[13,"Enter Word Decimal Value: "]
            SerIn RPin\CTSPin,RXBaudMode,[Dec InWord]
            SerOut TPin,TXBaudMode,[13,"Received: ",Dec InWord,13]

            SerOut TPin,TXBaudMode,[13,"Enter Byte&Word Dec Values: "]
            SerIn RPin\CTSPin,RXBaudMode,[Dec InByte, Dec InWord]
            SerOut TPin,TXBaudMode,[13,"Received: ",Dec InByte," ", |
                Dec InWord,13]

            SerOut TPin,TXBaudMode,[13,"Enter Indicated Hex Value: "]
            SerIn RPin\CTSPin,RXBaudMode,[iHex InByte]
            SerOut TPin,TXBaudMode,[13,"Received: ",iHex InByte,13]

            SerOut TPin,TXBaudMode,[13,"Enter Indicated Bin Value: "]
            SerIn RPin\CTSPin,RXBaudMode,[ibin InByte]
            SerOut TPin,TXBaudMode,[13,"Received: ",ibin InByte,13]

            SerOut TPin,TXBaudMode,[13,"Enter Signed Byte Value: "]
            SerIn RPin\CTSPin,RXBaudMode,[sdec InSByte]
            SerOut TPin,TXBaudMode,[13,"Received: ",sDec InSByte,13]


        GoTo Main
        End
```

Program 9-1G

Program 9-1G consists of several sections; each starts with a prompt identifying the type of variable to be entered, then a **SerIn** statement to read the variable, followed by outputting the entered value.

```
            SerOut TPin,TXBaudMode,[13,"Enter Byte Decimal Value: "]
            SerIn RPin\CTSPin,RXBaudMode,[Dec InByte]
            SerOut TPin,TXBaudMode,[13,"Received: ",Dec InByte,13]
```

A typical section is above; it prompts for a decimal value, and reads the input with a simple **Dec** modifier. The read value is then outputted with the same **Dec** modifier.

Let's see how these specific input sections interact with various input entries. The symbol <E> indicates the "Enter" key.

Prompt	Input Modifier	User Response	Output
Enter Byte Decimal Value:	[Dec InByte]	123<E>	Received: 123
		Xx123y<E>	Received: 123
Enter Byte Decimal Dec2 Value:	[Dec2 InByte]	12<E>	Received: 12
		Xxx33	Received: 12
Enter Word Decimal Value	[Dec InWord]	1234<E>	Received: 1234
		S%^4321c	Received: 4321
Enter Byte and Word Decimal Values:	[Dec InByte,Dec InWord]	11,22<E>	Received: 11 22
		123.4567<E>	Received: 123 4567
Enter Indicated Hex Value:	[iHex InByte]	$AB<E>	Received: $AB
		X$12	Received: $12
Enter Indicated Binary Value:	[ibin InByte]	%0101<E>	Received: %101
		%00001111<E>	Received: %1111
Enter Signed Byte Value:	[sdec InSByte]	−100<E>	Received: −100
		+123	Received: 123

To get a sense of how MBasic handles formatted inputs, run program 9-1F and try various possible responses.

Program 9-2A

So far, all our code has used software input and output routines. And, except for those provisions using flow control, the code in Programs 9-1A through 9-1G is equally usable with MBasic's hardware-based input and output routines, **HSerIn** and **HSerOut**. Now, however, we're going to look at input with the hardware-only **HSerStat** function.

We'll start with a simple activity; reading an input string and determining its length. However, instead of waiting for the user to enter data, Program 9-2A performs other functions and periodically checks to see if data has been received. When the user has finished entering the data, as reflected by a carriage return character, we count the number of characters received and output the length and the entered data.

```
;Constants
;-------------
MaxStrSize      Con     12        ;string length


;Variables
;------------
i               Var     Byte                    ;counter
j               Var     Byte                    ;counter
InStr           Var     Byte(MaxStrSize)        ;input string
InByte          Var     Byte                    ;input byte

EnableHSerial
SetHSerial H38400
HSerOut ["P-09-02A",13]
Pause 100

Main
        For i = 0 to 255        ;dummy procedure
                Pause 10
                HSerStat 3, CheckInput
                ReturnFromCI
        Next ;i
GoTo Main


CheckInput
        HSerIn [InByte]
        ;Echo back
        HSerOut [Str InByte\1]

        InStr(j) = InByte

        j = j+1
        If j = (MaxStrSize-1) Then
                j=0
        EndIf

        If InByte = 13 Then
                If j > 0 Then
                        j = j-1 ;get rid of CR
                EndIf
                GoSub CRReceived
        EndIf

        HSerStat 3,CheckInput

GoTo ReturnFromCI


CRReceived
;--------------
        HSerOut ["Chars Received: ",Dec j,13]
        If j > 0 Then
                HSerOut ["Complete Line Received: ",Str InStr\j,13]
```

```
              ELSE
                       HSerOut ["CR only",13]
              EndIf
              j = 0
       Return


       End
```

Let's look at the code.

```
       EnableHSerial
       SetHSerial H38400
```

First, we set the compiler instruction **EnableHSerial**, needed to cause the compiler to load the hardware serial support library. Next, we must set the desired baud rate, using **SetHSerial**. **SetHSerial** establishes the serial bit rate for both **HSerOut** and **HSerIn**. Since we have no need to change output speed, we set it once, during program initialization. Also, since the 16F87x family's USART is hardwired to pins C6 (TX out) and C7 (RX in), there are no input and output pins to define.

```
       Main
              For i = 0 to 255        ;dummy procedure
                       Pause 10
                       HSerStat 3, CheckInput
                       ReturnFromCI
              Next ;i
       GoTo Main
```

The main program loop is a dummy **For-Next** loop that simulates other activity, such as controlling a stepper motor, or monitoring temperature, or any other function suitable for a PIC programmed with MBasic.

The key element for our purpose is the statement **HSerStat 3, CheckInput**. As we learned earlier in this chapter, **HSerStat** is an undocumented function that, when called with a mode argument of 3, transfers program execution to the label **CheckInput** if there is data in the input buffer.

Both **HSerIn** and **HSerOut** have buffers. For 16F87x series PICs, the input and output buffers are each 47 bytes long. More importantly, however, is that *once the serial port is set up with the **SetHSerial** command, any data on the receive pin is automatically moved into the buffer.* If more than 47 bytes are received before the buffer is cleared, the buffer wraps around and newer data overwrites older data.

With this understanding, you should see a subtle, but clear, difference between **SerIn** and **HSerIn**: **SerIn** establishes the read procedure and sets a pin for input; data sent to that pin before **SerIn** is operating is lost. **HSerIn** does not establish anything; rather it checks the receive buffer opened by **SetHSerial** and reads that buffer. **SetHSerial** is the launching event, not **HSerIn**. Consequently, **HSerIn** may be executed at any time and retrieves whatever has arrived after **SetHSerial** has been executed, up to the maximum buffer size.

For many purposes, this difference between **SerIn** and **HSerIn** may be ignored. If all we do is send a user prompt message via **HSerOut** and then wait for a response, the difference between **SerIn** and **HSerIn** is immaterial. But, this difference is critical to our ability to quasi-multitask the PIC—that is, to execute unrelated code while data is being read over the serial port by the USART.

With this understanding of **HSerIn** and **HSerStat**, let's look at the **CheckInput** code. Remember, this code is executed only when there is at least one byte in the input buffer.

```
       CheckInput
              HSerIn [InByte]
              ;Echo back
              HSerOut [Str InByte\1]

              InStr(j) = InByte
```

First, we fetch one byte from the buffer with **HSerIn [InByte]**. To provide user feedback, we echo back this byte with a **HSerOut** procedure call. We then store this byte in the array **InStr**. (We initialized **j** by setting it to 0 in the initialization section of Program 9-2A and we'll see how it gets reset to 0 later.)

```
j = j+1
If j = (MaxStrSize-1) Then
        j=0
EndIf
```

We now update **j** and see if it exceeds the size of **InStr**. If it does, we reset **j** to 0.

```
If InByte = 13 Then
        If j > 0 Then
                j = j-1 ;get rid of CR
        EndIf
        GoSub CRReceived
EndIf
```

We next check to see if we have received an end-of-line (carriage return) character. If the character is a carriage return, we decrement **j** to prevent the carriage return from being considered in the length of **InStr**, and we execute the subroutine **CRReceived**.

```
HSerStat 3,CheckInput
```

```
GoTo ReturnFromCI
```

Finally, we see if additional bytes remain in the receive buffer through another **HSerStat 3, CheckInput** statement. If the buffer is not empty, we recycle through the **CheckInput** code until it is empty.

```
CRReceived
;--------------
        HSerOut ["Chars Received: ",Dec j,13]
        If j > 0 Then
                HSerOut ["Complete Line Received: ",Str InStr\j,13]
        ELSE
                HSerOut ["CR only",13]
        EndIf
        j = 0
Return
```

The subroutine **CRReceived** is mostly a dummy procedure; it sends a message that a complete line has been received and distinguishes between characters followed by a carriage return from a carriage return only.

We'll see in Program 9-3 how we may use this approach for more useful purposes. For now, run Program 9-2A and try differing responses to data entry. Here's one typical interaction.

Output of Program 9-2A	*User Interaction*
P-09-02A	
123 Chars Received: 3 Complete Line Received: 123	123<CR>
456 Chars Received: 3 Complete Line Received: 456	456<CR>
0123456789 Chars Received: 0 CR only	0123456789<CR>
012345678 Chars Received: 9 Complete Line Received: 012345678	012345678<CR>

Note that when we enter 11 characters (0...9 plus the carriage return) the **If j = (MaxStrSize-1)** conditional is true and **j** is reset to 0. Thus, the response is "Chars Received 0," exactly as we wish the program to behave. If we had entered 12 characters, 01234567891<CR>, for example, the response would be one character received, "1."

Program 9-2B

A common programming task is parsing input strings. Let's see how we might accomplish part of that task. We'll convert a numerical string of the form nnnn to a binary number and also we'll convert a string of the form we might use to represent a date or time, hh:mm:ss or dd/mm/yy to three separate numbers. We'll then use these routines in Program 9-3.

```
;Constants
;---------------
WorkLen Con             8


;Variables
;---------------
WorkStr         Var     Byte(WorkLen) ;working string for conversion
DecNum          Var     Long    ;output of string to decimal
i               Var     Byte    ;counter
k               Var     Byte    ;counter
NewVal          Var     Byte(3) ;for date/time items


;Initialization
;---------------
EnableHSerial
SetHSerial H38400

Main
;---------
        WorkStr = "1234 "
        GoSub StrToDecimal
        HSerOut ["Input String: ",Str WorkStr\WorkLen\" ",   |
        "  Value: ",Dec DecNum,13]

        WorkStr = "12:34:56 "
        GoSub StrToDecimal
        HSerOut [13,"Input String: ",Str WorkStr\WorkLen\" ", |
        "  Value: ",Dec DecNum,13]
        GoSub StrTo3Decimal
        HSerOut [Dec NewVal(0),"-", Dec NewVal(1),"-",   |
        Dec NewVal(2),13]

        WorkStr = "12/34/56 "
        GoSub StrToDecimal
        HSerOut [13,"Input String: ",Str WorkStr\WorkLen\" ", |
        "  Value: ",Dec DecNum,13]
        GoSub StrTo3Decimal
        HSerOut [Dec NewVal(0),"-", Dec NewVal(1),"-",   |
        Dec NewVal(2),13]
        Pause 100       ;don't forget pause before END to clear buffer
End

StrToDecimal
;-------------
        i = 0
        DecNum = 0
        While (WorkStr(i) >= "0") AND (WorkStr(i) <= "9")
                DecNum = DecNum * 10
                DecNum = DecNum + WorkStr(i) - "0"
                i = i+1
        WEND
Return

StrTo3Decimal
;-------------
        i = 0
        k = 0
        For k = 0 to 2
                NewVal(k) = 0
```

```
                   While (WorkStr(i) >= "0") AND (WorkStr(i) <= "9")
                         NewVal(k) = NewVal(k) * 10
                         NewVal(k) = NewVal(k) + WorkStr(i) - "0"
                         i = i+1
                   WEND
                   i = i+1
          Next ;k
   Return
   End
```

Program 9-2B sets up three preprepared test strings, "1234," "12:34:56" and "12/34/56" simulating user input. Each of these test strings is then submitted to one of two conversion subroutines, **StrToDecimal** or **StrTo3Decimal**. **StrToDecimal** assumes the input response is one decimal value, while **StrTo3Decimal** assumes the input is in either a time or date structure and it returns three decimal values, representing the three numbers separated by colons or slash symbols.

The converted single number value will be held in the type long variable **DecNum**, while the byte array **WorkStr** holds the simulated string input.

```
          WorkStr = "1234 "
          GoSub StrToDecimal
          HSerOut ["Input String: ",Str WorkStr\WorkLen\" ",   |
          " Value: ",Dec DecNum,13]
```

Let's see how subroutine **StrToDecimal** converts the string to a numerical value.

```
StrToDecimal
;--------------
          i = 0
          DecNum = 0
          While (WorkStr(i) >= "0") AND (WorkStr(i) <= "9")
                   DecNum = DecNum * 10
                   DecNum = DecNum + WorkStr(i) - "0"
                   i = i+1
          WEND
   Return
```

StrToDecimal is called with **WorkStr** containing "1234".

	Element Address									
	0	**1**	**2**	**3**	**4**	**5**	**6**	**7**	**8**	**9**
WorkStr Contents	"1"	"2"	"3"	"4"	0	0	0	0	0	0

We initialize the counter index **i** and the value holder, **DecNum** by setting both to 0.

The **While / Wend** loop steps through **WorkStr**, one element at a time, starting with **WorkStr(0)**.

We check this element to ensure it is a valid numerical character (between "0" and "9"). If it is a valid number, then we shift **DecNum** one decimal place to the left by multiplying it by 10. (The first pass through this loop, i = 0, so there is no shift to the left.) Then, we convert **WorkStr(i)** to its numerical value by subtracting the ASCII value of "0" from the character and add that value to **DecNum**. Let's go through this numerically.

i	WorkStr(i)	DecNum	DecNum=DecNum*10	DecNum=DecNum+ WorkStr(i)-"0"
0	"1"	0	0	0 + 1 = 1
1	"2"	1	10	10 + 2 = 12
2	"3"	12	120	120 + 3 = 123
3	"4"	123	1230	1230 + 4 = 1234

We use the same algorithm to convert the three-value input string, saving the numerical values in the three-element array **NewVal**.

```
StrTo3Decimal
;------------
          i = 0
          k = 0
          For k = 0 to 2
                    NewVal(k) = 0
                    While (WorkStr(i) >= "0") AND (WorkStr(i) <= "9")
                              NewVal(k) = NewVal(k) * 10
                              NewVal(k) = NewVal(k) + WorkStr(i) - "0"
                              i = i+1
                    WEND
                    i = i+1
          Next ;k
     Return
     End
```

The extension to three variables separated by a nonnumerical character should be obvious.

Program 9-3

We've concentrated on the basics in this book, and to emphasize algorithms and techniques many program use hard coded constants instead of prompting for user input information. As an exercise, therefore, we'll develop the skeleton of an interactive program that might be used to control the weather station and data logger developed in Chapter 28. The programs in Chapter 28 use hard coded time and date constants to set the DS1302 real time clock. Likewise the data save interval, the ground elevation barometer correction constant and other values are hard coded.

We'll approach Program 9-3 slightly differently than is our norm. Let's first look at the user's interaction with Program 9-3. And, you may wish to flip through Chapter 28 before going any further in this chapter. Understanding how Program 9-3 interacts with users will aid understanding the program logic. Many of the responses you'll see below are dummies, as the necessary measurements and computations to supply numerical values are contained in Program 28-2. In reviewing the response, user entered information appears after the prompt symbol >. Hence, >B? means that the user entered B? <CR>.

User Input and Interactive Response				*Comments*
P 09-3				
v 0.1				**Introductory screen. This**
>Commands – Letter & ? gets more info				**information is displayed when the program boots up. It is the short form of the help screen.**
? This Menu	A All Data B Barometer			
C Celsius Temp	D Date	E Elevation feet	F Fahrenheit Temp	
H Humidity	I Save Interval	M Mode		**User data is entered after the**
O Output	R Record	S Status		**prompt symbol >**
T Time	V Version X Stop Recording			
>				
>A?				**User enters A? and receives a**
A to output date/time/barometer/humidity/temp F/temp C				**better description of what the A prefix is for.**
>a				**Lower or upper case is the same.**
Status				**Entering A alone causes a status dump. In this case, the output is a dummy response "Status."**
>B?				**Command letter followed by ?**
B to output barometric pressure in in Hg & mbar				**gives expanded help**
>B				
Barometer inHg / mbar				**Dummy barometer response.**

User Input and Interactive Response	*Comments*
>C? C to output current temperature in degrees Celsius	**Command letter followed by ? gives expanded help**
>c Temp xx C	**Dummy Celsius temperature response. Note that the response was a lower case C.**
>D? D to output current date Dmm/dd/yy to set date	**D? gives a help sentence. Note that D has two options; D with nothing else gives the date, Dmm/dd/yy sets the date.**
>? Commands - Letter & ? gets more info	

? This Menu	A All Data B Barometer		
C Celsius Temp	D Date	E Elevation feet	F Fahrenheit Temp
H Humidity	I Save Interval	M Mode	
O Output	R Record	S Status	
T Time	V Version	X Stop Recording	

A ? by itself gives the summary help menu.

>E? E to output current elevation Exxxx to enter new elevation (feet AMSL)	**E? gives a help sentence. E is another double-purpose command; either read the current value or enter an updated value for ground elevation.**
>E	**No output for an E request at the present.**
>F? F to output current temperature in degrees Fahrenheit	**Command letter followed by ? gives expanded help**
>F Temp xx F	**Output a dummy report for the current temperature.**
>H? H to output relative humidity in %	**Command letter followed by ? gives expanded help**
>H Humidity	**Dummy humidity report.**
>I? I to set save interval Ixxxx in minutes	**Command letter followed by ? gives expanded help**
>I I to set save interval Ixxxx in minutes Log Interval	**Dummy response to command**
>M? M mode MF: record date/time/info MA: record info only	**Command letter followed by ? gives expanded help**
>M Mode	**Dummy response to command**
>O? O to output saved data to serial port Saved Data Output	**Command letter followed by ? gives expanded help**
>O O to output saved data to serial port Saved Data Output	**Dummy response to command**
>R? R to start saving data Start Recording	**Command letter followed by ? gives expanded help**

User Input and Interactive Response	Comments
>R R to start saving data Start Recording	Dummy response to command
>S? S to output status report	Command letter followed by ? gives expanded help
>S Status	Dummy response to command
>T? T to output current time Thh:mm:ss to set time 24 hour format only	Command letter followed by ? gives expanded help
>T	Dummy response to command
>V? v 0.1	Current software version is 0.1
>V v 0.1	Dummy response to command
>X? X to stop data recording	Command letter followed by ? gives expanded help
>X Stop Recording	Dummy response to command

The command syntax of Program 9-3 is:

Command	Name	Data Flow	Action
?	Help	To User	Display short form help menu.
A	All Data	To User	Displays all current weather information.
B	Barometer	To User	Current barometric pressure in inches Hg and hPa.
C	Celsius	To User	Display current temperature in degrees Celsius.
D	Date	Bidirectional	D to display current date. Ddd/mm/yy to set new data.
E	Elevation (feet)	Bidirectional	E to display current elevation (feet AMSL) setting. Exxxx to set elevation to xxxx feet AMSL.
F	Fahrenheit	To User	Display current temperature in degrees Fahrenheit.
H	Humidity	To User	Display current relative humidity in percentage.
I	Interval	Bidirectional	I to display current save interval (minutes). Ixx to set new save interval to xx minutes.
M	Mode	To User	Report is data being saved or not?
O	Output Dump	To User	Outputs the saved date/time/temp/humidity/barometer records from EEPROM.
R	Start Recording	To User	Start saving data to EEPROM.
S	Status	To User	Statistics on records saved, available space for new records.
T	Time		T to display current time. Thh:mm:ss to set the current time.
V	Version	To User	Display the current software version.
X	Stop Recording	To User	Stop saving data to EEPROM.
Any letter followed by a ? yields a help sentence for that command.			

Let's now look at the code for Program 9-3. We'll then go through the highlights; much of Program 9-3 we've already covered in this chapter.

One further change—Program 9-3 is very long. Rather than follow our normal structure of listing the program in full, followed by a section-by-section analysis, we'll skip the full listing. Program 9-3, like all other programs in this book, is available in an electronic copy in the associated CD-ROM. If you need a listing to follow while reading the analysis, you may print one from the program file.

Let's abstract the code into a pseudo-code form to clarify the program flow:

```
Main
        Execute a dummy loop to simulate other program code, such as that of
        Program 28-2

        Has any data been entered by the user into the serial input buffer? If yes,
        execute CheckInput
ReturnFromCheckInput :
GoTo Main


CheckInput
        Move a character in the input buffer into a string array InStr.
        Has a carriage return been received? If yes, GoSub CRReceived
        If more characters are waiting in the buffer, move them into InStr.
        GoTo ReturnFromCheckInput

CRReceived
        Convert InStr to upper case
        Check the first letter received.
        If it matches a command letter, branch to a subroutine
        to deal with that command letter.
Return

Supporting General Purpose Subroutines
        StrToDecimal—Converts a string to a number
        StrTo3Decimal—Converts dd/mm/yy or hh:mm:ss to numbers
        ContainsData—Determines if a string contains digits 0...9
        CopyToTemp—Copy one string to another string
```

We've already analyzed **CheckInput** as well as **StrToDecimal** and **StrTo3Decimal**. Only minor changes have been made to these subroutines from the earlier versions, so we won't reexamine this code.

Let's look at the new code in Program 9-3. We'll start with **CRReceived**. This subroutine is called only when a carriage return has been received. The contents of the input buffer are in the array **InStr**.

```
CRReceived
;---------
        For j = 0 to (MaxStrSize-1)
                If (InStr(j) < " ") OR (InStr(j) > "z") Then
                        InStr(j) = $0  ; non-printing chars
                EndIf
                ;Convert lower case to upper case
                If (InStr(j) >= "a") and (InStr(j) <= "z") Then
                                InStr(j) = InStr(j) - $20
                EndIf
        Next ;j
```

First, we scan the completed string for nonuseful characters and convert any lower case letters to upper case.

```
        ;how long is the valid part of the string?
        k = 0
        While InStr(k) >= " "
                k = k+1
        WEND
```

Next, we count the number of valid characters (0...9, A...Z and punctuation marks) contained in **InStr**.

```
        ;check for a CR only
        If k > 0 Then
                k = InStr(0) - "?"
                If k > 27 Then
                        k = 1
                EndIf
```

If the buffer holds something more than a carriage return, we then examine the first character of **InStr** (**InStr(0)**) to see if it is a valid command letter. We convert this letter to an ordinal digit by subtracting the

ASCII value for "?." Thus, if `InStr(0)` ="?," then `k=0`, if `InStr(0)` = "A," then `k=2`, etc. through "Z" corresponding to `k=27`. If this ordinal value is exceeded, we force `k=1`, thereby trapping one error source.

```
                Branch IndexArray(k),
[BR_Unk,BR_Ques,BR_Cent,BR_Date,BR_Fare,BR_Int,BR_Mode,  |
        BR_Out,BR_Rec,BR_Stat,BR_Time,BR_Xstop,BR_Vers,BR_Baro,  |
        BR_Hum,BR_All,BR_Elev]
        EndIf

        GoTo EndCmd
```

We then branch to separate code segments for each command letter. But, we don't branch directly on the value of `k`. Rather, we branch on the value of the byte table `IndexArray(k)`. Why would we do this? It adds complexity, that's for sure. But, it also makes it easier for us to add or delete command letters.

Let's look at `IndexArray`.

```
IndexArray  ByteTable 1,0,15,13,2,3,16,4,0,14,5,0,0,0,  |
        6,0,7,0,0,8,9,10,0,12,0,11,0,0
```

Suppose the input letter is C, corresponding to `k=4`. `IndexArray(4)` = 2. Our `Branch` instruction, therefore, executes `BR_Cent`. We've filled `IndexArray` with 0's for index values corresponding to letters that are not part of our command syntax. To add a new command syntax letter, all we have to do is add the next open digit to the corresponding place in `IndexArray` and insert a new branch label at the end in the `Branch` instruction. To add "Q" as a new command, therefore, we note the current highest value in `IndexArray` is 15. So, we'll revise `IndexArray(18)` to be 16, instead of 15. And, the last part of the Branch statement will be `[…,BR_All,BR_Elev,BR_QQ]`. This is at least slightly easier than trying to figure out where Q goes in `Branch`, and it lets us have a `Branch` statement that is fewer than 28 elements long, another advantage.

I've kept the code behind each `Branch` label as short as possible; each is a call to a specific subroutine, followed by a jump to the end of `CRReceived` subroutine. Breaking each command letter action into one or more subroutines modularizes the code, making it easier to write, debug and maintain.

```
        BR_Unk
                GoSub Sub_Unk
                GoTo EndCmd

        BR_Ques
                GoSub Sub_Ques
                GoTo EndCmd
        …

        EndCmd
                HSerStat 0
                j = 0
        Return
```

Let's now look at a couple of typical command-specific subroutines.

```
        Sub_Unk ;Unknown command
        ;--------
                HSerOut ["Unknown Command. Try Again",13]
                GoSub Sub_Ques
        Return
```

`Sub_Unk` is invoked for an unknown character—that is, a character not in the command syntax. It outputs a message to the user and then calls the subroutine `Sub_Ques`.

```
        Sub_Ques ;? Help request
        ;---------
                k = 0
                While Help1Str(k) <> 0
                        HSerOut [Str Help1Str(k) \1]
                        k = k+1
```

```
             WEND
             k = 0
             While Help2Str(k) <> 0
                     HSerOut [Str Help2Str(k) \1]
                     k = k+1
             WEND
                     k = 0
             While Help3Str(k) <> 0
                     HSerOut [Str Help3Str(k) \1]
                     k = k+1
             WEND
                     k = 0
             While Help4Str(k) <> 0
                     HSerOut [Str Help4Str(k) \1]
                     k = k+1
             WEND
                     k = 0
             While Help5Str(k) <> 0
                     HSerOut [Str Help5Str(k) \1]
                     k = k+1
             WEND
                     k = 0
             While Help6Str(k) <> 0
                     HSerOut [Str Help6Str(k) \1]
                     k = k+1
             WEND
             GoSub Sub_Prompt
      Return
```

Subroutine **Sub_Ques** is also reached by a command consisting of only a question mark; **Sub_Ques** displays six lines of abbreviated command syntax help. The six lines are stored in byte arrays **Help1Str** through **Help6Str**. Each of these strings is terminated with a $0 null. The byte arrays are sent out one byte at a time until the $0 is detected. (We could of course, used MBasic's internal end-of-line command modifier.)

```
      Sub_Prompt
      ;--------------
             HSerOut [">"]
      Return
```

Subroutine **Sub_Prompt** displays the ">" prompt symbol to the user.

Let's look at one of the numerical subroutines. Subroutine **Sub_Time** is invoked when the command letter is "T."

```
      Sub_Time
      ;----------
             If InStr(1) = "?" Then
                     k = 0
                     While TimeStr(k) <> 0
                             HSerOut [Str TimeStr(k) \1]
                             k = k+1
                     WEND
                     ELSE
                     GoSub ContainsData
             EndIf
```

First, we see if this is a request for help on using the "T" command. We determine this by seeing if **InStr(1) = "?"** and if true, we display the T-specific help string, contained in the byte array **TimeStr**. If not, we then see if **InStr** has numerical digits through a call to subroutine **ContainsData**. Subroutine **ContainsData** sets the variable **ContainsNum** to the number of characters ("0"…"9") received.

```
             If ContainsNum > 0 Then
                     GoSub CopyToTemp
                     GoSub StrTo3Decimal
             EndIf
```

If `ContainsNumbers` shows the presence of digits, we execute the subroutine `CopyToTemp` which copies the digits and digit separators from `InStr` to a working string `WorkStr` that is analyzed by `StrTo3Decimal`, which is identical with the `StrTo3Decimal` subroutine we analyzed in Program 9-2A.

```
        GoSub Sub_Prompt
    Return
```

We end by displaying the prompt.

The remainder of Program 9-3 consists of variations on these algorithms, and in order not to expand an already overly long chapter we will not conduct further analysis.

Ideas for Changes to Programs and Circuits

- The help strings consume significant program memory. Are there ways to make this more efficient? For example, suppose we adopt a 5-bit alphabet, sufficient to support (capital letters) A…Z and a few punctuations marks. We could introduce a shift character to reflect a shift to a second alphabet consisting of numbers and additional punctuation marks. (Yes, this has been done before; among other things, it's the Baudot code.) By bit packing (see Chapter 28), we can fit eight 5-bit symbols into five bytes. How else might we make fixed string store more efficient? Could we store common words, such as "the," "is," "enter," "number," "temperature" and the like in the PIC's on-board EEPROM, and give each word a byte-length index number. Sentences can then be constructed by stringing a series of word numbers together. How efficient would this be? How would you include word length information in the word number?
- Program 9-3 contains little error checking. How would you modify the input routines to ignore out-of-limits data? How much program space budget should be devoted to error checking versus required computation?
- How would you merge Programs 9-3 and 28-2? Does a 16F87x PIC have sufficient memory to run a fully interactive version of Program 28-2? How might the merged program be reduced in size?

References

[9-1] Dallas Semiconductor, *Fundamentals of RS-232 Serial Communications*, Application Note 83 (undated).

[9-2] The program Terminal (current release v1.9b) is available at: http://bray.velenje.cx/avr/terminal/.

[9-3] The program ComTest is available at: http://www.bb-elec.com/comtest.asp.

[9-4] Sipex Corporation, *SP231A/232A/233A/310A/312A Enhanced RS-232 Line Drivers/Receivers*, Document No. SP231ADS/01 (2000).

Interrupts and Timers in MBasic

What are interrupts? What are timers? How are they related? And, why should an MBasic programmer care?

Interrupts and Timers—Overview

What is an Interrupt?

An interrupt is an event that causes the current program sequence to temporarily halt, and for different code, called an "interrupt service routine" or ISR, to be executed. After the ISR completes, program flow returns to the point it was at before the interrupt occurred. You might think of an ISR as a subroutine that is called not by a **GoSub** in software, but rather an *event*. The event may be external, such as an input pin changing state from a 0 to a 1, or vice versa. Or, the event may be internal to the PIC, such as a timer reaching a defined count. It may even be a combination of external and internal events, such as where repeated external signals cause an internal counter to reach a preset value.

This chapter focuses on ISRs written purely in MBasic. As we'll learn, an ISR written in MBasic has one significant restriction—it will be executed only "between" other MBasic statements. In some cases, this latency is a serious drawback. For example, if our program is waiting for an input string in a **SerIn** statement, and an interrupt occurs, the ISR will not be executed until the **SerIn** statement completes, thereby negating the benefit of an interrupt. If a **Pause 1000** statement is executed, the MBasic ISR will not be executed until the **Pause 1000** statement completes. We'll learn in Chapter 15 to program the interrupt handler in assembler and have it execute essentially instantaneously when the interrupt triggering event occurs. But, we've learned in Chapter 6, for example, that we may successfully read a rotary encoder through an MBasic ISR far faster than you will be able to spin the encoder shaft by hand.

What are Timers?

Timers are internal, programmable, counters. The 16F87x/87xA family and many mid-range PICs have three hardware timers, timer 0, timer 1 and timer 2. These timers count input pulses until a certain number of pulses or "ticks" have been received. When this happens, an interrupt is triggered and specific interrupt handler code can be executed. These hardware modules are usually called "timers" instead of "counters" because their most often used purpose is to determine time intervals by counting pulses that occur at known intervals.

A timer's input may come from an internal oscillator, stabilized by an external crystal, or from the PIC's program clock, or from any external logic-level source. Some timers have a software-configurable prescaler, permitting input ticks to be divided by 1, 2, 4 or 8 and some have software-configurable post-scalers, permitting the output to be divided by defined factors up to 256. In the case of the 16F87x family, two timers have 8-bit registers, while one has a 16-bit register.

Interrupts

Let's start by looking at interrupts in more detail. We'll use the 16F87x family for this exercise. Not all PICs implement the same interrupts, so you should verify the capabilities of a PIC you may be considering before selecting it for your particular project.

Before jumping into the details, let's first learn a bit of terminology. In Microchip's reference documents, interrupts are identified by acronyms, while MBasic uses a slightly different name. The 16F87X has 14 possible interrupts, of which 11 are directly supported in MBasic:

Microchip Acronym	MBasic Identifier	Associated Setup Function	Description
ADIF	ADINT		Analog-to-digital conversion is completed.
BCLIF	BCLINT		Bus collision interrupt enable; not currently supported in MBasic.
CCP2IF	CCP2INT	SetCapture GetCapture SetCompare	Interrupt occurs when there is a capture/compare/period match with CCP module 2.
CCP1IF	CCP1INT	SetCapture GetCapture SetCompare	Interrupt occurs when there is a capture/compare/period match with CCP module 2.
EEIF	EEINT		EEPROM interrupt; occurs when writing to the on-board EEPROM is completed.
INTF	EXTINT	SetExtInt	Change of input pin state on RB0. Must separately set mode to select 0→1 or 1→0 transition for the triggering event.
PSPIF	None		Parallel slave port interrupt; not supported in MBasic.
RBIF	RBINT		Any pin RB4,RB5,RB6 or RB7 has changed state, in either the 0→1 or 1→0 direction.
RCIF	RCINT		Received a byte through the hardware USART. Note: When using HSERIN or HSEROUT functions RCIF is disabled.
SSPIF	SSPINT		Synchronous serial port interrupt; not supported in MBasic.
T0IF	TMR0INT	SetTmr0	Timer 0 overflow has occurred.
TMR1IF	TMR1INT	SetTmr1	Timer 1 overflow has occurred.
TMR2IF	TMR2INT	SetTmr2	Timer 2 overflow has occurred.
TXIF	TXINT		Finished transmitting a byte from the hardware USART. Note: When using HSERIN or HSEROUT functions RCIF is disabled.

The "F" at the end of Microchip's acronym signifies that reference is to the "interrupt flag" bit, which identifies source of the interrupt. Each of these 14 interrupts also has an associated "enable" bit, identified with the letter "E" at the end of the acronym, such as **INTE** which is the enable bit for the **INT** interrupt. We'll save further explanations until Chapter 15 as this information isn't necessary to use MBasic's interrupt procedures.

To program with an interrupt, we follow a five-element template:

1. With the **OnInterrupt** operator, link the interrupt source with the name of the associated interrupt service routine.
2. Set up any optional elements of the interrupt source, such as whether **EXTINT** is to be triggered on the 1→0 or the 0→1 transition, or presetting any counters.
3. Enable the interrupt.
4. Before the interrupt service routine, disable all interrupts. This step is necessary to prevent the interrupt service routine from being continually re-triggered and winding up in an endless loop.
5. The interrupt service routine itself.

Let's see how this works in practice.

Program 10-1

Configure your 2840 Development Board as shown in Figure 10-1. Switch SW1 is any normally open momentary contact switch. We'll use an interrupt to briefly flash the LED whenever SW1 is closed, even though the program is in the middle of mathematical calculations. Closing SW1 triggers an **EXTINT** (change in pin B0) interrupt and will invoke a simple ISR to flash the LED.

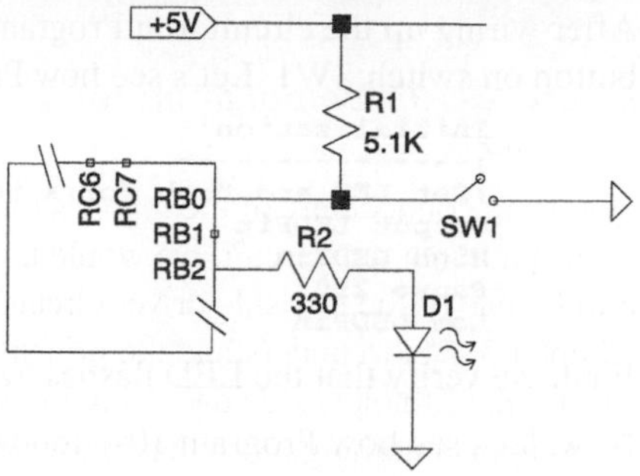

Figure 10-1: Program 10-1 connections.

```
;Program 10-1
;Elements of an interrupt program

;Constants
;---------
InputPin        Con     B0
LEDPin          Con     B2          ;LED connected


;Variables
;---------
i       Var     Byte
j       Var     Byte
k       Var     Word

Initialization
;-------------
;Set LED and flash for a test
Output LEDPin
High LEDPin
Pause 250
Low LEDPin

;[1] Link Interrupt to ISR
;-------------------------
OnInterrupt ExtInt,CloseSwitch

;[2] Setup any interrupt options
;-------------------------------
SetExtInt Ext_H2L

;[3] Enable the interrupt
;------------------------
Enable ExtInt


Main
        For i = 0 to 255
                For j = 0 to 255
                        k = i*j
                Next ;j
                Low LEDPin
        Next ;k
GoTo Main


;[4] Compliler directive to disable interrupts
;---------------------------------------------
Disable ExtInt

;[5] The Interrupt Service Routine
CloseSwitch
;-----------
        High LEDPin
        Pause 250
        Low LEDPin
Resume

End
```

After wiring up the circuit, run Program 10-1. You should see the LED light briefly whenever you press the button on switch SW1. Let's see how Program 10-1 works.

```
Initialization
;---------------
;Set LED and flash for a test
Output LEDPin
High LEDPin
Pause 250
Low LEDPin
```

First, we verify that the LED flashes by giving it a brief on/off at initialization.

Now, let's see how Program 10-1 follows our five-step template.

```
;[1] Link Interrupt to ISR
;-------------------------
OnInterrupt ExtInt,CloseSwitch
```

Step 1—The first step in the template is to associate the interrupt identifier with the ISR through the **OnInterrupt** procedure.

MBasic's **OnInterrupt** procedure takes two arguments:

```
OnInterrupt    InterruptSource, Label
```

InterruptSource—is the interrupt identifier from the table provided above. It identifies the source of the interrupt to be associated with the ISR. In Program 10-1, the interrupt source is **ExtInt**.

Label—The name of the location (the ISR) to which program execution jumps in when an **InterruptSource** interrupt occurs. You will normally structure the ISR similar to a subroutine, except instead of ending with a **Return**, an ISR terminates with a **Resume** statement. When the ISR finishes running, program execution returns to the point it was when the interrupt was triggered. In Program 10-1, the ISR's label is **CloseSwitch**.

```
;[2] Setup any interrupt options
;-------------------------------
SetExtInt Ext_H2L
```

Step 2—The second step is to set up any options for the interrupt. The interrupt we use, EXTINT, is set up with the **SetExtInt** procedure.

```
SetExtInt Mode
```

Mode—has two permitted values:

Ext_H2L—the interrupt triggers on a logic high to low transition; or

Ext_L2H—the interrupt triggers on a logic low to high transition.

Since SW1 is normally open, pin B0 will be held high by pull up resistor R1. When SW1 is closed, pin B0 will drop to logic 0. Hence we will trigger the interrupt on a high-to-low transition and accordingly use the statement **SetExtInt Ext_H2L** to set up **ExtInt**.

```
;[3] Enable the interrupt
;-------------------------
Enable ExtInt
```

Step 3—Next we activate the interrupt with the **Enable** function. Until an interrupt is enabled, it does not function; the interrupt event may occur but the ISR will not be executed.

```
Enable {InterruptSource}
```

If we use **Enable** without any interrupt source, all interrupts are enabled for all code compiled after the **Enable** function. If we add the **InterruptSource**, then only that particular interrupt source is enabled. Since we've used **Enable ExtInt**, only the **ExtInt** will be active.

```
Main
        For i = 0 to 255
            For j = 0 to 255
```

```
                      k = i*j
              Next ;j
              Low LEDPin
      Next ;k
GoTo Main
```

To simulate a program performing other tasks, we use two **For…Next** loops to compute the word length product of two byte variables.

```
;[4] Compliler directive to disable interrupts
;----------------------------------------------
Disable ExtInt
```

Step 4—Next, we disable the **ExtInt** using the **Disable** function.

```
Disable {InterruptSource}
```

Used without the **InterruptSource** argument, **Disable** is treated as a compiler directive, and it deactivates all interrupts for code compiled after the **Disable** statement until an **Enable** directive is reached.

This particular program will function equally well without the **Disable** command, but that's not generally true. In some cases, the ISR will get called over and over again, which will eventually cause MBasic's internal stack to overflow, thereby crashing your program. Hence, we'll always use the **Disable** function.

```
;[5] The Interrupt Service Routine
CloseSwitch
;-----------
          High LEDPin
          Pause 250
          Low LEDPin
Resume
```

Step 5—The final step is to write the ISR. Remember, the ISR ends with a **Resume** statement, not a **Return** statement. The ISR in this case flashes the LED on for 250 ms.

Program 10-1A

When you press the switch, the LED illuminates with no perceptible lag time to the human eye. This is because none of the MBasic statements being executed in the main program loop require more than a few hundred microseconds to execute. When the interrupt is triggered, any MBasic statement being executed continues to run and the ISR is not called until the then currently executing statement completes. The lag time between the interrupt action and start-of-execution for the ISR is often called *latency*. Let's modify Program 10-1 by adding a **Pause 1000** statement after each math calculation:

```
For i = 0 to 255
        For j = 0 to 255
                k = i*j
                Pause 1000
        Next ;j
        Low LEDPin
   Next ;k
GoTo Main
```

Save this as Program 10-1A. Run it and press the switch button. What happens? You press the switch button and after some perceptible delay, the LED illuminates. The delay between button press and LED illumination depends on which statement is executing where you press the button. If you are very fortunate, you press the button while one of the indexing statements or the product **i*j** is being calculated, and the LED appears to illuminate instantaneously. However, these statements require only in total a few hundred microseconds or less to execute. Accordingly 99.9% of the time when you press the button, the **Pause 1000** statement is being executed. The average latency you will see is therefore about 500 ms, and the worst-case is 1000 ms.

Program 10-1B

Let's see the effect of one more change. Delete the `Pause 1000` statement we added and reverse `ExtInt`'s trigger sense by changing the interrupt set up procedure to:

```
SetExtInt Ext_L2H
```

We're now trigger the interrupt on the low-to-high transition. We expect, therefore, to see the LED illuminate when we release the switch, not when we press it.

What happens? When I made this change, the LED illuminated twice; once when I pressed the switch and then again when I released the switch. What's going on here? It's our old friend, switch bounce, at work. Remember what we learned in Chapter 4? Mechanical switch contacts don't cleanly open and close; rather they may experience dozens of very short duration open/close periods until the contacts stabilize in the open or closed position. If you don't believe me, look at Figure 4-11 again. When we press the switch button, its contacts do not cleanly make or break. Rather SW1's contacts bounce causing RB0 to experience several high-low-high-low-high-low transitions until the contacts stabilize closed (low). Even though these bounces may seem short by the standards of human senses, they more than long enough to trigger an interrupt.

Why didn't we see multiple flashes when we ran Program 10-1 then? Most switch bounce occurs on make, not break. At least with the particular switch I used, the break action was clean, without bounce. Hence operating the switch with Program 10-1 produces only one flash, upon switch closure. Depending upon a switch to cleanly make or break is, however, not good practice and a practical design should include debounce measures.

Program 10-2

Let's look at the other Port B interrupt; **RBINT**, or interrupt on change of RB4, RB5, RB6 or RB7. A common use of **RBINT** is to read a rotary encoder, so let's see how this is done. Figure 10-2 shows the configuration required for Program 10-2. Figure 10-2 is identical with Figure 6-10 and, for that matter; Program 10-2 is identical with Program 6-3. If you've already read Chapter 6 and fully understand how Program 6-3 works,

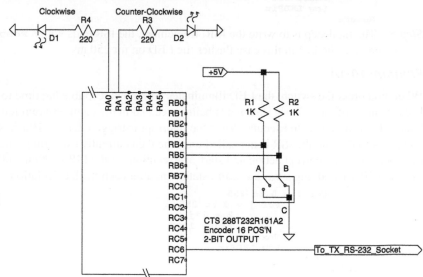

Figure 10-2: Program 10-2 connections.

you may skip our analysis of Program 10-2. We'll use the same CTS model 288T232R161A2 two-bit rotary encoder that we studied in Chapter 6.

Program 10-2 reads a relative rotary encoder and writes a step value to the serial port. Rotation clockwise increases the step value while counterclockwise rotation decreases it. Two LEDs show the rotation direction as well. The RBINT interrupt is executed whenever the value of pins RB4, RB5, RB6 or RB7 changes. A change on any pin, either 1→0 or 0→1 triggers an RBINT interrupt. As we learned in Chapter 6, every step rotation of the encoder shaft changes one output bit.

Direction	B4	B5	Direction
	0	0	
↑	1	0	↓
	1	1	
	0	1	

Hence, every time the encoder shaft rotates to a new position, B4 or B5 changes state, and an **RBINT** is triggered.

Since Program 10-2 duplicates Program 6-2, we will not print it again. If you need a listing to follow while reading the analysis, refer to Program 6-2 or print Program 10-2 from the CD-ROM program file.

We've extensively analyzed the polling version of this program in Chapter 6, so we'll concentrate only on its interrupt features.

```
OnInterrupt RBINT, ReadEncoder
Enable RBINT

Main
;--------------
GoTo Main
```

You should recognize the first three steps of our standardized interrupt template:

Step 1—Use **OnInterrupt** to link the occurrence of the **RBINT** to the ISR **ReadEncoder**.

Step 2—Set any options for the interrupt. **RBINT** has no options, so Step 2 does not exist.

Step 3—Enable the interrupt, via the **Enable** procedure.

Now, let's look at the remaining two steps.

```
Disable

ReadEncoder
      CurVal = (PortB & %00110000) >> 4
        If CurVal <> OldVal Then
      Branch CurVal, [S0,S1,S2,S3]
      EndIf
      AfterBranch
Resume
```

Step 4—We've disabled interrupts through the global compiler directive **Disable**, without any interrupt argument. This means that all interrupts are disabled for code below the **Disable** statement. In this simple program, that's acceptable. In a complex program with multiple interrupts enabled, you would likely wish to only disable the **RBINT** interrupt during the ISR, which you may accomplish by invoking **Disable RBINT**.

Step 5—is the ISR itself, which must conclude with a **Resume** statement.

Let's take a quick peak at how the ISR functions, and we'll discover one quirk of the **RBINT** interrupt.

```
ReadEncoder
      CurVal = (PortB & %00110000) >> 4
        If CurVal <> OldVal Then
      Branch CurVal, [S0,S1,S2,S3]
      EndIf
      AfterBranch
Resume
```

To read pins B4 and B5, we read all **PortB** pins; then mask all bits but those associated with B4 and B5 with the **& %00110000** operator and then shift the result four places to the right. The result is the equivalent of **%B5B4**, yielding a number that cycles from 0 through 3 as the encoder changes from one position to another. (Don't forget that the parentheses around the logical **AND** operation is required, as the **AND** operator is lower precedence than the shift ">>" operator.)

The quirk is that in order to determine whether or not to generate an interrupt, the PIC's hardware compares the status of pins B7...B4 with an internally stored copy of pins B7...B4 values. If the current value differs from the stored value, a mismatch is detected and the **RBINT** interrupt is issued. Somehow, therefore, we must get the PIC to discard the old stored comparison value and replace it with the new post-change value for pins B7...B4. There is no explicit assembler or MBasic instruction "refresh the comparison store." Rather, this is done automatically whenever Port B is read. In the **ReadEncoder** ISR, the statement **CurVal = (PortB & %00110000) >> 4** causes Port B to be read. In 99.9% of the code you will write, an **RBINT** ISR reads Port B and everything works exactly the way it's supposed to. However, you may run across the odd application where you don't care about the particular value of pins B7...B4. For example, you might connect these four pins to four safety switches so that if any are tripped the associated **RBINT** ISR shuts down a motor without ever reading Port B. In this case, the new switch settings will not be updated in the PIC's internal **RBINT** comparison store and the code may not work the way you think it should.

Timers

Figure 10-3 depicts a simplified, generic timer module. The 16F87x/87xA series PICs include three timer modules, each different from the other, but sharing certain common features illustrated in Figure 10-3. We'll go through the generic timer and then focus upon the 16F87x's three timer modules. The 16F87x's timers are representative of the timers found in other mid-range PICs, so what we learn in this chapter is not limited to the 16F87x/87xA family.

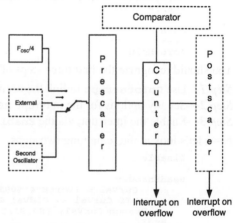

Let's first consider the term "timer." It means, in the dictionary sense, a device for measuring time, or a device that causes actions to be taken at certain times. A PIC's timers are capable, with software support, of both functions. But, of course, a PIC has no internal timepiece; rather it has either an internal or external oscillator. To determine the time that elapses between

Figure 10-3: Generic timer module.

events A and B, therefore requires us to (a) know the duration of one oscillator pulse, t_d, and (b) to count the number of oscillator pulses, N, that occur between events A and B. The elapsed time A-B is then Nt_d. Consequently, although often called a "timer," the literature more accurately refers to these modules and certain of their components as "counters." And, that's what they do; they count input events. If the input events occur at regular intervals, we may easily convert discrete counts to elapsed time. But, of course, we might just as easily use the counter to determine how many times a switch is closed.

Our generic timer consists of six components, and, of course, any particular timer may not have all of these components or all the options for any component:

Input signal—The counter section must have a logic level input signal. Three possible sources are available: (a) the PIC's clock signal; (b) an external signal applied to an input pin; or (c) a second internal oscillator with an external crystal. If the PIC's internal clock is used, the counter input signal is the clock frequency divided by four—that is, the PIC's instruction cycle time, usually called Fosc/4.

Prescaler—To provide more flexibility, the input signal runs through a prescaler. A prescaler is a predivider, for example, if the prescaler is set to divide by four, it takes four input cycles to produce one output cycle to the counter. The prescale divide ratio is software settable and includes a 1:1—that is, no extra division, setting.

Counter—The counter increments one count for every prescaler output pulse. Counters may be implemented as 8-bit or 16-bit devices. The counter is addressable and may be read or written to like any byte or word variable in MBasic. Writing to the counter is usually called *preloading*.

Comparator—A comparator is a register, which may be written to in MBasic. When the counter value equals the comparator value, the counter is reset to 0.

Postscaler—The output of the counter section may go to a postscaler. A postscaler works just like a prescaler, except its input is advanced one pulse every time the counter is reset to 0. The postscaler division ratio is also software settable and can be set to 1:1, effectively bypassing the postscaler.

Interrupt Source—Despite the differences among the 16F87x/87xA's three timers, the same condition—timer overflow—causes their associated interrupt to be issued. As we will see, overflow may include postscaler division.

Let's compare the 16F87x/87xA's three timers. Microchip identifies the three timers as timers 0, 1 and 2:

Parameter	Timer 0	Timer 1	Timer 2
MBasic Set-up Function	SetTmr0	SetTmr1	SetTmr2
MBasic Interrupt Name	TMR0INT	TMR1INT	TMR2INT
Input Clock Sources	• FOSC/4 • RA4 (external) • Watchdog Timer	• FOSC/4 • RC0 (external) • Second Oscillator (RC0-RC1)	• FOSC/4
Input Options	(When external input) • Increment on high-to-low • Increment on low-to-high	Synchronous or asynchronous mode (implicit synchronous for Fosc/4)	None
Prescaler?	Yes	Yes	Yes
Prescale Ratios	9 ratios, from 1:1 through 1:256	1, 2, 4 & 8:1	1, 4 & 16:1
Postscaler?	No	No	Yes
Prescale Ratios	N/A	N/A	16 ratios, from 1:1 to 16:1
Counter Register Length	8-bits	16-bits	8-bits
Counter Register Name (Read & Write)	TMR0	Tmr1H & Tmr1L for high and low bytes	TMR2
How to get "short count"	Preload TMR0	Preload Tmr1H & Tmr1L (But also see compare mode)	Load comparator with value; Timer 2 resets when count equals comparator
Turn Counter on/off	Timer 0 is always on; Don't enable interrupt if not needed	Off: SetTmr1 TMR1Off On: Any other constant Or: On: Tmr1on = %1 Off: Tmr1on = %0	Off: SetTmr2 TMR2Off On: Any other constant Or: On: Tmr2on = %1 Off: Tmr2on = %0
Interrupt Condition	Overflow (TMR0 goes from $FF to $00)	Overflow (TMR1H goes from $FF to $00)	Overflow of postscaler
Notes	Shares prescaler with watchdog timer	Can be used in conjunction with Compare CCP1 or CCP2 functions	For some releases of MBasic, Timer 2 constants are wrong

To select from among the various prescaler, postscaler and input source options, you use appropriate timer setup library constant, as MBasic provides, as the argument to the timer's setup function. There are dozens of setup constants, so we won't try and go through them, except as we may use a particular value in programming. Please refer to the MBasic User's Guide for the specific constant identifiers.

We may use timers with or without interrupts. We'll first look at an application without an interrupt.

Program 10-3

Let's use Timer 1 to measure the execution time of various MBasic statements. You may keep the same wiring connection as shown in Figure 10-2.

Our algorithm is simple; turn timer 1 on immediately before the statement(s) we wish to time; turn it off immediately after the statement(s) finish executing. We then read timer 1's counter and multiply it by the time per counter "tick."

```
;Program 10-3
;Use Timer 2 for execution time

;Constants
;----------
Enabled Con      %1
Stopped Con      %0

;Variables
;--------------
i                Var     Byte
j                Var     Byte
k                Var     Word
Value            Var     Word


;Initialization
;--------------
EnableHSerial
SetHSerial H115200
;Set for 8:1 prescale
;with 20 MHz this yields 1.6us
;per timer tick.
SetTmr1 Tmr1Int8


Main
        Tmr1on = Stopped
        Tmr1H = 0
        Tmr1L = 0
        Tmr1on = Enabled
    For i = 1 to 100
        j = i/2
        k = i*j
    Next
    Tmr1on = Stopped
    Value.LowByte = Tmr1L
    Value.HighByte = Tmr1H
    HSerOut ["Ticks: ",Dec Value,13]
    Pause 100
GoTo Main

End
```

Let's see how Program 10-3 works.

```
;Constants
;----------
Enabled Con      %1
Stopped Con      %0
```

For better readability, we define two bit constants to use with `Tmr1On`:

```
;Initialization
;--------------
EnableHSerial
SetHSerial H115200
;Set for 8:1 prescale
;with 20 MHz this yields 1.6us
;per timer tick.
SetTmr1 Tmr1Int8
```

We first must setup Timer 1. Since we wish to measure elapsed time, the most convenient input clock is the Fosc/4 source from the PIC's internal clock. I use a 20 MHz resonator in my development boards, so each Fosc/4 clock pulse (which we will call "ticks") represents 200 ns. One oscillator pulse is 50ns:

$$t_{osc} = \frac{1}{F_{osc}} = \frac{1}{20 \times 10^6} = 50 \times 10^{-9}$$

Hence Fosc/4 is four times one oscillator period, or 200 ns.

Which preselector division ratio shall we use? To select the division ratio, we must know, roughly, the time duration we wish to measure. We're going to measure 100 loops through a `For...Next` loop, with each loop having two integer math operations. Based on experience, we estimate each loop cycle will take about 750 µs. To measure 100 loop cycles requires us to measure 75 ms. Timer 1 has a 16-bit counter, so we can count a maximum of 65535 ticks. To count up to 75 ms, each tick's duration should be about 75ms/65535 (1.14 µs). If our prescale ratio is 4:1, the tick duration is 0.800 µs (4 times 200 ns), which is too short. An 8:1 prescale ratio gives 1.6 µs per tick, which meets our requirements.

```
Main
        Tmr1on = Stopped
        Tmr1H = 0
        Tmr1L = 0
```

We turn the timer off and clear both bytes of the counter.

```
        Tmr1on = Enabled
        For i = 1 to 100
                j = i/2
                k = i*j
        Next
        Tmr1on = Stopped
```

We turn the counter on, run 100 cycles through the loop, and then turn the counter off.

```
        Value.LowByte = Tmr1L
        Value.HighByte = Tmr1H
        HSerOut ["Ticks: ",Dec Value,13]
        Pause 100
GoTo Main
```

Now, we read the counter value and sent it through the serial interface. Note that we are able to read and write to Timer 1's counter just as if it was an MBasic variable. This, of course, is because there is no structural difference between a user-declared variable and the PIC's internal register. We don't even have to worry about declaring these variables, as MBasic's compiler already defines them for us.

In order to get the most information out of our timer, we'll run several modified versions of Program 10-3, commenting out various statements to determine individual execution times.

MBasic Code	Output	Analysis

`Tmr1on = Enabled`
`;For i = 1 to 100`
`;    j = i/2`
`;    k = i*j`
`;Next`
`Tmr1on = Stopped`
	Ticks: 25
	
	(40.0 µs)
		This time represents the overhead in starting and stopping Timer 1. We will subtract this overhead from the later readings to calculate accurate statement execution times.

MBasic Code	Output	Analysis
Tmr1on = Enabled For i = 1 to 100 ; j = i/2 ; k = i*j Next Tmr1on = Stopped	Ticks: 10846 (After subtracting overhead: 10821 ticks, or 17314 μs)	The For...Next loop, with no operations inside the loop, takes 17,314 μs to execute 100 times. Each loop cycle, therefore is 173 μs.
Tmr1on = Enabled For i = 1 to 100 j = i/2 ; k = i*j Next Tmr1on = Stopped	Ticks: 28539 (After subtracting the execution time for the bare loop: 17693 ticks, or 28309 μs)	The increased execution time by adding back in the j = i/2 statement is 28309μs. Since the j = i/2 statement is executed 100 times, the execution time for one j = i/2 statement is 283 μs.

Tmr1on = Enabled
For i = 1 to 100
    j = i/2
    k = i*j
Next
Tmr1on = Stopped
	Ticks: 42101
	(After subtracting the prior execution time: 13562 ticks, or 21699 μs)
		The increased execution time from adding back the k = i * j statement is 21699μs. Since the k = i * j statement is executed 100 times, the execution time for one k = i * j statement is 217 μs.

We now have determined an accurate execution time for each statement:

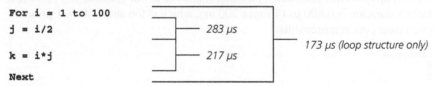

```
For i = 1 to 100
j = i/2                         283 μs
                                               173 μs (loop structure only)
k = i*j                         217 μs

Next
```

These execution time values are for MBasic version 5.2.1.1. Version 5.3.0.0 executes a few percent faster. For now, we'll ignore the errors that arise from the finite execution time of the statements turning timer 1 on and off. We'll see a way of compensating for this error in Chapter 14.

Program 10-4

Let's see how timers and interrupts work together. We'll reuse the circuit of Figure 10-2 but don't worry if you don't have a rotary encoder, as won't use it. Program 10-4 uses timer 1 to flash an LED briefly once every 100 ms, and timer 2 to illuminate a second LED, on for 1000 ms and off for 1000 ms.

```
;Program 10-4
;Two timers to flash LEDs

;Constants
;----------
Enabled        Con        %1
Stopped        Con        %0

;Timer 2 corrected constants
;---------------------------
T2Post1        Con        %00000000
T2Post2        Con        %00001000
T2Post3        Con        %00010000
T2Post4        Con        %00011000
T2Post5        Con        %00100000
T2Post6        Con        %00101000
T2Post7        Con        %00110000
T2Post8        Con        %00111000
T2Post9        Con        %01000000
T2Post10       Con        %01001000
T2Post11       Con        %01010000
T2Post12       Con        %01011000
```

```
T2Post13        Con     %01100000
T2Post14        Con     %01101000
T2Post15        Con     %01110000
T2Post16        Con     %01111000

T2Pre1          Con     %00000100
T2Pre4          Con     %00000101
T2Pre16         Con     %00000111

;Variables
;----------
PreLoad         Var     Word
i               Var     Byte

;Initialization
;--------------
SetTmr1 Tmr1Int8
;PreLoad based on measured duration
PreLoad = 3072
OnInterrupt Tmr1Int, Flash1

;Calculated for 1 second with *128
SetTmr2 T2Post16 + T2Pre16, 153
OnInterrupt Tmr2Int, Flash2

i = 0
Output A1
Low A1
Enable

Main
GoTo Main

Disable
Flash1 ;ISR for Timer 1
;------
        ;have to reload every time
        Tmr1on = Stopped
        Tmr1H = PreLoad.HighByte
        Tmr1L = PreLoad.LowByte
        Tmr1on = Enabled

        High A0
        Pauseus 300
        Low A0
Resume

Flash2 ;ISR for Timer 2
`-----
        ;adds an extra divide by 128
        ;no need to reset
        i = i+1
        PortA.Bit1 = i.Bit7
Resume
End
```

We'll analyze Program 10-4 functionally, not sequentially. Let's first look at timer 2.

```
;Calculated for 1 second with *128
SetTmr2 T2Post16 + T2Pre16, 153
OnInterrupt Tmr2Int, Flash2
```

Timer 2's setup function **SetTmr2** requires two arguments:

```
SetTmr2 Mode, Period
```

Mode—is a constant or variable that turns the counter off or turns it on and sets the prescale and postscale ratios. As we've seen, timer 2 has 16 possible postscale ratios, from 1:1 to 16:1, and three possible prescale ratios, 1:1, 4:1 and 16:1. Some earlier versions of MBasic have errors in the timer 2 mode library constants, so we define corrected constants in the **Constants** section of Program 10-4.

```
;Timer 2 corrected constants
;---------------------------
T2Post1      Con      %00000000
T2Post2      Con      %00001000
T2Post3      Con      %00010000
T2Post4      Con      %00011000
T2Post5      Con      %00100000
T2Post6      Con      %00101000
T2Post7      Con      %00110000
T2Post8      Con      %00111000
T2Post9      Con      %01000000
T2Post10     Con      %01001000
T2Post11     Con      %01010000
T2Post12     Con      %01011000
T2Post13     Con      %01100000
T2Post14     Con      %01101000
T2Post15     Con      %01110000
T2Post16     Con      %01111000

T2Pre1       Con      %00000100
T2Pre4       Con      %00000101
T2Pre16      Con      %00000111
```

We use these constants in pairs, added together. The **T2Postxx** series constants define the postscale divide ratio, from 1:1 (**T2Post1**) through 16:1 (**T2Post16**). The **T2Prexx** series constants define the prescale divide ratio, which is 1:1 (**T2Pre1**), 4:1 (**T2Pre4**) or 16:1 (**T2Pre16**). A timer 2 configuration constant is thus of the form **T2Postxx + T2Prexx**. (MBasic's library constant **Tmr2Off**, which disables timer 2, is correct.) If we wish to set the postscale ratio at 8:1 and the prescale ratio at 4:1, the required mode constant is **T2Post8 + T2Pre4**.

Period—Timer 2's module includes a comparator, which compares the counter's value with **Period**. When the counter value equals **Period**, the counter resets to 0. Since timer 2 has an 8-bit counter, **Period** is a byte length variable or constant. Once set, **Period** does not change when timer 2's counter resets. Remember that timer 2's interrupt, **Tmr2Int**, is triggered not upon counter reset, but rather upon postscaler overflow. Hence, timer 2's interrupt interval is:

$$t_2 = \frac{4 \bullet N_{PRE} \bullet N_{POST} \bullet N_{PERIOD}}{F_{OSC}}$$

t_2 is the time between interrupts
N_{PRE} is the prescale ratio
N_{POST} is the postscale ratio
N_{PERIOD} is **SetTmr2**'s **Period** argument

Let's see how we select N_{PRE}, N_{POST} and N_{PERIOD}. Our objective is to use timer 2 to flash an LED in a 1000 ms on, 1000 ms off cycle. What ratios provide this timing sequence? First, let's determine timer 2's absolute slowest interrupt period. The slowest interrupt period corresponds to the maximum prescale and postscale values, and the maximum period argument, 16:1, 16:1 and 255, respectively. Timer 2's maximum interrupt period, based upon the 20 MHz resonator I've installed in my 2840 Development Board, is:

$$t_2 = \frac{4 \times 16 \times 16 \times 255}{20 \times 10^6} = 13.056 \times 10^{-3}$$

Since timer 2's maximum interrupt period is only 13 ms, it's apparent we can't rely solely upon timer 2's interrupts to illuminate and extinguish the LED. Instead we use timer 2 interrupts to increment a variable and control the LED based upon that variable's value. The ISR called when timer 2 issues an interrupt is

Flash2:

```
        Flash2 ;ISR for Timer 2
            `-----
                    ;adds an extra divide by 128
                    ;no need to reset
                    i = i+1
                    PortA.Bit1 = i.Bit7
                    EndIf
            Resume
```

Instead of using an **If…Then…EndIf** test, we can use a faster process, by letting pin A1's value equal **i.Bit7**. **i.Bit7** has value **%0** until **i** equals 128 when it becomes **%1** and stays at **%1** until **i** rolls over and becomes zero. Hence the LED attached to A1 toggles between on and off every 128 times **Flash2** is called.

In order to make the on and off periods equal to 1 second, **Flash2** must be called 128 times per second, or every 7.8125 ms. In order for t_2 to be 7.8125 ms, timer 2's total division ratio must be:

$$N_{TOTAL} = \frac{t_2 \bullet F_{OSC}}{4} = \frac{7.8125 \times 10^{-3} \bullet 20 \times 10^6}{4} = 39063$$

N_{TOTAL} is the product $N_{PRE} * N_{POST} * N_{PERIOD}$

We'll set N_{PRE} and N_{POST} as 16. Hence, we determine N_{PERIOD}:

$$N_{PERIOD} = \frac{39063}{16 \bullet 16} = 152.59$$

Since **Period** must be a byte integer, we'll set **Period** at 153.

But, can we do better? Let's look at all legitimate combinations of N_{POST} and **Period**:

N_{POST}	Required Period	Integer Period	Resulting Interrupt Period (ms)	Error
1	2441.4063	2441	7.811	0.017%
…				
9	271.26736	271	7.805	0.099%
10	244.14063	244	7.808	0.058%
11	221.94602	221	7.779	0.426%
12	203.45052	203	7.795	0.221%
13	187.80048	187	7.779	0.426%
14	174.38616	174	7.795	0.221%
15	162.76042	162	7.776	0.467%
16	152.58789	152	7.782	0.385%

Since **Period** is a byte value it cannot exceed 255, which means that N_{POST} can't be less than 10, so I've omitted most of N_{POST} values resulting in illegal **Period** values. (The table calculates the zero error, floating point value for **Period**, and then determines the corresponding integer value. The error calculations are based upon the integer value.)

Our initial selection $N_{POST} = 16$ and **Period** = 152 isn't the best combination. We can reduce the timing error from 0.385% to 0.058%, by setting N_{POST} to 10 and **Period** to 244. The corresponding SetTmr2 statement is:

```
        SetTmr2 T2Post10 + T2Pre16, 244
```

For flashing an attention getting LED, this error improvement is immaterial, but many tasks demand greater accuracy.

Let's see how we use timer 1.

Our object is to briefly flash an LED 10 times a second, corresponding to an interrupt interval of 100 ms. We'll use $F_{OSC}/4$ as Timer 1's clock source, so each tick is 0.2 μs. We calculate the required division ratio as 100 ms/0.2 μs, or 500000:1. Since timer 1 has a 16-bit counter, its maximum division ratio is 65535:1, so we must use its prescaler to achieve 500000:1.

If we set timer 1's prescaler for 8:1 division, the required counter division ratio is 500000/8, or 62500. Setting timer 1's counter division ratio requires a different process than used with timer 2. We preload timer 1 with a starting value and timer 1 counts up from that value until it rolls over to $0000 and triggers the **Tmr1Int** interrupt. Unfortunately, the preload is lost with every rollover and must be re-entered after every interrupt. As we learned earlier, in MBasic we can both read and write the timer 1's 16-bit counter register, using the library declared byte variables **Tmr1H** (high byte) and **Tmr1L** (low byte). Hence, to preload the counter with the word preload value **PreLoad**, we set **Tmr1H=PreLoad.HighByte** and **Tmr1L=PreLoad.LowByte**.

The required preload value for timer 1 is thus:

$$N_{PRELOAD} = 65536 - N_{DIVISION}$$

$N_{PRELOAD}$ is the value to be preloaded into the timer 1's counter section
$N_{DIVISION}$ is the desired division ratio for timer 1's counter.

Since our desired division ratio is 62500, we calculate the preload value as 65536-62500, or 3036.

Let's look at the timer 1 related parts of Program 10-4.

```
SetTmr1 Tmr1Int8
;PreLoad based on measured duration
PreLoad = 3072
OnInterrupt Tmr1Int, Flash1
```

The ISR is **Flash1**:

```
Flash1 ;ISR for Timer 1
;------
        ;have to reload every time
        Tmr1on = Stopped
        Tmr1H = PreLoad.HighByte
        Tmr1L = PreLoad.LowByte
        Tmr1on = Enabled

        High A0
        Pauseus 300
        Low A0
    Resume
```

Why use a preload of 3072 when we calculated 3036 as the correct value? Look at **Flash1**. The first action stops timer 1 and reloads the counter registers **Tmr1H** and **Tmr1L** with the preload value. Then, we restart timer 1. These statements require time to execute, and during the execution time timer 1 is stopped. Hence, if we use our calculated preload value 3036, timer 1 will run slow and the resulting period will be longer than desired.

With a preload value of 3036, I measured the resulting period as 100.0693 ms. To arrive at the final value 3072, I kept increasing the preload value until the period was as close to 100 ms as possible. In Program 10-4, a preload of 3072 results in a measured period of 99.9980 ms. The difference between 3072 and 3036 corresponds to 57.6 μs. Some of the difference between our theoretical 3036 and the adjusted value may well correspond to error in the 20 MHz resonator in my 2840 Development Board.

What happens when interrupt requests collide? As we learned in Program 10-3, in MBasic an interrupt is only executed in between MBasic statements. However, since both **Flash1** and **Flash2** are compiled after the **Disable** compiler directive, during the execution of one ISR, the second interrupt is postponed.

Capture and Compare

The final interrupt procedures we examine are those related to the capture and compare functions. What are the capture and compare functions? By the way, "capture" and "compare" are two separate functions, not one function called "capture and compare."

These functions are part of the PIC's "CCP" hardware module, where the acronym CCP stands for "capture, compare and pulse width modulation functions. We'll see how the PWM module works in later chapters. The 16F87x/87xA series PICs have two CCP modules, CCP1 and CCP2, and either, or both, may be assigned to capture or compare functions. However, should you activate both CCP modules, be aware that the modules interact, a subject beyond the scope of an introductory text such as this one. More information is available in Microchip's 16F87x and 16F87xA data sheets, and in its Midrange Reference Manual. Module. CCP1 is tied to pin C2, whilE module CCP2 is tied to pin C1.

Capture—Capture mode "captures" or copies timer 1's counter value into the capture register **CCPR1** or **CCPR2** upon the occurrence of change in input pin state. Both are 16-bit registers comprised of two 8-bit halves, **CCPRnL** and **CCPRnH**, for the low and high bytes, respectively, where n is 1 or 2. The possible event options are:
- High to low transition
- Low to high transition
- Fourth low to high transition
- 16th low to high transition

When the selected event occurs, timer 1's value is copied and an interrupt is issued, **CCP1INT** if CCP module 1 is used or **CCP2INT** if CCP module 2 is used.

Compare—Compare mode makes timer 1 work something like timer 2; the value of timer 1's counter is continuously compared with the value loaded in either **CCPR1** or **CCPR2**. When the counter value matches the **CCPR1** or **CCPR2** value, one of four optional events occur:
- Pin C1 (CCP module 2) or C2 (CCP module 1) is driven high
- Pin C1 (CCP module 2) or C2 (CCP module 1) is driven low
- Pin C1 (CCP module 2) or C2 (CCP module 1) is unchanged
- Pin C1 (CCP module 2) or C2 (CCP module 1) is unchanged and timer 1 is reset to zero. If CCP module 2 is active, and if the analog-to-digital module is active, an A/D conversion is started.

And upon the match occurring, an interrupt is issued, **CCP1INT** if CCP module 1 is used or **CCP2INT** if CCP module 2 is used.

Capture

How might we use the capture interrupt? One possibility is to measure frequency or period. Suppose we have a pulse waveform connected to a CCP module's input pin and we set it to capture timer 1's value every positive transition:

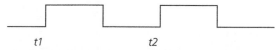

Timer values t1 and t2 permit us to calculate the waveform's period (t2 – t1) and its frequency, 1/(t2 – t1). We obviously must be concerned with timer 1 overflow when we compute t2 – t1.

Program 10-5

Program 10-5 reads the period and frequency of a square wave input signal applied to pin C1. Figure 10-4 shows the connection arrangement for Program 10-5.

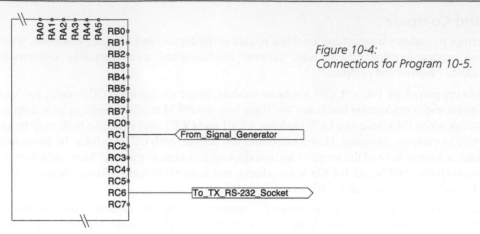

Figure 10-4:
Connections for Program 10-5.

Our approach in Program 10-5 may be summarized in pseudo-code as:

```
Set up Capture interrupt
Initialize timer 1 & select timer 1's prescaler
I = 0

Main Dummy Procedure

Interrupt Handler
        I = i+1
        If i is even then
                Read timer 1 and save in t1
        If i is odd then
                read timer 1 and save in t2
                gosub OutputProcedure
Resume
OutputProcedure
        Calculate t2-t1
        Adjust if t2-t1 is negative
        Period = timer 1 interval * (t2-t1)
        Frequency = 1/period
        Output frequency and period to serial port
Return
```

Here' the code implementing this pseudo-code.

```
;Program 10-5
;Capture operation

;Constants
;----------
UseCCP1         Con     %0      ;Pin C2
UseCCP2         Con     %1      ;Pin C1
usPerTick       Con      1.600

;Variables
;----------
t1              Var     Word
t2              Var     Word
i               Var     Byte
t1f             Var     Float
t2f             Var     Float
fPeriod         Var     Float
fFreq           Var     Float


;Initialization
;--------------
EnableHSerial
SetHSerial H115200
```

```
;set up for capture on pin C1
SetCapture UseCCP2,Capture1L2H
OnInterrupt CCP2Int,DumpPeriod

;turn timer 1 on
SetTmr1 Tmr1Int8
i = 0

Enable

Main
GoTo Main

Disable CCP2Int

DumpPeriod
;----------
        i = i+1
        If i.bit0 Then
                GetCapture USeCCP2,t1
        Else
                GetCapture USeCCP2,t2
                GoSub WriteOutput
        EndIf
Resume

WriteOutput
;-------------
        t1f = ToFloat t1
        t2f = ToFloat t2

        If (t2 < t1) Then
                t2f = t2f + 65536.0
        EndIf
        fPeriod = t2f - t1f

        HSerOut [Dec t1,9,Dec t2,9,Real FPeriod,9]
        fPeriod = fPeriod * usPerTick
        fFreq = 1000000.0 / fPeriod
        HSerOut [Real fPeriod," us",9,Real fFreq," Hz",13]

        Pause 1000
Return
End
```

We add three new constant definitions. The **UseCCP1** and **UseCCP2** constants select which CCP module is selected. The constant **usPerTick** is the duration in microseconds for each clock pulse, based on an 8:1 prescale ratio. With version 5.3.0.0 onward, we may define a floating point constant directly. (Earlier versions required the "FLOAT" prefix.).

```
;Constants
;----------
UseCCP1         Con     %0      ;Pin C2
UseCCP2         Con     %1      ;Pin C1
usPerTick       Con     1.600


;Initialization
;--------------
SetHSerial H115200

;set up for capture on pin C1
SetCapture UseCCP2,Capture1L2H
```

We initialize the capture interrupt using MBasic's **SetCapture** function:

```
SetCapture CCPPin, mode
```

CCPPin—is a constant or variable selecting which of the two CCP modules will be used. If **CCPPin = %0**, CCP module 1 (connected to pin C2 in the 16F87x/16F87xA series devices) will be used. If **CCPPin = %1**, CCP module 2 (connected to pin C1 in the 16F87x/16F87xA series devices) will be used. To make remembering module selection easier, we defined the two constants **UseCCP1** and **UseCCP2**.

Mode—is a variable or a constant defining which of five modes will be operational. MBasic's library mode constants are:

Library Constant Name	Functionality
CaptureOff	Disable capture module
Capture1H2L	Capture timer 1 on each high-to-low transition
Capture1L2H	Capture timer 1 on each low-to-high transition
Capture4L2H	Capture timer 1 every fourth low-to-high transition
Capture16L2H	Capture timer 1 every 16th low-to-high transition

I've arbitrarily used CCP module 2 and selected capture on every low-to-high transition.

```
OnInterrupt CCP2Int,DumpPeriod
```

We link the interrupt handler **DumpPeriod** to the **CCP2Int** using MBasic's **OnInterrupt** procedure.

```
;turn timer 1 on
SetTmr1 Tmr1Int8
```

In order to use the capture function, timer 1 must be turned on, which may be conveniently accomplished with MBasic's **SetTmr1** procedure. I've set timer 1's prescaler to 8:1. Why 8:1? As we learned earlier, with the prescaler set 8:1, we can measure up to 104 ms, corresponding to an input frequency just below 10 Hz. Based on using only MBasic for our interrupt handler, we expect execution time problems will limit our high frequency response, so it makes sense to concentrate on measuring lower frequencies.

```
i = 0
Enable
```

We use **i** as an odd/even determinant and initialize it to 0.

```
Main
GoTo Main
Disable CCP2Int
```

The main program loop is an empty **GoTo** statement.

Let's now look at our interrupt handler.

```
DumpPeriod
;----------
        i = i+1
        If i.bit0=0 Then
                GetCapture UseCCP2,t1
        Else
                GetCapture UseCCP2,t2
                GoSub WriteOutput
        EndIf
Resume
```

We use an **If i.bit0 Then** test to alternate between capturing data to **t1** and **t2**. Since **i** increases by one each time **DumpPeriod** is called, **i.bit0** alternates between **%0** and **%1**. By alternating between **t1** and **t2**, we capture one complete cycle of the incoming waveform. Upon the second capture, we call subroutine **WriteOutput** to compute the period and frequency of the waveform and write the result to the serial port.

```
WriteOutput
;--------------
        t1f = ToFloat t1
        t2f = ToFloat t2
```

We convert the two timer readings to floating point variables **t1f** and **t2f**.

```
If (t2 < t1) Then
        t2f = t2f + 65536.0
EndIf
```

It's possible that timer 1 has rolled over between the two captures. If so, the second timer value, **t2**, may be less than the first sample, **t1**. If this is the case, their floating point difference (**t2f-t1f**) will be negative. The rollover is functionally equivalent to carrying 65536 to the second timer value.

```
fPeriod = t2f - t1f
```

The period, measured in terms of counter ticks, is the difference between the second timer value and the first, or in terms of the floating point values, **t1f - t2f**.

```
HSerOut [Dec t1,9,Dec t2,9,Real FPeriod,9]
```

Next, we write the two timer readings and the difference to the serial port.

```
fPeriod = fPeriod * usPerTick
```

We convert **fperiod** from timer ticks to microseconds by multiplying by the number of microseconds per tick, held in the constant **usPerTick**.

```
fFreq = 1000000.0 / fPeriod
```

Frequency is defined as the reciprocal of the period, so we compute the floating point frequency **fFreq** as **1/fPeriod**. However, since **fPeriod** is in microseconds, we must multiply by 10^6 to yield the frequency in Hz.

```
HSerOut [Real fPeriod," us",9,Real fFreq," Hz",13]
Pause 1000
Return
```

Lastly, we write the period and frequency to the serial port. The **Pause 1000** limits updates to once every second and this value may be reduced should you desire more frequent updates.

With approximately a 10 Hz input signal from a function generator, we see a reasonably stable result.

3573	64940	61367.0000000000	98187.1875000000 µs	10.1846275329 Hz
60778	56609	61367.0000000000	98187.1875000000 µs	10.1846275329 Hz
52444	48275	61367.0000000000	98187.1875000000 µs	10.1846275329 Hz
44094	39923	61365.0000000000	98183.9921875000 µs	10.1849594116 Hz
35763	31591	61364.0000000000	98182.3906250000 µs	10.1851253509 Hz
27415	23245	61366.0000000000	98185.5937500000 µs	10.1847934722 Hz

I found that Program 10-5 worked well between 10 Hz and 125 Hz or so. We are constrained on the low frequency side by the counter length and our preselector ratio choice. Above 125 Hz, there isn't enough time for the interrupt and output procedures to complete their tasks and the results fail. It would be possible to measure higher input frequencies by capturing not every input pulse, but one for every four input cycles, or even one for every 16 input cycles, using the **SetCapture** options **Capture4L2H** or **Capture16L2H** mode library constants.

Ultimately, however, our upper frequency limit is constrained by the speed of program execution and the lower frequency limit by counter/preselector count maximum values.

You may be thinking "why not add code to the ISR to reset timer 1 to zero at the time of the t1 interrupt, thereby removing the t2 – t1 subtraction and its associated carry test"? Good question.

The capture function is executed independently, by dedicated hardware in the 16F87x/87xA device. Thus, without regard for whatever software may be executing, upon the occurrence of the defined trigger event timer 1's counter value is instantaneously captured into an internal register pair. At the same time, the capture interrupt flag is set. But MBasic will not act upon the interrupt flag and jump to the ISR until the current MBasic instruction completes. But any code to zero timer 1 cannot not execute until the ISR runs, so it is subject to the latency problem we've previously discussed. Consequently, resetting timer 1 to zero in the ISR to remove the t2-t1 computation introduces an unknown—and variable—time error compared with using the approach of Program 10-5. The function **GetCapture** reads the saved timer 1 value and is not time sensitive, so the latency issues related to Program 10-5 center around whether the second triggering event occurs before the first event processing completes.

However, we'll see in Chapter 15 that it's possible to remove almost all latency issues by writing the ISR in assembler using MBasic's **ISRASM** feature. If you choose to handle the ISR in assembler, zeroing timer 1 upon the first measurement is feasible, but still unnecessary.

Compare

The final interrupt we examine is tied to the "compare" element of the capture, compare and pulse width modulation module.

The compare function continuously monitors timer 1's counter value against the 16-bit value held in the **CCPRx** register. Each CCP module has one register, so it's possible to monitor timer 1's counter value against two different values, one held in CCP1's register and one held in CCP2's register. Upon a match, one of four possible user-selected events takes place:

- Pin C1 (CCP module 2) or C2 (CCP module 1) is driven high
- Pin C1 (CCP module 2) or C2 (CCP module 1) is driven low
- Pin C1 (CCP module 2) or C2 (CCP module 1) is unchanged
- Pin C1 (CCP module 2) or C2 (CCP module 1) is unchanged and timer 1 is reset to 0. Additionally, if CCP2 is active, and the analog-to-digital conversion module is enabled, an A/D conversion is initiated.

In all four cases, upon the match, a **CCPxInt** interrupt issues—**CCP1Int**, or **CCP2Int**, depending upon which CCP module is being used.

MBasic supports the compare operation with the function **SetCompare**, called with three arguments:

```
SetCompare CCPPin,Mode,CompareValue
```

CCPPin—is a constant or variable selecting which of the two CCP modules will be used. If **CCPPin = %0**, CCP module 1 (connected to pin C2 in the 16F87x/16F87xA series devices) will be used. If **CCPPin = %1**, CCP module 2 (connected to pin C1 in the 16F87x/16F87xA series devices) will be used. To make remembering module selection easier, we will define two corresponding constants **UseCCP1** and **UseCCP2**.

Mode—a constant or variable selecting the operational mode. Five library constants are available:

Library Constant Name	Functionality
CompareOff	*Disable compare module.*
CompareSetHigh	*Sets CCPx pin low upon match.*
CompareSetLow	*Sets CPPx pin high upon match.*
CompareInt	*Sets interrupt upon match, but no change in CCPx pin.*
CompareSpecial	*Reset timer 1 to 0 and if CCP2 is active and A/D is enabled, an A/D conversion is started.*

CompareValue—is a 16-bit (word) length constant or variable against which timer 1 is compared. In MBasic version 05.2.1.1 **SetCompare** has an error that reverses the byte order. The following "work around" may be employed until a compiler fix is available:

```
SetCompare CCPPin,Mode,CompareValue
CCPR1H = CompareValue.HighByte
CCPR1L = CompareValue.LowByte
```

Program 10-6

Let's see how we might use the compare feature. Configure your 2840 Development Board as shown in Figure 10-5.

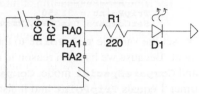

Figure 10-5: Circuit for Program 10-7.

Program 10-6 will use the interrupt upon compare feature to flash an LED. But, more importantly, Program 10-6 shows how we may use timer 1 to provide the same interrupt upon preset value functionality as we observed with timer 2. The advantage of using the 16-bit timer 1, instead of the 8-bit timer 2, of course is the far greater resolution timer 1 offers.

```
;Program 10-6
;Compare test

;Constants
;----------
UseCCP1         Con      %0      ;Pin C2
UseCCP2         Con      %1      ;Pin C1

;Variables
;---------
TripPoint       Var      Word


;Initialization
;--------------
TripPoint = 31250
SetCompare UseCCP1,CompareSpecial,TripPoint

;link interrupt with ISR
OnInterrupt CCP1Int, Flash1

;turn timer 1 on
SetTmr1 Tmr1Int8

Enable CCP1Int

Main
;-----
GoTo Main

Disable CCP1Int

Flash1 ;ISR for Timer 1
;------
        High A0
        Pauseus 300
        Low A0
Resume

End
```

We'll hold the compare value in the word variable **TripPoint**.

```
;Variables
;---------
TripPoint       Var      Word
```

Let's set the LED to flash 20 times/second, or one flash every 50 ms. Based on our earlier work, we know that durations in this range require us to set timer 1's preselector at 8:1. With a 20 MHz resonator, each tick is 1.6 µs. We now may calculate the count required for 50 ms:

$$N_{COMPARE} = \frac{50 \times 10^{-3}}{1.6 \times 10^{-6}} = 31250$$

```
;Initialization
;--------------
TripPoint = 31250
SetCompare UseCCP1,CompareSpecial,TripPoint
```

We set the variable **TripPoint** to the desired count, and use it in the **SetCompare CompareValue** argument. Because we have no reason to change the state of either pin C1 or C2, we will use the CCP1 module and **CompareSpecial** mode. **CompareSpecial** gives us what we desire; a **CCP1Int** interrupt whenever timer 1 equals **TripPoint** and it also automatically resets timer 1 to zero.

```
;link interrupt with ISR
OnInterrupt CCP1Int, Flash1
```

We associate the ISR, **Flash1**, with the **CCP1Int** interrupt using **OnInterrupt**.

```
;turn timer 1 on
SetTmr1 Tmr1Int8
Enable CCP1Int
```

Of course, don't forget to set up timer 1 and enable the interrupt. As decided, we activate timer 1 with 8:1 prescale.

```
Main
;-----
GoTo Main
```

Main is a dummy loop.

```
Disable CCP1Int
```

Remember to disable the interrupt before the interrupt handler. In this program, we only use the **CCP1Int** interrupt, so we need only disable it. This may not be the case in your programs, so disable any applicable interrupts. Be careful in this regard, as some MBasic procedures, such as **HSerOut** and **HSerIn** use interrupts and a blanket **Disable** statement may prevent these functions from executing.

```
Flash1 ;ISR for Timer 1
;------
        High A0
        Pauseus 300
        Low A0
Resume
```

Flash1 is identical with our earlier "flash the LED" interrupt handler.

How does it work? I found the period jumped between two values—49.9352 ms and 49.9142 ms, representing 65 µs and 86 µs short of the desired 50.000 ms period, respectively. (The period jumps between two values depending on where in the **Main...GoTo Main** cycle the interrupt fires.) Based on these measurements, **TripPoint**, should be increased by 47 to 31297. (This value is based upon an average error of 75 µs, corresponding to 47 ticks of 1.6 µs/tick.) As an experiment, I changed **TripPoint** to 31297 and observed the period jumping between 49.9981 ms and 50.0192 ms.

References

[10-1] A complete data sheet for most PICs comprises two elements; (a) a detailed "family" reference manual and (b) the particular device datasheet. MBasic supports only PICs from Microchip's "midrange" family and the associated PICmicro™ Mid-Range MCU Family Reference Manual may be downloaded at http://www.microchip.com/download/lit/suppdoc/refernce/midrange/33023a.pdf. This is a 688-page document, in almost mind numbing detail, but nonetheless is an essential reference to a complete understanding of PICs. For individual PIC family member datasheets, the easiest source is to go to http://www.microchip.com/1010/pline/picmicro/index.htm and select either the PIC12 or PIC16 group and from that link then select the individual PIC device. Of particular interest are the PIC16F87x and PIC16F87xA data sheets.

Analog-to-Digital Conversion

The world is analog, not digital, at least at scales perceptible by human senses. If you are asked the temperature, you might respond "it's 77 degrees." However the temperature doesn't abruptly jump from 77 to 78 degrees. Rather, with a sufficiently accurate thermometer, you might have said "it's 77.123 degrees," to which the questioner might respond, "no, according to my higher precision thermometer, it's 77.12345 degrees." In theory, until we start to reach quantum uncertainties, there is no limit to the precision with which we may state the temperature, or a voltage reading, or many other parameters human ingenuity is able to measure.

Programs you write using MBasic, however, deal with bits and bytes; bit, byte, word and long variables and constants are in discrete steps. A bit must be 0 or 1, not somewhere between 0 and 1. A byte variable can't have the value 128.3; it can be 128 or 129, but not something between. Even if we use MBasic's 32-bit floating point arithmetic package, looked at in sufficient detail we find it also jumps between finite, discrete steps, albeit small ones.

This chapter examines how MBasic translates between the analog and digital worlds in the analog-to-digital direction. We'll see in Chapter 16 how to accomplish the reverse process—digital-to-analog conversion.

We'll assume in this chapter that input signals to our A/D converter change slowly so that we may neglect anti-aliasing filters and other sampled system considerations. However, we'll touch upon Nyquist sampling rate limits in conjunction with Chapter 16 and the same concepts apply to A/D conversion. If you wish to learn more about digital sampling rates and frequency response, Reference [16-4] offers a very readable introduction.

Introduction to Analog-to-Digital Conversion

Suppose we have a crude A/D converter that has eight possible output values, with one digital output step per input volt. Thus, the possible output values are 0 volts, 1 volt…7 volts. Figure 11-1 illustrates this conversion process. Although our analog input voltage may assume any value—for example 3.123456789 volts—the A/D converter

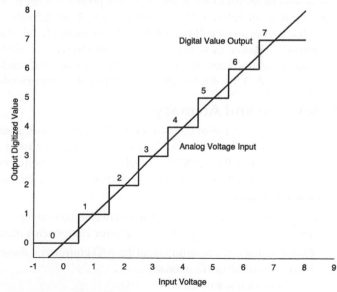

Figure 11-1: The A/D process.

forces this input value into one of the eight possible outputs, three in our example. This converter will report 3.4999 volts as three, and 3.5001 volts as four. In reality, the transition will have some jitter on either side of 3.5 volts. And, in a real A/D converter the transition point will not necessarily be the same for each voltage step. For example, the transition point between three and four may occur at 3.5200 volts, while the transition between four and five may occur at 4.4712 volts.

We may ask how does the A/D converter know that each step is one volt? Or, for that matter, how does it determine what voltage input corresponds to a zero output? The answer is found in the A/D's two voltage references; V_{REF+}, the positive reference voltage and V_{REF-}, the negative reference voltage. The voltage per step, V_{STEP} is simply:

$$V_{STEP} = \frac{\left(V_{REF+} - V_{REF-}\right)}{2^N}$$

N is the number of bits returned by the A/D converter.

Let's see how this works with our 3-bit converter. $2^3 = 8$ and we want 1 volt/step. We also will assume that V_{REF-} will be ground, or 0 volts. Hence, V_{REF+} must be 8 volts. There is a subtlety here, though. The conversion has only seven steps—0 to 1 volts, 1 to 2 volts … and 6 to 7 volts. So, it might appear that the reference voltage should be seven volts, not eight. Indeed, some A/D designs would require a 7 volt reference to yield 1 volt/step with a 3-bit converter—that is, the divisor in the equation would be $2^N - 1$ instead of 2^N. However, Microchip's A/D design requires 2^N as the divisor.

What about the negative reference? In most cases, we wish to measure a voltage with respect to a common reference point, generally referred to as "ground" or V_{SS} in our 2840 Development Board. V_{REF-} in this case will be that common reference point. But, there are excellent reasons to make the V_{REF-} connection at a point separate from the digital circuitry ground plane, primarily to avoid contaminating the input signal with the noise found on the digital ground plane. For example, a digital voltmeter may have the V_{REF-} point the negative binding post on the instrument front panel. See References [11-6], [11-7] and [11-8].

It may also be useful to work with V_{REF-} at a potential other than ground. Suppose the input signal to be digitized is never below 1.25 volts and is never above 2.5 volts. Simply setting V_{REF-} to 0 volts and V_{REF+} to 2.5 volts, throws away half the potential digital range. In this case, setting V_{REF-} to 1.25 volts will double the resolution. However, there may well be reasons to set V_{REF-} to 0 and to set V_{REF+} to something greater than 2.5 volts, accepting some resolution loss. Perhaps an error condition exists if the input drops below 1.25 volts or exceeds 2.5 volts, which may be detected via an extended measurement range.

Resolution and Accuracy

Resolution is a measure of the number of different values that are detectable by the A/D converter. For example, the 16F87x's 10-bit A/D converter has a resolution of 2^{10}, or 1024 possible output values. (Chapter 1 summarizes the A/D capabilities of selected PICs supported by MBasic.) Used with a positive reference voltage of 4.096 volts and a negative reference of zero volts, the voltage resolution is 4.096 volts/1024 steps, or 4.0 mV per step.

Accuracy is a measure of the difference between the true input voltage and the value reported by the A/D converter. Accuracy is limited by several error components including:

- Quantization error resulting from the A/D converter's finite resolution
- Errors within the A/D, such as:
  - o Integral linearity error
  - o Differential linearity error

- o Offset error
- o Gain error
- o Errors external to the A/D converter
- o Noise on the input signal or on the reference voltages
- o Errors in the reference voltages
- o Inadequate settling time
- o Excessive source impedance
- o Errors in any signal conditioning circuits

Estimating the overall accuracy of the A/D conversion process requires careful consideration of these error sources and others, both in magnitude and direction–that is, are the possible errors cumulative or offsetting? The error time frame must also be considered; are these short-term errors that may affect one measurement differently from the next, or are they long-term errors to which all measurements are more or less equally subject?

Ignoring those errors sources dependent upon the circuit layout and construction techniques and external factors, we can quantify several possible error sources in our design:

- Our design will use a widely available voltage reference device, a Linear Technology LT1634CCZ-4.096 voltage reference, with a specified error of ±0.2% over the temperature range 0...70°C. It is buffered by a Microchip MCP601 op-amp with ±2 mV offset voltage over the same temperature range. Since the MCP601 is configured as a unity gain buffer, this offset must be added directly to any error in the voltage reference itself. An error of 0.2% in the reference represents 8.2 mV, for a total error budget of ±10.2 mV, or ±0.25% of the 4.096 source.

Microchip's 16F87x data sheet provides the following error budget:

- Integral linearity error <±1 LSB
- Differential linearity error <±1 LSB
- Offset error <±2 LSB
- Gain error <±1 LSB

A worst-case estimate assumes all the internal A/D errors add in the same direction, yielding ±5 LSB. However, Microchip's *Midrange Reference Manual* claims the absolute error is less than ±1 LSB when V_{REF+} equals V_{DD}, noting that accuracy will "degrade as V_{DD} diverges from V_{REF+}." (Internal A/D errors are specified in terms of how many least significant bits, or LSB, may be in error. In our design, one LSB is 4 mV.) Our design has V_{REF+} within 1 volt of V_{DD}, so we are left in the uncomfortable position of wondering exactly how much the accuracy will degrade as a result of the 1 volt difference. However, spot measurements of the breadboard design against a Fluke 189 digital voltmeter rated at ±0.025% accuracy show that ±1 LSB remains a reasonable internal error figure for our 4.096 V reference.

Finally, there is a ±½ LSB quantization error inherent in the digitization process—even if everything else functions with zero error, in the worst case the analog input voltage may be ½ LSB below the reading, or ½ LSB above it. Our A/D measurement therefore has a possible error of ±1.5 LSB (6 mV), representing about ±0.2% of the reading.

Suppose the A/D output is 500 counts, corresponding to 2.000 volts. What is the possible error? Our voltage reference error budget contribution is 0.25% × 2.0000 volts = 5 mV. To this must we must add internal A/D errors of 4 mV (1 LSB) and inherent quantization error and quantization error corresponding to 2 mV (½ LSB). The total worst-case error is thus ±11 mV—although the A/D converter reports a value that corresponds to 2.000 volts, all we can truthfully say is that the actual input is likely to be between 1.989 volts and 2.011 volts.

MBasic's A/D Read Function—ADIN

MBasic provides two procedures to read an analog input, **ADIN** and **ADIN16**. Both procedures have identical calling sequences, so we'll look at **ADIN** first.

```
ADIN [PinIn],[ClockConstant],[SetupConstant],[ReadVariable]
```

These constants interact, and setting their values isn't always intuitive, so let's look at them in detail. For clarity, we'll discuss the parameters in a different order than they are used in **ADIN**.

Setup constant—the setup constant governs four A/D settings, which we'll identify as a…d:

(a-c) Internal Selection Switches (3 switches)

Figure 11-2 is a simplified version of the internal "wiring" within the 16F87x. (Similar information for other PICs appears in their data sheets.) In essence, we have three "switches" to set; (a) one to select internal or external positive reference voltage V_{REF+}, (b) one to select internal or external negative reference voltage V_{REF-} and (c) a multiposition selector switch to determine which of several possible pins connects to the A/D converter module input. But, before we connect a pin to the A/D converter, the pin must first be configured to be an analog input, instead of the default digital input or output.

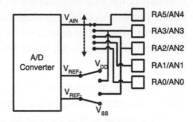

Figure 11-2: Simplified analog switching internals 16F876.

(d) Data Justification

ADIN returns a 10-bit result in **ReadValue** either left or right justified in a word variable. Suppose the A/D result is the 10-bit value (decimal) 750, or binary %1011101110. Our word variable is 16 bits wide, so this may be stored as right-justified with six leading 0's as %0000001011101110 or left justified with six trailing zeros as %1011101110000000. The later effectively multiplies the reading by 2^6 or 64.

Left Justified and Left Justified 10-Bit Values

High Byte								Low Byte								Just.	Hex	Decimal
0	0	0	0	0	0	1	0	1	1	1	0	1	1	1	0	Right	$2EE	750
1	0	1	1	1	0	1	1	1	0	0	0	0	0	0	0	Left	$BB80	48000

Almost always we will wish the 10-bit result to be right justified. Why have left justification? Suppose we need only an 8-bit A/D conversion. To obtain an 8-bit answer from the 10-bit returned value requires discarding the two least significant conversion bits and retaining the eight most significant bits, or the leftmost eight bits. This is easily accomplished by using the high byte of a left justified result, for example, **ReadValue.Byte1**.

Putting it All Together

Let's see how these parameters may be set. The table below presents the possible configuration values for both the 16F877 and 16F876 PICs (including "A" suffixed devices). The 16F876 has five possible analog input pins AN0…AN4, while the 16F877 has 8 possible analog inputs AN0…AN7. *Note that analog input AN4 corresponds to RA5, not to RA4.* (RA4's second function is to serve as the external input to the Timer 0 clock.) The 16F877's analog input pins AN5, AN6 and AN7 are found on inputs RE0, RE1 and RE2, respectively.

A/D Set-up Constants for 16F876/16F877/16F876A/16F877A

Hex for Left Just.	Hex for Right Just.	AN7 RE2	AN6 RE1	AN5 RE0	AN4 RA5	AN3 RA3	AN2 RA2	AN1 RA1	AN0 RA0	Vref+	Vref–	Chan/ Refs
										Volt Refs		
00	80	A	A	A	A	A	A	A	A	Vdd	Vss	8/0
01	81	A	A	A	A	Vref+	A	A	A	RA3	Vss	7/1
02	82	D	D	D	A	A	A	A	A	Vdd	Vss	5/0
03	83	D	D	D	A	Vref+	A	A	A	RA3	Vss	4/1
04	84	D	D	D	D	A	D	A	A	Vdd	Vss	3/0
05	85	D	D	D	D	Vref+	D	A	A	RA3	Vss	2/1
6/7	86/87	D	D	D	D	D	D	D	D	Vdd	Vss	0/0
08	88	A	A	A	A	Vref+	Vref–	A	A	RA3	RA2	6/2
09	89	D	D	A	A	A	A	A	A	Vdd	Vss	6/0
0A	8A	D	D	A	A	Vref+	A	A	A	RA3	Vss	5/1
0B	8B	D	D	A	A	Vref+	Vref–	A	A	RA3	RA2	4/2
0C	8C	D	D	D	A	Vref+	Vref–	A	A	RA3	RA2	3/2
0D	8D	D	D	D	D	Vref+	Vref–	A	A	RA3	RA2	2/2
0E	8E	D	D	D	D	D	D	D	A	Vdd	Vss	1/0
0F	8F	D	D	D	D	Vref+	Vref–	D	A	RA3	RA2	1/2

Shaded cells apply only to 16F877.

We'll start with a simple setup—V_{DD} as V_{REF+}, V_{SS} as V_{REF-}, analog voltage to be measured on pin A0, with right justified results. Selecting the conversion constant $8E meets our design requirements.

Clock constant—The A/D converter requires a certain minimum time to convert the analog input voltage to a digital value. Microchip divides the minimum time into two periods:

1. The "acquisition time," T_{ACQ}. During the acquisition time, the sample switch closes and the input voltage charges the A/D's internal 120 pF capacitance C_{HOLD}. T_{ACQ} is not user settable in MBasic.
2. The "analog-to-digital" conversion time, T_{AD}. During the conversion time, the sample switch opens and the stored voltage on C_{HOLD} is converted to a digital reading. The complete 10-digit A/D conversion requires 12 T_{AD} periods.

Starts when A/D input channel selected	Starts when A/D conversion initiated
Acquisition Time T_{ACQ}	*12 X A/D Conversion Time T_{AD}*
Hard-coded into the MBasic compiler—not settable by ADIN procedure	Define T_{AD} by selecting the Clock Constant in the ADIN procedure

ADIN and **ADIN16** require setting T_{AD} via the clock constant parameter, which has four possible values:

Clock Constant Selection

Clock Constant	Operation (Divisor)	Maximum PIC Frequency
0	$2T_{OSC}$	1.25 MHz
1	$8T_{OSC}$	5 MHz
2	$32T_{OSC}$	20 MHz
3	RC Clock	See Text

The minimum T_{AD} recommended by Microchip is 1.6 µs. With a 20 MHz clock, a divisor of 32 is necessary to yield 1.6 µs. The complete 10-bit conversion time would thus be 12 * 1.6 µs or 19.2 µs. **ADIN** executes in approximately 275 µs on a 16F877 with a 20 MHz clock. The difference between this time and the theoretical 19.2 µs (plus an acquisition time T_{ACQ} of typically 10–15 µS) is consumed partly in acquisition time but mostly in compiler overhead.

Regardless of the PIC's clock frequency, the 16F87x's internal RC oscillator can generate T_{AD} by setting the clock constant to three. The internal RC oscillator yields a nominal T_{AD} of 4 µs but Microchip specifies the expected duration between 2 µs to 6 µs, as component tolerance is not tightly controlled.

Input pin—The final parameter required by **ADIN** defines which pin is connected to the analog input source. (The parameter **SetupConstant** only establishes which pins may *may potentially* be used for analog input.) The **PinIn** constant may be selected from the following:

Input Pin Constant Selection

Analog Input Pin	Input Pin Constant	Notes
AN0/RA0	0	
AN1/RA1	1	
AN2/RA2	2	
AN3/RA3	3	
AN4/RA5	4	Note RA4 is not available for analog input
AN5/RE0	5	Not available on 16F876/876A
AN6/RE1	6	Not available on 16F876/876A
AN7/RE2	7	Not available on 16F876/876A

ADIN16

With version 5.3.0.0, MBasic added a new A/D input function, **ADIN16**.

```
ADIN16 [PinIn],[ClockConstant],[SetupConstant],[ReadVariable]
```

ADIN16 is identical with **ADIN**, with one difference—it conducts multiple A/D conversions and returns a quasi-16-bit result. How is it possible to get a 16-bit reading from a 10-bit converter? The answer is that by averaging multiple readings, it is possible to improve the converter's resolution. Averaging reduces some types of errors, such as noise pickup, but won't do anything to reduce other error sources, such as reference voltage error. **ADIN16** returns 65,535 possible values, so if we use a 4.096V reference, each voltage step is $4.096/2^{16}$, or 62.5 µV. You will have to decide whether **ADIN16** provides a useful improvement over **ADIN**, based upon your application and circuit layout.

Testing ADIN and ADIN16

Let's exercise **ADIN** and **ADIN16** with the simple test circuit of Figure 11-3 by reading the forward voltage across an LED with Program 11-1.

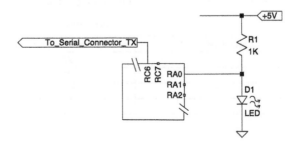

Figure 11-3: ADIN and ADIN16 test configuration.

Program 11-1

```
;Program 11-1
;Read A/D and output to serial

;Constants
;---------
Vdd            fCon              4.9508  ;Vdd used as reference
InPin          Con               A0            ;pin for input
ADSetUp Con              $8E            ;%10001110
Clk            Con               2             ;based on 20MHZ clock
VoltStep       fCon              Vdd / 1024.0 ;10 bit A/D =1024 states

;Variables
;---------
ADRaw          Var               Word
Volts          Var               Float

;Initialization
;--------------
EnableHSerial
SetHSerial H9600
Main
       ADin InPin,Clk,ADSetup,ADRaw
       HSerOut ["Raw: ",Dec ADRaw,9]
       Volts = (ToFloat ADRaw) * VoltStep
       HSerOut ["Real: ",Real6 Volts\3," V",13]
       Pause 100
GoTo Main
End
Let's look at our defined constants:
Vdd            fCon              4.9508  ;Vdd used as reference
InPin          Con               A0            ;pin for input
ADSetUp        Con               $8E           ;%10001110
Clk            Con               2             ;based on 20MHZ clock
VoltStep       fCon              Vdd / 1024.0 ;10 bit A/D =1024 states
```

For simplicity, we'll use the 2840 Development Board's +5 V regulated power supply V_{DD} as the reference voltage V_{REF+}. I measured V_{DD} as 4.9508 V on my board—your board will likely differ, but V_{DD} should be close to the nominal 5 V. We've already seen why **ADSetUp is $85**—it sets AN0 as an analog input, right justified output, $V_{REF+} = V_{DD}$ and $V_{REF.}=V_{SS}$. Likewise, because my 2840 Development Board has a 20 MHz resonator, we must select the clock constant, **CLK**, as 2 in order to assure the correct conversion time. We also define as a constant the voltage per A/D step. We do this letting the compiler perform the associated arithmetic via the division function:

```
       VoltStep          fCon Vdd / 1024.0.
```

For improved generality, I've used floating point notation in this calculation, so **VoltStep** is a floating point constant and must be defined using the **fCon** operator.

```
Main
       ADin InPin,Clk,ADSetup,ADRaw
       HSerOut ["Raw: ",Dec ADRaw,9]
       Volts = (ToFloat ADRaw) * VoltStep
       HSerOut ["Real: ",Real6 Volts\3," V",13]
       Pause 100
GoTo Main
```

The main program loop reads the analog voltage at pin AN0, and outputs both the raw reading **ADRaw** and the computed analog voltage **Volts** to the serial port. The computed analog value **Volts**, is, of course, the product of the voltage per step, **VoltStep**, and the number of steps, **Raw**. Since **Raw** is an integer, we must convert it to a floating point value via the **ToFloat** operator before using it in a floating point operation.

The result is a string of readings:

```
Raw: 451      Real: 2.180 V
Raw: 451      Real: 2.180 V
```

Raw: 451	Real: 2.180 V
Raw: 450	Real: 2.175 V
Raw: 451	Real: 2.180 V
Raw: 452	Real: 2.185 V
Raw: 451	Real: 2.180 V
Raw: 453	Real: 2.190 V
Raw: 451	Real: 2.180 V
Raw: 453	Real: 2.190 V

Using a Fluke 189 digital voltmeter with specified accuracy of ±0.025%, I measured the voltage on pin A0 as 2.1854 V, a 0.25% error from the most common output reading 2.180 V, and a close match to our error budget.

To see the difference **ADIN16** makes, modify Program 11-1 by substituting the following code for their counterparts in Program 11-1 and saving the result as Program 11-1A:

```
VoltStep        Con             Vdd / 65536.0           ;For 16 bit A/D
```

Since **ADIN16** is the functional equivalent of a 16-bit converter, the number of steps increases to 2^{16} or 65536 so we must modify **VoltStep** accordingly.

```
Main
        ADin16 InPin,Clk,ADSetup,ADRaw
        HSerOut ["Raw: ",Dec ADRaw,9]
        Volts = (ToFloat ADRaw) * VoltStep
        HSerOut ["Real: ",Real8 Volts\5," V",13]
        Pause 100
GoTo Main
```

The changes in **Main** are to substitute **ADIN16** for **ADIN** and to revise the output formatting to show more decimal places. The results are:

Raw: 28865	Real: 2.18058 V
Raw: 28870	Real: 2.18096 V
Raw: 28867	Real: 2.18073 V
Raw: 28865	Real: 2.18058 V
Raw: 28869	Real: 2.18089 V
Raw: 28853	Real: 2.17968 V
Raw: 28867	Real: 2.18073 V
Raw: 28865	Real: 2.18058 V
Raw: 28867	Real: 2.18073 V
Raw: 28868	Real: 2.18081 V
Raw: 28859	Real: 2.18013 V
Raw: 28864	Real: 2.18051 V
Raw: 28862	Real: 2.18036 V

The average of these readings is 2.180559V, 0.22% from the value I read with a Fluke 189 DVM, no real improvement over our **ADIN** 10-bit measurement.

Self-Contained DVM

Let's turn to a more ambitious project—building an automatic scaling 3-digit digital voltmeter. Figure 11-4 shows the display section of my breadboard prototype. There is not enough room on a 2840 Development Board for this project, so a second plugboard is required. Figure 11-5 shows the circuit we shall use.

The DVM uses the 7-segment LED driver circuit and software routine developed in Chapter 7, but instead of being driven by an up/down switch, the LEDs will display the voltage we measure at pin AN0. Before

Figure 11-4: 3-Digit PIC-based DVM.

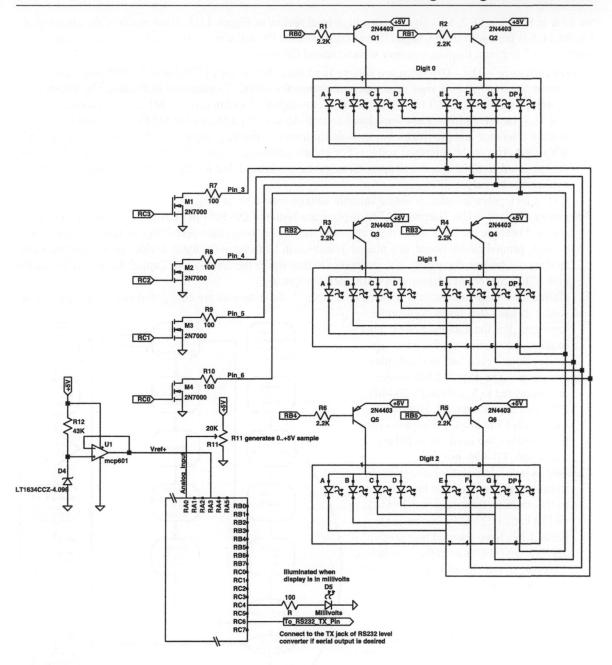

Figure 11-5: DVM schematic.

we look at Program 11-2, let's go through a high level review of Figure 11-5. Since much of the circuitry of Figure 11-5 is borrowed from Chapter 7, we'll discuss only the differences. If you are a bit hazy on how to multiplex a 7-segment display, you may wish to reread Chapter 7.

Voltage reference—The 4.096 reference voltage is a Linear Technology LT1634CCZ-4.096 precision source, accurate ±0.20% over the temperature range 0°C–70°C. To minimize drift caused by internal heating, I've biased the LT1634 at about 20 µA through a 43 k ohm resistor. Microchip recommends that V_{REF+} be driven from a low impedance source. Hence U1, a Microchip MCP601 op-amp buffers the voltage reference and provides a low impedance source to the V_{REF+} input. The MCP601 buffer adds ±2 mV offset error to the reference voltage. For greater accuracy, a Linear Technology LT1634A158-4.096 device, rated at ±0.05% and ±10 ppm/°C, might be considered, but it's three times as expensive and is surface mount only.

R11, a 20 k potentiometer, is used a variable voltage source for testing.

7-Segment displays—The 7-segment LED displays are Lumex LDS-M514RI-RA common anode devices selected because the leads are at right angles to the display body, thus permitting vertical mounting on a single printed circuit board, or a plug-in breadboard. Unfortunately, these devices have 0.050 inch pin spacing, so for breadboarding, it's convenient to first solder the display to a Capital Advanced Technologies 33117 Surfboard adapter with 0.100 inch pin spacing.

Unlike the 7-segment displays we used in Chapter 7, these devices are configured as two groups of four segments each (the seven digit stroke segments plus the decimal point). This requires the scanning software to separately address each half of the display.

Anode switch—The display LED anodes are connected to V_{DD} through 2N4403 PNP, just as in Chapter 7. However, since each LED display comprises two half-displays, we need six switching transistors, Q1–Q6, not three.

Cathode switch—In order for a segment to illuminate, its anode must be connected to V_{DD} through the associated 2N4403, and its cathode must be connected to ground. We have selected inexpensive 2N7000 MOSFETs as cathode switches. This part of our circuit is identical with that of Chapter 7.

If you reduce the current limiting resistor below 100 ohms, it may be necessary to power the 7-segment displays from a supply other than the one furnished with Basic Micro's development boards. Fortunately, 7805 regulators have built-in over temperature protection and should not be harmed by overloads.

Lumex rates the display at 150 mA maximum peak forward current with a corre-

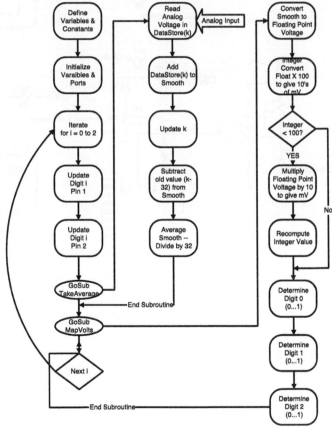

Figure 11-6: Program 11-2 structure.

sponding voltage drop of 2.5 volts. The total voltage drop across the 2N7000 and 2N4403 is approximately 300 mV. This leaves 2.2 volts across the current limiting resistor. For any desired segment current, the current limiting resistor is thus:

$$R = \frac{2200}{I}$$

Where:

  R is the resistor value in ohms
  I is the desired display current in mA

An additional stand-alone LED, D5, is driven directly from RC4 and illuminates when the measured voltage is below 1.00 volts to indicate that the display reads in millivolts.

Input limits—Microchip recommends that the voltage applied to an analog input pin should not exceed 0.3 V above V_{DD} nor be more than 0.3 V more negative than the PICs digital ground voltage. Inputs beyond these limits may increase errors, and, if not current limited, may damage the chip. Our simple circuit provides no protection against these problems. Also, our simple DVM circuit assumes the PIC's A/D input pin is driven from a low impedance source, so the time required to charge the internal sample-and-hold capacitor may be ignored.

Program 11-2 uses this circuit to display the DC voltage applied to pin AN0. Program 11-2's structure is illustrated in the flow chart of Figure 11-6.

Program 11-2

```
;Program 11-2
;Read 0...4.096 volts on Pin RA0 and display voltage reading
;on 3 LED digit readouts.

;Constants
;------------------------
AN0             Con         0           ;Use Pin RA0 for analog input
Clk             Con         2           ;set up A/D conversion clock
ADSetUp         Con         $85         ;ext Vref+; Gnd Vref-
AvgLength       Con         32          ;don't change
ShiftLen        Con         5           ;Shift to save time over a divide.
Vdd             fCon        4.096       ;precision voltage reference

;Variables
;------------------------
i               Var         Byte        ;i,j & k are
j               Var         Byte        ;counters used for various indexes
k               Var         Byte        ;in array and other access
Temp            Var         Byte        ;digit being displayed
Decode          Var         Byte(14)    ;patterns for LED display
DigitVal        Var         Byte(3)     ;digit for display
TempVolt        Var         Word        ;voltage reading
Temp10          Var         Byte        ;
DataStore       Var         Word(AvgLength);buffer array
Smooth          Var         Word        ;Averaged reading
RealVolt        Var         Float       ;Floating point volts
VoltStep        Var         Float       ;volts per A/D step
DpPosn          Var         Byte        ;decimal point location

;Initialization
;-----------------------------------
                ;ABCDEFG.                ;corresponding segments
                ;--------                ;valid for
Decode(0) = %11111100                    ;Lumex LDS-M514R1-RA LED
Decode(1) = %01100000                    ;
Decode(2) = %11011010                    ;LEDs are connected in two
Decode(3) = %11110010                    ;banks, unlike most displays
```

```
Decode(4)  = %01100110                ;Common anode
Decode(5)  = %10110110                 ;
Decode(6)  = %00111110                 ;
Decode(7)  = %11100000                 ;
Decode(8)  = %11111110                 ;
Decode(9)  = %11100110                 ;
Decode(10) = %00000001                 ;
Decode(11) = %10011100                 ;
Decode(12) = %11111010                 ;
Decode(13) = %00011100                 ;
DigitVal(0) = 0                        ;
DigitVal(1) = 0                        ;
DigitVal(2) = 0                        ;
TempVolt = 0                           ;

For i = B0 to B7              ;B0..B5 drive the digit halves.
     Output i                 ;[1-1 1-2][2-1 2-2][3-1 3-2]
     High i                   ;[   1   ][   2   ][   3   ]
Next                          ;so we scan 1-1,1-2,2-1 ... 3-2
;when illuminating the LEDs
                              ;3-1 = Digit 3,Pin 1
For i = C0 to C4              ;C0..C3 drive the diodes via 2N7000.
     Output i                 ;[Pin 3][Pin 4][Pin5 ][Pin 6]
     Low i                    ;[ RC0 ][ RC1 ][ RC2 ][ RC3 ]
Next
For i = 0 to (AvgLength-1)              ;Zero sampled values
     DataStore(i) = 0
Next
k=0                      ;initialize to index starting at zero
Smooth = 0               ;Used in Smooth = Smooth + X so must be 0
DPPosn = 0

EnableHSerial
SetHSerial      H1200         ;Initialize the serial port at 1200 b/sec
                              ;intentionally slow!!!
VoltStep = Vdd / 1024.0       ;For 10 bit A/D we have 1024 states


Main
;-------------------------------------------
;We scan the digits from left to right. For each digit we
;look at the A half then the B half.
;For each half, we use the corresponding nibble from Decode.

For i = 0 to 2                ;i is the digit, leftmost digit = Digit 0
     j = B0+(2*i)            ;j is the digit half, sequentially
0,1,2..5
                              ;Assume we are display the value 123
Temp = Decode(DigitVal(i))    ;which digit are we working ;on?
                              ;DigitVal() holds (1)(2)(3)

     ;Pin 1-half
     ;------------------------
     PortC.LowNib = Temp.HighNib          ;High nibble
                              ;holds pattern for Segments A,B,C & D

     Low j           ;to prevent ghosting order is important
     Pause 2         ;pause to illuminate LED. Larger = brighter
     High j          ;For a 20 MHz chip, 3 is max before flicker
                     ;is a problem. 2 provides cleaner display
     ;Pin 2-half
     ;------------------------
     j = j+1
     PortC.LowNib = Temp.LowNib           ;repeat for other half
     ;Low nibble holds pattern for Segments E,F,G & DP
     If i = DpPosn Then
          PortC.Bit0 = 1
     EndIf
```

```
            Low j            ;same approach as for Pin 1-half
            Pause 2          ;use this order to avoid ghosting
            High j
            GoSub TakeAvg    ;Get the average voltage value
            GoSub MapVolts   ;Computes individual digits
            Next
GoTo Main

TakeAvg
;----------------------------------------
            ;Subtract from head of buffer
Smooth = Smooth - DataStore(k)
            ADin AN0,Clk,ADSetup,DataStore(k)
            Smooth = Smooth + DataStore(k)
                    ;32 el circular averaging
                    ;buffer. This works as a low pass filter
            k = k + 1
            If k = (AvgLength) Then
                    k = 0
            EndIf

       TempVolt = Smooth >> ShiftLen                    ;Equal to / 32
Return

MapVolts
;----------------------
            RealVolt = VoltStep * (ToFloat TempVolt)
            ;RealVolt is the floating point voltage
            HSerOut [Real RealVolt,13]
            ;send it via serial if you want
            Low C4           ;C4 is tied to the millivolt flag
                             ;when high display is in millivolts
            DpPosn = 0
                    ;Assume volts, as the starting point
            TempVolt = ToInt (0.500 + (100.0 * RealVolt))
            ;round off in 10's of millivolts
                             ;the 0.5 is to get better rounding
            If (TempVolt < 100) Then
            ;If we are less than 1.00 volt we want
                    RealVolt = RealVolt * 10.0
            ;to display millivolts directly, so
                    DpPosn = 2
                    ;multiply by 10 (1,000 when considered
                    High C4
                             ;with the FMUL earlier) and light the
                    TempVolt = ToInt (0.500 + (100.0 * RealVolt))
            ;millivolt LED.
            EndIf
            DigitVal(0) = TempVolt / 100   ;display 0..512 divide by 2
            Temp10 = TempVolt - (DigitVal(0) * 100)     ;1st digit =
            DigitVal(1) = Temp10 / 10                   ;2nd digit
            DigitVal(2) = Temp10 - 10 * DigitVal(1)     ;3rd digit
       Return
       End
```

As usual, we first define constants and declare variables. The value choices for the constants related to setting up the A/D function **ADIN** will be discussed in conjunction with the subroutine **TakeAvg** analysis.).

```
AvgLength       Con         32      ;don't change
ShiftLen        Con         5       ;Shift to save time over a divide.
Vdd             fCon        4.096   ;precision voltage reference
```

The constant **AvgLength** defines the length of the circular buffer **DataStore** we use to hold individual voltage readings for averaging The constant **ShiftLen** defines the number of bits we shift right to equal a divide by 32 ($2^5 = 32$) when we compute the average voltage from the 32 element array **DataStore**. **Vdd** holds the value of the A/D convert module's reference voltage.

During the initialization section, we assign values to each LED segment mapping variable, `Decode()`, for example:

```
Decode(0) = %11111100          ;digit 0 display
```

We won't repeat all the digit-to-segment mapping variables, but show **De-code(0)** as an example. The primary difference between the digit-to-segment mapping of Chapter 7 and here relates to the split LED configuration of the Lumex devices used in this circuit. Since each half-digit requires controlling four 2N7000 drivers, a logical mapping structure is to hold a complete map for each displayed character in one byte, employing one nibble for each digit half. To hold definitions for multiple display digits, we use an array of bytes, **Decode()**. An illuminated segment is indicated by a high (logical 1) in the corresponding bit position. Consider **Decode(i)**. The index **i** corresponds to the desired digit; the segment mapping to display the digit "7," for example, is held in **Decode(7)**. The bit assignments are made as follows, where an X indicating an illuminated segment corresponds to a logical 1 in our mapping variable, based upon the segment identifiers of Figure 11-7:

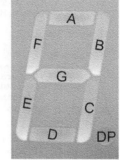

Figure 11-7: Segment identifiers for lumex LDS-M51RI-RA 7-segment LED display.

Displayed Digit to Segment Mapping
LUMEX LDS-M51RI-RA 7-Segment LED Display

	First half—Pin 1 Active (At +5V) Decode().Nib1				Second half—Pin 2 Active (At +5V) Decode().Nib0			
	Pin 3 Gnd	*Pin 4 Gnd*	*Pin 5 Gnd*	*Pin 6 Gnd*	*Pin 3 Gnd*	*Pin 4 Gnd*	*Pin 5 Gnd*	*Pin 6 Gnd*
Displayed Digit	*A*	*B*	*C*	*D*	*E*	*F*	*G*	*DP*
0	X	X	X	X	X	X		
1		X	X					
2	X	X		X	X		X	
3	X	X	X	X			X	
4		X	X			X	X	
5	X		X	X		X	X	
6			X	X	X	X	X	
7	X	X	X					
8	X	X	X	X	X	X	X	
9	X	X	X			X	X	
DP								X
C	X			X	X	X		
a	X	X	X	X	X		X	
L				X	X	X		
X indicates segment illuminated								

Thus, our assignment for **Decode(0) is %11111100** illuminating all segments except Segment G and the decimal point. A related decision is how to hold the numbers to be displayed. Suppose the voltage reading is 1.23 volts. Since we must display this one digit at a time (but scanned quickly so persistence of vision makes it look as if all digits are on all of the time), we allocate space for a three element byte array **DigitVal**. **DigitVal**, in our example, would hold **DigitVal(0) = 1**, **DigitVal(1) = 2** and **DigitVal(2) = 3**.

```
      For i = 0 to (AvgLength-1)          ;Zero sampled values
          DataStore(i) = 0
      Next
      k=0
```

```
Smooth = 0
DPPosn = 0

EnableHSerial
SetHSerial     H1200
```

The initialization section also sets the necessary Port B and Port C pins to be outputs and zeros variables. We also set the hardware serial output at 1200 bits/sec. If a serial output is not desired, omit this initialization.

```
VoltStep = Vdd / 1024.0          ;For 10 bit A/D we have 1024 states
```

To convert the A/D reading to volts, we calculate the volts per step. Since our design uses a 4.096 V reference, we calculate the voltage per step (4.0 mV) with a floating point division.

The main loop is endlessly repeated and sequentially scans the three display digits, left (digit 0) to right (digit 2). Within each display digit, pin 1 is first made active, followed by pin 2. The index variable **i** corresponds to the digit and has possible values 0...2. The variable **j** is a pin index that sequentially puts RB0...RB5 low, thereby placing V_{DD} on the segment display's common anode pins.

Recalling how we have stored the display mapping and the digit coding, the relationship between **i**, **j**, and output pins in Ports B and C is straightforward:

i	*j*	*Low on Pin*	*Port C.LowNibble Assignment*
0	*RB0 + 0*	*RB0*	*Decode(DigitValue(0)).HighNibble*
	RB0 + 1	*RB1*	*Decode(DigitValue(0)).LowNibble*
1	*RB0 + 2*	*RB2*	*Decode(DigitValue(1)).HighNibble*
	RB0 + 3	*RB3*	*Decode(DigitValue(1)).LowNibble*
2	*RB0 + 4*	*RB4*	*Decode(DigitValue(2)).HighNibble*
	RB0 + 5	*RB5*	*Decode(DigitValue(2)).LowNibble*

With this understanding, the logic flow of the main program loop is simple:

```
Main
;----------------------------------------------
For i = 0 to 2
 j =B0+(2*i)
Temp = Decode(DigitVal(i))
```

To save a bit of lookup time we use a new variable **Temp** instead of repeating **Decode(DigitVal(i))** which requires a double array index lookup each time it appears.

```
PortC.LowNib = Temp.HighNib
 Low j
Pause 2
High j
```

Since pins 3...6 of the display are controlled by RC0...RC3, we are concerned with only the PortC's low nibble. At this point in the **i** loop we are dealing with pin 1 of the display digit, so we must set segments A, B, C and D correctly. These control bits are held in the high nibble of **Decode()** array elements. Hence, we set **PortC.LowNib** to **Temp.HighNib**.

Having set the segments A...D correctly, we now apply +5 V to pin 1 of the display digit. To avoid a ghost image of incorrect segments, it's important to first set the segments first, and only then drop the RB pin to apply +5 V to the anodes. We keep the selected segments illuminated for 2 ms through a **pause 2** statement.

```
 j =j+1
PortC.LowNib = Temp.LowNib
If i = DpPosn Then
PortC.Bit0 = 1
End If
```

We now deal with pin 2 at V_{DD} by incrementing **j** and applying the same principles to segments E…DP attached to pin 2 of the display digit, except the segment map is contained in **Temp.LowNib**.

We treat the decimal point separately—its location is determined in subroutine **MapVolts** which sets **DpPosn** to 0 or 2, indicating which digit's decimal point is to be illuminated. The decimal point LED is connected to RC0, so we override the **PortC.Bit0** value contained in **Temp.LowNib** with a direct bit assignment.

```
GoSub TakeAvg
GoSub MapVolts
```

The analog voltage is read in subroutine **TakeAvg**. Subroutine **MapVolts** computes the decimal voltage and loads the digit values into the **DigitVal** array.

Subroutine **TakeAverage** reads the analog voltage, and calculates the average of the last 32 digital values. The resulting average is maintained in the variable **TempVolt**. Using the average value instead of an instantaneous reading reduces jitter and noise.

```
Take Avg
Smooth = Smooth - DataStore(k)
ADin AN0,Clk,ADSetup,DataStore(k)
```

(The statement **Smooth = Smooth - DataStore(k)** is discussed in connection with the averaging process below.)

We read the analog input voltage with the **ADIN** procedure. This requires setting three constants using the reasoning developed earlier in this chapter. Our A/D related constants are:

```
AN0         Con    0       ;Use Pin RA0 for analog input
Clk         Con    2       ;set up A/D conversion clock
ADSetUp     Con    $85     ;ext Vref+; Gnd Vref-
```

Since our 16F877 operates with a 20-MHz clock, and we do not wish to use the internal RC oscillator, our clock constant must be 2. Hence, we've defined **Clk** as 2.

We use an external voltage reference, V_{REF+}, and want RA0/AN0 to be our analog input pin. While experimenting with our breadboard layout, we prefer to use the internal ground for V_{REF-}, so we wish V_{REF-} to be assigned to V_{SS}. Finally, we also want the A/D data to be right justified. We wish the remaining dual-purpose pins to be available for digital I/O to the greatest extent possible. Hence, our setup constant must be $85, which results in the following analog pin assignments:

Pin	Function
AN0/RA0	Analog Input
AN1/RA1	Analog Input (Unused)
AN3/RA3	V_{REF+}

After the **ADIN** function executes, **DataStore(k)** has the returned A/D data, a 10-bit value ranging from 0…1023.

```
Smooth = Smooth + DataStore(k)
k = k + 1
If k = (AvgLength) Then
          k = 0
EndIf
TempVolt = Smooth >> ShiftLen
Smooth = Smooth - DataStore(k)
Return
```

The remainder of the subroutine implements a 32-sample moving average routine. **DataStore** is a circular buffer that stores new data in the next open position. When all 32 positions are filled with data (k = 31), the buffer index **k** wraps around to 0 and the process starts over again. We implement the average function in an efficient manner, as illustrated in Figure 11-8. A brute-force average algorithm would simply sum **DataStore** after every read and divide by 32:

```
Smooth = 0
For m = 0 to AvgLength
Smooth = Smooth + DataStore(m)
Next
TempVolt = Smooth / AvgLength
```

The brute-force algorithm computes 32 sums for every A/D read, a very inefficient process. A more elegant approach is to note that every time we complete an A/D read, 31 of the 32 values in **DataStore** are unchanged from the prior read; only the new A/D value has been added and it replaces the oldest reading—the **ADIN** command writes the *new* value over the *oldest* reading.

Hence, our strategy for an efficient average is:

```
Keep the sum of the prior 32 readings is in the variable Smooth
Subtract the oldest reading (held in DataStore(k))from Smooth
Take a new reading store it in DataStore(k) and add it to Smooth
Divide Smooth by the number of readings to obtain the average
```

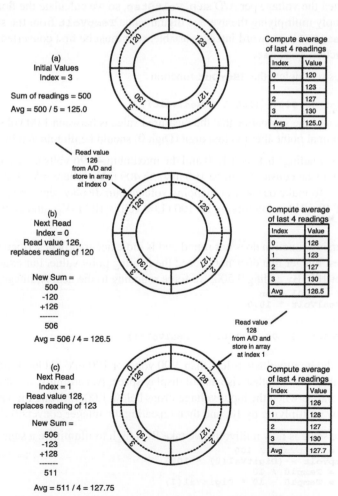

Figure 11-8: Implementing an efficient averaging algorithm.

227

Since we maintain the readings in a circular buffer array indexed by the variable *k* we must add a fifth, housekeeping, item to our strategy:

> *Update the array index k; if necessary execute a wrap around to zero*

One final point is that since instead of dividing by 32 to compute the average, we can simply execute a faster 5-bit shift procedure:

```
Temp Volt = Smooth >> ShiftLen
Temp Volt = Smooth / AvgLength
```

The two alternatives above produce identical results for our word length variables. Of course, the shift function only works if the average is taken over a length of the form 2^N.

The subroutine **MapVolts**, fills the array **DigitVal**, sets the location of the decimal point and activates or deactivates the millivolt indicator LED.

```
MapVolts
RealVolt = VoltStep * ToFloat TempVolt
HSerOut [Real RealVolt,13]
  Low C4
```

We previously determined the voltage per A/D step, **VoltStep**, so we calculate the floating point value of the input voltage by simply multiplying the average digital value **TempVolt** from the subroutine **TakeAvg** by **VoltStep**. Since **TempVolt** is a word integer variable, so it must be first converted to a floating point with the conversion function **ToFloat**.

If serial output is not desired, delete the **HSerOut** function.

```
DpPosn = 0
TempVolt = Int (0.500 + (100.0 * RealVolt))
```

We assume in this portion of the subroutine that the measured value is between 1.00 volts and the maximum 4.096 volts. If so, the decimal point after the first digit (Digit 0) should be illuminated by setting **DpPosn = 0**.

With three display digits, readings between 1.00 and the maximum 4.096 volts have a resolution of 10 mV. This voltage range can thus be considered to be $100\,\text{mV}_{10}$ to $409\,\text{mV}_{10}$ where mV_{10} means units of 10 mV, without loss of accuracy. To make our task of mapping voltage into display segments, therefore, we convert **RealVolt** into units of 10 mV by multiplying by 100 (1.00 volt is $100\,\text{mV}_{10}$) and converting the result to an integer.

MBasic's **ToInt** conversion function does not round and is more accurately described as a floor conversion, i.e., it returns the greatest integer that does not exceed the floating point value. For example, **ToInt(99.99)** returns the value 99, not 100. By adding 0.500, we force rounding to the nearest integer.

```
If (TempVolt < 100) Then
RealVolt = RealVolt * 10.0
DpPosn = 2
High C4
TempVolt = ToInt (0.500 + (100.0 * RealVolt))
EndIf
```

Suppose, however, that the input voltage is less than 1.00 volts, or $100\,\text{mV}_{10}$? Our 3 digit display will have a leading zero, and we throw away otherwise useful display space. Accordingly, we shift to displaying millivolts on a 0…999 mV range when the input voltage drops below 1.00. Shifting the value to millivolts from mV_{10} is accomplished by multiplying by 10, and then repeating the integer conversion and rounding process.

To alert the user that the scale is now millivolts, we take RC4 high to illuminate a sentinel LED.

```
DigitVal(0) = TempVolt / 100
Temp10 = TempVolt - (DigitVal(0) * 100)
DigitVal(1) = Temp10 / 10
DigitVal(2) = Temp10 - 10 * DigitVal(1)
Return
```

Whether the digits display millivolts or mV_{10} is immaterial to the display code in **Main**; it is simply concerned with illuminating the segments corresponding to the values in the **DigitVal** array.

Suppose the desired display is 234, representing 2.34 volts or 234 mV, depending on the location of the decimal point. **DigitVal(0)** should contain 2, **DigitVal(1)** should contain 3 and **DigitVal(2)** should contain 4. Thus, **DigitVal(0)** contains the 100's value, which we obtain by simply dividing **TempVolt** by 100. (Remember these are integer divisions.) To find the 10's value for **DigitVal(1)**, we subtract the 100's value from **TempVolt** and divide by 10. The same process is applied to find the units value for **DigitVal(2)**. Here's the numerical result of these steps, where the underscore indicates the results to be saved in the **DigitVal** array:

```
234/ 100 = 2          ;save 2 in DigitVal(0)
234- (2 * 100) = 34
34 / 10 = 3           ;save 3 in DigitVal(1)
34 - (3 * 10) = 4     ;save 4 in DigitVal(2)
```

Ideas for Modifications to Programs and Circuits

Although our voltmeter is usable in the breadboard format, no one would mistake it for a useful laboratory instrument. Here are a few ideas for additional experimentation:

- *High impedance input*—buffer the input with an op-amp follower. Figure 11-9 is a simplified depiction of the A/D input circuit in the 16F87x devices. Based on the circuit constants in Figure 11-9, what should the output impedance of the buffer? Is the buffer output impedance important if we only use **ADIN** or **ADIN16**, where we cannot control the precise A/D read time?

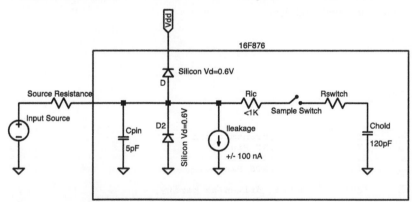

Figure 11-9: Simplified model of A/D input module.

- *Range selection input*—our voltmeter has a limited range. Higher voltages could be accommodated by a voltage divider input. Lower voltage ranges may require an amplified input. These can, of course, be combined with an input buffer amplifier.
- *Measure AC voltages*—add an op-amp precision rectifier after a buffered input. Read the peak voltage and add a calibration factor to the program to reflect the difference between average and RMS, in which almost all meters are calibrated.
- *Measure DC currents*—measure the voltage across a current shunt resistor through which the unknown current is passed. Is the design easier if you use both AN0 and AN1 analog voltage input pins? If you do use two input pins, how does the time delay between consecutive reads affect the accuracy?
- *Measure resistance*—add an ohmmeter function, by measuring the current through the unknown resistor and the voltage across the unknown. Calculate the resistance from Ohm's law.

References

[11-1] Mitra, S., et al. *Using the Analog-to-Digital Converter (A/D)*, Microchip Technology, Inc. Application Note AN546, Document No. DS00546E (1997).

[11-2] Bowling, Steve, *Understanding A/D Converter Performance Specifications*, Microchip Technology, Inc. Application Note AN693, Document No. DS00693A (2000).

[11-3] D'Souza, Stan, *Four Channel Digital Voltmeter with Display and Keyboard*, Microchip Technology, Inc. Application Note AN557, Document No. DS00557C (1997).

[11-4] *PICmicro™ Mid-Range MCU Family Reference Manual, Section 23—10-bit A/D Converter*, Microchip Technology, Inc., Document No. DS31023A (Preliminary-1997).

[11-5] PIC16F87X Data Sheet, Section 11 Analog-to-Digital (A/D) Converter Module, Microchip Technology, Inc., Document No. DS30292C (2001).

[11-6] Pease, Robert A., *Troubleshooting Analog Circuits, Chapter 10, The Analog-Digital Boundary,* Boston: Butterworth-Heinemann, 1991 paperback reprint 1993.

[11-7] Sauerwald, Mark, *Designing with High-Speed Analog-to-Digital Converters*, Application Note AD-01, National Semiconductor, Inc. (May 1998).

[11-8] Rempfer, William, *The Care and Feeding of High Performance ADCs: Get all the Bits You Paid For*, Application Note 71, Linear Technology (July 1997).

[11-9] Baker, Bonnie C., *Layout Tips for 12-bit A/D Converter Application,* Microchip Technology, Inc. Application Note AN688, Document No. DS00688B (1999).

Digital Temperature Sensors and Real-Time Clocks

We're going to look at two communications protocols, "1-wire" and "three-wire" (also known as SPI—Serial Peripheral Interface) for serial data exchange between a PIC and external sensors. We'll look at these protocols in the context of two specific devices, the DS18B20 12-bit temperature sensor and the DS1302 real time clock, but we'll see SPI again in later chapters. (By the way, don't confuse Dallas Semiconductor's DS18B20 sensor with its similar, cheaper, DS18S20 device. The DS18S20 is also a 1-wire temperature sensor, but with 9-bit resolution, yielding 0.5°C steps.)

In order to keep our discussion of manageable length, we'll briefly mention that these devices are but two examples from the world of sensors. One form or another of electronic sensor can measure almost any physical parameter of interest, directly or indirectly. Historically, sensors used an analog change in an electrically measurable parameter—resistance, capacitance or voltage being the most common—to measure a change in an underlying parameter, such as temperature, humidity, pressure, acceleration, or light intensity. Since the sensor output is analog, the resulting value must be read with an analog instrument, or converted to a digital value with an analog to digital converter.

Pure analog sensors are still widely used, but have been augmented with a built-in digital conversion, so the output can be directly read by a microcontroller, even though the microcontroller doesn't have a built-in ADC. (Don't be fooled by the digital output, however, as in almost every case the underlying physical sensor process remains analog.)

DS18B20 Temperature Sensor

One-Wire Protocol

Before delving into the specific temperature sensor we'll use, let's explore its 1-wire protocol. (If you want to skip this discussion and jump right into the program code, go ahead, as MBasic does an excellent job of encapsulating the nuts and bolts of 1-wire protocol into the **OWIn** and **OWOut** procedures.)

Maxim Integrated Products' Dallas Semiconductor division developed the "1-wire" protocol as an economical way of exchanging information between microprocessors and ancillary chips. It's called a 1-wire network because it uses one wire (plus ground) to communicate with the master microcontroller. One-wire devices are slaves and communicate only when so commanded by the master. Data is transmitted serially over the data line at an equivalent 14kb/s rate. Figure 12-1 shows how to connect 1-wire devices to a PIC. Dallas has a range of 1-wire devices that include memory, encryption and serial number generation, in addition to temperature sensing.

The 1-wire data protocol differs significantly from others we've seen, however. Although MBasic will hide the complexity for us, let's take a look at typical signaling in a 1-wire device as shown in Figure 12-2.

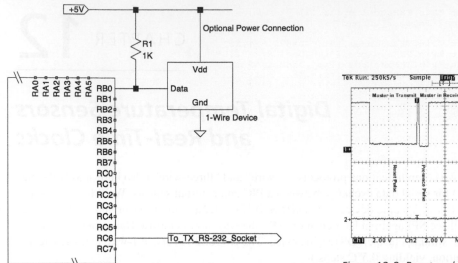

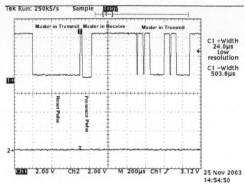

Figure 12-1: Typical 1-wire device connection.

Figure 12-2: Reset and presence pulses in 1-wire protocol.

Communications from the master PIC to slave 1-wire devices starts with the master outputting a 480 μs low reset pulse, following which the master switches to receive. The 1-wire device responds by taking the data line low sending a "presence" pulse, for approximately 150 μs. (We will use the presence pulse to detect whether or not a 1-wire device is connected.) Following the reset pulse, the master sends commands to the slave and the slave responds. Figure 12-3 shows the complete sequence; the reset/presence pulse, followed by a 1-byte interrogation request sent by the master, with an 8-byte reply by the slave device.

Let's take a closer look at how data bits are sent. Figure 12-4 shows the first byte transmitted by a DS18B20 responding to an interrogation request to send its device ID code. (The first byte is the family code, $28 (%00101000) in the case of the DS18B20, sent least significant bit first.) For each of the 8 bits to be received, the master switches to output mode, drops the data line low briefly (at least 1 μs), releases the data line and switches to input mode. The data line is then pulled high by the pull-up resistor, R1 in Figure 12-1. The slave 1-wire device then responds by either pulling the data line low (transmitting a 0) or leaving the data line high (transmitting a 1). The slave will maintain this status for 15 μs. The master must therefore read the data line status within 15 μs after it switches to input mode.

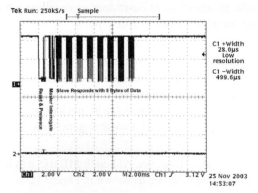

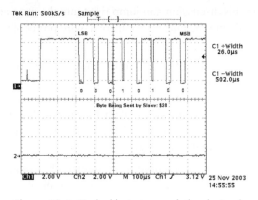

Figure 12-3: Reset – Interrogate – Response in 1-wire protocol.

Figure 12-4: Typical byte transmission in 1-wire protocol.

A master writes to the slave using a similar process, but with different timing. To write a 1 to the slave, the master pulls the bus low for 15 μs or less time and either releases the bus to go high (writes a 1 to the slave) or holds the bus low for a total time of 60 μs (writes a 0 to the slave).

Dallas refers to this technique as "time slot" coding—that is, the process of the master taking the data line low is seen as "issuing a time slot." The slave device then responds within the time slot with either a 0 or a 1, and the master then issues the next time slot (if any). When looking at the output on an oscilloscope, we see wide and narrow zero-level pulses, as reflected in Figure 12-4. A narrow pulse results if a 1 follows the time slot issuance from the slave; a wide pulse results when a 0 follows it.

An integral part of the 1-wire protocol is the ability to obtain operating power for the chips from the data line, without a separate power supply line, sometimes referred to as "parasitic" powering. One-wire devices have an internal capacitor that charges from the data line and power it when sending a 0 V signal. (If we look at it in detail, however, we'll see that parasitic power has its drawbacks, so all of our circuits use external power connections.)

One last comment and we'll see some real code. Each individual 1-wire part produced has its own unique 64-bit serial number and may be addressed either individually through the serial number, or globally, without reference to its serial number.

Reading a One-Wire Devices' Serial Number

Program 12-1 reads the serial number of a one-wire device connected with the circuit shown in Figure 12-1 and writes it to the serial output port. (When building the circuit don't forget the pull-up resistor!)

Program 12-1

```
;Program 12-1
;Read 1-wire device w/ READROM $33 Command
;Write ROM output (Device ID, SN & CheckSum)
;to serial output.
;Only works with 1 device attached!

;Variables
Temp            Var             Byte
i               Var             Byte

;Constants
ReadRom Con             $33

EnableHSerial
SetHSerial H19200

Main
        ;Issue universal READROM Command
        OWOut B0,1,Main,[ReadRom]

        ;Response is 8 bytes
        ;Device Family          1 Byte
        ;SN                     6 Bytes
        ;Check Sum              1 Byte
        For i = 0 to 7
                OWIn B0,0,[Temp]
                HSerOut [Hex Temp," "]
        Next
        ;Write  CR
        HSerOut [13]

        Pause 1000
GoTo Main
```

MBasic's support for 1-wire devices is contained in two procedures; **OWOut** for writing to the 1-wire device and **OWIn** for reading from the 1-wire device. Our first action is to instruct the 1-wire device to report its serial number, through the function: **OWOut B0,1,Main,[ReadRom]**. **OWOut** is invoked with three options:

Owout Pin,Mode,{NCLabel},[{Mods} Exp]

Pin—is the pin to which the data line is connected. **Pin** can be either a constant or a variable.

Mode—defines the specifics of the data transfer. Permissible **mode** values are:

Mode	Send Reset?	Data Mode	Speed
0	No	Byte	Low
1	Before Data	Byte	Low
2	After Data	Byte	Low
3	Before and After Data	Byte	Low
4	No	Bit	Low
5	Before Data	Bit	Low

(Dallas Semiconductor is expanding its 1-wire product line to include new devices with bit rates of several hundred kb/s and has a proposed an even higher 1 Mb/s rate. MBasic currently supports only the initial 14 kb/s speed 1-wire protocol.)

NCLabel—an optional address label to which execution will jump if no 1-wire chip is detected.

Mods—MBasic's standard command modifiers, for example, **Str**, **Real**, and so on.

Exp—is a variable or constant holding the value to be sent.

Figure 12-1 shows **B0** as the data pin. Commands written to a 1-wire device require a leading reset pulse, we'll use mode 1. For simplicity, we'll omit device connection error checking.

We'll look at 1-wire commands in more detail in connection with later programs, but for now we'll note that $33 is the command value for "read ROM," that is, send the serial number of the device. We declare a constant, **ReadRom**, and set its value to $33. Hence, our command is:

OWOut B0,1,Main,[ReadRom]

We now read the response from the 1-wire device with **OWIn** and write the output to the serial port, one byte at a time:

```
For i = 0 to 7
        OWIn B0,0,[Temp]
        HSerOut [Hex Temp," "]
Next
```

OWin follows the same syntax as **OWOut**:

OWIn Pin,Mode,{NCLabel},[{Mods} Var]

Pin, **Mode** and **NCLabel** are identical with their corresponding elements in **OWOut**, and require no further description. Since we are reading a value, of course, it must be read into a *variable*, not a constant.

Dallas Semiconductor's 1-wire specifications define device serial numbers as 8 bytes (64 bits) long, configured as:

8-bit CRC Code		48-bit Serial Number		8-bit Family Code	
MSB	LSB	MSB	LSB	MSB	LSB

The family code for 18B20 digital thermometer chips is $28. The CRC (cyclic redundancy code) is an error-checking feature, so that, should we desire, we may verify that the 56 bits of the family code and serial number have been correctly received and were not corrupted. We'll not further consider how CRCs are calculated, as it's a topic well beyond the level of this introductory book.

When we run the program, we see the following repetitive output:

28 4C D4 3E 0 0 0 D6
28 4C D4 3E 0 0 0 D6
28 4C D4 3E 0 0 0 D6
...

The digits 4C D4 3E 0 0 0 D6 are, of course, dependent upon the particular DS18B20 chip. I plugged in a second DS18B20 chip and found its serial number:

28 FE DA 3E 0 0 0 C1
28 FE DA 3E 0 0 0 C1
28 FE DA 3E 0 0 0 C1
...

In the output, $28 is the family number and $D6 (or $C1 in the second example) is the CRC. The center six bytes represent the serial number of the chip. But, there's a difference between the result and the serial number specification isn't there? The definition has the CRC sent first and the family code sent last. Yet, Program 12-1 displays the family code first, and the CRC last. The explanation is that 1-wire devices store the least significant byte at the lower address and the most significant byte at the higher address. Bytes are transmitted and received from lowest address to highest address. Hence, the net effect is the send/receive byte order is reversed from the data sheet description. This is more confusing to describe than to use; when we wish to address a particular device, we just repeat the byte order we read with Program 12-1.

Reading the Temperature

Program 12-1 will confirm that your circuit is properly wired up and the DS18B20 is functioning. Program 12-2 shows how we can read the temperature. (Remember the | symbol is a line continuation.)

```
;Program 12-2
;Reads a Dallas 1-wire Temperature sensor
;DS18B20. Assumes only ONE device is connected to the bus,
;since we use global address mode.

;Variables
;----------------------
Temp            Var     Word    ;holds raw binary temperature output
PlusFlag        Var     Byte    ;plus/minus flag
Centigrade      Var     Float   ;Floating point C temp
Fahrenheit      Var     Float   ;Floating point F temp
TH              Var     Byte    ;Lower alarm value
TL              Var     Byte    ;Upper alarm value
ConfigReg       Var     Byte    ;Temperature resolution

;Initialization
Init
;--------------
        EnableHSerial
        SetHSerial H19200

        ;$4E Write to RAM; dummy $FF to TH & TL
        ;$7F to ctrl reg for 12-bit
        OWOut B0,1,Init,[$CC,$4E,$FF,$FF,$7F]
        OWOut B0,1,Init,[$CC,$48]
        OWOut B0,1,Init,[$CC,$B8]

        ;Read scratchpad memory
        OWOut B0,1,Init,[$CC,$BE]
        Pause 1000
        ;Read output and check configuration
        OWin  B0,0,[Temp.Byte0,Temp.Byte1,TH,TL,ConfigReg]
        HSerOut ["Init OK  R1: ",BIN ConfigReg.Bit6 ,"  R0: " |
        ,BIN ConfigReg.Bit5,13]
```

```
Main
            ;Cause temp conversion to start
            OWOut B0,1,Main,[$CC,$44]

Wait
            ;read output and loop until conversion is done.
            ;conversion finished if writeback of 1, 0 if not done
            ;Note this only works with external power
            OWIn B0,0,[Temp]
            If Temp = 0 Then Wait

            OWOut B0,1,Main,[$CC,$BE]
            OWin  B0,0,[Temp.Byte0,Temp.Byte1,TH,TL,ConfigReg]
            ;Read TH,TL and ConfigReg, but don't use
            ;Temperature data returned as 2's complement for below 0.
            ;Hence, <0 is detected by a 1 in the returned highest bit.
            PlusFlag = 1
            If Temp.HighBit = 1 Then        ;check for < 0C
               ;Subtract from 0 to read below 0 value
               Temp = $0000-Temp
               PlusFlag = 0
            EndIf

            ;Returned in Centigrade steps of 0.0625 Deg C.
            Centigrade = (ToFloat Temp) * 0.0625
            If PlusFlag = 0 Then            ;below zero, multiply by -1
                  Centigrade = -Centigrade
            EndIf

            ;Standard conversion formula to get to F from C.
            Fahrenheit = (Centigrade * 1.8) + 32.0 ;deg F

            ;Write output. Note the \2 and \1 modifiers give number of
            ;places after decimal point
            HSerOut ["Temperature ",Real Fahrenheit \2," F " ,Real| Centigrade \1,"
            C", 13]

            Pause 2000
GoTo Main
```

Program 12-2 initializes the DS18B20 to 12-bit resolution mode, and then every 2 seconds reads the temperature and writes the value in Fahrenheit to the serial port.

Program 12-2 uses global addressing and will work only if there is one 1-wire device connected to the data bus.

After declaring the necessary variables, we initialize the DS18B20.

The DS18B20 has variable resolution, and can be programmed for 9, 10, 11 or 12-bit resolution. (The DS18S20 is a variant with only 9-bit resolution.) The trade-off for increased resolution is longer conversion time.

Resolution	Conversion Time	Resolution		R1	R0
9-bit	93.75 ms	0.5°C	0.900°F	0	0
10-bit	187.5 ms	0.25°C	0.450°F	0	1
11-bit	375 ms	0.125°C	0.225°F	1	0
12-bit	750 ms	0.0625°C	0.113°F	1	1

We'll use 12-bit resolution, the default configuration for DS18B20 devices. Two option bits, R1 and R0, in the configuration register control resolution, so we'll set both to 1 to establish 12-bit resolution. We should, of course, not confuse resolution and accuracy. Regardless of the resolution selected, the DS18B20's accuracy remains at ±0.5°C over the range −10°C through +85°C.

Configuration Register							
Bit 7	Bit 6	Bit 5	Bit 4	Bit 3	Bit 2	Bit 1	Bit 0
0	R1	R0	1	1	1	1	1

The configuration word to command 12-bit resolution is thus %01111111, or $7F. How, then, to write $7F to the configuration register?

The DS18B20's memory is organized as nine bytes of RAM and three bytes EEPROM. The configuration register is byte no. 4:

Address	Scratchpad Memory Contents		EEPROM
0	Temperature LSB (read only)		
1	Temperature MSB (read only)		
2	TH Register	←→	TH
3	TL Register	←→	TL
4	Configuration Register	←→	Configuration
5	Reserved ($FF)		
6	Reserved ($0C)		
7	Reserved ($10)		
8	CRC		

Our ability to selectively write to the scratchpad and EEPROM is limited. In fact, we have but four memory commands, plus two function commands:

DS18B20 Command Values			
Command	Description	Command Value	1-Wire Bus Activity
Convert Temperature	Initiate temperature reading	$44	Transmit conversion status (only with external power)
Read Scratchpad	Read all 9 scratchpad locations	$BE	DS18B20 transmits up to 9 bytes to master
Write Scratchpad	Write all 3 scratchpad locations, TH, TL and Configuration Register	$4E	Master transmits TH, TL and Configuration to DS18B20
Copy Scratchpad	Copies TH, TL and Configuration from the scratchpad to EEPROM	$48	None
Recall EEPROM	Copies TH, TL and Configuration from EEPROM to the scratchpad	$B8	DS18B20 transmits recall status to master
Read Power Supply	Informs master whether 18B20 is directly or parasitically powered	$B4	DS18B20 transmits power supply status to master

To write the resolution information to the configuration register and then to the EEPROM, we must write 3 bytes, TH, TL and the configuration byte. TH and TL are alarm trigger registers and store values for high (TH) and low (TL) temperature alarms. We shall not use either, so we will write a dummy value, $FF to both TH and TL.

One final bit of housekeeping and then we'll write our configuration bits. Before a 1-wire device may receive any commands, it must be addressed. We'll continue In Program 12-2 with global addressing, $CC. All 1-wire devices respond to a $CC address, so to avoid device conflict we may have only one 1-wire device connected to pin B0.

```
OWOut   B0,1,Init,[$CC,$4E,$FF,$FF,$7F]
OWOut   B0,1,Init,[$CC,$48]
OWOut   B0,1,Init,[$CC,$B8]
OWOut   B0,1,Init,[$CC,$BE]
Pause 1000
OWin    B0,0,[Temp.Byte0,Temp.Byte1,TH,TL,ConfigReg]
HSerOut ["Init OK  R1: ",BIN ConfigReg.Bit6 ," R0: " |
,BIN ConfigReg.Bit5,13]
```

The first four lines of code implement our strategy; we write **$FF, $FF, $7F** to the scratchpad with the **$4E** command. We then move scratchpad memory locations 2, 3 and 4 to EEPROM with a **$48** command. For our demonstration, we then copy the EEPROM back to scratchpad memory with a **$B8** command and, finally, read back the contents of the scratchpad with a **$BE** command. Note that in every case, we issue a reset and a global address (**$CC**) before the command byte.

We then read back the bytes sent after the **$BE** command with the **OWIn** procedure and write the values to the serial port. The response we see from the serial port should be:

Init OK R1: 1 R0: 1

Now, we initiate a temperature read.

```
Main
            ;Cause temp conversion to start
            OWOut B0,1,Main,[$CC,$44]

Wait

            ;read output and loop until conversion is done.
            ;conversion finished if writeback of 1, 0 if not done
            ;Note this only works with external power
            OWIn B0,0,[Temp]
            If Temp = 0 Then Wait
```

Temperature reads must start with the "convert temperature" **$44** command. Upon receipt of the $44, the DS18B20 starts the temperature reading process and upon completion stores the value in the scratchpad at locations 0 (least significant byte) and 1 (most significant byte). If externally powered, the DS18B20 will respond during the conversion process with a 0, changing to a 1 when completed. Hence, after initiating the convert temperature command, we keep reading the DS18B20 until it returns a 1, indicating temperature data is ready to be read.

We now are ready to read the temperature from the DS18B20's scratchpad memory. The data is returned in increasing address order, from 0 (least significant byte of temperature) to 4 (configuration register).

```
            OWOut B0,1,Main,[$CC,$BE]
            OWin  B0,0,[Temp.Byte0,Temp.Byte1,TH,TL,ConfigReg]
            ;Read TH,TL and ConfigReg, but don't use
```

This is the same code we used when reading the configuration register during initialization. Now, however, we discard all read data, save for **Temp.Byte0** and **Temp.Byte1**, as these contain our temperature reading.

The temperature data is returned in the following format:

Temp.Byte0	*Bit 7*	*Bit 6*	*Bit 5*	*Bit 4*	*Bit 3*	*Bit 2*	*Bit 1*	*Bit 0*
Least Significant Byte	2^3	2^2	2^1	2^0	2^{-1}	2^{-2}	2^{-3}	2^{-4}
Temp.Byte1	*Bit 15*	*Bit 14*	*Bit 13*	*Bit 12*	*Bit 11*	*Bit 10*	*Bit 9*	*Bit 8*
Most Significant Byte	S	S	S	S	S	2^6	2^5	2^4

A 1 at bit 0, for example, represents 1×2^{-4} degrees C, or 0.0625°C.

Since the range of temperatures reported by the DS18B20 extends below 0°C, the data includes a sign bit, identified as "S" in the table. If the temperature is above 0, S is 0; if below 0°C, S is 1. For temperatures below 0, we must subtract the reading from $0000 to obtain the number of degrees below 0°C. (The data is stored as "a 16-bit sign-extended two's complement number." Don't let this scare you; it's simple to make it useful.)

Here's how values around 0°C are reported by the DS18B20:

Temperature	Binary Output	Hexadecimal Output
+10.125°C	**0000 0000 1010 0010**	**$00A2**
+0.5°C	**0000 0000 0000 1000**	**$0008**
0°C	**0000 0000 0000 0000**	**$0000**
-0.5°C	**1111 1111 1111 1000**	**$FFF8**
-10.125°C	**1111 1111 0101 1110**	**$FF5E**

We read the two temperature bytes one at a time, storing them in the word variable **Temp**. After reading **Temp**, we check to see if the returned value is less than 0°C, as indicated by a 1 in at bit 15:

```
PlusFlag = 1
        If Temp.HighBit = 1 Then      ;check for < 0C
        ;Subtract from 0 to read below 0 value
        Temp = $0000-Temp
        PlusFlag = 0
        EndIf
```

If a 1 is found, then we remove the two's complement by subtracting the value from $0000. This gives the number of degrees below 0°C. (If you're not convinced, subtract it by hand, or use Window's accessory calculator in scientific view. For –10.125°C, $0000 – $FF5E = $00A2. Decimal ($A2) = 162. Each step is 0.0625°C, so 162 * 0.0625 = 10.125.)

```
;Returned in Centigrade steps of 0.0625 Deg C.
Centigrade = (ToFloat Temp) * 0.0625
If PlusFlag = 0 Then           ;below zero, multiply by -1
        Centigrade = -Centigrade
EndIf

;Standard conversion formula to get to F from C.
Fahrenheit = (Centigrade * 1.8) + 32.0 ;deg F
```

We now convert the returned raw temperature word (after any necessary below zero correction) to a floating point Celsius (Centigrade) value by multiplying it by the step size, 2^{-4}, or 0.0625 degrees C/step. (2^{-4} is the same as $1/2^4$, or 1/16, which is 0.0625.) Since **Temp** is an integer, we first convert it to a floating-point value with the **ToFloat** operator. We then multiply by the step size. (If you run the DS18B20 at less than 12-bit resolution, the output jumps in larger intervals, but each step remains 0.0625°C.)

If the returned value was less than 0°C, **PlusFlag** has been set to 0, so we must multiply the temperature by –1, which is accomplished by prefixing **Centigrade** with a minus sign.

Finally, we convert the floating point Celsius temperature to Fahrenheit.

```
;Write output. Note the \2 and \1 modifiers give number of
;places after decimal point
HSerOut ["Temperature ",Real Fahrenheit \2," F "  ,Real|
Centigrade \1," C", 13]
```

Now we write the temperature (in Celsius and Fahrenheit) to the serial port. Note our use of the undocumented \2 and \1 modifiers to display two and one digits following the decimal point. Here's a sample output from Program 12-2:

```
Init OK  R1: 1  R0: 1
Temperature 74.63 F 23.6 C
Temperature 74.63 F 23.6 C
Temperature 74.63 F 23.6 C
Temperature 74.63 F 23.6 C
Temperature 73.62 F 23.1 C
```

Reading Multiple Sensors on the Same Bus

A primary advantage of the 1-wire bus is that we may place many sensors on it, and selectively read each one. Let's try something simple, one indoor temperature sensor and one outdoor sensor, both connected to B0, as shown in Figure 12-5.

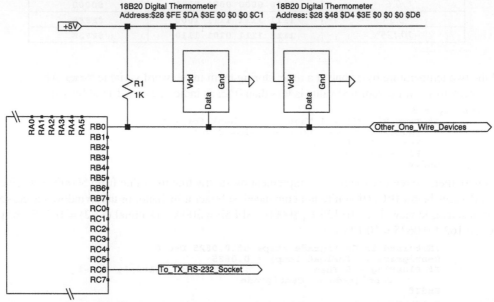

Figure 12-5: Multiple DS18B20 temperature sensors on one bus.

First, though, run Program 12-1 for each DS18B20 and note their serial numbers. The two sensors I had on hand identified themselves as:

28 4C D4 3E 0 0 0 D6
28 FE DA 3E 0 0 0 C1

I made a simple temporary outdoor sensor mount from a short length of PVC pipe and fittings, as shown in Figure 12-6. The sensor is at the top of the "T" fitting and I soldered its three pins to three conductors in a short length of CAT5 data cable. I stuffed the cable opening of the "T" with a few plastic shipping pellets for weatherproofing and brought the free end of the cable into my workshop. (Reference [12-7] discusses cable selection and distance issues and should be consulted for cable runs more than a couple dozen feet.) The other sensor I mounted on the breadboard for an indoor comparison reading.

Program 12-3 selectively reads the indoor and outdoor sensors and writes their readings to the serial port.

Figure 12-6: Temporary outdoor temperature sensor mounting.

Program 12-3

```
;Program 12-3 Read 2 DS18B20 temperature
;sensors connected to the same bus.
;Write the two temperatures to serial output
;The sensors are addressed by SN

;Variables
```

```
;------------------------
Temp            Var     Word    ;holds raw binary temperature output
PlusFlag        Var     Byte    ;plus/minus flag
Centigrade      Var     Float   ;Floating point C temp
Fahrenheit      Var     Float   ;Floating point F temp
TH              Var     Byte    ;Lower alarm value
TL              Var     Byte    ;Upper alarm value
ConfigReg       Var     Byte    ;Temperature resolution

; The full ROM SN of the two DS18B20 Sensor Chips
;Chip 1 $28 $FE $DA $3E $0 $0 $0 $C1
;Chip 2 $28 $48 $D4 $3E $0 $0 $0 $D6
;your sensors will have different serial numbers!

;Initialization
;----------------
        EnableHSerial
        SetHSerial H19200

;Initialize both sensors to 12-bit resolution.
;Since we have multiple sensors on same bus, we must use
;selective addressing $55 followed by 8-byte chip address
;Otherwise, routine is same as Program 12-2.
Init1
;----------------
        OWOut B0,1,Init1,[$55,$28,$FE,$DA, |
        $3E,$0,$0,$0,$C1,$4E,$FF,$FF,$7F]
        OWOut B0,1,Init1,[$55,$28,$FE,$DA,$3E,$0,$0,$0,$C1,$48]
        OWOut B0,1,Init1,[$55,$28,$FE,$DA,$3E,$0,$0,$0,$C1,$B8]
        OWOut B0,1,Init1,[$55,$28,$FE,$DA,$3E,$0,$0,$0,$C1,$BE]
        Pause 1000
        ;Read output and check configuration
        OWin  B0,0,[Temp.Byte0,Temp.Byte1,TH,TL,ConfigReg]
        HSerOut ["Init 1 OK  R1: ",BIN ConfigReg.Bit6 , |
        " R0: ",BIN ConfigReg.Bit5,13]

Init2
;----------------
        OWOut B0,1,Init2,[$55,$28,$48,$D4, |
        $3E,$0,$0,$0,$D6,$4E,$FF,$FF,$7F]
        OWOut B0,1,Init2,[$55,$28,$48,$D4,$3E,$0,$0,$0,$D6,$48]
        OWOut B0,1,Init2,[$55,$28,$48,$D4,$3E,$0,$0,$0,$D6,$B8]
        OWOut B0,1,Init2,[$55,$28,$48,$D4,$3E,$0,$0,$0,$D6,$BE]
        Pause 1000
        ;Read output and check configuration
        OWin  B0,0,[Temp.Byte0,Temp.Byte1,TH,TL,ConfigReg]
                HSerOut ["Init 2 OK  R1: ",BIN ConfigReg.Bit6 , |
        " R0  ",BIN ConfigReg.Bit5,13]

Main
        ;Selective start of conversion on Chip No. 1
        ;Again we start with $55,ChipAddress
        OWOut B0,1,Main,[$55,$28,$FE,$DA,$3E,$0,$0,$0,$C1,$44]

Wait
        ;Wait for good data
        OWIn B0,0,[Temp]
        If Temp = 0 Then Wait

        ;Selective read of data on Chip No. 1
        OWOut B0,1,Main,[$55,$28,$FE,$DA,$3E,$0,$0,$0,$C1,$BE]
        OWin  B0,0,[Temp.Byte0,Temp.Byte1]

        ;From raw data, compute Centigrade and Fahrenheit
        GoSub ComputeTemp

        HSerOut ["Sensor 1 "]
        ;Write the Degrees F and Degrees C to the serial port
        GoSub WriteTemp
```

```
;------------------------ Sensor No. 2 -----------------------
;Now we repeat but with the Sensor No. 2 Selective Address

Part2
        OWOut B0,1,Part2,[$55,$28,$48,$D4,$3E,$0,$0,$0,$D6,$44]

Wait2
        OWIn B0,0,[Temp]
        If Temp = 0 Then Wait2

        OWOut B0,1,Part2,[$55,$28,$48,$D4,$3E,$0,$0,$0,$D6,$BE]
        OWin  B0,0,[Temp.Byte0,Temp.Byte1]

        GoSub ComputeTemp

        HSerOut [$9,"Sensor 2 "]
        GoSub WriteTemp
        HSerOut [13]

        Pause 1000
GoTo Main

ComputeTemp             ;Subroutine accepts raw data and outputs Deg F and Deg C
;-------------

        ;Raw data is in 2's complement if less than 0C.
        ;Check high bit for < 0

        PlusFlag = 1
        If Temp.HighBit = 1 Then   ;check for < 0C
           Temp = $0000-Temp       ;Just subtract to convert
           PlusFlag = 0
        EndIf

        Centigrade = (ToFloat Temp) * 0.0625 ;step 0.0625 deg C
        If PlusFlag = 0 Then                        ;below zero
              Centigrade = fneg Centigrade   ;If < 0 add minus sign
        EndIf

        ;Standard conversion from Centigrade to Fahrenheit
        Fahrenheit = (Centigrade * 1.8) + 32.0
                        ;Centigrade to deg F
Return

WriteTemp
;--------------
        ;Write generic output to serial
        ;\2 and \1 fix the number of digits after .
        ;that are displayed
        HSerOut [Real Fahrenheit \2," F  ",Real Centigrade \1," C"]
Return
```

We'll concentrate on the changed parts of this program, as it largely duplicates what we learned in Program 12-2.

Since we have two DS18B20 sensors on the same bus, we must selectively address each one, instead of using the **$CC** universal address. We do this simply by replacing **$CC** with **$55**, followed by the 8-byte serial number of the device. Hence, to initialize the configuration register we have:

Universal address version:
```
        OWOut B0,1,Init,[$CC,$4E,$FF,$FF,$7F]
```

Selective address version for DS18B20 with SN $28 $FE $DA $3E $00 $00 $00 $C1:
```
        OWOut B0,1,Init1,[$55,$28,$FE,$DA,$3E,$0,$0,$0, |
        $C1,$4E,$FF,$FF,$7F]
```

(Remember the | is the line continuation symbol.)

Since we have two devices to initialize, we repeat the selective initialization twice, once for each device.

The main program loop likewise selectively addresses one sensor, calls the subroutine **ComputeTemp** to convert the raw temperature word to Centigrade and Fahrenheit values, and then calls a second subroutine, **WriteTemp**, to write the Centigrade and Fahrenheit values to the serial output port. (The **$9** character is a horizontal tab, used to space the results across the terminal screen.)

The two subroutines, **ComputeTemp** and **WriteTemp** track the embedded code in Program 12-2.

Here's the output of Program 12-3, taken before I assembled the outdoor sensor and moved it outside:

```
Init 1 OK  R1: 1  R0: 1
Init 2 OK  R1: 1  R0: 1
Sensor 1 74.74 F  23.7 C        Sensor 2 74.74 F  23.7 C
Sensor 1 74.74 F  23.7 C        Sensor 2 74.74 F  23.7 C
Sensor 1 73.84 F  23.2 C        Sensor 2 74.63 F  23.6 C
Sensor 1 74.97 F  23.8 C        Sensor 2 74.63 F  23.6 C
Sensor 1 80.26 F  26.8 C        Sensor 2 74.86 F  23.8 C  ← Finger on Sensor 1
Sensor 1 78.46 F  25.8 C        Sensor 2 74.86 F  23.8 C
Sensor 1 77.22 F  25.1 C        Sensor 2 79.02 F  26.1 C  ← Finger on Sensor 2
Sensor 1 76.32 F  24.6 C        Sensor 2 82.51 F  28.0 C
Sensor 1 75.76 F  24.3 C        Sensor 2 80.37 F  26.8 C
Sensor 1 75.64 F  24.2 C        Sensor 2 78.57 F  25.8 C
```

To see how fast the sensors react, I grasped each one between two fingers for a few seconds. (The measurement cycle is about 4 seconds per reading.) It took only one read cycle to see the temperature start to rise, and about five or six cycles to return to ambient.

DS1302 Real-Time Clock

Now that we've gotten the temperature out of the way, let's look at timekeeping. We could use our PIC as a clock, such as described in Reference [12-9]. But, we're going to unload the timekeeping function onto a dedicated special purpose chip from Dallas Semiconductor, the DS1302.

The DS1302 is a real time clock, requiring an external 32.678 KHz crystal. The DS1302 provides seconds, minutes, hours, day-of-the-week, date, month and year information, including leap year adjustment up to 2100. It also supports battery or capacitor backup, and includes an integrated trickle charger. The DS1302 communicates with the PIC via a three-wire serial connection, supported by MBasic's **ShiftIn** and **ShiftOut** procedures.

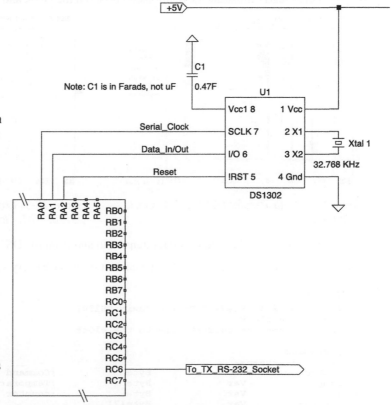

Figure 12-7: Connecting a DS1302 real-time clock.

We'll connect the DS1302 to our PIC using the circuit shown in Figure 12-7. A couple points in Figure 12-7 deserve amplification. First, we have chosen a very large capacitor, 0.47 F, to serve as our power supply backup. Once fully charged (it takes a little over 1 hour), the 0.47 F capacitor will keep the DS1302 running at least a day if the primary power is interrupted. Second, the DS1302's accuracy is dependent upon the accuracy of the 32.678 KHz crystal. The crystal's design capacitance should match the 6 pF specification of the DS1302. (If we desire precision timekeeping, there are many more things that will concern us, but those are beyond the scope of this chapter. See, for example, Reference [12-10].)

The DS1302 has three connections to the PIC:

SCLK (Serial Clock Input)—The clock determines when data bits are received and valid. Data is transferred to the DS1302 or read from the DS1302 when the clock changes state. Data is received by the DS1302 during the rising edge. Data is transmitted by the DS1302 on the falling edge of the clock, but are read by the PIC master on the rising edge. All clock pulses are generated by the PIC; the DS1302 only receives clock pulses and cannot generate them.

I/O (Data)—The data line establishes whether a 1 or a 0 is read at the appropriate clock edge. The data line is bi-directional; the DS1302 may send to the PIC or the PIC may send to the DS1302.

RST (Reset)—The reset line serves two purposes. First, when taken high it turns on the DS1302's control logic. Second, when taken from high to low, it terminates data transfer from the PIC to the DS1302. The PIC controls the reset line.

Figure 12-8 shows a sample data exchange between a PIC and a DS1302. A total of 5 bytes are exchanged. Figure 12-9 more clearly illustrate the relationship between the clock and data lines.

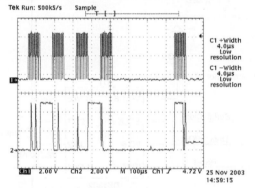

Figure 12-8: Data Transfer DS1302 Ch1: Clock; Ch2: Data.

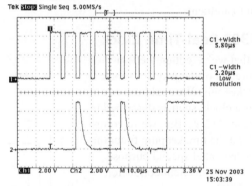

Figure 12-9: Expanded View of Data Transfer DS1302 Ch1: Clock; Ch2: Data

Fortunately, MBasic hides the details of data transfer to and from the DS1302.

Program 12-4 reads the DS1302 and writes the current time, once every second, to the serial port.

Program 12-4

```
;Program 12-4. Reads a Dallas/Maxim DS1302
;Real time clock.
;Also writes the current time to the clock

;Varibles
;--------------
RTCCmd          Var         Byte           ;Command byte
Temp            Var         Byte           ;Temporary variable
i               Var         Byte           ;counter
TimeData        Var         Byte(7)
OldSeconds      Var         Byte           ;last read seconds
```

```
;For convienence we alias the time array
Seconds        Var         TimeData(0)            ;Seconds
Minutes        Var         TimeData(1)            ;Minutes
Hours          Var         TimeData(2)            ;Hours
Date           Var         TimeData(3)            ;Day of Month
Month          Var         TimeData(4)            ;Month
Day            Var         TimeData(5)            ;Day of Week
Year           Var         TimeData(6)            ;Year

;Constants
;--------------                  () are DIP8 numbers
Clk            Con         A0             ;Clock pin (7)
Dta            Con         A1             ;Data pin  (6)
Reset          Con         A2             ;Reset pin (5)

CtrlReg        Con         %00111

;Initial load of Sec  Min   Hrs  Date MO   Day  YR
PreSet ByteTable $00, $53, $12, $24, $11, $02, $03
;These are in BCD, hence November (Month 11)
;is entered as $11, not 11.

;Command to select charger settings
Charger        Con              %01000
;1 diode, 2K charging resistance. See
;Figure 5 of DS1320 data sheet
Battery        Con              %10100101

;For Day and Month names, use the following. Since
;Sun=1 and Jan=1 we use nul at the beginning instead of
;subtracting 1.
DayName ByteTable "NulSunMonTueWedThuFriSatSun"
MoName  ByteTable "NulJanFebMarAprMayJunJulAugSepOctNovDec"


;Initialization
;--------------
        EnableHSerial
        SetHSerial H19200

        ;To set the clock we first disable write protection.
        ;Then send the clock set information, and then restore
        ;write protection.

        ;Disable DS1302 WriteProtection by 0 in Bit7
        Temp = $00
        RTCCmd = CtrlReg
        GoSub PreSetData

        Set the date and time from the Preset byte table
        For i = 0 to 6
                Temp = Preset(i)
                RTCCmd = i
                GoSub PreSetData
        Next

        ;Since we have a Super Cap backup, we want charger ON
        Temp = Battery
        RTCCmd = Charger
        GoSub PreSetData

        ;DS1302 Write protection back on
        Temp = $80
        RTCCmd = CtrlReg
        GoSub PreSetData

Loop
;----------
        ;Read the clock output. If seconds changed, write time
        GoSub ReadData
```

```
                    ;Write in form:
                    ;13:10:45 11/24/03  Mon 24-Nov-2003 to the serial port
                    ;New seconds, so write output
                    If OldSeconds <> Seconds Then
                    HSerOut [Dec2 (BCD2BIN Hours)\2,":",Dec2 |
                    (BCD2BIN Minutes)\2,":",Dec2 (BCD2BIN Seconds)\2]
                    HSerOut [" ",Dec2 (BCD2BIN Month)\2,"/",Dec2 |
                    (BCD2BIN Date)\2,"/",Dec2 (BCD2BIN Year)\2]
                    HSerOut ["   ",Str DayName(Day*3)\3," "]
                    HSerOut [Dec2 (BCD2BIN Date)\2,"-",Str |
                    MoName(3*BCD2BIN Month)\3,"-20",Dec2 (BCD2BIN Year)\2,13]
               EndIf
          GoTo Loop


          PreSetData              ;Routine to upload data to the DS1302
          ;-----------
                    High Reset
                    ;The %0\1,RTCCmd\5,%10\2 results in the equivelent of
                    ;sending 10 RTCCMD 0. Each element is sent in element order
                    ;but bit reversed, as required by the DS1320.
                    ShiftOut Dta,Clk,LSBFIRST,[%0\1,RTCCmd\5,%10\2,Temp\8]
                    Low Reset
          Return


          ReadData
          ;----------
                    ;Reset pin must be high for  read
                    High Reset
                    ;%10111111 is BURST SEND REGISTER command, so get 8 bytes
                    ;the last byte is the control register, which we don't use
                    ShiftOut Dta,Clk,LSBFirst,[%10111111\8]
                    OldSeconds = Seconds
                    ;Now actually read the burst send data.
                    ShiftIn Dta,Clk,LSBPRE,[Seconds\8,Minutes\8,Hours\8, |
                    Date\8,Month\8,Day\8,Year\8]
                    Low Reset
          Return
```

We've set up a seven-element data structure to hold time and date information from the DS1302:

```
          TimeData          Var                 Byte(7)

          ;For convienence we alias the time array
          Seconds Var               TimeData(0)              ;Seconds
          Minutes Var               TimeData(1)              ;Minutes
          Hours       Var               TimeData(2)              ;Hours
          Date        Var               TimeData(3)              ;Day of Month
          Month       Var               TimeData(4)              ;Month
          Day         Var               TimeData(5)              ;Day of Week
          Year        Var               TimeData(6)              ;Year
```

Clock and calendar data is held in the DS1302 in a 10-byte register structure:

Register Address	Contents
0	Seconds
1	Minutes
2	Hours
3	Date (Day of Month)
4	Month
5	Day of Week
6	Year
7	Control Register
8	Trickle Charger Control
9	Clock Burst Register

In addition, the DS1302 has 31 bytes of general-purpose RAM, at memory addresses 0…31

The index values of data array **TimeData** align with the DS1302's memory addresses. We've also aliased "user friendly" functional names for each element of **TimeData**.

```
;Constants
;--------------                 () are DIP8 numbers
Clk            Con         A0      ;Clock pin (7)
Dta            Con         A1      ;Data pin  (6)
Reset          Con         A2      ;Reset pin (5)
```

As usual, we define the three connection pins by declaring three custom constants, **Clk**, **Dta** and **Reset**.

```
;Initial load of Sec  Min  Hrs  Date MO   Day  YR
PreSet ByteTable $00, $53, $12, $24, $11, $02, $03
;These are in BCD, hence November (Month 11)
;is entered as $11, not 11.
```

At the first power-up—and whenever power is removed without the backup "super cap" in place—we must initialize the DS1302 with the current date, time and day of week. For clarity, we'll store the initialization information in a byte table.

The seven date, time and day of week values are stored in the DS1302 in binary coded decimal format. We've covered BCD in Chapter 7 and won't revisit it here, except to remind you that each nibble in a BCD byte is regarded as a separate digit, ranging from 0…9. Hence, the BCD representation of the number 28 (decimal) is $28 (decimal 40). The day of the week entry starts with Sunday = 1. For convenience in setting the clock, we'll store the byte table in the same order it will be sent to the DS1302—that is, in the DS1302's internal address order. Accordingly, 12:53:00, Monday, 24 November 2003 is stored in our byte table as **$00, $53, $12, $24, $11, $02, $03**.

```
;Command to select charger settings
Charger        Con         %01000
;1 diode, 2K charging resistance. See
;Figure 5 of DS1320 data sheet
Battery        Con         %10100101
```

Since we provide backup power, it's necessary to enable the internal charger circuitry in the DS1302, as the default status is disabled.

This is an appropriate time to look at how we exchange data with the DS1302. It's a straightforward two-step process:

- Send a command byte to the DS1302
- Send one or more bytes of data to the DS1302, or receive one or more bytes of data from the DS1302, if the command byte calls for reading data.

The DS1302's command byte structure is quite logical:

Bit 7	*Bit 6*	*Bit 5*	*Bit 4*	*Bit 3*	*Bit 2*	*Bit 1*	*Bit 0*
0 = Write Disabled	*0 = Register*						*0 = Write*
1 = Write Enable	*1 = RAM*	*5-bit representation of Register or RAM Address*					*1 = Read*

The 5-bit representations of the register or RAM address means that the leading three zeros are truncated from the normal 8-bit binary representation. For example, the normal byte-length binary representation of the digit 8 is **%00001000**. The 5-bit representation is **%01000**.

Bit 6 of the command byte determines whether we are addressing the DS1302's registers or its general-purpose RAM. (All of our communications will be with the registers, and we won't use the DS1302's RAM at all. Hence, Bit 6 will always be 0 in this chapter's programs.)

To set the value of the day of the month to the day-of-the-month register we send two bytes to the DS1302; first the command byte, followed by the value to be loaded into the DS1302. Suppose we are to set the day of the month to the 28th. The BCD representation of the 28th is \$28. Let's determine the command byte:

- Bit 7 must be %1, since we are writing data to the DS1302
- Bit 6 must be %0, since we are writing data to a register, not RAM.
- Bits 5…1 are the 5-bit representation of the address of the day of the month register. The day of the month register is number 3, so the 5-bit representation is %00011.
- Bit 0 must be %0, since we are writing data to the DS1302.

Hence, our command byte is:

Bit 7	Bit 6	Bit 5	Bit 4	Bit 3	Bit 2	Bit 1	Bit 0
1	0	0	0	0	1	1	0

This binary value corresponds to \$86. Hence, to set the day-of-the-month to the 28th, we must send the two byte sequence **\$86 \$28** to the DS1302.

One more housekeeping detail before we set the clock. Bit 7 of the control register is a master write protect bit. It must be set to 1 before any data may be written to either the DS1302's registers or RAM.

DS1302 Control Register							
Bit 7	Bit 6	Bit 5	Bit 4	Bit 3	Bit 2	Bit 1	Bit 0
0 = Master Write Enable 1 = Master Write Disable	0	0	0	0	0	0	0

The control register's address is 7, so its 5-bit binary address is %00111. We've predefined this value as a constant, **CtrlReg**, which we will use instead of %00111.

So, in pseudo-code, we'll use the following algorithm to set the clock values:

```
Take Reset pin high to permit writing
Set Bit 7 of the control register to 0 to enable write

Sequentially write the clock values:
For i = 0 (Seconds) to 6 (Years)
        Construct control byte
        Send control byte
        Send data byte from PreSet
Next i
Set Bit 7 of the control register to 1 protecting against
Further changes to the date and time.
Take Reset pin low to block writing
```

Here's the actual code:

```
;Disable DS1302 WriteProtection by 0 in Bit7
        Temp = $00
        RTCCmd = CtrlReg
        GoSub PreSetData

        Set the date and time from the Preset byte table
        For i = 0 to 6
                Temp = Preset(i)
                RTCCmd = i
                GoSub PreSetData
        Next

PreSetData                  ;Routine to upload data to the DS1302
;-----------
        High Reset
        ;The %0\1,RTCCmd\5,%10\2 results in the equivent of
        ;sending 10 RTCCMD 0. Each element is sent in element order
        ;but bit reversed, as required by the DS1320.
        ShiftOut Dta,Clk,LSBFIRST,[%0\1,RTCCmd\5,%10\2,Temp\8]
        Low Reset
Return
```

It tracks the pseudo-code quite closely. First, the master write protect bit is cleared, and then the seven time/date registers are set.

The only tricky part of this program fragment centers around:

```
ShiftOut Dta,Clk,LSBFIRST,[%0\1,RTCCmd\5,%10\2,Temp\8]
```

Let's look at **ShiftOut**. **ShiftOut** is invoked with three arguments:

Dta—A constant or variable that defines the data pin. (We declared **Dta** a constant equal to pin **A1**.)

Clk—A constant or variable that defines the clock pin. (We declared **Clk** a constant equal to pin **A0**.)

Mode—Mode is a value that defines the order in which the bits are sent, and the relationship between the clock (rise or fall) and data being valid, meaning is the data valid on clock rise or on clock fall? Beginning with version 5.3.0.0, MBasic permits **ShiftOut** to have one of 12 possible modes.

	Mode Constant	Bit Order	Clock / Data Relationship	Speed (Machine Cyles)	Speed (With 20 MHz clock)
High Speed (all new in 5.3.0.0)	FASTMSBPRE	MSB first	Data valid on leading edge	25	200 KHz (5 µs/bit)
	FASTLSBPRE	LSB first			
	FASTMSBPOST	MSB first	Data valid on falling edge		
	FASTLSPOST	LSB first			
Slow Speed (all new in 5.3.0.0)	SLOWMSBPRE	MSB first	Data valid on leading edge	100	50 KHz (20 µs/bit)
	SLOWLSBPRE	LSB first			
	SLOWMSBPOST	MSB first	Data valid on leading edge		
	SLOWLSBPOST	LSB first			
Normal Speed {} indicates backwards compatible for older versions	MSBPRE {MSBFIRST}	MSB first	Data valid on leading edge	50	100 KHz (10 µs/bit)
	LSBPRE {LSBFIRST}	LSB first			
	MSBPOST	MSB first	Data valid on leading edge		
	LSBPOST	LSB first			

The DS1302 requires medium speed, LSB-first bit order, and the bit sample before clock pulse speed. Hence, the mode we shall use is **LSBFIRST**, both for **ShiftIn** and **ShiftOut**.

Let's examine the information being sent, **[%0\1,RTCCmd\5,%10\2,Temp\8]**. We send four separate elements, **%0**, **RTCCmd**, **%10**, and **Temp**. Each element is sent in its order of appearance (left to right), but with its *bits* sent LSB first. Only the number of bits specified by the **\x** command is sent. Suppose we're clearing the master write protection bit. In this case, **RTCCmd** is set to **CtrlReg (%00111)** and **Temp** is set to $00. The control byte is sent as the composite of the first three elements, sent in *element order*, but with *bits reversed*:

Send Order							
%0	%00111					%10	
1	2	3	4	5	6	7	8
0	1	1	1	0	0	0	1

In sending order, therefore, the first three elements are sent as **%01110001**. The DS1302 receives eight bits in that order, and *reassembles it into the normal byte order* as **%10001110**. Recalling how the command byte is constructed, we see that the byte reassembled by the DS1302 instructs it that the next byte is to be stored in the command register. Since the next byte sent, **Temp**, is $00, the net result of **ShiftOut Dta,Clk,LSBFIRST, [%0\1,RTCCmd\5,%10\2,Temp\8]** is to write a **%00000000** into the command register, thus activating the master write enable.

With this understanding, we can now see how the following code fragment works.

```
For i = 0 to 6
        Temp = Preset(i)
        RTCCmd = i
        GoSub PreSetData
Next
```

The command byte is assembled in the subroutine **PreSetData**, just as we've gone through. Hence, the **Seconds** value that is contained at **Preset(0)** is written to the Seconds register (address 0 in the DS1302), the Minutes value that is contained at **Preset(1)** is written to the Minutes register (address 1 in the DS1302) and so forth.

```
;SInce we have a Super Cap backup, we want charger ON
Temp = Battery
RTCCmd = Charger
GoSub PreSetData
```

We use the same approach to enable the DS1302's built-in backup power supply charging circuit. We've predefined **Charger** as **%01000**, or **8**, the address of the charger control register. The charger options byte has multiple options:

Bit 7	Bit 7	Bit 5	Bit 4	Bit 3	Bit 2	Bit 1	Bit 0
Trickle Charger Control Bits (TCS)				Diode Control Bits		Resistor Control Bits	

The TCS bits determine whether the charger is active, or disabled. The default is disabled, and *only* the TCS bit pattern **%1010** enables the charger. Any other bit pattern in the TCS bits disables the charger.

To prevent the backup power from trying to run all the devices connected to the V_{CC} line, the DS1302 includes reverse current isolation diodes. Bits 3 and 2 determine whether one or two diodes will be in series. Two diodes increase the isolation, but add an extra 0.7V drop to the charging voltage.

Permitted values for bits 3 and 2 are:

Bit 3	Bit 2	Diode Connection
0	0	Trickle charger disabled, independently of TCS bits
0	1	One series diode selected
1	0	Two series diodes selected
1	1	Trickle charger disabled, independently of TCS bits

Finally, Bits 1 and 0 determine the charging current, by selecting the series resistor value:

Bit 1	Bit 0	Resistor Connection
0	0	None
0	1	Typically 2Kohm
1	0	Typically 4Kohm
1	1	Typically 8Kohm

We've selected TCS—enabled, one diode and 2 Kohm series resistance. Hence, our charger option byte is **%1010 01 01**, and we've defined the constant **Battery** accordingly:

```
Battery          Con              %10100101
```

To activate the charger we use the subroutine **PreSetData** to load the appropriate command byte (constructed in **PreSetData** from the 5-bit address **Charger**), followed by the **Battery** option byte.

A brief word on backup power is appropriate. At one time, our choice would have been limited to a choice of battery chemistry. We might have chosen from among a variety of primary (nonrechargeable) batteries, or secondary (rechargeable) batteries. But, recently supercapacitors—capacitors with values of 0.1 Farad or more—have appeared on the market. For low current drain devices, a supercap has many advantages over a battery, most importantly its much longer lifetime. The particular supercap we're using is a 0.47 F, 5 V device. It charges to essentially full capacity in about 4,000 seconds, or just over one hour, when we select a 2K charging resistor and one diode options in the DS1302 charger option byte. (Supercaps are known for high levels of dielectric absorption, so measurable charging current may be observed for hundreds, or thousands of hours.) The DS1302 consumes 0.3 µA operating current and will continue to operate until 2.0 V. Hence, neglecting internal losses in the supercap, the 0.47 F backup capacitor will allow the DS1302 to continue operating for approximately 3.6 million seconds—about 41 days after power is removed. In practice, the supercap's leakage current may be as much as 10 µA or more, which limits the backup operation to around one day, still more than adequate to carry operation over during normal power outages. If we require extended operation without mains power, then we should consider battery backup, perhaps a long life lithium primary battery.

Since the timekeeping continues during power outages, we need only set the clock once and thereafter may comment out the clock setting parts of Program 12-4.

Now that we've set the clock, let's read it and write the current time to the serial port. We've bundled all the read code into the subroutine **ReadData**, so let's look at it first.

```
ReadData
;----------
        ;Reset pin must be high for  read
        High Reset
        ;%10111111 is BURST SEND REGISTER command, so get 8 bytes
        ;the last byte is the control register, which we don't use
        ShiftOut Dta,Clk,LSBFirst,[%10111111\8]
        OldSeconds = Seconds
        ;Now actually read the burst send data.
        ShiftIn Dta,Clk,LSBPRE,[Seconds\8,Minutes\8,Hours\8, |
        Date\8,Month\8,Day\8,Year\8]
        Low Reset
Return
```

Large parts of this should look familiar.

Before doing anything with the DS1302, we must bring the reset pin high. Next, we send a command byte to the DS1302 using **ShiftOut**. The particular command byte we send is the "clock/calendar burst read" command, **%10111111**, and we don't have to construct it from a composite of three elements. The clock/calendar burst read command instructs the DS1302 to send the contents of the first eight registers in address order (starting with seconds and ending with the control register) sequentially, without further commands from the master PIC.

Following the clock/calendar burst read command, we read the first seven bytes with the **ShiftIn** procedure: **ShiftIn Dta,Clk,LSBPRE,[Seconds\8,Minutes\8,Hours\8, Date\8,Month\8,Day\8,Year\8]**. (**ShiftIn** is the receive version of **ShiftOut** and requires the same arguments and mode selection, so we won't elaborate on it.)

We read the first 7 bytes into the array **TimeData**, using the alias variables **Seconds**, **Minutes**, and so on that we earlier declared. Instead of using the DS1302's clock/calendar burst read function, we could have individually read the first seven registers into the elements of **TimeData()**, using code much like that found in **PreSetData**, that is, send a command byte to read register 0, then read the value into **TimeData(0)**, repeat for register 1, and so forth.

Now, let's look at how we call **ReadData** and what we do with the time information it returns.

```
Loop
;----------
          ;Read the clock output. If seconds changed, write time
          GoSub ReadData
          ;Write in form:
          ;13:10:45 11/24/03  Mon 24-Nov-2003 to the serial port
          ;New seconds, so write output
          If OldSeconds <> Seconds Then
          HSerOut [Dec2 (BCD2BIN Hours)\2,":",Dec2 |
          (BCD2BIN Minutes)\2,":",Dec2 (BCD2BIN Seconds)\2]
          HSerOut [" ",Dec2 (BCD2BIN Month)\2,"/",Dec2 |
          (BCD2BIN Date)\2,"/",Dec2 (BCD2BIN Year)\2]
          HSerOut ["    ",Str DayName(Day*3)\3," "]
          HSerOut [Dec2 (BCD2BIN Date)\2,"-",Str |
          MoName(3*BCD2BIN Month)\3,"-20",Dec2 (BCD2BIN Year)\2,13]
     EndIf
GoTo Loop
```

Our main program loop calls **ReadData** to obtain the current time and date. We then compare the current value of **Seconds** with the value last time we called **ReadData**, held in **OldSeconds**. If the value is different, we write the current time to the serial port with a series of **HserOut** calls and update the value of **OldSeconds**.

The **HserOut** calls are mostly straightforward, but a few points are worthy of note. Since the raw data is in BCD form, we use MBasic's **BCD2BIN** function to convert to the standard binary form with which we are accustomed. To give a month name and day-of-week name output, we earlier defined byte tables **DayName** and **MoName** (month name). We index into these arrays with the day-of-the-week number, or the month number. Since both the month and day-of-week start with 1, we added the dummy string "**Nul**" to the beginning of both **DayName** and **MoName**. Alternatively, we could have started the byte tables with "Sun" and "Jan" and subtracted 1 from the day-of-week number and month number.

Following is a sample of Program 12-4's output:

```
13:20:56  11/24/03   Mon  24-Nov-2003
13:20:57  11/24/03   Mon  24-Nov-2003
13:20:58  11/24/03   Mon  24-Nov-2003
13:20:59  11/24/03   Mon  24-Nov-2003
13:21:00  11/24/03   Mon  24-Nov-2003
13:21:01  11/24/03   Mon  24-Nov-2003
13:21:02  11/24/03   Mon  24-Nov-2003
```

Combination Date, Time and Temperature

Let's merge our program to read two temperature sensors, and add a DS1302 real-time-clock to yield a time and temperature logging output. Program 12-5 reads two DS18B20 temperature sensors and a DS1302 clock chip and writes the time and temperature readings to the serial port. Figure 12-10 shows the circuit hookup. Depending on the length of the cable to the remote DS18B20, it may be necessary to decrease the value of R1, the pull-up resistor. It can be as low as 220 ohms, if required.

Program 12-5

Program 12-5 is very long and is largely a composite of earlier programs in this chapter. Rather than follow our normal structure of listing the program in full, followed by a section-by-section analysis, we'll skip the full listing. Program 12-5, like all other programs in this book, is available in an electronic copy in the associated CD-ROM. If you need a listing to follow while reading the analysis, you may print one from the program file.

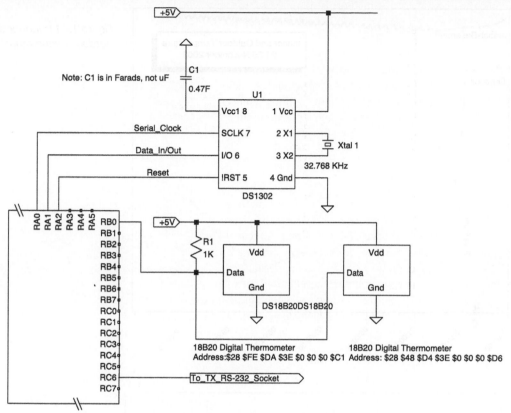

Figure 12-10: Circuit to read two temperature values and time of day.

Program 12-5 should, by now, require little analysis, as it simply merges programs we've exhaustively studied. Following is a sample of Program 12-5's output:

```
15:12:00 11/24/03   Sensor 1 73.62 F   Sensor 2 65.41 F
15:13:00 11/24/03   Sensor 1 73.73 F   Sensor 2 65.41 F
15:14:00 11/24/03   Sensor 1 73.62 F   Sensor 2 65.52 F
15:15:00 11/24/03   Sensor 1 73.51 F   Sensor 2 65.41 F
15:16:00 11/24/03   Sensor 1 73.39 F   Sensor 2 65.41 F
15:17:00 11/24/03   Sensor 1 73.51 F   Sensor 2 65.29 F
```

We've set the program to output a reading once every minute, but the output interval may be increased by changing the value of the constant **Interval**.

```
        Interval        Con            1         ;how often read data
```

Defining **Interval** as 5, for example, outputs a reading once every five minutes.

Figure's 12-11 and 12-12 show the indoor and outdoor temperature readings collected with this program over 24 hours. The abrupt drop in outdoor temperature coincided with the arrival of a cold front and rainstorm. The variation in indoor temperature (the data was collected in my basement workshop) results from the forced air furnace cycling off and on.

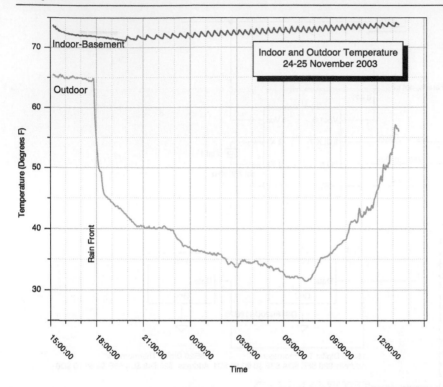

Figure 12-11: Indoor and outdoor temperature.

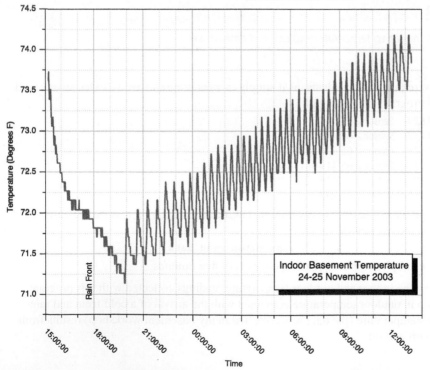

Figure 12-12: Indoor temperature.

Which Edge?

Serial clocked data is transferred on either the rising edge of the clock pulse, or the falling edge. By convention, clock idle is 0 V, so the leading edge is the transition from 0 to 1 and the falling edge is the transition from 1 to 0. Figure 12-13 shows both possibilities. Figures 12-14 and 12-15 show a bit with value 1 being transferred with these two options. The data being transferred is %10000000, MSB first.

Don't get confused when looking at Figures 12-14 and 15 because the data pulse is wider than the clock pulse. In Figure 12-15, for example, the data is valid on the falling edge of the first clock pulse, but it's also valid on the rising edge of the second clock pulse. But, the data is only read once for each clock pulse; i.e., data is always read on the rising edge or the falling edge, but not both. In Figure 12-15, data is read on the falling edge of each clock pulse; hence the data line's status during the rising edge is immaterial.

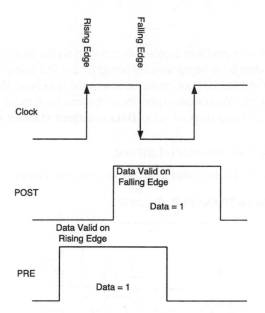

Figure 12-13: Rising and falling clock edge data transfer.

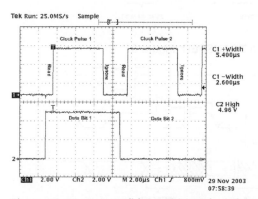

Figure 12-14: Data Valid on Rising Edge Ch1: Clock; Ch2: Data.

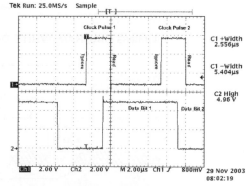

Figure 12-15: Data Valid on Falling Edge Ch1: Clock; Ch2: Data.

Bit Order

We have two choices to send the individual bits that comprise a byte; from most significant (B7) to least significant (B0), or from least significant to most significant.

Reading the Data Sheet

Let's look at the DS1302 data sheet to see where we find the clock edge and bit order information.

Let's start with bit order, as that's the simpler of the two. It's hard to be clearer than the DS1302 data sheet:

> ### DATA INPUT
>
> *Following the eight SCLK cycles that input a write command byte, a data byte is input on the rising edge of the next eight SCLK cycles. Additional SCLK cycles are ignored should they inadvertently occur.* **Data is input starting with bit 0.**
>
> ### DATA OUTPUT
>
> *Following the eight SCLK cycles that input a read command byte, a data byte is output on the falling edge of the next eight SCLK cycles. Note that the first data bit to be transmitted occurs on the first falling edge after the last bit of the command byte is written. Additional SCLK cycles retransmit the data bytes should they inadvertently occur so long as RST remains high. This operation permits continuous burst mode read capability. Also, the I/O pin is tri-stated upon each rising edge of SCLK.* **Data is output starting with bit 0.**

Since data is both input and output starting with bit 0, the bit order is LSB first.

To determine whether the data is to be read on the rising or falling clock edge, we examine the timing diagrams.

TIMING DIAGRAM: READ DATA TRANSFER Figure 5

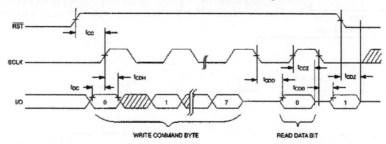

TIMING DIAGRAM: WRITE DATA TRANSFER Figure 6

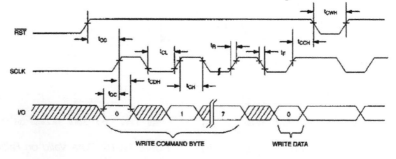

(Figures courtesy of Maxim Integrated Products, Inc. Reprinted with permission.)

With respect to the *DS1302* reading data (PIC writing data with `ShiftOut`), the data bits are shown to be read by the DS1302 on the rising edge of the clock. Hence we will use LSBPRE mode with `ShiftOut`.

With respect to the DS1302 writing data (PIC reading data with `ShiftIn`) the data is shown to be written immediately after the clock's falling edge and hence must be read by the PIC on the rising clock edge. Accordingly, we will use LSBPRE mode with `ShiftIn`.

Figures 12-16 through 12-19 show the difference among the four order and clocking options. In each case, the byte being sent is `%01000000`.

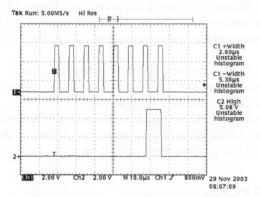

Figure 12-16: Mode: MSBPRE Ch1: Clock; Ch2: Data.

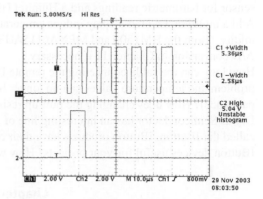

Figure 12-17: Mode: LSBPRE Ch1: Clock; Ch2: Data.

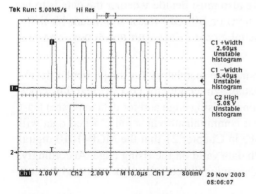

Figure 12-18: Mode: MSBPOST Ch1: Clock; Ch2: Data.

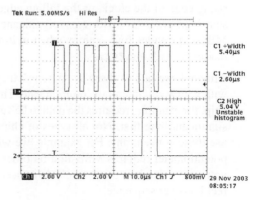

Figure 12-19: Mode: LSBPOST Ch1: Clock; Ch2: Data.

Predefined Constant	Value	Meaning		Sample Illustration
		Bit Order	Clock / Data Relationship	
MSBPRE (MSBFIRST)	0	MSB first	Data valid on leading edge	Figure 12-16
LSBPRE (LSBFIRST)	1	LSB first	Data valid on leading edge	Figure 12-17
MSBPOST	2	MSB first	Data valid on falling edge	Figure 12-18
LSBPOST	3	LSB first	Data valid on falling edge	Figure 12-19

Ideas for Modifications to Programs and Circuits

• The DS2401 is a 1-wire electronic serial number chip. Its sole function is to return a 64-bit unique serial number. How might this device be used? For example, it could be used as a security access device, or it could identify a calibrated printed circuit board. Or, it might be used to multiplex several switches on a single bus (each switch connects a different DS2401 to the bus. By reading the serial numbers, which switch is closed may be determined.) Try revising the circuits to read and use the unique serial number available from a dS2401.

• Program 12-5 is a start of a logging weather station. Add a Motorola MPXA6115AU absolute pressure sensor for barometric readings and a Humirel HM1500 humidity sensor to the circuit of Figure 12-10. Add a multiline LCD to display indoor temperature, outdoor temperature, barometric pressure and humidity. (Both the HM1500 and MPXA6115AU sensors require reading an analog voltage. Brush up on A/D conversion in Chapter 11 if necessary.)

• How long can the cable connecting the remote DS18B20 be before capacitive loading becomes a problem and temperature values can no longer be read? How does this vary with cable type (cable capacitance)? How does changing the pull-up resistor alter the distance performance?

• Dallas Semiconductor has an interesting line of 1-wire devices, packaged in stainless steel "buttons" called the IButton. IButton products have their own website, http://www.ibutton.com. Are any of the IButton devices useful for your projects? How would you go about interfacing them with a PIC?

Chapter 12 Sidebar
ShiftOut/ShiftIn – Rising or Falling Clock and Bit Order

When transferring serial clocked data we must decide whether the data is to be valid on the rising edge of the clock, or the falling edge. We also must decide whether the data is to be sent is in order B7...B0 (most significant bit first) or B0...B7 (least significant bit first). MBasic provides four options when writing serial clocked data with **ShiftOut**, or when reading it with **ShiftIn**.

Although we talk of these as "options" they in fact are mandatory for the particular device with which you intend to communicate. The Dallas Semiconductor DS1302 real time clock chip, for example, requires the data to be valid on the rising edge and for data to be sent least significant byte first. Not all types of chips that communicate via serial clocked data will use these conventions. For example, we will see in Chapter 21 that the MCP41010 digital potentiometer requires MSB-first bit order with data valid on the rising edge. Accordingly, you must delve into the device's data sheet to determine the correct clock edge and the correct bit order sequence.

References

[12-1] Motorola, Inc., *Sensor Device Data Book*, Publication DL200/D, Rev. 5 (January 2003).

[12-2] Dallas Semiconductor Corp. division of Maxim Integrated Products, Inc., *DS1302 Trickle Charge Time-keeping Chip*, (September 29, 2001).

[12-3] Dallas Semiconductor Corp. division of Maxim Integrated Products, Inc., *DS18B20 Programmable Resolution 1-Wire Digital Thermometer*, (January 5, 2002).

[12-4] Dallas Semiconductor Corp. division of Maxim Integrated Products, Inc., *DS2401 Silicon Serial Number*, (February 21, 2002).

[12-5] Dallas Semiconductor Corp. division of Maxim Integrated Products, Inc., *1-Wire Communication with a Microchip PICmicro Microcontroller*, (September 9, 2003).

[12-6] Awtrey, Dan, Transmitting *Data and Power over a One-Wire Bus, Sensors*, The Journal of Applied Sensing Technology (February 1997).

[12-7] Dallas Semiconductor Corp. division of Maxim Integrated Products, Inc., *MicroLAN Design Guide—Tech Brief 1*, (September 24, 2002).

[12-8] Dallas Semiconductor Corp. division of Maxim Integrated Products, Inc., *Printed Circuit Board Identification Using 1-Wire Products, Application Note 178*, (June 5, 2002).

[12-9] Richey, Rodger, *Yet Another Clock Featuring the PIC16C924*, AN649, Microchip Technology, Inc., Doc. No. DS00649A (1997).

[12-10] Dallas Semiconductor Corp. division of Maxim Integrated Products, Inc., *Crystal Considerations for Dallas Real-Time Clocks Application Note58*, (December 9, 2002).

Assembler 101

At the end of this chapter, you won't be an expert in assembler language programming. In fact, if this is your first introduction to PIC assembler, you won't even qualify as a novice. What I hope to do, however, is to illuminate a few of the more important assembler programming principles to prepare the way for Chapters 14 and 15 where we will learn assembler routines that do things otherwise impossible with MBasic and speed up certain operations.

The Basics

Let's start with the basics. To keep the discussion to a manageable length, we'll limit our discussion to the 16F876/877/876A/877A series devices. For more details on these devices, as well as other PICs, consult the sources identified in the References section.

What is Assembler?

As we learned in Chapter 2, the MBasic programs you write are compiled into an assembler format, linked with library functions by a linker ultimately converted into machine instruction executable by a PIC.

Assembler is a human compatible form of directly addressing a PIC's built-in instruction set. A PIC's instruction set—and mid-range PICs such as the 16F87x/87xA devices have only 35 instructions—works with data and hardware at the most elementary level. Typical assembler instructions involve moving bytes from one storage location to another, or setting and clearing bits. Even the most complex mid-range PIC instructions do nothing more than add or subtract 8-bit values. Consequently, one line of MBasic code may correspond to dozens, or even hundreds of lines of assembler code.

We may illustrate this difference by comparing MBasic, assembler and machine instructions to accomplish a very simple task: **A = 123**, where **A** is a byte variable.

MBasic	Assembler	Machine Code
A=123	movlw .123	11 00 00 01111011
	movwf A	00 00 00 11001000

It's easy to write **A=123**. It's more difficult to write the two assembler instructions and it's far more difficult and time consuming to manually construct the machine code instructions. Fortunately, we'll never have to deal with machine code; MBasic's integrated compiler and assembler will mask that complexity from us.

What do these two cryptic assembler instructions mean?

movlw .123—means to load the 8 bits corresponding to the decimal number 123 into the PIC's working register, W. The period symbol tells the assembler that the number is a decimal number, not a hexadecimal (written in the form 0xNN, or a binary number, written in the form b'NNNNNNNN'). In PIC assembler terminology, copying a value from one location to another is called a *move*. A constant is called a *literal* and the working register is called *w*. Hence, the assembler operation code name (usually called simply an opcode) is `movlw`, standing for <u>mov</u>e a <u>l</u>iteral to register <u>w</u>.

`movwf A`—means to copy the value currently in working register w into the file location identified by the value of the constant **A**. (We'll assume for the moment that **A** represents the storage register (file) at physical address $48.) The opcode name `movwf` is constructed from its actions; to <u>mov</u>e a value from register <u>w</u> to a <u>f</u>ile.

Not so fast, you're probably thinking. You just said **A** is a constant with a value of $48; but not one page earlier you said **A** is a variable that we are going to set to have the value 123. Make up your mind, which is it? The most accurate description of **A** is that it is neither a variable nor a constant, but rather it is an *address* at which a byte is stored. That *address* is static and is assigned—by the compiler, or by you if you write only in assembler—and does not change, so in that regard **A** is a constant. However, the *value* stored at *address* **A** can change. Hence, **A** can be considered a variable when we refer to the contents of the memory at address **A**, or a constant when we refer to the memory address itself.

Terminology

When starting out to learn something new, the hardest part is often understanding the terminology. The underlying idea may be simple, but it's described in unfamiliar words. Or, the words may be familiar, but they are used in a fashion that divorces them from their normal meaning. We've already thrown around the words, register, file, literal, move, opcode, and address, to mention but a few.

To understand the terminology, we must understand the PIC's architecture. First, remember what we learned in Chapter 1; the PIC follows the Harvard architecture and has separate *program* memory and *data* memory. Program memory holds—for mid-range PICs—a series of 14-bit machine code statements, similar to those we noted earlier. The data memory is 8-bits wide and holds not only variables of the type we use in MBasic and assembler, but also other variables controlling the PIC's functioning, its hardware and its features.

Data Memory

Register Files

Let's look first at how the data memory is organized in a 16F876/877/ 876A/877A device. Figure 13-1 provides a conceptual view of the data memory. The data memory is organized into four *banks*, with each bank having 128 addresses, giving a maximum of 512 unique addresses. Each address represents one of three types of 8-bit wide register files:

- *General-Purpose Register*—user RAM. The '87x chips we're discussing have 368 general-purpose registers.
- *Special Function Register*—a file that may be accessed by the user (in almost all cases, both read and write) but it is used internally by the PIC for a function. For example, all ports are special function registers, and thus we may both read from and write to, for example, PortB.
- *Unimplemented Address*—an address that is dead because it is not physically implemented. The '87x chips we're discussing have 15 dead addresses.

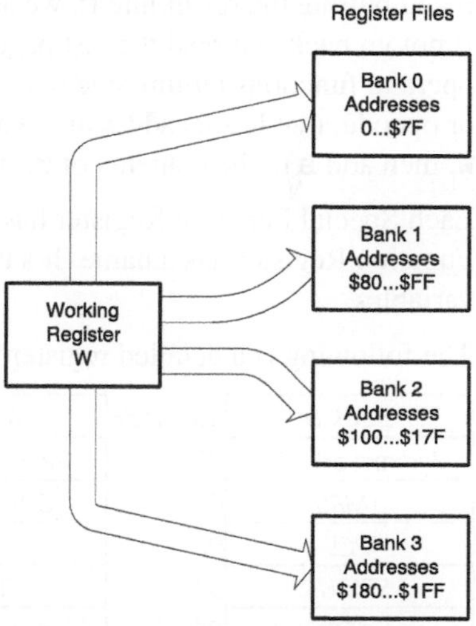

Figure 13-1: Conceptual view of 16F87x data memory.

Microchip calls these memory locations *register files*, or *registers*, or *files*. Don't get confused; whether we call it a register, or a file, it is still the location of an 8-bit information storage unit, identified by a numerical address in the range 0...511. (Banking makes the actual addressing a bit more complex, as we'll see later.)

The *working* or *W* register is special—it's an 8-bit memory location built into the PIC's central processing core and tied to the arithmetic and logic unit. It doesn't have an address in the same sense that the register files do; rather it's identified through opcodes as the source of data or the destination of a particular action.

Many logic and arithmetic function require two operands; such as addition, subtraction and logic operations. One operand may be a constant and the other a register value, or both may be register values. The arithmetic and logic unit portion of the PIC performs these functions and, as illustrated in Figure 13-2 , *one of the two input operands must be held in the* **w** *register*. This is a simple, but inviolate rule for mid-range PICs—if the opcode has two operands, one must be in the **w** register. Where the result of the operation is stored is determined by the opcode's destination bit and may be either the

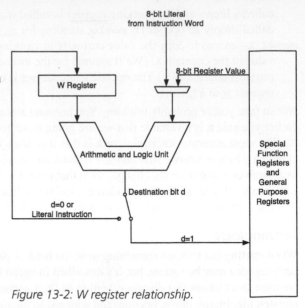

Figure 13-2: W register relationship.

w register or one of the operand registers. It can't be a third register—it's either the **w** register or one of the operand registers.

Suppose we have two files, **A** and **B** and that we wish to add their values together and store the result in file **C**. In MBasic, this operation is **C** = **A** + **B**. (We'll assume that somehow the letters **A**, **B** and **C** are linked to the addresses of three general-purpose registers. You do understand that when we talk about adding **A** to **B** and putting the result into **C**, we are referring to the contents of the memory locations **A**, **B** and **C**, right? If not go back and read the last pages again.) Addition is a two-operand function and, as we learned, two operand functions require one operand to be held in **w**. Consequently, there is no single machine instruction, or opcode, that lets us add **A** and **B** and store the result at **C**. Rather, we must first move the contents of **A** into **w**, then add **B** to the contents of **w**, storing the result in **w** and lastly move **w**'s contents to **C**.

Each Special Function Register has been given a name by Microchip and each bit within each Special Function Register has a name. It's up to us to assign names to General-Purpose Registers we use to hold variables.

The following is a detailed register file map for the 16F876/A/877/A devices.

Bank 0	Hex Adr	Bank 1	Hex Adr	Bank 2	Hex Adr	Bank 3	Hex Adr
Indirect addr.(*)	00	Indirect addr.(*)	80	Indirect addr.(*)	100	Indirect addr.(*)	180
TMR0	01	OPTION_REG	81	TMR0	101	OPTION_REG	181
PCL	02	PCL	82	PCL	102	PCL	182
STATUS	03	STATUS	83	STATUS	103	STATUS	183
FSR	04	FSR	84	FSR	104	FSR	184
PORTA	05	TRISA	85		105		185
PORTB	06	TRISB	86	PORTB	106	TRISB	186
PORTC	07	TRISC	87		107		187
PORTD(1)	08	TRISD(1)	88		108		188

Bank 0	Hex Adr	Bank 1	Hex Adr	Bank 2	Hex Adr	Bank 3	Hex Adr
PORTE(1)	09	TRISE(1)	89		109		189
PCLATH	0A	PCLATH	8A	PCLATH	10A	PCLATH	18A
INTCON	0B	INTCON	8B	INTCON	10B	INTCON	18B
PIR1	0C	PIE1	8C	EEDATA	10C	EECON1	18C
PIR2	0D	PIE2	8D	EEADR	10D	EECON2	18D
TMR1L	0E	PCON	8E	EEDATH	10E		18E
TMR1H	0F		8F	EEADRH	10F		18F
T1CON	10		90		110		190
TMR2	11	SSPCON2	91		111		191
T2CON	12	PR2	92		112		192
SSPBUF	13	SSPADD	93		113		193
SSPCON	14	SSPSTAT	94		114		194
CCPR1L	15		95		115		195
CCPR1H	16		96	General Purpose Register (RAM) 16 Bytes	116	General Purpose Register (RAM) 16 Bytes	196
CCP1CON	17		97		117		197
RCSTA	18	TXSTA	98		118		198
TXREG	19	SPBRG	99		119		199
RCREG	1A		9A		11A		19A
CCPR2L	1B		9B		11B		19B
CCPR2H	1C	CMCON	9C		11C		19C
CCP2CON	1D	CVRCON	9D		11D		19D
ADRESH	1E	ADRESL	9E		11E		19E
ADCON0	1F	ADCON1	9F		11F		19F
	20		A0		120		1A0
General-Purpose Register (RAM) 98 Bytes		General Purpose Register (RAM) 80 Bytes		General Purpose Register (RAM) 80 Bytes		General Purpose Register (RAM) 80 Bytes	
			EF		16F		1EF
		Accesses 70-7F	F0	Accesses 70-7F	170	Accesses 70-7F	1F0
	7F		FF		17F		1FF

* Not a physical register
(1) Not implemented in '876
(2) Reserved

This is an important table; so take some time to carefully look it over. You should see several things:

- Certain files are repeated in multiple banks. This permits us to access those files regardless of the current bank settings.
- Some General-Purpose Registers are marked "accesses." This represents multiple addressed shared memory; the same 16 bytes of memory are addressed at $70...$7D, $F0...$FF, $170...$17F and $1F0...$1FF. This permits quick access to certain memory locations regardless of the bank settings.
- It's full of strange names, like **INTCON**, **CCP1CON** and **SSPSTAT**. This book is not a comprehensive tutorial on assembler programming, and we don't have time to discuss these Special Purpose Registers

except when we find it necessary to use one. And, we'll leave those discussions to Chapters 14 and 15, and those other chapters were we incorporate assembler into our MBasic programs. Of course, detailed information is available on each Special Purpose Register in Microchip's data sheets.

Banking

Let's look at banks and banking. To address memory locations from 0...$1FF (0...511 decimal) requires 9 address bits. Opcodes that involve addressing files must have the file address embedded in the opcode. These opcodes have the following bit structure:

13	8	7	6	0
OPCODE		Destination	File Address	
6 bits		1 bit	7 bits	

For example, remember our **movwf A** opcode that translated into a machine instruction **00000011001000**? It turns out that the opcode for **movwf** is **0000001**, representing **000000** for the **mov** operation and **1** for the destination code.Since mid-range PICs have only a 14-bit instruction word, this leaves only 7 bits to hold **A**'s address, **1001000**. We've assigned A to represent storage address $48, and seven bits are adequate, since all higher bits are zeros. But, suppose **A**'s address is $196—a perfectly legitimate user file location, but one requiring all nine address bits.

We have a problem. The cleanest fix would have been for Microchip to use a 16-bit op code with 9 bits of address space, but that would have destroyed backwards compatibility with older devices as well as requiring Microchip to extensively rework its silicon layout, an expensive task. We are therefore forced to deal, for historical and compatibility reasons, with the mid-range PIC being able to address only file addresses in the range 0...127 directly in the opcode. Microchip's solution was to break the 9-bit file address into a 7-bit opcode *relative address* portion and a 2-bit *bank address* portion.

The bank address bits are held in the Special Function Register bits **RB1** and **RP0**. (These bits are bits 6 and 5 of the **status** register.) If we wish to address memory location $196, residing in Bank 3, therefore, we would first set **RB1** and **RB0** to **%1**. The rightmost seven bits of $196, or $16 represent the relative address. (Don't worry if this sounds complicated—the assembler hides much of this complexity from you.)

To simplify setting and clearing the bank bits, we may use an assembler *macro*, **BankSel**, invoked with the following syntax:

```
Banksel FileAddress
```

The term **FileAddress** is the file name. (Remember, the name of a file is defined as a number representing its address, so the assembler simply uses the address value.) A *macro* can be thought of as a "user defined" opcode. Instead of a real opcode that generates a specific machine instruction, the assembler expands a macro into one or more "real" opcodes. The macro's advantages are space savings and automating a repetitive task. In this case, the macro **BankSel** expands into instructions that set or clear **RB1** and **RB0** corresponding to the **FileAddress** value.

Suppose our variable name is **A** and its at address $196. (In assembler, the equate operation, **Equ**, is the equivalent to MBasic's constant assignment operator, **Con**.) How would we use **BankSel**?

```
A       Equ     0x196

BankSel A
movlw           .123
movwf           A
```

OK you're thinking I understand that the **BankSel** macro extracts the two leftmost bits from the value of **A**, but how does **movwf** get the rightmost seven bits without any special instruction? The answer is that since there are only seven bits of opcode space available for the address, only the rightmost seven bits of **A** are loaded into the opcode, without additional commands. (This simplistic construction may generate assembler warnings, so we'll learn a better way later in this chapter.)

Program Memory

In Chapter 1, we learned that the 16F876/877/876A/877A devices have 8K program memory. The *program counter* in these devices is 13 bits wide and 2^{13} is 8192, so it can address all 8K of program memory. The program counter is the part of the PIC's core that keeps track of which program instruction is currently executing and which program instruction should next be executed. If our assembler code uses all 8K sequentially, so that each program instruction address follows the next in strict numerical order, all is well.

However, when writing a program, it's common to have execution to jump from one location to another. You might write some code in a subroutine that you then *call* (the assembler equivalent of **GoSub**) from multiple locations within your main program. Or, program execution may need to jump around a data table stored in program memory. And, you may wish to repeat some code in the assembler equivalent of a **For...Next** or a **While...WEND** loop. This implies that our opcodes must include instructions to jump execution from the current program counter location to a different location.

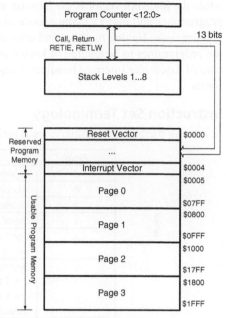

Figure 13-3: Conceptual view—program memory in 16F876/877 devices.

Do you see the problem? Our opcode structure must accommodate a series of jump instructions that say, in essence, stop executing the current code and transfer execution to a new program memory address. But our opcode only has 14-bits of space and a full jump address alone is 13 bits. This only leaves 1 bit for all jump opcode instructions, which is unsatisfactory to say the least. So, the engineers at Microchip decided to segment the program memory, and split the program counter into two registers. Figure 13-3 illustrates the program memory structure. Instead of calling them banks, the program memory segments are called *pages*. The 16F876/877 devices have four program memory pages, each 2K long.

There are two jump opcodes in mid-range PICs, the **call** and **goto** instructions. Both reserve 11 bits for the jump address:

13	11	10	0
OPCODE		*Execution Transfer Address*	
3 bits		11 bits	

Where then are the other two bits of the 13-bit jump address? Or, more pertinently, since our data structure only deals with eight bits, how do we access a 13-bit program control register? The answer is a bit complicated. The 13-bit program control register is split into two parts: the lower 8 bits are held in the **PCL** register, which may be read or written to, just like any other special purpose register. The upper 5 bits of the program control register are not directly accessible and may be reached only through the **PCLATH** register. We'll end our discussion of the **PCL** and **PCLATH** registers now, and return to them in the context of specific examples in Chapter 14.

The remaining parts of Figure 13-3 are the *stack* memory and the *reset* and *interrupt vectors*. The stack is dedicated memory where the PIC stores return addresses for interrupt and call operations. (This stack is internal to the PIC and is not the same as the general-purpose register file memory that MBasic requires for its stack.) This is transparent to the programmer and for the applications discussed in this book need not further concern you. The reset vector is the location where the program counter begins executing code after a reset, while the interrupt vector is the location where execution jumps to in the event of an interrupt. Normally the programmer will insert a **GoTo** opcode in those locations, pointing to the real reset code, or the real interrupt handler code. Having the reset and interrupt start point hardwired to $0000 and $0004, respectively, permits the programmer to be assured of known code execution point in the event of a reset or interrupt event. In the case of assembler interrupt handlers, Chapter 15 will show us that MBasic takes care of much of this work for us.

Instruction Set Terminology

In reading the opcode summary, we'll make extensive use of the following field descriptors.

Field	Descriptor Meaning
f	Register file address ($00...$7F)
W	Working Register (accumulator)
b	Bit address within an 8-bit file register
k	Literal field, constant data or label
x	Don't care location (0 or 1). The assembler generates code with 0.
d	Destination selection: d = 0 store result in W d = 1 store result in register f Default is d = 1
PC	Program Counter
TO	Time-out Bit
PD	Power-down Bit

One Special Purpose Register that we will deal with frequently when reviewing the opcodes is the **Status** register:

Bit 7	Bit 6	Bit 5	Bit 4	Bit 3	Bit 2	Bit 1	Bit 0
IRP	RP1	RP0	$\overline{TO}$	$\overline{PD}$	Z	DC	C

IRP—Register bank select bit for indirect addressing; covered in Chapter 14.

RP1 and RP0—As we previously saw, the **RP1** and **RP0** bits are the 9th and 8th register file address bits, respectively.

> 11 = Bank 3
> 10 = Bank 2
> 01 = Bank 1
> 00 = Bank 0

$\overline{TO}$—Time-out bit (the overscore indicates inverted status). We will not further discuss the **TO** bit.
$\overline{PD}$—Power-down bit (the overscore indicates inverted status). We will not further discuss the **PD** bit.
Z—Zero bit:

> 1 = the result of the arithmetic or logic operation is zero
> 0 = the result of the arithmetic or logic operation is nonzero

DC—Digit carry/$\overline{\text{Borrow}}$ bit (for **ADDWF**, **ADDLW**, **SUBLW** and **SUBWF** instructions). For borrow, the bit sense is reversed:

1 = A carry-out from the 4th low order bit of the result occurred

0 = No carry-out from the 4th low-order bit of the result

C—Carry/$\overline{\text{Borrow}}$ bit (for **ADDWF**, **ADDLW**, **SUBLW** and **SUBWF** instructions):

1 = A carry-out from the Most Significant bit of the result occurred

0 = No carry-out from the Most Significant bit of the result occurred.

For $\overline{\text{Borrow}}$, the bit sense is reversed. A subtraction is executed by adding the two's complement of the second operand. For rotate (**RRF**, **RRL**) instructions, this bit is loaded with the high (or low) order bit of the source register.

OpCodes

The remainder of this chapter is a summary of the 35 16F87x/87xA opcodes. Microchip's Midrange Reference Manual provides more information and should be consulted where necessary.

When reading the summary, remember that a "literal" is Microchip's name for a constant. A literal may be a number entered directly, such as .123 or as a named constant, linked to a numerical value with either MBasic's **Con** operation, or an **Equ** assignment inside an assembler module. The notation **Name** **<N>** indicates the Nth bit of file register **Name**. N may also be a name, such as **Status** **<C>** which means the bit named **C** of the file register named **Status**.

In providing examples, we will use Microchip's hexadecimal notation, **0xNN**, instead of Basic Micros's **$NN** identifier.

The field "cycles" indicates how many clock cycles ($F_{OSC}/4$) is required to complete the particular operation.

Standard Notes:

Three notes apply to several opcodes:

1. When an I/O register is modified as a function of itself; for example, **MOVF PORTB,1**, the value used will be that value present on the pins themselves. For example, if the data latch is **%1** for a pin configured as input and is driven low by an external device, the data will be written back with a **%0**.

2. If this instruction is executed on the **TMR0** register (and where applicable, d = **1**), the prescaler will be cleared if assigned to the **Timer0** module.

3. If Program Counter (PC) is modified, or a conditional test is true, the instruction requires two cycles. The second cycle is executed as a **NOP**.

OpCode	Operands	Description	Cycles	Affected	Notes
ADDLW	k	Add Literal and W	1	C, DC, Z	

Addlw adds a literal 'k' to the value currently in **w**. The result remains in **w**. The literal 'k' must be in the range 0...255. If the literal exceeds 255, only the lowest 8 bits will be used.

Example 1:

```
Addlw   .123    ;adds decimal 123 to the current contents of W
```

Remember that **w** is only 8-bits wide and that it is possible the result will overflow **w**. In the event of an overflow, **Status** **<C>** will be set. If the sum is zero (such as **w** holds 128, to which 128 is added) **Status** **<Z>** will be set. In this latter example, both **z** and **c** will be set.

Example 2:

Assume **A** is defined as a byte variable and its address is 0x30. The value stored at address 0x30 is .128 (0x80). Assume **w**'s initial value is 0x10.

```
Addlw   A
```

The result is 0x48, not 0x90. This is because we are adding the numerical value of **A**, which is its *address*, not the *value* of the contents at address **A**. Confusing **A**'s address with the value of the contents of the file at **A**'s address is a common error.

OpCode	Operands	Description	Cycles	Affected	Notes
ADDWF	f, d	Add W and f	1	C,DC,Z	1,2

Addwf adds the value currently in **w** to the value held in a file 'f'. The result is either placed in **w** (destination = **w**) or in the file 'f' (destination = **f**).

Example 1:

```
Addwf   A,f
```

Before the opcode is executed, assume the value at address **A** is .123 and that **w** holds .10. After execution, the value at address A will be .133 and **w**'s value remains at .10.

Example 2:

```
Addwf   A,w
```

Before the opcode is executed, assume the value at address **A** is .123 and that **w** holds .10. After execution, the value at address **A** remains as .123 and **w** will hold .133.

Since both **w** and register files are 8-bits wide, overflow is possible. Overflow will set **Status <C>**, and a zero result will set **Status <Z>**.

OpCode	Operands	Description	Cycles	Affected	Notes
ANDLW	k	AND Literal with W	1	Z	

Andlw performs a bit-by-bit logical AND of the value in **w** and the literal 'k.' The result is retained in the **w** register.

The AND function truth table is:

Input A	Input B	A AND B
0	0	0
0	1	0
1	0	0
1	1	1

The logical AND function is often used to "mask off" bits, as in the following example.

Example 1:

Assume **w** holds .123 before the instruction.

```
                    ;Before Instruction W: 01111011
        Andlw   31  ;                  31: 00011111
                                           --------
                    After W   00011011 = 27
```

We'll see in Chapter 14 how the AND function may be used to calculate the division remainder. It may also be used to test multiple bits to see if all are zero by using a mask literal with `%1` values in the places where the bits to be tested reside. If all bits coinciding with the `%1` mask bits are 0, the result will is 0 and `Status <C>` will be set.

OpCode	Operands	Description	Cycles	Affected	Notes
ANDWF	f, d	AND W with f	1	Z	1,2

`Andwf` performs a bit-by-bit logical AND of the value in `w` and the value held in a file 'f.' The result is either placed in `w` (destination = `w`) or in the file 'f' (destination = `f`).

The AND function truth table is:

Input A	Input B	A AND B
0	0	0
0	1	0
1	0	0
1	1	1

The logical AND function is often used to "mask off" bits, as in the following example.

Example 1:

Assume `w` holds .123 before the instruction and assume file `A` holds .31.

```
              ;Before Instruction W: 01111011 = .123
     Andwf  A,w  ;            31: 00011111 = .31
                          --------
              After W    00011011 = .27
```

The value stored at file address `A` remains .31.

Example 2:

Assume `w` holds .123 before the instruction and assume file `A` holds 31.

```
              ;Before Instruction W: 01111011 = .123
     Andwf  A,f  ;            31: 00011111 = .31
                          --------
              After f    00011011 = .27
```

The value in `w` remains .123.

OpCode	Operands	Description	Cycles	Affected	Notes
BCF	f, b	Bit Clear f	1		1,2

BCF clears (makes it a zero) bit 'b' of a file 'f'. Bit 'b' may be identified with a number (0...7) or a named constant in the range 0...7.

Example 1:

Assume PortB is set to be an output and we wish to set pin `B0` to low.

```
     Bcf    PortB,0
```

OpCode	Operands	Description	Cycles	Affected	Notes
BSF	f, b	Bit Set f	1		1,2

BSF sets (makes it a one) bit' b' of a file 'f'. Bit 'b' may be identified with a number (0…7) or a named constant in the range 0…7.

Example 1:

Assume PortB is set to be an output and we wish to set pin **B0** to high.

```
Bsf     PortB,0
```

OpCode	Operands	Description	Cycles	Affected	Notes
BTFSC	f, b	*Bit Test f, Skip if Clear*	1 (2)		3

Btfsc checks the status of bit 'b' in file 'f'. If bit 'b' is clear (equal to 0), the next opcode is skipped. If bit 'b' is set (equal to 1) the next opcode is executed.

Btfsc is a program flow control opcodes and is analogous to MBasic's **If…Then** test.

Example 1:

Suppose PortB is set as input and has a switch connected to pin **B0**, with a pull-up resistor. Depending upon whether the switch is set (switch open and **B0=%1**) or clear (switch closed and **B0=%0**) we wish to execute different code.

```
…
Btfsc   PortB,0
        GoTo SwitchOpenRoutine          ;SwitchOpenRoutine executed if no skip
…                                       ;code from here down is executed otherwise
```

The **Btfsc** opcode tests **PortB**'s bit **0** (pin **B0**) value. If the value is clear (**B0=0**), the next instruction (**GoTo SwitchOpenRoutine**) is skipped and code execution resumes with the statement immediately following the skipped instruction. If **PortB**, bit **0** is set (**B0=1**) the next instruction is <u>not</u> skipped so the **GoTo SwitchOpenRoutine** statement executes and program execution branches to the code beginning with the label **SwitchOpenRoutine**.

A common use for **btfsc** is to test the result of an arithmetic or logical operation and branch program flow based upon whether **Status <C>** or **Status <Z>** flags are set or clear.

OpCode	Operands	Description	Cycles	Affected	Notes
BTFSS	f, b	*Bit Test f, Skip if Set*	1 (2)		3

Btfss checks the status of bit 'b' in file' f'. If bit 'b' is set (equal to 1), the next opcode is skipped. If bit 'b' is clear (equal to 0) the next opcode is executed. **Btfss** is the mirror image of **btfsc**.

Btfsc is a program flow control opcodes and is analogous to MBasic's **If…Then** test.

Example 1:

Suppose PortB is set as input and has a switch connected to pin **B0**, with a pull-up resistor. Depending upon whether the switch is set (switch open and **B0=%1**) or clear (switch closed and **B0=%0**) we wish to execute different code.

```
…
Btfss   PortB,0
        GoTo SwitchClosedRoutine        ;SwitchClosedRoutine executed if no skip
…                                       ;code from here down is executed otherwise
```

The **Btfss** opcode tests **PortB**'s bit **0** (pin **B0**) value. If the value is set (**B0=1**), the next instruction (**GoTo SwitchClosedRoutine**) is skipped and code execution resumes with the statement immediately following the skipped instruction. If **PortB**, bit **0** is clear (**B0=0**) the next instruction is *not* skipped so the **GoTo**

SwitchClosedRoutine statement executes and program execution branches to the code beginning with the label **SwitchClosedRoutine**.

A common use for **btfss** is to test the result of an arithmetic or logical operation and branch program flow based upon whether **Status <C>** or **Status <Z>** flags are set or clear.

OpCode	Operands	Description	Cycles	Affected	Notes
CALL	k	*Call Subroutine*	2		

Call transfers program execution to a subroutine, similar to MBasic's **GoSub** function. Upon return from the subroutine, program execution resumes with the opcode immediately following the **Call** operator.

Example 1:

```
...
Call SubOne
...

SubOne
...        ;subroutine code goes here
Return
```

The operand for **Call** is a literal value, in the range 0...2047. However, in most instances, you will wish to use a label, such as **SubOne** in the example, and allow the assembler to resolve the label into a numerical argument for the **Call** opcode. As we learned earlier, the range of the call is one program page and if the jump distance is greater, the two extra page bits must be generated. This is an advanced topic and will not be further considered in this chapter.

OpCode	Operands	Description	Cycles	Affected	Notes
CLRF	f	*Clear f*	1	Z	2

Clrf clears, or sets to 0 all bits in file 'f'. Since the result is zero, **Status <Z>** is set by this operation.

Example 1:

Assume file **A** has been defined and that it holds the value .123.

```
              ;Before instruction A= .123 01111011
    Clrf    A ;After Clrf A        A= 0     00000000
```

OpCode	Operands	Description	Cycles	Affected	Notes
CLRW	–	*Clear W*	1	Z	

Clrw clears, or sets all bits to 0 in register **W**. Since the result is zero, **Status <Z>** is set by this operation.

Example 1:

Assume **W** holds the value .123.

```
              ;Before instruction W= .123 01111011
    Clrw      ;After Clrw A        W= 0     00000000
```

OpCode	Operands	Description	Cycles	Affected	Notes
CLRWDT	–	*Clear Watchdog Timer*	1	$\overline{TO},\overline{PD}$	

ClrWdt clears the watchdog timer. It also sets the $\overline{TO}$ and $\overline{PD}$ bits.

OpCode	Operands	Description	Cycles	Affected	Notes
COMF	f, d	Complement f	1	Z	1,2

Comf performs a one's complement on the contents of file 'f.' The result is either placed in **w** (destination = **w**) or in the file 'f' (destination = **f**). The one's complement operation replaces all 0's by one's and vice versa.

Example 1:

Assume file **A** has been defined and that it holds the value .123.

```
                  ;Before instruction A= .123 01111011
     Comf   A,f   ;After Comf A      A= .132 10000100
```

In 8-bit arithmetic used in **w** and **A**, the sum of **A** and its one's complement is 255. Note this is not the same as for two's complement, where the sum of **A** and its two's complement is 256.

OpCode	Operands	Description	Cycles	Affected	Notes
DECF	f, d	Decrement f	1	Z	1,2

Decf decrements by 1 the value in file 'f'. If, after decrementing, the result is 0, **Status <Z>** is set. The result is either placed in **w** (destination = **w**) or in the file 'f' (destination = **f**). **Decf** is the assembler analog of **A=A-1**.

Example 1:

Assume file **A** has been defined and that it holds the value .123.

```
                  ;Before instruction A= .123 01111011
     Decf   A,f   ;After Decf A      A= .122 01111010
```

The result of decrementing 0 is, of course .255.

Decf is useful in control loops, as it can be combined with a **btfss** or **btfsc** operator to provide the assembler analog of MBasic's **For...Next** loop.

OpCode	Operands	Description	Cycles	Affected	Notes
DECFSZ	f, d	Decrement f, Skip if 0	1(2)		1,2,3

Decfsz combines the decrement operation with a zero result test; the value in file 'f' is decremented by 1 and if the result is 0, the next instruction is skipped. The result is either placed in **w** (destination = **w**) or in the file 'f' (destination = **f**).

Example 1:

Assume file **A** has been defined and that it holds the value .123.

```
     Loop      ...          ;this code is executed 123 times
               Decfsz A,f
               GoTo   Loop
     ...
```

When program execution hits the **Decfsz** instruction, the then current value of **A** is decremented by 1 and the result tested. If the result is nonzero, the next instruction is executed and program flow jumps back to the label **Loop**. If the result is zero, the **GoTo Loop** statement is skipped over and program execution goes to the first opcode following the **GoTo**.

Example 1 is the assembler analog of:

```
A = 123
Repeat
    ...
    A=A-1
Until A=0
```

Note that the assembler code in Example 1 is executed 123 times, not 124 times because the test for zero is made at the bottom of the loop, *after* decrementing **A**.

OpCode	Operands	Description	Cycles	Affected	Notes
GOTO	*k*	*Go to Address*	*2*		

GoTo transfers program execution to a code sequence. Its effect is similar to MBasic's **GoTo** statement.

Example 1:

Assume file **A** has been defined and that it holds the value .123.

```
Loop    ...          ;this code is executed 123 times
    Decfsz  A,f
    GoTo    Loop
```

When A is nonzero, the **GoTo** statement is executed and program flow jumps to the address identified by the label **Loop**.

The operand for **GoTo** is a literal value, in the range 0…2047. However, in most instances, you will wish to use a label, such as **Loop** in the example, and allow the assembler to resolve the label into a numerical argument for the **GoTo** opcode. As we learned earlier, the range of the jump is one program page and if the jump distance is greater, the two extra page bits must be generated. This is an advanced topic and will not be further considered in this chapter.

OpCode	Operands	Description	Cycles	Affected	Notes
INCF	*f, d*	*Increment f*	*1*	*Z*	*1,2*

Incf increments by 1 the value in file 'f.' If, after incrementing, the result is 0, **Status <Z>** is set. The result is either placed in **W** (destination = **w**) or in the file 'f' (destination = **f**). **Incf** is the assembler analog of **A=A+1**.

Example 1:

Assume file **A** has been defined and that it holds the value .123.

```
                ;Before instruction A= .123 01111011
    Incf    A,f ;After Decf A       A= .124 01111100
```

The result of incrementing .255 is, of course .0 In this case, **Status <Z>** will be set.

Incf is useful in control loops, as it can be combined with a **btfss** or **btfsc** operator to provide the assembler analog of MBasic's **For…Next** loop or other control structures.

OpCode	Operands	Description	Cycles	Affected	Notes
INCFSZ	*f, d*	*Increment f, Skip if 0*	*1(2)*		*1,2,3*

Incfsz combines the increment operation with a zero result test; the value in file 'f' is incremented by 1 and if the result is 0, the next instruction is skipped. The result is either placed in **W** (destination = **w**) or in the file 'f' (destination = **f**).

Example 1:

Assume file **A** has been defined and that it holds the value .123.

```
Loop      ...          ;this code is executed 133 times not 123 times
          Incfsz A,f
          GoTo   Loop
          ...
```

When program execution hits the **Incfsz** instruction, the then current value of **A** is incremented by 1 and the result tested. If the result is nonzero, the next instruction is executed and program flow jumps back to the label **Loop**. If the result is zero, the **GoTo Loop** statement is skipped over and program execution goes to the first opcode following the **GoTo**.

Example 1 is the assembler analog of:

```
A = 123
Repeat
        ...
    A=A+1
Until A=0
```

A becomes 0, of course, due to rollover. Rollover occurs when **A** should be 256, but, since **A** only holds 8-bits it instead is zero. (256 is **%1 00000000** where the first **%1** is the 9th bit.) The number of times the loop code will execute is thus 256-123, or 133 times.

OpCode	Operands	Description	Cycles	Affected	Notes
IORLW	k	*Inclusive OR Literal with W*	1	Z	

Iorlw performs a bit-by-bit logical OR of the value in **W** and the literal 'k.' The result is retained in the **W** register.

The OR function truth table is:

Input A	Input B	A OR B
0	0	0
0	1	1
1	0	1
1	1	1

Example 1:

Assume **w** holds .123 before the instruction.

```
                  ;Before Instruction W: 01111011 = .123
    Iorlw   31    ;                  31: 00011111 = .31
                                        --------
                  After    W: 01111111 = 127
```

OpCode	Operands	Description	Cycles	Affected	Notes
IORWF	f, d	*Inclusive OR W with f*	1	Z	1,2

Iorwf performs a bit-by-bit logical OR of the value in **W** and the value held in a file 'f.' The result is either placed in **W** (destination = **w**) or in the file 'f' (destination = **f**).

The OR function truth table is:

Input A	Input B	A OR B
0	0	0
0	1	1
1	0	1
1	1	1

Example 1:

Assume **w** holds .123 before the instruction and assume file **A** holds .31.

```
                    ;Before Instruction W: 01111011 = .123
    Andwf   A,w     ;                     31: 00011111 = .31
                                             --------
                    After   W: 01111111 = .127
```

The value stored at file address **A** remains .31.

Example 2:

Assume **w** holds .123 before the instruction and assume file **A** holds 31.

```
                    ;Before Instruction W: 01111011
    Andwf   A,f     ;                     31: 00011111
                                             --------
                    After   f: 01111111 = 127
```

The value in **w** remains .123.

OpCode	Operands	Description	Cycles	Affected	Notes
MOVF	f, d	Move f	1	Z	1,2

Movf copies the value in file 'f' to either **w** or back to the same file 'f.' The result is either placed in **w** (destination = **w**) or in the source file 'f' (destination = **f**).

It may seem pointless to copy a file back to itself. However, copying the file back to itself sets the **Status <Z>** flag if the file value is 0, so it permits us to test the file 'f' for zero value. **Movf** is most often used to load **w** with the value of file 'f.'

Example 1:

Assume file **A** has been defined and that it holds the value .123.

```
    Movf    A,w
```

Register **w** will contain .123 after the operation is executed.

Remember—**movf** does not allow you to copy the contents of one file to another file. The copy is either to **w** or back to the source file.

OpCode	Operands	Description	Cycles	Affected	Notes
MOVLW	k	Move Literal to W	1		

Movlw loads register **w** with the value of the literal 'k' where k is in the range 0…255.

Example 1:

```
    Movlw   .123
```

After this operation is executed **w** contains the value .123.

OpCode	Operands	Description	Cycles	Affected	Notes
MOVWF	*f*	*Move W to f*	*1*		

Movwf copies the contents of register W to file 'f.' **Movwf** is how results in **w** are placed in a file 'f.'

Example 1:

Assume file **A** has been defined and that it holds an unknown value

```
Movlw   .123
Movwf   A
```

First, we load register **w** with the literal .123. Next, the **movwf** operation copies **w**'s value to file **A**. After these two operations complete, **A** holds the value .123.

OpCode	Operands	Description	Cycles	Affected	Notes
NOP	–	*No Operation*	*1*		

NOP is the "no operation" code. It consumes one cycle and is used to generate brief time delays such as to balance execution times where different code segments must execute in the identical time.

Example 1:

```
NOP
```

All register and file values are unchanged after the **NOP**.

OpCode	Operands	Description	Cycles	Affected	Notes
RETFIE	–	*Return from Interrupt*	*2*		

Retfie is used to terminate an interrupt handler.

Using **Retfie** is advanced topic and this opcode will not be further discussed. Our use of MBasic's assembler interrupt support will not require us to use the **Retfie** operation.

OpCode	Operands	Description	Cycles	Affected	Notes
RETLW	*k*	*Return with Literal in W*	*2*		

Retlw returns from a **GoTo** with a literal 'k' loaded in the **w** register.

Retlw is used most often for a data or look-up table, to return, and allows us to implement the assembler equivalent of MBasic's **ByteTable**, whereby we may access X(n), the value of the nth data entry.

Example 1:

Assume file **A** is declared and holds a value between 0...7.

```
        movf    A,w
        call    GetHalfPattern
...
GetHalfPattern
        addwf   PCL,f
        retlw   0x08
        retlw   0x0C
        retlw   0x04
        retlw   0x06
        retlw   0x02
        retlw   0x03
        retlw   0x01
        retlw   0x09
```

w is loaded with the index into the table. After the **Call** operation, **w** holds the indexed value. For example, if **A** holds .3, after the **Call** operation, **w** holds 0x6.

OpCode	Operands	Description	Cycles	Affected	Notes
RETURN	–	*Return from Subroutine*	2		

Return transfers program execution back to the **Call** statement at the end of a subroutine, similar to MBasic's **Return** operation. Upon return from the subroutine, program execution resumes with the opcode immediately following the **Call** operator.

Example 1:

```
        ...
        Call SubOne
        ...

    SubOne
        ...          ;subroutine code goes here
        Return
```

OpCode	Operands	Description	Cycles	Affected	Notes
RLF	f, d	*Rotate Left f through Carry*	1	C	1,2

RLF rotates the value held in file 'f' one bit to the left, through the carry flag. The result is either placed in **w** (destination = **w**) or in the source file 'f' (destination = **f**). This process re-circulates the bits.

Example 1:

Assume file **A** is declared and holds the value .179.

```
        RLF A,f
        RLF A,f
```

Status	Register A								
<C>	B7	B6	B5	B4	B3	B2	B1	B0	Comments
0	1	0	1	1	0	0	1	1	Before rotate command, A = 179
1	0	1	1	0	0	1	1	0	First shift to the left. A= 102
0	1	1	0	0	1	1	0	1	Second shift to left; note the 1 shifted through carry has now reappeared at B7. A= 205

If you wish a straight shift, not re-circulation, it is necessary to clear **Status** **<C>** after each **RLF**.

OpCode	Operands	Description	Cycles	Affected	Notes
RRF	f, d	*Rotate Right f through Carry*	1	C	1,2

RRF rotates the value held in file 'f' one bit to the right, through the carry flag. The result is either placed in **w** (destination = **w**) or in the source file 'f' (destination = **f**). This process re-circulates the bits.

Example 1:

Assume file **A** is declared and holds the value .123.

```
        RRF A,f
        RRF A,f
```

Status	Register A								Comments
<C>	B7	B6	B5	B4	B3	B2	B1	B0	
0	0	0	1	1	0	0	1	1	Before rotate command, A = 123
1	0	0	0	1	1	0	0	1	First shift to the right. A= 25
1	1	0	0	0	1	1	0	0	Second shift to right; note the 1 shifted through carry has now reappeared at B7. A= 140

If you wish a straight shift, not re-circulation, it is necessary to clear **Status <C>** after each **RRF**.

OpCode	Operands	Description	Cycles	Affected	Notes
SLEEP	–	Go into Standby mode	1	TO,PD	

Sleep places the PIC into standby mode. This is an advanced topic and will not be further discussed.

OpCode	Operands	Description	Cycles	Affected	Notes
SUBLW	k	Subtract W from Literal	1	C,DC,Z	

Sublw subtracts the **W** register from the literal 'k' using two's complement arithmetic. Note the order of subtraction is **W = k-W**.

Example 1:

Suppose file A is declared and holds the value .47

```
        movf A&0x7F,w           ;w = .47
        sublw  .100             ;w = (.100 - .47) = .53
```

Example 2:

Suppose file A is declared and holds the value .123

```
        movf A&0x7F,w           ;w = .123
        sublw  .100             ;w = (.100 - .123) = .233
```

Example 2 has the subtrahend larger than the minuend. The result may be calculated using the two's complement offset, .256:

Result = .256 + .100 - .123 = .233. Alternatively, the .256 may be viewed as a carry to the 9th bit.

OpCode	Operands	Description	Cycles	Affected	Notes
SUBWF	f, d	Subtract W from f	1	C,DC,Z	1,2

Subwf subtracts the value in register **W** from the value in file 'f' using two's complement arithmetic. The result is either placed in **W** (destination = **w**) or in the source file 'f' (destination = **f**). Note the order of subtraction is **Result = f-W**.

Example 1:

Suppose file **A** is declared and holds the value .100

```
                                ; A = .100
        movlw .47               ; W = .47
        subwf A,f               ; A = (.100-.47) = .53
```

Example 2:

Suppose file **A** is declared and holds the value .100

```
                          ; A = .100
        movlw .123        ; W = .123
        subwf A&0x7F,f ; A = (.100 - .123) = .233
```

Example 2 has the subtrahend larger than the minuend. The result may be calculated using the two's complement offset, .256:

Result = .256 + .100 − .123 = .233. Alternatively, the .256 may be viewed as a carry to the 9th bit.

OpCode	Operands	Description	Cycles	Affected	Notes
SWAPF	f, d	Swap nibbles in f	1		1,2

Swapf reverses the high and low nibble in file 'f.' The result is either placed in **w** (destination = **w**) or in the source file 'f' (destination = **f**).

Example 1:

Assume file **A** has been defined and that it holds the value .123.

```
                   ;Before Instruction A: 01111011 = .123
        Swapf   A,f    ;After Swapf        A: 10110111 = .183
```

OpCode	Operands	Description	Cycles	Affected	Notes
XORLW	k	Exclusive OR Literal with W	1	Z	

Xorlw performs a bit-by-bit logical XOR of the value in **w** with the literal 'k.' The result is retained in the **w** register.

Input A	Input B	A XOR B
0	0	0
0	1	1
1	0	1
1	1	0

Example 1:

Assume **w** holds .123 before the instruction.

```
                   ;Before Instruction W: 01111011 = .123
        Xorlw   31    ;                  31: 00011111 = .31
                                          --------
                        After   W: 01100100 = .100
```

OpCode	Operands	Description	Cycles	Affected	Notes
XORWF	f, d	Exclusive OR W with f	1	Z	1,2

Xorwf performs a bit-by-bit logical XOR of the value in **w** with the literal 'k.' The result is retained in the **w** register.

Input A	Input B	A XOR B
0	0	0
0	1	1
1	0	1
1	1	0

Example 1:

Assume the file **A** is declared and holds the value .123 and that **w** holds .31 before the instruction.

```
                 ;Before Instruction A: 01111011  = .123
     Xorwf   A,w   ;                   W: 00011111  = .31
                                          --------
                            After     W: 01100100  = .100
```

References

[13-1] A complete data sheet for most PICs comprises two elements; (a) a detailed "family" reference manual and (b) the particular device datasheet. MBasic supports only PICs from Microchip's "midrange" family and the associated PICmicro™ Mid-Range MCU Family Reference Manual may be downloaded at http://www.microchip.com/download/lit/suppdoc/refernce/midrange/33023.pdf. This is a 688-page document, in almost mind numbing detail, but nonetheless is an essential reference to a complete understanding of PICs. For individual PIC family member datasheets, the easiest source is to go to http://www.microchip.com/1010/pline/picmicro/index.htm and select either the PIC12 or PIC16 group and from that link then select the individual PIC device.

[13-2] Benson, David, *Easy PIC'n A Beginners Guide to Using PIC Microcontrollers version 3.1*, Square 1 Electronics, Kelseyville, CA (1999).

[13-3] Benson, David, *PIC'n Up the Pace PIC Microcontroller Applications Guide version 1.1*, Square 1 Electronics, Kelseyville, CA (1999).

[13-4] Predko, Myke, *Programming and Customizing PICmicro Microcontrollers, Second Ed.*, McGraw Hill, New York (2002).

[13-5] Predko, Myke, *PICMicro Microcontroller Pocket Reference*, McGraw Hill, New York (2001).

In-Line Assembler

Chapter 13 described the rudiments of assembler language, so it's now time to see how that knowledge may be used in MBasic. Our focus is not upon learning how to write complex assembler programs, but rather how to augment MBasic by adding a few lines of assembler here and there, making our programs run faster or access PIC features not otherwise available to us. This chapter provides a series of "bolt-in" replacements, simple assembler routines that may be directly substituted for MBasic statements. We will learn a process of stepwise refinement, where we replace MBasic statements one at a time, proceeding to the next statement only after verifying the earlier replacement.

We'll make frequent use of the word "macro" in this chapter, so let's first agree what a macro is. A "macro" in our context is a shortcut that is automatically expanded by the assembler into one or more real op codes.

Let's invent a macro for MBasic. MBasic doesn't allow user-defined macros, but we'll suppose it does. Our fictitious macro will save us from typing the full details to set up a **For…Next** loop. It invoked as: **@F v, s, f** where **v** is the variable name, **s** is the start value for the loop and **f** is the final value. Thus, our fictitious macro permits us to enter the line **@F i,0,10**. The compiler would then expand this macro shorthand into **For i = 0 to 10…Next** functionality.

If that were all the macro does, it would be an interesting way to save a few keystrokes, but not all that useful. There's nothing stopping us—at least in the realm of inventing features for MBasic that don't exist—from writing a second version of this macro, **@F2 i, 10** that expands into **For i = 0 to 2*10… Next**, where the blank template macro is **For <> = 0 to 2*<> Next** where the first <> is completed with the variable name, **i** in this example, and the second <> is completed with the second macro argument, **10** in this example. The point is that our macro can contain mathematical computations and constants. In many computer languages, macros can be multiline programs, so it may be more accurate to think of them as a special type of subroutine. And, in fact, the **@bank** macro we'll run into in this chapter is a multiline construct that includes an if…then test. We won't discuss the details of how macros are constructed, but you should understand the general concept of what a macro does. And, to confuse you even more, MBasic's in-line assembler macros have the same name as the assembler opcodes we studied in Chapter 13.

If you intend to do serious assembler programming, you'll need more information than imparted in this book. The references section recommends some useful publications.

Adding Assembler to MBasic Programs

MBasic provides four ways to add assembler code to your program:

Library macros—MBasic has library macros, for certain functions, such as **@high** for the standard function **high**, that automatically substitute quick assembler code for slower MBasic code. (We'll also see another library macro for assembler opcodes. We'll look at these when we examine how MBasic deals with general-purpose register file addressing.) Library macros are an undocumented feature of MBasic and are not covered in the current User's Guide.

Chapter 14

Assembler opcode macros—MBasic also has library macros for almost every assembler opcode so they may be used almost as if they were a standard BASIC language operator. These macros are not prefixed with the @ symbol, but rather have the same name as the associated opcode; for example, the direct macro for the assembler opcode **bcf** is **bcf**. (The call opcodes are not so defined and must be used through the **@call** and **@goto** macros, or via an **ASM** directives.)

ASM directive—We may package assembler code, grouped with curly braces {…}, following the compiler directive **ASM**. Code within the curly braces is passed directly to the assembler.

Interrupt handler—MBasic's special provisions for assembler interrupt code, **ISRASM**, is covered in Chapter 15.

Measuring Execution Speed

Since execution speed is the reason we substitute assembler, before we delve into how to mix assembler code with MBasic let's first develop a simple way to measure statement execution speed. As we learned in Chapter 10, almost every PIC includes at least one timer. Since we're working with the 16F87x/87xA series devices, we'll use the 16-bit timer 1. To time how long it takes for an MBasic statement or statements to execute, we'll use the concept developed in Program 10-3, but significantly improve upon its performance by using assembler start/stop routines. The only connection to the development board is to connect the PIC to the serial output as shown in Figure 14-1.

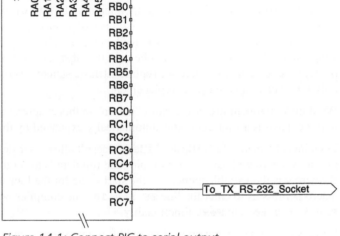

Figure 14-1: Connect PIC to serial output.

Our concept is, in pseudo-code:

```
Set Timer 1 to zero
Start Timer 1
Execute code to be profiled
Stop Timer 1

Read timer 1's value
```

Program 14-1 implements this pseudo-code.

Program 14-1

```
;Program 14-1
;Measure Execution Speed

;Constants
;----------
;For 20 MHz clock, Fosc/4 = 5 MHz
;so there are 5 ticks/usec or 0.2us per tick
usPerTick       fCon    0.2
;how many clock ticks for zero function
Latency         Con 4

;Variables
;----------
;holds the timer value
t2              Var         Word
Extime          Var         Float
A               Var         Byte
```

282

```
        ;Initialization
        ;--------------
        ;no prescaler
        SetTmr1 Tmr1Int1
        tmr1on = 1                    ;start timer  (T1CON, bit 0 [Bank 0] 1=on 0=off)
        EnableHSerial
        SetHSerial H115200

        Main
                ASM {   Call StartTime}
                ;statements to be timed go here
                ASM {   Call StopTime}
                GoSub DumpTime
                Pause 1000
        GoTo Main

        ASM
                {
        StartTime             ;reset timer 1 to 0
                bcf t1con,0   ;stop timer
                clrf tmr1h    ;clear high byte
                clrf tmr1l    ;clear low byte
                bsf t1con,0   ;turn timer on
                Return
        StopTime
                bcf t1con,0   ;stop timer 1
                Return
                }

        DumpTime
        ;--------
                t2.lowbyte = tmr1l
                t2.highbyte = tmr1h
                t2 = t2- Latency
                Extime = usPerTick * (Float t2)
                HSerOut [Real ExTime\2," us",13]
                Pause 10
        Return
        End
```

To maximize our measurement resolution, timer 1 runs with as short a tick interval as possible, that is, the prescaler set for 1:1. My 2840 Development Board has a 20 MHz resonator installed, so as we learned in Chapter 10, with the prescaler set for 1:1, timer 1 will receive one tick every Fosc/4, or once every 0.2 μs. Since we wish the result in microseconds, we define the conversion constant **usPerTick** as 0.2 so that when timer 1's count is multiplied by this constant, the product is elapsed time in microseconds.

```
        ;For 20 MHz clock, Fosc/4 = 5 MHz
        ;so there are 5 ticks/usec or 0.2us per tick
        usPerTick       fCon    0.2
```

We also remember from Chapter 10 that it isn't possible to start or stop a timer instantaneously; there will always be some latency in turning it on and off. We'll define a constant, **Latency**, to hold this dead time as we wish to subtract it from our measured data to provide an accurate time reading.

```
        ;how many clock ticks for zero function
        Latency         Con 4
```

We'll see how to determine the value for **Latency** later in this discussion.

```
        ;Initialization
        ;--------------
        ;no prescaler
        SetTmr1 Tmr1Int1
        tmr1on = 1                    ;start timer  (T1CON, bit 0 [Bank 0] 1=on 0=off)
        EnableHSerial
        SetHSerial H115200
```

The initialization sets timer 1 to the operating condition we desire—internal ticks at Fosc/4 and prescale ratio of 1:1.

```
Main
        ASM {   Call StartTime}
        ;statements to be timed go here
        ASM {   Call StopTime}
        GoSub DumpTime
        Pause 1000
GoTo Main
```

We've encapsulated the code to start and stop timer 1 into two assembler subroutines, **StartTime** and **StopTime**. The MBasic or assembler code we wish to time must be added between these two **Call** statements. Both are accessed by the assembler equivalent of **GoSub**, the **Call** opcode. Since these are assembler statements, we'll use the **ASM** compiler directive and enclose the assembler language portion of our code in curly braces. (We'll learn more about this later in this chapter; for now take it as a requirement.)

```
ASM
            {
StartTime               ;reset timer 1 to 0
        bcf t1con,0     ;stop timer
        clrf tmr1h  ;clear high byte
        clrf tmr1l  ;clear low byte
        bsf t1con,0 ;turn timer on
        Return
StopTime
        bcf t1con,0 ;stop timer 1
        Return
        }
```

You should see some familiar opcodes here.

This assembler module has two labels, **StartTime** and **StopTime** that, as their names suggest, start timer 1 and stop timer 1.

StartTime—First, we ensure timer 1 is stopped by setting to 0 its on/off control bit, held in the **t1con** register at bit 0. (We know the control register is named **t1con** and that the on/off control bit is bit 0 because we consulted the 16F87xA data sheet, Section 5.) We use the bit clear file, **bcf**, opcode to clear the control bit. Next, we reset timer 1's register files to 0 by though the **clrf** opcode. Since timer 1 is a 16-bit module, we separately clear its high and low bytes, **tmr1h** and **tmr1l**, respectively. Lastly, we turn timer 1 on by setting its control bit **t1con**, bit 0 to 1 through a **bsf** operation, then returning to the calling location via the **return** operator.

StopTime—All we do to stop timer 1 is to clear its control bit, **t1con**, bit 0, using the **bcf** operation.

```
DumpTime
;--------
        t2.lowbyte = tmr1l
        t2.highbyte = tmr1h
        t2 = t2- Latency
        Extime = usPerTick * (ToFloat t2)
        HSerOut [Real ExTime\2," us",13]
        Pause 10
Return
```

After we have stopped timer 1, we calculate the elapsed time in microseconds and send the value over the serial port with subroutine **DumpTime**.

We can determine the proper value for the constant **Latency** by experimentation. With the two ASM Call statements adjacent to each other without any intervening MBasic code, set **Latency** to 0 and run Program 14-1. It should return 0.8us, corresponding to four Fosc/4 instructions. You can then set **Latency** to 4 and re-run Program 14-1. This time, it should report 0 microseconds. (Or, of course, we may simply calculate the required **Latency** value based on the number of Fosc/4 intervals required to execute each instruction.)

Program 14-1A

Let's verify Program 14-1 provides meaningful timing information. Modify Program 14-1 by adding a byte arithmetic function and time it:

```
Main
        ASM {   Call StartTime}
        A=A+1 ;statements to be timed go here
        ASM {   Call StopTime}
        GoSub DumpTime
        Pause 1000
GoTo Main
```

We'll call this version Program 14-1A, and it has no other changes. (Program 14-1 declared **A** as a byte variable, but didn't do anything with it.) Run Program 14-1A and you should see the following result displayed in your terminal program:

87.59 us

The assignment operation **A=A+1** requires 87.6µs, with a 20 MHz oscillator. (Under version 5.2.1.1, the result was 96.99µs, so we see version 5.3.0.0 offers an noticeable improvement in execution time.)

Library Macros

This detour completed, let's start with the simplest way to add assembler to your MBasic programs, the library macro. In fact, it's so simple, it's transparent to you when programming. Library macros are defined for the following functions:

Standard Name	Macro Name
Pin Assignment Functions	
High	@High
Low	@Low
Input	@Input
Output	@Output
Timing Functions	
Pause	@MSDelay
Pauseus	@USDelay
Assembler Only	
BankSel	@Bank
Call	@Call
GoTo	@GoTo
Return	@Return
None	@TblJmp
None	@RateDelay

The macro name is usually formed by adding a "@" prefix to the MBasic normal name. The macro functions have one major limit compared with the standard MBasic version; their argument must be a constant.

In the case of the four pin assignment functions, the object of the macro is increased speed; for the two delay statements, the object is increased accuracy for short delays.

To see the difference between the standard and macro versions for the pin assignment functions, let's test them in Program 14-1. Replace the **Main...GoTo** Main code with the following and save it as Program 14-1B.

```
Main
        ASM {   Call StartTime}
        High A0
        ASM {   Call StopTime}
        HSerOut ["High A0",9]
        GoSub DumpTime
        Pause 1000
        ASM {   Call StartTime}
        @High A0
        ASM {   Call StopTime}
        HSerOut ["@High A0",9]
        GoSub DumpTime
        Pause 1000

        ASM {   Call StartTime}
        Low A0
        ASM {   Call StopTime}
        HSerOut ["Low A0",9]
        GoSub DumpTime
        Pause 1000
        ASM {   Call StartTime}
        @Low A0
        ASM {   Call StopTime}
        HSerOut ["@Low A0",9]
        GoSub DumpTime
        Pause 1000

        ASM {   Call StartTime}
        Input A0
        ASM {   Call StopTime}
        HSerOut ["Input A0",9]
        GoSub DumpTime
        Pause 1000
        ASM {   Call StartTime}
        @Input A0
        ASM {   Call StopTime}
        HSerOut ["@Input A0",9]
        GoSub DumpTime
        Pause 1000

        ASM {   Call StartTime}
        Output A0
        ASM {   Call StopTime}
        HSerOut ["Output A0",9]
        GoSub DumpTime
        Pause 1000
        ASM {   Call StartTime}
        @Output A0
        ASM {   Call StopTime}
        HSerOut ["@Output A0",9]
        GoSub DumpTime
        Pause 1000

GoTo Main
```

Standard MBasic v 5.3.0.0 (timing for v. 5.2.1.1)		MBasic Library Macros	
Name	*Execution Speed (μs)*	*Name*	*Execution Speed (μs)*
High	23.2 (28.6)	*@High*	0.8
Low	23.0 (28.4)	*@Low*	0.8
Input	22.6 (28.0)	*@Input*	0.6
Output	23.0 (28.4)	*@Output*	0.6

The four pin assignment procedures show remarkable improvement; from around 23 µs in standard MBasic v 5.3.0.0 to well under 1 µs in macro form.

The two macro delay functions, **@USDelay** and **@MSDelay** will not be further discussed, as they have perfectly serviceable MBasic forms, and the table jump macro, **@TblJmp**, is an advanced macro and will not be further analyzed.

Finally, you may have noticed the **ASM {Call StartTimer}** code is replaced by the **@Call** macro: **@call StartTimer**. The macro offers a cleaner representation and executes in identical time.

MBasic Variables

Before we discuss the library assembler macros and the ASM compiler directive, we must take yet another brief detour and examine the way MBasic passes variable information onto the assembler. As we learned in Chapter 2, the compiler translates MBasic statements into assembler commands, writing a file with the same name as your .BAS file, but with the extension .ASM. MBasic then runs assembler and linker programs (from v 5.3.0.0 onward, GPASM; earlier versions used Microchip's MPASM v. 3 assembler and linker) that merges the directly translated assembler code with MBasic's library functions, and assembles the composite code into a form suitable for loading into the PIC by a loader program. Figure 14-2 illustrates this process.

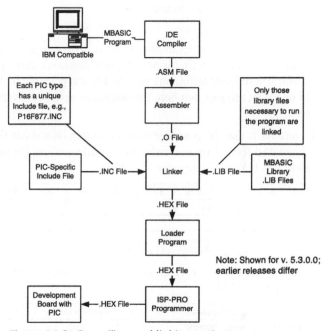

Figure 14-2: Compiling and linking actions.

We learned in Chapter 13 that the 16F87x/87xA's register addresses are in the range 0…511, and require a 9-bit address. We also learned that the 9-bit register address is broken into a 7-bit part that may be used as an opcode argument to directly address a register in the current bank, with the two higher order bits being "bank selection bits" that select between one of four register banks. If this sounds unfamiliar to you, take a moment and reread Chapter 13, as it's essential that you grasp how the 16F87x/87xA addresses data registers.

Let's see how MBasic's compiler identifies variable addresses. Remember, of course, that the variable we define thusly as **A Var Byte** is transformed into a register address, i.e., a number. In the assembler output of the compiler, therefore, this statement is linked to a specific register assignment, using the assembler equate directive.

Program 14-2

Let's see how this works in a simple program that does nothing more than defines variables:

```
;Variables
;------------
AByte       Var       Byte
AWord       Var       Word
ALong       Var       Long
```

```
ABit1          Var          Bit
ABit2          Var          Bit
ByteArray      Var          Byte(8)
WordArray      Var          Word(4)
Clear

End
```

Let's now look at the pertinent portions of compiler's assembler output, found in Program 14-2.ASM. About the middle of Program 14-2.ASM we find the variable assignments:

```
ALONG equ 0AA0h
AWORD equ 06A4h
WORDARRAY equ 06A6h
ABYTE equ 02AEh
BYTEARRAY equ 02AFh
ABIT1 equ 08B7h
ABIT2 equ 018B7h
```

The variable **Abyte**, for example, is stored in register address **02AEh** (remember, the assembler uses a **h** suffix instead of a **$** prefix for hexadecimal notation), corresponding to decimal 686. Wait just a minute, you say. You just finished telling me that the highest possible register address is decimal 512, since it must be held in a 9-bit number, but here the assembler assigned a variable to register address 686. What's going on here?

The answer is that the compiler requires certain information concerning the type of variable, whether it is an array, and so forth. Consequently, the compiler assigns the variable a 16-bit identifier, adding seven compiler control bits to the 9-bit register file address:

B15	B14	B13	B12	B11	B10	B9	B8	B7	B6	B5	B4	B3	B2	B1	B0
AM	Bit/Nib Index			Variable Type			Bank No.			7-bit File Address for OpCodes					
							9-bit File Address								

AM = Signed Variable Marker
B14 also identifies signed/unsigned variables

The three variable type bits are assigned according to the following arrangement, as found near the top of Program 14-2.ASM:

ASM Equates	Bit/Nib Index	Variable Type Bit Assignment	Comments
_FLOATTYPE equ 0E00h	000	111	(new in 5.3.0.0)
_SLONGTYPE equ 4A00h	100	101	(new in 5.3.0.0)
_LONGTYPE equ 0A00h	000	101	
_SWORDTYPE equ 4600h	100	011	
_SBYTETYPE equ 4200h	100	001	
_WORDTYPE equ 0600h	000	011	
_BYTETYPE equ 0200h	000	001	
_NIBTYPE equ 0400h	000	010	
_BITTYPE equ 0800h	000	100	

The bits/nib index runs from 0...7 and applies where more than one nibble or more than one bit variable is allocated. The compiler attempts to efficiently use memory space, so up to 8 bit-length variables will be packed into one byte, and up to two nibbles will be packed into one byte. Word and long variables are stored in consecutive bytes with the lowest order byte occupying the lowest memory address. These type identifiers may change with new compiler releases.

Where the variable is a multibyte type, such as a type word, type long or an array of bytes, words or longs, the compiler assigned address is the "base" address, i.e., the location of the lowest byte in the collection of registers that are assigned to hold the variable. In an array of byte, for example, **Barray Var Byte(5)**, the address represents the location of **BArray(0)**. In the case of a word variable, for example, **AWord Var Word**, the assigned address represents **AWord.Byte0**. The compiler will not split a multibyte type variable across different banks, however. Arrays may occupy more than one bank, but within the array no variable will be split across banks.

Assembler Macros versus ASM Compiler Directive

Why have we taken this detour to examine the details of how MBasic's compiler allocates variables to specific register files? The answer is that it is necessary to understand the difference between the two following alternate assembly structures:

ASM Construct	In Line Macro Construct
AByte Var Byte	AByte Var Byte

ASM
{
Clrf AByte
}
	clrf AByte

Both accomplish the same task; setting the byte variable **AByte** to zero. Program 14-3 incorporates these two constructs.

Program 14-3

```
;Variables
;-----------

AByte   Var             Byte

Clear

Main
        ASM
        {
        Clrf AByte
        }
        clrf AByte
End
```

What happens when we run Program 14-3? The compiler reports warning messages:

Warnings/Errors/Messages:
Warning[219] C:\PROGRA~1\MBasic~1\PROGRAMS\CHAP14~1\PR22DF~1.ASM 140 : Invalid RAM location specified.
Message[302] Message[302] C:\PROGRA~1\MBasic~1\PROGRAMS\CHAP14~1\PR22DF~1.ASM 140 : Register in operand not in bank 0. Ensure bank bits are correct.

These warning messages relate not to the MBasic source code, but rather to problems the assembler found with the compiler's output. To determine where the offending line is, open the associated assembler file, Program 14-3.ASM and go to line 140. It is:

CLRF ABYTE (the one inside the **ASM** braces { })

By now, you should understand why the assembler objected to this operation; the argument for the **clrf** operator must be a 7-bit register address in the range 0…127, not a 16-bit address constructed to also provide compiler-specific information. No wonder, therefore, that the assembler warns us that the address for **ABYTE** is not in the expected range.

We must, therefore "trim off" these excess compiler bits to extract the 7-bit register address in a suitable form to use as a variable address with an assembler operator. The easiest way to trim the high order bits is to take advantage of the assembler program's ability to perform bitwise logical operations with arguments and use the logical AND function with a masking variable. (We've seen how the AND function works as a mask when studying the **ANDLW** opcode in Chapter 13.). Accordingly, we may extract the 7-bit register address from the 16-bit compiler-generated version with the following operation:

```
7-Bit Version = 16-Bit Version AND %1111111
```

To implement this with assembler, we would simply write:

```
CLRF ABYTE&0x7F
```

Where the ampersand "**&**" is the AND operator and **0x7F** is the hexadecimal form of **%1111111**.

Let's try an amended version of Program 14-3.

Program 14-3A

```
;Variables
;-----------
AByte   Var             Byte

Clear

Main
        ASM
        {
        Clrf AByte&0x7F
        }
        clrf AByte
End
```

The compiler reports no errors. But, is the program correct? Unfortunately, it isn't, as **AByte** is in Bank 1. We'll see how we know this and how to fix the banking bits when we re-examine Program 14-3A later in this chapter.

Why did not the macro clear operation, **clrf Abyte**, generate a warning message? After all, it has the same 16-bit version of **Abyte**'s address as did the pure assembler opcode inside the **ASM {…}** section. *The answer is that the macro **clrf**, as with all the other assembler-named library macros defined in MBasic automatically incorporate the **&0x7F** function. However, these macros do not incorporate automatic banking selection.*

It also turns out that MBasic has yet another set of library macros that automatically include address trimming. These macros may be used only inside the **ASM {…}** section and are identified as **_@opcode**. For example, **_@clrf** is the address trimming macro version of the assembler opcode **clrf**, and it may be used only inside the **ASM {…}** section.

One additional point is that although using the 16-bit address form generates "Warnings/Error Messages," in fact the code will assemble and function correctly (assuming, of course, that you have correctly set any required banking bits!) as the assembler ignores all bits beyond the seven legitimate address bits. And, it's possible to turn off these error messages by an assembler directive.

Using MBasic variables in assembler routines is a bit confusing to the beginner, particularly since macros and the actual opcodes share the same name. The following table summarizes the options in using MBasic variable addresses in assembler code:

If you write the assembler code in ASM {...} form

- *Trim the extra bits off with the &0x7F operator:*

  clrf 0x7F&Abyte

- *Use the assembler macro version (prefixed with the _@).*
  This macro automatically trims the extra bits.

  _@clrf AByte

- *Don't trim the extra bits and either ignore the warning*
  messages or suppress the warning messages by adding the
  following assembler directive before your ASM {...} code:

```
ASM
{
errorlevel -219
errorlevel -302
}
```

If you use inline code without the ASM {...} directive

- *Use the MBasic macro with the same name as the assembler*
  opcode. The macro automatically trims the extra bits.

  clrf AByte

Bank Selection

Now that we've learned how to make the 7-bit register address available to our assembler code, how do we deal with the bank selection? For the purpose of this discussion, I'll assume you are dealing only with byte variables, not byte arrays, nor any other variable type. (We'll see how to access some non-byte variables later in this chapter.)

We have several options:

Inspect the Assembler Code and Determine All Variables are in Bank 0

Let's deal with the simplest case first. After declaring all the variables you intend to use in your program, compile it and inspect the associated ASM file. Using the techniques we learned manually determine which bank is associated with each variable you plan to use in assembler code. MBasic sets the banking bits to Bank 0 when it hands over control to your assembly code, so if all the variables and special purpose registers you plan to use in your assembler routine are in bank 0, you don't have to worry about bank switching.

To determine into which bank MBasic has placed a variable, convert the variable's hexadecimal address to binary and examine bits 8 and 7; if they are both zero, the variable is in bank 0. To convert from hexadecimal to binary you may use the calculator program included Microsoft Windows. (It's found under Accessories. Don't forget that you have to switch it to scientific mode to get hexadecimal to binary conversion features.) Or, you may keep the Microsoft calculator in hexadecimal mode, enter the variable's address and AND the address with $1FF. Then shift the calculator display to binary. Bits 8 and 7 give the bank number. Since the calculator blanks leading 0's, the display may show fewer than nine bits.

Manually inspecting the result is an acceptable process only for simple programs that you don't intend to modify. It's too easy to add a variable or two and shift a variable you are using into a different bank without realizing it. Or if you do realize it, you are forced to search your work and add bank selection code with the risk that you miss making the change in a place or two.

If we look at the assembly output for Program 14-3A, we see the compiler (v 5.3.0.0) has made the following memory allocation:

ABYTE equ 02A0h

Converting $02A0 to binary, we find (after adding spaces for clarity) 1 01 010 0000. This tells us the variable **Abyte** is assigned space in Bank %01 (decimal 1), at relative address %0100000 (decimal 32). **Abyte**'s 9-bit address is %01 010 0000, or $A0. (This 9-bit address can also be computed by ANDing the equate value found in the ASM listing with $1FF.)

MBasic changed its variable assignment procedure, beginning with version 5.3.0.0. Prior versions by default reserved the first available 20 words (40 bytes) of memory for the compiler's internal stack. (Larger or smaller stack sizes could be set by through the STACK command.) User variables commenced with the 41st byte of RAM. Since the 16F87x family has 98 bytes of RAM in bank 0, 48 bytes were available for assignment to user variables. Beginning with version 5.3.0.0, MBasic's default is to reserve 100% of bank 0 memory for its internal stack requirements. Thus, all user variables in version 5.3.0.0 will be in banks 1…3, unless you override the default stack setting to claim less than 100% of bank 0.

If you decide to override the default stack setting, you might try **STACK = 10** as a starting point, which reserves 40 bytes for stack memory, freeing up 28 bytes in bank 0 for user variables. (In 5.3.0.0, the stack size is specified in increments of 4 bytes, a change from 5.2.1.1 where the stack size is specified in increments of 2 bytes.) If you see random errors or freezes in your code, you may need to increase the stack reservation.

In the 16F87x family, MBasic's compiler makes a number of standard memory assignments in bank 0 for internal use. You will seldom need to use these reserved variables, but it is useful to understand how MBasic carves up bank 0 memory.

RAM Address (hex)	MBasic Reservation Name	Comment
$20	_WORK	
$21	_WORK2	4 bytes for working variables
$22	_WORK3	
$23	_WORK4	
$24	__ENHWORK	
$25	__ENHWORK2	
$26	__ENHWORK3	
$27	__ENHWORK4	8 bytes for extended working variables
$28	__ENHWORK5	
$29	__ENHWORK6	
$2A	__ENHWORK7	
$2B	__ENHWORK8	
$2C	STACK start	The stack uses memory locations starting at $2C. The amount of memory it uses is determined by the compiler directive STACK = N, where 4*N bytes are reserved for stack memory.
...	...	
...	...	
$2C + 4*N	STACK end	Maximum stack size 68 bytes (N=17)
	Possible User Variable Memory	If stack is not set at maximum
$70	_PC	
$71	_PCH	
$72	_PCU	
$73	_PCSP	
$74	_SP	
$75	_CMD and _DEBUGWORK	16 bytes for temporary storage
$76	_CMD2_ and _DEBUGWORK2	
$77	_CMD3 and _DEBUGWORK3	Note: These addresses are "ACCESS ADDRESSES" and are reachable in any bank via the variables 7-bit offset address without regard for the bank selection bits
$78	_CMD4 and _DEBUGWORK4	
$79	_CMD5 and _DEBUGWORK5	
$7A	_INTSTATE	
$7B	_INTSTATEH	
$7C	_fsr_temp	
$7D	_pclath_temp	
$7E	_status_temp	
$7F	_w_temp	

Force a Bank Selection with BankSel

The assembler macro, **BankSel**, automatically decodes the variable address and generates assembler code setting the correct bank selection bits.

Banksel usage changed slightly beginning with MBasic version 5.3.0.0. MBasic before 5.3.0.0 used Microchip's MPASM assembler, in which **BankSel** expects a 9-bit argument, i.e., an address in the range 0...511. To provide this argument, we must mask out the upper seven bits of the 16-bit compiler version variable address. We may accomplish this by using the same logical AND process we used to mask all but the lower seven bits for opcode arguments. In this case, however, our mask variable will be **%111111111**, or **0x1FF**:

BankSel   Abyte&0x1FF

The following example shows **BankSel**'s usage in MBasic before version 5.3.0.0:

```
;versions earlier than 5.3.0.0
ASM
{
banksel AByte&0x1FF
Clrf AByte&0x7F
}
```

With version 5.3.0.0, MBasic switched to the GPASM assembler, which requires the variable to be used without trimming to the range 0.511:

```
;version 5.3.0.0 onward
ASM
{
banksel Abyte
Clrf AByte&0x7F
}
```

If you use the **&01FF** operator with the **banksel** macro in version 5.3.0.0 and later, you will receive an "illegal label" error message and the code will not assemble.

Make an Optional Bank Selection with The _@Bank or the @Bank Macros

The disadvantage of forcing a bank selection with the **BankSel** macro is that it adds bank selection code whether it's required or not. If the bank selector bits are already set for bank 0, and **Abyte** is in bank 0, the operation **BankSel Abyte** adds unnecessary code to the program. MBasic has defined two assembler macros that test whether a changed bank selection is necessary and generates code only if the new variable is not in the same bank as the old variable.

For use inside the **ASM {...}** construction, the macro is **_@bank**, with syntax:

_@bank OldVar,NewVar

OldVar is the last variable used

NewVar is the new variable, to be tested against **OldVar** and a bank selection performed if necessary. The following example shows **_@bank**'s usage:

```
Main
ASM
{_@bank 0,AByte
Clrf AByte&0x7F
_@bank,Abyte,0
}
```

The first time we invoke **_@Bank**, the **OldVar** argument is 0, because upon entering an **ASM {...}** structure, MBasic sets the bank selector to 0. The **NewVar** argument is **Abyte**, and because the macro takes care of trimming unnecessary bits, we require no masking operation. We then access the variable **AByte**, and use **_@Bank** a second time to restore the bank selector to 0, as MBasic expects the bank selector to be set to 0 when exiting an **ASM {...}** structure.

For use in inline code, MBasic defines the macro **@bank**, with syntax identical to that of _@bank:

```
@bank OldVar,NewVar
```

The following example shows **@bank**'s usage:

```
Main
        @bank 0,AByte
        clrf AByte
End
```

The macro takes care of trimming unnecessary bits from the 16-bit address.

Macros in ISRASM

Before version 5.3.0.0, MBasic's assembler macros, that is, all macros prefixed with _**@** such as _@bank, are unusable inside an **ISRASM** routine. This is because the macros are defined after the **ISRASM** code and hence are not recognized by the assembler. Chapter 15 provides more detail on this limitation. Fortunately, this restriction is removed in 5.3.0.0.

Page Selection

We also learned that the 16F87x/87xA devices are limited to a 2048 location jump page restriction for **GoTo** and **Call** operations. MBasic has defined some macros to assist in making jumps outside a page (sometimes called long jumps or long calls) but these are advanced topics and will not be directly dealt with in this introductory level book.

_@RateDelay Macro

Let's close out this section by looking at the one of the two remaining special assembler macros, _**@RateDelay**. (The last macro, **@TblJmp**, we'll leave for a future book dealing with more advanced subjects.)

_**@RateDelay** is intended to help program loops that must provide an output at a defined periodic rate, but it also works as a stand-alone delay function not associated with a loop. Its syntax is:

```
_@RateDelay Rate, Offset
```

Rate—is a constant that defines how many times per second the loop is to cycle. If you wish the loop to cycle 5000 times per second, set **Rate** at 5000.

Offset—is a constant defining how many instruction cycles are consumed by code in the rest of the loop. For example, the following assembler loop has five instruction cycles:

```
loop:
    @ratedelay 1000,5
    movfw   temp        ;1
    decf    temp,w      ;1
    movwf   temp        ;1
    goto    loop        ;2
```

(The number of instruction cycles each assembler instruction consumes can be found in either Microchip's Midrange Reference Manual, or the specific data sheet for the device you are using.)

Let's look at how **@ratedelay** might be used in your code. This code fragment assumes **i** is declared as a byte variable. It puts a train of 10 brief positive pulses out on pin B0 at the rate of 5000 pulses per second, waits 1 second and repeats the pulse train.

```
Main
        i = 10 ;times to repeat loop
        ASM
        {
Loop:
        ;set for 5000 reps/sec, or 200us/rep
        _@ratedelay 5000,6
        bsf     PortB,0         ;1
```

```
NOP                     ;1  delay for pulse width
bcf       PortB,0       ;1
_@decfsz          i,f   ;1
GoTo Loop               ;2
NOP     ;provides one machine cycle to make it exact
}
Pause 1000
GoTo Main
```

The extra **NOP** after the loop compensates for the difference in the **decfsz** instruction, which is one (no skip) or two (skip) instruction cycles. When combined with the **GoTo** loop, the instruction pair executes in either three instruction cycles (**GoTo** loop executed) or two instruction cycles (**i=0**, so the **GoTo** is skipped). Hence, we are short one instruction cycle at the end, which we make up for by the **NOP**.

Bolt-In Assembler Functions

The remainder of this chapter will focus upon specific code examples that you may use to replace certain MBasic operations. Our focus is upon simple replacements and a process of stepwise refinement. We're not, for example, going to write an assembler replacement for general-purpose multiplication or division. But, we will show some fast ways to divide by a binary power, such as 2, 4, 8… or to obtain the remainder function after such a division. With a bit of care and planning, most if not all bottleneck functions can be converted to an assembler routine based upon code in this chapter.

Stepwise Refinement

The process of stepwise refinement is to write and debug the program in MBasic and then determine whether greater speed is required to meet your objectives. If increased speed is necessary, then we start replacing small elements of MBasic—perhaps as small as one line of code—with assembler routines. After each replacement is tested and debugged, we again determine if our speed objectives are met. If not, we repeat the process until either we've run out of code to convert or our target is met.

Chapter 19 provides a detailed example of applying stepwise refinement to a stepper motor control program. The 100% MBasic control loop of Program 19-1 permits a maximum motor speed of 55.8 RPM. As we progressively replaced MBasic statements with assembler routines, the maximum motor speed increased to 317 RPM, a 5.7:1 speed up. Program 19-2 shows the completed replacement. At this point, our performance objectives had been achieved.

Here's the main loop of Program 19-1:

```
Main
;----
For i = 0 to CyCLen-1
        j = (i+8) / 16
        k = i//32
        n = (i+16)//32
        HPWM 0,PWMPeriod,Cosine(k)
        HPWM 1,PWMPeriod,Cosine(n)
        MPort = FullStep(j)
Next
```

You may wish to take a break from this chapter and sneak a peak at Chapter 19, but the following table shows the effectiveness of each replacement:

Refinement Step	MBasic Code Replaced	Maximum Motor Speed
0	None – 100% MBasic	55.8 RPM
1	j = (i+8) / 16	71.6 RPM
2	k = i//32	95.8 RPM

(Continued)

295

Refinement Step	MBasic Code Replaced	Maximum Motor Speed
3	`n = (i+16)//32`	158.2 RPM
4	`MPort = FullStep(j)`	185.2 RPM
5	`HPWM 0,PWMPeriod,Cosine(k)`	231 RPM
6	`HPWM 1,PWMPeriod,Cosine(n)`	317 RPM

I retained the main **For i…Next** loop in MBasic, as we had more than met our performance objective through replacing the six statements. The order of replacement roughly followed the order of complexity; the easiest statements were the ones first replaced. You also will note that the division, remainder and addition values are all powers of 2. That is not an accident as it simplified the assembler routines.

Let's now turn to "bolt in" assembler routines. The term "bolt-in" means that you should, with minimal work, be able to replace an MBasic statement with the assembler code. In essence, you unbolt the MBasic code and bolt in the assembler code. Of course, some adjustments may be necessary, and you must verify the result.

To verify the operation of the code, and the speed timing, we'll use Program 14-4 as the template.

```
;Program 14-4 template
Stack = 10
;Constants
;----------
;For 20 MHz clock, Fosc/4 = 5 MHz
;so there are 5 ticks/usec or 0.2us per tick
usPerTick       fCon    0.2
;how many clock ticks for zero function
Latency         Con     4

ACon    Con     2
BCon    Con     8
CCon    Con     16

;Variables
;----------
;holds the timer value
t2              Var     Word
Extime          Var     Float

AByte           Var     Byte
BByte           Var     Byte
CByte           Var     Byte

AWord           Var     Word
BWord           Var     Word
CWord           Var     Word


;Initialization
;--------------
;no prescaler
SetTmr1 Tmr1Int1
tmr1on = 1                  ;start timer   (T1CON, bit 0 [Bank 0] 1=on 0=off)
EnableHSerial
SetHSerial H115200

Main
        @Call StartTime
        @Call StopTime
        HSerOut ["Null",9]
        GoSub DumpTime
        Pause 1000

GoTo Main

ASM
```

296

```
              {
StartTime                    ;reset timer 1 to 0
        bcf t1con,0          ;stop timer
        clrf tmr1h           ;clear high byte
        clrf tmr1l           ;clear low byte
        bsf t1con,0          ;turn timer on
        Return
StopTime
        bcf t1con,0          ;stop timer 1
        Return
              }

DumpTime
;--------
        t2.lowbyte = tmr1l
        t2.highbyte = tmr1h
        t2 = t2- Latency
        Extime = usPerTick * (ToFloat t2)
        HSerOut [Real ExTime\2," us",13]
        Pause 25
Return
End
```

This is our earlier timing program, to which we've added three constants, three byte variables and three word variables to use in our test code.

```
ACon     Con            2
BCon     Con            8
CCon     Con            16
AByte           Var     Byte
BByte           Var     Byte
CByte           Var     Byte
AWord           Var     Word
BWord           Var     Word
CWord           Var     Word
```

All execution time data is taken using a 20 MHz resonator with the compiler optimization switch set to "speed." The times shown are for version 5.3.0.0, with times for version 5.2.1.1 shown in parentheses. Since all variables in the test programs are in bank 0, the execution time for the assembler version does not include bank switching time. I've ensured these variables are in bank 0 by limiting the stack size via a **Stack = 10** compiler directive. However, for generality, I've included bank switching code in the samples. If required, bank switching adds 0.2 µs execution time for each switch operation, assuming a 20 MHz resonator.

Finally, I will show only the in-line macro code version of the assembler algorithm. Based upon what you've learned in this chapter, you should be able to rewrite the code using the **ASM {…}** convention, should you so desire.

There's another difference between using an in-line macro or _@macro compared with using pure assembler inside an **ASM {…}** construction. Using pure assembler inside **ASM {…}** permits the assembler to insert a default value for missing destination parameters. If you forget to add the destination parameter in the corresponding macro, you will receive an error message. This is actually a benefit, as the default destination assigned by the assembler may not be the one you intended.

MBasic sets the default radix for assembler literals as decimal. However, it is not good practice to rely upon default radix settings, and you should always explicitly identify any literals you use, for example:

- .123 for decimal
- 0x123 for hexadecimal
- b'010101010' for binary

MBasic Code	MBasic Time (µs)	Assembler Time (µs)
AByte = 0	36.0 (38.2)	0.2
Restrictions: Byte variable Program 14-100		

We use the **clrf** operation to set all bits of the variable to zero.

```
Main
        ;MBasic
        ;----------------
        AByte = 99
        @Call StartTime
        AByte = 0
        @Call StopTime
        HSerOut [Dec AByte,9,"MBasic",9]
        GoSub DumpTime
        Pause 1000

        ;Assembler In-Line Macro
        ;-----------------------
        AByte = 99
        @Call StartTime
        @Bank 0,AByte
        clrf AByte
        @Bank AByte,0
        @Call StopTime
        HSerOut [Dec AByte,9,"Assembler",9]
        GoSub DumpTime
        Pause 1000
GoTo Main
```

MBasic Code	MBasic Time (µs)	Assembler Time (µs)
AWord = 0	37.6 (38.8)	0.4
Restrictions: Word variable Program 14-101		

We use the **clrf** operation to set all bits of the variable to zero. Since the variable is type word, we must separately clear both the low byte (at address **AWord**) and the high byte (at address **AWord+1**).

```
Main
        ;MBasic
        ;----------------
        AWord = 9999
        @Call StartTime
        AWord = 0
        @Call StopTime
        HSerOut [Dec AWord,9,"MBasic",9]
        GoSub DumpTime
        Pause 1000

        ;Assembler In-Line Macro
        ;-----------------------
        AWord = 9999
        @Call StartTime
        @Bank 0,AWord
        clrf AWord
        clrf AWord+1
        @Bank AWord,0
        @Call StopTime
        HSerOut [Dec AWord,9,"Assembler",9]
        GoSub DumpTime
        Pause 1000
GoTo Main
```

MBasic Code	MBasic Time (μs)	Assembler Time (μs)
AByte = Constant	38.8 (40.8)	0.4
Restrictions: Byte variable Program 14-102		

Loading a constant to a variable requires a two-step move. First, copy the constant into **w** and second, copy the value of **w** to the register file. The sample code shows a named constant, but a literal may be directly substituted—that is, **movlw** .123.

```
Main
          ;MBasic
          ;----------------
          @Call StartTime
          AByte = CCon
          @Call StopTime
          HSerOut [Dec AByte,9,"MBasic",9]
          GoSub DumpTime
          Pause 1000

          ;Assembler In-Line Macro
          ;-----------------------
          @Call StartTime
          @Bank 0,AByte
          movlw   CCon
          movwf AByte
          @Bank AByte,0
          @Call StopTime
          HSerOut [Dec AByte,9,"Assembler",9]
          GoSub DumpTime
          Pause 1000
GoTo Main
```

MBasic Code	MBasic Time (μs)	Assembler Time (μs)
AWord = Constant	40.4 (45.0)	0.8
Restrictions: Word variable—may be used only inside ASM{...} construction Program 14-103		

Loading a constant into a word variable requires us to apply the technique used to load a constant into a byte variable, but to apply it twice, once to the low byte and once to the high byte. It is, of course, necessary to copy the low byte of the constant to the low byte of the variable and the high byte of the constant to the high byte of the variable. We may use the assembler function **high** and **low** to select the high and low bytes of the constant. The code also works if a literal, such as .1234 is used instead of a defined constant such as **CCcon**.

This program uses the **ASM {...}** construction, necessary because MBasic does not define in-line macros for the assembler mathematical operations **high** and **low**, as these keywords are already used for other functions. Hence, the in-line version of this code fails and should not be used.

```
Main
          ;MBasic
          ;----------------
          @Call StartTime
          AWord = CCon
          @Call StopTime
          HSerOut [Dec AWord,9,"MBasic",9]
          GoSub DumpTime
          Pause 1000

          ;Assembler In-Line Macro
          ;-----------------------
          AWord = 1111
          @Call StartTime
```

299

```
@Bank 0,AWord
ASM
{
movlw Low CCon
_@movwf AWord
movlw High CCon
_@movwf AWord+1
}
@Bank AWord,0
@Call StopTime
HSerOut [Dec AWord,9,"Assembler",9]
GoSub DumpTime
Pause 1000

GoTo Main
```

MBasic Code	MBasic Time (μs)	Assembler Time (μs)
AByte = Abyte + 1	90.0 (97.0)	0.2
Restrictions: Byte variable Program 14-104		

Incrementing a byte variable is a task directly suited to the **incf** operator. Don't forget that the result of the increment should be stored back in the variable, so the destination should be 'f'. The demonstration program 14-104 also sets **AByte** and **Bbyte** to 0 in the initialization stage, and separately increments both variables, to make the demonstration clearer.

```
Main
        ;MBasic
        ;----------------
        @Call StartTime
        BByte = BByte+1
        @Call StopTime
        HSerOut [Dec BByte,9,"MBasic",9]
        GoSub DumpTime
        Pause 1000

        ;Assembler In-Line Macro
        ;-----------------------

        @Call StartTime
        @Bank 0,AByte
        incf AByte,f
        @Bank AByte,0
        @Call StopTime
        HSerOut [Dec AByte,9,"Assembler",9]
        GoSub DumpTime
        Pause 1000
GoTo Main
```

MBasic Code	MBasic Time (μs)	Assembler Time (μs)
AWord = AWord + 1	93.0 (100.0)	0.6
Restrictions: Word variable Program 14-105		

To increment a word variable, we increment its low byte and test the **Status <Z>** bit to see if the low byte is zero. If it is, it means the low byte has rolled over from 255 to 000 and hence the high byte must be incremented. The test condition is implemented in a **btfsc** test; if **Z** is clear (no rollover) skip incrementing the high byte. If **Z** is set (rollover) then increment the high byte. We need not change banks when testing **Z** as the **Status** register is reachable from all four banks.

(If desired, this sequence can be extended for a 4-byte type long variable.)

```
Main
        ;MBasic
        ;----------------
        @Call StartTime
        BWord = BWord+1
        @Call StopTime
        HSerOut [Dec BWord,9,"MBasic",9]
        GoSub DumpTime
        Pause 1000

        ;Assembler In-Line Macro
        ;-----------------------

        @Call StartTime
        @Bank 0,AWord
        incf AWord,f
        btfsc Status,Z
        incf AWord+1,f
        @Bank AWord,0
        @Call StopTime
        HSerOut [Dec AWord,9,"Assembler",9]
        GoSub DumpTime
        Pause 1000
    GoTo Main
```

MBasic Code	MBasic Time (µs)	Assembler Time (µs)
AByte = AByte – 1	94.4 (101.2)	0.2
Restrictions: Byte variable Program 14-106		

Decrementing a byte variable is a task directly suited to the **decf** operator. Don't forget that the result of the decrement should be stored back in the variable, so the destination should be 'f'.

```
Main
        ;MBasic
        ;----------------
        @Call StartTime
        BByte = BByte-1
        @Call StopTime
        HSerOut [Dec BByte,9,"MBasic",9]
        GoSub DumpTime
        Pause 1000

        ;Assembler In-Line Macro
        ;-----------------------

        @Call StartTime
        @Bank 0,AByte
        decf AByte,f
        @Bank AByte,0
        @Call StopTime
        HSerOut [Dec AByte,9,"Assembler",9]
        GoSub DumpTime
        Pause 1000
    GoTo Main
```

MBasic Code	MBasic Time (µs)	Assembler Time (µs)
AWord = AWord – 1	96.4 (104.2)	0.8
Restrictions: Word variable Program 14-107		

To decrement a word variable, we use the same principle of decrementing the low byte and high byte separately, but the order of decrements is critical. Before decrementing, we test the low byte by copying it back to itself using the **movf** operator, with the destination set to 'f.' This operation sets the **Status <Z>** bit if the low byte is zero. We test for **Z** with the **btfsc** operator and if **Z** is set (low byte is zero) we decrement the high byte. If not, we skip the high byte decrement. Then, regardless of whether we have decremented the high byte or not, we decrement the low byte. Decrements use the **decf** command, with a destination of 'f' to place the result back in the variable.

```
Main
            ;MBasic
            ;----------------
            @Call StartTime
            BWord = BWord-1
            @Call StopTime
            HSerOut [Dec BWord,9,"MBasic",9]
            GoSub DumpTime
            Pause 1000


            ;Assembler In-Line Macro
            ;-----------------------

            @Call StartTime
            @Bank 0,AWord
            movf AWord,f
            btfsc Status,Z
            decf AWord+1,f
            decf AWord,f
            @Bank AWord,0
            @Call StopTime
            HSerOut [Dec AWord,9,"Assembler",9]
            GoSub DumpTime
            Pause 1000
    GoTo Main
```

MBasic Code	MBasic Time (µs)	Assembler Time (µs)
AByte = BByte + CByte	98.0 (105.4)	0.8
Restrictions: Byte variable		
Program 14-108		

Forming the sum **AByte = BByte + CByte** in assembler may be accomplished in four steps:

1. Copy **BByte**'s value to **W**;
2. Copy **W** to **AByte** (at this point, **AByte=BByte**);
3. Copy **CByte** to **W**;
4. Add **W** to **AByte** and store result in **AByte** (at this point **AByte = AByte + CByte**).

The addition function in step 4 uses the **addwf** opcode. Note that since we use three register files, we must ensure that, if necessary, the correct bank bits are set before each read or write.

If the desired operation is **AByte = AByte + CByte**, only steps 3 and 4 need be used.

```
Main
            ;MBasic
            ;----------------
            @Call StartTime
            AByte = BByte + CByte
            @Call StopTime
            HSerOut [Dec AByte,9,"MBasic",9]
            GoSub DumpTime
            Pause 1000


            ;Assembler In-Line Macro
            ;-----------------------
```

```
            AByte = 0
            @Call StartTime
            @Bank 0,BByte
            movf BByte,w
            @Bank BByte,AByte
            movwf AByte
            @Bank AByte,CByte
            movf CByte,w
            @Bank CByte,AByte
            addwf AByte,f
            @Bank AByte,0
            @Call StopTime
            HSerOut [Dec AByte,9,"Assembler",9]
            GoSub DumpTime
            Pause 1000
      GoTo Main
```

MBasic Code	MBasic Time (µs)	Assembler Time (µs)
AWord = BWord + CWord	102.4 (109.8)	2.00
Restrictions: Word variable Program 14-109		

Forming the sum **AWord = BWord + CWord** in assembler may be accomplished in nine steps:

1. Copy **BWord** low byte to **w**;
2. Copy **w** to **AWord** low byte (at this point **AWord.LowByte = BWord.LowByte**);
3. Copy **BWord** high byte to **w**;
4. Copy **w** to **AWord** high byte (at this point **AWord = BWord**);
5. Copy **CWord** high byte to **w**;
6. Add **w** to **AWord** high byte, saving result in **AWord** high byte;
7. Copy **CWord** low byte to **w**;
8. Add **w** to **AWord** low byte, saving result in **AWord** low byte;
9. If the result of step 8 is a carry, increment **AWord** high byte.

We use the **addwf** operator to perform the addition in steps 6 and 8, and test the carry bit, **Status <C>**, in step 9 using a **btfsc** test. If the carry bit is set, increment **AWord** high byte with the **incf** operation; if it is clear skip the increment. By performing the addition in high-byte-first order, our code is somewhat simplified. Each time we read or write the register files, we must consider the bank status, so this code makes frequent use of the **@bank** macro.

Remember, in assembler a word variable has its low byte at the base address and its high byte at its base address +1.

```
      Main
            ;MBasic
            ;----------------
            @Call StartTime
            AWord = BWord + CWord
            @Call StopTime
            HSerOut [Dec AWord,9,"MBasic",9]
            GoSub DumpTime
            Pause 1000

            ;Assembler In-Line Macro
            ;-----------------------
            AWord = 0
            @Call StartTime
            @Bank 0,BWord
            movf BWord,w
            @Bank BWord,AWord
            movwf AWord
```

```
            @Bank AWord+1,BWord+1
            movf BWord+1,w
            @Bank BWord+1,AWord+1
            movwf AWord+1

            @bank AWord+1,CWord+1
            movf  CWord+1,w
            @bank CWord+1,AWord+1
            addwf AWord+1,f

            @Bank AWord+1,CWord
            movf  CWord,w
            @Bank CWord,AWord
            addwf AWord,f
            btfsc Status,C
            incf  AWord+1,f

            @Bank AWord+1,0
            @Call StopTime
            HSerOut [Dec AWord,9,"Assembler",9]
            GoSub DumpTime
            Pause 1000
      GoTo Main
```

MBasic Code	MBasic Time (µs)	Assembler Time (µs)
AByte = AByte – Constant	93.4 (101.2)	0.4
Restrictions: Byte variable		
Program 14-110		

To subtract a constant from a byte variable, we use the **subwf** operation. This requires a two step process:

1. Copy the subtrahend to **W** using the **movlw** opcode;
2. Subtract **W** from register file **AByte** using the **subwf** operation, saving the result in file **AByte**.

If you wish to subtract a number, such as 123, directly instead of using a defined constant, the line **movlw CCon** is replaced by **movlw .123**.

```
      Main
            ;MBasic
            ;----------------
            AByte = 222
            @Call StartTime
            AByte = AByte - CCon
            @Call StopTime
            HSerOut [Dec AByte,9,"MBasic",9]
            GoSub DumpTime
            Pause 1000

            ;Assembler In-Line Macro
            ;-----------------------
            AByte = 222
            @Call StartTime
            movlw CCon
            @Bank 0,AByte
            subwf AByte,f
            @Bank AByte,0
            @Call StopTime
            HSerOut [Dec AByte,9,"Assembler",9]
            GoSub DumpTime
            Pause 1000
      GoTo Main
```

304

MBasic Code	MBasic Time (µs)	Assembler Time (µs)
AWord = AWord – Constant	81.8 (104.2)	0.8
Restrictions: Word variable Program 14-111		

To subtract a constant from a word variable, we apply the **subwf** operator to both the high and low bytes and adjust for a carry, if necessary. This requires six steps:

1. Copy the high byte of the constant into **W**;
2. Subtract **W** (high byte of constant) from **AWord** high byte, saving the result in **AWord** high byte;
3. Copy the low byte of the constant into **W**;
4. Subtract **W** (low byte of constant) from **AWord** low byte, saving the result in **AWord** low byte;
5. If the result of step 4 sets the carry flag, **Status <C>**, then decrement **AWord** high byte.

If you wish to subtract a number, such as (decimal) 123, directly instead of using a defined constant, the line **movlw High CCon** is replaced by **movlw High .123** and the line **movlw Low CCon** is replaced by **movlw Low .123**.

As with subtraction of variables, we proceed in the high-byte-first order to simplify the carry test.

```
Main
            ;MBasic
            ;----------------
            AWord = 2222
            @Call StartTime
            AWord = AWord - CCon
            @Call StopTime
            HSerOut [Dec AWord,9,"MBasic",9]
            GoSub DumpTime
            Pause 1000

            ;Assembler In-Line Macro
            ;-----------------------
            AWord = 2222
            @Call StartTime
            ASM
            {
            _@movlw High CCon
            _@Bank 0,AWord
            _@subwf AWord+1,f
            _@movlw Low CCon
            _@subwf AWord,f
            _@btfss Status,C
            _@decf AWord+1,f
            _@Bank AWord,0
            }
            @Call StopTime
            HSerOut [Dec AWord,9,"Assembler",9]
            GoSub DumpTime
            Pause 1000
    GoTo Main
```

MBasic Code	MBasic Time (µs)	Assembler Time (µs)
Swap AByte and BByte	81.8 (65.6)	0.8
Restrictions: Byte variable Program 14-112		

It's occasionally necessary to swap two variables. MBasic has a special function **Swap** to perform this task.

To avoid using an intermediate storage register file, we use an arithmetic trick from Reference [14-5] requiring four steps:

1. Copy **AByte** to **W**;
2. Subtract **W** from **BByte**, store result in **W**. (**W = BByte − AByte**);
3. Add **AByte** to **W**, store the result in **AByte**. (**AByte = ~~AByte~~ + (BByte − ~~AByte)~~**, or **AByte = BByte**);
4. Subtract **W** from **BByte** and store the result in **BByte**. (**BByte = BByte − (BByte −AByte)**, or **BByte = ~~BByte~~ − ~~BByte~~ + AByte**, so **BByte = AByte**).

This algorithm requires only four Fosc/4 clock periods to complete, and is thus faster than the usual approach to swapping a variable without using an intermediate variable, using three consecutive XOR operations, which requires six clock periods:

```
movf   BByte,w
xorwf  AByte,f
movf   AByte,w
xorwf  BByte,f
movf   BByte,w
xorwf  AByte,f
```

The previous XOR example omits bank selection.

```
Main
        ;MBasic
        ;----------------
        AByte = ACon
        BByte = BCon
        @Call StartTime
        Swap AByte,BByte
        @Call StopTime
        HSerOut [Dec AByte,9,Dec BByte,9,"MBasic",9]
        GoSub DumpTime
        Pause 1000

        ;Assembler In-Line Macro
        ;------------------------
        @Call StartTime
        @Bank 0,AByte
        movf   AByte,w
        @Bank AByte,BByte
        subwf  BByte,w
        @Bank BByte,AByte
        addwf  AByte,f
        @Bank AByte,BByte
        subwf  BByte,f
        @Bank BByte,0
        @Call StopTime
        HSerOut [Dec AByte,9,Dec BByte,9,"Assembler",9]
        GoSub DumpTime
        Pause 1000
GoTo Main
```

MBasic Code	MBasic Time (µs)	Assembler Time (µs)
Remainder (Special)	266.8 (277.8)	0.8
Restrictions: Byte variable; Divisor is of form 2^N Program 14-113		

The remainder operation, **/ /** in MBasic, may be easily implemented with a masking operation where the divisor is a power of 2; for example, 2, 4 … 128. We mask off all higher order bits with an appropriate AND mask, (2^N-1). For example, suppose the divisor is 32, so N = 5. The mask is 32 − 1, or 31. We see how this works by examining 31 in binary: %00011111; so it masks off the three high bits. This is the functional equivalent of the remainder after dividing by 32.

The remainder function is often used to keep a variable within a certain range. Examples of this can be seen in Program 19-1 and 19-2, where a **For i = 0 to 63 … Next** loop has two inner that variables must cycle from 0…31 within the outer loop. These sub-loop variables, **k** and **n** are easily obtained, for example, **k = i//32**.

To compute **CByte = AByte//Divisor** where **Divisor** has the form 2^N, the algorithm has three steps:
1.   Copy the dividend (**AByte**) to **w**;
2.   AND **w** with the divisor minus 1 ($2^N - 1$);
3.   Copy **w** (which holds the remainder) to **CByte**.

The example uses a divisor of 32 (2^5) and $2^5 - 1$ is 31.

```
Main
For AByte = 0 to 255
        ;MBasic
        ;----------------
        @Call StartTime
        BByte = AByte // 32
        @Call StopTime
        HSerOut [Dec BByte,9,"MBasic",9]
        GoSub DumpTime
        Pause 1000

        ;Assembler In-Line Macro
        ;-----------------------
        @Call StartTime
        @Bank 0,AByte
        movf AByte,w
        andlw (.32-.1)
        @bank AByte,CByte
        movwf CByte
        @Bank CByte,0
        @Call StopTime
        HSerOut [Dec CByte,9,"Assembler",9]
        GoSub DumpTime
        Pause 1000
Next
GoTo Main
```

MBasic Code	MBasic Time (µs)	Assembler Time (µs)
Division (Special)	266.8 (277.8)	6.2*

Restrictions: Byte variable; Divisor is of form 2^N
* = time for / 16; other divisors will be different
Requires auxiliary byte variable Temp
Program 14-114

If the divisor is of the form 2^N, we may perform division by shifting right N bits. Since the assembler rotate operation shifts through carry, we must zero out the carry bit.

The operation to be duplicated in assembler is **CByte = AByte / 16**. The simplicity of the algorithm is masked by the complexity introduced in dealing with bank selection:
1.   The divisor must be of the form 2^N. Since the divisor in the example is 16, N = 4. Copy N into a temporary byte variable **Temp**. This is done by a **movlw** opcode to get N into **w** followed by a **movf** operation to copy N into the file **Temp**.
2.   Copy **AByte** into **CByte**, using the two-step copy process; copy **AByte** to **w** and then copy **w** to **CByte**.
3.   Shift **CByte** one bit to the right with the **RRF** operator. This rotation puts the bit 0 into the carry bit, **Status <C>**.
4.   Clear the carry bit **Status<C>** to convert the rotation into a shift and loss operation.

5. Decrement `Temp`.

6. Test `Temp` to see if it is zero, indicating we have completed N steps. If `Temp` is not zero, go to Step 3. We use the `decfsz` operator to combine steps 5 and the test part of Step 6 into one operation.

Because we cannot be sure that `Temp` and `CByte` are in the same bank, we must make an auxiliary jump to `FixUp`, where we add bank selection code. After the bank selection is completed, we go back to `Shift` (Step 3 in the above description).

Since `GoTo` is an MBasic function, the in-line macro for the assembler opcode `GoTo` is `@GoTo`.

We use "short" `GoTo` calls here, which means that all code (including the associated MBasic code) must be within the same 2K program page. Calls outside the current program page are an advanced topic and will not be considered in this book.

```
          Main
          For AByte = 0 to 255
                  ;MBasic
                  ;----------------
                  @Call StartTime
                  BByte = AByte / 16
                  @Call StopTime
                  HSerOut [Dec BByte,9,"MBasic",9]
                  GoSub DumpTime
                  Pause 1000

                  ;Assembler In-Line Macro
                  ;-----------------------
                  @Call StartTime

                  movlw           0x4
                  movwf           Temp&0x7F                ;temp=4
                  @Bank           0,AByte
                  movf            AByte,w
                  @Bank           AByte,CByte
                  movwf           CByte

          Shift
                  RRF             CByte,f ;/2 but this is rotate, not shift
                  bcf             STATUS,C        ;clear the carry flag
                  @bank           CByte,Temp
                  decfsz  Temp,f          ;temp=temp-1
                  @GoTo   FixUp           ;while Temp > 0 GoTo Shift
                  @GoTo           EndIt
          FixUp
                  @bank           temp,CByte
                  @GoTo           Shift
          EndIt
                  @Call StopTime
                  HSerOut [Dec CByte,9,"Assembler",9]
                  GoSub DumpTime
                  Pause 1000
          Next
          GoTo Main
```

MBasic Code	MBasic Time (µs)	Assembler Time (µs)
Set / Clear Port Bit	26.6 (31.2) Set 26.6 (31.0) Clear	0.2 Set 0.2 Clear
Restrictions: Bit must be a constant Program 14-115		

The `bfs` and `bfc` opcodes provide a direct way to set or clear an output pin.

The example shows clearing B0 (MBasic command `low B0`). To set B0 (MBasic command `High B0`) replace the `bcf` operation with the `bsf` function. Ports A...E are all in Bank 0, so if the port bits are set or cleared without further assembler code, the **@bank** macros are unnecessary. Don't forget to set the desired pin to output mode. This may be conveniently done in MBasic; for example, **Output B0**.

```
Main
            ;MBasic
            ;----------------
            @Call StartTime
            Low B0
            @Call StopTime
            HSerOut ["MBasic",9]
            GoSub DumpTime
            Pause 1000

            ;Assembler In-Line Macro
            ;------------------------
            @Call StartTime

            @Bank 0,PortB
            bcf   PortB,0
            @Bank PortB,0

            @Call StopTime
            HSerOut ["Assembler",9]
            GoSub DumpTime
            Pause 1000
     GoTo Main
```

MBasic Code	MBasic Time (μs)	Assembler Time (μs)
Set / Clear Port Bit and Set to Output	26.6 (31.2) Set 26.6 (31.0) Clear	0.8 Set 0.8 Clear
Restrictions: Bit must be a constant Program 14-116		

If you wish to also set the port to output direction in assembler, it is necessary to clear the associated bit in the **TRIS** file register. **TRISA** corresponds to **PortA**, **TRISB** to **PortB**, etc. **PortB** pin 0 corresponds to **TRISB** bit 0, etc. Setting the pin to input requires setting the associated TRIS bit to **%1**; setting the pin to output requires setting the associated TRIS bit to **%0**.

To set **B0** (MBasic command **High B0**) replace the **bcf** operation with the **bsf** function.

This is a case where the two file registers we address are in different banks, so the bank selection macro causes code to be generated, accounting for the extra two Fosc/4 cycles consumed above the time required for the two **bcf** operators.

Remember, you can't use MBasic's library pin constants, such as **B0**, here. Rather, you must identify pins by digits 0...7. (Or, you can let the assembler do the math for you; for example, to reference pin **B6**, use (**B6-B0**) as the bit identifier. The assembler will perform the subtraction, coming up with the correct number.)

```
Main
            ;MBasic
            ;----------------
            @Call StartTime
            Low B0
            @Call StopTime
            HSerOut ["MBasic",9]
            GoSub DumpTime
            Pause 1000

            ;Assembler In-Line Macro
            ;------------------------
            @Call StartTime
```

```
          @Bank 0,TRISB
          bcf      TRISB,0
          @Bank TRISB,PortB
          bcf  PortB,0
          @Bank PortB,0

          @Call StopTime
          HSerOut ["Assembler",9]
          GoSub DumpTime
          Pause 1000
     GoTo Main
```

MBasic Code	MBasic Time (µs)	Assembler Time (µs)
For ... Next Loop	164.8 (183.2)*	1.4
Restrictions:		
Loop Variable must be a byte		
Step is +1 or –1		
Step is a constant		
* Time stated is for one empty loop iteration		
Program 14-117		

We'll structure a simple **For...Next** loop, but by now you should see how to extend it to use variables for step and the start/finish bounds.

The loop example implements an assembler version of:

```
     For Temp = ACon to CCon
     ;MBasic code here
     Next
```

Our assembler algorithm has four steps:

1. Copy **ACon** to **W** using the **movlw** operator and copy **W** to **Temp** using the **movf** operator. The result is **Temp = ACon**;
2. Insert the assembler or MBasic code that resides within the loop;
3. Copy **Temp** to **W** with the **movf** operator and subtract **W** from **CCon** using the **subwf** opcode;
4. If the result of step 3 is zero (**Status <Z>** is set) then we are finished and are ready to execute the next code sequence. This is accomplished via a **btfsc Status,Z** test. If **Z** is set, the next instruction, the **GoTo EndIt** instruction executes and the code goes onto the next sequence. If **Z** is not set (**Temp <> CCon**) increment **Temp** through the **Incf** operator. After incrementing **Temp**, then **GoTo** step 2.

Since **GoTo** is an MBasic function, the macro for the assembler opcode **GoTo** is **@GoTo**.

We use "short" **GoTo** calls here, which means that all code (including the associated MBasic code) must be within the same 2K program page. Calls outside the current program page are an advanced topic and will not be considered in this book.

```
          For Temp = ACon to CCon Step -1
          ;MBasic code here
          Next
```

If we wish to reverse step—that is, Step = –1 in MBasic code, replace the **incf Temp,f** instruction with **decf Temp,f**.

```
     Main
          ;MBasic
          ;----------------
          BByte = 0
          @Call StartTime
          For Temp = ACon to CCon
```

```
                    ;MBasic code here
                    Next
                    @Call StopTime
                    HSerOut [Dec BByte,9,"MBasic",9]
                    GoSub DumpTime
                    Pause 1000

                    ;Assembler In-Line Macro
                    ;-----------------------
                    BByte = 0
                    @Call StartTime
                    movlw ACon
                    @Bank 0,Temp
                    movwf Temp
          Loop
                    ;assembler code here
                    movf    temp,w
                    sublw   CCon
                    btfsc   Status,Z
                    @GoTo   EndIt
                    incf    Temp,f
                    @GoTo   Loop
          EndIt
                    @bank Temp,0
                    @Call StopTime
                    HSerOut [Dec BByte,9,"Assembler",9]
                    GoSub DumpTime
                    Pause 1000
          GoTo Main
```

MBasic Code	MBasic Time (µs)	Assembler Time (µs)
Byte Table	74.0 (76.6)	2.6

Restrictions:
Table size <256 elements
Table only byte elements
Index into table byte variable
Program 14-118

The two most complicated assembler modules we will concern ourselves with involve indexing into data and variables. This module provides a faster alternative to MBasic's ByteTable function, while a later module will show how to read from and write to an array of byte variables.

The algorithm we use is based upon References [14-4] and [14-7] and implements an assembler equivalent of:

```
AByte = FullStepTable(Temp)
```

Where **FullStepTable** is the functional equivalent of a **ByteTable** data array.

Let's consider a high level view of how our assembler code works:

1. The table entry number (0...7 in our sample table) is in the byte variable **Temp**. The data table itself starts at label **FullStepTable**.
2. Each element in **FullStepTable** is a **ret1w** operator, followed by a literal, the element value corresponding with that offset into the table. The **ret1w** operator executes a return from call with its literal argument loaded into **w**.
3. The act of accessing the table can be thought of as a computed **GoTo**. A **call** operator results in program execution jumping to the program code at position **FullStepTable + Temp**. The op-code at **FullStepTable + Temp** is a **ret1w** operator that causes program execution to jump back to the statement immediately following the **Call** operation.

In turning this high level description into functioning code, we have to deal with a couple of PIC architectural issues:

- As we learned in Chapter 13, the program counter in the 16F87x/87xA devices is 13 bits wide and we have direct access only to the program counter's low 8-bits, contained in the PCL register. Suppose that the address of the data table is such that the PCL register's value is .250 and that we are indexing 10 positions into the table. The required PCL address is .250 + .10 = .260. However, since the PCL is only an 8-bit register, the result, .260, overflows and wraps around to give an incorrect result.
- As we also learned in Chapter 13, **Call** and **GoTo** instructions only use 11 address bits, and hence are restricted to 2K page boundaries. Suppose the data table is in a different page? If so, we must increment or decrement the page counter accordingly.

The sample code provided takes care of both of these architectural issues.

I'm not going to provide a detailed line-by-line analysis of this code. If you are interested in learning more about how it works, consult References [14-4] and [14-7].

To modify this code for your own use, change:

- the table name (**FullStepTable**)
- the index variable (**Temp**)
- the variable the returned table value is placed in (**AByte**)

The rest of the code can remain as shown. One final note—because we use the high and low assembler operators, that part of the calling code must be contained in an **ASM {…}** structure. For consistency, I'm placed the entire calling code sequence in an **ASM {…}** structure. The table itself may either reside inside an **ASM {…}** construction, or you may use in-line code.

```
ACon    Con             0
CCon    Con             7

BTable ByteTable $C,$6,$3,$9,$C,$6,$3,$9

Main
        ;MBasic
        ;-----------------
BByte = 0
          For Temp = ACon to CCon
          @Call StartTime
                    AByte = BTable(Temp)
          @Call StopTime

          HSerOut [Dec AByte,9,"MBasic",9]
          GoSub DumpTime
          Pause 1000
        Next

        ;Assembler In-Line Macro
        ;-----------------------
BByte = 0
For Temp = ACon to CCon
          @Call StartTime
          ASM
          {
          _@movlw HIGH FullStepTable
          _@movwf PCLATH
          _@movlw LOW (FullStepTable+1)
          _@bank  0,Temp
          _@addwf Temp,w
          _@btfsc STATUS,C
          _@incf  PCLATH,f
          _@call  FullStepTable
```

```
                    _@bank  Temp,AByte
                    _@movwf AByte
                    _@bank  AByte,0
                    }
          @Call StopTime
          HSerOut [Dec AByte,9,"Assembler",9]
          GoSub DumpTime
          Pause 1000
          Next
       GoTo Main


    FullStepTable
          movwf PCL
          retlw 0xc
          retlw 0x6
          retlw 0x3
          retlw 0x9
          retlw 0xc
          retlw 0x6
          retlw 0x3
          retlw 0x9
```

MBasic Code	MBasic Time (µs)	Assembler Time (µs)
Byte Variable Array Read and Write	82.8 (87.4) (Read) 82.8 (87.4) (Write)	2.0 (Read) 2.0 (Write)
Restrictions: Array of bytes Array cannot span banks Program 14-119-A (Read), 14-119B (Write)		

Our final module shows how to read from and write to an array of byte variables. As with the table read, we'll confine our description to a relatively high level analysis.

Array access involves indirect memory addressing—instead of addressing a register file with a specific address, the register file to be addressed is defined as computed location. This is, of course, exactly how an array variable is addressed; the array index is the offset into the array. Suppose we define an array of bytes **ByteArray**. The register file pointed to by **ByteArray(N)** has an address of **ByteArray(0)**'s address plus **N**. In fact, in MBasic we can reference the base address **ByteArray(0)** simply as **ByteArray**. (This discussion assumes that the array does not span register banks; MBasic supports multibank arrays, but our assembler code does not.)

Midrange PICs have two special features to facilitate indirect memory addressing, the indirect file register, or **indf**, register and the file selection register or **FSR**. The **indf** register has an address of $0, $80, $100 and $180 so it may be reached with an address of 0, regardless of the bank selection bits. The **indf** register is a fictitious register; writing to **indf** actually writes to a real register at the address corresponding to the value in **FSR**. Likewise, reading from **indf** actually reads from the real register at the address corresponding to the value in **FSR**.

We must pay attention to the mid-range PIC architecture. **FSR** is an 8-bit register, so this permits us to access memory addresses in the range 0…255, that is, banks 0 and 1 in the 16F98x/87xA devices. The 9th address bit, necessary to address all four banks, is held in **Status <7>**, named the **IRP** bit. If the **IRP** bit is 0, we may address banks 0 and 1; it the **IRP** bit is 1, we address banks 2 and 3. The banking bit setting, held in **RP1:RP0** does not apply when using indirect addressing through the **FSR/indf** process.

Assuming our byte array is **ByteArray**, and that the index variable into **ByteArray** is **Temp**, (accesses are thus via the equivalent of **ByteArray(Temp)**) our strategy for indirectly addressing memory is:

1. Load the starting address (**ByteArray**) into **W** via a **movlw** operation. **ByteArray** is a 16-bit variable, since it is in the form of an MBasic compiler-compatible variable address. The **movlw** operation only copies the lower 8 bits, which is exactly our desire.

2. Add the offset (**Temp**) to **W** with an **addwf** operation, saving the result in **W**. Then copy **W** to **FSR** via a **movwf** operator. At this point, **FSR** holds the 8-bit address of the subject register file.

3. To put the **IRP** bit into a known state, we clear it with a **bcf Status,IRP** operation.

4. To determine if **ByteArray** is in banks 2 or 3, we need to determine if the 9th bit of **ByteArray**'s 16-bit MBasic compiler-compatible address is set or clear. This we accomplish by loading the high byte of **ByteArray** into **W** using the **movlw HIGH ByteArray** operation, using the assembler to handle the high byte conversion. As we've seen earlier, the keyword conflict between the assembler operator **HIGH** and MBasic's **High** function require this part of the code to be within an **ASM {...}** segment. After loading the high byte of **ByteArray**'s 16-bit address into **W**, the 9th bit is now **W** bit 0. We test **W** bit 0 by a logical AND of **W** and %00000001 to mask off all but bit 0. If the result is zero, the 9th bit is zero and **Status <Z>** is set. If the result is %1, **Status <Z>** is not set. This permits us to test for bit 9 = 0 with a **btfss Status<Z>** test, immediately followed by a **bsf Status, IRP** operation to set the **IRP** bit to reflect banks 2 and 3. If **Z** is set, the **bsf Status, IRP** operation is skipped and **IRP** remains cleared. If Z is clear (bit 9 = 1), the **bsf Status, IRP** executes and the bank 2/3 bit is set.

5. We now either write to **INDF** or read from **INDF**, as **INDF** is correctly pointing to the register file address **ByteArray+Temp**.

This algorithm requires the array to be completely contained within one bank, as we do not check for discontinuous addressing. For both read and write demonstration programs, we make the following changes to our template program:

- Modify the constant definitions:

```
ACon    Con         0
CCon    Con         7
```

- Define a new byte array:

```
ByteArray  Var        Byte(CCon)
```

- Initialize the byte array to hold some arbitrary values:

```
For Temp = 0 to CCon
        ByteArray(Temp)= Temp + 10
Next
```

The code for reading from the byte array (**AByte = ByteArray(Temp)**):

```
Main
        ;MBasic
        ;----------------
        BByte = 0
                For Temp = ACon to CCon
                @Call StartTime
                        AByte = ByteArray(Temp)
                @Call StopTime

                HSerOut [Dec Temp,9,Dec AByte,9,"MBasic",9]
                GoSub DumpTime
                Pause 1000
        Next
```

```
              ;Assembler In-Line Macro
              ;-----------------------

              For Temp = ACon to CCon
                      @Call StartTime
                      @Bank   0,ByteArray
                      movlw   ByteArray
                      @Bank   ByteArray,Temp
                      addwf   Temp,0
                      movwf   FSR
                      bcf     Status,IRP
                      ASM {   movlw   HIGH ByteArray}
                      andlw   0x1
                      btfss   Status,Z
                      bsf     Status,IRP
                      @Bank   ByteArray,AByte
                      movf    INDF,0
                      movwf   AByte
                      @Bank   AByte,0
                      @Call StopTime
                      HSerOut [Dec Temp,9,Dec AByte,9,"Assembler",9]
                      GoSub DumpTime
                      Pause 1000
          Next
      GoTo Main
```

The code for writing to the bye array (**ByteArray(Temp)=AByte**):

```
      Main
              ;MBasic
              ;-----------------
              BByte = 0
                      For Temp = ACon to CCon
                      AByte = Temp + 20
                      @Call StartTime
                              ByteArray(Temp) = AByte
                      @Call StopTime

                      HSerOut [Dec Temp,9,Dec ByteArray(Temp),9,"MBasic",9]
                      GoSub DumpTime
                      Pause 1000
              Next

              ;Assembler In-Line Macro
              ;-----------------------

              For Temp = ACon to CCon

                      AByte = Temp + 30
                      @Call StartTime
                      @Bank   0,ByteArray
                      movlw   ByteArray
                      @Bank   ByteArray,Temp
                      addwf   Temp,w
                      movwf   FSR
                      bcf     Status,IRP
                      ASM {   movlw   HIGH ByteArray}
                      andlw   0x1
                      btfss   Status,Z
                      bsf     Status,IRP
                      @Bank   ByteArray,AByte
                      movf    AByte,w
                      movwf   INDF
                      @Bank   AByte,0
                      @Call StopTime
                      HSerOut [Dec Temp,9,  |
      Dec ByteArray(Temp),9,"Assembler",9]
                      GoSub DumpTime
                      Pause 1000
          Next
      GoTo Main
```

References

[14-1] A complete data sheet for most PICs comprises two elements; (a) a detailed "family" reference manual and (b) the particular device datasheet. MBasic supports only PICs from Microchip's "midrange" family and the associated PICmicro™ Mid-Range MCU Family Reference Manual may be downloaded at http://www.microchip.com/download/lit/suppdoc/refernce/midrange/33023a.pdf. This is a 688-page document, in almost mind-numbing detail, but nonetheless is an essential reference to a complete understanding of PICs. For individual PIC family member datasheets, the easiest source is to go to http://www.microchip.com/1010/pline/picmicro/index.htm and select either the PIC12 or PIC16 group and from that link then select the individual PIC device.

[14-2] Benson, David, *Easy PIC'n A Beginners Guide to Using PIC Microcontrollers version 3.1*, Square 1 Electronics, Kelseyville, CA (1999).

[14-3] Benson, David, *PIC'n Up the Pace PIC Microcontroller Applications Guide version 1.1*, Square 1 Electronics, Kelseyville, CA (1999).

[14-4] Predko, Myke, *Programming and Customizing PICmicro Microcontrollers, Second Ed.*, McGraw Hill, New York (2002).

[14-5] Predko, Myke, *PICMicro Microcontroller Pocket Reference*, McGraw Hill, New York (2001).

[14-6] Microchip Technology, Inc., *MPASM User's Guide with MPLink and MPLib*. Available for download at Microchip's website, http://www.microchip.com under Development Tools, MPLAB. MPLAB is a large program that contains, among other things, MPASM and its documentation. MPASM is currently in version 6.x, and although version 3 is the one used in MBasic prior to v.5.3.0.0, the current MPASM User's Guide will be of importance to you in learning to program in assembler. GPASM, used beginning with MBasic v.5.3.0.0. A guide to GPASM may be downloaded at: http://gputils.sourceforge.net/gputils.pdf. Since GPASM is broadly compatible with MPASM, MPASM documentation is useful as well.

[14-7] D'Souza, Stan, *Implementing a Table Read, AN556*, Microchip Technologies, Inc., Doc. No. DS00556E (2000).

Interrupt Handlers and Timers in Assembler

We seen how timers and interrupts work in MBasic (Chapter 10) and learned a bit about assembler programming (Chapter 13) and how to mix assembler with MBasic (Chapter 14). It's time now to fuse these three strands into one—writing an assembler interrupt service routine. We'll see that the assembler ISR has several advantages over its MBasic cousin; perhaps most importantly, it solves the latency problems we observed in Chapter 10. Its primary disadvantage is that it must be written solely in assembler. That's not to say that the assembler ISR can't communicate with the rest of the MBasic program—because we'll see how to do exactly that—but it does mean that we can't seamless alternate in-line assembler and MBasic code inside the ISR, as we were able to in Chapter 14.

Chapter 10 offers the following definition of an interrupt:

> *An interrupt is an event that causes the current program sequence to temporarily halt, and for different code, called an "interrupt service routine" or ISR, to be executed. When the ISR is completed, program flow returns to the point it was at before the interrupt occurred. You might think of an ISR as a subroutine that is called not by software, but rather an event. The event may be an external, such as an input pin changing state from a 0 to a 1, or vice versa. Or, the event may be internal to the PIC, such as a timer reaching a defined count. It may even be a combination of external and internal events, such as where repeated external signals cause an internal counter to reach a preset value.*

ISRASM – MBasic's Gateway to Assembler Interrupt Service Routines

There's good news and bad news about writing an assembler ISR in MBasic. The good news is that MBasic includes a powerful aid to writing an assembler ISR—the **ISRASM** function. The bad news is that **ISRASM** is undocumented in the MBasic User's Guide. Before we delve into the **ISRASM** function, however, let's step back and look at how a PIC deals with an interrupt. And, since we must deal with interrupts in assembler, we'll need to examine some details that MBasic's **OnInterrupt** procedure automatically handles for us and therefore we omitted from Chapter 10.

We'll base our discussion upon PICs of the 16F87x/87xA series. Other PICs may not implement all the interrupts we discuss or may implement others. All midrange PICs, however, follow the same interrupt process, so what you learn here will be directly transferable to other midrange devices. When programming in assembler, read Microchip's Midrange Reference Manual and the specific data sheet for the device you will be using. It may seem impenetrable the first time you read it, but don't give up. You can do a great deal with MBasic without ever opening Microchip's documents; that is not the case when we start dealing directly with the PIC's hardware and internal firmware in an **ISRASM** procedure.

Chapter 10 developed a five-element template for MBasic interrupt handling. We'll recast that template for our assembly-based ISR:

1. Set up any options for the interrupt, such as timer preselectors, timer source signals, direction of input change, and so forth.

2. Clear the interrupt flag. (The interrupt flags are cleared as part of processor initialization, so in many cases you can skip this step.)
3. Enable:
   a. The specific interrupt;
   b. Enable global interrupts (**GIE**); and
   c. Enable peripheral interrupts (**PEIE**) if the specific interrupt is a member of the peripheral interrupt family
4. Write the interrupt service handler inside the **ISRASM** wrapper. Include:
   a. Check the source of interrupt to see if it is the one your code expects. If not, pass code execution to the end of the **ISRASM** wrapper. (You can, of course, check for multiple interrupt sources and branch to appropriate code routines based upon the particular calling interrupt.
   b. If the interrupt type is not self-clearing, clear the interrupt bit in your assembler code.
   c. Then add your code to take whatever actions you wish based on the interrupt.

Because they are so closely related, we'll treat steps 1, 2 and 3 together.

Every interrupt has one dedicated control bit and one status bit, perhaps more often called a "status flag" rather than a status bit. Some interrupts have a third control element; turning the underlying device on or off.

Control Bit—The control, or "enable," bit turns the interrupt on or off. When the control bit is set to **%0**, the interrupt is *disabled*; when it is set to **%1**, the interrupt is *enabled*. Once enabled, the interrupt remain enabled until the control bit is cleared.

Status Bit—The status bit, or "status flag," is set to **%1** to indicate that the condition causing the interrupt has occurred. The user is responsible for clearing the status bit for many interrupts.

Device Control—Some interrupt-generating internal modules, such as the timers, have a third control bit to turn the module on or off. For certain modules, turning the module on sets the interrupt flag, whether or not the interrupt is enabled. We'll deal with device control only in the context of specific examples.

In addition to the control bit associated with each interrupt, some interrupts are grouped into the *peripheral interrupt* family which may be collectively disabled or enabled. In order for a member of this interrupt family to function, both the PI family must be enabled and the particular interrupt must be enabled. Finally, there is a global interrupt enable bit that must be set for any interrupt to work.

Where are these various control and status bits? They are in various special purpose registers, associated with the interrupt hardware or software. The details of how each interrupt works, and how it is controlled, can be found in Microchip's reference documents. But let's summarize all 14 interrupts found in the 16F87x/ 87xA series PICs.

MBasic Identifier	Description	Peripheral Family?	Control		Status	
			Register	*Bit*	*Register*	*Flag*
ADINT	Analog-to-digital conversion is completed.	Yes	PIE1	ADIE	PIR1	ADIF *
BCLINT	Bus collision interrupt enable; not supported in MBasic	Yes	PIE2	BCLIE	PIR2	BCLIF *
CCP2INT	Interrupt occurs when there is a capture/compare/period match with CCP module 2	Yes	PIE2	CCP2IE	PIR2	CCP2IF
CCP1INT	Interrupt occurs when there is a capture/compare/period match with CCP module 2	Yes	PIE1	CCP1IE	PIR1	CCP1IF

(Continued)

MBasic Identifier	Description	Peripheral Family?	Control		Status	
			Register	Bit	Register	Flag
EEINT	EEPROM interrupt; occurs when writing to the on-board EEPROM is completed.	Yes	PIE2	EEIE	PIR2	EEIF
EXTINT	Change of input pin state on RB0. Must separately set mode to select 0→1 or 1→0 transition for the triggering event	No	INTCON	INTE	INTCON	INTF
None	Parallel slave port interrupt; not supported in MBasic	Yes	PIE1	PSPIE	PIR1	PSPIF *
RBINT	Any pin RB4,RB5,RB6 or RB7 has changed state, in either the 0→1 or 1→0 direction.	No	INTCON	RBIE	INTCON	RBIF **
RCINT	Received a byte through the hardware USART. Note: When using HSERIN or HSEROUT functions RCIF is disabled.	Yes	PIE1	RCIE	PIR1	RCIF *
SSPINT	Synchronous serial port interrupt; not supported in MBasic	Yes	PIE1	SSPIE	PIR1	SSPIF
TMR0INT	Timer 0 overflow has occurred	No	INTCON	T0IE	INTCON	T0IF
TMR1INT	Timer 1 overflow has occurred	Yes	PIE1	TMR1IE	PIR1	TMR1IF
TMR2INT	Timer 2 overflow has occurred	Yes	PIE1	TMR2IE	PIR1	TMR2IF
TXINT	Finished transmitting a byte from the hardware USART. Note: When using HSERIN or HSEROUT functions RCIF is disabled.	Yes	PIE1	TXIE	PIR1	TXIF *

* Indicates a flag that is automatically cleared by the PIC; all other flags require software clearing in the interrupt service routine.
** In addition to clearing the RBIF flag in software, you must read the PortB register to reload the internal comparison latch.

After reading this table, you should see several patterns:

- The enable and interrupt flag bits share a common naming system, with the enable bit acronym ending in "E" for *enable* while the interrupt flag bit ends in an "F" for *flag*.
- Only the oldest interrupts (oldest in the sense that they have been in the PIC architecture since the earliest PICs were designed) have their enable and flag bits in the interrupt control register, **INTCON**.
- Interrupts associated with "newer" features are all members of the peripheral interrupt family and have their enable bits in two peripheral interrupt enable (**PIE**) registers, **PIE1** and **PIE2**. Their corresponding flag bits are held in two peripheral interrupt registers (**PIR**), **PIR1** and **PIR2**. In fact, although not seen in the table, the register and bit sequence align between the **PIE** and **PIR** registers, for example, **PIE1** bit 0 is the enable bit for timer 1, while **PIR1** bit 0 is the interrupt flag for timer 1.
- The 7-bit register addresses align between the **PIE** and **PIR** registers, with the **PIR** registers in bank 0 and the **PIR** registers in bank 1. For example, **PIR1**'s relative address is **$0C**, in bank 0, while **PIE1**'s relative address is **$0C** in bank 1.

In addition, we have two higher level interrupt control bits:

Function	Register	Bit
Global interrupt enable bit—controls all interrupts	INTCON	GIE
Peripheral family interrupt enable bit—controls all interrupts that are members of the peripheral family	INTCON	PEIE

We have several choices in how to enable and disable interrupts and set any associated options:

- *Set options using MBasic commands exactly as we learned in Chapter 10*—This is the simplest approach as it encapsulates the details into a simple MBasic procedure or two:
  - o Set up the interrupt using the appropriate MBasic setup function, such as `SetExtInt`, or `SetTmr0`, etc. For example we set up the external (**RB0**) interrupt to trigger on the 1→0 transition with the following MBasic statement:

    ```
    SetExtInt Ext_H2L
    ```
  - o You cannot enable or disable the interrupt using MBasic's `Enable` and `Disable` functions, however. These functions work only if you also establish an interrupt handler using the `OnInterrupt` function, which conflicts with an `ISRASM` routine. Rather, you must use one of the three approaches discussed below to enable and disable the interrupt.

- *Set or clear the setup and enable bits from MBasic*—All of the control and flag bit identifiers shown in the interrupt table are also library defined MBasic bit variables and may be written to or read from like any other variable. The associated special purpose file registers are also library defined byte variables. To enable the external interrupt (**RB0**) either of the following two MBasic statements may be used:

  ```
  Inte = %0
  INTCON.Bit4 = %0
  ```

  To determine that the `Inte` flag corresponds to `Intcon.bit4`, you must consult the 16F87x or 16F87xA datasheet. You may also use this approach to set up interrupt options. For example, the choice between **RB0** interrupt occurring on the 0→1 or 1→0 direction is determined by the control bit `IntEdg`, which is `Option_Reg.Bit6`. From the data sheet, we determine that triggering the interrupt upon a 1→0 transition (falling edge) corresponds to `IntEdg` being cleared. This gives us three ways in MBasic to configure the external (**RB0**) interrupt:

  ```
  SetExtInt Ext_H2L
  IntEdg = %0
  Option_Reg.Bit6 = %0
  ```

- *Use in-line assembler to set or clear enable and setup bits*—From the data sheet for the device you are using, identify the relevant enable and option control registers and bits. Then, set or clear these as, appropriate, using the in-line assembler macros `bsf` and `bcf`. For example, to set up and enable the external (**RB0**) interrupt, we may use the following in-line code (assuming, of course, that the prior bank is bank 0):

  ```
  @bank 0,Option_Reg
  bcf Option_Reg,6          ;same as SetExtInt Ext_H2L
  bcf Option_Reg,IntEdg     ;can use named const IntEdg

  bsf Intcon,4              ;enable the interrupt
  bsf Intcon,Inte           ;can use named constant Inte
  ```

- *Use `ASM {...}` assembler code to set or clear enable bits*—Using the same approach as for in-line assembler code, we identify the relevant enable and option control registers and bits. We then set or clear these, as appropriate, using assembler the opcodes `bsf` and `bcf`, or their macro versions `_@bsf` and `_@bcf`.

  ```
  ASM
  {
          _@bank 0,Option_Reg
          bcf Option_Reg,6          ;same as SetExtInt Ext_H2L
          bcf Option_Reg,IntEdg     ;use named constant instead
          bsf Intcon,4              ;enable the interrupt
          bsf Intcon,Inte           ;can used named constant

  }

  ASM
  {
  ```

```
            _@bank 0,Option_Reg
            _@bcf Option_Reg,6           ; same as SetExtInt Ext_H2L
            _@bcf Option_Reg,IntEdg      ;use named constant
            _@bsf Intcon,4               ;enable the interrupt
            _@bsf Intcon,Inte            ;use named constant
      }
```

Checking the memory map in Chapter 13, or the Microchip data sheet, we determine that **Option_Reg** is in bank 1, and **Intcon** is duplicated across all four banks, so it may be addressed from any bank setting. If we chose to write to **Option_Reg** with either in-line assembler macros or from within an **ASM {...}** block, we must ensure that the proper bank bits are set.

I've shown two ways to identify the bit arguments; by looking up the bit number in Microchip's documentation, or using the named library constant:

```
            bsf Intcon,4                ;enable the interrupt by number
            bsf Intcon,Inte             ;or (preferable) used named constant
```

Both of these assembler statements perform identical tasks, but using the named constant, **Inte**, makes the code more human readable, and is preferred over the numerical form. You'll find the named constants for all special purpose registers and bits in Microchip's device-specific data sheet. Interrupt enable and flag related register and bit names also appear in the summary table earlier in this chapter.

Since the interrupt flag is cleared at processor reset, manually clearing the flag is not always necessary, but you may wish to do it as part of interrupt setup and initialization. It's good programming practice to clear the associated interrupt flag for all the interrupts you intend to use, particularly if the interrupt has been responding to inputs during initialization, so that its flag status is unknown.

We have the same three options to clear the interrupt flag:

- From MBasic using the library variable identical to the interrupt flag name. For example, to clear the **INTF** flag associated with the **INTE** interrupt, we could use the following assignment:

      INTF = %0

- Using in-line assembler macros. We first must determine the address and bit number of the flag we wish to clear. For the external (**RB0**) interrupt, Microchip's data sheet informs us that the interrupt flag is at **INTCON <1>**, or **INTCON <INTF>** We check **INTCON**'s bank and discover that it is aliased at all four banks, so we need not concern ourselves with bank setting. The following bit clear operation can be used:

      Bcf   Intcon,Intf

- Use the **ASM {...}** construction, either with assembler macros, or with pure assembler opcodes:

```
      ASM
      {
      _@bcf     Intcon,Intf    ;macro version
      Bcf       Intcon,Intf    ;assembler opcode version
      }
```

What Happens When the Interrupt Fires?

After laboring in the details of special purpose registers and files, let's step back a moment and look at what happens when an interrupt event occurs. In MBasic, as we learned in Chapter 10, the interrupt handler is executed at the completion of the MBasic statement executing at the instant the interrupt occurs. We demonstrated the resulting latency could be dozens of microseconds, or even much longer, depending upon the particular MBasic statement. We also learned how to associate an interrupt handler with the source of the interrupt, so that the compiler automatically determined the source of the interrupt and routes control to the appropriate handler.

From the prospective of the MBasic programmer, the assembler-based **ISRASM** handler works quite differently. When an interrupt event occurs, two things happen:

- The flag bit associated with the interrupt is set.
- The processor stops executing the current code within, at most, two Fosc/4 instruction intervals and transfers program execution to whatever assembler code resides at code address $0004, the location of the *interrupt vector*. This jump may occur in the middle of an MBasic instruction. At the end of the **ISRASM** routine, program flow returns to the MBasic code seamlessly and without loss of information.

Regardless of the interrupt source, all interrupts cause the code at $0004 to be executed. (Some more complex microprocessors dispatch different interrupt sources to code at different addresses; not so with the mid-range PIC's we're using, however.) It's up to you, the programmer, to write code to determine which interrupt sent execution to your **ISRASM** routine and what to do if the interrupt isn't from the source you expected. (MBasic, for example, uses the **TXIE** and **RCIE** hardware serial module interrupts in sending and receiving serial data via **HSerOut** and **HSerIn**. If your program uses these functions, your ISRASM code must pass the **TXIE** and **RCIE** interrupts through without change, or else the **HSerIn** or **HSerOut** functions may fail to correctly perform.)

Why is location $0004 called an *interrupt vector*? The term comes from the fact that upon an interrupt event occurring, program execution jumps, or is vectored to, the program code at location $0004. Usually, the program code at location $0004 is a **GoTo** statement that transfers program execution to a different memory location where the real interrupt service routine resides.

There's a bit of housekeeping that must occur in order for program execution to seamlessly return to MBasic processing after interruption in the middle of a MBasic statement. For example, we must save the **W, Status, PCLath** and **FSR** registers and then jump to the **ISRASM** module. Upon completion of the **ISRASM** code, we must restore them to the value they had before the interrupt. These processes are often called *context saving* and *context restoration* or, generally, *context switching*. The good news is that MBasic's **ISRASM** automatically performs context switching. We may have to account for the extra delay caused by context switching if our interrupt is engaged critical time measurement activity, but by in large context switching is invisible to you when writing **ISRASM** code.

Another twist that you must consider is what happens if another interrupt occurs in the middle of your ISRASM code? Your **ISRASM** code isn't likely to be reentrant, so you will wish to disable all interrupts at the outset of the **ISRASM** code. Don't forget to turn them back on at the end of the **ISRASM** code.

Special Considerations for ISRASM Code

Before we start looking at sample code, there's one final point you should be aware of. In versions prior to 5.3.0.0, **ISRASM** procedures do not work with MBasic's pre-defined in-line macros, or assembler macros. This is a by-product of the MBasic's compiler order of library linkage that was modified in version 5.3.0.0. In these earlier versions, MBasic's macro library follows ISRASM code assembly, and hence the assembler doesn't see the macro definitions.

In order to maintain compatibility with earlier releases, the ISRASM code in this book will not use any pre-defined macro. (You could write your own macros, of course, but if you know enough to do that, you don't need this book to tell you how.) Our coding therefore incorporates the following features:

- We will not use any **@** prefixed macros, or **_@** prefixed macros.
- We will not use our time saving optional bank selection macro, **@bank**. Instead, if we are uncertain of which bank a particular register is in and whether we need to change banks, we must force a bank selection using the macro defined inside the assembler, **BankSel**. The **banksel** macro is called with the following syntax:

```
Banksel <register file name or value>
```

Banksel expects a 9-bit argument. It generates the code to change the bank bits to point at the bank corresponding to the 9-bit address. The bank selection code will always be generated, even if the bank bits were already set correctly before the **banksel** macro.

● In order to avoid a string of error messages, we will strip the excess compiler bits from register file definitions.

  o To obtain a 7-bit register address for opcode arguments, we trim with the following operation:
  RegisterName & 0x7F

  o In MBasic versions earlier than 5.3.0.0 which use Microchip's MPASM assembler, it is necessary to obtain a 9-bit register address for **banksel**, which we do by we trimming the excess bits with the following operation: **RegisterName &0x1FF**. The GPASM assembler used with version 5.3.0.0 onward does require the excess bits to be trimmed via the **&0x1FF** mask. Rather, the variable name must be used directly, for example, **banksel RegisterName**. Using a **&0x1FF** mask causes the assembler to generate an "invalid label" error message, preventing your from compiling.

Program Examples

Let's start with an **ISRASM** version of Program 10-1. Configure your 2840 Development Board as shown in Figure 15-1. As you may recall, Program 10-1 monitors the status of pin **B0** through the external interrupt. When the switch SW1 is closed, **B0** goes low and causes the external interrupt to be issued. The external interrupt flag, **INTF** is set and execution jumps to our **ISRASM** code. We'll take the same action in Program 15-1 as we did in Program 10-1; we'll briefly flash the LED connected to pin **B2** every time SW1 is closed.

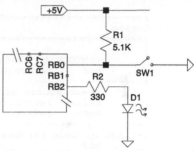

*Figure 15-1: Program 15-1 Connection].
As you may recall*

Program 15-1

```
Stack = 10
;Constants
;---------
Dly             Con             255

;Variables
;---------
i               Var             Byte


;MBasic Initialization
;---------------------
Input B0        ;Switch on B0
Output B2       ;LED on B2
High B2         ;test that it works
Pause 250
Low B2

i = Dly         ;must use variable or it won't be allocated


;........................
;Interrupt Initialization
;........................

;=========================
;(A) set up the interrupt
;=========================

;(1) MBasic function
;----------------
```

```
        SetExtInt Ext_H2L

        ;==========================
        ;(B) enable the interrupt
        ;==========================

        ;(1) Write to enable bit in MBasic
        Inte = %1              ;use named bit variable


        ;===================================
        ;(C) Enable PEIE (If necessary) & GIE
        ;===================================

        ;(1) Write to enable bit in MBasic
        ;(PEIE not necessary for this interrupt
        ;but it's shown for completeness)
        PEIE = %1              ;use named bit variable
        GIE = %1


        ;=============================================================
        israsm
        ;=============================================================
        {
                ;INTCON is shadowed in all 4 banks, no need to banksel
                bcf     Intcon,GIE     ;disable interrupts
        ;------------------
                ;test source of interrupt and if INTF not set, go to end
                btfss   Intcon,Intf        ;if not INTF then jump to the end
                GoTo    EndProc
        ;------------------
        ;Following code executed only for INTF
                bcf             Intcon,intf    ;clear the interrupt flag
                BankSel PortB
                ;light the LED on B2
                bsf             PortB,2
                ;pause for a brief time with delay loop
                ;loads value of Dly into i and decrements until zero
                movlw   Dly     ;w=Dly
                BankSel i       ;get right bank-no mask is allowed!
                movwf   i&0x7F  ;i = Dly
                ;simple delay loop= 10 * Dly value
                ;measured in Fosc/4 cycles
        LoopIt
                NOP     ;total delay is 10 Fosc/4 cycles
                NOP     ;for one pass through the loop
                NOP
                NOP
                NOP
                NOP
                decfsz          i&0x7F,f
                GoTo    LoopIt
                banksel PortB               ;back to PortB's bank
                bcf             PortB,2 ;turn LED off

        ;end of ISRASM
        ;------------------
        EndProc
        ;------------------
                bsf             IntCon,GIE      ;enable interrupts
                BankSel 0                       ;point at Bank 0 for MBasic
        }

        Main
        ;------
        GoTo Main

        End
```

When you close SW1, you should see the LED briefly illuminate.

Let's go through the code.

```
Dly             Con         255
i               Var         Byte
```

We define one constant, **Dly** and one variable **i**. We use both to create a delay between turning the LED on and turning it off.

```
;MBasic Initialization
;----------------------
Input B0        ;Switch on B0
Output B2       ;LED on B2
High B2         ;test that it works
Pause 250
Low B2
```

We set up **B0** as an input—the default is input, but it's good practice to explicitly define the pin's input/output status. We also define **B2** as an output and briefly flash the LED to verify that it's properly connected.

```
i = Dly         ;must use variable or it won't be allocated
```

As we learned in Chapters 13 and 14, MBasic is an optimizing compiler and will not allocate storage for variables that are unused within MBasic code. One way to efficiently accomplish this is the **Clear** procedure, which sets all variables to zero, thereby forcing the compiler to allocate space for all declared variables. We'll just use **i** in an assignment statement, thereby ensuring the compiler allocates space for it.

Now we go through the multistep template we developed earlier.

```
;Interrupt Initialization
;(A) set up the interrupt
;(1) MBasic function
;----------------
SetExtInt Ext_H2L
```

Since we only initialize the interrupt once, there's no benefit from using in-line or other assembler code. Hence, we'll use the MBasic **SetExtInt** function to set the **RB0** interrupt to fire on the high-to-low transitions.

```
;(B) enable the interrupt
;(1) Write to enable bit in MBasic
Inte = %1               ;use named bit variable
```

The next step in our interrupt template is to enable the interrupt. Again, since we only initialize the interrupt once, there's no benefit from using anything other than MBasic.

```
;(C) Enable PEIE (If necessary) & GIE
;(1) Write to enable bit in MBasic
;(PEIE not necessary for this interrupt
;but it's shown for completeness)
PEIE = %1               ;use named bit variable
GIE = %1
```

Next we set the global interrupt enable bit, **GIE**. As we learned earlier, the **GIE** is a master switch that must be set for *any* interrupt to work. While we are at it, we'll also set the peripheral interrupt bit, **PEIE**, a second master switch required to be set before any *peripheral* interrupt will trigger. The particular interrupt we use, the external or **RB0**, interrupt is not a member of the peripheral interrupt family; so enabling the **PEIE** isn't necessary. But, we'll enable it to demonstrate how it is done.

We'll skip the step of clearing the external interrupt flag, **INTF** as it isn't necessary for this simple program. If you wish to do so, clearing the flag may be accomplished by adding the MBasic statement **INTF=%0**.

Now we reach the **ISRASM** itself. Note that the **ISRASM** is followed by curly braces to contain all the interrupt routine's assembler code. It isn't necessary to add the compiler instruction **ASM**.

I like to place the **ISRASM** code near the top of the program, after variables and constants are declared and after initialization. This is, however, a personal preference and you may place the **ISRASM** code anywhere outside program flow, such as at the bottom of the program, even after the **End** statement.

```
israsm
{
        ;INTCON is shadowed in all 4 banks, no need to banksel
        bcf     Intcon,GIE      ;disable interrupts
```

First thing we disable all interrupts by opening the master interrupt switch—clearing the global interrupt enable bit. We don't want our code to be re-triggered by another interrupt before it has completed executing.

```
        ;test source of interrupt and if INTF not set, go to end
        btfss   Intcon,Intf             ;if not INTF then jump to the end
        GoTo    EndProc
```

The next step—and it should not be postponed until later in the **ISRASM** code—is to determine the interrupt source and terminate the **ISRASM** routine if the interrupt is not from a source we wish to handle in the **ISRASM**. In this case, with only one interrupt to handle, we test to see if the associated flag bit, **Intf**, is set. If the **Intf** flag is clear (not set) we transfer execution, via the **GoTo** statement, to code starting at the label **EndProc**, which we will find completes one small housekeeping function and then ends the **ISRASM**. (All 14 possible interrupt flags are in registers located in Bank 0, and we know MBasic hands over control to the **ISRASM** in bank 0, so we do not have to worry about bank selection.)

```
        ;Following code executed only for INTF
        bcf             Intcon,intf     ;clear the interrupt flag
```

Next, we clear the interrupt flag, so that the interrupt doesn't go into an endless loop.

```
        BankSel PortB
        ;light the LED on B2
        bsf             PortB,2
```

We turn the LED on by setting pin **B2** high via the **bsf** operation, performed on PortB, bit 2. We know that PortB is in bank 0, so the **banksel PortB** operation is unnecessary, but I included it as a reminder that we must always be conscious of banking.

```
        ;pause for a brief time with delay loop
        ;loads value of Dly into i and decrements until zero
        movlw   Dly             ;w=Dly
        BankSel i               ;get right bank
        movwf   i&0x7F          ;i = Dly
```

We'll use a simple delay loop to pause long enough for us to see the LED illuminate. First, we copy the number of repeat cycles in the delay loop, held in the constant **Dly**, into the counter variable **i**. We've seen how this is accomplished in Chapters 13 and 14. Note we use the **Banksel i** macro to set the banking bits.

```
        ;simple delay loop= 10 * Dly value
        ;measured in Fosc/4 cycles
LoopIt
        NOP     ;total delay is 10 Fosc/4 cycles
        NOP     ;for one pass through the loop
        NOP
        NOP
        NOP
        NOP
        NOP
        decfsz  i&0x7F,f
        GoTo    LoopIt
```

The loop structure uses the **decfsz** operation, which is similar to the **For...Next** assembler template discussed in Chapter 14. The **decfsz** operator decrements the register file **i** and tests **i** for zero. If **i** is not zero, the statement immediately following the **decfsz** operator executes; if **i** is zero, that statement is skipped and execution branches to the second statement following the **decfsz** operator. This structure causes the **LoopIt** statement to be repeated **Dly** times. The **decfsz** and **GoTo** operations consume a combined total of three instruction cycles, so to provide more delay, we add seven no-operation instructions, for a total of ten instruction cycles delay. One instruction cycle is Fosc/4, or for the 20 MHz clock, 0.200 μs. Ten such instruction cycles is 2 μs, and when repeated 255 times results in a total delay of 5100 μs.

```
banksel        PortB                ;back to PortB's bank
bcf            PortB,2              ;turn LED off
```

After the delay, we turn the LED off by taking pin **B2** low.

```
;end of ISRASM
EndProc
        bsf            IntCon,GIE       ;enable interrupts
}
```

At the end of the **ISRASM**—whether we reach it by a valid interrupt and completion of the delay loop or by a nonexpected interrupt and the jump to **EndProc**, all we need do to wrap up the **ISRASM** is to reactivate the master interrupt switch by setting **GIE** and ensure that the banking bits are selected for bank 0, which MBasic expects. We've earlier used **banksel portB** to set the banking bits, and since **PortB** is in bank 0, we need no additional bank selection to meet MBasic's re-entry requirements.

That's it. Pretty simple, isn't it? MBasic hides the context switching from us, permitting us to concentrate on useful code to deal with the interrupt.

```
Main
;------
GoTo Main

End
```

Finally, have an endless **Main...GoTo Main** loop to keep the MBasic program elements exercised.

Run program 15-1 and press switch SW1. With no delay perceptible to your eye, the LED will briefly flash. So far, no major improvement over Program 10-1, right? In either program, you perceive no delay between switch closure and LED illumination.

Let's modify Program 15-1 by changing the following:

```
Main
;------
        Pause 1000
GoTo Main
```

We'll name the amended code Program 15-1-A. Recompile and run it. Press the button and what do you see? No change from Program 15-1; the LED flashes with no perceptible delay as soon as you operate SW1. This is quite different behavior than we observed after adding delay to Program 10-1. There, in what became Program 10-1-A, we saw noticeable and variable delay— up to one second—between SW1's operation and LED illumination. The extended delay in Program 10-1-A results from executing interrupts only between MBasic statements; in contrast, the near instantaneous response in Program 15-1-A is a product of the **ISRASM** executing after any of the dozens or hundreds of machine instructions that comprise any single MBasic statement.

Program 15-2

Since our program can have but one **ISRASM** procedure, how do we accommodate multiple interrupts? Let's look at a simple way to support multiple interrupts. Add a second LED and current limiting resistor to your 2840 Development Board, as shown in Figure 15-2. We'll use timer 1 to flash the LED connected to pin B1 with approximately 128 ms on / 128 ms off periods. We'll also test for change of status on pin B0 with the external interrupt and, when the

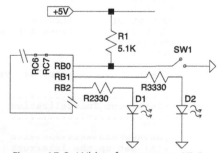

Figure 15-2: Wiring for program 15-2.

switch connected to pin B0 closes, we'll flash the LED on pin B2 using a third interrupt, timer 2, to control the flash duration. Thus, Program 15-2 must handle three interrupts.

```
;Program 15-2
;multiple interrupts

;Constants
;---------
TicksPerms      Con        $FFFF-5000       ;for timer 1

LED2            Con                  2 ;LED on Pin B2
LED1            Con                  1 ;LED on Pin B1

;In case MBasic's errors in Timer 2 constants have not been
;fixed, we supply our own
T2Post1         Con                  %00000000
T2Post2         Con                  %00001000
T2Post3         Con                  %00010000
T2Post4         Con                  %00011000
T2Post5         Con                  %00100000
T2Post6         Con                  %00101000
T2Post7         Con                  %00110000
T2Post8         Con                  %00111000
T2Post9         Con                  %01000000
T2Post10        Con                  %01001000
T2Post11        Con                  %01010000
T2Post12        Con                  %01011000
T2Post13        Con                  %01100000
T2Post14        Con                  %01101000
T2Post15        Con                  %01110000
T2Post16        Con                  %01111000

T2Pre1          Con                  %00000100
T2Pre4          Con                  %00000101
T2Pre16         Con                  %00000111

;we select 8us per comparision unit, so we need
;a total pre*post of 40:1.
Tmr2SetUp               Con     T2Post10 + P2Pre4
;for 1 ms on, the period must be 125
Tmr2Period              Con     125

;Variables
;---------
i               Var                  Byte

;MBasic Initialization
;---------------------
Input B0        ;Switch on B0
Output B2       ;LED on B2
High B2         ;test that it works
Pause 250
Low B2

Output B1       ;LED on B1
High B1         ;test that it works
Pause 250
Low B1

Clear           ;must use variable or it won't be allocated

;......................
;Interrupt Initialization
;......................

;========================
;(A) set up the interrupt
;========================
```

```
;(1) MBasic function
;---------------
SetExtInt Ext_H2L       ;the B0 interrupt
SetTmr1 Tmr1Int1        ;setup timer 1, prescale=1, internal

;set timer 2 for 10:1 post & 4:1 pre & 125 divide so
;that we get a 1000us period
SetTmr2 Tmr2Setup,Tmr2Period

Tmr2On = %0     ;turn timer 2 off until we need it


;=========================
;(B) enable the interrupt
;=========================


;(1) Write to enable bit in MBasic
Inte = %1               ;enable timer 1 interrupt
Tmr1IE = %1             ;timer 1 interrupt on
Tmr2IE = %1             ;timer 2 interrupt on

;==============================
;(C) Clear the interrupt flags
;==============================
intf = %0
Tmr1If = %0
Tmr2IF = %0


;===================================
;(D) Enable PEIE (If necessary) & GIE
;===================================


;(1) Write to enable bit in MBasic
;PEIE is necessary for Tmr1 interrupt
PEIE = %1               ;use named bit variable
GIE = %1                ;globals


;==============================================================
israsm
;==============================================================
{
        ;INTCON is shadowed in all 4 banks, no need to banksel
        ;MBasic sets bank=0 going into israsm
        bcf     Intcon,GIE      ;disable interrupts
;------------------
        ;test source of interrupt and if INTF not set, go to end
        btfsc           Intcon,Intf     ;if not INTF then jump to the end
        Goto            IntfProc        ;if set execute IntfProc

        btfsc           PIR1,Tmr2IF     ;test for timer 2 interrupt
        GoTo            Tmr2Proc        ;if source of interrupt is timer 2

        btfsc           PIR1,Tmr1IF     ;test for timer 1 interrupt
        GoTo            Tmr1Proc        ;if source of interrupt is timer 1
                                        ;all other cases - go to end
        GoTo            EndProc

;------------------
IntfProc
;------------------
;Following code executed only for INTF
        bcf             Intcon,intf     ;clear the interrupt flag
        bsf             PortB,LED2      ;turn B2 LED on
        ;both PIR1 and T2Con are in bank 0, no need to change
        bcf             PIR1,Tmr2IF     ;clear timer 2 interrupt flag
        ;clear timer 2
        clrf            Tmr2            ;clear timer 2's main counter
        ;write to T2con also clears pre & post scale counters
        bsf             T2Con,Tmr2On    ;turn timer 2 on
        GoTo            EndProc
```

```
;-------------------
Tmr2Proc
;-------------------
;following code executed only if Timer 2 interrupt
;called with bank = 0
        bcf             PortB,LED2      ;turn B2 LED off
        ;both PIR1 and T2Con are in bank 0, no need to change
        bcf             PIR1,Tmr2IF ;clear timer 2 interrupt flag
        bcf             T2Con,Tmr2On ;turn timer 2 off
        GoTo            EndProc


;-------------------
Tmr1Proc
;-------------------
;following code executed only if Timer 1 interrupt
;called with bank = 0
        bcf             PIR1,Tmr1IF ;clear timer 1 interrupt flag

        ;timer 1 must be reset every time it rolls over.
        ;high byte first-avoids rollover during
        ;the time it takes to reset the counter.

        movlw           High TicksPerms         ;put high byte of load into W
        ;timer spr's are in Bank 0 no need for banksel
        movwf           TMR1H                   ;move into timer register
        movlw           Low TicksPerms          ;now for the low byte
        movwf           TMR1L                   ;move into timer register

        BankSel I                               ;where is i?
        incf            i&0x7F,f                ;i=i+1
        btfss           i&0x7F,7                ;use bit 7 as /128 flash
        Goto            TurnOff
        BankSel PortB
        bsf             PortB,LED1
        GoTo            EndProc
TurnOff
        BankSel PortB
        bcf             PortB,LED1
        GoTo            EndProc

;end of ISRASM
;-------------------
EndProc
;-------------------
        bsf             IntCon,GIE              ;enable interrupts
}

Main
;------
        Pause 1000
GoTo Main

End
```

Let's start our analysis of Program 15-2 with a pseudo-code view of our **ISRASM** structure. For clarity, we'll use MBasic type pseudo-code instead of assembler operators.

```
Initialization
        Setup Timer 2 to provide 1ms period
        Setup Timer 1 to provide 1ms period
        Setup Timer 1, Timer 2 and External Interrupts

ISRASM
        Determine source of Interrupt:

        If External (Button pushed)
                Clear Timer 2 interrupt flag
Turn LED on B2 on
                Reset Timer 2
                Turn Timer 2 on
```

```
        If Timer 2
                Turn LED on B2 off
                Turn Timer 2 off

        If Timer 1
                Reload timer 1 pre-set values
                i = i+1
                LED on B1 = i.Bit7

        If none of the above, exit ISRASM

    MBasic dummy routine:
    Main
    Pause 1000 ms
    GoTo Main
```

Why set timers 1 and 2 for 1 ms? And, why complicate things with a second timer, when the delay process of Program 15-1 works?

Let's take the second question first. Remember, the first thing we do in the **ISRASM** procedure is disable interrupts. Hence, to reduce latency problems caused by interrupts firing but not being recognized during the **ISRASM** procedure, we should keep the **ISRASM** code as short as feasible. Instead of stopping interrupts during the delay period we used in Program 15-1, we'll offload the delay to timer 2—a hardware module that consumes no software resources. Thus, other than a few microseconds it takes execute the minimum needed code, our program is free to respond to any other interrupts.

I selected 1ms for timer 2 because it yields an easily viewed but still brief flash from an LED. The exact on-time isn't critical, so experiment if you wish. I wanted to flash the second LED at a rate that makes a noticeable on/off sequence. And, I wanted minimum housekeeping to alternate the LED between on and off, so I planed to use a trick we've seen before—increment a byte variable counter and make the LED pin track the counter's high bit. This gives 128 periods off, followed by 128 periods on. If we increment the counter once every 1ms, the result is a visually pleasant 128 ms off/128 ms on cycle.

We've learned in Chapter 10 how to set timers 1 and 2 to produce these intervals. With a 20 MHz resonator in my 2840 Development Board, it requires a total count of 5000 to produce interrupts once every one millisecond. For timer 2, we achieve the 5000 count with a prescale value of 4, a postscale value of 10 and a counter period of 125. For timer 1, we may set the prescale ratio as 1:1 and pre-load its counter register with **$FFFF-5000**. If you don't understand how I arrived at these numbers, go back and reread Chapter 10. We'll cheat a bit and not worry about counter dead time, since our purpose here is to demonstrate the rudiments of the **ISRASM** procedure and a few microseconds error isn't important. To see how to correct for dead time when setting timer 1, study Program 19-3.

```
    TicksPerms      Con     $FFFF-5000      ;for timer 1
    LED2            Con     2               ;LED on Pin B2
    LED1            Con     1               ;LED on Pin B1
    Tmr2SetUp       Con     T2Post10 + P2Pre4
    Tmr2Period      Con     125
```

Based on these calculations, we define the three timer-related constants shown above. To improve program readability, we define two constants LED2 and LED1 to reference **PortB**'s bit numbers corresponding to the LEDs on pins **B2** (**LED2**) and **B1** (**LED1**).

Let's concentrate on the changes from Program 15-1 in interrupt setup.

```
    ;Interrupt Initialization
    ;(A) set up the interrupt
    ;(1) MBasic function
    ;---------------
    SetExtInt Ext_H2L      ;the B0 interrupt
    SetTmr1 Tmr1Int1       ;setup timer 1, prescale=1, internal
```

```
;set timer 2 for 10:1 post & 4:1 pre & 125 divide so
;that we get a 1000us period
SetTmr2 Tmr2Setup,Tmr2Period

Tmr2On = %0     ;turn timer 2 off until we need it
```

We've added setup code for timers 1 and 2. Remember our use for timer 2 is as a "one shot" action; it is turned on when the external interrupt fires, cycles once and is turned off until the next time the external interrupt fires. Hence, we set up timer 2, but turn if off as it isn't needed yet.

```
;(B) enable the interrupt
;(1) Write to enable bit in MBasic
Inte = %1               ;enable timer 1 interrupt
Tmr1IE = %1    ;timer 1 interrupt on
Tmr2IE = %1             ;timer 2 interrupt on
```

We use MBasic to enable the interrupts. There is no advantage to enabling the interrupts in assembler.

```
;(C) Clear the interrupt flags
intf = %0
Tmr1If = %0
Tmr2IF = %0
```

To demonstrate how it's accomplished in MBasic, we've added a flag clearing section to our initialization code.

```
;(D) Enable PEIE (If necessary) & GIE
;(1) Write to enable bit in MBasic
;PEIE is necessary for Tmr1 interrupt
PEIE = %1               ;use named bit variable
GIE = %1                ;globals
```

At the end of the initialization section, we enable global and peripheral interrupts. Both timers 1 and 2 are members of the peripheral interrupt family, so we must enable peripheral interrupts by setting the **PEIE** bit.

```
israsm
{
        ;INTCON is shadowed in all 4 banks, no need to banksel
        ;MBasic sets bank=0 going into israsm
        bcf     Intcon,GIE      ;disable interrupts
```

As in Program 15-1, the first order of business in the **ISRASM** procedure is to disable interrupts to avoid multiple triggers. We use the same "master switch" approach discussed in our analysis of Program 15-1.

```
;test source of interrupt and if INTF not set, go to end
        btfsc           Intcon,Intf     ;if not INTF then jump to the end
        Goto            IntfProc        ;if set execute IntfProc

        btfsc           PIR1,Tmr2IF     ;test for timer 2 interrupt
        GoTo            Tmr2Proc        ;if source of interrupt is timer 2

        btfsc           PIR1,Tmr1IF     ;test for timer 1 interrupt
        GoTo            Tmr1Proc        ;if source of interrupt is timer 1
                                        ;all other cases - go to end

        GoTo            EndProc
```

Our interrupt handler must recognize and deal with three interrupts—external (**RB0**), timer 1 and timer 2. We test each of the three associated interrupt flags with a **btfsc** operation; if the bit is set, execution branches to the associated interrupt handler code via a **GoTo**, otherwise (the flag bit is clear) the test proceeds to the next **btfsc** comparison. If the interrupt flag has not been found after the last test, the interrupt source must be one other than the three we intend, so execution jumps to the end of the **ISRASM** code via the **GoTo End-Proc** operation.

The **IntfProc** code is executed if the interrupt source is the external interrupt (**RB0**). MBasic sets the bank selection bits to bank 0 going into **ISRASM**, and our flag test sequence does not alter the banking bits, so we are assured entry to each of the three interrupt-specific sections in bank 0.

```
IntfProc
;------------------
;Following code executed only for INTF
        bcf             Intcon,intf     ;clear the interrupt flag
        bsf             PortB,LED2      ;turn B2 LED on
        ;both PIR1 and T2Con are in bank 0, no need to change
        bcf             PIR1,Tmr2IF     ;clear timer 2 interrupt flag
        ;clear timer 2
        clrf            Tmr2            ;clear timer 2's main counter
        ;write to T2con also clears pre & post scale counters
        bsf             T2Con,Tmr2On    ;turn timer 2 on
        GoTo            EndProc
```

We first clear the **intf** interrupt flag to avoid multiple triggers of **ISRASM**. We next set PortB pin B2 high to illuminate the LED. Now, we must initialize and start timer 2 so that we may use it to turn the LED off after 1ms. Before we do anything else, we clear timer 2's interrupt flag. This flag should already be clear, but as a matter of caution, we'll clear it anyway. Next we initialize timer 2, which requires us to clear its 8-bit counter register, **Tmr2**, and its prescaler and postscaler. We clear **Tmr2** with a **clrf** operation. (When we write to timer 2's control register to turn timer 2 on, the prescaler and postscaler are automatically reset, so they need no explicit clearing operation.) The final step is to turn timer 2 on, which we do by setting **T2Con** **<Tmr2On>** high. Before writing the code, we verified, of course, that the **PIR1**, **Tmr2** and **T2Con** special purpose registers are all in bank 0, so no **banksel** operation is necessary.

```
Tmr2Proc
;-------------------
;following code executed only if Timer 2 interrupt
;called with bank = 0
        bcf             PortB,LED2      ;turn B2 LED off
        ;both PIR1 and T2Con are in bank 0, no need to change
        bcf             PIR1,Tmr2IF     ;clear timer 2 interrupt flag
        bcf             T2Con,Tmr2On    ;turn timer 2 off
        GoTo            EndProc
```

When the timer 2 interrupt is detected, we turn off the LED by taking PortB, pin **B2** low through the **bcf** **PortB,LED2** operation. Since we use timer 2 as a "one-shot" event, we clear timer 2's interrupt flag, **Tmr2If**, and shut timer 2 down by clearing the **Tmr2On** bit. Until restarted following an external interrupt, timer 2 remains off. **PIR1** and **T2CON** registers both are in bank 0, so we need not alter the banking bit selection.

```
Tmr1Proc
;-------------------
;following code executed only if Timer 1 interrupt
;called with bank = 0
        bcf             PIR1,Tmr1IF ;clear timer 1 interrupt flag

        ;timer 1 must be reset every time it rolls over.
        ;high byte first-avoids rollover during
        ;the time it takes to reset the counter.

        movlw           High TicksPerms         ;put high byte of load into W
        ;timer spr's are in Bank 0 no need for banksel
        movwf           TMR1H                   ;move into timer register
        movlw           Low TicksPerms ;now for the low byte
        movwf           TMR1L                   ;move into timer register

        BankSel i&                              ;where is i?
        incf            i&0x7F,f                ;i=i+1
        btfss           i&0x7F,7                ;use bit 7 as /128 flash
        Goto            TurnOff
        BankSel PortB
        bsf             PortB,LED1
        GoTo            EndProc
```

Timer 1 requires a bit more complex treatment. First, of course, we clear timer 1's interrupt flag, **Tmr1IF**.

As you may recall from Chapter 10, to make timer 1 reset (and cause an interrupt) at our target number of counts, we preload the counter register with $FFFF-target value. Timer 1 counts up and when it rolls over at $0000, it resets. But, we must reload timer 1 with this preset value every time it rolls over, or else the counter resumes from $0000, not the desired preloaded value. We preload the counter high byte first, to avoid having the lower counter byte rollover during the time it takes to accomplish the preload.

After preloading timer 1, we then increment the on/off counter, **i**. Our strategy is to use **i** to both count the number of 1ms interrupts and to serve as a status holder to determine when to turn the pin **B1** LED on and off. We use **i.bit7** to control the LED; if **i.bit7** is high, we turn the LED on by setting **PortB, LED1** high, otherwise we clear **PortB, LED1**.

Because we are not sure in which bank **i** resides, we add **banksel** operations. (Since we have not overridden MBasic's default stack setting in Program 15-3, and since we only have one variable, in fact, we know **i** is in bank 1. But, these circumstances may not always be true, so we'll use a **banksel** operator to ensure the correct banking.)

```
TurnOff
        BankSel PortB
        bcf             PortB,LED1
        GoTo            EndProc
```

Since the turn-off process requires more code than we can get into one line, we jump to the label **TurnOff** when the **i.bit7** is low.

```
;end of ISRASM
EndProc
        bsf             IntCon,GIE              ;enable interrupts
}
```

We wrap up the **ISRASM** procedure by re-enabling the master interrupt switch, **GIE**.

When running Program 15-2, you should see the LED connected to pin B1 flashing about four times a second and the LED at pin B2 should flash once for 1ms every time SW1 is pressed.

What happens if timers 1 and 2 fire their interrupts simultaneously? Or what happens if all three interrupts fire simultaneously? If so, the **ISRASM** will be executed once for each interrupt that has fired, with the interrupts dealt with one at a time by the **ISRASM** code, in the order established by the test/jump statements at the beginning of the ISRASM procedure. If simultaneous interrupts are possible, and if handling simultaneous interrupts in a specific order is important, then your **ISRASM** code should be structured to service the interrupts in the desired order.

Rather than extending this chapter to provide additional abstract examples of interrupt handling, you should examine the specific uses we've made of **ISRASM** routines in:

- Chapter 16—to generate a synthesized sine wave by direct digital synthesis;
- Chapter 19—to control stepper motors; and
- Chapter 25—to read bar codes.

References

[15-1] MBasic's ISRASM procedure is documented only in comments posted in Basic Micro's User's Forum, which may be reached at http://forums.basicmicro.net/. Search for "israsm" to review the available information.

[15-2] With respect to assembler programming in general, the references provided at Chapters 13 and 14 may be consulted.

Digital-to-Analog Conversion

Digital-to-analog conversion (DAC) is the process of converting a numeric value to an analog parameter, often a voltage. The voltage may represent an alerting "beep" tone driving a loudspeaker, or it may, through an amplifier, control the speed of a DC motor, or it may be used for wide range of other applications. The voltage may have a 1:1 relationship to the digital value (numerical value 123 generates 123 volts) or it may scaled (numerical value 123 generates 1.23 volts), level shifted (digital value 123 generates 523 volts; digital 124 generates 524 volts) or otherwise linearly or nonlinearly transformed. The output may change with time, as the digital value changes with time. Finally, the digital value may be user controlled or it may be generated by the PIC in response to an external sensor.

We will explore MBasic's digital-to-analog abilities by building a direct digital synthesis sine wave generator. Our first approach, entirely in MBasic, has a maximum output frequency just under 1 KHz. By coding time critical elements in assembler, we'll see usable results at more than 10 kHz. We will also compare the DDS design with other D/A approaches available with MBasic.

Introduction to Digital-to-Analog Conversion

Let's examine a simple digital-to-analog voltage converter, the DAC equivalent of the 3-bit ADC we examined in Chapter 11. Figure 16-1 illustrates the relationship between digital input and output for this DAC. Comparing Figure 11-1 with Figure 16-1 shows—and it should be no great surprise—digital-to-analog conversion is the inverse of the analog-to-digital process. Each input value results in a specific output voltage. Since the digital input can only have discrete values, 0, 1, 2…7, the output voltage likewise assumes distinct values.

The DAC has the same need for positive and negative reference voltages V_{REF+} and V_{REF-} as the ADC and the formula we developed in Chapter 11 for calculating the ADC step size applies to a DAC equally well:

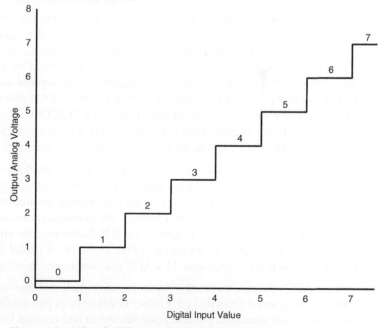

Figure 16-1: The digital-to-analog conversion process.

$$V_{STEP} = \frac{\left(V_{REF+} - V_{REF-}\right)}{2^N}$$

N is the number of input bits used by the DAC.

Let's apply this to our hypothetical 3-bit DAC. Suppose we wish to generate an output voltage that steps from 0 to 3.5 volts in 0.5 volt steps. Since our minimum voltage is to be 0, we will set V_{REF-} as ground. V_{REF+} is thus required to be 4.0 volts, since V_{STEP} is desired to be 0.5 volts. The output voltage is thus 0.5 multiplied by the input value.

Remember, though, that our hypothetical 3-bit DAC only has 0..7 as valid inputs. Hence, the maximum voltage it can output is 3.5 volts, not 4.0, the reference voltage. If we set the input to 8 (binary 1000) our 3-bit DAC, being connected to only the rightmost three bits (000), will obligingly output 0 V, not the desired 4.0 V.

As in the case of the ADC, we select V_{REF+} and V_{REF-} to meet our overall circuit design goals, consist, of course, with the physical limits of the DAC. Many DACs permit V_{REF-} to be negative with respect to ground, a useful option if we wish our waveform to swing to either side of ground. If we set V_{REF+} at +5 V and V_{REF-} at –5V, for example, we can output a waveform on either side of ground with a peak-to-peak voltage of nearly 10 V.

Although we usually use a stable DC reference voltage, many DACs permit audio frequency or time varying DC voltage references. In this case, the DAC functions as a programmable attenuator. (Inexpensive devices known as "digital potentiometers" are available to perform that function and are discussed in Chapter 21.)

Resolution – Accuracy and Signal-to-Noise Ratio

Let's look in detail at a particular DAC, Texas Instruments' model TLC7528 parallel loading dual 8-bit DAC. This inexpensive ($3.30 in single lot quantities, at the time of this writing) device is available in a 20-pin DIP package suitable for plugboard construction. It uses the common R-2R ladder design and may be operated either as a current-mode or voltage-mode output device.

Resolution is a measure of the number of different output levels that DAC can generate. The TLC7528 is an 8-bit device and hence has 2^8 or 256 possible output values. If we use +5 volts as the positive reference voltage for the DAC and 0 volts (ground) as the negative reference, each output will differ from the adjacent step by 5.00 V/256 or 19.53 mV. As was the case with the ADC, if we wish our output to have convenient step intervals, we should select a reference voltage that is an appropriate multiple of the device span. The LT1634CCZ-4.096 precision reference we used in Chapter 11's digital voltmeter design, for example, yields a convenient 16 mV step interval when used with a TLC7528 DAC. To avoid undue complexity and since our purpose is to illustrate principles, not to build a production circuit with accurate amplitude control, out design uses the 2840 Development Board's V_{DD} supply as the reference voltage.

Accuracy is a measure of the difference between the DAC's output and the desired value. Although we will not build it, let's briefly examine a hypothetical TLC7528 circuit with a precision LT1634CCZ-4.096 voltage reference, followed up by an op-amp buffer and a switched capacitor low-pass filter. V_{STEP} is 4.096 V/256, or 16.0 mV. Suppose we wish to generate 1.0000 V. We immediately see one problem; if we set the DAC input to 62, it outputs 0.992 V, while 63 generates 1.008 V. Either way, we are 8 mV from our target. Unfortunately, this quantization error is not the end of our problems. A partial list of potential error sources includes many of the suspects seen in Chapter 11's ADC discussion, which we won't repeat here.

As with our ADC analysis, we must consider the magnitude and direction of these error sources. Are they short-term or long term? Does one error offset in whole or in part another error, or do they directly add? Ignoring those error sources dependent upon the layout and external factors, we can identify several possible error sources in our hypothetical design:

- *Reference voltage error*—The LT1634CCZ-4.096 has a specified error of ±0.2% over the temperature range 0...70°C. An error of 0.2% in the reference represents 8.2 mV.
- *Linearity error*—Texas Instruments' quotes the TLC7528 DAC as not to exceed ±½ LSB in current mode or ±1 LSB in voltage mode.
- *Output buffer amplifier error*—The voltage output buffer amplifier, a LF411 op-amp configured as a voltage follower, adds ±0.5 mV offset voltage error.
- *Quantization error*—inherently ±½ LSB in our device and design.
- *Errors in the low-pass filter*—Many active low-pass filters have DC amplitude errors, in addition to the expected frequency and transient based errors
- *AC errors in the low-pass filter*—The AC response of the low-pass filters is not perfectly flat in the passband and transient waveforms may experience overshoot and undershoot from the filter as well.

A worst-case error analysis (where all errors are in the same direction) yields a maximum error into the low-pass filter is ±1 LSB (DAC linearity) ±0.5 mV (LF411 offset) ±0.2% (reference error) ± ½ LSB mV (quantization error). Since 1 LSB is 16 mV, we can restate the general error as ±0.2%, ±24.5 mV, or where the desired voltage is an integral number of voltage steps ±0.2%, ±16.5 mV. If we wish to generate 2.000 volts, therefore, the expected error magnitude is 0.002 * 2.000 + 16.5 mV = 20.5 mV. If we were designing a production instrument, of course, a comprehensive error and stability analysis of the entire circuit, including the low-pass filter, is required. Our purpose here is limited to highlighting a few DAC accuracy concerns.

Signal-to-Noise Ratio

Inherent in quantization—both A/D and D/A—is noise. The relationship between the maximum signal to noise ratio and the number of bits in the sample is:

$$SNR_{dB} = 6.02N + 1.76$$

N is the number of bits of amplitude level used in the sample

SNR_{dB} is the signal to noise ratio in decibels.

The TLC7528 DAC is an 8-bit device, so the maximum signal to noise ratio we may expect is 49.92 dB. This figure may be misleading as quantization noise is spread over the entire Nyquist sampling range of 0 to one-half the sampling frequency. If we generate a 1 KHz waveform, using a 100 KHz sample frequency, and if the DAC is followed by a 2 KHz low-pass filter, quantization noise—appearing over the range 0-50 KHz—may reduced by as much as 14 dB measured at the low-pass filter output, yielding an expected SNR approaching 65 dB.

Henry Nyquist and his Sampling Theorem

When analyzing the digital voltmeter in Chapter 11 we more-or-less implicitly assumed the voltage to be measured was constant, or at least changed very slowly with time. It wasn't necessary, therefore, to delve into what happens when we digitize a time variant signal. Our DAC project would be dull indeed if all it did were to put out a constant voltage. Since we have chosen to generate a time varying signal—in particular a sine wave—we must briefly examine a theorem that underpins much of modern electronics, Nyquist's sampling theorem. We shall skip over all but the most essential detail but the interested reader can consult Reference [16-4] for a more detailed, but highly readable, exploration of the A/D and D/A processes.

Nyquist discovered—in 1927, two decades before the transistor was invented—that in order to accurately reproduce an analog signal, it must be digitally sampled at a rate twice the highest frequency to be reproduced. *Or, in the context of a DAC, the highest frequency that may be generated without error is one-half the sampling rate.*

Suppose our DDS program sends 1,000 updated sine calculations (samples) to the DAC every second. In this case, Nyquist proved that we cannot accurately reproduce a frequency exceeding (1000 samples/sec)/2, or 500 Hz. (To allow for imperfections in the low-pass filter, we likely would not trust this design beyond about 400 Hz.) This fundamental sample rate versus output frequency limitation drives our quest to make DDS programs execute in the minimum possible time.

In addition to observing Nyquist's sampling rate, the number of bits that quantify the analog signal is important; the more bits (greater resolution), the more accurately the signal amplitude may be described. Suppose, for example, we have a 1-bit system, with V_{STEP} of 1 volt. The only possible output waveforms are (a) nothing or (b) one with a peak value of 1 volt. If our target output is 0.5 volts, we clearly are in trouble. (One-bit systems can be made to provide much greater resolution than this simplified discussion indicates—we'll see an example later in this chapter when we look at MBasic's **PWM** function.)

Our minimal DAC configuration is illustrated in Figure 16-2; a PIC to generate numerical values defining the desired waveform, a DAC to convert those values to analog voltages and, finally, a low-pass filter with a cut off frequency one-half the DAC sampling rate, or lower.

The low-pass filter is a critical element in the analog conversion process as it suppresses artifacts caused by the stepped nature of the DAC's output. It must be chosen with care. If we reproduce only a sinusoidal output, at one fixed frequency or slowly varying frequencies,

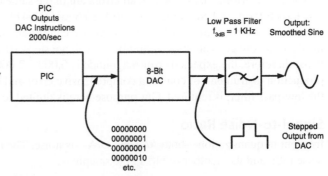

Figure 16-2: DAC configuration.

a filter offering excellent attenuation versus frequency characteristics (but with poor transient response), such as a Chebyshev or Cauer (elliptical) design is preferred. If we wish to generate an arbitrary waveform with fast rise and fall times, our filter needs good transient response (but which comes with relatively poorer attenuation versus frequency characteristics) such as a Bessel or linear phase design. Since we will demonstrate both sinusoidal and arbitrary waveforms, our design uses Butterworth filters as a compromise. Butterworth filters offer reasonable transient response, and acceptable attenuation versus frequency performance in the stop band. Considerable caution must be applied when choosing a low-pass filter to follow a DAC—this is one area where a "if a little bit of stop band attenuation is good, a lot must be even better" approach can cause trouble.

For outputs below 1 KHz, we use a LTC 1062 fifth-order switched capacitor low-pass filter with a cut-off frequency of 1 KHz and, for outputs with a maximum frequency between 1 KHz and 10 KHz, a second-order active RC low-pass design with a 10 KHz cut-off. The 10 KHz filter is implemented as a multiple feedback active RC filter and is inverting.

Although our sine wave generator design focuses upon direct digital synthesis with an external DAC, in fact we have a rich choice of methods to generate sinusoidal analog output signals with a PIC:

- Generate a square wave with MBasic's **SOUND** or **PULSOUT** procedures and convert it to a sine waveform by passing it through signal processing circuitry.
- Use MBasic's **PWM** or **FREQOUT** procedure to generate a pulse width modulated waveform, and convert the PWM output to an analog signal with a low-pass filter.
- Use a specialized PIC, such as the 16C781 and 16C782, with a built-in DAC. MBasic, however, does not currently support these devices.

- Use the hardware PWM built into many PICs, including the 16F87x family. (MBasic's `PWM` procedure is implemented in software only and does not use the PIC's hardware PWM, supported by the `HWPM` function, used in Chapter 24 for motor speed control.)
- Finally, our preferred solution; use an external DAC to directly convert a digital value to a defined voltage level. Depending on the desired waveform purity, additional signal filtering may not be required.

After completing the DDS design, we will examine these other options.

DAC Circuit Design

Figure 16-3 shows the DAC-PIC connection. The TLC7528 DAC is a parallel-load device and connects to PIC's pins RB0...RB7. Since we only use one of the two DACs in the TLC7528 and do not share the RB0...RB7 bus with any other device, we hard connect the Chip Select (pin 15), DAC A/B (pin 6) and Write Enable (pin 16) to ground.

The TLC7528 operates in either "current mode" or "voltage mode," that is, where the DAC control byte input sets the output *current* or output *voltage*. In current mode, an op-amp current-to-voltage converter

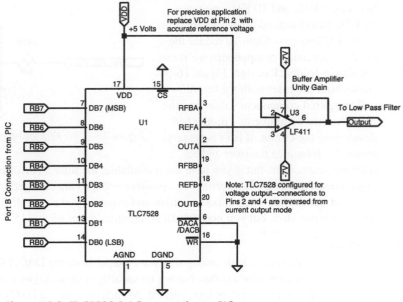

Figure 16-3: TLC7528 DAC connection to PIC.

usually follows the DAC. We'll use a simpler approach and operate the DAC in voltage mode and buffer its output with a unity gain LF411 op-amp. We'll used the 2840 Development Board's V_{DD} as our reference voltage. The LF411 buffer output connects to two parallel low-pass filters:

1 KHz low-pass—We use an LTC1062 switched capacitor fifth-order Butterworth design as shown in Figure 16-4. For our purpose, we can treat the LTC1062 as a "black box" that implements a filter. Our design follows the LTC1062 data sheet which you should read before starting construction. I've selected a 100:1 $F_{CLOCK}:F_{3dB}$ ratio by

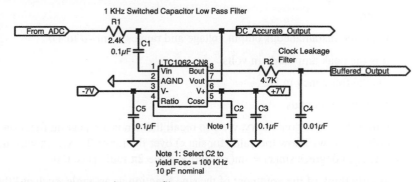

Figure 16-4: 1 KHz low-pass filter.

connecting the ratio pin (pin 4) to +7 V. Accordingly, for a 1 KHz cutoff, we need a 100 KHz clock. Rather than add an external clock source we use the LTC1062's internal timer, trimming it to 100 KHz with capacitor, C2. C2 is nominally 12 pF, but must be selected to yield 100 KHz with your particular chip and breadboard layout. Select C2 by measuring F_{CLOCK} with a counter or oscilloscope at pin 5 or based on the filter's frequency response. If you measure F_{CLOCK} at pin 5, a low capacitance probe is necessary, as even a 10X probe has enough shunt capacitance to appreciably shift F_{CLOCK}'s frequency.

The LTC1062 has two outputs; pin 7 is DC accurate but has no buffering, while pin 8 is buffered, but is subject to offset errors in the internal amplifier. The LTC 1062 is a mixed digital/analog device so we follow it with a simple RC filter to suppress clock feed through.

10 KHz low-pass— For output frequencies between 1 KHz and 10 KHz, we use a 10 KHz cutoff second order Butterworth active RC low-pass filter. To reduce the need for precision components, we'll use a multiple feedback design. Figure 16-5 shows both provides both the theoretically required component values and the 5% tolerance components I used when developing the circuit. It isn't necessary to use 1% tolerance resistors and 5%

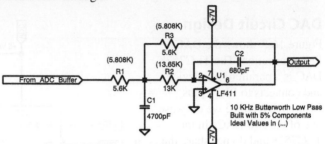

Figure 16-5: 10 KHz low-pass filter.

tolerance capacitors, but if you have them available, the filter's performance will be closer to the design objectives. The design is inverting, i.e., positive voltages are changed to negative and vice versa. Finally, note that our design requires both positive and negative power supplies. I've shown ±7 V supply voltages in Figure 16-5, but you may substitute anything from ±5 to ±15 V.

What is DDS?

Our objective is to produce a high quality sine wave output from the DAC. DDS, our preferred way to meet this objective, generates a waveform (often, but not necessarily, a sinusoid) by outputting the precomputed voltage value of the waveform at a constant time interval. At each time step, a DAC converts the numeric value to a corresponding voltage. For this reason, a DDS is also known as a "numerically controlled oscillator."

First, a quick review of elementary trigonometry and sinusoidal waveforms. For any time t (in seconds) the instantaneous voltage V_{OUT} of a sinusoidal waveform is:

$$v_{out} = a \times \sin(2\pi f t)$$

where

v_{out} is the instantaneous output voltage and ranges from $-a$ to $+a$

a is the peak amplitude in volts

f is the frequency in Hz

t is time in seconds

From high school trigonometry, we also recall that sin is a repeating function—that is, $\sin(x) = \sin(x + 2\pi)$. Accordingly, if we know the value for $\sin(x)$ over the range $0...x...2\pi$ we can find $\sin(y)$ for any positive value y, or in degrees $\sin(x) = \sin(x + 360°)$, since 2π radians is $360°$.

We usually think of the argument of the sine function as an angle, such as "the sine of $45°$ is $0.707...$" The term $2\pi ft$ is thus represents the phase (angle, in radians) of the waveform at time t. During a period of time equal to $1/f$, the waveform phase angle goes through 2π radians, or $360°$. It is often far more convenient to refer to the value of a sine wave at a specific phase angle, rather than at some specific time in microseconds or milliseconds.

Consider a very simple DDS system using a 3-bit DAC and an output range of 0…7 V. We'll calculate our sine wave target every 45 degrees and store in array SX() the associated numeric values the DAC is to output at 0, 45, 90…360 degrees. Thus, by stepping SX(i), i = 0 to 8, the DAC outputs one complete sine cycle. (Since sin(360°) equals sin(0°), SX(8) = SX(0). Therefore, we may simply wrap the index variable i around 7 so there is no need for an array element SX(8) or for any greater value of i. Since the maximum value of the sine function is 1, we will increase the resolution of its digital representation by multiplying it by 3. (We can always use a 1:3 voltage divider on the DAC's output if we desire a 1 volt maximum.) Finally, we bias the stored value upward by adding 4, so that we do not have to worry about outputting negative voltages. The SX(1) value (45°) for example is computed as 3*Sin(45°) + 4 = 6.121. Since SX() must be an integer 0…7, we round to 6. SX() holds the following values:

Index (i)	Angle (Degrees)	Sin (Angle)	Array Sx(i)
0	0	0.000	4
1	45	0.707	6
2	90	1.000	7
3	135	0.707	6
4	180	0.000	4
5	225	−0.707	2
6	270	−1.000	1
7	315	−0.707	2

Figure 16-6 plots SX() and we see that with as few as eight samples, the result closely resembles a continuous sinusoidal when drawn with a smooth connecting curve, mathematically equivalent to passing it through a low-pass filter.

Suppose that our elementary DDS program outputs one SX() value every 0.100 seconds, and that we desire to synthesize a 3 Hz sine wave. One complete sine cycle (360°) requires 8 index steps. Hence, for three complete cycles, we require 24 index steps or 1080°. Since our sample rate is 0.1 steps/second, we output SX() at the rate of 0.1 × 24, or 2.4 indexes per clock step, that is, SX(0), SX(2.4), SX(4.8), and so on. This is equivalent to 1080° × 0.10, or 108.0° phase advance per clock step. Obviously we can't output SX(2.4); it must be either SX(2) or SX(3). Figure 16-7 illustrates this process. We may represent SX() as a circular array, since SX(8) = SX(0), SX(9) = SX(1), and so on. Imagine that every 0.100 second clock step, the arrow advances clockwise 108.0°. The array index value *i* to which the arrow points is the value used in outputting SX(i). Figure 16-8 shows the indexes selected for time steps 0, 1, 2 and 3, while the following table shows the complete first second of output.

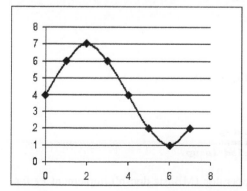

Figure 16-6: Sinusoid formed from eight data points.

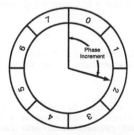

Figure 16-7: Phase wheel.

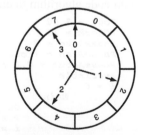

Figure 16-8: Phase wheel after three steps.

Time (Seconds)	Time Step	Phase Angle (Degrees)	(Phase Accumulator) Perfect Index	Wrap Around Index	Integer SX(i) Index	Index Error	DAC Output
0.000	0	0	0	0	0	0	4
0.100	1	108	2.4	2.4	2	0.4	7
0.200	2	216	4.8	4.8	4	0.8	4
0.300	3	324	7.2	7.2	7	0.2	2
0.400	4	432	9.6	1.6	1	0.6	6
0.500	5	540	12	4	4	0	4
0.600	6	648	14.4	6.4	6	0.4	1
0.700	7	756	16.8	0.8	0	0.8	4
0.800	8	864	19.2	3.2	3	0.2	6
0.900	9	972	21.6	5.6	5	0.6	2
1.000	10	1080	24	0	0	0	4

Plotting the DAC output with a smooth curve we see a 3 Hz sinusoidal waveform. As Figure 16-9 shows, even with coarse voltage and phase steps the DDS output after passing through the mathematical equivalent of a low-pass filter looks like the sinusoid we expect. (Points in the graph are connected by spline interpolation which can be shown to be related to low-pass filtering in the time domain.)

In our hypothetical, the phase step size is 108° or 2.4 units. The total phase is held in the *phase accumulator*. The relationship among the various parameters we have discussed is:

$$\phi_{STEP} = \frac{F_{out}2^N}{F_{Clock}}$$

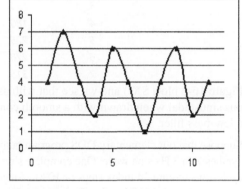

Figure 16-9: Simulated 3 Hz output.

ϕ_{STEP} is the phase step in index units.

F_{OUT} is the output frequency in Hz

F_{CLOCK} is the number of output samples per second

N is the number of bits required to hold the index to SX().

A pseudo-code algorithm to output our 3 Hz sine wave would be:

```
Fout = 3.00
FClock = 10
IndexSize = 8
PhaseStep = (Fout * IndexSize)/FClock
PhaseAccum = 0.0 ;phase accumulator is a floating point
;variable
i = 0
        Main
                Output to the DAC SX(i)
                PhaseAccum = PhaseAccum + PhaseStep
                i = INT (PhaseAccum MOD IndexSize)
                Pause until the full 100 ms period is up.
        GoTo Main
```

The MOD or "modulus" function is MBasic's **//** or remainder operator. (MBasic doesn't implement **//** for floating point numbers, so the real code gets messier than the pseudo-code indicates.)

This algorithm, although correct from a computational prospective, has one major practical problem—floating point arithmetic operations are slow in a PIC and unfortunately appear inside the main loop. Microchip did not design its midrange PIC products to be mathematical powerhouses nor to be digital signal processing devices. It's a testament to the inherent power in a PIC and to Basic Micro's MBasic compiler that we can obtain useful results at all this approach. Let's look at working code implementing this simple algorithm.

Program 16-1

```
;Program 16-1
;Constants
;-----------------------
Interval                Con         515         ;uSecs for output
StepsPerCycle           Con         32          ;steps per 1 complete cycle

;Variables
;-----------------------
i               Var         Byte        ;counter variable
OutIndex        Var         Byte        ;lookup index
Sx              Var         Byte(32)    ;holds sine table
Accum           Var         Word        ;phase accumulator
Freq            Var         Float       ;desired frequency
FStep           Var         Float       ;real phase step
PhaseAdder      Var         Word        ;steps added
Ftemp           Var         Float       ;debugging variable

;Initialization
;-----------------------
Clear
;Preload sine lookup table
For i = 0 to 31
        Sx(i) = Sin(i*8)+128
Next
;PortB is connected to DAC
For i = B0 to B7
        Output i
        Low i
Next
;C0 used for o'scope timing
Output C0
Low C0

;we hard code in the desired output freq
;in a real program, this might be
;entered by RS232, keypad or switches
Freq = 60.0     ;Hz
;calculate the number of steps per second
FStep = ToFloat (StepsPerCycle) * (Freq * ToFloat Interval)
FStep = FStep / 1000000.0
FTemp = (FStep * 256.0)
PhaseAdder = ToInt FTemp

Main
;----------
GoSub OutToPort
GoTo Main

OutToPort
;----------
        ;470 uS execution time not counting overhead
        bsf     PortC,0
        ;advance phase accumulator
        Accum = Accum + PhaseAdder
        ;look only at "integer" part of Accum
        ;and MOD 32 the "integer" part
        OutIndex = Accum.HighByte // 32
        PortB = Sx(OutIndex)
        bcf     PortC,0
Return
End
```

Program 16-1 follows the pseudo-code with some important speed-up factors:

SX array—The array **sx** is 32 elements long, and its values are filled through MBasic's **SIN** function. MBasic's **SIN** function's argument is 0...255, i.e, it divides the circle into 256 angular units and it returns signed byte values from –127...128. Hence, when filling **SX(i)** over the range 0...31, we multiply **i*8** and add 127 to shift the output to 0...255.

Calculate the phase advance—We first calculate the floating point phase advance.

```
FStep = ToFloat (StepsPerCycle) * (Freq * ToFloat Interval)
FStep = FStep / 1000000.0
```

This computation follows the definition of phase advance developed earlier. Rather than trying to make Program 16-1 output sample values at a pre-determined interval, we write the code and accept whatever time interval results—we determine **Interval** by measuring the execution time of the main program loop in microseconds and then define it as a constant:

```
Interval            Con     515           ;uSecs for output
```

You can use the execution timer routine seen in Program 14-1 among others, or you can measure the output frequency and adjust **Interval** until the measured frequency matches the target frequency. I measured **Interval** as 515 µs, but your time may differ, as this value is based on a prerelease version of MBasic's version 5.3.0.0 compiler. Since **Interval** is in microseconds, it is necessary to divide by 1,000,000 to convert to seconds. And, since frequency is 1/interval instead of dividing by F_{clock}, we multiply by $1/F_{clock}$, a mathematically identical operation.

After these two lines of code have been executed, **FSTEP** equals 0.98880.

Integer conversion—A standard floating point phase accumulator incurs a significant speed penalty. For improved execution speed, we'll keep the phase accumulator in a word (16-bit) variable and work with only integer arithmetic inside time sensitive program loops. To do this, we consider the 16-bit phase accumulator variable **Accum** to consist of two 8-bit halves:

Accum.HighByte	Accum.LowByte
Integer Value	Fractional Value

The fractional component is in units of 1/256. We convert the floating point accumulator value to this 16-bit integer/fraction form by multiplying **FSTEP*256** and converting the product to an integer.

```
FTemp = (FStep * 256.0)
PhaseAdder = ToInt FTemp
```

In this case, **PhaseAdder** contains **Int(0.98880 * 256.0)**, or 253, representing 0 + 253/256. If **Fstep** had been greater than 1, say 1.23, the result would have been 314. However, **PhaseAdder**'s low byte would still contain the 1/256 fractional value approximating 0.23. (Not convinced? 314-256 = 58; **Int(0.23 * 256) = 58.**) In this case, we would understand **PhaseAdder**'s contents to be 1+58/256, or approximately 1.23.

If the phase accumulator had five decimal digits resolution, after the second step it would contain 0.98880 + 0.98880, or 1.9776. Our 16-bit phase accumulator **Accum** will contain (253 + 253)/256, or 1.9766, a 0.05% error.

Generation loop—Here we generate the next output index by adding the phase step to the current phase accumulator value.

```
Accum = Accum + PhaseAdder
```

The rollover as **Accum** exceeds 65535 is not of concern since we use relative phase.

```
;look only at "integer" part of Accum
;and MOD 32 the "integer" part
Out Index = Accum.HighByte // 32
PortB = Sx(OutIndex)
```

Since **SX(i)** has only 32 elements, we must wrap the "integer" part of **Accum**, that is, **Accum.HighByte**, around 32. This is conveniently done through remainder division with the **// 32** operator. (At the cost of a more complex example, we could have made the phase accumulator more accurate by making the fractional part the lower 11 bits of **Accum**, and the integer part the upper 5 bits, thus preserving the full 16 bits for useful data. As it is, the phase accumulator is effectively 13 bits wide, not 16.)

Figure 16-10 shows the DAC output (Ch 1) and the output following the 1 KHz low-pass filter (Ch 2). The result is a quite acceptable 60 Hz waveform.

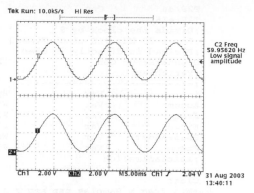

Figure 16-10: 60 Hz output from Program 16-1. Chan 1: DAC Output; Chan 2: Following 1 KHz low-pass filter.

The problems with Program 16-1 center on execution speed. The output interval is 515 µs, corresponding to Fclock = 1.942 KHz. Nyquist's sampling rule would thus limit our maximum output frequency to Fclock/2 or 970 Hz. And, MBasic can't be depended upon to generate jitterless timing in a **GoTo** or **GoSub** loop, with ±2 µs variations in timing observed. For many purposes, a couple microsecond jitter in loop timing is immaterial, but in DDS, clock jitter translates into frequency jitter and unwanted noise modulation and sidebands.

Program 16-2 ameliorates our speed concerns by coding all time sensitive elements in assembler and relying upon a timer-driven interrupt clock to establish the sample interval. Using an ISRASM interrupt procedure does not totally remove jitter, but reduces it to less than 1µs. And, with a 20 MHz oscillator we are able to operate with a 80 KHz Fclock, permitting up to 40 KHz output. We'll also use a 256 element waveform table.

Program 16-2

```
;Program 16-2
;Program Simple Phase Accum-7.bas
;Demonstrate the principles behind
;a DDS sine wave system using a PIC
;Constants
;-------------------------
Interval        Con     20              ;uSecs for output
StepsPerCycle   Con     256             ;steps per 1 complete cycle
;Interval between steps in us. Can't be less than 13
;because delay in ISR overhead & ISR procedure
T2Post1         Con                     %00000000
T2Post2         Con                     %00001000
T2Post3         Con                     %00010000
T2Post4         Con                     %00011000
T2Post5         Con                     %00100000
T2Post6         Con                     %00101000
T2Post7         Con                     %00110000
T2Post8         Con                     %00111000
T2Post9         Con                     %01000000
T2Post10        Con                     %01001000
T2Post11        Con                     %01010000
T2Post12        Con                     %01011000
T2Post13        Con                     %01100000
T2Post14        Con                     %01101000
T2Post15        Con                     %01110000
T2Post16        Con                     %01111000
;Note that bit 3 is Tmr2On and is set
;to be ON
T2Pre1          Con                     %00000100
```

```
T2Pre4           Con                  %00000101
T2Pre16          Con                  %00000111
;Divide the clock by 5
;For a 20 MHz PIC clock Timer2
;Preload is directly in uSec
CounterSetup Con                 T2Post5 + T2Pre1
;Variables
;-----------------------
i                Var      Byte        ;counter variable
Accum            Var      Word        ;phase accumulator
Freq             Var      Float       ;desired frequency
FStep            Var      Float       ;real phase step
PhaseAdder       Var      Word        ;steps added
Ftemp            Var      Float       ;debugging variable
;Initialization
;-----------------------
Clear
;Min freq = Samples per sec / Accum Size
;= 50000 / 2**16 = Min Freq
;Minimum frequency 0.762 Hz for 20us steps
;Defult generation frequency
Freq = 1000.0   ;Hz
;PortB is connected to DAC
For i = B0 to B7
        Output i
        Low i
Next
;C0 used for o'scope timing
Output C0
Low C0
;C1 used for program interrupt for external
;data input to be added if desired
Input C1

;set up frequency related elements
GoSub SetUp
;Timer 2 Setup
;----------------------------------------------------------------
'Don't forget the -1 here, as Counter2 starts at 00
SetTmr2 CounterSetup, Interval - 1
peie = 1                ;peripheral interrupt enable (INTCON, bit 6 [Bank 0/1/2/3])
gie = 1                 ;global interrupt enable (INTCON, bit 7 [Bank 0/1/2/3])
tmr2ie = 1              ;Enable Tmr2Int
;Enable Tmr2Int


;================================================================
israsm
;================================================================
{
        bcf     PIR1,tmr2if     ;reset timer 2 interrupt flag
        bsf     PortC,0         ;flag for o'scope timing
        ;advance phase accumulator
        ;Accum = Accum + PhaseAdder
        ;----------------------------
        ;16 bit add from PICMicro Microcontroller Pocket Reference
        ;
        banksel       PhaseAdder
        movf    (PhaseAdder&0x7F)+1,w          ;add the high bytes
        banksel       Accum
        addwf   (Accum&0x7F)+1,f
        banksel       PhaseAdder
        movf    (PhaseAdder&0x7F),w            ;add the low bytes
        banksel       Accum
        addwf   (Accum&0x7F),f
        btfsc   STATUS,C                       ;add 1 to high bytes if carry
        incf    (Accum&0x7F)+1,f
        banksel       PortC
        bcf     PortC,0                 ;for o'scope timing
        ;look only at "integer" part of Accum
```

```
                ;PortB = Sx(OutIndex)
        movlw           HIGH GetSine        ;w=highbyte table location
        movwf           PCLATH              ;PCLATH has high byte of table
        movlw           LOW (GetSine+1)     ;first actual data point
        banksel  Accum
        addwf           (Accum&0x7F)+1,w    ;offset adr of datapoint
        btfsc           STATUS,C
        Incf            PCLATH,f
        call            GetSine             ;what is sine value
        banksel  PortB
        movwf           PortB               ;output it to PortB
        GoTo            EndRoutine

GetSine
        movwf PCL
    retlw 128
    retlw 131
    retlw 134
<< note - remainder of GetSine table omitted from this listing >>
<< see Program 16-2.bas for all 256 table values >>
EndRoutine
        banksel PortC
        ;bcf     PortC,0
        }
;----------
Main
;----------
        If PortC.Bit1 = 0 Then
                tmr2on = 0
                GoSub GetNewFreq
                GoSub SetUp
        ELSE
                tmr2on = 1
        EndIf
GoTo Main
SetUp
;----------------
        ;we hard code in the desired output freq
        ;in a real program, this might be
        ;entered by RS232, keypad or switches
        ;calculate the number of steps per second
        FStep = ToFloat (StepsPerCycle) * (Freq * ToFloat Interval)
        FStep = FStep / 1000000.0

        FTemp = (FStep * 256.0) + 0.50
        PhaseAdder = ToInt FTemp
Return
GetNewFreq
;----------------
;Subroutine to get the frequency from user keypad, switches,
;RS-232, etc can be placed here.just put dummy in now
        Freq = 440.0    ; Hz
Return
End
```

Constants—As usual, we start defining necessary constants. The equivalent of **SX(i)** is 256 elements long, so **StepsPerCycle** is defined as 256. The interrupt is to be called via timer 2 every 20 μs, the value to which **Interval** is defined.

Some versions of MBasic have erroneous definitions of certain of the timer 2 predefined constants, so we define our own timer 2 constants. Timer 2's clock source runs at Fosc/4, or 5 MHz for a 20 MHz master clock. By setting the postscaler at 5, each timer count corresponds to 1μs.

Variables—We define a word variable **Accum** as the phase accumulator and a word variable **PhaseAdder** to hold the phase step interval. Three floating point variables, **Freq** (frequency), **Fstep** (floating point phase step) and **Ftemp** (a temporary variable) are declared, as floating point variables.

Initialization—Program 16-2 has no provision for frequency entry other than the value hard coded in at initialization, although a hook is provided to a possible frequency entry procedure. Change the `Freq = 1000.0` line as desired.

In DDS, the output frequency F_{OUTPUT} is constrained to be an integer multiple of the minimum frequency step.

$$F_{STEP} = \frac{F_{CLOCK}}{2^N}$$

N is the number of bits in the phase accumulator, 16 in this program

F_{CLOCK} is the sample rate, 50,000 Hz for a 20 μs interval

Hence,

$$F_{OUTPUT} = kF_{STEP}$$

Where $k = 1, 2 \ldots 2^{N-1}$

For a 20 μs sample interval and a 16-bit accumulator, F_{STEP} is 50,000 / 65536, or 0.76294 Hz. By increasing `Interval`, i.e., slowing the output clock, the frequency step may be reduced. However, the maximum possible frequency will correspondingly be changed. If, for example, we decrease F_{CLOCK} to 20 KHz by setting `Interval Con 50`, we may set the frequency output in steps of 0.3 Hz. However, the maximum output frequency will be reduced to 10 KHz.

We calculate the phase step in the subroutine `SetUp` and place this initialization code in a subroutine so that it may be called as necessary for as the frequency output is changed.

Subroutine SetUp—`SetUp` replicates the corresponding code in Program 16-1.

Timer 2 Setup—This code enables all interrupts (GIE) and peripheral interrupts (PEIE), of which timer 2 interrupt is a subset. See Chapter 15, "Interrupt Handling and Timers in Assembly." Finally, timer 2 is enabled.

ISRASM—The code in the interrupt service routine is an assembler version of the code found in Program 16-1, with limited changes. The corresponding MBasic statements appear as comments. PortC, bit 0 is used to measure the execution speed of certain assembler statements by an external oscilloscope and associated bit set and bit clear statements (`bsf/bcf PortC,0`) may be omitted if timing information is of no concern.

Instead of storing the sine look-up values in an MBasic array such as `SX(i)` in Program 16-1, we have implemented a standard "jump table" as discussed at Chapter 14. Jump table access loads the `w` register with the desired index `i` and program execution jumps to the `i`th `retlw` statement following the start of the table. The `retlw` statement then returns the associated literal value in the `w` register. We then send the returned literal value to Port B where it is converted to a voltage by the DAC. This process is the functional equivalent of `PortB = SX(i)` but executes perhaps 100 times faster. The index `i`, of course, is actually the high byte of `Accum`.

We have implemented the sine table in the simplest form using all 256 elements but it could be reduced to 64 elements through two trigonometric identities:

- $\sin(x)$ where x is in the range $90° \ldots 180°$ equals $\sin(180° - x)$.
- $\sin(x)$ where x is in the range $180° \ldots 360°$ equals $-\sin(x - 180°)$.

The extra computations associated with these identities would obscure the principle however and we have ample code space.

Main—**Main** is a simple endless loop, with a check of Port C, bit 1. If Port C, bit 1 is low, program execution jumps to the subroutine **GetNewFreq**. This serves as an example of how a subroutine permitting the user to enter a new frequency might be added to Program 16-2. Following new frequency entry, the new phase advance is calculated by calling subroutine **SetUp**.

Subroutine GetNewFreq—At the present, **GetNewFreq** is a dummy subroutine. It simply substitutes a new pre-determined frequency for the one established at the initialization section of the program. To be useful, this subroutine could, for example, read a set of thumb-wheel switches, keypad presses, or an RS-232 input, to allow the user to select a new output frequency.

Putting It Through Its Paces

I tested Program 16-2 for output at nominal frequencies of 100 Hz, 1000 Hz and 10000 Hz. The figures in this chapter are for an interval of 10 µs under the MBasic 5.2.1.1 version of Program 16-2. I've modified this chapter's ISRASM interrupt handler code to be less dependent upon bank 0 register access, and thus more version 5.3.0.0 friendly, but the added bank selection overhead increases the minimum interval in Program 16-2 to about 15 µs. For the version 5.2.1.1 code, with a 10 µs output interval, the frequency increment is 1.526 Hz, and the corresponding actual output frequencies are 100.7 Hz, 999.4 Hz and 10000.6 Hz, subject to the residual error from the PIC's 20 MHz resonator.

I've captured two oscilloscope waveforms for each output frequency, along with spectrum analysis plots. Channel 1 is the raw DAC output, while channel 2 is the output of the low-pass filter, the LTC1062 in the case of 100 Hz and 1 KHz nominal frequencies and the two-pole active low-pass in the case of 10 KHz nominal frequency.

For comparison, Figures 16-11 and 16-12 show a 1 KHz sinusoidal generated by an analog function generator. Although the analog oscillator provides a low noise output, it suffers from drift, tuning precision and other problems that have lead to increasing DDS use. Even the simple breadboard DDS device in this chapter is far more stable than the analog function generator. Whether the DDS's unwanted spurious sidebands, harmonics and noise are acceptable depends on the application. And, we should not forget that a commercial DDS chips have 32 bit or larger phase accumulators that significantly reduce many of the artifacts seen in our simple design.

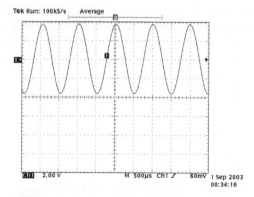

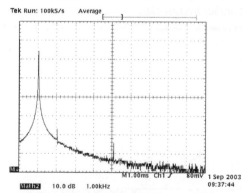

Figures 16-11: Reference 1 KHz sine waveform from analog function generator.

Figure16-12: Reference spectrum view 1 KHz sine waveform from analog function generator.

Note: The following graphics have no explicit text reference other than the previous page.

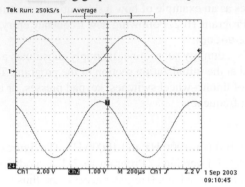

Figure 16-13: 1 KHz DDS Output. Ch1: DAC Output; Ch2: following 1 KHz low-pass filter.

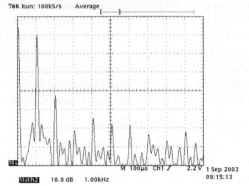

Figure 16-14: 1 KHz DDS. Spectrum analysis of DAC output.

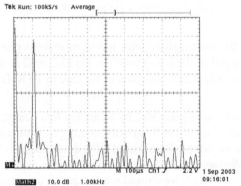

Figure 16-15: 1 KHz DDS. Spectrum Analysis of DAC output following low-pass filter. Note filter feed-through in the breadboard layout is seen at higher frequencies.

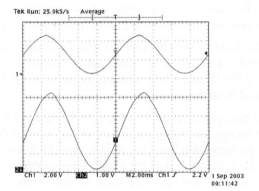

Figure 16-16: 100 Hz DDS. Ch1: DAC Output; Ch2: following 1 KHz low-pass filter.

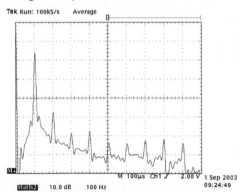

Figure 16-17: 100 Hz DDS. Spectrum analysis of DAC output.

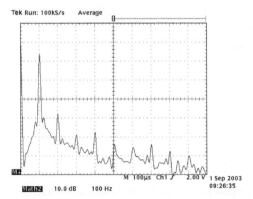

Figure 16-18: 100 Hz DDS. Spectrum snalysis of 1 KHz low-pass filter output. Since the 100 Hz fundamental is well below the filter cut-off frequency, proportional harmonic suppression is much less than seen for the 1 KHz DDS signal.

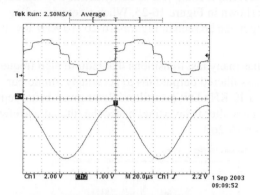

Figure 16-19: 10 KHz DDS. Ch1: DAC Output; Ch2: Following 10 KHz low-pass filter.

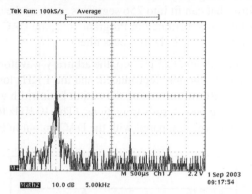

Figure 16-20: 100 Hz DDS. Spectrum analysis of DAC output.

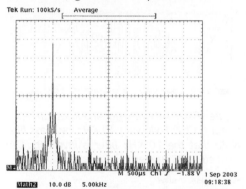

Figure 16-21: 10 KHz DDS. Spectrum Analysis of 10 KHz low-pass filter output. This low-pass is a second order filter and accordingly rolls off slower with frequency than the fifth-order 1 KHz low-pass.

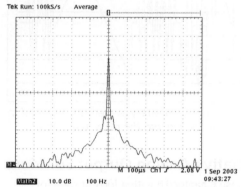

Figure 16-22: 1 KHz DDS output following 1 KHz low-pass filter. Spectrum analysis centered on signal. Close in sidebands are approximately −32 dBc.

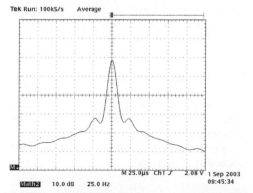

Figure 16-23: Expanded view of 1 KHz DDS output following 1 KHz low-pass filter. Spectrum analysis centered on signal. Close in sidebands are approximately ±20 Hz from the 1 KHz signal.

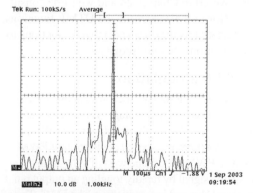

Figure 16-24: 10 KHz DDS output following 10 KHz low-pass filter. Spectrum analysis centered on signal.

Instead of using the jump table to define a sinusoidal waveform, it could just as easily define any arbitrary waveform that can fit into 256 segments, such as the one shown in Figure 16-25. When a corresponding `retlw` table is substituted in Program 16-2, the DAC output can be seen at Figures 16-26 (10 KHz low-pass filter) and 16-27 (1 KHz low-pass filter).

When clocked out at 25 µs/step, the arbitrary waveform has many fast transitions that require high frequency response be accurately reproduced, and the 1 KHz low-pass filtered output fails to accurately mimic our desired waveform. Increasing the output bandwidth with a 10 KHz low-pass filter allows much better reproduction of the waveform's transitions. (The 10 KHz low-pass filter is inverting; hence the filtered waveform is "upside down" compared with the DAC output in Figure 16-26.)

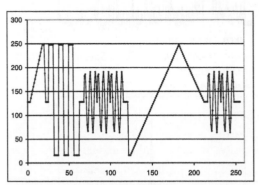

Figure 16-25: Arbitrary waveform designed in Excel and loaded into the retlw jump table.

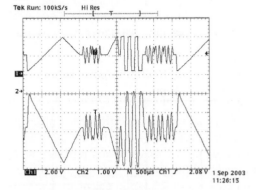

Figure 16-26: Arbitrary signal output. Ch1: DAC Output; Ch2: Following 10 KHz low-pass filter. Note that the 10 KHz filter is inverting. DAC clocked at 25 µs/step.

Alternative Analog Output Solutions

We'll close Chapter 16 with a look at some of MBasic's built-in waveform output functions—**PWM**, **FreqOut**, **Sound**, and **PulsOut**. We'll then write a simple program to test each of these four functions. (MBasic's hardware pulse width modulation function, **HWPM**, is covered in Chapter 24 and won't be considered here. Nor will we examine the DTMF functions, **DTMF** and **DTMF2** as they are specialized functions to generate touchtone dialing frequencies.)

Pulse Width Modulation

MBasic's **PWM** and **FREQOUT** procedures allow a PIC to produce an analog output voltage without an external DAC through pulse width modulation, or PWM, techniques. **FREQOUT** is a special application of PWM that outputs a PWM-based sine wave. PWM simulates an analog signal by rapidly toggling an output pin

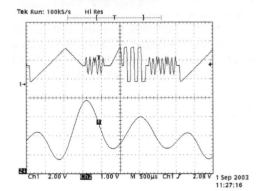

Figure 16-27: Arbitrary signal output. Ch1: DAC output; Ch2: Following 1 KHz low-pass filter. Note the 1 KHz filter masks all the high frequency components of the arbitrary waveform. DAC clocked at 25 µs/step.

between 0 and V_{DD} and is essentially a 1-bit DAC. The off/on output is averaged by a low-pass filter—even a simple RC network—into a smoothed output where the individual pulses are inconsequential. For a higher voltage, longer V_{DD} pulses are sent; for a lower voltage shorter 0 pulses are sent. The "wider" and "shorter" pulses are produced by consolidating multiple short pulses.

MBasic's **PWM** procedure has the following invocation:

PWM pin, duty, cycles

Pin is a constant specifying the output pin to be used.

Duty is a constant or variable specifying the desired output level with a permissible value between 0 and 255. 0 represents approximately 0 volts, 255 is approximately V_{DD}.

Cycles is a constant or variable (32-bit) specifying the duration of the output in microseconds from about 24 µs to 2^{32} µs, permitting a maximum duration of approximately 4295 seconds.

Figures 16-28, 16-29 and 16-30 show the PWM output (Ch1 is the raw PWM output from the PIC's C1 pin; Ch 2 is the PWM output following a 4.7 k/0.033 µF RC low-pass filter as shown in Figure 16-35) for **Duty** values of 240, 128 and 16 respectively. Each individual pulse is 4 µs duration. (Figure 16-29 has an expanded time axis compared with the other two figures.) As the ratio of V_{DD} pulses to 0 V pulses change, the filtered output voltage tracks quite well. We see linearity problems, however, near either end of the output scale, as a low on the PIC's output pin is 0.6 V, not 0 V and a high is $V_{DD} - 0.7$ V, not V_{DD}. If absolutely necessary, these figures could be improved through outboard circuitry.

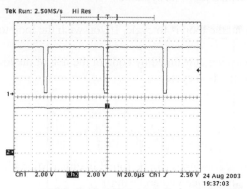

Figure 16-28: PWM 240 Output. Ch1: PWM Output unfiltered; Ch2: PWM following RC filter.

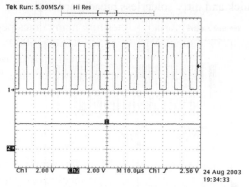

Figure 16-29: PWM 128 Output. Ch1: PWM Output unfiltered; Ch2: PWM following RC filter.

To produce a time varying analog output, therefore, we could use the concepts developed for the external DAC and instead use the **PWM** procedure. To produce a sawtooth ramp output, clocked out at one voltage step per millisecond, for example, we might use the following code:

```
For i = 16 to 240
        PWM C1, i, 1000
Next
```

Although we could use this concept in conjunction with Program 16-1 to produce a PWM-based sine wave output, it's much easier to take advantage of MBasic's **FREQOUT** procedure as it performs exactly that function. **FreqOut** is invoked as:

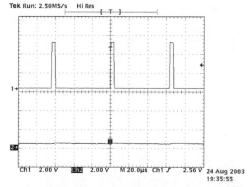

Figure 16-30: PWM 16 Output. Ch1: PWM Output unfiltered; Ch2: PWM following RC filter.

`FREQOUT Pin, Duration, Freq1,{Freq2}`

Pin is a constant specifying the output pin to be used.

Duration is a constant or variable (32-bit) specifying the duration of the output for each frequency argument, in units of 0.5 microseconds from about 24 units to 2^{32}units, permitting a maximum duration of approximately 2147 seconds.

Freq1 is a constant or variable specifying the frequency in Hz of the output sine wave from 0 to 32767 Hz.

Freq2 is a constant or variable specifying the frequency in Hz of an optional second simultaneous output sine wave from 0 to 32767 Hz..

Figure 16-31 shows the raw **PWM** output (Ch 1) and the RC filtered (Ch 2) 1 KHz sine generated by **FREQOUT C1, 65535, 1000**. The simple RC filter takes out most, but not all the high frequency component of the individual PWM pulses.

Comparing the spectrum of the raw PWM output shown in Figure 16-32 with that of the RC filtered output of Figure 16-33 we see the RC filter yields a significant reduction in broadband noise and harmonics. Of course, a well designed printed circuit board with a good ground plane will produce much less noise than our quick and dirty solderless breadboard trial.

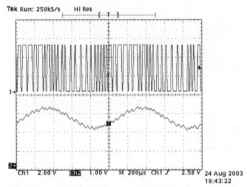

Figure 16-31: FreqOut 1000 Hz. Ch1: Raw PWM output; Ch2: RC Filtered output.

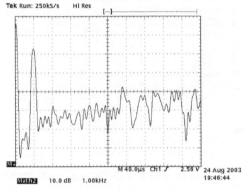

Figure 16-32: 1 KHz FreqOut Spectrum analysis raw PWM output.

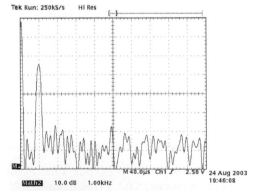

Figure 16-33: 1 KHz FreqOut Spectrum analysis PWM output following RC low-pass filter.

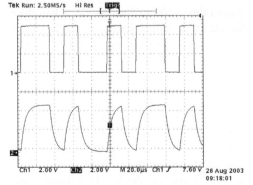

Figure 16-34: FreqOut 25 KHz. Ch1: Raw PWM output; Ch2: RC filtered output.

As we try to produce higher frequency outputs with **FreqOut**, however, the individual 4μs pulses become a greater proportion of the desired waveform. At 1 KHz, each sine wave output cycle comprises 250 4μs-wide PWM pulses. A 25 KHz output, however, each output sine wave is constructed from only 10 PWM pulses with the result shown in Figure 16-34. (For 25 KHz output the RC filter values are 4.7 k and 0.001 μF.).

If our requirements are not overly critical, **FreqOut** and **PWM** should not be rejected. Combined with a high quality low-pass filter, such as the LTC1062 fifth-order device, quite decent output quality can be achieved. If lesser quality can be accepted, a very simple RC filter can be used, or one op-amp might be configured as a two-pole low-pass filter. And, only one output pin is required to support **FreqOut** and **PWM**.

Calculating RC Filter Values

Figure 16-35 shows the simple RC low-pass filter we will use. Usually we design so that the 3 dB cut off frequency of the RC filter equals the highest frequency waveform to be generated. From elementary circuit analysis, we know the –3 dB frequency of an RC network is:

$$f_{3dB} = \frac{1}{2\pi RC}$$

f is the frequency in Hz
R is the resistance in ohms
C is the capacitance in farads

We arbitrarily select *R* as 4.7 k, and calculate C for a f_{3dB} of 1000 Hz:

$$C = \frac{1}{2\pi f_{3dB} R} = \frac{1}{2 \times 3.14159 \times 1000 \times 4700} = 33.9 \times 10^{-9} \ F \text{ or } 0.0339 \mu F$$

The nearest standard value is 0.033 μF or 33 nF. Figure 16-36 illustrates the frequency response of this RC filter.

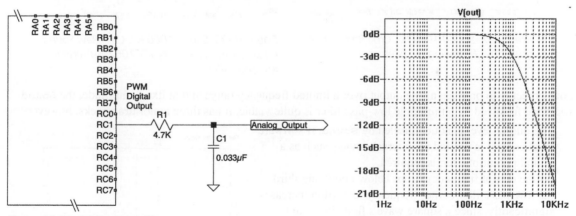

Figure 16-35: RC low-pass filter.   Figure 16-36: RC filter frequency response.

Sound and PulsOut Solutions

The last two analog output procedures we consider are MBasic's **PulsOut** and **Sound** procedures. As with the relationship between **PWM** and **FreqOut**, **PulsOut** is a generic procedure and **Sound** prepackages the code necessary to produce an audio frequency range output using **PulsOut** techniques. We will concentrate upon the **Sound** procedure.

Sound generates one or a series of outputs and is invoked as:

> **Sound Pin, [duration₁\note₁, duration₂\note₂, … durationₙ\noteₙ]**

Pin is a constant specifying the output pin to be used.

Duration_x is a constant or variable (32-bit) specifying the duration of the output for x^{th} frequency argument, in microseconds from about 24 μs to 2^{32} μs, permitting a maximum duration of approximately 4295 seconds.

Note_x is a constant or variable specifying the frequency in Hz from 0 to 32767 for the xth output. A 0 **note** indicates no output.

Unfortunately, **Sound** generates a square wave at the frequency **Note**. An RC filter doesn't come close to transforming the square wave into a sine, as illustrated in Figure 16-37. The best that can be said for the RC filtered waveform seen in Ch. 2 of the figure is that it is better than the unfiltered waveform. If we feed the square wave into an LTC1062 fifth-order low-pass filter, of course, we obtain a high quality, clean sine output, as shown in Figure 16-38.

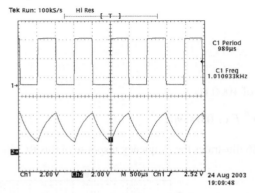

Figure 16-37: Sound 1000 Hz. Ch1: Raw Sound output; Ch2: Following RC filter.

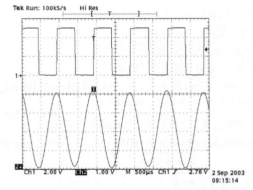

Figure 16-38: Sound 1000 Hz. Ch1: Raw Sound output; Ch2: Following LTC1062 Butterworth low-pass filter.

If our object is to produce a sine output over a limited frequency range and at fixed amplitude, the **Sound** procedure followed by a multipole low-pass filter is quite viable. It has three major drawbacks, however:

- The output level is not controlled by **Sound**. If software level control is desired an external device, such as a digital potentiometer, is required.

- As the output frequency drops below about one-third the low-pass filter corner frequency distortion increases significantly, since a square wave's first significant harmonic is the third. Thus, a wide range frequency generator will require either a tracking or switch-selected low-pass filter, or acceptance of high distortion for lower frequency signals. See, for example, Figure 16-39 showing the result of a 250 Hz **Sound** procedure with a 1 KHz low-pass filter.

- If we are going to the trouble of adding an expensive low-pass filter, why not opt for a PWM output using the **FreqOut** procedure?

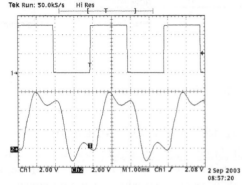

Figure 16-39: Sound 250 Hz. Ch1: Raw Sound Output; Ch 2: Following LTC062 fifth-order Butterworth low-pass filter.

These problems limit the usefulness of the **Sound** procedure to alert tones or other uncritical applications, or to those applications where a square wave is desired.

Sample Program

Program 16-3 illustrates how these alternative analog output procedures are used. By changing the definition of **SoundMode**, the program will branch to one of the four alternative outputs. All outputs are at pin C1. All branches output the frequency **freq**, except **PWMProc**, which outputs a static PWM signal with average level defined by **PwmValue**. Since **PulsOut** only defines the duration of the high level output, The **PulsOutProc** uses the **Pauseus** procedure to produce an output signal with equal duration high and low outputs. The low output duration also includes the overhead associated with the **GoTo** and **Branch** control elements, so we subtract an offset of 170 µs to the **Pauseus** procedure. If the high and low portions of the output are not of equal duration, significant distortion of sine output results. Compare Figures 16-40 and 16-41.

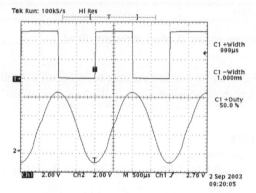

Figure16-40: PulsOut 1 KHz with equal high/low periods. Ch1: Raw Sound Output; Ch 2: Following LTC062 fifth-order Butterworth low-pass filter.

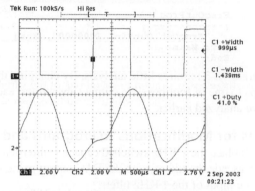

Figure 16-41: PulsOut 1 KHz with equal 59% low/41% low. Ch1: Raw Sound Output; Ch 2: Following LTC062 fifth-order Butterworth low-pass filter.

Program 16-3

```
;Program 16-3
;Constants
;--------------------
;change this to select
;generation method

;Change SoundMode to use desired
;method of generating output
SoundMode       Con         1
OutPin          Con         C1
Freq            Con         1000
Period          Con         1000000 / Freq ;period in us
HalfPeriod      Con         500000/ Freq
PWMValue        Con         128     ;DC Avg output desired

;Initialize
;-----------
Output OutPin
Low OutPin
Main
;--------
Branch SoundMode, [SoundProc,PulsOutProc,PWMProc, FreqOutProc]

SoundProc       ;Uses Sound procedure
;---------
```

```
            Sound OutPin,[1000000\Freq]
      GoTo Main

      PulsOutProc      ;Uses PulsOut procedure
      ;-----------
            PulsOut OutPin, HalfPeriod
            ;now pause for same time
            ;beause the pause time includes the GoTo
            ;we must subtract 170uS to allow for
            ;the GoTo overhead
            PauseUs HalfPeriod-170
      GoTo Main

      PWMProc
      ;-----------
            ;Output a steady-state PWM value
            PWM OutPin, PwmValue, 65535
      GoTo Main

      FreqOutProc
      ;-----------
          FreqOut OutPin, 10000000, Freq    ;remove extra 0 when FreqOut fixed
      GoTo Main
      End;
```

Program 16-3 requires little analysis. By changing the value of the constant **SoundMode**, program execution jumps to one of four possible output subroutines, each of which demonstrates one of the four procedures we've analyzed earlier.

Ideas for Modifications to Programs and Circuits

* Replace the 1 KHz switched capacitor filter with an active low-pass filter.
* Cascade the filters, with the DAC output going first to the 10 KHz LPF, and the output of that filter to the input of the 1 KHz filter.
* Add the ability of the user to input the desired frequency.
* Use the second DAC output as an amplitude input to the first DAC. This will allow the output level of the DAC No. 1 to be under program control. Apply a different frequency to each DAC and observe the output is an amplitude modulated signal.
* Allow the user to set the level via switches or other input devices.
* Use two DACs to output two different frequencies and combine them.
* Experiment more with arbitrary waveforms
* Have F_{CLOCK} change proportionally to frequency
* Add a precision voltage reference.
* Shift between two frequencies depending upon the status of an input pin. This is a modulation technique known as frequency shift keying. Shift the phase of a single frequency output (phase shift keying).

References

[16-1] Brokaw, Paul A., *Analog Signal-Handling for High Speed and Accuracy*, Analog Devices, Inc. Application Note AN-342 (undated).

[16-2] Buchanan, David, *Choosing DACs for Direct Digital Synthesis*, Analog Devices, Inc. Application Note AN-237 (undated).

[16-3] *TLC7528C, TLC7528E, TLC7528I Dual 8-Bit Multiplying Digital-to-Analog Converters*, Texas Instruments, Inc., Document No. SLAS062C (Rev. Sept 2000).

[16-4] Smith, Steven W. *The Scientist and Engineer's Guide to Digital Signal Processing*, California Technical Publishing (1999). Note: This book is available for free downloading on the Internet at http://www. DSPguide.com.

[16-5] Zverev, Anatol I. *Handbook of Filter Synthesis* , John Wiley & Sons (1967).

[16-6] Lancaster, Don, *Active-Filter Cookbook 2nd edition,* Butterworth-Heinemann, (1996).

[16-7] *LTC 1062 5th Order Lowpass Filter,* Linear Technology, Inc., (undated).

[16-8] *MAX280/MXL1062 5th Ordeer, Zero DC Error, Lowpass Filter,* Maxim Integrated Products (August 1995) (Note: both Linear Technology and Maxim data sheets must be consulted to obtain a complete view of LTC1062 performance issues.)

[16-9] *Fundamentals of Sampled Data Systems*, Analog Devices, Inc., Application Note AN-282 (undated).

DTMF Tone Decoding and Telephone Interface

We've looked at state-of-the-art high-speed inter-IC communications protocols, but we'll see there's still life in the 1960's touch-tone signaling technology. (touch-tone was a registered service mark of AT&T, but is now a public domain term.) We'll see how to decode touch-tone signals with a special purpose decoder IC and build an automatic remote control answering machine to receive touch-tone commands.

What is Touch-Tone Signaling?

Touch-tone is an example of dual-tone multifrequency (DTMF) signaling. As the name suggests, touch-tone involves simultaneously sending two tones from a set of multiple tones. As shown in Figure 17-1, each row and each column in a standard keypad is assigned a unique audio tone frequency. (Only a few specialty telephones have column 4 keys.) When you press a button, its row and column frequencies are simultaneously sent. Figure 17-2 shows the resulting envelope waveform, and Figure 17-3 shows the corresponding spectrum display. Figure 17-3 shows the spectrum of the "*" key, involving simultaneous transmission of 941 Hz and 1209 Hz tones.

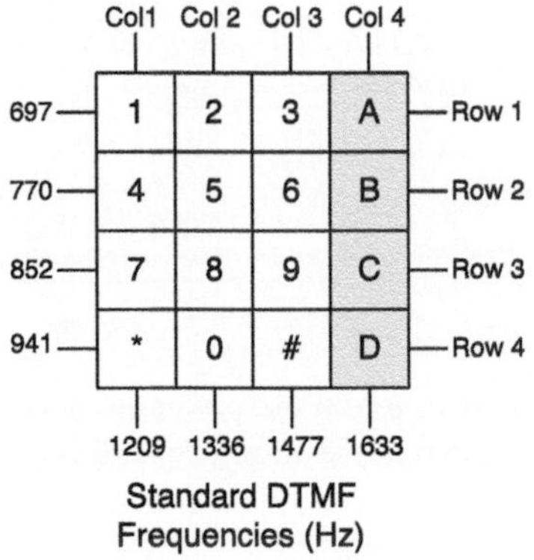

Touch-tone is not the only example of DTMF used within the telephone network. At one time, the Bell System extensively used "inter-office" DTMF—another 4 × 4 matrix system but with tone frequencies incompatible with touch-tone—to set up calls within its network. Inter-office multifrequency signaling—along with other "in-band" signaling—however, turned out to be vulnerable to hacking and has largely been replaced

	Col1	Col 2	Col 3	Col 4	
697	1	2	3	A	Row 1
770	4	5	6	B	Row 2
852	7	8	9	C	Row 3
941	*	0	#	D	Row 4
	1209	1336	1477	1633	

Standard DTMF Frequencies (Hz)

Figure 17-1: Standard Touch-tone frequencies.

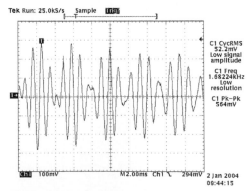

Figure 7-2: Time domain display of touch-tone signal.

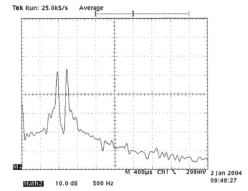

Figure 17-3: Frequency domain display of touch-tone signal.

by more efficient and more secure common channel digital signaling technology, such as Signal System No. 7. (Inter-office MF signaling, or simply MF signaling as it is commonly called, is still used on some analog trunks, and to transfer caller-ID information to some 911 call centers.)

Generating Touch-Tone Signals

To develop and test the touch-tone decoder hardware and software, you'll need a source of test signals. You could use a second PIC to generate test touch-tone signals, via the **DTMFout** or **DTMFOut2** functions. Or, you could purchase a new touch-tone generator pad. But, if you have an old unused Western Electric touch-tone telephone around, it's easy to remove the dial and set it up as a touch-tone generator. (Even though a touch-tone pad isn't round and doesn't rotate, it's still a dial in telephone terminology.) Figure 17-4 shows

how to connect a common Western Electric / Lucent touch-tone dial to function as a stand-alone touch-tone generator. It's necessary to power the dial with at least 9 V, and I usually run my 1969-vintage dial with 18 to 20 V. The dial only draws current when a key is pressed, so a pair of 9 V transistor radio batteries is a suitable power source. Don't forget the current limiting resistor, R1. Without it, you may destroy the dial. In order to reduce the output level to a more usable level, we use a voltage divider, R2 and R3. The output of the pad alone is around 0 dBm, and the R2/R3 voltage divider reduces

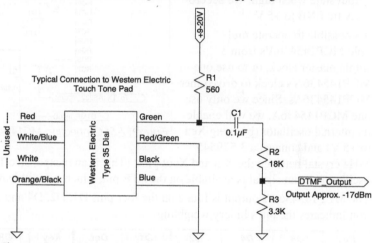

Figure 17-4: Wiring a Touch-Tone dial.

it to approximately –17 dBm. (If you are not familiar with measuring output levels in dBm, Chapter 21 has a brief refresher on decibels.) This chapter's dBm references are not rigorous, as we are seldom measuring voltage or power levels at the standard 600-ohm audio reference impedance level.

Decoding a Touch-Tone Signal

Decoding a touch-tone signal requires us to determine which tone pairs are being sent at any given instant and to discover from the brief no-tone period when one digit stops and a second digit begins. The good news is that it's possible to do this with a PIC in software alone. The bad news is that you can't do it in MBasic; it requires many lines of tightly written assembler code. Since our objective is to introduce you to MBasic with a smattering of assembler here and there, we'll use a special purpose touch-tone decoder IC.

Perhaps the two most readily available decoder ICs are Motorola's MC145436A (and the earlier MC145436) and Mitel's MT8870 devices. Motorola recently discontinued the MC145436 but it's still available from some distributors. The MT8870 and its clones have similar performance specifications to the MC145436A but are not quite as easy to use. We'll use an MC145436A, but an MT8870 can be substituted with only minor changes in the circuit.

Figure 17-5 shows how we may connect an MC145436A to a PIC. Let's review a few nonobvious aspects of Figure 17-5.

The MCP145436A operates from a single +5 V supply. It includes an interval $V_{DD}/2$ bias generator so it's necessary couple the touch-tone input to its analog signal input pin, "Ain" via a blocking capacitor, C1.

The enable pin, ENB, sets the four output pins to a high impedance tri-state when low and to a normal output state when high. We accordingly tie ENB to +5 V.

It's possible to operate multiple MCP145436A's from a single master clock, or to use one MCP145436A's clock to drive other MCP145436As. Since we only use one MCP145436A, we will enable its internal oscillator (by tying Xen to +5 V) and connect a 3.579545

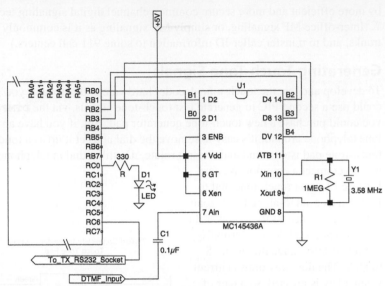

Figure 17-5: Connecting MC145436A to PIC.

MHz crystal between the Xin and Xout pins. (This is an inexpensive TV color burst crystal.) A buffered oscillator output signal is available on the ATB pin, which we will leave unconnected.

The MCP145436A's output is found on the four pins D1, D2, D4 and D8. The following numeric designation indicates the pin's binary weighting.

Key	D8	D4	D2	D1	Dec.	Key	D8	D4	D2	D1	Dec.
0	1	0	1	0	10	8	1	0	0	0	8
1	0	0	0	1	1	9	1	0	0	1	9
2	0	0	1	0	2	*	1	0	1	1	11
3	0	0	1	1	3	#	1	1	0	0	12
4	0	1	0	0	4	A	1	1	0	1	13
5	0	1	0	1	5	B	1	1	1	0	14
6	0	1	1	0	6	C	1	1	1	1	15
7	0	1	1	1	7	D	0	0	0	0	0

Keys 1…9 decode directly from their numerical value; the 0 key, however, decodes as 10. The * and # keys decode as 11 and12, respectively. We've connected these four data output pins to the PIC's **B0…B3** pins so that we may read the MC145436A's decoded output as **PortB.LowNib**.

How then, do we know when one key press ends and a second begins? And, how do we prevent one tone from triggering an output? (If you simultaneously press two keys in a row or in a column, the single tone corresponding to that row or column is generated.) The answer to these questions is that the DV (digit valid) output pin goes high only when the MC145436A decodes a valid tone pair. A single tone or tone pairs outside the ±2.5% frequency tolerance window will not cause the DV pin to go high. We've connected the DV output to the PIC's **B4** pin.

A final issue concerns how long a valid tone pair must be received in order to be properly decoded and the DV pin to go high. Here we have two options; if we tie the GT (guard time) pin to +5 V, we require longer time for tone detection. If the pin is left disconnected or directly grounded, the detection time is shortened.

GP Pin Status	Detection Time		Release Time	
	Minimum	*Maximum*	*Minimum*	*Maximum*
+5V	32	50	18	30
Ground	22	40	28	40

Operating with GT tied to +5 V is better at rejecting voice while GP at ground is better if the signal has noise and drop-outs. For our application, there is little or no difference between the two options, so we'll arbitrarily tie GT to +5 V. Manual key presses yield typical tone durations of about 100 ms with a 100 ms pause between keys, so either state works with manual sending.

Program 17-1

Let's see, then, how to read the MC145436A when connected in the circuit of 17-5. Don't forget, of course, to connect a source of touch-tone signals to the "DTMF_Input" point in Figure 17-5. The MC145436A's maximum permitted signal level is –2 dBm (615 mV RMS) and its minimum guaranteed signal level is –35 dBm (13.8 mV RMS). Our target level is –17 dBm, or 109 mV RMS. Don't worry about getting this exactly right; anywhere from 50 mV to 200 mV is close enough.

Program 17-1

```
;Program 17-1
;Reads a Motorola MC145436A DTMF
;Decoder chip. D1...D8 connected to
;B0...B3 to read value on lower nibble
;B4 connected to Valid Data pin on decoder

;Constants
;------------
;Maps BCD against key cap caption
KeyName ByteTable "D1234567890*#ABC"

;Variables
;-----------
BCDOut          Var          Byte

;Initialization
;----------------
TRISB = $FF
EnableHSerial
SetHSerial H115200
HSerOut [13,"P 17-1",13,13]

Main
;----
        ;Check to see if valid output
        BCDOut = PortB
        ;If nothing, go back and wait some more
        If BCDOut.Bit4 = 0 Then
                GoTo Main
        EndIf
        ;Valid output; write the result
        HSerOut [iHex BCDOut.Nib0," ",  |
        Str KeyName(BCDOut.Nib0) \1,13]

        ;Wait for reset of Valid Data pin
Waiting
        Repeat
                BCDOut = PortB
        Until BCDOut.Bit4 = 0

GoTo Main
End
```

To map the MC145436's decoded value to the ASCII string representation of key name, we use the byte table **KeyName**:

```
KeyName ByteTable "D1234567890*#ABC"
```

A decoded value of 12, for example returns **KeyName(12)**, or the "#" character.

```
;Check to see if valid output
BCDOut = PortB
;If nothing, go back and wait some more
If BCDOut.Bit4 = 0 Then
        GoTo Main
```

The code is simple; we keep checking the status of the MC145436A's DV pin. When it goes high, we know a valid touch-tone key sequence has been received. To save a bit of time, we read all eight PortB pins and copy the value into the variable **BCDOut**. (**BCDOut** isn't named for Binary Coded Decimal, because the MC145436A doesn't have a binary coded decimal output. Rather, it's my mnemonic for binary code data.) We then check the status of the DV pin by testing **BCDOut.Bit4**. If it's 0, we know a valid digit has not been received, so we loop back and read **BCDOut** again, waiting for it to read 1, indicating receipt of a valid digit.

```
;Valid output; write the result
HSerOut [iHex BCDOut.Nib0," ",  |
Str KeyName(BCDOut.Nib0) \1,13]
```

When a valid digit has been received, we write the information to the serial port.

```
                ;Wait for reset of Valid Data pin
Waiting
        Repeat
                BCDOut = PortB
        Until BCDOut.Bit4 = 0

GoTo Main
```

Finally, to avoid multiple readings of the same key press, we wait until the DV pin goes low, indicating the current key press has ended. When DV goes low, we loop back to **Main** and await another key press. Here's the output from Program 17-1 resulting from pressing the touch-tone keys from left to right, top to bottom. The first digit is the hex value output of the MC145436A while the second character is the corresponding string from **KeyName**. Since I used a 12-button dial, the special touch-tone characters A, B, C and D do not appear in the decoded output.

P 17-1

```
$1  1
$2  2
$3  3
$4  4
$5  5
$6  6
$7  7
$8  8
$9  9
$B  *
$A  0
$C  #
```

Program 17-2

Since it's so easy to read the touch-tone output, let's do something slightly more useful than reading single digits. We'll read a string of digits and use these to latch a PIC's output pin high or low, to simulate driving an external circuit. The circuit is the same we one we developed for Program 17-1.

Program 17-2

```
;Program 17-2
;Reads a Motorola MC145436A DTMF
;Decoder chip. D1...D8 connected to
;B0...B3 to read value on lower nibble
;B4 connected to Valid Data pin on decoder
;We will decode several multi-digit strings \
;for function selection/deselection
;Can be used for remote control

;Constants
;---------
;Maps BCD against key cap caption
KeyName        ByteTable "D1234567890*#ABC"
Fcn1           Con    $123456   ;code for function 1
Fcn2           Con    $654321   ;code for function 2
CodeLen        Con    6         ;digits in code
Timeout        Con    3         ;number of false attempts before timeout

;Variables
;---------
BCDOut         Var    Byte                    ;PortB value
FcnInput       Var    Long
DigitCount     Var    Byte                    ;how many digits received?
DigitVal       Var    BCDOut.Nib0             ;output nibble
InDigits       Var    Byte(CodeLen+2)         ;holds received bcd
i              Var    Byte                    ;counter for work
CodeTries      Var    Byte

;Initialization
;--------------
Clear
Low C0         ;LED is on C0 turned on/off via code
TRISB = $FF    ;B to input
EnableHSerial
SetHSerial H115200
HSerOut [13,"P 17-2",13,13]

Main
;----
        ;Check to see if valid output
        BCDOut = PortB
        ;If nothing, go back and wait some more
        If BCDOut.Bit4 = 0 Then
                GoTo Main
        EndIf

        ;Wait for ValidDigit to clear
        Repeat
                BCDOut = PortB
        Until BCDOut.Bit4 = 0

        ;Reset if character is "*"
        If DigitVal = 11 Then
                DigitCount = 0
        EndIf

        ;Store received digit in InDIgits()
        InDigits(DigitCount) = DigitVal

        ;For demonstration ReportInfo dumps data
        GoSub ReportInfo

        ;Valid start and end keys in the right places?
        ;If so, maybe we have a good input code
        If (InDigits(0) = 11) AND (InDigits(CodeLen+1) = 12) Then
                If DigitCount = (CodeLen+1) Then
                        GoSub CheckCode
                EndIf
        EndIf
```

```
        ;Ready for next digit
        DigitCount = DigitCount + 1

        ;Can't let it overflow the array
        If DigitCount = (CodeLen+2) Then
                DigitCount = 0
        EndIf

        ;To prevent brute force attack on code, freeze from new
        ;code reception for a minute or two after several attempts
        ;at code breakin
        If CodeTries = TimeOut Then
                HSerOut ["Attempted Code Breakin!!",13]
                ;Insert long pause, keep it short for development
                Pause 10
        EndIf

GoTo Main

ReportInfo
;-----------
        ;Report info for program development purposes
        HSerOut ["DCnt: ",Dec DigitCount,9]
        For i = 0 to DigitCount
                HSerOut [Dec i,"/",Dec InDigits(i)," "]
        Next
        HSerOut [13]
Return


CheckCode
;-----------
        ;Convert the array into a Long value
        FcnInput = 0
        ;Add each nibble at right and shift left for
        ;the next nibble. Don't shift at the end
        For i = 1 to CodeLen
                FcnInput.Nib0 = InDigits(i)
                If i <> CodeLen Then
                        FcnInput = FcnInput << 4
                EndIf
        Next

        ;Report for development
        HSerOut ["FcnInput: ",iHex FcnInput,13]

        CodeTries = CodeTries + 1

        ;Check to see if valid code. If so branch to
        ;appropriate function subroutine
        If FcnInput = Fcn1 Then
                GoSub Function1
                CodeTries = 0
        EndIf

        If FcnInput = Fcn2 Then
                GoSub Function2
                CodeTries = 0
        EndIf
Return

Function1
;----------
        ;Insert code to perform the desired task
        HSerOut ["Function 1",13]
        ;Turn LED on
        High C0
Return
```

```
Function2
;---------
          ;Insert code to perform the desired task
          HSerOut ["Function 2",13]
          ;Turn LED off
          Low C0
Return
End
```

We'll assume, until we reach Program 17-3, that the touch-tone signal arrives at the MC145436A's analog input pin by direct connection to a dial, such as shown in Figure 17-4.

We'll start by defining two multicharacter digits:

```
Fcn1              Con     $123456 ;code for function 1
Fcn2              Con     $654321 ;code for function 2
```

Function 1 turns the LED connected to **C0** on; Function 2 turns it off. Note that we've defined these two digits as hex. This lets us take advantage of the way a hexadecimal number is stored; each digit is stored in a separate nibble. (This also means if we wish to have use the 0 key in the password, it must be stored as **$A**.) Thus, by looking at **Fcn1** a nibble at a time, we can access the decimal numbers 1, 2, 3, 4, 5, 6. If we tried storing our function code in decimal as **Fcn Con 123456**, it would be stored as **$1E240**, not what we intended. Alternatively, we could store the function codes in RAM byte variable arrays, such as **Fcn1(6)**, with each element of **Fcn1()** holding one of function code 1's digits. But, this would be wasteful of RAM.

```
CodeLen Con       6         ;digits in code
Timeout Con       3         ;number of false attempts before timeout
```

We've also defined the number of digits in the two codes, and the number of attempts the user gets before being locked out of entering the codes. (Since we store a decoded value in a type long variable, **CodeLen** cannot exceed 8.)

To simplify our algorithm, we'll require that valid function codes be entered as *NNNNNN#, that is, you start with the * key, enter the function code digits followed by the # key.

We reuse the "wait for a valid digit" code from Program 17-1. Once a valid digit has been received, we add it to the array **InDigits(DigitCount)** with the following code.

```
          ;Reset if character is "*"
          If DigitVal = 11 Then
                    DigitCount = 0
          EndIf

          ;Store received digit in InDIgits()
          InDigits(DigitCount) = DigitVal

          ;For demonstration ReportInfo dumps data
          GoSub ReportInfo
```

We use the * key (read as decimal value 11) to reset **DigitCount**, so we are guaranteed of a fixed starting point when we evaluate the received key strokes to see if they match **Fcn1** or **Fcn2**.

```
          ;Valid start and end keys in the right places?
          ;If so, maybe we have a good input code
          If (InDigits(0) = 11) AND (InDigits(CodeLen+1) = 12) Then
                    If DigitCount = (CodeLen+1) Then
                              GoSub CheckCode
                    EndIf
          EndIf
```

We then check to see if we have a potentially valid code sequence—a * (value 11) at **InDigits(0)** and a # (value 12) at **InDigits(CodeLength+1)**. If so, we then execute subroutine **CheckCode** to see if the potentially valid entry matches a real function code. (We'll look at subroutine **CheckCode** after we finish the rest of the main program loop.)

```
;Ready for next digit
DigitCount = DigitCount + 1

;Can't let it overflow the array
If DigitCount = (CodeLen+2) Then
        DigitCount = 0
EndIf
```

We next bump up the **DigitCount** to be ready for the next received keystroke, and check if we have reached the maximum permitted **DigitCount** value. If so, we reset to zero.

```
;To prevent brute force attack on code, freeze from new
;code reception for a minute or two after several attempts
;at code breakin
If CodeTries = TimeOut Then
        HSerOut ["Attempted Code Breakin!!",13]
        ;Insert long pause, keep it short for development
        Pause 10
EndIf

GoTo Main
```

Finally, we check to an attempted break-in. If the correct function codes have not been received in a required number of attempts, we lock out further input for a time.

Let's now look at subroutine **CheckCode**.

```
CheckCode
;-----------
        ;Convert the array into a Long value
        FcnInput = 0
        ;Add each nibble at right and shift left for
        ;the next nibble. Don't shift at the end
        For i = 1 to CodeLen
                FcnInput.Nib0 = InDigits(i)
                If i <> CodeLen Then
                        FcnInput = FcnInput << 4
                EndIf
        Next
```

To compare the received keystrokes with the function codes, we place the keystroke values in a type **Long** variable **FcnInput**. We do this one nibble at a time, adding each keystroke value to the **FcnInput.Nib0**. After each nibble is entered, we shift **FcnInput** four bits to the left, to make room for the next keystroke value. (We don't shift after the last digit is entered, because there is no next value to add.) We could have simply hard-coded this conversion:

```
FcnInput.Nib0 = InDigits(1)
FcnInput.Nib1 = InDigits(2)
Etc.
```

But, using a loop and shift procedure allows us to change the code length by altering the value of **CodeLen** without mucking about in the rest of the code.

```
HSerOut ["FcnInput: ",iHex FcnInput,13]
CodeTries = CodeTries + 1
```

We reach subroutine **CheckCode** only if the entry appears to be a valid function code, so it's appropriate to increase the **CodeTries** counter. (Remember, we use **CodeTries** to detect attempted break-ins.)

```
;Check to see if valid code. If so branch to
;appropriate function subroutine
If FcnInput = Fcn1 Then
        GoSub Function1
        CodeTries = 0
EndIf

If FcnInput = Fcn2 Then
        GoSub Function2
        CodeTries = 0
```

```
            EndIf
      Return
```

Having put the keystrokes into **FcnInput**, it's a simple matter of comparing **FcnInput** to the pre-defined function codes **Fcn1** and **Fcn2**. If there's a match, then we execute code specific to that function, and reset **CodeTries** to 0, since we have received a valid function code.

```
      Function1
      ;---------
            ;Insert code to perform the desired task
            HSerOut ["Function 1",13]
            ;Turn LED on
            High C0
      Return

      Function2
      ;---------
            ;Insert code to perform the desired task
            HSerOut ["Function 2",13]
            ;Turn LED off
            Low C0
      Return
```

The functions associated with the **Fcn1** and **Fcn2** codes turn on or off the LED attached to pin **C0**.

As with Program 17-1, to help understand and develop the code, we output information to the serial port. Each line reports the current digit count (**DCnt**) and then the contents of **InDigits(i)** in the form index/value for **i = 0 to DigitCount**. Here's a typical interaction report, with my comments added in **bold**.

P 17-2

```
DCnt: 0 0/1
DCnt: 1 0/1 1/2
DCnt: 2 0/1 1/2 2/3
DCnt: 3 0/1 1/2 2/3 3/4
DCnt: 4 0/1 1/2 2/3 3/4 4/5
DCnt: 5 0/1 1/2 2/3 3/4 4/5 5/6
DCnt: 6 0/1 1/2 2/3 3/4 4/5 5/6 6/7
DCnt: 7 0/1 1/2 2/3 3/4 4/5 5/6 6/7 7/8    Keyed in too many digits, so DigitCount resets to 0
DCnt: 0 0/9
DCnt: 0 0/11
DCnt: 1 0/11 1/10
DCnt: 2 0/11 1/10 2/12
DCnt: 0 0/11                                Start a function code with the * key
DCnt: 1 0/11 1/1                            1 key
DCnt: 2 0/11 1/1 2/2                        2 key
DCnt: 3 0/11 1/1 2/2 3/3                    3 key
DCnt: 4 0/11 1/1 2/2 3/3 4/4                4 key
DCnt: 5 0/11 1/1 2/2 3/3 4/4 5/5           5 key
DCnt: 6 0/11 1/1 2/2 3/3 4/4 5/5 6/6      6 key
DCnt: 7 0/11 1/1 2/2 3/3 4/4 5/5 6/6 7/12  End the function code with the # key
FcnInput: $123456                          Keypress input received as $123456
Function 1                                 Function 1 executed based on correct reception of Fcn1
DCnt: 0 0/11                               Same but Function 2 to be executed
DCnt: 1 0/11 1/6
DCnt: 2 0/11 1/6 2/5
DCnt: 3 0/11 1/6 2/5 3/4
DCnt: 4 0/11 1/6 2/5 3/4 4/3
```

```
DCnt: 5 0/11 1/6 2/5 3/4 4/3 5/2
DCnt: 6 0/11 1/6 2/5 3/4 4/3 5/2 6/1
DCnt: 7 0/11 1/6 2/5 3/4 4/3 5/2 6/1 7/12
FcnInput: $654321                         Keypress input received as $654321
Function 2                                Function 2 executed based on correct reception of Fcn2
DCnt: 0 0/11
DCnt: 1 0/11 1/1                          Now we test some false function codes to verify the
DCnt: 2 0/11 1/1 2/4                      program will reject incorrect entries
DCnt: 3 0/11 1/1 2/4 3/7
DCnt: 4 0/11 1/1 2/4 3/7 4/2
DCnt: 5 0/11 1/1 2/4 3/7 4/2 5/5
DCnt: 6 0/11 1/1 2/4 3/7 4/2 5/5 6/8
DCnt: 7 0/11 1/1 2/4 3/7 4/2 5/5 6/8 7/12
FcnInput: $147258                         Keypress input received as $147258
DCnt: 0 0/11                              Does not correspond to any known function code
DCnt: 1 0/11 1/7
DCnt: 2 0/11 1/7 2/8
DCnt: 3 0/11 1/7 2/8 3/9
DCnt: 4 0/11 1/7 2/8 3/9 4/9
DCnt: 5 0/11 1/7 2/8 3/9 4/9 5/8
DCnt: 6 0/11 1/7 2/8 3/9 4/9 5/8 6/7
DCnt: 7 0/11 1/7 2/8 3/9 4/9 5/8 6/7 7/12  Second attempt at break-in
FcnInput: $789987                         Keypress input received as $789987
DCnt: 0 0/11                              Does not correspond to any known function code
DCnt: 1 0/11 1/6
DCnt: 2 0/11 1/6 2/5
DCnt: 3 0/11 1/6 2/5 3/4
DCnt: 4 0/11 1/6 2/5 3/4 4/6
DCnt: 5 0/11 1/6 2/5 3/4 4/6 5/5
DCnt: 6 0/11 1/6 2/5 3/4 4/6 5/5 6/4
DCnt: 7 0/11 1/6 2/5 3/4 4/6 5/5 6/4 7/12  Third attempt at break-in
FcnInput: $654654                         Keypress input received as $654654
Attempted Code Breakin!!                  Attempted break-in detected after three invalid attempts
```

Program 17-3

Now that we can read multikeystroke entries and take action based on their value, let's see how we might build upon Program 17-2 and use the telephone network to remote control devices in your home or office. Program 17-3 allows you to dial a telephone number and have it automatically answered. If you enter the correct password, you can then send four-digit commands by touch-tone. The four-digit command structure is patterned after the X10 home automation protocol, covered in a later chapter. To keep Program 17-3 at a manageable length, we'll omit the touch-tone-to-X10 transmission elements. First, however, we'll take a side trip and examine the analog telephone network subscriber connection.

Figure 17-6 provides a conceptual overview of the traditional single-party telephone connection and Figure 17-7 provides a similar picture of a typical subscriber instrument. We'll look at nonelectronic subscriber instruments, but the same functionality—implemented with electronic circuitry, not electromechanical components—is found in modern subscriber sets. And, of course, a similar electronic revolution has replaced electromechanical crossbar and step-by-step switches in the central office with computer-controlled equipment. But, the functionality is the same and it's easier to understand the electromechanical technology. There are two components we will examine; signaling (answering and placing a call) and communications (transmission and reception of voice frequency signals).

We'll first consider signaling in the local loop. From Figure 17-7, it should be apparent that when the subscriber instrument is *on-hook*, no DC current flows through the *subscriber loop*.

Answering a call—When ringing voltage (A 20 Hz sinusoidal waveform, typically 100 V RMS) is applied at the central office, AC current flows through the ringer, which causes its armature to strike a bell gong. To answer, you pick up the handset, which closes the *hookswitch*, thereby permitting DC current to flow through the subscriber loop. The inductors L1 and L2 shown in Figure 17-6 are actually relay coils, and thus their associated contacts are pulled in by the DC current flow, informing the central office that the subscriber has picked up the line and has thus answered the call. Ringing voltage is disconnected and conversation may commence.

Originating a call—When you lift the handset, the hookswitch closes and permits DC current to flow through the subscriber loop. The inductors L1 and L2 shown in Figure 17-6 are actually relay coils, and thus their associated contacts are pulled in by the DC current flow, informing the central office that the subscriber has picked up the

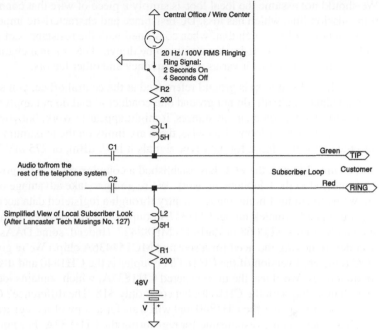

Figure 17-6: Simplified view of subscriber loop.

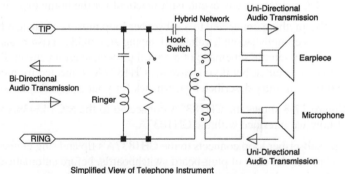

Figure 17-7: Simplified view of subscriber instrument.
(Shown on-hook.)

line. Since there is no ring signal being applied, the central office knows you are attempting to originate a call, and thus connects the line to a dial tone generator and a touch-tone digit receiver.

Communications over the subscriber loop requires a completed connection, that is, there must be DC current flow through the local loop sufficient for the central office to recognize supervision of the call and cut through the audio path. The subscriber loop supports both talk and receive audio signals over a single pair of wires. At the central office, and at the subscriber end, however, we must have separate audio paths for talk and receive, or else we will have undesired feedback and an unusable connection. (In telephone terminology, a circuit with separate transmit and receive paths is a "four-wire" system.) A two-wire to four-wire converter is called a *hybrid*, and historically was constructed as a multiwinding transformer. It's now usually done with several op-amps in a subtraction configuration where a local copy of the transmit signal is subtracted from the receive signal.

We should not assume the local loop is simply a piece of wire that connects you with the central office. It's a transmission line, which has loss, DC resistance and characteristic impedance. The DC resistance is usually a few hundred ohms, such that, when combined with the resistances of relay coils in the central office, typically 15 to 30 mA of off-hook current can be drawn. The nominal characteristic impedance of the local loop is 600 ohms, but in fact it varies with frequency and other factors.

Although the local loop is ground referenced at the central office, you should never treat it as anything other than a balanced circuit; do not ground one conductor, and do not apply signals between ground and one conductor. Under some circumstances, it might appear to work, but you run the risk of coupling your signals into other subscriber loops. Likewise, there are limits on the maximum signal level that you may transmit into the subscriber loop; for data-type signals it is –9 dBm, or 275 mV RMS. Do not exceed this level.

In the United States, the FCC has established a complex set of rules governing equipment that may be directly connected to the telephone network. We're going to take advantage of a provision in the FCC's rules that allows us to connect home made circuitry through a registered data access arrangement (DAA). The DAA we'll use is a Cermetek model CH1837A coupler. Similar devices are available from other manufactures, such as Xecom's XE0068 or Zarlink's MH88422. (Indeed, some DAAs include a touch-tone encoder and decoder, removing the need for a separate MC154436A chip.) We're going to cheat a little bit, though. The FCC registered version of the CH1837A coupler is the CH1840 and it sells for around $60 each in single lot quantities. We'll use the un-registered CH1837A, which contains identical circuitry, and is pin-for-pin interchangeable with the CH1840, but costs only $18. The difference? Cermetek has gone through the FCC paperwork to register the CH1840 and will transfer a copy of its registration to you with each CH1840 you purchase, but you are responsible for registering the CH1837A. For home made one-off experimental work, we'll use the unregistered CH1837A and follow good engineering practices, but not file any paperwork with the FCC, which, in any event, isn't practical for the home experimenter.

Lastly, anything connected to a subscriber loop must be protected from over-voltage and over-current. Usually this is accomplished with voltage clamping zinc oxide varistors and fuses. Cermetek's CH1837A data sheet provides a sample schematic of a typical protective arrangement. I've omitted these elements from the schematic to avoid clutter, but if you leave the CH1837A connected for more than brief tests, you must add the varistor and fuse circuitry described by Cermetek for your own safety as well as to protect the telephone network.

Figure 17-8 shows the CH1837A connected to the MC145436A touch-tone decoder and the PIC. Let's look at how we interface with the CH1837A.

The subscriber loop connects to the CH1837A's tip and ring connections. (The terms *tip* and *ring*, by the way, go back to the days of plug-board switchboards, before automatic dialing, when one side of the loop connected to the tip of the plug and the other side connected to the ring connection. The sleeve of the plug was ground.)

When ringing voltage is detected, the CH1837A's ringing indicator (RI) pin goes low. The RI pin's response is fast enough to follow the individual cycles of the 20 Hz ringing voltage, so we add a 4.7 µF capacitor to provide a single pulse for every ring interval. (Ringing voltage is applied for two seconds, followed by four seconds of silence.) Figure 17-9 shows RI's response during ringing. The individual cycles of ringing voltage are bridged over, so RI stays low during the entire ringing voltage interval. We've connected RI to the PIC's **C2** pin so we may read RI's status.

To take the CH1837A *off-hook*, that is, to answer a ring, or to seize the line to originate a call, you take the OFFHK pin high. A low on OFFHK places the CH1837A in the *on-hook* mode. We've connected OFFHK to the PIC's **C0** pin so we may place the CH1837A in either the *on-hook* or *off-hook* mode via software control. To disconnect a call, we return to *on-hook* mode.

The CH1837A performs the hybrid function, giving us independent transmit and receive audio paths. The net loss in either direction is nominally 0 dB, so the signal level applied to the CH1837A's transmit port

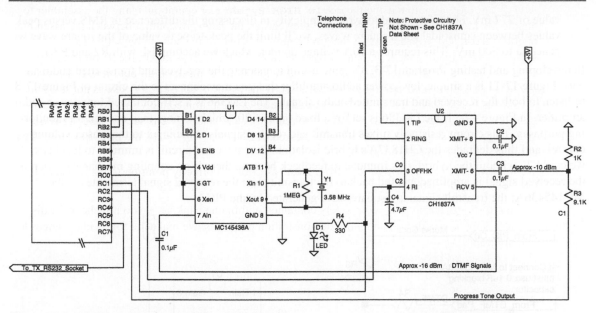

Figure 17-8: Telephone interface to PIC and TT decoder.

appears without loss on the tip and ring connections and the receive signal level on the tip and ring connections appear without loss at the CH1837A's receive port.

- *The receive signal*, i.e., the touch-tone signal coming in from the remote user, appears at the RCV pin. It is a single-ended signal, that is, it is with respect to ground, DC biased at approximately $V_{cc}/2$. Hence, we couple out of RCV with a 0.1 μF blocking capacitor. Based on the telephone company's nominal loss plans, we expect the received touch-tone signal to be in the –10 dBm to –20 dBm range. Figure 17-10 shows a typical touch-tone signal (–16 dBm) as seen at the CH1837A's RX pin.

- *The transmit signal*, i.e., the audio prompt signals we will send to the remote user, is applied *differentially* to the XMIT+ and XMIT- pins. Both of these pins are internally biased at approximately $V_{cc}/2$. Since it's more convenient for the PIC to generate a singled-ended signal, we'll connect the XMIT+ input to ground through a 0.1 μF blocking capacitor and connect the XMIT- input to the PIC's **c1** pin. Again, we use a 0.1 μF blocking capacitor to avoid disrupting the internal $V_{cc}/2$ bias. And, since the raw output we generate is well over the –9 dBm transmit limit, we use R2 and R3 as a resistive voltage divider. (We generate the response tone as a square wave on **C0**. **C0** swings from +5 V to 0 V, or peak-to-peak 5 V. A sinusoidal waveform at –9 dBm has a peak-to-peak

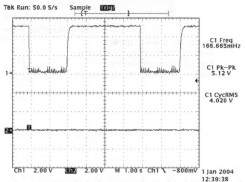

Figure 17-9: RI pin with 4.7 μF filter during ringing.

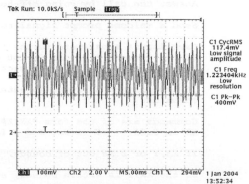

Figure 17-10: Touch-Tone signal as received at RX pin.

value of 777 mV. For safety—and to avoid complexity in discussing the difference in RMS versus peak values between sinusoidal and a square waves, we'll limit the peak-to-peak value of the square wave we generate to 500 mV. This requires a 10:1 voltage divider, which we accomplish with R2 and R3.)

In developing and testing Program 17-3, it's quite useful to listen to the received and transmitted audio signals. Figure 17-11 is a simple, low-power audio amplifier that you may connect to the circuit of Figure 17-8 to listen to both the received and transmitted audio signals. The LM386 is a self-contained low-power audio amplifier. In Figure 17-11, the LM386 is set for a fixed gain of 20, which is more than enough for monitoring purposes. This circuit resistively mixes transmit and receive signals at the input to the master volume control, and thus degrades the CH1837A's hybrid isolation. But, since our circuit is immune to feedback, we could do without the hybrid. (It's immune to feedback because there is no coupling mechanism whereby the received signal is amplified and fed back to the transmit side; the received signal terminates in the MC145436A; the transmit signal is a square wave generated by the PIC.)

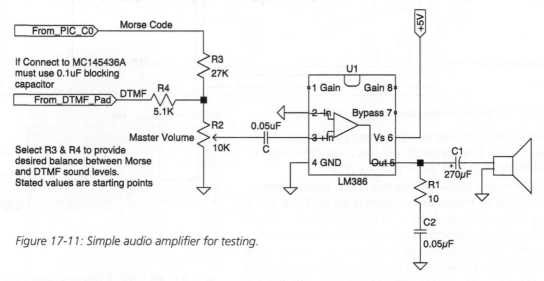

Figure 17-11: Simple audio amplifier for testing.

Before we look at the code for Program 17-3, let's summarize in pseudo-code what we expect it to accomplish.

```
Wait on Hook for Ring.
Count rings and after a preset number of rings,
Answer the call by going off-hook

Send a welcome prompt signal to the user

User inputs a password in the form *NNNNNN#
If the correct password is not entered after a preset number of attempts, send a
response message and disconnect by going on-hook

If the correct password is entered, acknowledge it and allow the user to enter
commands.

Commands are of the form *XXYZ# where
XX is 0...15    Y is 0...1 and Z is 0...9

If XX,Y or Z is out of bounds, send an appropriate error message to the user.

If the XXYZ is within bounds, echo it back to the user

Disconnect if *9999# code is received, or if there are no key presses for a certain
interval
```

We must decide how to communicate with the user. Unfortunately, simple speech synthesizer chips are no longer readily available, and those chips that are available are too complex for beginner's projects. So, we are left with a simple tone or beep type response. Because I've been a ham radio operator since I was a teenager, many years ago, I've made Program 17-3 respond with Morse code prompts and messages. If you don't know Morse code, I'll suggest a simple beep routine alternative. (It's not that difficult to learn a few characters of Morse. When I was a graduate student in electrical engineering at Wayne State University in the late 1960's, I worked on a research project with the Detroit Police Department and spent quite a bit of time at the DPD's Belle Isle mobile radio repair shop. About half the officers at the shop were hams and instead of a voice PA system to announce "telephone call for Joe on line 1," they used an audio oscillator and Morse key to send the initials of the person being called. It didn't take long for non-hams to learn enough Morse to identify their initials when heard over the sound system. To show how far the technology has come, it was only in the mid-60s that the regional police Morse code emergency radio network was abandoned.)

Program 17-3

Program 17-3 is very long. Rather than follow our normal structure of listing the program in full, followed by a section-by-section analysis, we'll skip the full listing. Program 17-3, like all other programs in this book, is available in an electronic copy in the associated CD-ROM. If you need a listing to follow while reading the analysis, you may print one from the program file.

We'll look at our constants, variables and Morse code routines after we work through the main elements of Program 17-3.

As usual, the first order of business after defining constants and declaring variables is to initialize variables and establish pins and input or output.

```
Initialization
;---------------
TRISB = $FF             ;B to input
TickCtr = 0             ;time passage
RingCount = 0           ;rings received
EnableHSerial
SetHSerial H115200
SetTmr1 TMR1Int8        ;With 20 MHz clock this yields
                        ;one tick about every 105mS
OnInterrupt Tmr1Int, UpDateTicks
```

It's important that our program release the subscriber loop automatically if a user forgets to enter the disconnect code, or if the remote control line is called in error or by a telemarketer. We keep track of elapsed time after answering with an interrupt service routine, **UpDateTicks**. Approximately every 0.1 second, **UpDateTicks** increments the variable **TickCtr** so we can determine elapsed time by the value of **TickCtr**. The foregoing code initializes **TickCtr** and establishes the entry call for the interrupt handler. (Timer 1 is a 16-bit timer, and we've set it to operate with a 8:1 prescaler with its input being the instruction clock. Hence, with a 20 MHz clock rate, timer 1's interrupt service routine is called once every $(8 \times 65536 / 5 \times 10^6)$ or 0.1049 seconds. See Chapter 10 for more information on timers and interrupt service routines.)

```
;Serial output for development; remove if un-necessary
HSerOut [13,"P 17-3",13,13]
DigitCount = 0
PWOK = Bad ;no password yet

Input    RIPin          ;Ring indicator
Output   OffHook        ;pin to control connection
Low      OffHook        ;Ready to receive a call
Enable   Tmr1Int        ;start counting ticks
```

Next, we set the pins connected to the CH1837A's RI and OffHook pins as input and output, respectively. (We aliased the constants **RIPin** to **C2** and **OffHook** to **C0** in the definitions file.) And, lastly, we start timer 1's interrupt process.

```
;Initialization Morse Code
;--------------------------
ELtime = 1200 / WPM   ;From PARIS = 50 Elements = 1 Word
Low TonePin
```

Now we calculate the base length of time for the shortest element of a Morse code symbol (a single dot). Morse code speed is universally defined in "words per minute" and the "standard word" "PARIS" is comprised of 50 elements, including inter-character, inter-letter and inter-word spaces. We'll spend a more time on the Morse code aspects of Program 17-3 later.

```
Main
;------
            ;Return from WaitForRing only after defined number
            ;of rings received
            GoSub WaitForRing

            ;Go off-hook when ring stops
            ;Wait for ringing to stop
            GoSub AnswerPhone

            ;At this stage, answered call and waiting for user TT input
            ;Allow certain time to enter valid password and user data.
            ;Timer1 checks for idle condition
            TickCtr = 0
            Enable Tmr1Int
```

We start the main portion of the program by waiting for a pre-determined number of rings, following which occurrence we place the CH1837A off-hook. The functional code for these two activities appears in the subroutines **WaitForRing** and **AnswerPhone**.

```
WaitForRing
;-----------
Waiting
            ;Waiting for ring
            If RI = %1 Then
                    GoTo Waiting
            EndIf

            TickCtr = 0

            For i = 0 to 99 : Pause 10 : Next
```

(In declaring variables, we have aliased **RI** to **PortC.Bit2**.) When ringing voltage is detected by the CH1837A, it drops RI low. Hence we continuously test RI's status until it is low. Upon RI going low, we reset **TickCtr** and wait for 1 second. (Because MBasic executes interrupt service routines only between BASIC statements, we've used 100 delays of 10 ms instead of one 1000 ms delay. This way, the interrupt service routine will continue executing without more than 10 ms latency.)

```
            If RI = %1 Then
                    GoTo Waiting
            EndIf
```

One second after detecting ringing, we check to see if RI is still low. If it is, we likely have a valid ring, and not just a momentary voltage spike on the subscriber loop.

```
            ;Wait 2 seconds
            For i = 0 to 199 : Pause 10 : Next

            RingCount = RingCount + 1
```

We then wait two seconds, at which time we should be in the four-second window between ringing voltage pulses. We update the ring count.

```
            If TickCtr > (RingVal * 60) Then
                    RingCount = 0
            EndIf

            If RingCount < RingVal Then    ;Ring limit reached?
```

```
                GoTo Waiting
        EndIf

        RingCount = 0
    Return
```

We need a way of resetting **RingCount** to avoid the case of several isolated rings, or line disturbance, being considered a valid answer criteria. We do this by resetting **RingCount** if the requisite number of rings (**RingVal**) has not been reached within 6 seconds for each expected ring. (**TickCtr** increments 10 times/second.)

```
;Answer phone when ring stops
AnswerPhone
;-------------
AnswerOnQuiet
        If RI = %0 Then
                GoTo AnswerOnQuiet
        EndIf

        ;Ringing stopped - now answer by taking OffHook HIGH
        High OffHook
        TickCtr = 0

        ;Send welcome message
        GoSub Welcome
        PWOK = Bad
    Return
```

Subroutine **AnswerPhone** does exactly that; when invoked, it waits until there is no ringing voltage detected and then takes **OffHook** high, which causes the CH1837A to go off-hook and signal the central office that the call has been answered. (We've aliased **OffHook** to **C0**.) Note also we reset **TickCtr** to 0, as we must monitor the connected call for inactivity and disconnect if appropriate. Subroutine **Welcome** sends "?" by Morse code as a welcome message.

```
;After auto answer send welcome msg
;-------
Welcome
;-------
        Pause 2000      ;Wait 2 seconds
        Message = "?["
        Message(2) = 0
        GoSub SendMsg
    Return
```

We wait 2 seconds before sending the message as the ring tone indicator heard by the calling party is not necessarily in phase with the ringing voltage on the called line. If you are not implementing Morse code, then substitute a routine that sends, for example, two short pips after a 2-second delay.

Before looking at the rest of the main program loop, let's look at the idle timeout feature we've implemented. The main program loop looks something like this when stripped of detail:

```
Main
<<Wait for ring and answer if valid rings>>
CheckForTimeout
        If TickCtr > TimeOutVal Then
                GoSub GoodBye   ;If timed out, disconnect
                GoTo Main       ;Wait for new call
        EndIf
<< Check for key presses--if Key Pressed reset TickCtr to 0 >>
GoTo CheckForTimeout
```

Every time a key is pressed and a valid touch-tone signal is received, **TickCtr** is reset to 0. Thus, while the line is idle (speech is regarded as idle, as it is unlikely to be accepted as a valid touch-tone key press by the MC145436A decoder) **TickCtr** keeps incrementing, approximately 10 times every second. If **TickCtr** exceeds the timeout constant **TimeOutVal**, subroutine **GoodBye** is executed and program execution returns to **Main**, where it awaits the next ring sequence.

```
                ;Exit with SK and disconnect
                GoodBye
                ;-------------
                        Message = "\"   ;Send SK
                                Message(1) = 0
                                GoSub SendMsg
                                Pause 2000        ;no signal
                                Low OffHook       ;Disconnect
                Return
```

Subroutine **GoodBye** sends a disconnect message, the Morse code sign for "end of communications"—the letters SK sent without a letter space—waits two seconds and then instructs the CH1837A to disconnect by taking **OffHook** low. The 2-second wait follows FCC rules requiring 2 silent seconds preceding disconnect. Again, you could substitute a goodbye beep signal, or simply silently disconnect the call.

We'll now go back to the main program loop.

```
                CheckForTimeout

                        If TickCtr > TimeOutVal Then
                                GoSub GoodBye             ;If timed out, disconnect
                                GoTo Main                 ;Wait for new call
                        EndIf

                        ;Wait for a keypress
                        BCDOut = PortB
                        ;If nothing, go back and wait some more
                        If PortB.Bit4 = 0 Then ;ValidDigit = 0 -> no keypress
                                GoTo CheckForTimeOut
                        EndIf

                        ;Wait for ValidDigit to clear
                        Repeat
                                BCDOut = PortB
                        Until BCDOut.Bit4 = 0

                        ;Reset if character is "*"
                        If DigitVal = 11 Then
                                DigitCount = 0
                        EndIf

                        ;Store received digit in InDIgits()
                        InDigits(DigitCount) = DigitVal

                        ;For demonstration ReportInfo dumps data
                        GoSub ReportInfo
```

The above code is, by in large, lifted directly from Program 17-2 and requires little discussion; wait for a valid digit to be received, and store its numerical value in the array **InDigits(DigitCount)**. Reset **DigitCount** when a * key is pressed.

```
                        If (InDigits(0) = 11) AND (InDigits(CodeLen+1) = 12) Then
                                If DigitCount = (CodeLen+1) Then
                                        GoSub CheckCode
                                EndIf
                EndIf
```

Likewise, the above code to determine if a key press sequence meets the format for being a valid password is lifted directly from Program 17-2. The subroutine **CheckCode** also follows its namesake in Program 17-2.

```
                CheckCode
                ;-----------
                        ;Remember 0 is reported as 10
                        ;we need numerical so must zero it
                        For i = 1 to CodeLen
                                If InDigits(i) = 10 Then
                                        InDigits(i) = 0
                                EndIf
                        Next
```

```
                        ;Convert the array into a Long value
                        FcnInput = 0
                        ;Add each nibble at right and shift left for
                        ;the next nibble. Don't shift at the end
                        For i = 1 to CodeLen
                                FcnInput.Nib0 = InDigits(i)
                                If i <> CodeLen Then
                                        FcnInput = FcnInput << 4
                                EndIf
                        Next

                        ;Report for development
                        HSerOut ["FcnInput: ",iHex FcnInput,13]

                        ;Attempted password entries
                        CodeTries = CodeTries + 1

                        ;Check to see if valid code. If so branch to
                        ;appropriate function subroutine
                        If FcnInput = Password Then
                                GoSub PasswordOK
                                CodeTries = 0
                        EndIf

                        ;If bad PW, send ?. Don't send on last try because we
                        ;send different message
                        If (CodeTries > 0) AND (CodeTries < Attempts) Then
                                Message = "[?["
                                Message(3) = 0
                                GoSub SendMsg
                        EndIf
        Return
```

If the received password entry is invalid, we sent a Morse "?" and increase the counter **CodeTries**. You could substitute a routine that sends a series of short pips to indicate a bad password.

```
        PasswordOK
        ;---------
                        TickCtr = 0      ;Reset the timer
                        PWOK = Good      ;;Good password
                        ;Insert code to perform the desired task
                        HSerOut ["Password OK",13]
                        Message = "OK["
                        Message(3) = 0
                        GoSub SendMsg
        Return
```

If the password entry is valid, "OK" is sent by Morse to the calling party. The sentinel variable **PWOK** is set to the constant **Good**, thereby indicating a valid password has been received.

Returning to the main program loop, we next check to see if the key sequence matches that of a function command.

```
                        If (InDigits(0) =11) AND (InDigits(CommandLen+1) =12) AND |
                        PWOK = Good) AND (DigitCount = 5) Then
                                GoSub FunctionCheck
                                DigitCount = 0
                        EndIf
```

Remember that function commands are four digits long. Hence we check for a * (value 11) at **InDigits(0)** and a # (value 12) at **InDigits(5)**. If the key sequence matches the function command template, we evaluate the key sequence at subroutine **FunctionCheck**.

```
        FunctionCheck
        ;------------
                        ;Remember 0 is reported as 10
                        ;we need numerical so must zero it
                        For i = 1 to 4
                            If InDigits(i) = 10 Then
```

```
              InDigits(i) = 0
        EndIf
Next
```

First, we modify **InDigits()** to hold the decimal values of the keys. Since only the 0 key differs from the decimal value of the key face label, we only need convert any array element with a value of 10 to 0.

Remember, function commands are of the form XXYZ, where XX is the address, Y is the application number and Z is the application value. The variables associated with these are XX: **FcnAdr**, Y: **AppNo** and Z:**AppVal**.

```
FcnAdr = 10 * InDigits(1) + InDigits(2)

If FcnAdr = 99 Then
        RingCount = 0
        TickCtr = 0
        CodeTries = 0
        GoSub GoodBye
        GoTo Main
EndIf
```

We use *99NN# (where NN is any digit 0…9) as an exit command, so first check for the exit code. If found, we use the subroutine **GoodBye** to perform the disconnect sequence.

```
If FcnAdr > 15 Then
        Message = "ADR?["
        Message(5) = 0
        GoSub SendMsg
        FcnAdr = BadData
EndIf
```

Valid X10 addresses are in the range 0…15, so we check for valid addresses. If the address is bad, we set the **FcnAdr** to a **BadData** flag value and send an error message by Morse. You may, of course, substitute beeps for Morse.

```
AppnNo = InDigits(3)

If AppnNo > 1 Then
        Message = "APP?["
        Message(5) = 0
        GoSub SendMsg
        AppnNo = BadData
EndIf
```

We do the same thing for the application number, which is valid only for 0 and 1.

```
AppnVal = InDigits(4)

If AppnVal > 9 Then
        Message = "VAL?["
        Message(5) = 0
        GoSub SendMsg
        AppnVal = BadData
EndIf
```

We then check the application value for validity, and send an error message if it is in error.

```
If (FcnAdr <> BadData) AND (AppnNo <> BadData) AND |
(AppnVal <> BadData) Then
    Message(0) = InDigits(1) + "0"
    Message(1) = InDigits(2) + "0"
    Message(2) = ","
    Message(3) = InDigits(3) + "0"
    Message(4) = ","
    Message(5) = InDigits(4) + "0"
    Message(6) = "["
    Message(7) = "O"
    Message(8) = "K"
    Message(9) = 0
    GoSub SendMsg
    ;Insert code to actually execute function
EndIf
```

```
        TickCtr = 0
Return
```

If all three elements of the function code are correct, we echo the code back via Morse. After sending the Morse message, we insert code to send the function value via an X10 transceiver to X10 receiving modules. Of course, there's nothing stopping you from using other control approaches, such as turning on or off a relay or triac via setting or clearing a PIC output pin in response to a four-digit function code.

The remainder of the main program loop performs some housekeeping activities and checks for attempted password break-ins.

```
;Ready for next digit
        DigitCount = DigitCount + 1

        ;Can't let it overflow the array
        If DigitCount = (CodeLen+2) Then
                DigitCount = 0
        EndIf

        ;To prevent brute force attack on password disconnect after
        ;defined number of attempts. Could delay re-answer
        If CodeTries = Attempts Then
                HSerOut ["Attempted Code Breakin!!",13]
                Message = "BAD[PW["
                Message(8) = 0
                CodeTries = 0
                GoSub SendMsg
                GoSub GoodBye
                GoTo Main
        EndIf

GoTo CheckForTimeout
```

If the correct password has not been entered in **Attempts** tries (set at 3), an error message is sent in Morse and the telephone connection is broken. For additional security, you could add a delay loop that prevents answering another call until after a considerable time, perhaps five or ten minutes has passed.

We've now gone through the main program and its subroutines, so we'll take a quick look at the variables and constants.

```
;Constants
;------------
;Maps BCD against key cap caption
KeyName ByteTable "D1234567890*#ABC"
PassWord        Con     $012345 ;Master Password
CodeLen         Con     6       ;digits in code <<-- Do not use 4
Attempts        Con     3       ;number of false attempts before timeout
RIPin           Con     C2      ;!Ring Indicator - goes low on Ring
OffHook         Con     C0      ;Off/On Hook. Take high to receive
TimeOutVal      Con     250     ;Time to wait for valid password in ;seconds/10
CommandLen      Con     4       ;command length
RingVal         Con     3       ;How many rings for answer >>NTE 4!!<<
Bad             Con     0       ;for password status
Good            Con     1       ;for password status
BadData         Con     98      ;Bad address

;Morse Code Constants
;--------------------
;Encode Morse characters as di-bits in a Word:
;       - = 11
;       . = 10
;Space = 01
;End   = 00

;Numbers 0...9 & Some Specials
;-------------- ,      -    .      /    0    1    2    3    4
;5      6    7    8    9
Nums WordTable |
```

```
    $FAF0,$EAC0,$BBB0,$EB80,$FFC0,$BFC0,$AFC0,$ABC0,$AAC0,$AA80, |
    $EA80,$FA80,$FE80,$FF80

;Letters A...Z & Some Specials
;-------------- ?     @    A    B    C    D    E    F
  G    H    I    J    K    L    M    N    O
Alpha WordTable |
$AFA0,$BB80,$B000,$EA00,$EE00,$E800,$8000,$AE00,$F800,$AA00, |
$A000,$BF00,$EC00,$BA00,$F000,$E000,$FC00, |
$BE00,$FB00,$B800,$A800,$C000,$AC00,$AB00,$BC00,$EB00,$EF00, |
$FA00,$4000,$ABB0,$AB80
;P    Q    R    S    T    U    V    W    X    Y    Z
;[space \SK  ]SN


WPM           Con      30      ;speed in WPM
TonePin       Con      C1      ;Pin for beep/Morse output
Dash          Con      %11     ;DiBit for Dash
Dot           Con      %10     ;DiBit for Dot
WSpaceCh      Con      %01     ;DiBit for Word Space
EndCh         Con      %00     ;end of Morse dibits
```

The constraints of page length render some of the embedded comments in the reproduced code a bit cryptic. Let's look at what are likely the most confusing parts of the constant definitions, those related to the Morse code sending. Morse code is comprised not only of "dots" and "dashes" but, equally importantly, spaces. The duration of dots, dashes and spaces are defined with respect to the duration of a single dot. The relationship is:

Element	Relative Length
Dot	1
Dash	3
Inter-Element Space	1
Inter-letter Space	3
Inter-Word Space	7

Morse is a variable length code, where the most common letters are represented by the shortest combination of elements. E, for example, is the most common letter in the English alphabet and is assigned the shortest possible Morse character, a single dot, and can be transmitted in the interval of two dot lengths—one for the dot, one for the space following the dot. Q, in contrast, is an infrequent letter is English and is assigned a long symbol, dash-dash-dot-dash, and requires 14 dot lengths to send—three dashes at four dot lengths each (three for the dash plus one for the following space) and one dot of two dot lengths (one for the dot, one for the space).

Our initial idea might be to use binary to represent the Morse characters, a 0 for dots and a 1 for dashes. In this code, E is `%0` and Q is `%1101`. This is perfectly practical, as far as it goes, but how do we know when the elements comprising a character end? We must have an additional piece of information—the length of each dot/dash sequence. The longest Morse characters we'll use are the punctuation marks, with six elements, such as the comma: dash-dash-dot-dot-dash-dash. It takes three bits to store the numbers 0...7, so we could therefore represent a Morse character using nine bits, organized as [3 bit length information] [up to 6 bits of dot/dash sequences]. Although no letter or number in Morse has more than five elements, punctuation marks do. Hence we must either use a 16-bit word for storage of each character, or use a more complex storage scheme, such as byte packing, or bytes for alphanumeric characters and words for all others.

I decided to use a different approach, and encode the elements of each Morse character with 2 bits, or a "di-bit":

Element	Di-Bit	Decimal
Dash	11	3
Dot	10	2
Word Space	01	1
End of Character	00	0

We still must use 16 bits to hold each character, but di-bit encoding gives us a logical approach. Let's see how we encode E and Q, and the comma:

	Nibble 3		Nibble 2		Nibble 1		Nibble 0	
E	$8		$0		$0		$0	
	10		00		00		00	
	00		00		00		00	
Q	$F		B		$0		$0	
	11	11	10	11	00	00	00	00
,	$F		$A		$F		$0	
	11	11	10	10	11	11	00	00

Thus, the hex values for E, Q and comma are **$8000**, **$FB00** and **$FAF0**, respectively.

Finally, we must decide in which order to place the encoded words. We wish to be able to easily index into one or two word table(s) to retrieve the encoded character. Examining the ASCII collating sequence shows that although we could use one word table, it would be quite long. (The ASCII collating sequence can be found at MBasic User's Guide, Appendix C.) We'll therefore have two tables, one for numbers 0…9 and one for letters A…Z. In looking over the ASCII collating sequence, it makes sense to include the comma, period, dash and slant (fraction sign) in their order immediately preceding 0, and to precede "A" with "?" and "@," with the thought that the "@" sign may be used for a special character. Likewise, we tack on the characters "]", "\" and "]" after Z. We use these ASCII characters for special purpose "prosign" Morse characters. (Prosign characters are two letters sent as a single Morse character, that is, the inter-letter spacing is reduced to an inter-element space. The prosigns have special meanings.)

ASCII	Morse Character	Meaning
@	AR	End transmission
[	None	Word Space
\	SK	End of contact
]	SN	Understood

To maintain continuity in our analysis, let's look at the Morse routines.

```
;Following subroutines send Morse code.
;Replace with beeps if you don't want Morse outputs
;----------
SendMsg            ;Main subroutine- Send message held in Message()
;----------        ;Must use upper case. Use [ for space.
                   ;Message() ends with a 0
        ixx = 0
        While Message(ixx) > 0          ;are we at the end of message?

                ;check for numeric & special sequence, if so
                ;get the offset value
                If Message(ixx) >= "?" Then
                        x = Alpha(Message(ixx)-"?")
                EndIf

                ;Is character in range comma to 9?
                ;If so get its offset value
                If (Message(ixx) >= ",") AND (Message(ixx) <= "9") |
                Then
                        x = Nums(Message(ixx)-",")
                EndIf

                ;Get DiBit going into send loop
                DiBit = 2 * x.Bit15 + X.Bit14
                While DiBit > EndCh    ;are we at the end?
```

```
                                ;What is the DiBit? Use Subs so can't Branch
                                If DiBit = Dot  Then : GoSub DitOut   : EndIf
                                If DiBit = Dash Then : GoSub DahOut   : EndIf
                                If DiBit = WSpaceCh Then : GoSub WspaceOut |
                                : EndIf
                                ;Get the next DiBit
                                x = x << 2
                                DiBit = 2 * x.Bit15 + X.Bit14
                                ;If the next dibit is End, then
                                ;add a Character Space
                                If DiBit = EndCh Then: GoSub CSpaceOut: EndIf
                        WEND ;individual element of a character
                        ixx = ixx+1
                WEND ;Message(ixx) loop
        Return

        ;Subroutines below send dot/dash or space
        ;Standard Morse is dot/element space = 1
        ;--------
        DitOut
        ;--------
                Sound TonePin, [1000*ELtime\440]
                Pause ElTime
        Return

        ;Dash is 3xdot plus one space
        ;------
        DahOut
        ;------
                Sound TonePin, [3000*ElTIme\440]
                Pause ElTime
        Return


        ;Character space is 3x element time
        ;already have 1 space after element
        ;--------
        CSpaceOut
        ;--------
                Pause 2*ElTime
        Return

        ;Word space is 7x element time
        ;Already have 3 spaces automatically added
        ;by end of character
        ;------
        WSpaceOut
        ;------
                Pause 4*ElTime
        Return
```

The main subroutine is **SendMessage**. It assumes that the text to be sent has been loaded into the byte array **Message()** and that after the last character to be sent **Message()** contains a **0**, not the ASCII symbol for "0," but **$0**.

```
        SendMsg
                ixx = 0
                While Message(ixx) > 0          ;are we at the end of message?

                        If Message(ixx) >= "?" Then
                                x = Alpha(Message(ixx)-"?")
                        EndIf

                        If (Message(ixx) >= ",") AND (Message(ixx) <= "9") |
                        Then
                                x = Nums(Message(ixx)-",")
                        EndIf
```

Since we have put our encoded Morse word values into a word table, in ASCII collation order, to obtain the encoded value for any letter, we simply index into the word table using the letter to be sent, minus the ASCII value of the character at the base of the table. Since the start of the table is "?" for the alphabetic letters and "," for the numeric group, the offset is found by subtracting "?" or "," from the letter to be sent. We use the word variable **x** to hold the encoded Morse definition.

We then peel off each di-bit until we reach the end of character di-bit, **%00**.

```
              DiBit = 2 * x.Bit15 + X.Bit14
              While DiBit > EndCh
                      If DiBit = Dot   Then : GoSub DitOut   : EndIf
                      If DiBit = Dash Then : GoSub DahOut   : EndIf
                      If DiBit = WSpaceCh Then : GoSub WspaceOut |
                      : EndIf
                      x = x << 2
                      DiBit = 2 * x.Bit15 + X.Bit14
                      If DiBit = EndCh Then: GoSub CSpaceOut: EndIf
              WEND
              ixx = ixx+1
      WEND ;Message(ixx) loop
Return
```

(We defined constants **Dot** (**%10**), **Dash** (**%11**), **WspaceCh** (**%01**) and **EndCh** (**%00**) in the constants definition section.) Based upon the value of each di-bit, we go to an according subroutine; send a dot (**DitOut**), send a dash (**DahOut**) or send a word space (**WspaceOut**.)

When the **While/Wend** loop reads the end-of-character di-bit **EndCh**, the complete character has been sent and **SendMessage** checks if additional characters remain in **Message()**.

The remaining subroutines in our Morse section send dots, dashes, character space or word space.

```
;Subroutines below send dot/dash or space
;Standard Morse is dot/element space = 1
;--------
DitOut
;--------
        Sound TonePin, [1000*ELtime\440]
        Pause ElTime
Return

;Dash is 3xdot plus one space
;------
DahOut
;------
        Sound TonePin, [3000*ElTIme\440]
        Pause ElTime
Return

;Character space is 3x element time
;already have 1 space after element
;--------
CSpaceOut
;--------
        Pause 2*ElTime
Return

;Word space is 7x element time
;Already have 3 spaces automatically added
;by end of character
;------
WSpaceOut
;------
        Pause 4*ElTime
Return
```

All timing is tied to the basic unit, element time, `ElTime`, calculated from the desired words per minute speed. I've selected a tone pitch of 440 Hz, as it's my personal preference. Many people prefer their Morse to be received with a pitch of about 800 Hz. Since MBasic's `Sound` function requires its duration argument in microseconds, and `ElTime` is in milliseconds, we multiply by 1000 to convert ms to μs.

And, in case you don't have it, here's the Morse "cheat sheet."

0	_ _ _ _ _	M	_ _	
1	. _ _ _ _	N	_ .	
2	. . _ _ _	O	_ _ _	
3	. . . _ _	P	. _ _ .	
4	. . . . _	Q	_ _ . _	
5	. . . . .	R	. _ .	
6	_	S	. . .	
7	_ _ . . .	T	_	
8	_ _ _ . .	U	. . _	
9	_ _ _ _ .	V	. . . _	
A	. _	W	. _ _	
B	_ . . .	X	_ . . _	
C	_ . _ .	Y	_ . _ _	
D	_ . .	Z	_ _ . .	
E	.	?	. . _ _ . .	
F	. . _ .	"/"	_ . . _ .	
G	_ _ .	.	. _ . _ . _	
H	. . . .	,	_ _ . . _ _	
I	. .	=	_ . . . _	
J	. _ _ _	SK	. . . _ . _	
K	_ . _	AR	. _ . _ .	
L	. _ . .	SN	. . . _ .	

We now return to the variable declarations.

```
;Variables
;-----------
BCDOut          Var     Byte                    ;PortB value
FcnInput        Var     Long
DigitCount      Var     Byte                    ;how many digits received?
DigitVal        Var     BCDOut.Nib0             ;output nibble
InDigits        Var     Byte(CodeLen+2)         ;holds received bcd
i               Var     Byte                    ;counter for work
CodeTries       Var     Byte                    ;How many tries for password
RI              Var     PortC.Bit2              ;Ring Indicator
TickCtr         Var     Byte                    ;Timer tick counter (0.1sec/tick)
RingCount       Var     Byte                    ;How many rings received
PWOK            Var     Byte                    ;Password rcvd = 1 Not rcvd = 0

;Command is FcnAdr + AppnNo + AppnVal
FcnAdr          Var     Byte    ;2 digits
AppnNo          Var     Byte    ;1 digit
AppnVal         Var     Byte    ;1 digit

;Morse Code Variables
;--------------------
ELtime          Var     Word    ;Element length in milliseconds
ixx             Var     Byte    ;Loop
```

```
DiBit          Var     Byte    ;element variable
x              Var     Word    ;temp variable
Message        Var     Byte(32) ;message to be sent
```

Program 17-3 is far from a finished product. For example, it's important that a software glitch, perhaps caused by a momentary power interruption, not result in the CH1837A seizing the phone line and not releasing it. Such a feature might make use of the watchdog timer in addition to employing MBasic's **ONBOR** brownout reset command.

Ideas for Modifications to Programs and Circuits

- Although we've used the CH1837A to automatically answer a call, it can just as easily originate a call. Taking its OFFHK pin high will cause the serving central office to send dial tone and wait for a call. If you send touch-tone signals through the CH1837A's TX port, they will be treated just as manual dialing. How might you use automatic call origination? (Remember, MBasic has two functions to generate touch-tone signals, **DTMFOut** and **DTMFOut2**.)
    - o Develop a remote temperature alarm. Use a DS18B20 temperature sensor to sense temperature and automatically call a programmed number if the temperature drops below a defined limit. What practical considerations enter into such a program? How does the calling program know that dial tone has been placed on the line? How many times should it attempt to make a call?
    - o Make an automatic dialer to work in conjunction with a conventional telephone. How do you connect the telephone and CH1837A?
- The CH1837A may serve as the heart of a speakerphone. A small amplifier, such as the LM386 for received audio, a microphone and amplifier for transmitted audio and a PIC for dialing and control complete the design. How would you assemble a speakerphone? How would you prevent feedback between the speaker and microphone, since the hybrid cancellation is not perfect?
- Suppose you wish to send slow bit rate information using inexpensive Family Radio Service low power UHF radio transceivers. For example, you might wish to remotely telemeter water level at a creek. Is this an appropriate use of touch-tone signaling? How might you sense stream level and send it back?
- How would you add a LCD display to Program 17-3? What local control features would you add?

References

[17-1] Motorola, Inc., *MC145436A Low-Power Dual Tone Multiple Frequency Receiver* (1996).

[17-2] Mitel Networks Corp., *MT8870D/MT8870D-1 Integrated DTMF Receiver,* (undated).

[17-3] Cermetek Microelectronics, Inc., *CH1837A/7F/8A Data Access Arrangement Module V34.bis High Speed DAA Module* (2002).

[17-4] Hedley, David, *Connecting a PICmicro Microcontroller to a Standard Analog Telephone Line AN854*, Microchip Technologies, Inc., Document No. DS00854A (2002).

[17-5] Hulous, Joel & Morel, Patrice, *Application Note Phone Remote System, AN488/0695*, SGS-Thompson Microelectronics (1995).

[17-6] Deosthali, Amey, et al., A Low-Complexity ITU-Compliant Dual Tone Multiple Frequency Detector, IEEE Trans. On Signal Processing (Oct. 1999). (Implemented with a 16C711.)

[17-7] Philips Semiconductors, *Application Note Data access arrangement (DAA) AN812* (2003).

[17-8] Lancaster, Don, Tech Musings No. 127, (August 1998). Downloadable at http://tinaja.com/glib/muse127.pdf.

[17-9] Microchip Technologies, Inc., DTMF Decoding on PIC18, Document DTF18, presented at Microchip's 2003 Master's Conference; available at http://techtrain.microchip.com/masters2003/.

[17-10] Constantinescu Radu, *DTMF Remote control – A software DTMF decoder for PIC 16F87X*, including associated assembler code, available at http://www.geocities.com/constantinescuradu/content/dtmf.htm. Implements a software DTMF decoder, relay controller and LCD display in some 2200 lines of assembler code.

[17-11] AT&T Company, Long Lines Department, *Principles of Electricity applied to Telephone and Telegraph Work* (1953). Probably best known as "the Green Book," used as training text by the Bell System for many years. If you are at all interested in how the classical telephone system was built, search the used bookstores for a copy.

[17-12] AT&T Company, *Network Planning Division, Notes on the Network*, (1980). Originally published as *Notes on Distance Dialing*, and subsequently published under other names, such as *Bellcore Notes on the Network* and, most recently, *Telcordia Notes on the Network*. The current pricetag of $575 for a printed copy of *Telcordia Notes on the Network*, however, is difficult to justify unless you are involved in the telecom business.

CHAPTER 18

External Memory

There are times when the internal RAM and EEPROM memory in a PIC just isn't enough. You might wish to save sensor readings for later analysis, or perhaps your program displays long text strings that can't be squeezed into the available memory. In this chapter, we'll look at two types of external memory—slow serial EEPROM devices and fast parallel static RAM (SRAM) chips.

I²C-Bus Devices

Serial memory chip are available in several interface protocols. Since we've already studied 1-wire and SPI (3-wire) serial protocols in other contexts, to learn something new we'll limit our examination of external memory chips to those employing the "Inter IC" bus, or I²C, originally developed by Philips Semiconductors, but widely adopted by other manufactures. Although we'll study it in the context of external memory, many other chips, such as LCD controllers, analog-to-digital converters and frequency synthesizers are I²C bus controlled.

The I²C shares elements of both the 1-wire and SPI protocols, but, as might be expected, adds its own flavors. Since MBasic's **I2Cout** and **I2Cin** functions encapsulate the details of I²C communications, we'll satisfy ourselves with an overview of the protocol. I²C is a two-wire interface, with separate clock and data lines. Originally the data speed was set at a maximum of 100 Kbits/s, now called the "standard" mode, but newer devices are rated at 400 Kbit/s, or "fast" mode or up to 3.4 Mbit/s, the "high-speed" mode. **I2Cout** and **I2Cin** support only standard mode data transfers, but all higher speed devices remain compatible with standard mode data speeds. At any instant, one device on the I²C bus is the master and it communicates with a slave device. The master device generates the clock, and initiates and terminates data transfer. During data transfer, the master and slave devices alternate transmitting and receiving data.

As with the SPI protocol, several I²C devices may be paralleled from a single pair of clock and data lines, but individual chip select lines are not required. Rather, individual I²C devices are selected by a combination of hardware address and command byte family ID code. We'll see how this works when we parallel two 24LC256 memory chips.

The I²C standards contemplate multiple microcontrollers on one bus, including possible inter-controller communications, i.e., microcontrollers may be either slaves or masters and may alternate that role. And, a bus may have multiple masters. Master microcontrollers communicate only with slave microcontrollers; master-to-master communications is not allowed. MBasic's **I2Cout** and **I2Cin** procedures support only master operation and hence cannot be used for inter-PIC communications.

Finally, we'll see that although the I²C protocol is well standardized, memory chip manufactures have not always approached chip operation consistently.

24LC16B 16Kbit EEPROM

Let's look a simple application, reading and writing to a 16 kbit EEPROM chip, a Microchip 24LC16B. By the way, the standard identification system for generic serial EEPROM chips starts with the number 24, followed by an identifying letter or two indicating its manufacturing process and possibily its voltage rating, ending with digits indicating the number of kilobits of data capacity. Hence a 24LC16B is a low (L) voltage capable (2.5V to 5.5V) memory chip manufactured with CMOS (C) technology, and with a data capacity of 16 kbits. (Microchip's 24AA16 is an extended voltage range chip, operating down to 1.8V.) Since we will read and write in 8-bit bytes, the 24LC16B's storage capacity is perhaps more usefully stated as 2 Kbytes.

Figure 18-1 shows how we connect the 24LC16B to a PIC. Don't forget the pull up resistor! The three "chip address" pins, A0, A1 and A2 are not used on the 24LC16B and hence are left unconnected. (We'll see these in use in connection with Program 18-2.) The generic connections for 24-family devices are:

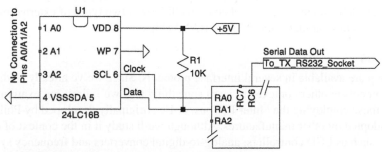

Figure 18-1: Connecting a 24LC16B I2C EEPROM chip.

Pin Name	8-pin DIP Pin Number	Function and Comments
A0	1	Chip address bit A0 in most 24-family devices; unused in 24LC16B
A1	2	Chip address bit A1 in most 24-family devices; unused in 24LC16B
A2	3	Chip address bit A2 in most 24-family devices; unused in 24LC16B
V_{ss}	4	Ground (Identified as Gnd by some manufacturers)
SDA	5	Serial Data. This is an open drain pin and must have a pull-up resistor. 10K is a suitable value for Standard Speed data transfer.
SCL	6	Serial Clock
WP	7	Write Protect (active high). When high, the device is read-only; for read and write must be low.
V_{DD}	8	Plus power supply; +5V in our breadboard designs. (Identified as V_{cc} by some manufacturers.

We've arbitrarily decided to use the PIC's **A0** pin as the data (SDA) connection, and its **A1** pin as the clock (SCL) connection.

Program 18-1 shows how we write to and read from the 24LC16B.

Program 18-1

```
;Program 18-1 to read/write
;16Kbit 24LC16B Serial EEPROM
;I2C communications
;Demonstrate random read/write

;Constants
;---------
Clk             Con     A0      ;clock is on A0
```

```
Dta             Con     A1         ;data on A1
                                   ;have to add the page select to Ctrl
                                   ;8-bit address word, so ends in 0
RCtrl   Con             %10100000        ;control word starting point
Display Con             1 ;set to 0 to block writing to serial port


;Variables
;----------
PSel            Var     Byte    ;Page selector
CtrlB           Var     Byte    ;control byte
i               Var             Word            ;index
j               Var             Byte            ;byte value to/from 25LC16B
InStr           Var             Byte(12)        ;text string

;Initialization
;--------------
;Communicate with outside world via serial
EnableHSerial
SetHSerial H19200

Main
        ;Demonstrate random access byte at a time
        ;16LC16B is organized as 8 pages of 256 bytes.
        ;The 8 pages are addressed in the control byte
        For i = 0 to 2047
                ;CtrlByte page address is Bits Bit3/Bit2/Bit1
                CtrlB = (i.HighByte * 2) + RCtrl
                ;To write something distinctive to RAM
                ;we write the index plus the offset
                j = i.LowByte+i.HighByte
                ;Write to addressed memory
                I2COut Dta,Clk,ErrorMsg,CtrlB,i.LowByte,[j]
                Pause 4

                If Display > 0 Then
                        HSerOut ["Write  i: ",Dec i,  |
                        " Data: ",Dec j,13]
                        Pause 25        ;needed if write output
                EndIf
        Next

        j= 0

        ;Now read it back
        For i = 0 to 2047
                ;Set up the control byte in the same way
                CtrlB = (i.HighByte * 2) + RCtrl
                I2CIn Dta,Clk,ErrorMsg,CtrlB,i.LowByte,[j]
                If Display > 0 Then
                        HSerOut ["Read  i: ",Dec i," Data: ",Dec j,13]
                        Pause 25                ;needed if write output
                EndIf
        Next

        ;Now, a totally random write and read, to prove it
        ;to the doubtful
        i = 1111
        CtrlB = (i.HighByte * 2) + RCtrl
        j = 77
        HSerOut ["Custom Write  i: ",Dec i," Data: ",Dec j,13]
        Pause 50
        I2COut Dta,Clk,ErrorMsg,CtrlB,i.LowByte,[j]
        j = 0
        ;If pause < 4, I2COut hangs up
        Pause 4
        I2CIn Dta,Clk,ErrorMsg,CtrlB,i.LowByte,[j]
        HSerOut ["Custom Read  i: ",Dec i," Data: ",Dec j,13]
        Pause 50
```

```
TextWrite:

;How, let's try text read/write
i = 0
InStr = "xxxxxxxxxxxx"
CtrlB = (i.HighByte * 2) + RCtrl
;Write to addressed memory
I2COut Dta,Clk,ErrorMsg,CtrlB,i.LowByte,["HelloWorld!"]
HSerOut ["Write  i: ",Dec i,"  Data: ",Str InStr\12,13]
Pause 50

i = 0
;Set up the control byte in the same way
CtrlB = (i.HighByte * 2) + RCtrl
I2CIn Dta,Clk,ErrorMsg,CtrlB,i.LowByte,[Str InStr\12]
HSerOut ["Read  i: ",Dec i," Data: ",Str InStr\12,13]
;delay needed to let HSerOut flush before End!
Pause 50
End
;GoTo Main

ErrorMsg
        HSerOut ["Error!"]
        Pause 50
End
```

Let's first look at MBasic's I²C support functions, **I2Cout** and **I2Cin** as we use it in Program 18-1.

I2COut Dta,Clk,ErrorMsg,CtrlB,i.LowByte,[j]

I2Cout is invoked with six arguments:

Dta—A constant or variable defining the pin used for the data line. Since we wish to use pin **A1** as the data line, we earlier defined a constant **Dta** as **A1**.

Clk—A constant or variable defining the pin used for the clock line. Since we wish to use pin **A0** as the clock line, we earlier defined a constant **Clk** as **A0**.

ErrorMsg—An optional program label to which execution will jump in the event of an error communicating with the slave device. Program 18-1 has a brief error routine that writes an alert message to the serial port.

CtrlB—The control byte that selects the chip and certain options, as we will see below.

i—An optional memory address to which to write. If the memory address is omitted, the external memory chip uses its internal address counter to write to the next address. The internal address counter increments after each byte is written. In most cases, however, we will wish to write and read at specific addresses and hence will use an explicit memory address. As we'll see later, **i** is a word variable and 24LC16B's requires a byte-length memory address, so we use **i**'s low order byte.

[Argument]—The specific data to be written to the EEPROM appears in the square brackets [ ], where MBasic's standard modifier/expression rules apply. My experiments, however, show that the data to be written must be in the form of individual bytes, such as **[Byte1,Byte2,Byte3...]** or an array of bytes such as **["Hello World!"]**. I was unable to successfully write or read word or long variables without breaking them up—that is, **[WordVar]** fails, but **[WordVar.Byte1, WordVar.Byte0]** works.

Let's look at the control byte. It's comprised of three elements:

Device Family Code				Chip Addressing Information			8/16 Bit
Bit 7	Bit 6	Bit 5	Bit 4	Bit 3	Bit 2	Bit 1	Bit 0

Device Family Code—The upper nibble of the control byte defines the family type. The EEPROM chips we are interested in are "Group A" devices, so their family code is **$A**, or **%1010**. The 16 possible family codes are as follows:

Group	Bit 7	Bit 6	Bit 5	Bit 4	Device Group Types
0	0	0	0	0	*General call*
1	0	0	0	1	*Digital receivers & demodulators*
2	0	0	1	0	*FM and TV related*
3	0	0	1	1	*CD-ROM and audio related*
4	0	1	0	0	*MPEG, speech, DAC, DTMF, pagers*
5	0	1	0	1	*Unidentified by Philips*
6	0	1	1	0	*Computer monitor devices*
7	0	1	1	1	*LCD controllers, video controllers*
8	1	0	0	0	*Audio and TV processors and decoders*
9	1	0	0	1	*TV processors*
A	1	0	1	0	*EEPROM*
B	1	0	1	1	*Audio and TV processors and decoders*
C	1	1	0	0	*Radio and TV RF devices, frequency synthesizers*
D	1	1	0	1	*Miscellaneous, including clock/calendar, mobile telephone, satellite.*
E	1	1	1	0	*Special purpose*
F	1	1	1	1	*Reserved*

Some newer I²C devices use a 10-bit control sequence, but `I2Cin` and `I2Cout` do not support this mode.

8/16-bit addressing—We'll take the remaining two elements out of order. Bit 0 determines whether the EEPROM address information sent immediately following the control byte is a single 8-bit byte (256 possible addresses) or a 16-bit word (65536 possible addresses). If Bit 0 is 0, the address is sent as an 8-bit byte; if Bit 0 is 1, the address is sent as a 16-bit word.

If you look at the control byte definition in a data sheet for an I²C device, you will find that Bit 0 is defined as the "read/write" or R/W bit, and that it determines the direction of data transfer; if 0, the master sends and if 1, the slave sends. *However, MBasic, for reasons of compatibility with other versions of PIC BASIC, automatically inserts the correct R/W bit and instead uses Bit 0 of the control byte for address length information.* It's easy to get confused, so this is another case where attention to detail is critical.

To determine whether the 24LC16B requires 8-bit or 16-bit addressing, we consult its data sheet. We see that the memory is organized as "eight blocks of 256 × 8-bit memory," and that the byte write and byte read instructions show an 8-bit address. Hence, we set Bit 0 as 0 in preparing the control byte.

Addressing Information—Perhaps the most confusing aspects of using serial EEPROM devices is understanding how the three chip address bits in the control byte are used. We'll see each of the four EEPROMs models examined in this chapter has a different chip address bit configuration.

Device Type	Bit 3	Bit 2	Bit 1	Comments
24LC16B	B2	B1	B0	*B2, B1 and B0 are block selection bits; they select which of eight 256-byte memory blocks is active. [Requires 8-bit memory addressing within each block.]*
AT24C256	0	A1	A0	*A1 and A0 select the chip address match of Pins A0 and A1. [Memory is organized as one 32K-byte block; no block selection bit. Requires 16-bit addressing.]*
AT24C512	0	A1	A0	*A1 and A0 select the chip address match of Pins A0 and A1. [Memory is organized as one 64K-byte block; no block selection bit. Requires 16-bit addressing.]*

(Continued)

Device Type	Bit 3	Bit 2	Bit 1	Comments
24FC515	B0	A1	A0	B0 selects which of two 32K-byte memory blocks is active; A1 and A0 select the chip address match of Pins A0 and A1. [Requires 16-bit memory addressing within each block.]

As we learned earlier, the 24LC16B's 2048 bytes of memory is organized as eight 256-byte blocks. The three chip address bits define which block is active. To write to block 2 (**%010**), for example, the control byte will be **%10100100**. If we are to write the value 99 to memory cell no. 34 (starting from 0) in block 2, our command will be **I2Cout Dta, Clk, ErrorMsg, %10100100, 34, [99]**.

We rather would have a flat, or linear, address space, that we might address from 0…2047, instead of worrying about blocks and offset in each block. Using **i** as the word-length address from 0…2047, we can construct the control byte, **CtrlB**, as follows:

```
CtrlB = (i.HighByte * 2) + RCtrl
```

Since **i** is a word variable, **i.LowByte** is simply the 0…255 individual block offset. The high byte, **i.HighByte**, runs from 0…7 as **i** goes from 0…2047. We therefore insert the low three bits of **i.HighByte** into P2/P1/P0 of **CtrlB**, by multiplying **i.HighByte** by 2 (functionally equivalent to shifting the bits one place to the left) and adding the result to **CtrlB**. We have defined the constant **RCtrl** as **%10100000** in Program 18-1, adding it to (**i.HighByte * 2**) yields the complete control word.

With this understanding, let's examine the write loop in Program 18-1:

```
For i = 0 to 2047
        CtrlB = (i.HighByte * 2) + RCtrl
        j = i.LowByte+i.HighByte
        I2COut Dta,Clk,ErrorMsg,CtrlB,i.LowByte,[j]
        Pause 4

        If Display > 0 Then
                HSerOut ["Write  i: ",Dec i,  |
                " Data: ",Dec j,13]
                Pause 25        ;needed if write output
        EndIf
Next
```

Our demonstration program writes an arbitrary byte value to each of the 2048 memory locations in the 24LC16B. Rather than write the same value, we'll write unique value, **j**, where **j = i.LowByte+i.HighByte** to each memory cell. Each memory cell thus contains the sum of its 0…255 index plus the block number. Since memory cells are byte length, **j** rolls over to 0 when **i.LowByte+i.HighByte** exceeds 255. If, for example, **i.LowByte+i.HighByte** is 257, **j** holds the value 2. With this understanding, the write command should be clear:

```
I2COut Dta,Clk,ErrorMsg,CtrlB,i.LowByte,[j]
Pause 4
```

EEPROMs require time to write to memory, with the devices we use in this chapter having 5 ms maximum write cycle duration. We accordingly pause for 4 ms, with the remaining cycle time being the overhead associated with the remainder of the code in the **For i** write loop. It's important not to forget the write cycle time when writing to EEPROM.

The remainder of the write loop writes, if desired, **i** and **j** to the serial port, where the progress may be observed.

After writing the data, we next read it back.

```
;Now read it back
For i = 0 to 2047
        ;Set up the control byte in the same way
        CtrlB = (i.HighByte * 2) + RCtrl
```

```
                    I2CIn Dta,Clk,ErrorMsg,CtrlB,i.LowByte,[j]
                    If Display > 0 Then
                            HSerOut ["Read i: ",Dec i,"  Data: ",Dec j,13]
                            Pause 25        ;needed if write output
            Next    EndIf
```

The code for reading the 24LC16B tracks the write code, except that we use MBasic's I²C read procedure, **I2Cin**:

```
        I2CIn Dta,Clk,ErrorMsg,CtrlB,i.LowByte,[j]
```

Since **I2Cin** is an exact analog of **I2Cout**, with the same arguments we will not further discuss it, except to add a reminder that Bit 0 in **CtrlB** is the 8/16-bit switch, and is not a read/write selector bit. MBasic automatically adds the appropriate read/write selector bit. Note also that no delay is required after reading the EEPROM. After the data is read we optionally write it to the serial port.

We've so far demonstrate byte-at-a-time reading and writing, but **I2Cout** and **I2Cin** support multiple byte writes and reads. Let's see how this works with a byte array string.

```
        ;How, let's try text read/write
        i = 0
        InStr = "xxxxxxxxxxx"
        CtrlB = (i.HighByte * 2) + RCtrl
        ;Write to addressed memory
        I2COut Dta,Clk,ErrorMsg,CtrlB,i.LowByte,["Hello World!"]
        HSerOut ["Write  i: ",Dec i,"  Data: ",Str InStr\12,13]
        Pause 50

        i = 0
        ;Set up the control byte in the same way
        CtrlB = (i.HighByte * 2) + RCtrl
        I2CIn Dta,Clk,ErrorMsg,CtrlB,i.LowByte,[Str InStr\12]
        HSerOut ["Read  i: ",Dec i," Data: ",Str InStr\12,13]
        ;delay needed to let HSerOut flush before End!
        Pause 50
```

We've arbitrarily written the text string at memory address 0, but almost any address in the range 0…2047 could have been used. Of course, we can't start writing 12 bytes at 2047 and expect them to be saved. More subtly, however, when writing in this fashion we cannot span a block boundary. Let's look at an example. Suppose we start writing at **i=253**. This writes the first three characters of **"Hello World!"** at block 0, locations 253, 254 and 255, but the remaining characters will <u>not</u> be written at Block 1, locations 0…8. Rather, since the block select bits do not change from 000 during the multibyte **I2Cout** procedure, the remaining nine characters will be saved at block <u>0</u>, locations 0…8. Likewise, we cannot read across the 24LC16B's block boundaries in one **I2Cin** statement. If we are concerned with spanning block boundaries, we must revert to a system such as that seen earlier in the program, writing and reading one byte at a time, with each byte having its own **I2Cout** statement and its own control byte.

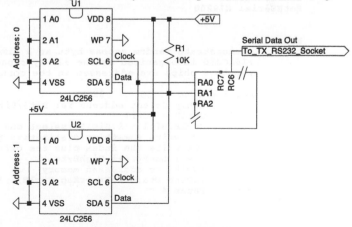

Multiple 24LC256 256 Kbit EEPROMs

We'll now turn our attention to paralleling multiple serial memory chips. Figure 18-2

Figure 18-2: Multiple 24LC256 Serial EEPROMs.

shows two parallel 24LC256 256Kbit serial memory devices. Figure 18-2 follows Figure 18-1, except for the three chip address pins, A0, A1 and A2. The 24LC256 data sheet shows that A0, A1 and A2 are all active address pins.

By selectively setting A0, A1 and A2 at either ground or V_{DD}, we assign each 24LC256 a unique chip address. Since we have three chip address pins, we have 2^3, or eight, possible unique addresses, 0...7. We've assigned U1's address as %000 by grounding A2, A1 and A0. By setting U2's A0 pin to V_{DD}, we've set its address at %001, or 1. One last point; don't forget the 10K pull-up resistor on the data line.

Program 18-2 exercises the pair of 24LC256 EEPROMs just as we did the 24LC16B. However, since the 24LC256 is set up as a single 32 Kbyte array, with 16-bit addressing, we must make some adjustments to the code from Program 18-1.

Program 18-2

```
;Program 18-2 to read/write
;selectively to two 256Kbit 24LC256 Serial EEPROM
;I2C communications ;Demonstrate random read/write

;Constants
;---------
Clk           Con     A0        ;clock is on A0
Dta           Con     A1        ;data on A1
                                ;have to add the page select to Ctrl
                                ;16-bit address word, so ends in 1
RCtrl         Con     %10100001 ;control word starting point
Display       Con     1         ;set to 0 to block writing to serial port
Adr0          Con     0         ;Address of 24LC256 No. 1 A0=A1=A2=0
Adr1          Con     1         ;Address of 24LC256 No. 2 A0=1 A1=A2=0


;Variables
;---------
CtrlB         Var     Byte      ;control byte
i             Var     Word      ;index
j             Var     Byte      ;byte value to/from 25LC256
InStr         Var     Byte(12)  ;text string


;Initialization
;--------------
;Communicate with outside world via serial
EnableHSerial
SetHSerial H19200


Main
      ;Demonstrate random access byte at a time
      ;16LC256 is organized as one 32K byte page.
      ;Up to 8 chips are addressed in the control byte

      For i = 0 to 65535     ;$FFFE
            ;Chip Select address is: Bit3/Bit2/Bit1
            CtrlB = RCtrl
            CtrlB.Bit1 = i.Bit15 ;select one of two 32K chips
            ;To write something distinctive to RAM
            ;we write the index plus the offset
            j = i.LowByte+i.HighByte
            ;Write to addressed memory
            I2COut Dta,Clk,ErrorMsg,CtrlB,i,[j]
            Pause 4
```

```
                If Display > 0 Then
                        HSerOut ["Write  i: ",Dec i," Data: ",Dec j,13]
                        Pause 25          ;needed if write output
                EndIf
        Next

        j= 0
        Pause 50
        HSerOut ["End of Write",13]
        Pause 50

        ;Now read it back
        For i = 0 to 65535
                ;Set up the control byte in the same way
                ;Chip Select address is: Bit3/Bit2/Bit1
                CtrlB = RCtrl
                CtrlB.Bit1 = i.Bit15 ;select one of two 32K chips
                I2CIn Dta,Clk,ErrorMsg,CtrlB,i,[j]
                If Display > 0 Then
                        HSerOut ["Read  i: ",Dec i,"  Data: ",Dec j,13]
                        Pause 25          ;needed if write output
                EndIf
        Next

        ;Prove access is truely random, read address 12345.
        ;Should have (12345-12288) = 57 plus offset.
        ;12288-12543 has offset 48, so we expect to read 105.
        i = 12345
        CtrlB = RCtrl
        CtrlB.Bit1 = i.Bit15 ;select one of two 32K chips
        I2CIn Dta,Clk,ErrorMsg,CtrlB,i,[j]
        HSerOut [13,"Custom Read  i: ",Dec i,"  Data: ",Dec j,13]
        Pause 50

        ;How, let's try text read/write
        i = 45000
        InStr = "xxxxxxxxxxxx"
        CtrlB = RCtrl
        CtrlB.Bit1 = i.Bit15 ;select one of two 32K chips
        ;Write to addressed memory
        I2COut Dta,Clk,ErrorMsg,CtrlB,i,["Hello World!"]
        HSerOut ["Write  i: ",Dec i,"  Data: ",Str InStr\12,13]
        Pause 50

        i = 45000
        ;Set up the control byte in the same way
        CtrlB = RCtrl
        CtrlB.Bit1 = i.Bit15 ;select one of two 32K chips
        I2CIn Dta,Clk,ErrorMsg,CtrlB,i,[Str InStr\12]
        HSerOut ["Read  i: ",Dec i," Data: ",Str InStr\12,13]
        ;delay needed to let HSerOut flush before End!
        Pause 50
End

ErrorMsg
        HSerOut ["Error!"]
        Pause 50
End
```

Since Program 18-2 largely duplicates Program 18-1 we'll concentrate on the differences. Although we have two separate 32 Kbyte memories, we'll address them as if we had one 64 Kbyte memory.

We have wired U1 and U2 with chip addresses 0 and 1, respectively. Further, the 24LC256's data sheet states that the memory is organized as a single $32K \times 8$-bit block and that consequently 15 memory address bits are required. (We actually write 16 bits, but the 24LC256 ignores the leading bit.) Based on this information, we can define the control bytes to address U1 and U2.

Fig 18-2 Chip	Device Family Code				Chip Addressing Information			8/16 Bit
	Bit 7	Bit 6	Bit 5	Bit 4	Bit 3 (A2)	Bit 2 (A1)	Bit 1 (A0)	Bit 0
U1	1	0	1	0	0	0	0	1
U2	1	0	1	0	0	0	1	1

As in Program 18-1, we would like a seamless linear address space 0...65535 so we don't have to worry about individual chip addressing. Let's see how we accomplish this.

```
For i = 0 to 65535      ;$FFFF
        CtrlB = RCtrl
        CtrlB.Bit1 = i.Bit15 ;select one of two 32K chips
        j = i.LowByte+i.HighByte
        I2COut Dta,Clk,ErrorMsg,CtrlB,i,[j]
        Pause 4
Next
```

For clarity, the above code fragment is stripped of comments and serial write commands. We start by noting that earlier in the program we defined the constant **RCtrl** to hold the baseline control byte. (Since we must use 16-byte addressing, **RCtrl.Bit0** is 1, unlike the case in Program 18-1.)

```
RCtrl        Con        %10100001
```

We then set the byte variable **CtrlB** to **RCtrl**.

```
CtrlB = RCtrl
```

Now we insert the chip address bits into **CtrlB**, as necessary.

```
CtrlB.Bit1 = i.Bit15
```

By defining the two chip addresses as 0 and 1, we greatly simplify setting the control byte's address bits. When **i <= 32767**, **i.Bit0** is 0; when **i > 32767**, **i.Bit0** is 1. Hence, to automatically switch between U1 and U2 whenever **i** crosses over the 32K boundary all we need to is to copy **i.Bit0** to the control byte's A0 position, **CtrlB.Bit1**. If we had assigned U1 and U2 different addresses, such as 3 and 5, respectively, we would need a more complicated assignment scheme, likely involving an **If i.Bit15 = 0 Then ...** construction. It's more elegant, not to mention yielding faster executing code, to wire the chip addresses as 0 and 1 and make a simple bit assignment.

If we use four 24LC256 memory chips, for example, and wire their addresses as 0, 1, 2 and 3, this simple bit assignment is easily extended. In this case, of course, the index **i** must be a type **long** variable, and it's necessary to assign two bits (**i.Bit15** and **i.Bit16**) to **CtrlB**, but the concept is identical.

```
j = i.LowByte+i.HighByte
```

As for Program 18-1, we construct a unique byte **j** to be written to memory by summing **i.LowByte** and **i.HighByte**.

```
I2COut Dta,Clk,ErrorMsg,CtrlB,i,[j]
```

Finally, we write **j** to EEPROM, using the **I2Cout** procedure. Note, however, we use all 16 bits of **i** in the **I2Cout** procedure, and we do not force **i.Bit15** to 0 to accommodate the each 24LC256's 32K address space limit. We can get away with this time saving measure because the 24LC256's data sheet defines this as a "don't care" bit, ignored when the chip internally decodes the received 16 address bits.

The remainder of Program 18-2 is identical with Program 18-1, save for those changes relating to the addressing. Before we look at the next memory chip variation, however, we again remind ourselves that we cannot continuously read or write across chip boundaries with a multiple-byte **I2Cin** or **I2Cout** statement.

One final note applying to MBasic versions earlier than 5.3.0.0; **I2Cin** and **I2Cout** processes freeze when addressing memory 65535. This problem is fixed in version 5.3.0.0 onward.

24LC515

Let's look at the 24LC515 EE-PROM next. It is a 64K × 8-bit device, organized as two banks of 32K × 8-bit memory. As usual, we start by examining the data sheet, and find the usual three chip address pins, A0, A1 and A2 have a peculiarity; A0 and A1 are normal address pins; A2 is a "nonconfigurable Chip Select" which must be tied to V_{DD} for the 24LC515 to function. Accordingly, as shown at

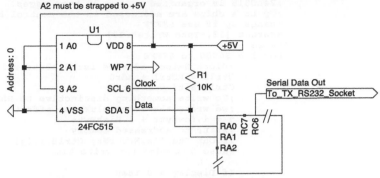

Figure 18-3: Connecting a 24LC515 EEPROM Device.

A2 is strapped to V_{DD}, while A0 and A1 are strapped to ground, yielding a chip address of 0.

Finally, we learn from the data sheet how to select between the two 32K × 8-bit banks in the 24LC515.

Device Family Code				Chip Addressing Information			8/16 Bit
Bit 7	*Bit 6*	*Bit 5*	*Bit 4*	*Bit 3*	*Bit 2*	*Bit 1*	*Bit 0*
1	0	1	0	B1	A1	A0	1

Reminiscent of the 24LC16B, the 24LC515 uses bit 3 of the control byte to select between the two 32K banks. Our Program 18-2B to exercise the 24LC515 will accordingly resemble Program 18-1's bank selection.

Program 18-2A

```
;Program 18-2 to read/write
;selectively to 512Kbit 24FC515 Serial EEPROM
;I2C communications ;Demonstrate random read/write

;Constants
;---------
Clk             Con     A0      ;clock is on A0
Dta             Con     A1      ;data on A1
                ;have to add the page select to Ctrl
                ;16-bit address word, so ends in 1
RCtrl           Con     %10100001       ;control word starting point
Display         Con     0 ;set to 0 to block writing values to serial port
Adr0            Con     0 ;Address of 24FC515 A0=A1=0
                ;IMPORTANT-Pin 3 MUST = +5V 24FC515 differs from
                ;24FC512 in this regard!
                ;Also 24FC515 has a bank selection bit
                ;that must be set to jump between 32K banks


;Variables
;----------
CtrlB           Var     Byte    ;control byte
i               Var     Word    ;index
j               Var     Byte    ;byte value to/from 25FC512
ShouldBe        Var     Byte    ;calculated read value


;Initialization
;--------------
;Communicate with outside world via serial
EnableHSerial
SetHSerial H19200


        ;Demonstrate random access byte at a time
```

```
                        ;24LC515 is organized as two 32 byte pages.
                        ;Up to 4 chips are addressed in the control byte
                        ;hangs up if use $FFFF
                        HSerOut [13,"Into Write",13]
                        Pause 50
                        For i = $0000 to $FFFE
                                ;Block Select address is: Bit3 to select 32K blocks
                                CtrlB = RCtrl + Adr0
                                CtrlB.Bit3 = i.Bit15
                                ;To write something distinctive to RAM
                                ;we write the index plus the offset
                                j = i.LowByte + i.HighByte
                                ;Write to addressed memory
                                I2COut Dta,Clk,ErrorMsg,CtrlB,i,[j]
                                Pause 5 ;needed for write time

                                If Display > 0 Then
                                        HSerOut ["Write i: ",Dec i," Data: ",Dec j,13]
                                        Pause 25      ;needed if write output
                                EndIf

                                If (i//256) = 0 Then
                                        HSerOut ["Write  i: ",Dec i," Data: ",Dec j,13]
                                        Pause 25      ;needed if write output
                                EndIf
                        Next


                        j= 0
                        Pause 50
                        HSerOut ["End of Write",13]
                        Pause 50

                        ;Now read it back
                        For i = $0000 to $FFFE
                                ;Set up the control byte in the same way
                                ;Chip Select address is: Bit2/Bit1
                                CtrlB = RCtrl + Adr0
                                ;Bit 3 is a bank selection switch
                                CtrlB.Bit3 = i.Bit15
                                ShouldBe = i.LowByte + i.HighByte
                                I2CIn Dta,Clk,ErrorMsg,CtrlB,i,[j]

                                If Display > 0 Then
                                        HSerOut ["Read  i: ",Dec i,"  Data: ",Dec j,13]
                                        Pause 25       ;needed if write output
                                EndIf

                                If (i//256) = 0 Then
                                        HSerOut ["Read  i: ",Dec i,"  Data: ",Dec j,13]
                                        Pause 25        ;needed if write output
                                EndIf

                                If j <> ShouldBe Then
                                        HSerOut ["Read  i: ",Dec i,"  Data: ",Dec j]
                                        HSerOut [" <-- Error!",13]
                                        Pause 25
                                EndIf
                        Next
                        HSerOut ["End of Read",13]
                        Pause 50

        End


ErrorMsg
        HSerOut ["Error!"]
        Pause 50
End
```

Let's look at the write code, again stripped of comments and serial output statements.

```
For i = $0000 to $FFFF
        CtrlB = RCtrl + Adr0
        CtrlB.Bit3 = i.Bit15
        j = i.LowByte + i.HighByte
        I2COut Dta,Clk,ErrorMsg,CtrlB,i,[j]
        Pause 5 ;needed for write time
Next
```

Instead of using decimal notation for the **i** loop limits, Program 18-2B uses hexadecimal. Hexadecimal notation is clearer, once you are familiar with it, as the 64K address boundary is **$FFFF**.

```
CtrlB = RCtrl + Adr0
CtrlB.Bit3 = i.Bit15
```

We construct the control byte and address bank selection just as we did in Program 18-1, except that here we need deal only with one bank selection bit. Since we have hard wired the chip select address to **%00**, and we have only one 24LC515 device, the chip selection bits A0 and A1 are set to 0.

```
j = i.LowByte + i.HighByte
I2COut Dta,Clk,ErrorMsg,CtrlB,i,[j]
Pause 5 ;needed for write time
```

We compute a byte to write, **j**, exactly as before, and the **I2Cout** statement is identical. Again, the most significant bit of **i** is ignored by the 24LC515 and we need not force it to zero.

The read loop is similar, except we pre-compute the value we expect to read from the memory and send an error message over the serial output should the expected value not match what is read from the EEPROM.

AT24C512

As we should always do, we first check Atmel's data sheet for the AT24C512. It is a 64K × 8-bit device, but unlike the 24FC515, it is organized as one contiguous address space, without pages. (Don't get confused when the data sheet says internally the AT24C512 is organized as 512 pages of 128 bytes each. From the prospective of reading and writing to the device from MBasic, it appears to have a

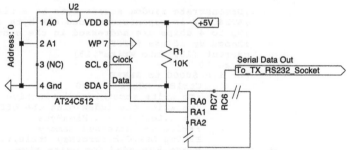

Figure 18-4: Connecting an AT24C512 EEPROM Device.

linear 64Kbyte address space and we need not concern ourselves with the specific chip construction.) Chip address pins and bits A0 and A1 are active; what is normally chip address pin A2 is unused. Hence, if we needed extra memory, we could parallel up to four AT24C512 devices for a total of 256K × 8-bit capacity. As reflected in Figure 18-4 we'll strap A0 and A1 to ground, setting the chip address at **%00**. Our control byte is thus:

Device Family Code				Chip Addressing Information			8/16 Bit
Bit 7	*Bit 6*	*Bit 5*	*Bit 4*	*Bit 3*	*Bit 2*	*Bit 1*	*Bit 0*
1	0	1	0	X	0	0	1

Bit 3 is "don't care," so we'll set it at zero.

Program 18-2B performs the same sample write, read and check for error functions for the AT24C512 that Program 18-2A does for the 24FC515 chip.

Program 18-2B

```
;Program 18-2B to read/write
;selectively to 512Kbit AT24C512 Serial EEPROM
;I2C communications ;Demonstrate random read/write

;Constants
;---------
Clk          Con     A0        ;clock is on A0
Dta          Con     A1        ;data on A1
                               ;have to add the page select to Ctrl
                               ;16-bit address word, so ends in 1
RCtrl        Con     %10100001 ;control word starting point
Display      Con     0 ;set to 0 to block writing values to serial port
Adr0         Con     0 ;Address of AT24C512 A0=A1=0
                        ;has full 64K write address space


;Variables
;---------
CtrlB        Var     Byte      ;control byte
i            Var     Word      ;index
j            Var     Byte      ;byte value to/from AT24C512
ShouldBe     Var     Byte      ;calculated read value


;Initialization
;--------------
;Communicate with outside world via serial
EnableHSerial
SetHSerial H19200

        ;Demonstrate random access byte at a time
        ;AT24C512 is organized as one 64K byte space
        ;Up to 4 chips are addressed in the control byte
        ;hangs up if use $FFFF
        HSerOut [13,"Into Write",13]
        Pause 50
        For i = $0000 to $FFFE
                CtrlB = RCtrl + Adr0
                ;To write something distinctive to RAM
                ;we write the index plus the offset
                j = i.LowByte + i.HighByte
                ;Write to addressed memory
                I2COut Dta,Clk,ErrorMsg,CtrlB,i,[j]
                Pause 5 ;needed for write time

                If Display > 0 Then
                        HSerOut ["Write i: ",Dec i,"  Data: ",Dec j,13]
                        Pause 25        ;needed if write output
                EndIf

                If (i//256) = 0 Then
                        HSerOut ["Write i: ",Dec i,"  Data: ",Dec j,13]
                        Pause 25        ;needed if write output
                EndIf
        Next

        j= 0
        Pause 50
        HSerOut ["End of Write",13]
        Pause 50

        ;Now read it back
        For i = $0000 to $FFFE
                ;Set up the control byte in the same way
                CtrlB = RCtrl + Adr0
                ShouldBe = i.LowByte + i.HighByte
                I2CIn Dta,Clk,ErrorMsg,CtrlB,i,[j]
```

```
                    If Display > 0 Then
                            HSerOut ["Read  i: ",Dec i,"  Data: ",Dec j,13]
                            Pause 25        ;needed if write output
                    EndIf

                    If (i//256) = 0 Then
                            HSerOut ["Read  i: ",Dec i,"  Data: ",Dec j,13]
                            Pause 25        ;needed if write output
                    EndIf

                    If j <> ShouldBe Then
                            HSerOut ["Read  i: ",Dec i,"  Data: ",Dec j]
                            HSerOut [" <-- Error!",13]
                            Pause 25
                    EndIf
            Next
            HSerOut ["End of Read",13]
            Pause 50

    End

    ErrorMsg
            HSerOut ["Error!"]
            Pause 50
    End
```

Again, let's look at a simplified version of the write loop.

```
            For i = $0000 to $FFFE
                    CtrlB = RCtrl + Adr0
                    j = i.LowByte + i.HighByte
                    ;Write to addressed memory
                    I2COut Dta,Clk,ErrorMsg,CtrlB,i,[j]
                    Pause 5 ;needed for write time
            Next
```

Since the AT24C512 is set up as a flat 64K memory, we have no page or chip boundary concerns; we can simply read and write with a full 16-bit address without worry. Of course, we could expand to a maximum of four AT24C512 chips on the I²C bus, in which case we would modify our program along the lines suggested in connection with Program 18-2.

Practical Use of External EEPROM

Our demonstration programs writing and reading arbitrary bytes in EEPROM are interesting, so far as they go, but are not all that useful by themselves. Why would we want a bunch of extra EEPROM memory hanging on our PIC?

Saving data for later retrieval—We'll see this in detail in Chapter 28, but perhaps the most common use of external EEPROM is to save captured data for later retrieval. For example, a roadside traffic counter might save hourly lane axle count totals, along with clock data. After capturing data for a week or two, the counter is collected and the stored data is transferred to a PC for analysis.

String data—It's possible to program the external EEPROM with a series of string messages and add the pre-programmed EEPROM to your circuit. (If you have the correct adapter, Basic Micro's PIC programmer also will program many I²C EEPROM devices. Otherwise, you could write a simple program using a PIC to do nothing but load your string data into an EEPROM.)

Memory swap—It's possible to indirectly extend PIC's internal RAM by swapping data between RAM and EEPROM. Suppose, for example, you program must generate values for two 100-element byte arrays A and B, and store the word-length products of the two arrays in a 10,000 element array X. Leaving aside the question of whether a PIC is the best vehicle for such mathematically intensive computations, we can use an external AT24C512 EEPROM to hold both the individual arrays, and their product.

Let's take a look at how we might accomplish this. In pseudo-code, our algorithm is:

```
Create array A(i) for i=0...99
Save array A in EEPROM
Create array B(j) for j=0...99
Save array B in EEPROM
For i = 0 to 99
        Retreive A(i) from EEPROM
        For j = 0 to 99
                Retreive B(j) from EEPROM
                X = A(i)*B(j)
                Store X in EEPROM
        Next j
Next i
```

Program 18-3 contains functioning code.

Program 18-3

```
;Program 18-3
;Shows how to use external EEPROM
;for a 100x100 element array product
;Use a single AT24C512 64K EEPROM
;Address hardwired to %00
;
;Constants
;----------
RCtrl           Con     %10100001       ;control word starting point
                                        ;in this case can use directly
iStart          Con     0               ;memory starting point for i Array
jStart          Con     100             ;memory starting point for j array
xStart          Con     200             ;memory starting point for X array
Clk             Con     A0              ;clock is on A0
Dta             Con     A1              ;data on A1


;Variables
;----------
i               Var     Byte            ;Array A index
j               Var     Byte            ;Array B index
k               Var     Word            ;Array X index
Prod            Var     Word            ;Product A(i)*B(j)
RTemp           Var     Word            ;Random variable
Ti              Var     Byte            ;temporary i array
Tj              Var     Byte            ;temporary j array
ShouldBe        Var     Word            ;A*B calculated


;Initialization
;--------------
EnableHSerial
SetHSerial H19200

;Main
;-------
RTemp = 77

;Create array A and write to EEPROM
;----------------------------------
HSerOut [13,13,"*** Write ***",13]
For i = 0 to 99
        ;fill array A with random numbers
        RTemp = Random RTemp
        Ti = RTemp.LowByte
        ;write A array to EEPROM
        I2COut Dta,Clk,ErrorMsg,RCtrl,i+iStart,[Ti]
        Pause 5
Next ;i
```

```
;Now create array B and fill with randoms
;------------------------------------------
For j = 0 to 99
        RTemp = Random RTemp
        Tj = RTemp.LowByte
        ;write B array to EEPROM
        I2COut Dta,Clk,ErrorMsg,RCtrl,j+jStart,[Tj]
        Pause 5
Next ;j

;Now multiply X=A*B
;-------------------
For i = 0 to 99
        I2CIn Dta,Clk,ErrorMsg,RCtrl,i+iStart,[Ti]
        For j = 0 to 99
                I2CIn Dta,Clk,ErrorMsg,RCtrl,j+jStart,[Tj]
                Prod = Ti* Tj
                ;calculate storage index for product
                k = i*200 + (j*2) + XStart
                'write the product to EEPROM
                I2COut Dta,Clk,ErrorMsg, RCtrl,k, |
                [Prod.HighByte,Prod.LowByte]
                Pause 5

                GoSub WriteInfo
        Next ;j
Next ;i

;Read EEPROM
;---------------
HSerOut ["*** Read ***",13]
For i = 0 to 99
        ;Read the A(i) values
        I2CIn Dta,Clk,ErrorMsg,RCtrl,i+iStart,[Ti]
        For j = 0 to 99
                ;Read the B(j) values
                I2CIn Dta,Clk,ErrorMsg,RCtrl,j+jStart,[Tj]

                ShouldBe = Ti* Tj

                ;calculate storage index for product
                k = i*200 + (j*2) + XStart
                'write the product to EEPROM
                I2Cin Dta,Clk,ErrorMsg,RCtrl,k, |
                [Prod.HighByte,Prod.LowByte]

                If ShouldBe <> Prod Then
                        HSerOut ["*** ERROR! ***",9]
                EndIf

                GoSub WriteInfo
        Next ;j
Next ;i

End

WriteInfo
;--------
        HSerOut ["A(",Dec i,"): ",Dec Ti," * B(",Dec j,"): ",Dec Tj,9," A*B= ",Dec
Prod,13]
        Pause 25
Return

ErrorMsg
;--------
        HserOut ["Error!"]
End
```

Before we write the first line of code, we must devise a memory allocation plan. We have three arrays to save, **A(0...99)** and **B(0...99)**, both byte length, and **X(0...99,0...99)**, word length.

We'll use a simple allocation plan for arrays **A** and **B**:

AT24C512 Memory Address	0	1	2	3	...	98	99
Array A value	A(0)	A(1)	A(2)	A(3)	A(...)	A(98)	A(99)
AT24C512 Memory Address	100	101	102	103	...	198	199
Array B value	B(0)	B(1)	B(2)	B(3)	B(...)	B(98)	B(99)

Hence, we'll reserve the first 200 bytes of space in the AT24C512 for **A** and **B**.

We'll start the product array **X** at address 200. Since **X(i,j)** is a word-length variable, we must reserve two bytes for each element.

AT24C512 Memory Address	200	201	202	203	...	20198	20199
Array X value	X(0).Byte1	X(0).Byte0	X(1).Byte1	X(1).Byte0	X(...)	X(9999).Byte1	X(9999).Byte0

The resulting memory map is:

Let's see how we go about calculating the memory indexes to read and write these memory locations.

- Memory Address for **A()** is simple enough; we index into memory with **i**.
- Memory Address for **B()** is **j + 100**.
- Memory Address for **X()** starts at 200, but jumps in two byte steps, to accommodate the word length variable. We'll define the starting point as a constant, **XStart**. Array **X(0...9999)** can be more conveniently thought of as a two-dimensional array **X(0...99,0...99)**, or **X(i,j)**. For each **i**, therefore, we reserve 200 bytes and step with **j**, but in intervals of two bytes. Thus, **k = i*200 + (j*2) + Xstart**, where **k** is the memory address for **X(i,j).Byte1**.

After mapping memory usage, the code is straightforward.

```
;Create array A and write to EEPROM
;-----------------------------------
HSerOut [13,13,"*** Write ***",13]
For i = 0 to 99
        ;fill array A with random numbers
        RTemp = Random RTemp
        Ti = RTemp.LowByte
        ;write A array to EEPROM
        I2COut Dta,Clk,ErrorMsg,RCtrl,i+iStart,[Ti]
        Pause 5
Next ;i
```

```
;Now create array B and fill with randoms
;---------------------------------------
For j = 0 to 99
        RTemp = Random RTemp
        Tj = RTemp.LowByte
        ;write B array to EEPROM
        I2COut Dta,Clk,ErrorMsg,RCtrl,j+jStart,[Tj]
        Pause 5
    Next ;j
```

The above code fragment constructs the arrays **A()** and **B()**, fills them with random values, using MBasic's **Random** function, and writes the arrays to EEPROM at the memory addresses we developed. Of course, in a real program these arrays would be filled with measured or calculated data of more utility than our random number assignments.

For increased generality, we previously defined the starting points for **i** (**iStart**) and **j** (**jStart**) memory segments. In accordance with our memory plan, **iStart** is defined as 0, and **jStart** is defined as 100

```
;Now multiply X=A*B
;------------------
For i = 0 to 99
        I2CIn Dta,Clk,ErrorMsg,RCtrl,i+iStart,[Ti]
        For j = 0 to 99
                I2CIn Dta,Clk,ErrorMsg,RCtrl,j+jStart,[Tj]
                Prod = Ti* Tj
                ;calculate storage index for product
                k = i*200 + (j*2) + XStart
                'write the product to EEPROM
                I2COut Dta,Clk,ErrorMsg, RCtrl,k, |
                [Prod.HighByte,Prod.LowByte]
                Pause 5

                GoSub WriteInfo
        Next ;j
    Next ;i
```

Now that we have the arrays A and B defined, we are in a position to calculate the product of each element of A and B. We do this by reading the AT24C512 for the elements. **Ti** is **A(i)** and **Tj** is **B(j)** via nested **For** loops. The product **A(i)*B(j)** is simply **Ti*Tj**. We store this product, which we assign to the word variable **Prod**, in the AT24C512 at location **k**. We store **Prod.HighByte** and **Prod.LowByte** sequentially in one **I2Cout** procedure call.

The subroutine **WriteInfo** outputs i,j **Ti**, **Tj** and **Prod** to the serial data port.

```
;Read EEPROM
;----------------
HSerOut ["*** Read ***",13]
For i = 0 to 99
        ;Read the A(i) values
        I2CIn Dta,Clk,ErrorMsg,RCtrl,i+iStart,[Ti]
        For j = 0 to 99
                ;Read the B(j) values
                I2CIn Dta,Clk,ErrorMsg,RCtrl,j+jStart,[Tj]

                ShouldBe = Ti* Tj

                ;calculate storage index for product
                k = i*200 + (j*2) + XStart
                'write the product to EEPROM
                I2Cin Dta,Clk,ErrorMsg,RCtrl,k, |
                [Prod.HighByte,Prod.LowByte]

                If ShouldBe <> Prod Then
                        HSerOut ["*** ERROR! ***",9]
                EndIf

                GoSub WriteInfo
```

```
        Next ;j
   Next ;i
```

Finally, we verify that the data has been correctly computed and saved, by repeating the product calculation, saving it as the word variable **ShouldBe**, and comparing that with the value retrieved from the AT24C512. Should the retrieved value differ from the newly computed value, an error flag is written from the serial port.

Here's a sample of the serial port output:

```
*** Write ***
A(0): 155 * B(0): 150      A*B= 23250
A(0): 155 * B(1): 44       A*B= 6820
A(0): 155 * B(2): 89       A*B= 13795
A(0): 155 * B(3): 178      A*B= 27590
A(0): 155 * B(4): 101      A*B= 15655
A(0): 155 * B(5): 203      A*B= 31465
A(0): 155 * B(6): 150      A*B= 23250
```

<omitted output>

```
*** Read ***
A(0): 155 * B(0): 150      A*B= 23250
A(0): 155 * B(1): 44       A*B= 6820
A(0): 155 * B(2): 89       A*B= 13795
A(0): 155 * B(3): 178      A*B= 27590
A(0): 155 * B(4): 101      A*B= 15655
A(0): 155 * B(5): 203      A*B= 31465
A(0): 155 * B(6): 150      A*B= 23250
```

A final reminder; if we try writing and reading a word variable, such as:

```
    I2COut Dta,Clk,ErrorMsg, RCtrl,k,[Prod]
    I2Cin Dta,Clk,ErrorMsg, RCtrl,k,[Prod]
```

only the **Prod.LowByte** is actually stored and read. **I2Cout** and **I2Cin** function correctly with individual byte or byte-array arguments, not multibyte types, such as **Word**, **Long** or **Real**.

Parallel Access Memory

Serial EEPROM memory is compact, inexpensive and consumes only two PIC pins, often a scarce resource. But, it's slow—writing to EEPROM requires 5ms for the devices we've examined in this chapter. Even though no pause is required, reading a byte from EEPROM at standard speed requires several hundred microseconds.

Why would we want faster access? Suppose we wish to capture a fast changing analog input signal with the A/D converter, store several kilobytes of consecutive A/D values and then analyze the results post-capture. Perhaps we wish to do a fast Fourier transform (FFT) on the data, or have the PIC take action based on some other mathematical analysis of the data. If we must capture more than a couple hundred bytes of A/D readings, the PIC's onboard RAM is inadequate, so we must look to external memory. From Chapter 11, we know that an A/D read in MBasic consumes less than 300 µs, but if it takes 5ms to save each result to serial EEPROM memory, our A/D capture-save rate drops markedly. We'll see that even if we keep our programming to only MBasic, we can save a byte value in SRAM in about 300 µs, a huge improvement over writing to serial EEPROM. And, if we aren't afraid to mix a bit of assembler into our MBasic programming, we can shave the SRAM-save time to under 10 µs, thus making the A/D conversion routine the limiting factor in the A/D capture-save cycle.

For faster access, we might look at serial memory that uses static RAM, not EEPROM, technology. SRAM can be written to in nanoseconds, without the long wait for EEPROM writing. SRAM memory retains information without periodic refreshing so long as power is maintained. Unlike EEPROM, however, SRAM will not retain information once power is removed. And, unfortunately there are not many serial SRAM devices available, and those that are available carry price tags significantly above those of serial EEPROM.

If we require inexpensive fast SRAM access, unfortunately, we are often left with only parallel access devices. And, parallel devices

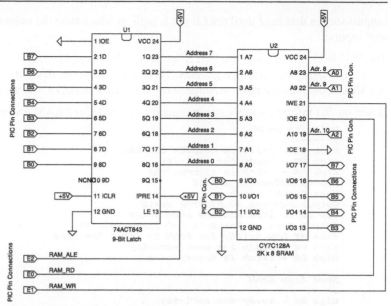

Figure 18-5: Parallel access DRAM.

consume PIC pins like there is no tomorrow. Let's look at a simple application of parallel SRAM with Cypress Semiconductor's CY7C128A 2K × 8-bit SRAM device. Figure 18-5 shows the strategy for connecting a parallel SRAM memory chip to a PIC. Even with a latch to share address and data pins on Port B, the PIC still must dedicate 14 pins to communications with the CY7C128A.

Before we look at the code associated with the CY7C128A, let's look at the two active devices we'll use. (This overview is no substitute for a thorough review of the data sheet.)

The CY7C128A memory is a 2K × 8-bit SRAM with 11 address inputs, A0…A10 and eight bidirectional output pins, I/O0…I/O7, and three control pins: chip enable ($\overline{CE}$), output enable ($\overline{OE}$) and write enable ($\overline{WE}$). (These three pins are active low, normally shown by a horizontal bar over the acronym.)

The I/O pins are bidirectional; much like a PIC they may be switched between input and output functions. In addition, the I/O pins may be placed in a third, high impedance, state where they are effectively disconnected from the outside world.

Function	$\overline{CE}$	$\overline{WE}$	$\overline{OE}$
Write Data to SRAM	Low	Set up Address and Data to be Written with WE High. Take WE Low to Write Data to SRAM	High during relevant time
Read Data from SRAM	Low	High during relevant time	Set up Address and go into Read With OE High and Take Low to Output Valid Data

The 74ACT843 is a 9-bit latch. (We only use 8-bits, but I had a 74ACT843 on hand and used it. You may substitute an 8-bit latch, such as 74ACT574, if you wish.) A latch performs the function that its name suggests; if the latch enable (LE) pin is high, the each output follows its associated input, but when the LE is pulled low, the outputs are "latched" into the state they were in on the falling edge of the LE waveform. The outputs stay in that state until the LE pin is high, at which time the outputs go back to tracking their associated inputs.

The '843 latch allows us to use Port B for both addressing (A0…A7) the SRAM and for data transfer to and from the SRAM, thus saving eight pins. This means, however, that our code must ensure that the Port B and the SRAM I/O pins are never simultaneously in output mode, or else we risk damage to the devices.

Based on this information, we can prepare pseudo-code for SRAM access:

```
Write to SRAM
;-------------
High OE - tristate outputs of SRAM
High WE - ready for activity
High LE - latch is transparent
Set up Addresses A8…A10
Make PortB Output
Output Address A0…A7 from PortB
Low LE to lock address into place
Output Data on PortB to I/O lines IO0…IO7
Take WE Low Briefly for SRAM to input the data
High WE - ready for next activity
High LE - latch is transparent

Read from SRAM
;-------------
High WE - ready for activity
High OE - tristate outputs of SRAM
Set up Addresses A8…A10
Make PortB Output
Output Address A0…A7 from PortB
Low LE to lock address into place
Make PortB Input
Low OE - put I/O on line now avoiding clash
Read PortB, thereby Reading Data on I/O lines IO0…IO7
High OE - back to tristate for SRAM outputs
High LE - latch is transparent
```

This circuit presses the limits of what is feasible to simulate on a plug-board style breadboard. The CY7C128A is a high-speed chip, with the one I used having an access time of 25 ns and is accordingly sensitive to ground bounce and other artifacts that might be ignored by slower speed logic. (You can purchase CY7C128A versions with as little as 15 ns access time.) The space on a 2840 development board is marginal at best so I built the latch and SRAM elements on a larger, separately-powered plug-board, and used a 6" 14-pin DIP-to-DIP jumper cable to connect it to the 2840 board. I found lead dress and jumper wire routing was critical and that I could cause random read or write errors by changing certain jumper wires' physical position or routing.

One final point; I used a 16F877A for this project, but you could substitute a 16F876 if desired. If so, substitute C0…C2 for the E0…E2 connections.

Program 18-4 writes data to the SRAM and reads it back. Any discrepancy between the value that should have been written and the as-read value sends an error message over the serial port.

Program 18-4

```
;Program 18-4. Minimalist
;R/W to CY7C128A 2Kx8 SRAM
;Uses SN74ABT843 Latch
;See Microchip TB011


;Constants
;----------------
RAM_ALE Con     E2      ;address latch enable on SN74ABT843 latch
RAM_WR Con      E1      ;Write Enable
```

```
RAM_RD          Con     E0      ;Output Enable
Display         Con     0       ;0 blocks writing to serial port


;Variables
;--------------
RamAdr          Var     Word    ;RAM read/write address
RamData         Var     Byte    ;data to be read or written
i               Var     Word    ;Loop variable
ShouldBe        Var     Byte    ;Read back value should be


;Initialization
;--------------
TRISB = %00000000       ;Dual Purpose In/Out
TRISA = %00000000       ;Use A0..A3 as outputs
TRISE = %00000000       ;

;Communicate with outside world via serial
EnableHSerial
SetHSerial H19200
High RAM_RD
High RAM_WR
High RAM_ALE


Main
        For i = 0 to 2047
                RamAdr = i
                RamData = i.LowByte + i.HighByte
                GoSub WriteRam

                If Display > 0 Then
                        HSerOut ["i: ",Dec i," Write: ",Dec RamData,13]
                        pause 25
                EndIf
        Next

        HSerOut [13,13]
        For i = 0 to 2047
                RamAdr = i
                GoSub ReadRam

                If Display > 0 Then
                        HSerOut ["i: ",Dec i," Read: ",Dec RamData,13]
                        Pause 25
                EndIf

                ShouldBe = i.LowByte  + i.HighByte
                If (RamData <> ShouldBe) Then
                        HSerOut ["i: ",Dec i,"  Read: ",iHex RamData, |
                        "  Should Be: ",iHex ShouldBe," <-- Error!",13]
                        Pause 500
                EndIf
        Next
        HSerOut ["Endit",13]
        Pause 30
End


WriteRAM
;--------
        ;Subroutine executes in 320 uSec, not counting call/return
        PortA.LowNib = RamAdr.Nib2      ;get high 3 bits of address
        TRISB = $00                      back to output
        PortB = RamAdr.LowByte
        Low RAM_ALE
        PortB = RamData                 ;write the data
        PulsOut RAM_WR,5                ;accomplish the write to RAM
        High RAM_ALE
```

411

```
        Return


        ReadRAM
        ;--------
                ;Subroutine executes in 350 uSec, not counting call/return
                ;PortB and PortA are in output mode
                PortA.LowNib = RamAdr.Nib2 ;get high 3 bits of address
                TRISB = $00
                PortB = RAMAdr.LowByte
                Low RAM_ALE
                TRISB = $FF             ;PIC pins now inputs input
                Low RAM_RD              ;enable output drivers in RAM
                RamData = PortB         ;read the data
                High RAM_RD             ;tristate to disable output drivers
                High RAM_ALE
        Return


        End
```

Following Microchip's Application Note TB 11, we start by giving the three control pins aliases that are more understandable:

```
        RAM_ALE Con     E2      ;address latch enable on SN74ABT843 latch
        RAM_WR  Con     E1      ;Write Enable
        RAM_RD  Con     E0      ;Output Enable used for Read RAM
```

The main **For i** loops should be familiar, as they largely reproduce the concept we developed for serial EEPROM testing; fill the memory with test bytes generated by summing the two address bytes, read the memory and compare the read value with the calculated value and write an error message if the two values differ. Hence, our analysis will jump to the two subroutines that perform the heavy lifting in Program 18-4, **WriteRAM** and **ReadRAM**.

Both **WriteRam** and **ReadRam** track the pseudo-code quite closely.

Pseudo-Code	WriteRam Subroutine
Write to SRAM ;------------- High OE - tristate outputs of SRAM High WE - ready for activity High LE - latch is transparent Set up Addresses A8…A10 Make PortB Output Output Address A0…A7 from PortB Low LE to lock address into place Output Data on PortB to I/O lines IO0…IO7 Take WE Low Briefly for SRAM to input the data High WE - ready for next activity High LE - latch is transparent	WriteRAM ;-------- PortA.LowNib = RamAdr.Nib2 TRISB = $00 PortB = RamAdr.LowByte Low RAM_ALE PortB = RamData PulsOut RAM_WR,5 High RAM_ALE Return

We can omit several of the control line settings that were shown in the pseudo-code for clarity because they are implicitly dealt with. For example, **RAM_ALE** is always high going into **WriteRam**, so it isn't necessary to explicitly set it high again. Similarly, MBasic's command **PulsOut** leaves **RAM_WR** in the state (high) it was in before being called, and inserts a brief low pulse, so resetting **RAM_WR** high is unnecessary.

We see the same parallelism in **ReadRam**:

Pseudo-Code	ReadRam Subroutine

`Read from SRAM`
`;------------`
`High WE - ready for activity`
`High OE - tristate outputs of SRAM`
`Set up Addresses A8…A10`
`Make PortB Output`
`Output Address A0…A7 from PortB`
`Low LE to lock address into place`
`Make PortB Input`
`Low OE - put I/O on line now avoiding clash`
`Read PortB, thereby Reading Data on I/O`
`lines IO0…IO7`
`High OE - back to tristate for SRAM outputs`
`High LE - latch is transparent`
	`ReadRAM`
	`;--------`
	
	`PortA.LowNib = RamAdr.Nib2`
	`TRISB = $00`
	`PortB = RAMAdr.LowByte`
	`Low RAM_ALE`
	`TRISB = $FF`
	`Low RAM_RD`
	
	`RamData = PortB`
	
	`High RAM_RD`
	
	`High RAM_ALE`
	
	`Return`

Again, we've been able to omit the two entrance set-up commands in the pseudo-code because the status is known when the subroutine is called.

After all this work in making nanosecond SRAM work with a PIC, you may wonder why expend so much effort if it still takes 300µs or so to read or write a byte. Well, we can knock 290 µs off that time by recoding **WriteRAM** and **ReadRAM** in assembler, resulting in execution times of 10–11 µs.

Our approach to assembler programming in Chapters 13 and 14 emphasized a process of "stepwise substitution," whereby we develop and debug a program in pure MBasic and then, for time-critical program elements, write an in-line assembler replacement, one statement at a time. (We leave complicated things, such as serial output, floating point routines and the like in MBasic.) After replacing one MBasic statement, we debug and test the new chunk of assembler code and when functioning, move to the next MBasic statement. When completed, we may merge the various individual in-line assembler code chunks into a single longer in-line segment, or leave them in place.

Program 18-5 shows how we might replace the MBasic code in both **WriteRAM** and **ReadRAM** with assembler chunks. I've intentionally not merged the code into a single long in-line assembler segment so that you might see how each MBasic statement is simply rendered into a few lines of assembler. However, I've removed the intermediate **BankSel** commands, necessary when exiting assembler and returning to MBasic. (GPASM will report error messages when assembling the instruction **banksel 0**. So, we substitute a known bank 0 register, such as **PortA**—that is, **banksel PortA**.)

Program 18-5

The listing for Program 18-5 is available in the associated CD-ROM and as a space saving measure will not be shown.

```
;PortA.LowNib = RamAdr.Nib2    ;get high three address bits
ASM
{
BankSel RamAdr1
movf    (RamAdr+1)&0x7F,w
BankSel PortA
movwf   PortA
}
```

We cheat a bit in writing the three high-bit addresses A8…A10 to Port A. The MBasic code writes only to **PortA.LowNib**, but since we use Port A for no other purpose, the assembler code simply writes **RamAdr.HighByte** to **PortA**. In accessing a word variable declared in MBasic, its high byte is the base address plus 1; hence in assembler we use **(RamAdr+1)** as the equivalent of MBasic's **RamAdr.HighByte**. In order to avoid operator precedence issues, we force **RamAdr+1** to be executed first by enclosing it in parentheses. The assembler code copies **RamAdr+1** to the PIC's W register and then copies the W register contents to **PortA**, yielding the equivalent of **PortA=RamAdr.HighByte**. GPASM does not permit us to perform **(RamAdr+1)** as an argument for **BankSel**, so we have created **RamAdr1** as an alias for **RamAdr.HighByte** in the MBasic variable definitions.

```
;TRISB = $00          ; back to output
ASM
{
BankSel TRISB
Clrf          TRISB
}
```

Setting a file to zero is accomplished quickly and efficiently in assembler by clearing it.

```
;PortB = RamAdr.LowByte
ASM
{
BankSel RamAdr
movf RamAdr&0x7F,w
BankSel PortB
movwf PortB
}
```

We write **RamAdr.LowByte** to **PortB** in the same fashion we did with the **RamAdr.HighByte** and **PortA**; copy **RamAdr** to the **W** register and then copy the **W** register to **PortB**. Since **RamAdr**'s address is the low byte, we need not do anything special to load **RamAdr.LowByte** into **W**.

```
;Low RAM_ALE
ASM
{
BankSel PortE
bcf            PortE,2
}
```

We drop **RAM_ALE** to low by clearing its defined PortE bit, thereby latching the low order address values A0…A7.

```
;PortB = RamData          ;write the data
ASM
{
BankSel RamData
movf    RamData&0x7F,w
BankSel PortB
movwf   PortB
}
```

Copying the data byte, **RamData** to **PortB** is identical with how we moved the low byte of the address to PortB.

```
;PulsOut RAM_WR,5      ;accomplish the write to RAM
ASM
{
BankSel PortE
bcf            PortE,1
NOP
NOP
NOP
NOP
bsf            PortE,1
}
```

To generate a brief zero-level pulse on **RAM_WR**, we clear its associated bit in **PortE** with a **bcf** operation. We follow with four **NOP**s for a brief delay (0.8 μs) and then restore **RAM_WR** to high by setting the associated bit with a **bsf** operation. The delay we used in MBasic was 5 μs, far longer than required by the CY7C128A. (MBasic's **PulsOut** function has a minimum period of 4 μs.)

```
              ;High RAM_ALE
              ASM
              {
              BankSel PortE
              bsf             PortE,2
              }
      Return
```

Finally, we return **RAM_ALE** to high by setting the associated bit in PortE, thereby enabling the '843 latch's output pins to again follow input values.

For completeness, we'll step through **ReadRam**.

```
      ReadRAM
      ;--------
              ;PortA.LowNib = RamAdr.Nib2 ;get high 3 bits of address
              ASM
              {
              BankSel RamAdr1
              movf    (RamAdr+1)&0x7F,w
              BankSel PortA
              movwf   PortA
              }
```

Same code as in **WriteRam**.

```
              ;TRISB = $00
              ASM
              {
              BankSel TRISB
              Clrf    TRISB
              }
```

Same code as in **WriteRam**.

```
              ;PortB = RAMAdr.LowByte
              ASM
              {
              BankSel RamAdr
              movf RamAdr&0x7F,w
              BankSel PortB
              movwf PortB
              }
```

Same code as in **WriteRam**.

```
              ;Low RAM_ALE
              ASM
              {
              BankSel PortE
              bcf             PortE,2
              }
```

Same code as in **WriteRam**.

```
              ;TRISB = $FF   ;PIC pins now inputs
              ASM
              {
              BankSel TRISB
              movlw   0xFF
              movwf   TRISB
              }
```

In **WriteRam**, **PortB** was outputting data to the SRAM so **TRISB** was kept at **$00**; however, here we must read data from the SRAM. Hence, we set **PortB** to input by setting **TRISB** to **$FF**. We accomplish this by loading register W with **$FF** and copying register W to **TRISB**.

```
;Low RAM_RD      ;enable output drivers in RAM
ASM
{
BankSel PortE
bcf             PortE,0
}
```

We drop **Ram_RD** low by clearing its associated bit on **PortE**.

```
;RamData = PortB ;read the data
ASM
{
BankSel PortB
movf    PortB,w
BankSel RamData
movwf   RamData&0x7F
}
```

At this time, the SRAM I/O pins have shifted to output and contain the byte value to be read by **PortB** and then transferred to **RamData**. We first copy **PortB**'s value to the **W** register and then copy the **W** register's value to **RamData**.

```
;High RAM_RD     ;tristate to disable output drivers
ASM
{
BankSel PortE
bsf             PortE,0
}
```

To tri-state the SRAM's output drivers, we simply bring the **RAM_RD** pin high by setting its associated **PortE** pin through a **bsf** operation.

```
;High RAM_ALE
ASM
{
BankSel PortE
bsf             PortE,2
BankSel PortB   ;switches to Bank 0 for MBasic
}
Return
```

We finish with the same code as in **WriteRam** to re-enable the '843 latch.

Perhaps some of these steps could be accomplished more elegantly in assembler, but we've opted for clarity over elegance.

References

[18-1] Philips Semiconductors, *The I2C-Bus Specification Version 2.1, January 2000*, Document No. 9398 393 40011 (2000).

[18-2] Microchip Technology, Inc., *24AA16/24LC16B 16K I²C Serial EEPROM*, Document No. DS21703C (2003).

[18-3] Microchip Technology, Inc., *24AA256/24LC256/24FC256 256K I²C CMOS Serial EEPROM*, Document No. DS21203K (2003).

[18-4] Microchip Technology, Inc., *24AA515/24LC515/24FC515 512K I²C CMOS Serial EEPROM*, Document No. DS21673C (2003).

[18-5] Atmel Corp., *2-wire Serial EEPROM 512K (65536 x 8) AT24C512*, Rev. 1116J-SEEPR-7/03 (2003)

[18-6] Texas Instruments Inc., *SN54ABT843, SN74ABT843 9-Bit Bus-Interface D-Type Latches With 3-State Outputs*, (1997).

[18-7] Cypress Semiconductor Corp., *CY7C128A 2K x 8 Static RAM*, Document No. 38-05028 (Aug. 24, 2001).

[18-8] Evans, Rick, Using *SRAM with a PIC16CXXX, TB011*, Microchip Technology, Inc., Document No. DS91011A (1997).

[18-9] Atmel Corp., *Interfacing AT24CXX Serial EEPROMs with AT89CX051 Microcontrollers*, Rev. 0507D-05/01 (2001)

[18-10] Philips Semiconductors, *I2C-bus allocation table General*, (Mar. 3, 1997)..

CHAPTER **19**

Advanced Stepper Motors

Chapter 8 introduces the stepper motor, and shows how to drive both unipolar and bipolar steppers in full step, half step and wave modes. It also covers the fundamentals of motor speed up through L/R drive methods.

This chapter covers two more advanced topics; microstepping and a general-purpose serial-controlled stepper driver program. Both topics are illustrated with programs that blend MBasic with assembler language for time-critical routines. If you haven't read Chapters 8, 10, 13, 14 and 15 yet, now is a good time to do so.

One final note; this chapter focuses on bipolar stepper motors, but as you know from Chapter 8, unipolar motors work with same program by simply changing the driving hardware from a bipolar H-bridge to one aimed at unipolar motors. And, you may even use the same bipolar H-bridge driver to drive a unipolar motor, operating in bipolar mode.

Microstepping

Introduction

Let's go back to the conceptual bipolar stepper motor introduced in Chapter 8. As illustrated in Figure 19-1 if the currents through windings A-C and B-D are equal, the rotor orients itself at an angle halfway between the two stator poles, in our simple four steps per revolution motor, at 45 degrees. We'll say our imaginary motor has 1 ampere flowing through both windings A-C and B-D. Suppose we maintain windings B-D at 1 ampere, but reduce A-C's current to 0.5 ampere. Intuitively, you should see that the rotor will shift to being more closely aligned with winding B-D. If we maintain windings B-D at 1 ampere and reduce the current through winding A-C further, to say 0.25 ampere, the rotor will move even more towards alignment with winding B-D's poles.

Thus, by holding the current constant in one winding pair and reducing it in the other pair, we can position the rotor, at least in theory, at any arbitrary angle. The thing that makes a stepper motor what it is—the ability to move in precise increments or steps—can be negated, turning the stepper into a continuously positional vari-

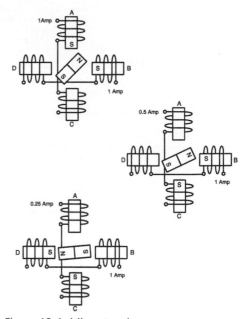

Figure 19-1: Microstepping.

able motor. In practice, there are advantages in varying the current to provide a binary power of intermediate steps, for example, 8, 16 or 32. Programs 19-1 and 19-2 use 16 intermediate steps. If applied to a 48-step motor, such as the PF35-48 we use in Chapter 8, the motor has 48 × 1 6, or 768 microsteps per revolution. Applied to a 200-step/revolution motor, the result is 3,200 microsteps per revolution, corresponding to 0.112° per microstep.

Is this too good to be true—that we can take a stepper motor and position the shaft within 6.75 arc-minutes? The devil, as usual, is in the details, and in the case of microstepping, the details are getting the current ratios correct. But, even if the current ratio isn't quite accurate, microstepping provides another major advantage by replacing the abrupt on/off current waveform we observed in Chapter 8 with a smooth, continuously varying one. This improves motor efficiency and smoothes the step transitions. The current error, moreover, applies only to intermediate microsteps—full step positions are identical to those of a conventional full step controller.

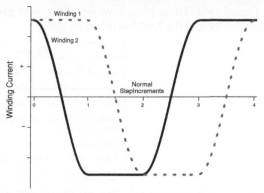

Figure 19-2: Current Distribution for Microstepping.

Figure 19-2 illustrates the theoretically desired current distribution. The target current waveform holds the current in one winding pair constant while the current in the second winding pair is varied following a cosine wave shape, repeating the process with the second winding pair held constant while the first varies in a cosine pattern. Figure 19-2 simulates the cosine distribution.

Drive Circuit

How do we efficiently generate these current waveforms? One approach would be a variable constant current supply under PIC control through a D/A converter. However, an active constant current source is beyond the complexity appropriate for an introductory book such as this one. Hence, we'll use something a bit simpler, but not as good. As we learned in Chapter 8, when voltage is applied to a stepper winding, the resulting current is a function both of time (the L/R time constant), the applied voltage and series resistance of the motor winding plus any external rise time improvement resistors. If the incremental motor steps occur relatively slowly so that we may disregard the L/R component of the current response, the winding current is proportional to the applied voltage. So, we'll ramp the motor supply voltage up and down, and trust that the motor current will follow the supply voltage. As we'll see when we examine the performance of our simple circuit, this trust is not misplaced; although the current waveform isn't always what we desire, it's adequate to demonstrate the principles behind microstepping.

Let's take another look at our SN754410 bipolar driver circuit from Chapter 8, reproduced as Figure 19-3. What we need is a way of easily and efficiently independently varying the voltage across each motor winding. Fortunately, there is such a way. The SN754410 has two extra "enable" pins connected directly to the +5 V line

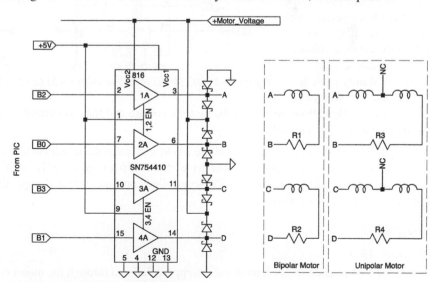

Figure 19-3: Bipolar control circuit.

in Figure 19-3. These pins, however, allow us to gate the voltage applied to the motor off and on, as we see when we look at the SN754410's truth table:

Inputs		Output
A	**Enable**	**Y**
H	H	H
L	H	L
X	L	Z
H= high level		
L=low level		
X= irrelevant (could be either H or L)		
Z= high impedance (off) state		

The "A" connections are the bridge driver inputs, pins 2, 7, 10 and 15 in Figure 19-3. The "Y" connections are the outputs, pins 3, 6, 11 and 14 in Figure 19-3. By taking the enable pin high or low, we gate the output off or on. (When the output is in a high impedance state, it is effectively an open circuit.) Each enable pin controls one set of windings when the SN754410 is configured according to Figure 19-3.

We've looked at pulse width modulation in general in Chapter 16, and in Chapters 24 and 26 at the hardware pulse width modulation (PWM) module found in most PICs. As we'll see later in this chapter when we look at Program 19-1, we use the PWM modules to drive the enable input of the SN754410, with the circuit shown in Figure 19-4. If the PWM module operates with a duty cycle of 50%, we reduce the average voltage across the motor by 50%. If the PWM duty cycle is 1%, the average voltage is

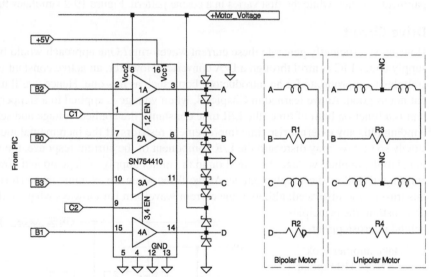

Figure 19-4: SN754410 H-Bridge With PWM Control for Microstepping.

reduced to 1%. This allows us to vary the motor's voltage and thus its current through a completely digital mechanism under PIC control.

We could extend this concept further and use PWM to control the motor current directly, not the voltage, constructing what is known as a "chopper drive" circuit. However, the SN754410 offers no easy way to sense individual winding current and a PIC-controlled constant current chopper drive is more complex than appropriate for our discussion level.

Programs

Program 19-1

Before we look at code to implement microstepping, let's go through the motor control sequence. We'll break each step into 16 microsteps. Our high level strategy is to hold the current through one winding pair constant while ramping the current through the other winding pair, following the cosine function:

Full Steps	Microsteps	Windings A-C	Windings B-D
1	0...15	Cosine Ramp from + to –	Hold Constant +
2	16...31	Hold Constant –	Cosine Ramp from + to –
3	32...47	Cosine Ramp from – to +	Hold Constant –
4	48...63	Hold Constant +	Cosine Ramp – to +

Let's dig into this in detail. We'll rearrange the bit order from the normal B3...B0 sequence so that the motor sign bits match the winding connections to the SN754410; Windings A–C are driven via PIC pins B2–B0 and windings B–D via PIC pins B3–B1 in Figure 19-4.

Full Step	Winding A-C P	Winding A-C S	Winding B-D P	Winding B-D S	μstep	Motor Sign Bits B3	B1	B2	B0	Binary Value
1a	+	R		C	0...6	1	0	1	0	12
		A	+	O	7	1	0	1	0	12
2	–	M	+	N	8...14	0	1	1	0	6
		P		T	15	0	1	1	0	6
	–	C	+	R	16...22	0	1	1	0	6
		O		A	23	0	1	1	0	6
3	–	N	–	M	24...30	0	1	0	1	3
		T		P	31	0	1	0	1	3
	–	R	–	C	32...38	0	1	0	1	3
		A		O	39	0	1	0	1	3
4	+	M	–	N	40...46	1	0	0	1	9
		P		T	47	1	0	0	1	9
	+	C	–	R	48...54	1	0	0	1	9
		O		A	55	1	0	0	1	9
1b	+	N	+	M	56...62	1	0	1	0	12
		T		P	63	1	0	1	0	12

P=polarity of current into windings
S=status;
   RAMP—current change in cosine relationship
   CONT—constant current
Underline indicates changed value; intermediate steps follow correspondingly

This is a complicated table, but since it's key to understanding our code, we'll study it carefully. The first five columns expand the summary table, with two "winding" columns showing the polarity of the current (P), and its status (S), which may be either constant (CONT) or ramping (RAMP). The "full step" and "microstep" columns show the full step number and its relationship to the microstep increments. The motor sign bits and binary value columns are directly repeated from the full step patterns developed in Chapter 8, that is, step 1 corresponds to %1010, or decimal 12, step 2 corresponds to %0110 or binary 6, and so on. If this representation is unclear to you, take a break and reread Chapter 8.

I've broken full step 1 into "1a" and "1b" halves to better show how the pattern repeats. But, there's a more important reason for breaking the steps this way. If we examine the polarity reversals during current ramping, we see that the breakpoints from positive to negative are exactly aligned with the full step transitions. This means that we may use the PIC's motor control bits B3...B0 to drive our SN754410 motor controller and let the SN75440 take care of polarity, while we pulse width modulate the corresponding enable pins to control the current level. In other words, we may use the structure of the a full step motor control program, such as those of Chapter 8 and overlay on top of it two PWM-based cosine ramp waveforms, one for each winding. Of course, we must ensure the ramp waveforms are synchronized with the full step switching, but that's an easy task.

Programs 19-1 through 19-3 use the circuit shown in Figure 19-4. Using the methodology developed in Chapter 8, determine the appropriate values of series resistance and motor voltage for your stepper motor. The oscilloscope illustrations in this chapter are of a PF35-48 48-step unipolar wound motor, with no external series resistance and 8 V into the SN754410. And, to remind you again, don't use the 2840 Development Board's +5 V supply to drive your stepper—it can't safely supply enough current to power anything but the smallest steppers and it's marginal for that.

With this understanding, let's look at Program 19-1.

```
;Program 19-1 Microstep
;in MBasic

;Constants
;---------------------
PWMPeriod       Con     255     ;PWM width 20MHz clock = 12.75us
                                ;correspond to 78 KHz
DefaultDC       Con     128     ;default duty cycle 50%
MPin            Con     B0      ;starting pin for motor driver
CycLen          Con     64      ;micro-cycles per repetition

;Cosine lookup table for 180 degrees
;We add the steady-state part to simplify the stepping code

Cosine  ByteTable 252,243,224,196,161,120, |
                  74,25,25,74,120,161,196,224,243,252, |
                  255,255,255,255,255,255, |
                  255,255,255,255,255,255,255,255,255,255

FullStep  ByteTable 12,6,3,9,12,6,3,9 ;j is 0...4

;Variables
;------------------
i               Var     Byte    ;overall counter
j               Var     Byte    ;full step index
k               Var     Byte    ;cos index for w1
n               Var     Byte    ;cos index for w2
MPort           Var     PortB.Nib0      ;output port pins

;Initialization
;-----------------

        HPWM 0,PWMPeriod,DefaultDC
        HPWM 1,PWMPeriod,DefaultDC

        ;set motor port to output
        For j = MPin to MPin+3
                Output j
        Next
        ;all windings off
        MPort = 0

Main
;----
For i = 0 to CyCLen-1
        j = (i+8) / 16
        k = i//32
        n = (i+16)//32
        HPWM 0,PWMPeriod,Cosine(k)
        HPWM 1,PWMPeriod,Cosine(n)
        MPort = FullStep(j)
Next
```

```
GoTo Main

End
```

We'll start by looking at the constants.

```
;Constants
;----------------------
PWMPeriod      Con     255     ;PWM width 20MHz clock = 12.75us
                               ;correspond to 78 KHz
DefaultDC      Con     128     ;default duty cycle 50%
```

MBasic's **HPWM** function takes period and duty cycle arguments between 0 and 16383. For reasons that will become apparent when we study Program 19-2, I've set the period at 255, the largest possible byte value. The period does not change, so I've defined a constant, **PWMPeriod**, to hold this value. Chapter 24 discusses the **HPWM** function extensively and if you are not familiar with it, a brief detour to Chapter 24 may be in order.

MBasic supports hardware PWM through the function **HPWM**, with syntax:

HPWM CCPx, Period, Duty

CCPx—identifies the PWM module to use. The 16F87x devices, for example, have two hardware PWM modules, one hardwired to pin **C1** and the other hardwired to pin **C2**. **CCPx** may be a constant or a variable, and has two permissible values, 0 and 1:

CCPx value	PWM Generator Used	16F87x Pin Connection
0	CCP Module 1	C2
1	CCP Module 2	C1

In some printings, MBasic's User's Guide incorrectly reverses the **CCPx** values required to select the CCP modules. Incidentally, the term CCP is used because the hardware modules may be configured to perform three functions, capture, compare and PWM, or CCP for short.

Period and Duty—These parameters are interrelated, so we'll look at them together. We'll use waveform B to illustrate the terminology used in **HPWM**.

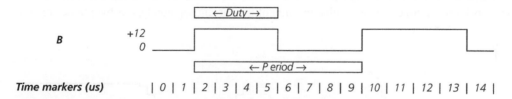

The **period** might more clearly be referred to as the "repetition period," i.e., the time between successive repetitions of the output waveform, measured from leading edge to leading edge. In the example, the period is 8 μs. **Duty** is the time that the output waveform is at logical high, 4 μs in the example. A commonly used term is "duty cycle," which is the ratio of high to low periods. In the example, the duty cycle is 4 μs/8 μs or 0.50. The value is often expressed as a percentage, so our example has a duty cycle of 50%. (Occasionally duty cycle is used to express the percentage of time low, not high, so watch the context in which the term is used.) *The average DC output of the PWM waveform is the duty cycle.* We've set the period at 255, so by varying **duty** from 0...255, we vary the average value of the DC output from 0 to 100% of the maximum.

As we learn in Chapter 24, the hardware PWM module's period is set by the relationship:

$$t_p = \frac{N}{f_{osc}}$$

t_p is the pulse period (duration) in seconds

N is the period argument to the MBasic function **HPWM**, 255 in Program 19-1

f_{osc} is the PIC's oscillator clock frequency, 20 MHz in my case

$$t_p = \frac{255}{20 \times 10^6} = 12.75 \times 10^{-6}$$

The pulse frequency shouldn't be so low as to cause acoustic noise, nor so high as to cause excessive loss. A period of 12.75 µs corresponds to 78 KHz, which is a bit high, but still within the acceptable range.

We initialize the duty cycle so that the PWM output is approximately a square wave, with a 50% duty cycle, thus requiring **HPWM**'s duty cycle argument to equal 128. Accordingly, we define a constant, **DefaultDC**, equal to 128, to use in initializing the PWM modules.

```
MPin        Con     B0      ;starting pin for motor driver
CycLen      Con     64      ;micro-cycles per rep
```

MPin is the starting pin for the drive pattern; **CycLen** is the number of microsteps before the microstep cycle repeats itself.

Although the full repetition period is 64 microsteps, looking at windings A-C we see that the second 32 microsteps repeat the first 32, except for the direction of current flow. Likewise, we see that the current flow in windings B-D has an identical pattern to windings A-C, except that B-D is delayed by 16 microsteps. These symmetries suggest that we create a byte table of length 32, starting with a length 16 cosine ramp, followed by a length 16 constant trailer. Since the PWM period argument is 255, the duty cycle ranges from 0 (no output) to 255 (maximum output). We therefore scale the maximum and minimum cosine values to range from 0...255.

```
Cosine  ByteTable    252,243,224,196,161,120, |
                     74,25,25,74,120,161,196,224,243,252, |
                     255,255,255,255,255,255, |
                     255,255,255,255,255,255,255,255,255,255
```

Let's see how we constructed this table. Figure 19-5 is a view of the cosine transition current waveform objective, quantized in 16 steps. Since we determine the sign of the current separately, our byte table contains

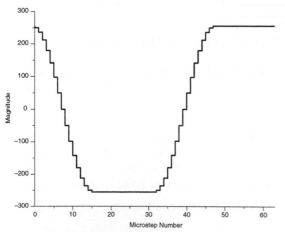

Figure 19-5: Winding A-C Current waveform.

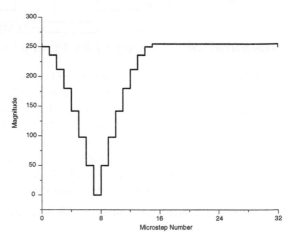

Figure 19-6: Desired cosine byte table values.

only the magnitude of the cosine function, and, because of symmetry, it need be only length 32. Figure 19-6 shows the resulting desired byte table values. Mathematically, the first eight entries may be calculated as:

$$BT(N) = 255 \times \cos\left(\frac{(N+0.5) \times 360}{32}\right)$$

N is the microstep number, 0...7. We add one-half step to N to get a better approximation to the average value of the cosine function, since we evaluate it with only eight steps.

255 Is the multiplier necessary to scale the cosine function value. Over the range N = 0...7, the value of cosine ranges from +1 to 0; hence we multiply it by 255 to rescale its range to 0...255

cos is the trigonometric cosine function, with the argument stated in degrees

Using this relationship, we calculate the first eight entries in the table. The next eight entries are identical to the first eight, but in reverse order. The final 16 entries are constant, set at the maximum output, 255.

```
FullStep  ByteTable 12,6,3,9,12,6,3,9 ;j is 0...4
```

We also require access to the full step pattern, so as to accomplish the polarity reversals. The byte table **FullStep**, copied from Chapter 8, provides this access.

```
;Initialization
;--------------------

        HPWM  0,PWMPeriod,DefaultDC
        HPWM  1,PWMPeriod,DefaultDC

        ;set motor port to output
        For j = MPin to MPin+3
                Output j
        Next
        ;all windings off
        MPort = 0
```

At initialization, we turn both PWM modules on and set the period and duty cycle at the default values, previously defined as 256 and 128 respectively. We also set the output pins to output mode, but made low to unenergize the motor windings.

```
Main
;----
For i = 0 to CyCLen-1
        j = (i+8) / 16
        k = i//32
        n = (i+16)//32
        HPWM 0,PWMPeriod,Cosine(k)
        HPWM 1,PWMPeriod,Cosine(n)
        MPort = FullStep(j)
Next

        GoTo Main
```

The main program relies on four index variables, **i**, **j**, **k** and **n**:

i—runs from 0...63 and represents one complete microstep cycle.

j—indexes into the byte table **FullStep()**. In our earlier table overview, **FullStep(j)** returns the "binary value" value. To accomplish this, **j** steps from 0...4, with each **j** increment amounting to 16 microsteps. Since the first full step is divided into two half steps, we add one half of 16 microsteps as an offset. Hence, **j=(i+8)/16**.

k—indexes into the cosine array **cosine()** for windings A-C. Based on our earlier table, **k** must run 0...31 and then repeat 0...31 as **i** runs 0...63.

n—indexes into the cosine array **cosine()** for windings B-D. Based on our earlier table, n tracks k, but with an offset of 16 microsteps. Hence, we compute **n = (i+16)//32**.

After each complete microstep cycle is repeated, the `GoTo Main` restarts the cycle.

Let's see how these four variables and the two byte tables fit into our summary:

Full Step	Winding A-C		Winding B-D		μstep	Motor Sign Bits				Binary Value
	P	S	P	S	i	B3	B1	B2	B0	
	*	Cosine(k)	*	Cosine(n)	i			FullStep(j)		
1a	+	R		C	0...6	1	0	1	0	12
		A		O	7	1	0	1	0	12
2	-	M	+	N	8...14	0	1	1	0	6
		P		T	15	0	1	1	0	6
	-	C	+	R	16...22	0	1	1	0	6
		O		A	23	0	1	1	0	6
3	-	N	-	M	24...30	0	1	0	1	3
		T		P	31	0	1	0	1	3
	-	R	-	C	32...38	0	1	0	1	3
		A		O	39	0	1	0	1	3
4	+	M	-	N	40...46	1	0	0	1	9
		P		T	47	1	0	0	1	9
	+	C	-	R	48...54	1	0	0	1	9
		O		A	55	1	0	0	1	9
1b	+	N	+	M	56...62	1	0	1	0	12
		T		P	63	1	0	1	0	12

P=polarity of current into windings
S=status;
    RAMP—current change in cosine relationship
    CONT—constant current
Underline indicates changed value; intermediate steps follow correspondingly
* = the sign is taken care of automatically by FullStep(j)

There's a lot going on in six lines of code, so let's unroll the loop and see exactly what happens as we go through one complete cycle of 64 microsteps.

i	J	FW(J)	B3	B1	B2	B0	k	Cosine (k)	n	Cosine (n)	Comments
0	0	12	1	0	1	0	0	252	16	255	AC=+ BD=+
1	0	12	1	0	1	0	1	243	17	255	
2	0	12	1	0	1	0	2	224	18	255	
3	0	12	1	0	1	0	3	196	19	255	
4	0	12	1	0	1	0	4	161	20	255	
5	0	12	1	0	1	0	5	120	21	255	
6	0	12	1	0	1	0	6	74	22	255	
7	0	12	1	0	1	0	7	25	23	255	
8	1	6	0	1	1	0	8	25	24	255	Reversal: AC= - / BD = +
9	1	6	0	1	1	0	9	74	25	255	
10	1	6	0	1	1	0	10	120	26	255	
11	1	6	0	1	1	0	11	161	27	255	
12	1	6	0	1	1	0	12	195	28	255	

(Continued)

i	J	FW(J)	B3	B1	B2	B0	k	Cosine (k)	n	Cosine (n)	Comments
13	1	6	0	1	1	0	13	224	29	255	
14	1	6	0	1	1	0	14	243	30	255	
15	1	6	0	1	1	0	15	252	31	255	
16	1	6	0	1	1	0	16	255	0	252	
17	1	6	0	1	1	0	17	255	1	243	
18	1	6	0	1	1	0	18	255	2	224	
19	1	6	0	1	1	0	19	255	3	196	
20	1	6	0	1	1	0	20	255	4	161	
21	1	6	0	1	1	0	21	255	5	120	
22	1	6	0	1	1	0	22	255	6	74	
23	1	6	0	1	1	0	23	255	7	25	
24	2	3	0	1	0	1	24	255	8	25	Reversal: AC= - / BD = -
25	2	3	0	1	0	1	25	255	9	74	
26	2	3	0	1	0	1	26	255	10	120	
27	2	3	0	1	0	1	27	255	11	161	
28	2	3	0	1	0	1	28	255	12	195	
29	2	3	0	1	0	1	29	255	13	224	
30	2	3	0	1	0	1	30	255	14	243	
31	2	3	0	1	0	1	31	255	15	252	
32	2	3	0	1	0	1	0	252	16	255	
33	2	3	0	1	0	1	1	243	17	255	
34	2	3	0	1	0	1	2	224	18	255	
35	2	3	0	1	0	1	3	196	19	255	
36	2	3	0	1	0	1	4	161	20	255	
37	2	3	0	1	0	1	5	120	21	255	
38	2	3	0	1	0	1	6	74	22	255	
39	2	3	0	1	0	1	7	25	23	255	
40	3	9	1	0	0	1	8	25	24	255	Reversal AC= + BD = -
41	3	9	1	0	0	1	9	74	25	255	
42	3	9	1	0	0	1	10	120	26	255	
43	3	9	1	0	0	1	11	161	27	255	
44	3	9	1	0	0	1	12	195	28	255	
45	3	9	1	0	0	1	13	224	29	255	
46	3	9	1	0	0	1	14	243	30	255	
47	3	9	1	0	0	1	15	252	31	255	
48	3	9	1	0	0	1	16	255	0	252	
49	3	9	1	0	0	1	17	255	1	243	
50	3	9	1	0	0	1	18	255	2	224	
51	3	9	1	0	0	1	19	255	3	196	
52	3	9	1	0	0	1	20	255	4	161	
53	3	9	1	0	0	1	21	255	5	120	
54	3	9	1	0	0	1	22	255	6	74	
55	3	9	1	0	0	1	23	255	7	25	
56	4	12	1	0	1	0	24	255	8	25	Reversal: AC= + / BD = +

(Continued)

i	J	FW(J)	B3	B1	B2	B0	k	Cosine (k)	n	Cosine (n)	Comments
57	4	12	1	0	1	0	25	255	9	74	
58	4	12	1	0	1	0	26	255	10	120	
59	4	12	1	0	1	0	27	255	11	161	
60	4	12	1	0	1	0	28	255	12	195	
61	4	12	1	0	1	0	29	255	13	224	
62	4	12	1	0	1	0	30	255	14	243	
63	4	12	1	0	1	0	31	255	15	252	

Let's see how Program 19-1 works. It's easier to illustrate Program 19-1's waveforms with a resistor instead of stepper motor windings, so I substituted an 8 ohm, 5 watt resistor for the stepper motor winding connected between the SN754410's pins 3 and 6 for Figures 19-7 through 19-9. Figure 19-7 shows the enable voltage (pin 1) to the SN754410 and the resulting current (pin 3). Both match our desired target quite well; the current swings smoothly between its maximum positive and negative values. I've expanded the timebase in Figure 19-8 so you may see that the current and voltage are indeed quantized in discrete values.

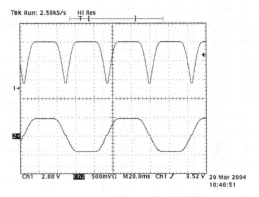

Figure 19-7: Program 19-1 Output; Ch1: SN754410 Pin 1; Ch2: Current from SN754410 Pin 3.

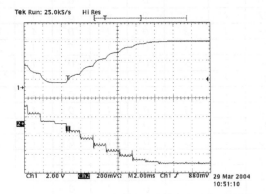

Figure 19-8: Expanded View of Program 19-1 Output; Ch1: SN754410 Pin 1; Ch2: Current from SN754410 Pin 3.

But, just a minute you're thinking. Figures 19-7 and 8 show the SN754410's enable input driven by an analog voltage, not a logic-level PWM output. Good catch, but what you see is a digital sampling oscilloscope artifact. As we saw in Chapters 11 and 16, any analog-to-digital conversion only accurately represents the input waveform up to frequencies of one-half of the sample rate. Figure 19-8 is taken at a 25 kilo-samples/sec, permitting accurate reproduction of input signals up to only about 12.5 KHz. What Channel 1 shows in both Figures 19-7 and 8 is the *average* voltage output of the HWPM module. If we further expand the time base to 25 mega-samples/sec, as shown in Figure 19-9, we see exactly what should be there; a PWM digital output that switches between logical 0 and 1.

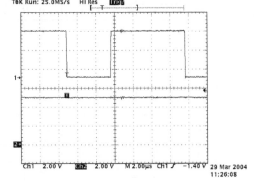

Figure 19-9: Ch1: SN754410 Pin 1; Ch2: Unused.

Now that we've verified Program 19-1 does what is expected, connect a stepper motor to the SN754410. It runs, but very slowly compared with Chapter 8's programs. Measuring the duration for two current reversals, (64 microsteps) we find that one microstep requires about 1410µs. Since it takes 768 microsteps for one revolution of my PF35-48 stepper, we calculate 1.08 seconds for each motor shaft revolution, about 55.4 RPM.

Program 19-2

Program 19-2 answers the question "can we improve Program 19-1 to a more useful speed"? The answer is yes, we can increase the speed about six-fold by replacing all MBasic statements within the **For i = 0 to CycLen-1...Next** loop by assembler routines. We'll use the stepwise refinement approach developed in Chapter 14—replace one MBasic statement with an assembler routine and verify its operation, repeating the process with different MBasic statements until our speed objective is met. And, we'll make liberal use of the "bolt-in" MBasic-to-assembler replacements in Chapter 14.

There's little point in providing six different versions of Program 19-2, each with one more MBasic statement converted to assembler. Instead, we'll show the "as converted" program, making clear which MBasic statement is replaced by assembler code. Of course, this is not the way Program 19-2 evolved. As we learned in Chapter 14, it's much easier to replace the statements one-at-a-time and go to the next replacement only after the first is debugged and verified.

```
;Program 19-2 Microstep
;in MBasic with assembler speedups

STACK = 10        ;need space in Bank 0 for variables

;Constants
;------------------------
PWMPeriod      Con      255      ;PWM width 20MHz clock = 12.75us
                                 ;correspond to 78 KHz
DefaultDC      Con      128      ;default duty cycle 50%
MPin           Con      B0       ;starting pin for motor driver
MDelay  Con    2000     ;microseconds delay per step
CycLen  Con    64       ;micro-cycles per rev
Zero           Con      0        ;for banking

;Variables
;--------------------
i              Var      Byte     ;overall counter
j              Var      Byte     ;full step index
k              Var      Byte     ;cos index for w1
n              Var      Byte     ;cos index for w2
l              Var      Byte     ;ASM counter index
x              Var      Byte     ;temp
MPort          Var      PortB.Nib0      ;output port pins

;Cosine lookup table for 180 degrees
;We add the steady-state part to simplify the stepping code
;Only comment out these two byte tables after converting to ;assembly code,as they
are needed during the piecewise BASIC-to-;Assembler conversions.

;Cosine        ByteTable        252,243,224,196,161,120,74,25,
;                               25,74,120,161,196,224,243,252,
;                               255,255,255,255,255,255,255,255,
;                               255,255,255,255,255,255,255,255
;
;FullStep       ByteTable 12,6,3,9,12,6,3,9 ;j is 0...4

;Initialization
;------------------
        HPWM 0,PWMPeriod,DefaultDC
        HPWM 1,PWMPeriod,DefaultDC

        ;set motor port to output
        For j = MPin to MPin+3
```

```
                Output j
        Next
        ;all windings off
        MPort = 0
        l = 0
        i = 0
        k = 0
        n = 0
        x = 0

Main
;----
For i = 0 to CyCLen-1

        ;j = (i+8) / 16
        ASM
        {
        movf      i&0x7F,w          ;i+8
        addlw     0x8
        movwf     j&0x7F            ;j = i+8
        movlw     0x4
        movwf     l&0x7F            ;L=4

Shift16
        RRF               j&0x7F,f          ;/2 but this is rotate, not shift
        bcf               STATUS,C          ;clear the carry flag
        decfsz            l&0x7F,f          ;L=L-1
        GoTo              Shift16           ;while L > 0 GoTo Shift
        }

        ;k = i//32
        ASM
        {
        movf      i&0x7F,w
        andlw     b'00011111'       ;masks out high 3 bits = //32
        movwf     k&0x7F
        }

        ;n = (i+16)//32
        ASM
        {
        movf      i&0x7F,w          ;i+8
        addlw     .16
        andlw     b'00011111'       ;masks out high 3 bits = //32
        movwf     n&0x7F            ;j = i+8
        }

        ;HPWM 0,PWMPeriod,Cosine(k)      ;Pin C2
        ;HPWM 1,PWMPeriod,Cosine(n)      ;Pin C1

        ASM
        {
        ;PWM No. 1 on Pin C2
        ;----------------------
        movlw             HIGH CosineTable       ;w=high byte table location
        movwf             PCLATH                 ;PCLATH has high byte
        movlw             LOW (CosineTable+1)    ;first data point
        addwf             k&0x7F,w               ;offset address of datapoint
        btfsc             STATUS,C
        incf              PCLATH,f
        call              CosineTable            ;what is duty cycle value
        movwf             x&0x7F                 ;Store in x

        ;now get two LSB and move to CCP1CON home
        ;two lowest bits of duty cycle are held in CCP1CON<5> & <4>
        ;presumptive clear CCP1CON<4>

        bcf                       CCP1CON,4
        ;If DutyCycle.Bit0 = 1 then set CCP1CON<4>=1
        btfsc             x&0x7F,0
```

```
        bsf              CCP1CON,4
        ;Presumptive clear of CCP1CON<5>
        bcf              CCP1CON,5
        ;If DutyCycle.Bit1 = 1 then Set CCP1CON<5>=1
        btfsc     x&0x7F,1
        bsf              CCP1CON,4

        ;now we move DutyCycle to CCPR1L. But first must
        ;mask off lower two bits and shift right 2 bits
        movlw            b'11111100'
        andwf     x&0x7F,f          ;mask in place
        rrf       x&0x7F,f          ;right 1 bit
        rrf       x&0x7F,f          ;right 1 more bit
        movf      x&0x7F,w          ;copy it to w
        movwf            CCPR1L ;now insert into CCPR1L

        ;PWM No. 2 on Pin C1
        ;----------------------
        movlw            HIGH CosineTable      ;w=high byte table location
        movwf            PCLATH                ;PCLATH = high byte of table
        movlw            LOW (CosineTable+1)   ;first data point
        addwf     n&0x7F,w                     ;offset address of datapoint
        btfsc     STATUS,C
        incf      PCLATH,f
        call             CosineTable           ;what is duty cycle value
        movwf            x&0x7F                ;Store in x

        ;now get two LSB and move to CCP2CON home
        ;two lowest bits of duty cycle are held in CCP2CON<5> & <4>
        ;presumptive clear CCP2CON<4>
        bcf              CCP2CON,4
        ;If DutyCycle.Bit0 = 1 then set CCP2CON<4>=1
        btfsc     x&0x7F,0
        bsf              CCP2CON,4

        ;Presumptive clear of CCP2CON<5>
        bcf              CCP2CON,5
        ;If DutyCycle.Bit1 = 1 then Set CCP2CON<5>=1
        btfsc     x&0x7F,1
        bsf              CCP2CON,4

        ;now we move DutyCycle to CCPR2L. But first must
        ;mask off lower two bits and shift right 2 bits
        movlw            b'11111100'
        andwf     x&0x7F,f          ;mask in place
        rrf       x&0x7F,f          ;right 1 bit
        rrf       x&0x7F,f          ;right 1 more bit
        movf      x&0x7F,w          ;copy it to w
        movwf            CCPR2L     ;now insert into CCPR1L
        GoTo             EndCosRoutine

CosineTable
    movwf PCL
    retlw 0xfc
    retlw 0xf3
    retlw 0xe0
    retlw 0xc4
    retlw 0xa1
    retlw 0x78
    retlw 0x4a
    retlw 0x19
    retlw 0x19
    retlw 0x4a
    retlw 0x78
    retlw 0xa1
    retlw 0xc4
    retlw 0xe0
    retlw 0xf3
    retlw 0xfc
    retlw 0xff
```

```
        retlw 0xff
        retlw 0xff
        retlw 0xff
        retlw 0xff
        retlw 0xff
        retlw 0xff
        retlw 0xff
        retlw 0xff
        retlw 0xff
        retlw 0xff
        retlw 0xff
        retlw 0xff
        retlw 0xff
        retlw 0xff
EndCosRoutine
    }


;MPort = FullStep(j)
ASM
{
    movlw       HIGH FullStepTable          ;w=high byte table location
    movwf           PCLATH                  ;PCLATH =high byte of table
    movlw           LOW (FullStepTable+1)   ;first data point
    addwf           j&0x7F,w                ;offset address of datapoint
    btfsc           STATUS,C
    incf            PCLATH,f
    call            FullStepTable           ;what is step value
    movwf           MPort&0x7F              ;output it to PortB
    GoTo        EndRoutine

FullStepTable
    movwf PCL
    retlw 0xc
    retlw 0x6
    retlw 0x3
    retlw 0x9
    retlw 0xc
    retlw 0x6
    retlw 0x3
    retlw 0x9
EndRoutine
        BankSel Zero
    }

    ;For max speed, remove the following statement
        Pauseus MDelay
Next
GoTo Main
End
```

The constants and variable declaration section tracks Program 19-1, with four exceptions:

```
MDelay Con     2000     ;microseconds delay per step

l              Var   Byte   ;ASM counter index
x              Var   Byte   ;temp
Zero           Con   0      ;for banking
```

Since Program 19-2 executes a microstep in 246µs, in many applications it may be necessary to slow it down. Hence, we've added a **pauseus** statement to the microstep control loop; **MDelay** is the value of the delay, in microseconds. I've made this a constant to simplify our programming, but you may wish to make the delay time a user-settable variable. The assembler program GPASM, used with MBasic 5.3.0.0 onward, will not accept a numerical argument for **banksel**; hence we define the symbolic constant **zero** as 0 thus permitting us to set the bank selector bits to 0 using the operation **banksel zero**. The operation **banksel 0**, permissible in MPASM used in MBasic's earlier versions is not accepted by GPASM's current release.

Two additional byte variables, **l** and **x** are used in the assembler routines.

```
HPWM 0,PWMPeriod,DefaultDC
HPWM 1,PWMPeriod,DefaultDC

;set motor port to output
For j = MPin to MPin+3
        Output j
Next
;all windings off
MPort = 0
l = 0
i = 0
k = 0
n = 0
x = 0
```

The only initialization change adds "variable = 0" statements for all variables only used in an assembler routine. As we learned in Chapter 14, MBasic's compiler allocates space for a variable only if it's used in an MBasic statement. To force the compiler to allocate space, we either use the variable (such as using **j** in the **For...Next** loop) for some beneficial purpose, or we set it to some value, 0 in this case, in a dummy statement. (The variable **i** is used in a **For..Next** loop, so it does not require a dummy assignment in the initialization section.)

Before we dig into the code, two reminders:

* Our assembler program does not use **banksel** or otherwise set the banking bits, except once at the end of the **asm** code. This is because all the variables, ports and registers we access are in Bank 0, and when MBasic passes control to an in-line assembler routine, the banking bits are already set to bank 0. Chapters 13 and 14 explain how to verify whether bank selection is necessary. In Program 14-2, we force the compiler to reserve space in bank 0 for our variables by manually setting the maximum stack size to 40 bytes via the **STACK = 10** instruction.

* To remove spurious assembler warnings, I've trimmed off the compiler control bits by "anding" variables with $7F, using the operator **&0x7F**. (Remember assembler uses the prefix **0x** instead of **$** to indicate hexadecimal.) To load the content of memory location **i** into the PIC's **w** register, therefore, we write **movf i&0x7F,w**. Correct code will be generated if the compiler control bits are not masked off, as would be the case if this statement were written **movf i,w**, but the assembler will issue warning messages such as the following:

Warning[219] C:\DOCUME~1\SMITH\MYDOCU~1\MEASUR~1\EXPERI~1\CH19SE~1\PROGRAMS\PROGRA~2.ASM 197 : Invalid RAM location specified.

Message[302] C:\DOCUME~1\SMITH\MYDOCU~1\MEASUR~1\EXPERI~1\CH19SE~1\PROGRAMS\PROGRA~2.ASM 197 : Register in operand not in bank 0. Ensure that bank bits are correct.

Speedup 1: We convert **j = (i+8) / 16** to assembler. This conversion saves 319 μs over the pure-MBasic code, and increases the maximum speed of my PF35-48 stepper from 55.4 RPM under pure-MBasic code to 71.6 RPM.

When programming a multipart mathematical expression, we must function as a "human parser" and break the expression into a logical evaluation order. We'll follow the standard order of precedence rules for arithmetic, and evaluate the expression in two steps:

1. Calculate the value **i+8**. Store this value in **j**.
2. Calculate **j/16** and store the answer in **j**.

```
;j = (i+8) / 16
ASM
{
movf    i&0x7F,w        ;i+8
```

```
        addlw   0x8
        movwf   j&0x7F          ;j = i+8
```

We know that **i** is always in the range 0…63, so **i+8** cannot exceed 71. Hence, we have no worry about overflowing byte length arithmetic and may use a simple algorithm for addition. We first copy the value of **i** into the working register **w** with the **movf** operator. We then use the operator **addlw** to add the value of a constant **8** (a "literal" in Microchip terminology) to **w**, retaining the sum **i+8** in **w**. Lastly, we copy the value of the **w** register to the variable **j** by the operator **movwf**.

```
        movlw   0x4
        movwf   l&0x7F                          ;L=4

Shift16
        RRF         j&0x7F,f        ;/2 but this is rotate, not shift
        bcf         STATUS,C        ;clear the carry flag
        decfsz      l&0x7F,f        ;L=L-1
        GoTo        Shift16         ;while L > 0 GoTo Shift
        }
```

We now divide **j** by 16. Since 16 is an exact binary power, 2^4, integer division is simply bit-shifting **j** four bits to the right. (It wasn't accidental that we decided to use 16 microsteps per full step instead of 10 or 12; by keeping this relationship as an exact binary power value, we simplify the division algorithm considerably.)

To repeat the shift four times, we use the loop operator, **decfsz**. First, we load a temporary variable **l** with the number of times we wish the loop to be executed, four in this case. Since the mid-range PIC has no direct "load a constant to memory" operator, we first load 4 into the **w** register with the **movlw** operator, and then copy **w**'s value into **l** with the **movwf** operator.

The operator **decfsz f,d** decrements the value of the file **f** and returns the decremented value to destination **d**. In our case, the file is variable is **l** and we wish the result to be returned to **l**, so the destination option will be **f**. Consequently, the assembler statement (including the compiler trim provision) is **decfsz l&0x7F,f**. If **l=0**, the next instruction is skipped. If **l<>0**, the next instruction is executed. Hence, the code structure using **decfsz** to repeat a group of assembler instructions **N** times, in pseudo-assembler, is:

```
        Set a temporary variable l to N
Loop1
            Assembler code to be repeated goes here
            decfsz l,f
            GoTo Loop1
```

Note that the decrement is performed before the comparison. Hence to execute a code sequence four times, the counter variable must be preloaded with 4. After decrementing and comparing, the first three times through the loop, the result will be nonzero, so the operation **GoTo Loop1** will be executed. After the fourth pass through the loop, **l** equals 1. The **decfsz** operator then decrements **l** again, at which point it is zero. The test for zero is now true, so the **GoTo Loop1** statement is skipped and execution proceeds to the next statement after the **GoTo**. The code within the loop has been executed exactly four times, as we desired.

Microchip's mid-range PICs implement bit shifting as the assembler instructions **RLF** and **RRF**, standing for rotate left through carry flag and rotate right through carry flag. This is not quite binary division. Let's see how **RRF** works. Suppose we wish to divide the value 51 by 4. (I've highlighted the movement of the carry bit in bold.)

					Register				
C	*B7*	*B6*	*B5*	*B4*	*B3*	*B2*	*B1*	*B0*	*Comments*
0	0	0	1	1	0	0	1	**1**	*Value = 123*
1	0	0	0	1	1	0	0	1	*First shift to the right; equivalent to / 2; result is25*
1	**1**	0	0	0	1	1	0	0	*Second shift to right; note the 1 shifted through carry has now reappeared at B7. The decimal value is now 140, not 12 as desired.*

What happened is as we shift one bit to the right, the **%1** we shifted out moves to the carry bit in the PIC's status register. With the next shift, the carry bit moves to the register's B7 bit. This rotate through carry might be more properly named "re-circulation" as the bits we shift out don't disappear, but rather are re-circulated back through the carry flag bit back to the high order bit in the register. To prevent these unwanted bits from being re-circulated, we adopt a strategy of shift, followed by clearing the carry bit. Repeating our calculations for dividing 51 by 4, we see that correct answer is produced.

				Register					
C	*B7*	*B6*	*B5*	*B4*	*B3*	*B2*	*B1*	*B0*	*Comments*
0	*0*	*0*	*1*	*1*	*0*	*0*	*1*	*1*	*Value = 123*
1	*0*	*0*	*0*	*1*	*1*	*0*	*0*	*1*	*First shift to the right; equivalent to / 2; result is25*
0	*0*	*0*	*0*	*1*	*1*	*0*	*0*	*1*	*Now clear the carry bit*
1	*0*	*0*	*0*	*0*	*1*	*1*	*0*	*0*	*Second shift. The result is 12, as desired*

To clear the carry flag, we use the bit clear operator, **bcf**, with file **Status**, where the carry flag bit is held at bit **C**. Hence, our two-operator rotate and clear sequence is implemented as:

```
RRF        j&0x7F,f        ;/2 but this is rotate, not shift
bcf        STATUS,C        ;clear the carry flag
```

Speedup 2: Here we convert **k = i//32** to assembler. This conversion saves 276µs, and increases the maximum speed of my PF35-48 stepper to 95.8 RPM.

```
;k = i//32
ASM
{
movf     i&0x7F,w
andlw    b'00011111'       ;masks out high 3 bits = //32
movwf    k&0x7F
}
```

MBasic's **//** operator is the modulus or remainder function. Here we may take advantage of another convenient feature of using an exact binary power as the divisor. If the divisor is of the form 2^N, the modulus can be obtained by masking out higher order bits of the dividend, retaining only the bottom N bits. And, we may simply mask off bits by the bitwise AND operator. Let's see how this works in calculating **i//32** supposing **i=51**.

B7	*B6*	*B5*	*B4*	*B3*	*B2*	*B1*	*B0*	*Comment*
0	*0*	*1*	*1*	*0*	*0*	*1*	*1*	*=51*
0	*0*	*0*	*1*	*0*	*0*	*1*	*1*	*Right 5 bits only = 19*

Logical AND to perform the bit masking

0	*0*	*1*	*1*	*0*	*0*	*1*	*1*	*=51*
0	*0*	*0*	*1*	*1*	*1*	*1*	*1*	*=(32-1)*
0	*0*	*0*	*1*	*0*	*0*	*1*	*1*	*51 AND 31 = 19*

To implement this algorithm, our assembler code copies the value of **i** into the **w** register, masks out the upper three bits by **AND**ing **w** with the literal value **%00011111** and copies the result into **k**.

Speedup 3: Here we convert **n = (i+16)//32** to assembler. This conversion saves 320µs, and increases the maximum speed of my PF35-48 stepper to 158.2 RPM.

We've already done both parts of this conversion in Speedups one and two. Since **i+16** can never be greater than 79, the result fits in a single byte and we need not worry about overflow in the addition. Applying the laws of mathematical precedence, we evaluate this expression as:

1. Copy the contents of **i** to register **w** and add 16
2. Mask out the highest three bits in register **w**, thereby yielding the remainder after dividing by 32. Copy the contents of **w** to **n**.

```
;n = (i+16)//32
ASM
{
   movf    i&0x7F,w          ;i+8
   addlw   .16
   andlw   b'00011111'       ;masks out high 3 bits = //32
   movwf   n&0x7F            ;j = i+8
}
```

The above code implements this strategy, using the same techniques we used earlier.

Speedup 4: Here we convert **MPort = FullStep(j)** to assembler. This conversion saves 72μs, and increases the maximum speed of my PF35-48 stepper to 185.2 RPM.

The assembler code implements a standard jump table. Jump tables are discussed at Chapter 14 and the table here follows the template developed there. In essence, we load the **w** register with the desired index **j** and program execution jumps to the **j**[th] **retlw** statement following the start of the table.

The **retlw** statement then returns the associated literal value in the **w** register that we then send to Port B.

```
;MPort = FullStep(j)
ASM
{
   movlw              HIGH FullStepTable       ;w=high byte table loc'n
   movwf              PCLATH                   ;PCLATH =high byte of table
   movlw              LOW (FullStepTable+1)    ;first data point
```

To get program execution to jump to the desired **retlw** statement, we load the starting address of the **retlw** sequence (identified by the label **FullStepTable** here) into the program counter and add the offset to retrieve the specific item. Mid-range PICs—including the 16F877A we use here—have a 13-bit program counter, so we can't load a complete address in one step with the PIC's 8-bit wide data memory. And, the upper five bits of the program counter are only indirectly addressable via the **PCLATH** register. Hence, we load the starting address of the jump table in two pieces; the high byte of the address is loaded via the **PCLATH** register, while the low address byte is directly entered into **PCL**, which holds the low 8 bits of the program counter.

```
   addwf        j&0x7F,w               ;offset address of datapoint
   btfsc        STATUS,C
   incf         PCLATH,f
   call         FullStepTable          ;what is step value
```

We then add the offset address to **PCL** to obtain the full 13-bit address of the desired jump table entry. If the result of adding the offset address to the lower 8 bits of the program counter is greater than 255, the overflow sets the carry flag. If the carry flag is set, the statement following the **btfsc STATUS, C** is executed and we add **%1** to **PCLATH**, thereby bumping up the high order five bits by one to represent the carry. If the carry flag is not set (no carry), the **incf PCLath, f** statement is skipped. The actual jump is executed through the **call** operator.

```
   movwf              MPort&0x7F               ;output it to PortB
   GoTo               EndRoutine
```

Execution returns to the statement after the **call**, and with the indexed value stored in **w**. We then copy **w**'s value to the output port through the **movwf** operator. Lastly, we jump over the data table via a **GoTo** that transfers execution to the end of this routine.

```
FullStepTable
   movwf PCL
   retlw 0xc
   retlw 0x6
```

```
        retlw 0x3
        retlw 0x9
        retlw 0xc
        retlw 0x6
        retlw 0x3
        retlw 0x9
EndRoutine
```

Speedup 5: Here we convert **HPWM 0, PWMPeriod,Cosine(k)** to assembler. This conversion saves 85µs, and increases the maximum speed of my PF35-48 stepper to 231 RPM.

```
ASM
{
;PWM No. 1 on Pin C2
;----------------------
movlw           HIGH CosineTable        ;w=high byte table location
movwf           PCLATH          ;PCLATH has high byte
movlw           LOW (CosineTable+1)     ;first data point
addwf           k&0x7F,w                ;offset address of datapoint
btfsc           STATUS,C
incf            PCLATH,f
call            CosineTable             ;what is duty cycle value
movwf           x&0x7F                              ;Store in x
```

The above code is another jump table lookup, performing the same function, and with the same algorithm, we used in connection with the **FullStepTable** lookup. The only differences are that we index based on the variable **k** and the lookup table is **CosineTable**. Hence, we will not further analyze this part of the code.

The returned value from **CosineTable** is stored in the temporary variable **x**. We will now see how the value of **x** is copied into the PWM duty cycle control registers. The duty cycle value is 10 bits wide, organized as 2-bit and 8-bit segments:

10-bit Duty Cycle Value (B9...B0)									
B9	B8	B7	B6	B5	B4	B3	B2	B1	B0

CCPR1L								CCP1CON	
B7	B6	B5	B4	B3	B2	B1	B0	B5	B4

The most significant eight duty cycle bits are contained in the **CCPR1L** register, while the two least significant bits are stored at bits 5 and 4 in the **CCP1CON** register. Although we wish to write an 8-bit duty cycle value, we must split it into a 6-bit and a 2-bit segment and place the values in **CCPR1L** and **CCP1CON**. After inserting the value of **x** into the duty cycle control registers, the result will be the following, where x.7 means bit no. 7 of the value of our temporary duty cycle holder **x**, x.6 means bit no. 6 of **x**, etc.:

CCPR1L								CCP1CON	
B7	B6	B5	B4	B3	B2	B1	B0	B5	B4
0	0	x.7	x.6	x.5	x.4	x.3	x.2	x.1	x.0

```
bcf                     CCP1CON,4
;If DutyCycle.Bit0 = 1 then set CCP1CON<4>=1
btfsc           x&0x7F,0
bsf             CCP1CON,4
```

Midrange PICs do not have a "copy bit" operation, so to move the a bit from one location to another, we follow a two-step process:

1. Set the destination bit to **%0** using the bit clear file **bcf** operator.
2. Check the source bit using the bit test file skip if clear **btfsc** operator. If the source bit is **%0**, we are finished. If it is **%1**, set the destination bit to **%1** using the bit set file **bsf** operator.

The above code implements this algorithm for bit x.0.

```
;Presumptive clear of CCP1CON<5>
bcf                 CCP1CON,5
;If DutyCycle.Bit1 = 1 then Set CCP1CON<5>=1
btfsc           x&0x7F,1
bsf                     CCP1CON,4
```

These three lines implement the same algorithm for bit x.1.

Now, we must move **x**'s six high order bits to **CCPR1L**, right justifying the six bits in the process. We could use the rotate through carry function and zero out the carry flag, as when did when dividing by 16 earlier. We'll use an alternative approach and force the lowest two order bits of **x** to zero by **AND**ing with **%11111100** and then rotate.

```
;now we move DutyCycle to CCPR1L. But first must
;mask off lower two bits and shift right 2 bits
movlw           b'11111100'
andwf           x&0x7F,f        ;mask in place
```

To accomplish this, we move the mask literal **%11111100** to the **w** register and then **AND** the **w** register's value with **x**, storing the result in **x**, with the above two operations.

```
rrf             x&0x7F,f        ;right 1 bit
rrf             x&0x7F,f        ;right 1 more bit
```

Since the rightmost two bits are zero, we may rotate two places to the right without worry that **%1**'s will be shifted out and return through the carry flag. After these two rotations, the six bits left are right justified in **x**.

```
movf            x&0x7F,w        ;copy it to w
movwf           CCPR1L          ;now insert into CCPR1L
```

The last step remaining is to copy **x**'s value to the **CCPR1L** register, which we do in the usual way by first copying **x**'s value to the **w** register and then copying the **w** register to the destination, **CCPR1L**.

Speedup 6: Here we convert **HPWM 1, PWMPeriod,Cosine(n)** to assembler. This conversion saves 91μs, and increases the maximum speed of my PF35-48 stepper to 317 RPM.

This code duplicates the code in Speedup 5, except it is for second hardware PWM module. The only changes are in the index into the **CosineTable** (**n** instead of **k**) and the destination registers for the duty cycle values (**CCPR2L** instead of **CCPR1L** and **CCP2CON** instead of **CCP1CON**). There is no need, therefore, to further analyze Speedup 6.

After making these six replacements, we have replaced almost all the code inside the control loop:

```
Main
        For i = 0 to CyCLen-1
        ...
        Next
Goto Main
```

The only remaining MBasic code is an optional **Pauseus** statement to provide speed control. While we could replace the **For...Next** loop, or even the **Main...GoTo Main** loop, with assembler code, we won't. By keeping these control structures in MBasic, we permit easier control or modification by other MBasic statements. We'll see this concept in more detail in Program 19-3.

Let's take a look at the output of Program 19-2. Figure 19-10 shows the output from both hardware PWM modules. As we discussed before, these look like analog waveforms because the digital sampling oscilloscope sample rate is much slower than the PWM rate. The cosine ramp waveforms are out-of-phase by one-half waveform, which is correct.

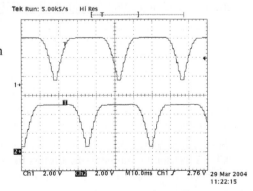

Figure 19-10: Output of Hardware PWM Modules Ch1: PIC C2; Ch2: PIC C1.

Figures 19-11 through 19-15 show the hardware PWM output and the associated motor current under Program 19-2 with added delays of 0, 500, 1000, 1500 and 2000µs per microstep, respectively. The current waveforms depart quite markedly from the resistive waveforms, particular as the step speed increases. These figures were taken with 8V drive and no additional series resistance, and the deviation is caused by the relatively slow L/R time constant for this drive arrangement compared with the slew rate of the applied average motor voltage. Even with the divergence from the desired current waveform, the motor follows microsteps reasonably well, particularly at slower speed.

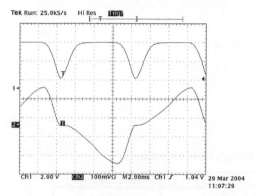

Fig 19-11: Ch-19-K.TIF Caption: Output of Program 19-2, No Delay; Ch1: HPWM Output; Ch2: PF45-48 Motor Current.

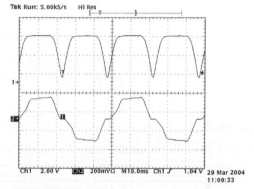

Figure 19-12: Output of Program 19-2, 500 µs Delay; Ch1: HPWM Output; Ch2: PF45-48 Motor Current.

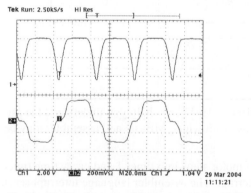

Figure 19-13: Output of Program 19-2, 1000 µs Delay; Ch1: HPWM Output; Ch2: PF45-48 Motor Current.

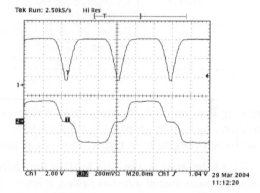

Figure 19-14: Output of Program 19-2, 1500 µs Delay; Ch1: HPWM Output; Ch2: PF45-48 Motor Current.

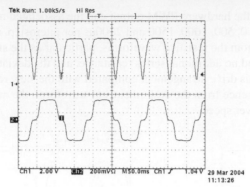

Figure 19-15: Output of Program 19-2, 2000 μs Delay; Ch1: HPWM Output; Ch2: PF45-48 Motor Current.

Program 19-3

Program 19-3 allows us to serially control a stepper motor through eight commands:

Command	Function	Status Query?
U12345	Set delay between motor steps at 12345 us	U?
DF or DR	Set direction Forward (DF) or Reverse (DR)	D?
S	Status report of parameters	—
MF or MH	Set mode to half-step (MH) or full step (MF)	M?
T12345	Set target position to 12345 motor steps; stop when reached	T?
#	Execute; start motor rotating	—
$\overline{F}$ or $\overline{L}$	Stop motor, release current to free the motor shaft ($\overline{F}$) or keep current flowing to lock motor in place ($\overline{L}$)	—
Z	Zero the motor position. Z? reports the current motor position.	Z?

When reading Program 19-3, the flow charts of Figures 19-16 and 19-17 may help you understand the program structure. As with Program 19-2, it demonstrates the ease with which MBasic and assembler may be mixed, with each working to its respective strong points.

Program 19-3 is interrupt driven—we load the time between motor steps into timer 1 and every time timer 1 determines the programmed time has passed, it issues an interrupt. The assembler language interrupt service routine ISRASM is immediately executed and advances the motor one step, if it has not previously reached its target position. It also updates the location counter and resets timer 1. These operations are transparent to any MBasic statement that might be executing when the interrupt service routine is called.

Program 19-3 works with a bipolar motor using the circuit of Figure 19-3, and with unipolar motors using any unipolar driver circuits shown in Chapter 8. However, don't forget to connect the PIC's hardware serial input and output pins to the RS-232 interface port. Figure 19-18 shows these connections added to an SN754410 bipolar driver circuit. Figure 19-19 shows how you may use the SN754410 in pure unipolar mode. You may also use this circuit for any of other stepper driver programs in this chapter, or those in Chapter 8.

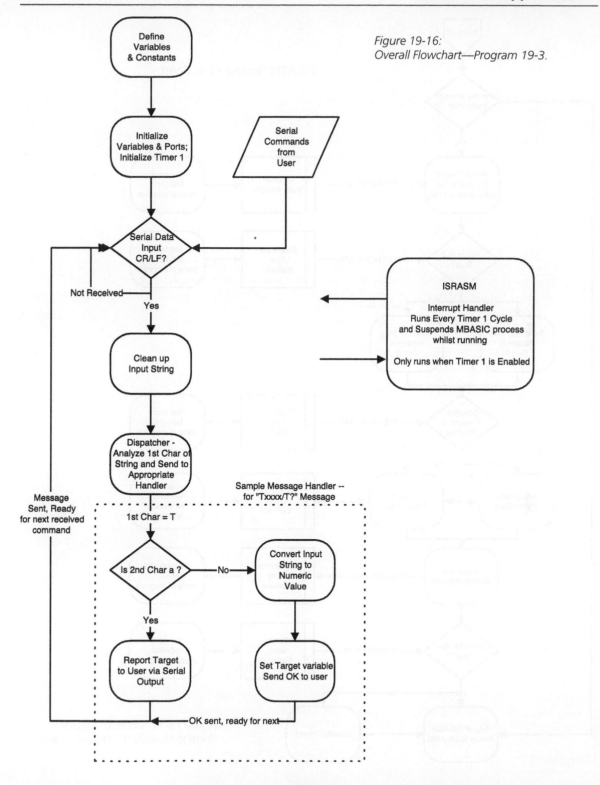

Figure 19-16:
Overall Flowchart—Program 19-3.

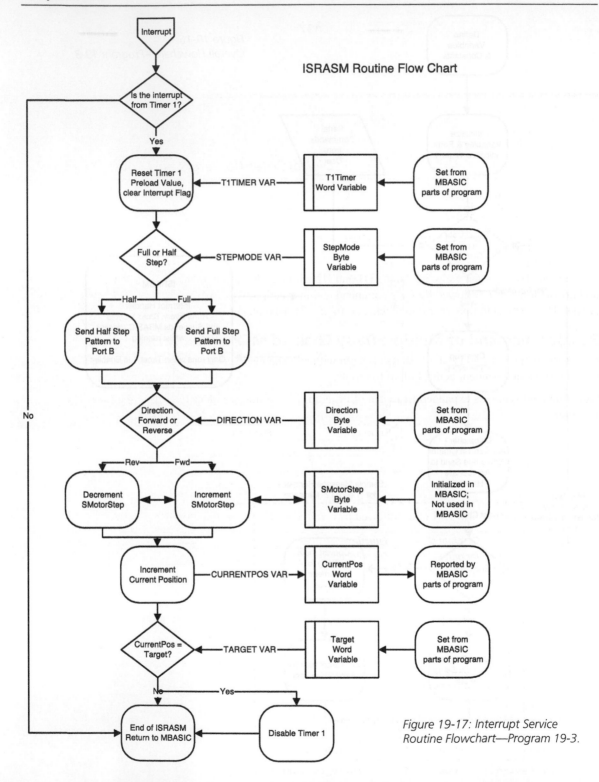

Figure 19-17: Interrupt Service Routine Flowchart—Program 19-3.

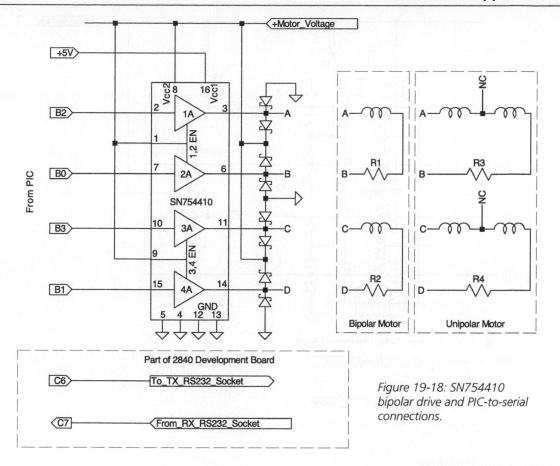

Figure 19-18: SN754410 bipolar drive and PIC-to-serial connections.

Since Program 19-3 is quite lengthy, and since we've seen almost all its concepts elsewhere, we'll not provide a printed listing in the text and we also depart from our normal form of analysis and will not reproduce the program in sections, followed by a detailed analysis, except in a few cases where justified. Please refer to the CD-ROM electronic copy of Program 19-3.

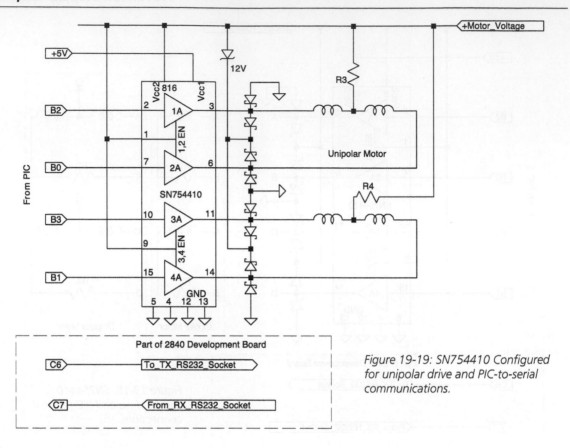

Figure 19-19: SN754410 Configured for unipolar drive and PIC-to-serial communications.

Constants and Variables

At the outset, we define all needed variables and constants. Certain variables are used in both MBasic and assembler routines. As we reminded you in the case of Program 19-2, MBasic is an optimizing compiler and a variable defined in the normal way in MBasic but used only in an assembler routine, will not be allocated space. The **Clear** command sets all MBasic variables to zero and ensures that the compiler defines and allocates all variables, even if used only in an assembler routine, such as **SmotorStep**.

For convenience, we define constants **FullStep**, **HalfStep**, **Fwd** and **Rev** so that comparisons may be written in a more meaningful fashion. For example, the following two forms are equal, but one is clearly more meaningful to the reader.

```
If StepMode = 0 Then …

If StepMode = FullStep Then …
```

We also decided to declare the two position variables, **CurrentPos** and **Target** as type **word**, thus limiting their range to 0 … 65,535. For a stepper motor with 400 half steps per revolution, using word variables permits counting to just over 163 revolutions. Defining these two variables as type **long** would provide a much greater counting range, but would have complicated the assembler portions of the program considerably.

Initialization

We use **Clear** to set all variables to 0 and ensure the MBasic compiler allocates space for each variable. I've set the serial input speed at 9600 b/s. This is an arbitrary choice and can be changed to other standard values as desired. Since the USART implements a buffer, high data speeds are not be a problem. Next, Port B, pins **B0 … B3** are defined as outputs and initialized low. Port C, pin **C0** is also defined as an output. I use **C0** to measure execution times by connecting it to an oscilloscope. At the start of a particular routine **C0** is set high, and dropped low at the end of the routine, so the positive pulse width measured on the oscilloscope provides the execution time for the code bracketed by the **C0** high/low operations.

The short assembly routine we use to get the PIC's clock speed is interesting:

```
ASM
{
    movlw   (_MHz/.1000000)     ;_MHz is aconstant
    Banksel Fosc
    Movwf   Fosc&0x7F   ;likewise for command
    Banksel 0           ;Compiler expects Bank 0 on exit
}
```

In order to calculate the Timer 1 preset value to achieve a defined delay in microseconds, we must know the clock speed. Beginning in version 5.3.0.0, the clock speed is defined at the top of an MBasic program as part of the initialization string. This initialization string is normally hidden by MBasic's editor, but may be seen by examining your .bas file with a text editor, such as Notepad:

```
;%CONFIG% 16F877A $1312d00 $3f3a $0 $0  $0  $0  $0  $0  $0  $0
$0  $0  $0  $0  $0  $0  $0  $0
```

I've highlighted the clock value by enclosing it in a box. This number is the value you enter from MBasic's configuration dialog box as the oscillator frequency in Hz. This initialization string is from a program for a 16F877A PIC, with a 20 MHz clock (20000000 Hz). 20,000,000 in hexadecimal is $1312D00.

When your program is compiled, the clock speed value is converted into a constant appearing in the assembler file as:

```
_MHZ equ 01312D00h
```

The four lines of assembler code above copy the **_MHZ** value into the byte variable **Fosc**, accessible to MBasic. We use the assembler to perform the arithmetic to convert the frequency in Hz to MHz, that is, to divide by 1,000,000.

We now are in a position to calculate the number of nanoseconds per timer 1 "tick"—4000/Fosc. The factor 4 comes about because each timer tick is 4 oscillator cycles, while the factor 1000 converts microseconds to nanoseconds. We then initialize timer 1, enable its interrupt, but do not turn timer 1 on, as its interrupts are not yet required.

ISRASM

We'll postpone discussion of the interrupt service routine until later in this chapter.

Wait for Input String and Clean Up Received Data

The program assumes the user will send a series of command strings, with each string terminated by a carriage return or carriage return / line feed combination. (Don't forget to connect the PIC's hardware transmit and receive pins to the RS-232 socket on your 2840 Development Board.)

```
HSerIn [Str TempStore\ArraySize\$D]
```

The program pauses at the above statement and waits for a command string. Upon receipt of 10 characters (recall that we defined **ArraySize** as **10**) or upon receipt of a carriage return (Hex value **$D**) the received string is stored in **TempStore** and execution passes to the next program statement.

The length of incoming string is calculated and stored in variable **Ctr**.

The subroutine **CleanUp** is called to remove any control characters that might have been included in the input string and convert lower case alphabetic characters to upper case to remove case sensitivity. The cleaned string is copied into **WorkArray** and **TempStore** is zeroed out and is ready for the next analysis.

Analyze and Dispatch

The cleaned up command string, now contained in **WorkArray**, is analyzed in subroutine **Dispatcher**. The first character of the command string, in **WorkArray(0)** is analyzed and control is passed to one of a number of subroutines based upon **WorkArray(0)**.

Following is an overview of the command sequence:

Command	Function	Status Query?
U12345	Set delay between motor steps at 12345 us	U?
DF or DR	Set direction Forward (DF) or Reverse (DR)	D?
S	Status report of parameters	—
MF or MH	Set mode to half-step (MH) or full step (MF)	M?
T12345	Set target position to 12345 motor steps; stop when reached	T?
#	Execute; start motor rotating	—
$\overline{F}$ or $\overline{L}$	Stop motor, release current ($\overline{F}$) or keep current flowing to lock motor in place ($\overline{L}$)	—
Z	Zero the motor position. Note Z? reports the current motor position.	Z?

Each subroutine looks at the remaining characters (if any) in **WorkArray** and executes the appropriate functions.

Many subroutines call the subroutine **Convert** to get the numeric value associated with the command string. Consider, for example, the set delay function U. A command string beginning with U will be evaluated by the subroutine **GetUSec**.

First, **GetUSec** checks **WorkArray(1)**. If this character is a question mark "?" the command string calls for a parameter report. Using the **HserOut** function, **GetUSec** reports the current delay value.

If **WorkArray(1)** is not a question mark, **WorkArray** is evaluated by the subroutine **Convert**, where it is converted to a **Long** variable, **TempLong**. If **TempLong** exceeds the maximum 16-bit value, 65535, an error flag is set and an error message sent back to the user. If the converted value is within the permissible range, it is stored in **TempWord**.

GetUSec then must compute the proper timer 1 load setting to yield the desired interval between motor steps. First, a raw count value is computed, which is the count load value assuming there is no lost time in resetting the timer once an interrupt is triggered. However, some time is always lost due to the interrupt calling procedure and also due to elapsed time between starting the interrupt service routine and the code that actually resets the timer. This lost time is a constant (unless the early portions of the ISRASM routine are modified) and is held in the constant **T1Offset**. To arrive at the final timer preload, **T1Timer**, **T1Offset** is added to the raw count value. (Timer 1 counts up, so by adding an offset, the actual elapsed time is reduced, thereby making up for time lost in calling overhead.) If your PIC does not use a 20 MHz clock, you may find it necessary to change **T1Offset**.

ISRASM Portion Program Structure

General

As we've mentioned in Chapter 15, MBasic handles the complicated "stuff" both going into and returning from an Assembler Interrupt Service Routine (ISR). It saves the necessary information in the MBasic stack and restores the context at the end of the ISR, thereby freeing the user to concentrate on the functionality of the interrupt code.

Variables defined in MBasic are available in the ISR, but with some twists. MBasic adds extra bits to its variable address locations that are used by the compiler. When we are reading or writing these address locations in assembler, we trim these extra bits off. We also must consider in which bank MBasic has stored each variable. Fortunately, these tasks are simpler than they sound.

Bank Selection and Address Issues—Before reading or writing a variable we must ensure that the correct bank selection bits are set. In an ISR, we are use the pre-defined **BankSel** assembler macro. Thus, before we move data in or out of the byte length variable **StepMode** we will first execute the **BankSel** command:

```
BankSel        StepMode
```

When moving or modifying data in the variable, we trim off the extra compiler control bits:

```
movf    StepMode&0x7F,w        ;put StepMode into w
```

Here we are copying the value of the variable **StepMode** into the 16F876's working register, **w**.

At the end of the ISR, we will execute a **BankSel** command, as MBasic expects bank 0 to be selected when exiting the ISR.

Many times we wish to access a word variable. This is straightforward—the low order byte is addressed at the base address, while the high order byte is at the base address plus one. This is illustrated below in dealing with the word variable **CurrentPos**.

```
BankSel        CurrentPos
Incf           CurrentPos&0x7F,f
btfss          STATUS, Z
GoTo           CompareWithTarget
BankSel        CurrentPos
incf           (CurrentPos+1)&0x7F,f
```

The first two instructions deal with the low byte of **CurrentPos**; the last two instructions deal with the high byte of **CurrentPos**. The high byte is addressed as **CurrentPos+1**. (Remember that we are dealing here with the *address* of **CurrentPos**, not its *value*. Thus, one reads the instruction **incf** as to add one to the value stored at *address* **CurrentPos**.) Since MBasic's compiler will not break a word variable across bank boundaries, both **CurrentPos** and **CurrentPos+1** are in the same bank and may be set via **banksel CurrentPos**. (The GPASM assembler will not accept **BankSel (CurrentPos+1)** as a legitimate argument.)

Check the Interrupt Type

In addition to interrupts generated by timer 1, other interrupts may be triggered. In particular, **HSerIn** causes an interrupt upon character receipt. We do not wish to interfere with MBasic's handling of these interrupts, so the first thing we check is whether the timer 1 interrupt flag, bit **tmr1if** of register **PIR1**, is set. If it is set, we have a timer 1 interrupt; if it is clear the interrupt came from another source and program flow immediately jumps to the end so that the interrupt service routine established by the MBasic compiler may service the interrupt call.

Reset Timer 1

After clearing the **tmr1if** interrupt flag (to prevent multiple calls of the same interrupt) we must reset the timer 1 preload. This routine follows Microchip's recommended instruction sequence, setting the high byte first, followed by the low byte.

Check Mode and Output Bits

Next, the ISR outputs the bit pattern to Port B, pins **B0...B3**. We have the option of either half steps or full steps. It takes eight half steps for a complete pattern repeat, but only four full steps. Rather than use two different counters, one ranging from **0...3** and the other from **0...7**, we instead just double-up the full-step pattern and use a single **0...7** counter **SMotorStep**. This is the same approach developed in Chapter 8. Note the full step repeat.

SmotorStep	Half Step B0	B1	B2	B3	Full Step B0	B1	B2	B3
0	1	0	0	0	1	0	0	1
1	1	1	0	0	1	1	0	0
2	0	1	0	0	0	1	1	0
3	0	1	1	0	0	0	1	1
4	0	0	1	0	1	0	0	1
5	0	0	1	1	1	1	0	0
6	0	0	0	1	0	1	1	0
7	1	0	0	1	0	0	1	1

It's necessary, of course, to output either the full or half step patterns, depending upon the status of step mode variable, **StepMode**. We accomplish this by using either a full-step (**GetFullPattern**) or half-step subroutine (**GetHalfPattern**), as appropriate. The subroutine jumps to the address represented by **SmotorStep** plus the entry point, implemented by the **addwf PCL,f** instruction. The **retlw 0xnn** command then returns the literal value **nn** in the **W** register. The **W** register then is copied to Port B.

Since the pattern subroutines immediately follow their calling command, and this is a short assembler program, we need not worry about page boundaries in the subroutine call. Caution should be exercised in general, of course, to ensure that page boundaries are not inadvertently crossed without proper consideration. (Compare our jump table calling procedure in Program 19-2 which does consider page boundaries and the one we use here, which does not.)

Determine Direction and Update SmotorStep

We use **SmotorStep** as the index into either the half-step or full-step patterns. **SmotorStep** must run from **0...7**.

If we are moving in the forward direction, we increment **SmotorStep**, so the index runs 0, 1, ... 7. If we are moving in the reverse direction, we decrement **SmotorStep**, so the index runs 7, 6, ... 0.

We accomplish this by a simple check of the direction variable **Direction**, and use either an increment (**incf**) or a decrement (**decf**) instruction on **SMotorStep**, as appropriate.

Following incrementing or decrementing **SmotorStep**, we must ensure it does not go outside the range 0...7. Since 0...7 represents the range of the last the last three bits of the byte variable, we can simply mask off the higher 5 bits with the **AND 0x7** command. This gives a three-bit counter that is forced to range between 0 and 7.

Count Completed Steps

We also desire to keep track of how many step instructions have been executed, so they may be compared against the target position. Completed instructions are stored in **CurrentPos**, a word variable.

Because **CurrentPos** is a word variable, we cannot use one simple **incf** instruction to count. Rather, we deal separately with the low and high order bytes that comprise **CurrentPos**.

First, we increment the low order byte and check it for zero. If it is zero, we have counted past 255 and therefore must roll over one count to the high order byte.

Check Against Target

Finally, we must check the current motor position against the target position. If the match, we will turn off timer 1 and stop the motor.

In MBasic this would be a simple **IF CurrentPos = Target Then StopMotor** statement. Assembler requires us to get a bit closer to the nuts and bolts of these two variables, however.

The comparison follows the algorithm of reference [19-2] for a 16-bit comparison— compare the high order bytes; if they are equal, compare the low order bytes. If both the high and low order bytes are equal, the two 16-bit variables are equal.

Finally, if **CurrentPos** is identical with **Target**, then we clear the **Tmr1On** bit, thereby disabling Timer 1 and stopping the motor.

Command Structure

Commands are case insensitive—that is, u is the same as U. Where a numeric command is to be sent up to 5 digits may be used. Do not add commas. After entering a command, the response will be the string sent back OK.

Command Type	Specific Command	Description
Motor Speed	**U?**	*Report the current interval in microseconds between motor steps. Response will be:* `Delay 12345` *Where 12345 will be replaced by the actual delay in microseconds*
	Unnnnn	*Set the interval between pulses. Nnnnn is in microseconds. The maximum interval is dependent upon the clock speed of the PIC. For a 20 MHz clock, the maximum interval is 13,100 µs (13.1 ms).*
		The minimum clock interval is also dependent upon the clock speed, but for almost any stepper motor, the limiting factor will be the stepper motor itself. (Since we are entering the time between pulses, the shorter the time, the faster the motor will step.)
		To calculate the required pulse interval for a desired motor RPM, use the following formula:
		$$u = \dfrac{1,000,000}{\dfrac{N \times RPM}{60}}$$
		where:
		u *is the interval in microseconds* ***RPM*** *is the desired revolutions per minute* ***N*** *is the number of steps per revolution, for the stepping mode in which the motor will be operated. Half-step operation doubles the number of steps per revolution, for example, in full step mode for a particular motor, N = 48, but in half-step mode N = 96.*

Command Type	Specific Command	Description
		Example: *RPM = 60* *N = 48 (full step) / 96 (half step). We will operate in half step mode.* $$u = \frac{1,000,000}{\frac{96 \times 60}{60}} = \frac{1,000,000}{96} = 10417$$ *To set this speed and mode, the following commands would be used:* *MH (set to half-step mode)* *U10417 (sets step interval at 10417 uSeconds)*

Motor Direction	D?	*Report the current motor direction. Response will be either:*
		`Forward`
		`Reverse`
		Note that the directions forward and reverse are arbitrary and depend upon the motor connections.

	DF DR	*Set the motor to Direction Forward* *Set the motor to Direction Reverse* *You will receive an OK response after a successful DF/DR command.*

Status	S	*Outputs a status report of various parameters:*
		`Variable Dump`
		`Fosc  nn`
		`Delay nnnnn uSec`
		`Timer Status n  T1Timer " nnnnn`
		`Cur Posn nnnnn`
		`Target Posn nnnnn`
		`<Half Step><Full Step>`
		`<Forward><Reverse>`
		
		***Fosc** is the oscillator frequency in MHz as set in the MHZ= line*
		***Delay** is the time interval in microseconds between motor steps*
		***Timer** Status is 0 or 1, 0=off and 1=on.*
		***T1Timer** is the numeric value of the Timer 1 count load*
		***Cur Posn** is the number of motor steps executed*
		***Target** is the number of motor steps to be executed to cause a motor stop*
		***Half Step** or **Full Step** indicates the current mode*
		***Forward** or **Reverse** indicates the current direction.*
		
		*Note that **Cur Posn** is as of the time the data is sent back and that the motor will continue to step.*

Step Mode	M?	*Report the current step mode. Response will be either:*
		`Half Step`
		`Full Step`

	MH MF	*Set Mode to Half Step* *Set Mode to Full Step* *You will receive an OK response after a successful MH/MF command.*

Command Type	Specific Command	Description
Set Target Position	*T?*	*Report on the current target position. The response is:* `Target Posn nnnnn` *nnnnn will be the target step count.*
	Tnnnnn	*Set the Target position. When the nnnnn motor pulses have been generated, the motor stops. nnnnn is in units of motor steps. If, for example, the motor has 98 steps/revolution and 10 complete revolutions are desired, the command would be:* `T980` *Note that the motor will not start stepping until an Execute command has been received.* *Target is an unsigned Word, so the maximum value is 65536.* *Target is an relative value, i.e., it will execute an additionalTarget steps based upon the current position at the time the Target command is received. If absolute position is required, i.e., from some fixed zero, the user is responsible for keeping track of absolute positions and translating them to relative Target values.*
Execute (Start Rotation)	*#*	*Based upon the entered parameters, start the motor running.* *Note that only minimal error checking has been implemented in the sample software, so the user is responsible for ensuring that reasonable parameters have been entered.* *You will receive an OK response after a successful # command.*
Stop	*$\bar{F}$*	*Stops the motor and de-energizes the motor windings, so the motor shaft is Free.*
	$\bar{L}$	*Stops the motor with the last stepping pattern in place, so the motor windings are energized and the motor shaft is Locked.* *The $\bar{F}$ and $\bar{L}$ commands retain the Current Position. Hence a # will resume from the current position. If the user desires to reset the Current Position, a zero (Z) command may be issued.* *You will receive an OK response after a successful Stop command.*
Zero (Zero Current Motor Position and Report Positon)	*Z?*	*Reports the current motor position in completed steps:* `Cur Posn nnnnn` *Where nnnn is completed motor steps. If the motor is running, the reported position will vary somewhat from the actual position at the time the string is received due to internal program and serial transmission time lags.*
	Z	*Sets the current motor position to Zero.* *While it is possible to zero the motor position while the motor is running, the user should remember that there will be a time lag between sending the motor command and the internal execution of the zeroing. In most cases, you will wish to stop the motor before sending a zero command.* *You will receive an OK response after a successful Z command.*

Ideas for Changes to Programs and Circuits

* Merge the concepts in Programs 19-2 and 19-3, along with the work of Chapter 8 to give the user the option of four modes: full step, half step, wave and microstep.
* Modify Program 19-2 to allow the user to select different size microsteps, such as 4, 8, 16 or 32 microsteps per full step. This will involve rewriting some of the assembly code to replace constants with variables.
* Modify Program 19-2 to increase user interaction, not necessarily to the extent of Program 19-3, but beyond entering hard coded values at program time. Which parameters should be user settable? How do you process user inputs (most of the time for which is spent waiting for keystrokes) without being too much of a processing time burden in the time-sensitive main program loop?

References

[19-1] Condit, Reston, *Stepper Motor Control Using the PIC16F684, AN 906*, Microchip Technology, Inc., Doc. No. DS00906A (2004).

[19-2] Predko, Myke, *Programming and Customizing PICmicro Microcontrollers*, McGraw-Hill/TAB Electronics; 2nd edition (December, 2000).

[19-3] Predko, Myke, *PICmicro Microcontroller Pocket Reference*, McGraw-Hill/TAB Electronics, Pocket edition (November 2000).

[19-4] Texas Instruments, *SN754410 Quadruple Half-H Driver Datasheet*, Doc. No. SLRS007B, (Nov. 1995).

X-10 Home Automation

In the late 1970's, a small Scottish company, Pico Electronics, joined with BSR, famed for its high fidelity turntables, to produce inexpensive home automation equipment. Pico Electronics' deceptively simple idea was to use a home's power wiring to communicate from a controller to small remote control modules. The result was the BSR System X10, originally sold at Radio Shack and Sears. BSR is no longer in business, but its X-10 business is carried on by X10 USA. (The Company's name is "X10" but the technology is identified as "X-10.") Many manufacturers have licensed X-10 technology, and a wide variety of remote switches, dimmers and other devices are available.

MBasic's **Xin** and **Xout** functions provide built-in support for both transmitting and receiving X-10 signals. This chapter also covers nontraditional use of X-10 technology, such as telemetering temperature readings.

How X-10 Works

Introduction

The technology behind X-10 is to send brief bursts of 120 KHz signals over the power lines, timing the bursts to coincide with the 60 Hz power line zero-voltage crossings. At the zero voltage point the power line noise level is at a minimum and the 120 KHz signal has the best chance to be received by a remote control module. We'll go through X-10's protocol by starting at the top and working down to the bit level.

At the highest level, X-10 messages are comprised of three elements:

House Code	Unit Code	Function Code

Let's look at each element in more detail.

House Code—There are 16 possible house codes, conventionally identified as A…P. For convenience when using **Xin** and **Xout**, MBasic provides library constants of the form **X_A..X_P**.

Unit Code—There are 16 possible unit codes, conventionally identified as 1…16. For convenience when using **Xin** and **Xout**, MBasic provides library constants of the form **X_1...X_16**.

Function Code—There are 16 possible function codes, identified by their operations, such as "On," "Off" and "Dim." For convenience when using **Xin** and **Xout**, MBasic provides library constants of the form **X_On**, **X_Off** and **X_Dim**, etc.

Controlled modules are addressed by both house and unit codes. Figure 20-1 shows the setting switch for a typical X-10 control module. By rotating the switches, the module's address is set to the desired house and unit codes. Figure 20-1 shows an address set to "B-8," which means that it will only respond to function codes prefixed with house code B, unit code 8. (Some modules are designed to also accept "all call" signals,

Figure 20-1: House and unit code setting switches,

where only the house code must match the module setting. In this case the instruction "B Off" would turn off all modules set to house code B, regardless of the unit code setting.) The module shown in Figure 20-1 will turn off upon receipt of the code sequence corresponding to "B 8 OFF" and will ignore the code sequences "C 8 OFF" and "B 7 OFF."

Since there are 16 house codes and 16 unit codes, we have a maximum of 256 possible unique addresses. Modules recognize only certain function codes associated with their purpose. For example, the "appliance control" module is limited to turning loads on and off; it accordingly recognizes only on and off function commands. In addition to off and on, a light dimmer module also recognizes the dim and bright commands.

The remainder of this section delves into the TW523 transceiver and the details of X-10 transmission. You may treat the TW523 as a black box and skip over the X-10 theory, should you prefer.

TW523 X-10 Transceiver

MBasic's **Xin** and **Xout** functions assume the PIC is connected to an X-10 transmitter, or, in the case of **Xin**, an X-10 transceiver. We'll use a TW523 transceiver for this chapter's programs. If you are interested in only transmitting X-10 commands, the transmit-only PL513 device is slightly cheaper. I recently paid $24 for a TW523, while a PL513 cost $16 from the same supplier. And, although the device I bought was sold as a "TW523," it turned out to carry the part number PSC05. Regardless of the number, the X-10 Pro PSC05 I purchased is pin-compatible with a TW523. When I refer to the X-10 transceiver, I'll call it a TW523, but not all manufacturers use this part number to identify their X-10 transceiver module.

Let's take a quick look at what's inside a TW523. Figures 20-2 and 20-3 show both a TW523 and a PAM02 on/off appliance control module. The TW523 is at the left in both photos and may be identified by the RJ11 4-pin modular connector. Figure 20-4 provides a simplified block diagram of the TW523. (Reference [20-1] includes a component-level schematic for both the TW523 and PL513 devices.)

Figures 20-2: TW523 X-10 transceiver and PAM02 appliance control module, front view.

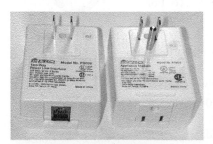

Figure 20-3: TW523 X-10 transceiver and PAM02 appliance control module, back view.

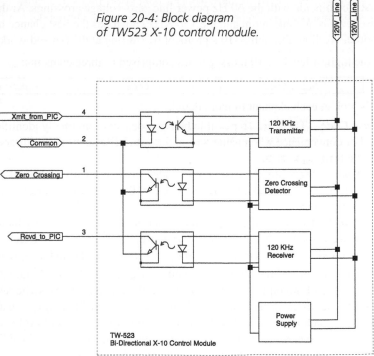

Figure 20-4: Block diagram of TW523 X-10 control module.

The TW523 uses opto-couplers to isolate its input and output circuits from the 120 V power line. However, the TW523's internal circuitry is *not* isolated from the 120 V power line and you should not succumb to the temptation to disassemble the device and start probing around with your oscilloscope. We've seen opto-couplers in Chapters 3 and 4. If you need a refresher on how an opto-coupler works, take a moment and read Chapters 3 and 4.

We'll go into the purpose of each of the TW523's three input/output lines later in this chapter when take a detail look at **Xin** and **Xout**. For now, we'll content ourselves by noting that it has a zero-crossing detector that toggles between logic 1 and logic 0 whenever the 120 V line changes sign; a transmit input line for sending data and a receive output line for incoming data.

The X-10 Protocol

Continuing in the top-down direction, let's look at the bit level data flow in the X-10 protocol. A binary 1 is represented as a 1 ms duration burst of 120 KHz signal, while a binary 0 is represented as an absence of 120 KHz signal. Each burst (or absence of burst) occurs within 200 μs following a 60 Hz zero crossing. Figure 20-5 shows a single logic 1 burst. There are three 1ms bursts, separated by 2.78 ms because in a three-phase power system there are three zero crossings, each separated from the other by 60° corresponding to 2.78 ms in a 60 Hz system. Figure 20-6 shows the relationship between zero crossings in three-phase power. You're probably thinking "three phase power? I don't have three phase power in my house." And, you are correct. However, X-10's designers wanted their product to include industrial and commercial users, where three-phase power is the norm. And, in some European countries three phase residential service is common.

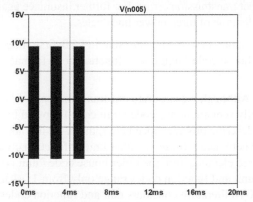

Figure 20-5: Logic 1 bursts.

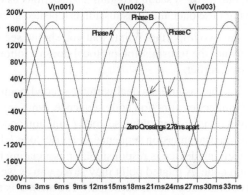

Figure 20-6: Zero crossings in three-phase power system.

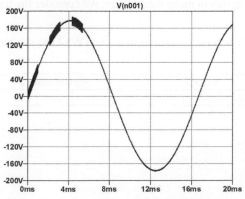

Figure 20-7: Transmitting a 120 KHz burst.

Next, you may be thinking, "even if I had three phase power, why transmit at all three zero crossings?" The answer is that the X-10 transmitter and receiver may not be powered from the same phase. By transmitting the burst once for each possible zero crossing, the protocol ensures that the *receiving* device is guaranteed to receive a signal at *its* zero crossing, regardless of when the *transmitter's* zero crossing occurs. And, at the local receiver's zero crossing, the interference and noise on the power line will be at a minimum. Figure 20-7 provides a conceptual view of how the 120 KHz tone burst appears when transmitted over a 60 Hz power line. The power line has the sum of the 60 Hz power and the superimposed 120 KHz tone bursts. For clarity, I've shown the 120 KHz bursts as about 20 V peak-to-peak, but the actual 120 KHz signal level found on the power line is much less, typically a few hundred millivolts.

The TW523 automatically manages receiving multiple bursts and returns one pulse—coinciding with the local zero crossing—for each received burst triplet. When transmitting through the TW523, MBasic's **xout** function incorporates timing and multiburst requirements without additional intervention. Hence, we won't further concern ourselves with the fact that the X-10 standard calls for bursts to be sent in triplets.

Let's look at how X-10 commands are sent. (The complement of a bit is shown by an overscore. Transmitted bits are shown in italics.) We'll send the house code A, followed by the unit code 2.

Start Code	House Code								Data										
	H1	$\overline{H1}$	H2	$\overline{H2}$	H4	$\overline{H4}$	H8	$\overline{H8}$	D1	$\overline{D1}$	D2	$\overline{D2}$	D4	$\overline{D4}$	D8	$\overline{D8}$	D16	$\overline{D16}$
1 1 1 0	*0*	*1*	*1*	*0*	*1*	*0*	*0*	*1*	*1*		*1*		*1*		*0*		*0*	
1 1 1 0	*0*		*1*		*1*		*0*		*1*		*1*		*1*		*0*		*0*	
SC=1110	House Code = "A" (0110)								Unit Code = 2 (11100)										

X-10 data is sent in three consecutive blocks; each block comprising a 4-bit start block, an 8-bit house block and a 10-bit data block. (The start code is always 1110.) Except for the start bits, each bit is sent twice consecutively, first in its normal or true form, followed immediately by its complement, or inverted form. A binary 1 is thus sent as 10 and a binary 0 is sent as 01. This doubles the transmission length of the house code and unit codes, but, as we'll see in Program 20-7, permits us to detect transmission errors. For further insurance against receiving false data, the transmission is repeated after an idle time of three power line cycles.

There are two important points that you should see in this table; first, the bits are sent in LSB-first order, and second, the transmission protocol doesn't follow the house code / unit code / function code format we discussed earlier.

Since it's the easier, let's deal with the second point first. Although we structure our transmission in a house code / unit code / function code sequence, what are actually sent are house code / unit code followed by a second transmission of house code / function code. To describe this sequence better, the X-10 literature defines the transmission sequence as house code / data, where data represents either unit or function codes. The receiving device is responsible for associating the three elements into the order house code / unit code / function code. A related point may strike you as well; if unit and function codes range only from 1...16, why use five bits when four is sufficient? The answer is that X-10 literature considers unit and function codes as a single group of five-bit binary data values, running from 0...31, with values from 0...15 being unit codes and values from 16...31 being function codes. We'll look at this in more detail when we consider the transmitted bit order.

Thus, a complete transmission of a house code / unit code / function code requires the TW523 to send the following sequence:

Block 1				Repeat Block 1				Block 2				Repeat Block 2		
SC	HC	UC	IDL	SC	HC	UC	IDL	SC	HC	FC	IDL	SC	HC	FC

SC= Start Code
HC=House Code
UC=Unit Code
FC=Function Code
IDL=Idle Time

Fortunately, the TW523 and MBasic's **Xin** and **Xout** functions cooperate to hide this complexity from us.

Let's now look at the second curiosity; bit order. By convention, Microchip's documentation (and virtually all of the rest of the microcontroller and microcomputer industry) shows binary values most significant bit first. (This is the way, of course, we normally write number, e.g., 1234 is in most significant digit first order.) Thus, the binary representation of decimal 1 is **%00000001**. If, on the other hand, we decided to display bits starting with the least significant bit first, **%00000001** represents decimal 128, not 1. When transmitting a byte serially, there's nothing that makes sending it LSB first or MSB first preferable. In fact, the PC serial interface sends bits LSB first. What is important is consistency; nothing but trouble will ensue if we transmit bits LSB first, but the receiving program interprets the received data as being MSB first or vice versa. In the case of serial transmission to or from a PC to a PIC, both the transmitting and receiving ends understand the data is sent LSB first, so there is no confusion.

As long as you stick to MBasic's library constants for house, unit and function codes, you need not concern yourself with bit order and values. The library constant **x_A**, for example, produces the correct bit sequence for house code "A" when used as an argument in **Xin** or **Xout**. However, when reading "official" X-10 documents, you should note that binary values are presented in LSB first order. The following table cross-references MBasic's library constants with the names and bit sequences followed in X-10 documents.

House Code	X-10 Documentation		MBasic		
	TW523 LSB-first Order	TW-523 Value (Decimal)	Library Constant Name	Library Constant Value	
				Decimal	Hex
A	0110	6	X_A	6	6
B	1110	7	X_B	14	E
C	0010	4	X_C	2	2
D	1010	5	X_D	10	A
E	0001	8	X_E	1	1
F	1001	9	X_F	9	9
G	0101	10	X_G	5	5
H	1101	11	X_H	13	D
I	0111	14	X_I	7	7
J	1111	15	X_J	15	F
K	0011	12	X_K	3	3
L	1011	13	X_L	11	B
M	0000	0	X_M	0	0
N	1000	1	X_N	8	8
O	0100	2	X_O	4	4
P	1100	3	X_P	12	C

Unit Code / Function Name	X-10 Documentation						MBasic		
	TW523 LSB-first Order						Library Constant Name	Library Constant Value	
	D1	D2	D4	D8	D16	Decimal Value		Decimal	Hex
1	0	1	1	0	0	6	X_1	12	C
2	1	1	1	0	0	7	X_2	28	1C
3	0	0	1	0	0	4	X_3	4	4
4	1	0	1	0	0	5	X_4	20	14
5	0	0	0	1	0	8	X_5	2	2
6	1	0	0	1	0	9	X_6	18	12
7	0	1	0	1	0	10	X_7	10	A
8	1	1	0	1	0	11	X_8	26	1A
9	0	1	1	1	0	14	X_9	14	E
10	1	1	1	1	0	15	X_10	30	1E
11	0	0	1	1	0	12	X_11	6	6
12	1	0	1	1	0	13	X_12	22	16
13	0	0	0	0	0	0	X_13	0	0
14	1	0	0	0	0	1	X_14	16	10
15	0	1	0	0	0	2	X_15	8	8
16	1	1	0	0	0	3	X_16	24	18
All Units Off	0	0	0	0	1	16	X_Units_On *	1	1
All Lights On	0	0	0	1	1	24	X_Lights_On	3	3
On	0	0	1	0	1	20	X_On	5	5
Off	0	0	1	1	1	28	X_Off	7	7
Dim	0	1	0	0	1	18	X_Dim	9	9
Bright	0	1	0	1	1	26	X_Bright	11	B
All Lights Off	0	1	1	0	1	22	X_Lights_Off	13	D
Extended Code	0	1	1	1	1	30	X_Hail	15	F
Hail Request	1	0	0	0	1	17	N/A	17	11
Hail Acknowledge	1	0	0	1	1	25	N/A	19	13
Extended Data (Analog)	1	1	0	0	1	19	N/A	25	19
Status=on	1	1	0	1	1	27	X_Status_On	27	1B
Status=off	1	1	1	0	1	23	X_Status_Off	29	1D
Status Request	1	1	1	1	1	31	X_Status_Request	31	1F
Pre-Set Dim 0	1	0	1	0	1	21	N/A	21	15
Pre-Set Dim 1	1	0	1	1	1	29	N/A	23	17

* This constant appears to be mis-named in MBasic, as the function is identified in X-10 literature as "off" not "on."

X-10's designers worked with LSB-first bit order when preparing the encoding scheme, as is apparent from the fact that the data codes separate nicely into 0…15 (unit codes) and 16…31 (function code) groups. When these binary values are reversed, this clean grouping is broken.

The **xin** and **xout** sections of the MBasic User's Guide provide binary equivalents for the compiler library function and unit constants. These binary equivalents are, however, in error, and do not correspond to the actual binary values of the named constants. Instead, the binary information presented consists of the binary equivalent (MSB first order) of the TW523 numerical values. If these binary values are used in **Xin** or **Xout** instead of the library constant names, incorrect codes will be sent.

To clarify this explanation, let's look at **x_1**:

MBasic User's Guide					X-10 Documentation	
Unit Constant	Constant as Stated in User's Guide	Value of Binary Constant (MSB first; Decimal Value)	Actual Value of X_1 Constant in MBasic Library (Decimal)	Binary Representation of 12 (Decimal)	TW523 Value (Decimal) of Unit 1 Code	TW523 Representation of Unit 1 Code (LSB First)
X_1	%00110	6	12	%01100	6	01100

For example, to send the code for Unit 1, the bits must be sent in the following order 0-1-1-0-0. If these bits are held in a MBasic byte variable and are sent left-to-right, the MBasic byte value must be decimal 12 ($C). In fact, the value of the constant **x_1** is $C or decimal 12. However, the bit value shown in the User's Guide is **%00110**. This corresponds to decimal 6 ($6), not the actual value of **x_1**. This is a documentation error, and the named constants work correctly when used with MBasic's Xin and Xout functions.

Programs

Let's now look at some code implementing what we've learned about X-10. Connect your TW523 as shown in Figure 20-8. An oscilloscope connected to RB0 will display the zero crossing output waveform, as shown in Figure 20-9.

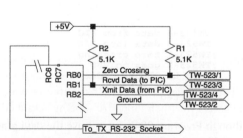

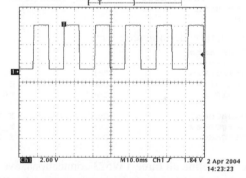

Figure 20-8: Configuration for program 20-1.

Figure 20-9: Zero Crossing Waveform; Ch1: Pin B0; Ch2: Not Used.

To verify Program 20-1's functionality, I used two X-10 Pro model PAM02 appliance modules, set to codes K8 and J8. I plugged both appliance modules into the same power distribution strip as the TW523 module. An appliance module implements only on/off functionality, and in the case of the PAM02 modules, the switching element is a mechanical relay.

Program 20-1 commands the appliance modules to turn on, and, after a brief pause, to turn off. It tests both house/unit/function addressing and house/function "all call" modes.

Program 20-1

```
;Program 20-01. Verify transmit
;and control with minimal program

;Constants
```

```
;------------
DataOutPin              Con             B2      ;TW523-4 data from PIC
DataInPin               Con             B1      ;TW523-3 data into PIC
ZeroXPin                Con             B0      ;TW523-1 zero crossing
                                                ;TW523-2 goes to VSS

HouseCode1              Con             X_K     ;test two house codes
HouseCode2              Con             X_J     ;these are arbitrary
UnitCode                Con             X_8     ;selected

Main
        ;By omitting the Unit Code we control all
        ;units with the same HouseCode
        XOut DataOutPin\ZeroXPin,HouseCode1,[X_On]
        Pause 1000
        XOut DataOutPin\ZeroXPin,HouseCode1,[X_Off]
        Pause 1000

        ;Add the Unit Code to address specific controller
        XOut DataOutPin\ZeroXPin,HouseCode1,[UnitCode,X_On]
        Pause 1000
        XOut DataOutPin\ZeroXPin,HouseCode1,[UnitCode,X_Off]
        Pause 1000

        XOut DataOutPin\ZeroXPin,HouseCode2,[X_On] ;X_On
        Pause 1000
        XOut DataOutPin\ZeroXPin,HouseCode2,[X_Off] ;X_Off
        Pause 1000

        ;Add the Unit Code to address specific controller
        XOut DataOutPin\ZeroXPin,HouseCode2,[UnitCode,X_On]
        Pause 1000
        XOut DataOutPin\ZeroXPin,HouseCode2,[UnitCode,X_Off]
        Pause 1000
GoTo Main
End
```

As usual, we start by defining constants.

```
;Constants
;------------
DataOutPin              Con             B2      ;TW523-4 data from PIC
DataInPin               Con             B1      ;TW523-3 data into PIC
ZeroXPin                Con             B0      ;TW523-1 zero crossing
                                                ;TW523-2 goes to VSS
HouseCode1              Con             X_K     ;test two house codes
HouseCode2              Con             X_J     ;these are arbitrary
UnitCode                Con             X_8     ;selected
```

In this case, we define all three TW523 connections, even though Program 20-1 doesn't use the data pin. I've arbitrarily set the two appliance modules at codes K8 and J8, and defined three corresponding constants, **HouseCode1**, **HouseCode2** and **UnitCode**.

Although it receives frequent mention earlier in this chapter, this is our first official introduction to MBasic's function **Xout**. Let's see how we use **Xout** to transmit X-10 codes.

Xout DataPin\ZeroPin,House,[{Unit},{Modifiers},KeyCode]

DataPin—is a constant or variable identifying the data output pin. This pin connects to the TW523's pin 4. No pull-up resistor is required. As seen in Figure 20-4, the TW523's transmit data line is connected to the LED section of an opto-isolator. In Figure 20-4 and elsewhere in this chapter, I've shown the PIC's data pin driving the LED without a current limiting series resistor. This is acceptable for brief experiments; but for permanent construction I recommend a series resistor of 100 to 220 ohms to limit the LED's drive current. The TW523 data sheet [20-1] sample interface circuit shows 100 ohms series resistance and should be consulted for more detailed design recommendations.

ZeroPin—is a constant or variable identifying the zero crossing input pin. This pin connects to the TW523's pin 1. Don't forget this pin requires a pull-up resistor. Figure 20-8 shows a 5.1K, but any value in the range 3.3K through 10K is acceptable.

House—is a constant or variable holding the house code. The house code must be in the range 0…15. MBasic has defined library constants **X_A…X_P** that provide the proper house code value corresponding to house codes A…P.

Unit—is a constant or variable holding the unit code. MBasic has defined library constants **X_1…X_16** that provide the proper unit code values corresponding to unit codes 1…16.

KeyCode—is a constant or variable holding the key code. MBasic has defined library constants for certain key codes, such as **X_On** and **X_Off**. For key codes without a library constant, values provided earlier in this chapter may be used.

Xout with both a unit code and key code argument sends the sequence <start><house code><unit code> followed by <start>><house code><function code>. (For simplicity, I'm omitting the fact, as we learned earlier, that each sequence is sent twice.) If the unit code argument is omitted, the transmitted sequence is <start><house code><function code>. In other words, **Xout** sends the sequence <start><house code><argument N> for all arguments 1…N within the square brackets.

```
Main
          ;By omitting the Unit Code we control all
          ;units with the same HouseCode
          XOut DataOutPin\ZeroXPin,HouseCode1,[X_On]
          Pause 1000
          XOut DataOutPin\ZeroXPin,HouseCode1,<X_Off]
          Pause 1000
```

The first call to **Xout** sends the "all call" version, which causes all receivers with house codes equal to **HouseCode1** to switch to "on" mode. After a 1 second pause, we repeat the all call transmission, but this time with an "off" command.

```
          ;Add the Unit Code to address specific controller
          XOut DataOutPin\ZeroXPin,HouseCode1,<UnitCode,X_On]
          Pause 1000
          XOut DataOutPin\ZeroXPin,HouseCode1,<UnitCode,X_Off]
          Pause 1000
```

Next, we repeat the on/off sequence, but with both a house code and a unit code address.

```
          XOut DataOutPin\ZeroXPin,HouseCode2,<X_On]  ;X_On
          Pause 1000
          XOut DataOutPin\ZeroXPin,HouseCode2,<X_Off]  ;X_Off
          Pause 1000

          ;Add the Unit Code to address specific controller
          XOut DataOutPin\ZeroXPin,HouseCode2,<UnitCode,X_On]
          Pause 1000
          XOut DataOutPin\ZeroXPin,HouseCode2,<UnitCode,X_Off]
          Pause 1000
GoTo Main
```

We then repeat this sequence, but for the second appliance module. (Your module may or may not respond to an "all call" sequence. My PAM02 modules do not recognize "all call" commands.)

Let's look at the PIC's data output pin while sending commands. Figure 20-10 shows part of an output sequence. A logic high on Channel 2 represents an X-10 logic high. As we learned earlier, each high is sent three times, at intervals representing 60 degrees phase at 60 Hz. Figure 20-11 expands one output triplet to show how the burst structure fits the 60 Hz zero crossings.

I've captioned the logic levels in Figure 20-12. Let's see if we can determine what these particular bits mean. The visible part of the data stream in Figure 20-12 is 010010110101. We know the X-10 protocol require bit pairs of either 10 or 01 (except for the start signal) to be transmitted. Hence, the only permissible patterns of

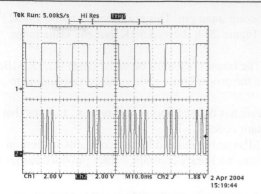

Figure 20-10: Xout Output Ch1: Zero Crossing Reference; Ch2: Data Output.

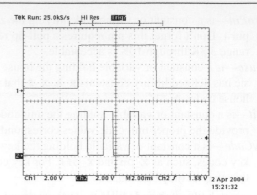

Figure 20-11: Expanded View of Data Output Ch1: Zero Crossing Reference; Ch2: Data Output.

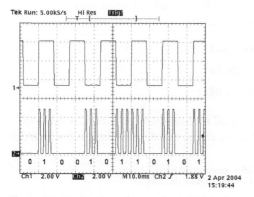

Figure 20-12: Captioned Xout Output Ch1: Zero Crossing Reference; Ch2: Data Output,

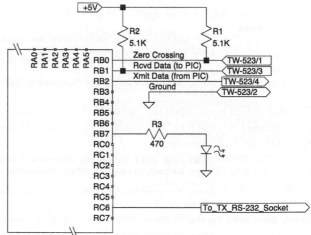

Figure 20-13: Circuit for Program 20-2.

11 or 00 must occur as parts of 0110 or adjacent 1001 patterns. The only possible bit pattern sequence that meets this requirement is 0 10 01 01 10 10 1. We can add the missing elements at the two edges and conclude the actual sequence is 10 10 01 01 10 10 10. Or, dropping the redundant complement bits, the sent bits that generated this oscilloscope waveform are 1100111.

Program 20-2

Next, let's cycle through all the possible transmit codes. This lets us test the false code rejection ability of an X-10 module. (I ran Program 20-2 for two days and observed no false triggers.) And, more importantly, Program 20-2 serves as a test transmitter when we develop receiving programs 20-3 and 20-7.

Figure 20-13 shows the connection arrangement. We've added an LED to show progress and a jumper between the PIC's hardware serial output port and the 2840 Development Board's RS-232 socket.

```
;Program 20-02. Verify transmit
;and control with minimal program

;Constants
;-------------
```

```
DataOutPin      Con     B2      ;TW523-4 data from PIC
DataInPin       Con     B1      ;TW523-3 data into PIC
ZeroXPin        Con     B0      ;TW523-1 zero crossing
                                ;TW523-2 goes to VSS
LEDPin          Con     B7      ;LED to flash to show status

KeyMax          Con     15      ;16 legit key codes
KeyMin          Con     0

UnitMin         Con     0       ;16 legit unit codes
UnitMax         Con     15

HouseMin        Con     0       ;16 legit house codes
HouseMax        Con     15

;Because the codes are not in numerical sequence we construct
;a table for each to get the order correct.
XUnit   ByteTable X_1,X_2,X_3,X_4,X_5,X_6,X_7,X_8,X_9,X_10, |
                  X_11,X_12,X_13,X_14,X_15,X_16

Xhouse ByteTable X_A,X_B,X_C,X_D,X_E,X_F,X_G,X_H,X_I,X_J,X_K, |
                 X_L,X_M,X_N,X_O,X_P

;Use all keys, in the order given in TW523 interface documents

XKey            ByteTable 1,3,5,7,9,11,13,15,17,19,21,23,25,27,29,31


;Each entry = 13 bytes; index (N*13)
;In the order given in TW523 interface documents
KeyName ByteTable "All Units Off", |          ;1
                                "All Lghts On ", |          ;3
                                "On           ",  |          ;5
                                "Off          ",  |          ;7
                                "Dim          ",  |          ;9
                                "Bright       ",  |          ;11
                                "All Lghts Off",  |          ;13
                                "Extended Code",  |          ;15
                                "Hail Request ",  |          ;17
                                "Hail Acknldge",  |          ;19
                                "Pre-Set Dim 0",  |          ;21
                                "Status=Off   ",  |          ;23
                                "Extended Data",  |          ;25
                                "Status=On    ",  |          ;27
                                "Pre-set Dim 1",  |          ;29
                                "Status Rqst  "              ;31

;Variables
;-------------
uc      Var     Byte    ;unit code
kc      Var     Byte    ;key code
hc      Var     Byte    ;house code

;Initialization
;----------------
        EnableHSerial
        SetHSerial H115200
        HSerOut["P20-02",13,13]

        ;dump function codes and values & names
        HSerOut ["Function Codes",13]
        For kc = KeyMin to KeyMax
                HSerOut [Dec kc,9,"Code: ",Dec XKey(kc), |
                9,iBin XKey(kc)\5,9, |
                Str KeyName((kc)*13)\13,13]
        Next
        HSerOut[13]
        Pause 1000

        ;Dump unit code values
```

```
                    HSerOut ["Unit Codes",13]
                    For uc = UnitMin to UnitMax
                            HSerOut [Dec uc,9,iHex XUnit(uc)\2,   |
                            9,iBin XUnit(uc)\5,13]
                    Next
                    HSerOut [13]

                    ;Dump house codes
                    HSerOut ["House Codes",13]
                    For hc = HouseMin to HouseMax
                            HSerOut [Dec hc,9,iHex XHouse(hc),   |
                            9,iBin XHouse(hc)\4,13]
                    Next
                    HSerOut [13,13]
                    Output LEDPin
                    GoSub Flash

        Main
                    ;Loop through all possible house/unit/function codes
                    For hc = HouseMin to HouseMax
                        For uc = UnitMin to UnitMax
                            ;Send ALL to house code when house code changes
                            If uc = 0 Then
                                For kc = KeyMin to KeyMax
                                        XOut DataOutPin\ZeroXPin,  |
                                        XHouse(hc),[XKey(kc)]
                                        HSerOut [("A"+hc),"-ALL",9]
                                        HSerOut [Str KeyName((kc)*13)\13,  |
                                        13]
                                Next
                            EndIf
                            For kc = KeyMin to KeyMax
                                    ;to get these in
                                    ;order A-01 All Units off, etc.
                                    ;we use the index tablesw
                                    XOut DataOutPin\ZeroXPin,  |
                                    XHouse(hc),[Xunit(uc),XKey(kc)]
                                    ;Output to serial port to know
                                    ;what's being sent
                                    HSerOut [("A"+hc),"-"]
                                    HSerOut [Dec uc+1\2,9]
                                    HSerOut [Str KeyName((kc)*13)\13,13]
                            Next
                            ;Visual indication for users
                            GoSub Flash
                        Next
                        GoSub Flash
                        GoSub Flash
                    Next
                    GoSub Flash
                    GoSub Flash
                    GoSub Flash
        GoTo Main

        ;Subroutine flashes the LED once for
        ;250 ms.
        Flash
        ;------
                    High LEDPin
                    Pause 250
                    Low LEDPin
                    Pause 250
        Return

        End
```

We'll start by looking at Program 20-2's data structure. Our objective is to transmit all possible house, unit and function codes, but in a logical order. We wish to start with A-1-All Units Off and end with P-16-Status Request. As we've seen earlier in this chapter, the MBasic library constant values are not in an order that permits us to, for example, use a loop structure `For HouseCode = X_A to X_P … Next`. Similar issues exist for the unit codes and function codes. Hence, we return to a data structure we've seen before; we create a byte table to hold the underlying values, organized in the order we wish them to be used.

```
XUnit   ByteTable X_1,X_2,X_3,X_4,X_5,X_6,X_7,X_8,X_9,X_10, |
                  X_11,X_12,X_13,X_14,X_15,X_16
```

The byte table `XUnit`, for example, is comprised of the unit code library constants `X_1` through `X_16`, in the order we wish them to be sent. These library constants are easily accessed by a loop construct For `i = 0 to 15 UnitCode = Xunit(i) … Next`.

```
Xhouse ByteTable X_A,X_B,X_C,X_D,X_E,X_F,X_G,X_H,X_I,X_J,X_K, |
                 X_L,X_M,X_N,X_O,X_P
```

```
XKey         ByteTable 1,3,5,7,9,11,13,15,17,19,21,23,25,27,29,31
```

We modify this approach slightly for the function codes. MBasic provides library constants for 11 of the 16 possible function codes. So, instead of mixing library constant identifiers, such as `X_On` with numerical values, the function byte table `Xkey` uses only numerical values. The values 1, 3, 5… are from the consolidated table developed earlier in this chapter.

We learned earlier in this chapter that unit codes and function codes are subsets of a unified 5-bit data set. And, if the numerical values of those codes are evaluated in the bit order sent, unit codes are in the range 0…15 and function codes are in the range 16…31. If we look at the numerical sequence we use, however, the function key values are odd integers in the range 1…31. Why the discrepancy? `Xout` sends its arguments in left-to-right order, i.e., MSB-first. But the X-10's developers established its command structure assuming data is sent LSB-first. A bit of time with the tables presented earlier demonstrates that the reversed bit order causes the odd/even interleaving effect.

Since unit codes are all even integers and function codes are all odd integers, couldn't we simplify matters considerably, skipping both the unit and function byte tables, by a construct such as:

```
For UnitCode = 0 to 30 Step 2
    For FunctionCode = 1 to 31 Step 2
        Send UnitCode and Function Code
    Next FunctionCode
Next UnitCode
```

The answer is "yes we could." But, we cannot use this construct and have the unit codes come out in the sequence we desire. The first three unit code *values* sent would be 0, 2 and 4. These values correspond to unit *addresses* 13, 5 and 3.

```
KeyName ByteTable    "All Units Off",      ;1
                     "All Lghts On ",      ;3
                     "On          ",       ;5
                     "Off         ",       ;7
                     "Dim         ",       ;9
                     "Bright      ",       ;11
                     "All Lghts Off",      ;13
                     "Extended Code",      ;15
                     "Hail Request ",      ;17
                     "Hail Acknldge",      ;19
                     "Pre-Set Dim 0",      ;21
                     "Status=Off   ",      ;23
                     "Extended Data",      ;25
                     "Status=On    ",      ;27
                     "Pre-set Dim 1",      ;29
                     "Status Rqst  "       ;31
```

To identify the function code being transmitted, we construct an additional byte table holding the name function code's command name. For convenience, I've made each command name 13 bytes long. This allows us to index into the table and write the string to the serial port with via the command `str`

KeyName((kc)*13)\13 where **kc** is the key code index, 0…15. The command names in **KeyName** are in the same order as the function code values are in **Xkey**.

```
;Variables
;-------------
uc        Var        Byte        ;unit code
kc        Var        Byte        ;key code
hc        Var        Byte        ;house code
```

The three index variables are **uc**, **kc** and **hc**, for unit, key (function) and house codes, respectively.

The initialization code largely concerns itself with writing the byte tables to the serial port so we may verify the data entry against the values contained in the tables earlier in this chapter. The initialization section of Program 20-2 is straightforward and needs no detailed analysis.

Let's now turn to the main program loop.

```
Main
            ;Loop through all possible house/unit/function codes
            For hc = HouseMin to HouseMax
                For uc = UnitMin to UnitMax
```

Our program structure is three nested **For … Next** loops. The outer two loops are the house code and unit code loops. We've defined constants **HouseMin** (0), **HouseMax** (15), **UnitMin** (0) and **UnitMax** (0) to serve as the loop ranges.

```
                ;Send ALL to house code when house code changes
                If uc = 0 Then
                    For kc = KeyMin to KeyMax
                            XOut DataOutPin\ZeroXPin, |
                            XHouse(hc),[XKey(kc)]
                            HSerOut [("A"+hc),"-ALL",9]
                            HSerOut [Str KeyName((kc)*13)\13, |
                            13]
                    Next
                EndIf
```

For completeness, we also wish to send the "all call" commands, i.e., sequences of the type <house code><function code>, omitting the unit code. We'll send the all call commands before starting the unit code sequence 1…16. The above code sequence checks for a new house code (which we identify by the unit code loop index **uc** being zero) and when **uc=0**, we cycle through all 16 possible function code combinations with the **For kc = KeyMin to KeyMax** loop. The **Xout** statement is **XOut DataOutPin\ZeroXPin,XHouse(hc),[XKey(kc)]** which omits the unit identifier. And, of course, we use the index variables **hc** and **kc** to index into the byte tables **Xhouse** and **Xkey**, for the reasons earlier developed.

```
                For kc = KeyMin to KeyMax
                        ;to get these in
                        ;order A-01 All Units off, etc.
                        ;we use the index tablesw
                        XOut DataOutPin\ZeroXPin, |
                        XHouse(hc),[Xunit(uc),XKey(kc)]
                        ;Output to serial port to know
                        ;what's being sent
                        HSerOut [("A"+hc),"-"]
                        HSerOut [Dec uc+1\2,9]
                        HSerOut [Str KeyName((kc)*13)\13,13]
                Next
```

After getting the all call sequence out of the way, we loop through all 16 possible function codes, this time sending the sequence <house code><unit code><function code>, as can be seen when we examine the **Xout** statement: **XOut DataOutPin\ZeroXPin, XHouse(hc), [Xunit(uc), XKey(kc)]**.

```
                        ;Visual indication for users
                        GoSub Flash
                Next
            GoSub Flash
```

```
                        GoSub Flash
                Next
                GoSub Flash
                GoSub Flash
                GoSub Flash
        GoTo Main
```

Finally, we close the various **For...Next** loops with calls to subroutine **Flash**. **Flash** blinks an LED to provide a visual cue of the program's progress and needs no detailed analysis.

The serial output of Program 20-2 is shown below. To save space, I've reformatted the output into two columns.

P20-02

Function Codes

0	Code: 1	%00001	All Units Off
1	Code: 3	%00011	All Lghts On
2	Code: 5	%00101	On
3	Code: 7	%00111	Off
4	Code: 9	%01001	Dim
5	Code: 11	%01011	Bright
6	Code: 13	%01101	All Lghts Off
7	Code: 15	%01111	Extended Code
8	Code: 17	%10001	Hail Request
9	Code: 19	%10011	Hail Acknldge
10	Code: 21	%10101	Pre-Set Dim 0
11	Code: 23	%10111	Status=Off
12	Code: 25	%11001	Extended Data
13	Code: 27	%11011	Status=On
14	Code: 29	%11101	Pre-set Dim 1
15	Code: 31	%11111	Status Rqst

Unit Codes

0	$0C	%01100
1	$1C	%11100
2	$04	%00100
3	$14	%10100
4	$02	%00010
5	$12	%10010
6	$0A	%01010
7	$1A	%11010
8	$0E	%01110
9	$1E	%11110
10	$06	%00110
11	$16	%10110
12	$00	%00000
13	$10	%10000
14	$08	%01000
15	$18	%11000

House Codes

0	$6	%0110
1	$E	%1110
2	$2	%0010
3	$A	%1010
4	$1	%0001
5	$9	%1001
6	$5	%0101
7	$D	%1101
8	$7	%0111
9	$F	%1111
10	$3	%0011
11	$B	%1011
12	$0	%0000
13	$8	%1000
14	$4	%0100
15	$C	%1100
A-ALL	All Units Off	
A-ALL	All Lghts On	
A-ALL	On	
A-ALL	Off	
A-ALL	Dim	
A-ALL	Bright	
A-ALL	All Lghts Off	
A-ALL	Extended Code	
A-ALL	Hail Request	
A-ALL	Hail Acknldge	
A-ALL	Pre-Set Dim 0	
A-ALL	Status=Off	
A-ALL	Extended Data	
A-ALL	Status=On	
A-ALL	Pre-set Dim 1	
A-ALL	Status Rqst	
A-01	All Units Off	
A-01	All Lghts On	
A-01	On	
A-01	Off	
A-01	Dim	

Program 20-2's output continues through P-16 Status Request and then begins again with A-ALL All Units Off.

Program 20-3

Let's now try receiving X-10 signals. This requires two 2840 Development Boards, as well as two TW523s, or one TW523 transceiver and one PL513 transmitter. Transmit and receive Development Boards should be connected as illustrated in Figure 20-13. The transmit Development Board should be running Program 20-2.

```
;Program 20-03
;Read specific house code info
;Demonstrates XIN

;Constants
;------------
DataOutPin              Con             B2      ;TW523-4 data from PIC
DataInPin               Con             B1      ;TW523-3 data into PIC
ZeroXPin                Con             B0      ;TW523-1 zero crossing
                                                ;TW523-2 goes to VSS
HouseCode               Con             X_K     ;selected code to listen for

;Variables
;------------
RcvdCode                Var             Byte ;will hold received info


;Initialization
;----------------
EnableHSerial
SetHSerial H115200


Main
;-----
        ;Each transmission that has a unit code and
        ;a function code will give two returns from
        ;XIN. One for the unit code and one for fcn code
        XIN DataInPin\ZeroXPin,HouseCode,[RcvdCode]
        ;Echo out to serial port
        HSerOut ["Rcvd: ",Dec RcvdCode,13]
GoTo Main

End
```

We use the same pin connections and pin identifier constants as in Program 20-1 and 20-1. The main program loop waits for data to be received, stores the received data byte in **RcvdCode** and sends **RcvdCode** out the serial port.

```
Main
;-----
        XIN DataInPin\ZeroXPin,HouseCode,[RcvdCode]
        ;Echo out to serial port
        HSerOut ["Rcvd: ",Dec RcvdCode,13]
GoTo Main
```

After each byte is received, execution returns to **Main** to await the next received data element.

Before we see Program 20-3's output, let's look at MBasic's **Xin** procedure.

Xin DataPin\ZeroPin,House,{TimeoutLabel, TimeoutCount}, [{{Modifier}, Variable]

DataPin—is a constant or variable identifying the data input pin. This pin connects to the TW523's pin 3.
   Don't forget the required pull-up resistor.

ZeroPin—is a constant or variable identifying the zero crossing input pin. This pin connects to the TW523's
   pin 1. Don't forget this pin requires a pull-up resistor. Figure 20-8 shows a 5.1K, but any value in the
   range 3.3K through 10K is acceptable.

TimeoutLabel and TimeoutCount—are optional parameters. If **TimeoutCount** commands are received without a match for **House**, execution will jump to the program label **TimeoutLabel**.

House—is a constant or variable holding the house code. The house code must be in the range 0...15. MBasic has defined library constants **X_A...X_P** that provide the proper house code value corresponding to house codes A...P.

Variable—is a byte length variable holding the received 5-bit data value.

Xin only returns a value for **Variable** when the received X-10 sequence <house code><data> has a house code that matches **House**. If **House** does not match the received <house code> sequence, **Xin** continues to monitor the input data stream and does not complete execution.

Let's compare Program 20-3's output with the transmissions of Program 20-2.

Program 20-02 Sending		Program 20-03 Receiving
K-07	Pre-Set Dim 0	Rcvd: 10 Rcvd: 21
K-07	Status=Off	Rcvd: 10 Rcvd: 23
K-07	Extended Data	Rcvd: 10 Rcvd: 25
K-07	Status=On	Rcvd: 10 Rcvd: 27
K-07	Pre-set Dim 1	Rcvd: 10 Rcvd: 29
K-07	Status Rqst	Rcvd: 10 Rcvd: 31
K-08	All Units Off	Rcvd: 26 Rcvd: 1
K-08	All Lghts On	Rcvd: 26 Rcvd: 3
K-08	On	Rcvd: 26 Rcvd: 5
K-08	Off	Rcvd: 26 Rcvd: 7
K-08	Dim	Rcvd: 26 Rcvd: 9
K-08	Bright	Rcvd: 26 Rcvd: 11

If we check our earlier data table, we see the value of **X_7**, is 10 and **X_8**'s value is 26. These values match the decoded value reported by Program 20-3. Likewise, we can check the function code values, for example **X_Off**, and determine that it is 7, which matches the decoded value reported by Program 20-3.

Program 20-4

As explained at the outset of this chapter, our focus is not on using X-10 to control lights or appliances, but rather on data transmission via the X-10 interface. If you wish to build a X-10 house controller, by the time you compete this book you should have learned enough to design, construct and program a controller that meets your specific requirements.

Let's first verify our ability to transmit and receive unrestricted data variables (but restricted to values in the range 0...31) through Program 20-4.

```
;Program 20-04. Verify transmit
;and control with minimal program
```

```
;Constants
;------------
DataOutPin          Con         B2      ;TW523-4 data from PIC
DataInPin           Con         B1      ;TW523-3 data into PIC
ZeroXPin            Con         B0      ;TW523-1 zero crossing
                                        ;TW523-2 goes to VSS
HouseCode           Con         X_J

;Variables
;------------
i                   Var         Byte

Main
        For i = 0 to 31
                XOut DataOutPin\ZeroXPin,HouseCode,[i]
                Pause 1000
        Next
GoTo Main

        End
```

Program 20-4 sends the values 0…31 using house code **x_J**. Its purpose is to demonstrate that Program 20-3 correctly receives these values.

Load Program 20-4 into the transmitting Development Board and examine the output from Program 20-3. You should see the following output (reformatted to four columns to save space):

Rcvd: 0	Rcvd: 8	Rcvd: 16	Rcvd: 24
Rcvd: 1	Rcvd: 9	Rcvd: 17	Rcvd: 25
Rcvd: 2	Rcvd: 10	Rcvd: 18	Rcvd: 26
Rcvd: 3	Rcvd: 11	Rcvd: 19	Rcvd: 27
Rcvd: 4	Rcvd: 12	Rcvd: 20	Rcvd: 28
Rcvd: 5	Rcvd: 13	Rcvd: 21	Rcvd: 29
Rcvd: 6	Rcvd: 14	Rcvd: 22	Rcvd: 30
Rcvd: 7	Rcvd: 15	Rcvd: 23	Rcvd: 31

Program 20-5

Let's try sending something a bit more adventuresome—the current temperature. Reconfigure your transmitting 2840 Development Board by adding a Dallas Semiconductor DS18B20 digital temperature sensor as shown in Figure 20-14.

We've extensively covered the DS18B20 in Chapter 12 so if you've skipped ahead without reading Chapter 12 first, you might wish to read it now. In short, reading a DS18B20 temperature sensor returns a 12-bit temperature reading (in Celsius) over a one-wire interface. MBasic supports the one-wire interface with the **OWin** and **OWout** procedures for reading from and writing to one-wire devices.

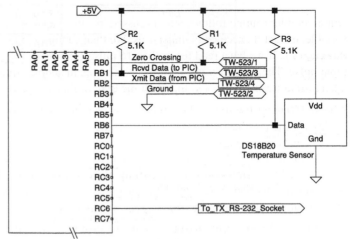

Figure 20-14: X-10 temperature transmitting circuit.

Program 20-5 reads the ambient temperature from the DS18B20 and transmits the value over an X-10 link to a receiving program. We'll test Program 20-5's functionality with Program 20-3, and then write Program 20-6 to properly manage and display received temperature data.

```
;Program 20-05.
;Reads a Dallas 1-wire Temperature sensor
;DS18B20. Assumes only ONE device is connected to the bus,
;since we use global address mode. Sends the temp data
;over the X10 interface for reading remotely.

;Constants
;------------
DataOutPin              Con         B2      ;TW523-4 data from PIC
DataInPin               Con         B1      ;TW523-3 data into PIC
ZeroXPin                Con         B0      ;TW523-1 zero crossing
                                            ;TW523-2 goes to VSS
SensorPin               Con         B6      ;DS18B20 to this pin

;Data is sent as : <Start Data> <Channel No>
;<Data Elements 1...N> <End Data>
;Data elements are 0...9, +. -, .
StartData               Con         10      ;first element
EndData                 Con         11      ;last element
DecimalPt               Con         12      ;decimal point
Minus                   Con         13      ;minus sign
Plus                    Con         14      ;plus sign
Channel0                Con         16      ;data channels 0
Channel1                Con         17      ;data channel 1
                                            ;Number = Channel + 16


;Reads a Dallas 1-wire Temperature sensor
;DS18B20. Assumes only ONE device is connected to the bus,
;since we use global address mode.

HouseCode               Con         X_J ;arbitrary selection

;Variables
;-----------
i               Var         Byte
Temp            Var         Word    ;holds raw binary temperature output
PlusFlag        Var         Byte    ;plus/minus flag
Centigrade      Var         Float   ;Floating point C temp
Fahrenheit      Var         Float   ;Floating point F temp
TH              Var         Byte    ;Lower alarm value
TL              Var         Byte    ;Upper alarm value
ConfigReg       Var         Byte    ;Temperature resolution
FString Var                 Byte(10) ;holds degrees F in string
OutByte Var                 Byte    ;output to X10 transmitter
Channel Var                 Byte    ;Data Channel


;Initialization
Init
;--------------
        EnableHSerial
        SetHSerial H115200

        ;$4E Write to RAM; dummy $FF to TH & TL,
        ;$7F to ctrl reg for 12-bit
        OWOut SensorPin,1,Init,[$CC,$4E,$FF,$FF,$7F]
        ;Write from scratchpad to EEPROM
        OWOut SensorPin,1,Init,[$CC,$48]
        ;read back from EEPROM to scratchpad
        OWOut SensorPin,1,Init,[$CC,$B8]
        ;Read scratchpad memory
        OWOut SensorPin,1,Init,[$CC,$BE]
        Pause 1000
        ;Read output and check configuration
        OWin  SensorPin,0,[Temp.Byte0,Temp.Byte1,TH,TL,ConfigReg]
```

```
            HSerOut ["Init OK  R1: ",BIN ConfigReg.Bit6 ,"  R0: ", |
            BIN ConfigReg.Bit5,13]

Main
            ;Cause temp conversion to start
            OWOut SensorPin,1,Main,[$CC,$44]

Wait
            ;read output and loop until conversion is done.
            ;conversion done through a writeback of 1,0 if not done
            ;Note this only works with external power
            OWIn SensorPin,0,[Temp]
            If Temp = 0 Then Wait

            OWOut SensorPin,1,Main,[$CC,$BE]
            OWin  SensorPin,0,[Temp.Byte0,Temp.Byte1,TH,TL,ConfigReg]
            ;read TH,TL and ConfigReg, we won't use this information
            ;Temperature data returned as 2's complement for below 0.
            ;Hence, below 0=1 in the returned highest bit.
            PlusFlag = 1
            If Temp.HighBit = 1 Then          ;check for < 0C
               ;Subtract from 0 to read below 0 value
               Temp = $0000-Temp
               PlusFlag = 0
            EndIf

            ;Returned value is in Centigrade steps of 0.0625 Degrees C.
            Centigrade = (ToFloat Temp) * 0.0625
            If PlusFlag = 0 Then   ;below zero, multiply by -1
                    Centigrade = -Centigrade
            EndIf

            ;Standard conversion formula to get to F from C.
            Fahrenheit = (Centigrade * 1.8) + 32.0 ;convert to F

            ;Write output. Note the \2 and \1 modifiers
            ;give number of places after decimal point
            HSerOut ["Temperature ",Real Fahrenheit \2," F "  , |
            Real Centigrade \1," C", 13]
            ;Use Modifier as Function to convert to string
            Fstring = Real Fahrenheit \2
            HserOut ["Sending: ",9,Str FString\10\0,13]
            Channel = Channel0
            GoSub SendData

            Fstring = Real Centigrade \2
            HserOut ["Sending: ",9,Str FString\10\0,13]
            Channel = Channel1
            GoSub SendData

GoTo Main

;String to be sent is in FString
;Channel no is in Channel
SendData
;------------
;now we send  the string via X10 transmitter
        ;start with StartData and Channel No.
        XOut DataOutPin\ZeroXPin,HouseCode,[StartData,Channel]
        Pause 100
        i = 0
        ;strip the string apart and look for
        ;- and . and numbers. Encode these
        While (FString(i) <> 0) AND (i<10)     ;printing char
                If FString(i) = "-" Then
                        OutByte = Minus
                EndIf
                If FString(i) = "." Then
                        OutByte = DecimalPt
                EndIf
```

472

```
                        If (FString(i) >= "0") AND (FString(i) <="9") Then
                               OutByte = FString(i) - "0"
                        EndIf
                        ;now send the encoded byte
                        XOut DataOutPin\ZeroXPin,HouseCode,[OutByte]
                        i = i+1
                        Pause 50
                WEND
                ;all data gone, now wrap it up with end data
                XOut DataOutPin\ZeroXPin,HouseCode,[EndData]
                ;zero FString for the next data value
                For i = 0 to 9
                        FString(i) = 0
                Next
                Pause 1000
        Return
        End
```

Before plunging into Program 20-5's code, let's consider how we might go about transmitting generic data over an X-10 interface. The X-10 protocol limits us to five data bits. We could split every transmitted byte into its high and low nibbles and transmit each nibble separately. Or, if we wish to make maximum advantage of X-10's five-byte transmission space, we could pack five bytes of data into eight X-10 messages.

Program 20-5 takes a different approach. First, convert the numerical data to be transmitted to a string representation. Then, transmit a coded version of the string. And, to make the transmission protocol more general, we'll structure it as a 16-channel telemetry system. The structure we use is:

Start Data	Channel Code	Data Element 1	Data Element 2	...	Data Element N	End Data

The values for these elements are defined in the **Constants** section of Program 20-5:

```
;Data elements are 0...9, +. -, .
StartData               Con             10      ;first element
EndData                 Con             11      ;last element
DecimalPt               Con             12      ;decimal point
Minus                   Con             13      ;minus sign
Plus                    Con             14      ;plus sign
Channel0                Con             16      ;data channels 0
Channel1                Con             17     ;data channel 1
                                                ;Number = Channel + 16
```

The string digits 0...9 are represented by data values 0...9. The decimal point, plus and minus signs have data values 12, 14 and 13, respectively. The channel number runs from 16...31, representing data channels 0...16. If you would rather have 32 data channels, let the channel numbers run from 0...31.

Each element of this structure is sent separately, with a preceding house code, so that the actual transmission sequence is <house code> <start code> <house code> <channel code> <house code> <data element 1> ... <house code> <data element N> <house code> <end code>. If you wish to use the X-10 protocol only for data transmission, and have no X-10 receiving modules, you could combine the house code and data elements into a single 9-bit data space and send a full byte with every **Xout** command. However, this will require you to write a general-purpose input routine, such as the one we present in Program 20-7, as **Xin** will not return both the house and data elements.

Now, let's look at the initialization section

```
;Initialization
Init
;--------------
        EnableHSerial
        SetHSerial H115200

        ;$4E Write to RAM; dummy $FF to TH & TL,
        ;$7F to ctrl reg for 12-bit
```

```
OWOut SensorPin,1,Init,[$CC,$4E,$FF,$FF,$7F]
;Write from scratchpad to EEPROM
OWOut SensorPin,1,Init,[$CC,$48]
;read back from EEPROM to scratchpad
OWOut SensorPin,1,Init,[$CC,$B8]
;Read scratchpad memory
OWOut SensorPin,1,Init,[$CC,$BE]
Pause 1000
;Read output and check configuration
OWin   SensorPin,0,[Temp.Byte0,Temp.Byte1,TH,TL,ConfigReg]
HSerOut ["Init OK  R1: ",BIN ConfigReg.Bit6 ,"  R0: ", |
BIN ConfigReg.Bit5,13]
```

The initialization code, other than setting the hardware serial output speed, sets the DS18B20 for 12-bit accuracy. This initialization section duplicates the code of Program 12-2 and the analysis there will not be repeated here.

We'll step through the DS18B20 related code quickly, which is cloned from Program 12-2.

```
Main
        ;Cause temp conversion to start
        OWOut SensorPin,1,Main,[$CC,$44]

Wait
        ;read output and loop until conversion is done.
        ;conversion done through a writeback of 1,0 if not done
        ;Note this only works with external power
        OWIn SensorPin,0,[Temp]
        If Temp = 0 Then Wait
```

The above code initiates a temperature reading and then loops until the temperature reading is completed.

```
OWOut SensorPin,1,Main,[$CC,$BE]
OWin   SensorPin,0,[Temp.Byte0,Temp.Byte1,TH,TL,ConfigReg]
;read TH,TL and ConfigReg, we won't use this information
;Temperature data returned as 2's complement for below 0.
;Hence, below 0=1 in the returned highest bit.
PlusFlag = 1
If Temp.HighBit = 1 Then        ;check for < 0C
   ;Subtract from 0 to read below 0 value
   Temp = $0000-Temp
   PlusFlag = 0
EndIf
```

After the temperature reading is complete, the above code reads the 12-bit result, and, if the result is below zero Celsius, decomplements the data and clears a plus/minus flag, **PlusFlag**.

```
;Returned value is in Centigrade steps of 0.0625 Degrees C.
Centigrade = (ToFloat Temp) * 0.0625
If PlusFlag = 0 Then   ;below zero, multiply by -1
        Centigrade = -Centigrade
EndIf

;Standard conversion formula to get to F from C.
Fahrenheit = (Centigrade * 1.8) + 32.0 ;convert to F
```

Now, the raw temperature reading is converted to floating point Celsius and Fahrenheit values, held in the variables **Centigrade** and **Fahrenheit**.

```
;Write output. Note the \2 and \1 modifiers
;give number of places after decimal point
HSerOut ["Temperature ",Real Fahrenheit \2," F " , |
Real Centigrade \1," C", 13]
;Use Modifier as Function to convert to string
Fstring = Real Fahrenheit \2
HserOut ["Sending: ",9,Str FString\10\0,13]
```

To convert the floating point **Fahrenheit** to a string, **Fstring**, we use the **Real** modifier as a function. MBasic's ability to use a modifier as a conversion function is undocumented in its User's Guide. At this point, **Fstring** holds a string representation of the floating point **Fahrenheit**. If, for example, **Fahrenheit** holds the value 74.123456, after **Fstring = Real Fahrenheit \2**, the variable **Fstring** equals "74.12."

```
            Channel = Channel0
            GoSub SendData
```

We call subroutine **SendData** to output **Fstring** over the X-10 interface. We set **Channel = Channel0** to send **Fstring** with a channel identifier of Channel 0.

```
            Fstring = Real Centigrade \2
            HserOut ["Sending: ",9,Str FString\10\0,13]
            Channel = Channel1
            GoSub SendData
     GoTo Main
```

We repeat the process with the Celsius temperature floating point variable **Centigrade**, but send it with a channel identifier of Channel 1.

Let's examine the subroutine **SendData**. **SendData** requires the output string to be held in the variable **Fstring**, and that **Fstring**'s characters are restricted to the digits 0…9, decimal point, plus and minus signs. The channel number is set in the variable **Channel**.

```
     ;String to be sent is in FString
     ;Channel no is in Channel
     SendData
     ;------------
     ;now we send  the string via X10 transmitter
            ;start with StartData and Channel No.
            XOut DataOutPin\ZeroXPin,HouseCode,[StartData,Channel]
            Pause 100
            i = 0
```

We know the channel number so we can send the start data code and the channel number. I added a 100 ms pause for debugging purposes, but it isn't necessary and may be removed if you desire.

```
            ;strip the string apart and look for
            ;- and . and numbers. Encode these
            While (FString(i) <> 0) AND (i<10)     ;printing char
                    If FString(i) = "-" Then
                            OutByte = Minus
                    EndIf
                    If FString(i) = "." Then
                            OutByte = DecimalPt
                    EndIf
                    If (FString(i) >= "0") AND (FString(i) <="9") Then
                            OutByte = FString(i) - "0"
                    EndIf
                    ;now send the encoded byte
                    XOut DataOutPin\ZeroXPin,HouseCode,[OutByte]
                    i = i+1
                    Pause 50
            WEND
            ;all data gone, now wrap it up with end data
            XOut DataOutPin\ZeroXPin,HouseCode,[EndData]
            ;zero FString for the next data value
            For i = 0 to 9
                    FString(i) = 0
            Next
            Pause 1000
     Return
```

The main body of **SendData** loops through **Fstring** so long as it has data, or the length of **Fstring**, whichever occurs first. A **While … WEND** loop provides an appropriate control structure, as it permits checking the end conditions at the top of the control loop.

Each character in **Fstring** is tested to see if it is a special character, i.e., decimal point, or a plus or minus sign. If it is, **OutByte** is set to the appropriate character code. If the character is a number character "0"…"9" **OutByte** is set to the numerical value 0…9 by subtracting the value of the zero character "0." **OutByte** is then sent via the **Xout** procedure.

After all characters in **Fstring** have been sent, the **EndData** code is sent and **Fstring** zeroed to be ready for the next string to be sent.

Let's see how Program 20-3 receives the sent data.

Program 20-05 Sending	Program 20-03 Receiving	Comments
Temperature 71.37 F 21.8 C	Rcvd: 10	Start Code
Sending: 71.37	Rcvd: 16	F Data is on Channel 0
Sending: 21.87	Rcvd: 7	7
	Rcvd: 1	1
	Rcvd: 12	.
	Rcvd: 3	3
	Rcvd: 7	7
	Rcvd: 11	End Code
	Rcvd: 10	Start Code
	Rcvd: 17	C Data is on Channel 1
	Rcvd: 2	2
	Rcvd: 1	1
	Rcvd: 12	.
	Rcvd: 8	8
	Rcvd: 7	7
	Rcvd: 11	End Code

Program 20-6

Program 20-3's output verifies that the temperature data is correctly encoded and set, so we may now concentrate on an improved display of received data. Program 20-6 uses the same receiving configuration as Program 20-3.

```
;Program 20-06 demonstrates reading the remote
;telemetered data from Program 20-05. We read
;Channel 0 and display the temp in Deg F.

;Constants
;------------
DataOutPin          Con          B2      ;TW523-4 data from PIC
DataInPin           Con          B1      ;TW523-3 data into PIC
ZeroXPin            Con          B0      ;TW523-1 zero crossing
                                         ;TW523-2 goes to VSS
HouseCode           Con          X_J     ;Unused control code

;Data is sent as : <Start Data> <Channel No>
;<Data Elements 1...N> <End Data>
;Data elements are 0...9, +. -, .
StartData           Con          10      ;first element
EndData             Con          11      ;last element
DecimalPt           Con          12      ;decimal point
Minus               Con          13      ;minus sign
Plus                Con          14      ;plus sign
Channel0            Con          16      ;data channels 0...15
                                         ;Number = Channel + 16
;reserved length for received string
RXStrLen            Con          12      ;arbitrary selection

;status of main loop
```

```
Waiting                  Con           0      ;waiting for input
ValidStart               Con           1      ;has rcvd "StartData"
ValidEnd                 Con           2      ;has rcvd "EndData"
Unknown                  Con           3      ;Unknown status

;We have only two telemetry channels now
;We'll send Deg F on Channel 0 and Deg C on
;Channel 1
ChannelUnits    ByteTable "Deg F","Deg C"

;Variables
;-----------
RcvdCode        Var     Byte      ;5-bit received data
i               Var     Byte      ;loop counter
j               Var     Byte      ;loop counter
RcvdData        Var     Byte(RxStrLen) ;holds incoming RcvdCodes
RcvdStr         Var     Byte(RxStrLen) ;decoded string
InStatus        Var     Byte      ;holds status of main loop
ChannelNo       Var     Byte      ;telemetry channel number

;Initialization
;---------------
EnableHSerial
SetHSerial H115200
;set to null
For i = 0 to (RXStrLen-1)
        RcvdStr = 0
Next
;don't know where status is
InStatus = Unknown
i = 0

Main
;-----
        ;read data input-must match HouseCode
        XIN DataInPin\ZeroXPin,HouseCode,[RcvdCode]
        ;remove comments to see raw data
        ;HSerOut ["Rcvd: ",Dec RcvdCode,13]
        ;If we have start byte set status
        If RcvdCode = StartData Then
                InStatus = ValidStart
        EndIf
        If RcvdCode = EndData Then
                InStatus = ValidEnd
        EndIf
        If InStatus = ValidStart Then
                RcvdData(i) = RcvdCode
                i = i+1
                If i = RXStrLen Then
                        i = 0
                EndIf
        EndIf
        If InStatus = ValidEnd Then
                For j = 0 to (RXStrLen-1)
                        RcvdStr(j) = 0
                Next

                ChannelNo = RcvdData(1)-16

                For j = 0 to (i-3)
                        RcvdStr(j) = RcvdData(j+2)+"0"
                        If RcvdData(j+2) = DecimalPt Then
                                RcvdStr(j) = "."
                        EndIf
                        If RcvdData(j+2) = Plus Then
                                RcvdStr(j) = "+"
                        EndIf
                        If RcvdData(j+2) = Minus Then
                                RcvdStr(j) = "-"
                        EndIf
```

```
                              Next
                              HSerOut ["Chan: ",9,Dec ChannelNo,9]
                              HSerOut [Str RcvdStr\RXStrLen-1\0,9]
                              HSerOut [Str ChannelUnits(ChannelNo*5)\5,13]
                              InStatus = Unknown
                              i = 0
                      EndIf
              GoTo Main
              End
```

The constant and initialization section of Program 20-6 is similar to Program 20-5 and will not be analyzed.

Our approach to reading the incoming data stream is to begin collecting data when the start code is received, and stop collecting data when the end code is received. We control program flow with a status variable, **In-Status**. In pseudo-code our algorithm is:

```
              Initialize InStatus = Unknown

      MainStart
      Read incoming character
      Is it Start? If so then InStatus = ValidStart
      Is it End? If so then InStatus = EndData

      If InStatus=ValidStart then:
      Store the incoming character in the array RcvdData(i)
      Increment (i)

      If InStatus=EndData then:
              Convert RcvdData to numerical value and display
              Reset i to 0

      Goto MainStart and read the next character
```

Until we receive the start character, nothing is stored; the program continues to loop through reading the incoming character. When a start character is received, the state variable **InStatus** latches to **ValidStart** condition and we store each received character in the array **RcvdData** so long as **InStatus** remains in **Va-lidStart** condition. When the end character is received, **InStatus** changes to a new status, **ValidEnd**, and remains in **ValidEnd** state until a new start character is received. At the time the status changes from **Valid-Start** to **ValidEnd**, we convert the saved data in **RcvdData** to the desired format and display the results.

Let's see how we might implement this algorithm.

```
              InStatus = Unknown
              i = 0
```

We initialize **InStatus** as **unknown** and set the counter variable to **i** to **0**. This provides a known starting point; since **InStatus** is neither **ValidStart** nor **ValidEnd**, none of the activities associated with the latter two states will be executed. Instead, we await either a start or end code.

```
      Main
      ;-----
              ;read data input-must match HouseCode
              XIN DataInPin\ZeroXPin,HouseCode,[RcvdCode]
              ;remove comments to see raw data
              ;HSerOut ["Rcvd: ",Dec RcvdCode,13]
              ;If we have start byte set status

              If RcvdCode = StartData Then
                      InStatus = ValidStart
              EndIf

              If RcvdCode = EndData Then
                      InStatus = ValidEnd
              EndIf
```

We read the incoming data and change **InStatus** based on the received data. Remember, **InStatus** toggles between **ValidStart** or **ValidEnd** status, and it stays in one value until it flips to the opposite value.

```
If InStatus = ValidStart Then
        RcvdData(i) = RcvdCode
        i = i+1
        If i = RXStrLen Then
                i = 0
        EndIf
EndIf
```

If the data stream is valid, as evidenced by receipt of the start code, and **InStatus = ValidStart**, we save the received character in **RcvdData(i)** and increment **i**. We also check for possible **i** overflow and reset (**i=0**) if necessary to prevent overflow.

```
If InStatus = ValidEnd Then
```

At this point, **RcvdData** holds:

i=	0	1	2	3		i-1	i
RcvdData(i)=	Start Data	Channel Code	Data Element 1	Data Element 2	...	Data Element N	End Data

```
For j = 0 to (RXStrLen-1)
        RcvdStr(j) = 0
Next
```

We next initialize a new array variable **RcvdStr** by filling it with null (0) characters.

```
ChannelNo = RcvdData(1)-16
```

The channel number is held as the second element in **RcvdData**, with a bias of 16. We convert it to an integer value 0...15 by subtracting 16.

```
For j = 0 to (i-3)
        RcvdStr(j) = RcvdData(j+2)+"0"
        If RcvdData(j+2) = DecimalPt Then
                RcvdStr(j) = "."
        EndIf
        If RcvdData(j+2) = Plus Then
                RcvdStr(j) = "+"
        EndIf
        If RcvdData(j+2) = Minus Then
                RcvdStr(j) = "-"
        EndIf
Next
```

Next, we copy the data element bytes from **RcvdData** to **RcvdStr**. As we've seen, the first valid data element is **RcvdData(2)** and the last valid data element is **RcvdData(i-1)**. Our objective is to copy only the data element bytes from **RcvdData** to **RcvdStr**, and left justify the copied data in **RcvdStr**. The above code accomplishes this task.

We also convert the data byte values to a string character by adding **"0"** and checking for the three special characters, decimal point and the plus and minus signs. If any of these special characters are found, we insert the appropriate string symbol.

```
HSerOut ["Chan: ",9,Dec ChannelNo,9]
HSerOut [Str RcvdStr\RXStrLen-1\0,9]
HSerOut [Str ChannelUnits(ChannelNo*5)\5,13]
InStatus = Unknown
i = 0
        EndIf
GoTo Main
```

Finally, we write the decoded string to the serial port. To display the appropriate dimension name, we earlier defined dimensional constants for Channels 0 and 1:

```
ChannelUnits   ByteTable "Deg F","Deg C"
```

We index into the byte table **ChannelUnits** by **ChannelNo*5**. Hence, Channel 0 returns "Deg F" while Channel 1 returns "Deg C."

Following is the output from Program 20-5 compared with the received data from Program 20-6:

Program 20-5 Sending	Program 20-6 Receiving			
Temperature 70.58 F 21.4 C				
Sending: 70.58	*Chan:*	*0*	*70.58*	*Deg F*
Sending: 21.43	*Chan:*	*1*	*21.43*	*Deg C*
Temperature 70.69 F 21.5 C				
Sending: 70.69	*Chan:*	*0*	*70.69*	*Deg F*
Sending: 21.50	*Chan:*	*1*	*21.50*	*Deg C*
Temperature 70.69 F 21.5 C				
Sending: 70.69	*Chan:*	*0*	*70.69*	*Deg F*
Sending: 21.50	*Chan:*	*1*	*21.50*	*Deg C*
Temperature 70.81 F 21.5 C				
Sending: 70.81	*Chan:*	*0*	*70.81*	*Deg F*
Sending: 21.56	*Chan:*	*1*	*21.56*	*Deg C*
Temperature 70.81 F 21.5 C				
Sending: 70.81	*Chan:*	*0*	*70.81*	*Deg F*
Sending: 21.56	*Chan:*	*1*	*21.56*	*Deg C*
Temperature 70.81 F 21.5 C				
Sending: 70.81	*Chan:*	*0*	*70.81*	*Deg F*
Sending: 21.56	*Chan:*	*1*	*21.56*	*Deg C*
Temperature 70.81 F 21.5 C				
Sending: 70.81	*Chan:*	*0*	*70.81*	*Deg F*
Sending: 21.56	*Chan:*	*1*	*21.56*	*Deg C*
Temperature 70.92 F 21.6 C				
Sending: 70.92	*Chan:*	*0*	*70.92*	*Deg F*
Sending: 21.62	*Chan:*	*1*	*21.62*	*Deg C*
Temperature 70.81 F 21.5 C				
Sending: 70.81	*Chan:*	*0*	*70.81*	*Deg F*
Sending: 21.56	*Chan:*	*1*	*21.56*	*Deg C*

Program 20-7

Our final program is an "X-10 code sniffer" that monitors the power line and displays the last four X-10 commands on an LCD display. It also outputs received commands over the RS-232 port. It builds on the algorithm of Program 20-6, but with a software version of `Xin`.

Program 20-7 is very long. Rather than follow our normal structure of listing the program in full, followed by a section-by-section analysis, we'll skip the full listing. Program 20-7, like all other programs in this book, is available in an electronic copy in the associated CD-ROM. If you need a listing to follow while reading the analysis, you may print one from the program file.

Figure 20-15 shows the circuit I used with Program 20-7. Although I used a 20 × 4 LCD display, only simple changes are necessary to Program 20-7 should your display have fewer lines or characters.

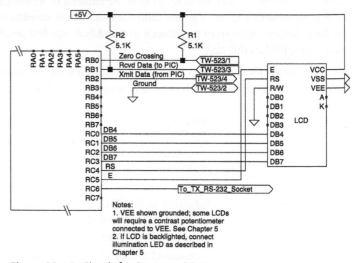

Figure 20-15: Circuit for Program 20-7.

To illustrate that we may successfully work with either LSB-first or MSB-first bit order, we'll use LSB-first order in Program 20-7.

Although we will use the same byte table approach as in Program 20-2, the table orders are different, to reflect the LSB-first approach. We also define a new byte table **PowerTable** to assist us in calculating the value of received data in LSB-first order.

If you examine the earlier comparison table, you will see that values in **HouseTable**, **UnitTable** and **Key-Name** are in the sequence of the TW523 LSB-first documentation.

```
PowerTable      ByteTable 1,2,4,8,16

HouseTable      ByteTable "MNOPCDABEFGHKLIJ"

UnitTable       ByteTable 13,14,15,16,3,4,1,2,5,6,7,8,11,12,9,10

KeyName         ByteTable                   "All Units Off",
                                            "Hail Request ",
                                            "Dim          ",
                                            "Ext Dta-Analg",
                                            "On           ",
                                            "Pre-Set Dim 0",
                                            "All Lghts Off",
                                            "Status=Off   ",
                                            "All Lghts On ",
                                            "Hail Acknldge",
                                            "Bright       ",
                                            "Status=On    ",
                                            "Off          ",
                                            "Pre-Set Dim 1",
                                            "Extended Code",
                                            "Status Rqst  "
```

The initialization section initializes the LCD display:

```
;Initialization
;---------------
        Pause 500               ;Allows the LCD to initialize
        ;Clear the LCD
        LCDWRITE RegSel\Clk, LCDNib, [INITLCD1,INITLCD2, |
            TWOLINE, CLEAR, HOME, SCRBLK]
        LineNo = 1
        LCDWrite RegSel\Clk,LCDNib,[ScrRAM+LineOFf(LineNo)]
        LCDWrite RegSel\Clk,LCDNib,["P-20-07"]
        Pause 2000
        LCDWRITE RegSel\Clk, LCDNib, [INITLCD1,INITLCD2, |
            TWOLINE, CLEAR, HOME, SCRBLK]
```

Since we've previously covered LCD initialization in Chapter 5, we will not review the above initialization code, save to note that we defined the byte table **LineOff**:

```
LineOff ByteTable $0,$0,$40,$14,$54
```

LineOff(LineNo) returns the LCD RAM offset to position the text to start at the left edge for lines 1...4 (**LineNo** = 1...4). **LineOff(0)** returns a dummy value, 0, as legitimate line numbers do not include 0.

```
        EnableHSerial
        SetHSerial H115200
        HSerOut ["P-20-07",13]
        Pause 500
        Input B0        ;read zero crossing
        Input B1        ;read data

        OldXPin = %0
        j = 0
        i = 0
        RXStatus = Waiting
```

The remainder of the initialization section establishes initial values for several variables. In particular, we use the same discrete state structure in Program 20-7 that we did in Program 20-6. For the reasons discussed earlier, therefore, we initialize the control variable **RXStatus** to **Waiting**.

Our strategy in Program 20-7 mimics Program 20-6 is key respects. Instead of reading a sequence of bytes, instead here we must read a sequence of bits from the TW523. Each bit must be read immediately after the zero crossing pin changes state. Let's look at our algorithm in pseudo-code.

```
Read Zero Crossing pin
If Zero Crossing Pin changes state (0→1) or (1→0) Then
        Read Data Pin
Invert Data Pin Value (TW523 output has inverted logic)

Count number of consecutive 1's received
If three consecutive 1's received the:
        it is a start signal
        Set RXStatus to ValidStart

        If RXStatus=ValidStart, save data pin value in RcvdBit(i)
        Increment i and test value of i
                If i<RXBitLen then go back to zero crossing status check
                        and wait for more bits
                If i=RXBitLen, then last bit of sequence is received

If RXBitLen bits have been received,then:
        Change RXStatus to Waiting
        Decode the data held in RcvdBit()
Go back and wait for next zero crossing
```

Let's see how we implement this algorithm.

```
ReadXPin
;--------
        ;read zero-crossing pin and wait for a change in value
        If XPin = OldXPin Then
                GoTo ReadXPin
        EndIf
```

We read the value of the zero crossing pin and detect a change in state by comparing the currently read value (**Xpin**) with the last read value, held in **OldRXPin**. Until the zero crossing pin changes state, the program remains in a tight loop, continuing to read and compare the zero crossing pin value.

```
        ;Have a zero crossing, so read the data pin value
        OldXPin = XPin
        InVal = InPin
        ;TW-523 has +5V = LOGIC 0, so invert by
        ;adding %1.
        InVal = InVal+1%
```

When the zero crossing pin changes state, we update the old pin status (**OldXPin**) and read the data pin. Figure 20-16 shows the received data output from a TW523. The TW523's output is inverted logic; logic 1 in the PIC world is represented as a low TW523 output. To invert the data pin reading, we add %1 to the result. (%1+%1=0; %0+%1=%1) As Figure 20-16 shows, the TW523 maintains its logic 1 (low voltage) output commencing immediately upon zero crossing and holds the value for approximately 1 ms. As we learned earlier, the TW523 consolidates the multiple tone burst transmissions required by the X-10 protocol into one logic output, coincident with the zero crossing.

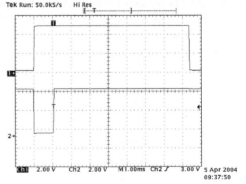

Figure 20-16: TW523 Receiving; Ch1: Zero Crossing Pin; Ch2: Data Pin.

```
;X10 start bit pattern is 1110. This is a unique
;pattern. To detect it, we count consecutive 1's
;via j=j+1. Reset j=0 if read an input 0
If InVal = %1 Then
        j = j + 1
Else
        j = 0
EndIf
```

As we learned earlier in this chapter, the X-10 protocol starts each data sequence with a unique starting code **%1110**. To detect this starting code, we count consecutive logic 1s, resetting the counter **j** to zero whenever logic 0 is received.

```
;Given the Value/Inverted Value format of X10
;we only have 111 pattern at start code. Receipt of
;111 gives j=3
If j=3 Then
        RXStatus = ValidStart   ;status=start
        i = 0
        j = 0
EndIf
```

Upon receiving **%111**, we assume a valid start code is received. (Actually, we should wait and verify that the next bit is a **%0**, but we'll take care of that when we verify the bit complements.) When received, we toggle the **RXStatus** to **ValidStart** and begin capturing the received bits.

```
;After a valid start, we capture the bit data in
;the array RcvdBit. Store received bit in RcvdBit(i)
;and bump up the counter (i).
If RXStatus = ValidStart Then
        RcvdBit(i) = InVal
        i = i+1
        If i = RXBitLen Then            ;check for exact receipt
                RXStatus = Waiting      ;if overflow restart
                GoSub DumpOutput        ;and dump output
        EndIf
EndIf
GoTo ReadXPin
```

We know that a valid sequence—and the bit array **RcvdBit**— will contain 20 bits:

Start		House Code								Data									
S2	S3	H1	$\overline{H1}$	H2	$\overline{H2}$	H4	$\overline{H4}$	H8	$\overline{H8}$	D1	$\overline{D1}$	D2	$\overline{D2}$	D4	$\overline{D4}$	D8	$\overline{D8}$	D16	$\overline{D16}$
1	0	x	$\overline{x}$	x	$\overline{x}$	x	$\overline{x}$	x	$\overline{x}$	x	$\overline{x}$	x	$\overline{x}$	x	$\overline{x}$	x	$\overline{x}$	x	$\overline{x}$
0	1	2	3	4	5	6	7	8	9	10	11	12	13	14	15	16	17	18	19

S2 and S3 are the last two of the four start bits
H are house code bits
D are data bits
The overscore indicates bit complement
x Indicates either 1 or 0

The last two start bits are included because we start saving received bits in **RcvdBit** upon the third consecutive **%1**. This means we save the last 2 bits of the 4-bit start sequence.

When we have received 20 bits, we call subroutine **DumpOutput** to update the LCD and also write the decoded information to the serial port.

```
DumpOutput
;---------
        If DataDump = %1 Then
                For k = 0 to RXBitLen-1
                        If (k = StartHouse) OR (k = StartData) Then
                                HSerOut ["/ "]
                        EndIf
                        HSerOut[Bin RcvdBit(k)," "]
                Next
```

```
                              HSerOut[13]
              EndIf
```

By setting the constant **DataDump** (**DataDump Con %1**), the bits held in **RcvdBit** will be sent to the serial port. This can be useful when debugging.

```
              HouseCode = 0
              DataCode = 0
              ValidDecode = Good
              ;Data is transmitted as <Bit> - <NOT Bit> so we can perform
              ;parity check to see if data is corrupted.
              For k = 2 to RXBitLen-1 Step 2
                      ;Valid codes are %1 %0 and %0 %1.
                      ;These evaluate as %1 in XOR
                      ;invalid codes are %1 %1 and %0 %0.
                      ;These evaluate as %0 in XOR
                      ;If bad, then set ValidDecode error flag
                      If RcvdBit(k) XOR RcvdBit(k+1) = 0 Then
                              ValidDecode = Bad
                              HSerOut ["Bad Decode at k= ",Dec k\2, |
                              "  Bits are ",iBin RcvdBit(k)," ", |
                              iBin RcvdBit(k+1),13]
                      EndIf
              Next
```

We now compare the received bits and their complements to verify that the received data is uncorrupted. In reality, this check is redundant, as the TW523's circuit also makes this comparison. But, we'll use it as a way of demonstrating how to make the complement comparison.

The heart of the code is the exclusive or, or XOR, function which has the following logic table:

		Input 1	
Input 2		1	0
	1	0	1
	0	1	0

If input 1 equals input 2, the output is 0; if input 1 and input 2 are complements of each other, the output is 1. Hence, to test two adjacent bits to verify that they are complements, we use the test **If RcvdBit(k) XOR RcvdBit(k+1) = 0 Then ValidDecode = Bad**.

If they are complements, then the XOR test returns 1, and the sentinel variable **ValidDecode** remains equal to **Good**. If **ValidDecode** equals **Bad**, then an appropriate error message is emitted.

```
              If ValidDecode = Good Then
              ;We evaluate in LSB order.
                      For k = 0 to 3
                              HouseCode = HouseCode + |
                                  RcvdBit(2+k*2) * PowerTable(k)
                      Next
                      For k = 0 to 4
                              DataCode = DataCode + |
                                  RcvdBit(10+k*2) * PowerTable(k)
                      Next
```

We now have known good bits, and compute the house code and data code values, based upon LSB first data transmission.

```
              ;If DataCode < 16 then it is a unit code
              If DataCode < 16 Then
                      ;HSerOut [Str HouseTable(HouseCode)\1,"-", |
                          Dec UnitTable(DataCode)\2,"  "]
                      UnitVal = UnitTable(DataCode)
              EndIf
```

By evaluating the data as LSB first, we can use the 0...15 and 16...31 division between unit codes and function codes. This division is replaced by an odd/even separation when the data code is evaluated MSB first.

```
                              ;now evaluate data code.
                              If DataCode >= 16 Then
                                      HSerOut [Str HouseTable(HouseCode)\1,"-", |
                                          Dec UnitVal\2," "]
                                      HSerOut [Str KeyName((DataCode-16)*13)\13,13]
                                      GoSub WriteToLCD
                                      UnitVal = 99 ;99 = Fcn Only, no Unit Code.
                              EndIf
                      EndIf
          Return
```

We send the received data out the serial port, formatting it into a house code – unit code and function name form. Displaying a unit code of 99 identifies "all call" transmissions.

Finally, we update the LCD.

```
          WriteToLCD
          ;----------
                      ;Now put the same info to the LCD.
                      LCDWrite RegSel\Clk,LCDNib,[ScrRAM+LineOFf(LineNo)]
                      LCDWrite RegSel\Clk,LCDNib,[Dec LineNo," ", |
                          Str HouseTable(HouseCode)\1,"-", |
                          Dec UnitVal\2," ",Str KeyName((DataCode-16)*13)\13]
                      ;Add a line number
                      LineNo = (LineNo + 1)
                      If LineNo = 5 Then
                              LineNo = 1
                      EndIf
          Return
```

We write the decoded data to the next available line, recycling to line 1 after four lines are written.

Figure 20-17 shows the LCD display when receiving commands sent by Program 20-2.

The following is a representative sample of the serial output of Program 20-07 when receiving commands sent by Program 20-2.

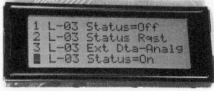

Figure 20-17: LCD output displaying Program 20-7 results.

Transmitting Program 20-02 Output		Receiving Program 20-07 Output	
L-16	Pre-Set Dim 0	L-16	Pre-Set Dim 0
L-16	Status=Off	L-16	Pre-Set Dim 1
L-16	Extended Data	L-16	Ext Dta-Analg
L-16	Status=On	L-16	Status=On
L-16	Pre-set Dim 1	L-16	Status=Off
L-16	Status Rqst	L-16	Status Rqst
M-ALL	All Units Off	M-99	All Units Off
M-ALL	All Lghts On	M-99	All Lghts On
M-ALL	On	M-99	On
M-ALL	Off	M-99	Off
M-ALL	Dim	M-99	Dim
M-ALL	Bright	M-99	Bright
M-ALL	All Lghts Off	M-99	All Lghts Off
M-ALL	Extended Code	M-99	Extended Code
M-ALL	Hail Request	M-99	Hail Request
M-ALL	Hail Acknldge	M-99	Hail Acknldge
M-ALL	Pre-Set Dim 0	M-99	Pre-Set Dim 1
M-ALL	Status=Off	M-99	Pre-Set Dim 1
M-ALL	Extended Data	M-99	Ext Dta-Analg
M-ALL	Status=On	M-99	Status=On
M-ALL	Pre-set Dim 1	M-99	Status=Off
M-ALL	Status Rqst	M-99	Status Rqst
M-01	All Units Off	M-01	All Units Off
M-01	All Lghts On	M-01	All Lghts On
M-01	On	M-01	On
M-01	Off	M-01	Off

Range Problems

During the course of experimenting with two TW523's, I discovered that high quality power strips, identified as "filtered" or "protected" contain interference reducing filter components that attenuate the 120 KHz bursts and therefore reduce the possible range. For maximum range, plug both the TW523 and any receiving modules directly into the wall socket and don't use a power strip.

A second range problem often observed with X-10 is due to the way power is distributed inside a residence. Figure 20-18 shows a typical US residential wiring arrangement. Units A and B are X-10 receivers. The TW523 is on one leg of the power distribution line, and its signals can cross over to the other leg only by passing through appliances that are connected to the 240 V line, that is, to both hot circuits. In many cases, these appliances will leak enough 120 KHz energy through, even when turned off, to permit decoding with modules connected on either 120 V leg. My house had insufficient leakage between the two power lines to work. If you experience this problem you can purchase an X-10 signal bridge that couples the 120 KHz burst signals between both sides of your house wiring.

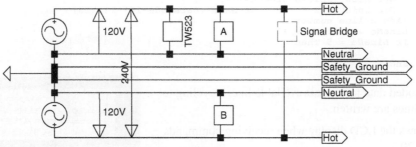

Figure 20-18: X-10 Power distribution issues.

References

[20-1] Rye, Dave, *Technical Note, The X-10 POWERHOUSE Power Line Interface Model # PL513 and Two-Way Power Line Interface Model #TW523, Rev. 2.4*, X-10 Powerhouse (undated). Available for download at http://www.x10.com/support/support_manuals.htm.

[20-2] Davidson, Ken, *The X-10 TW523 Two-Way Power Line Interface*, Circuit Cellar Ink, No. 5, September/ October 1988. Available for download at http://www.circuitcellar.com/library/print/hcs-pdf/5-Davidson. pdf.

[20-3] Davidson, Ken, Power-Line Based Computer Control, Circuit Cellar Ink, No. 3, May/June 1988. Available for download at http://www.circuitcellar.com/library/print/hcs-pdf/3-Davidson.pdf.

[20-4] Moews, Paul and Moews, David, The X-10 Spy; Making X-10 Signals Visible, Circuit Cellar Ink, No. 63, October 1995. Available for download at http://www.circuitcellar.com/library/print/hcs-pdf/63-moews.pdf.

Digital Potentiometers and Controllable Filter

Digital potentiometers (usually called a "pot" in electronics speak) are a solid-state replacement for mechanically adjustable variable resistors historically used for volume, balance and tone controls in consumer electronics, among many other applications. Instead of responding to shaft rotation, digital pots change their value—in microseconds—based upon a received command message.

Digital pots are offered in a variety of control interfaces, including 1-wire, contact closure (up/down steps), parallel, increment/decrement, SPI (3-wire) and others. Some even include nonvolatile memory to retain their last setting when power is removed. We'll look at Microchip's MCP41xxx/42xxx family of digital pots, with an SPI interface. Other manufacturers, such as Maxim (including its Dallas Semiconductor division), have wider product ranges, and you should review their offerings if you think digital potentiometers may be useful in your particular project.

First, let's clear up the terminology. The device we are talking about is known as a potentiometer, a variable resistor, a rheostat or a volume control, among many other terms. In reality, these terms can be boiled down into two possible connection arrangements, as shown in Figure 21-1. The *potentiometer* arrangement uses all three connections and is usually configured as a variable voltage divider. The *variable resistor* or *rheostat* arrangement uses two connections (the wiper—the variable connecting piece—is connected to one end of the fixed winding, but to the external circuit only two connections are seen) and operates as a variable resistance. We'll use the term "pots" as the generic name for these devices, even though they may be connected, in certain applications, as variable resistors.

One major difference between mechanical and digital pots is resolution. Mechanical pots vary continuously with shaft rotation, or in the case of some wirewound designs, have very small jumps in value with rotation. Digital pots, in contrast, are more accurately though of as a string of resistors with an electronic switch, as illustrated in Figure 21-2. Consequently, their value moves in steps. The MCP42xxx/41xxx devices have 256 steps. Rx is residual resistance, and Rw can be thought of as the resistance of the "wiper" connection. For Microchip's MCP41xxx/42xxx devices,

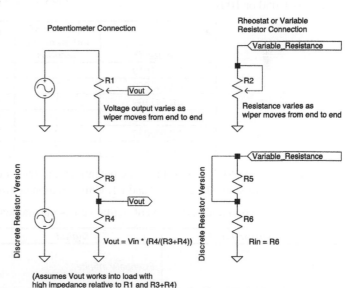

Figure 21-1: Potentiometer versus variable resistor (Rheostat) connection.

the parasitic resistance from terminal A includes one resistance step as well, that is, the wiper will never connect directly to terminal A.

For the MCP41xxx/42xxx devices, the resistance between terminals A, B and the wiper and the command step D_n (neglecting parasitic resistance Rx) is:

$$R_{BW} = R_{AB}\frac{D_n}{256} + R_W$$

$$R_{AW} = R_{AB}\frac{256 - D_n}{256} + R_W$$

where:

D_n is the command step, with possible values 0…255.

R_{BW} is the resistance between terminals B and the wiper

R_{AW} is the resistance between terminals A and the wiper

R_{AB} is the resistance between terminals A and B.

The MCP41xxx/42xxx family consists of six devices, with the device ID identifying the number of sections (1 or 2) and the nominal resistance (in Kohm, 010, 050 or 100).

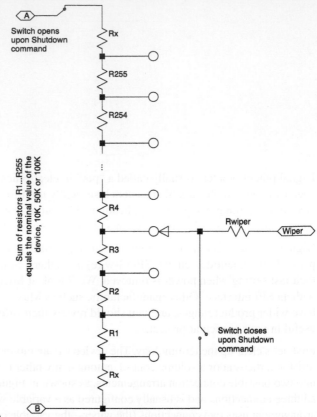

Figure 21-2: Conceptual model of digital potentiometer.

Device ID	Number of Independent Sections	Resistance R_{AB}
MCP41010	1	10Kohm
MCP41050	1	50Kohm
MCP41100	1	100Kohm
MCP42010	2	10Kohm
MCP42050	2	50Kohm
MCP42100	2	100Kohm

Single devices, type MCP41xxx, have one potentiometer. Dual devices, type MCP42xxx, have two independent potentiometers, identified as P0 and P1, with common control logic and power supply.

Let's look at the good and bad points of digital versus mechanical pots.

Comparison Point	Digital	Mechanical
Lifetime	Essentially infinite.	Limited by number of shaft rotations and construction. Ranges from a few hundred cycles to hundreds of thousands of cycles.

(Continued)

Comparison Point	Digital	Mechanical
Reliability; environmental	Excellent when operated within specifications; hermetically sealed.	Good, but can become noisy and some construction types can become contaminated when operated in harsh environments.
Resistance ranges available	Limited; very high and very low values not available.	Wide range of values available from 1 ohm to 10 megohms.
Power handling	Very limited; dissipation limited to a few milliwatts; maximum permitted current typically 1 mA.	Wide range available, including power rheostats rated into the Kilowatt range.
Speed of response	Excellent; microsecond or faster.	Limited by mechanical considerations.
Remote control ability	Excellent, small package with direct microprocessor connection.	Can be motor driven, but large volume results.
Linearity	Very good, well under 1%.	High linearity (0.1%) available, but at a premium price.
Voltage Rating	Can't exceed the supply voltages (V_{DD} and V_{SS}) of the device.	Limited only by the insulation used.
Frequency Response	Less than 1 MHz for 10K; less than 145 KHz for 100K devices.	Limited by stray capacitance and inductance; can easily exceed 10 MHz for carbon or conductive plastic devices.
Distortion and Intermodulation	Not specified; but Rw is a nonlinear function of applied voltage.	Essentially zero.
Shock and vibration resistance	Excellent.	Moderate; may require shaft locks to prevent inadvertent rotation.
Temperature Coefficient	Approx. ±800 PPM/°C.	Can be as good as ±100 PPM/°C in premium units.
Resolution	Typically 100, 256, 512 or 1024 steps.	Theoretically unlimited in most designs, but in practice limited by friction and backlash. Wirewound single turn pots have jumps as the wiper moves from wire to wire.
Taper Available	Linear.	Linear, log, reverse log.
Nominal Resistance Accuracy	Typically ±20–30%.	Typically ±10–25%.

Now, it's time to look at some code.

We've used the three-wire, or SPI, protocol to read the DS1302 real-time clock in Chapter 12, so we won't repeat the introductory material appearing in that chapter.

Getting Started with an MCP41010

Our initial test configuration uses a MCP41010, a single section 10K device, connected as illustrated in Figure 21-3. Program 21-1 exercises the potentiometer by running it through all possible 256 states.

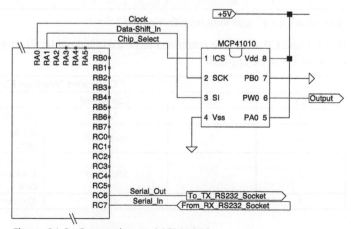

Figure 21-3: Connecting an MCP41010.

Program 21-1

```
;Program 21-1
;Control a MCP41010
;Digital potentiometer

;Constants
;-----------
Clk             Con     A0      ;Clock
Dta             Con     A1      ;Data
CSel            Con     A2      ;ChipSelect


Cmd_NOP         Con     %00000000
Cmd_Write       Con     %00010011
Cmd_ShutDn      Con     %00100011

;Variablesw
;-----------
i       Var             Byte    ;loop var

;Initialization
;--------------
Output CSel
Low CSel

Main
        For i = 0 to 255
                Low CSel
                ShiftOut Dta,Clk,MSBPre,[Cmd_Write\8,i\8]
                High CSel
        Next
        GoTo Main
End
```

Similar to the DS1302 real-time clock, the command syntax for the MCP42xxx/41xxx chips is a command byte, followed by an instruction byte.

The key statement in Program 21-1 is **ShiftOut Dta,Clk,MSBPre,[Cmd_Write\8,i\8]**. It sends two sequential bytes to the MCP41010; the constant **Cmd_Write**, followed by the step instruction, **i**.

The possible command bytes are illustrated below.

MCP42xxx/41xxx Command Byte							
Bit 7	Bit 6	Bit 5	Bit 4	Bit 3	Bit 2	Bit 1	Bit 0
X	X	C1	C0	X	X	P1	P0

X is "don't care" and may be either a 0 or 1.

Bits C1 and C0 select the command to be executed, while P1 and P0 select the potentiometer to which the command is to be applied.

Command Selection Bits			
C1	C0	Command	Command Summary
0	0	None	No command executed.
0	1	Write Data	Write the data contained in the instruction byte to the potentiometer(s) selected by the selection bits.
1	0	Shutdown	Potentiometer(s) selected by the selection bits enter shutdown mode. Data bits for this mode are "don't care."
1	1	None	No command executed.

Potentiometer Selection Bits		
P1	*P0*	*Potentiometer Selections*
0	0	*Dummy code; neither potentiometer is affected.*
0	1	*Command executed on P0.*
1	0	*Command executed on P1.*
1	1	*Command executed on both P0 and P1.*

Since we are using a single section device, the potentiometer selection bits P0 and P1 are "don't care." We'll set them to **%11**.

When dealing with an MCP41xxx device, we have only two useful commands; write a step value and shut down, and a third dummy command "do nothing" that we'll later see is necessary when we daisy chain multiple MCP42xxx devices:

```
Cmd_NOP        Con     %00000000
Cmd_Write      Con     %00010011
Cmd_ShutDn     Con     %00100011
```

We have accordingly defined the three possible command bytes as constants.

Let's take a quick look at what these command bytes do.

Cmd_NOP—A "no operation" or dummy command. When received, the device takes no action. All previous values are retained without change.

Cmd_Write—Instructs the device to set the tap setting to the immediately following byte value.

Cmd_ShutDn—Instructs the device to enter shutdown mode. As illustrated in Figure 21-2, upon a shutdown instruction, terminal A is disconnected from the resistive divider string and the wiper connection is connected directly to terminal B.

One final point regarding our use of the **ShiftOut** statement concerns selecting the proper mode constant. As we saw in Chapter 12, the DS1302's read and write protocol is LSB first, and bit sample before the clock pulse. Hence we use the mode **LSBPRE** or its alias **LSBFIRST**. The MCP42xxx/41xxx's data sheet (Section 5.8) tells us that data is sent MSB first and that the data should be set before the clock pulse. Hence, the correct mode for a **ShiftOut** statement is **MSBPRE**. If you don't understand why this is the case, compare the data transfer diagrams for the DS1302 and the MCP42xxx/41xxx devices from their respective data sheets.

A final reminder of SPI functionality is that data may be written only when the MCP42xxx/41xxx's chip select pin is high. The command and data then becomes active when the chip select pin drops low.

The data byte sets the wiper at the position defined by the byte value. Let's see how that works in practice.

Suppose we send the command byte **Cmd_Write** followed by the instruction byte, D_n, 123. When the chip select line goes high, the pot wiper goes to position 123. Since the nominal resistance R_{AB} for a MCP41010 chip is 10Kohm, we can calculate the resulting nominal resistance between the wiper and terminals A and B:

$$R_{BW} = R_{AB}\frac{D_n}{256} + R_W = 10 \times 10^3 \frac{123}{256} + 52 = 4,857\Omega$$

$$R_{AW} = R_{AB}\frac{256 - D_n}{256} + R_W = 10 \times 10^3 \frac{256 - 123}{256} + 52 = 5,247\Omega$$

We've used the word "nominal" as the 10K ohm R_{AB} value may be as low as 8K ohm or as high as 12K ohm and remain within specification. The good news is that we are less often concerned with the precise values R_{AB}, R_{AW} or R_{BW} than we are with the accuracy of the relative steps. And in this regard the MCP42xxx/41xxx devices are excellent, with a typical relative accuracy error of ±0.25 LSB.

Our circuit shown in Figure 21-3 connects V_{DD} to terminal A, ground (V_{SS}) to terminal B and uses the wiper as the output connection. Since V_{DD} is 5 V, this gives an unloaded current through the resistor string R_{AB} of 0.5 mA, well within its power and current rating.

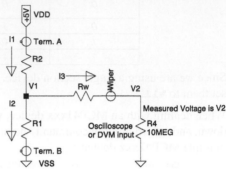

If we attach the wiper to a high impedance device, such as a digital voltmeter, or an oscilloscope probe, the voltage output at the wiper will be directly proportional to the instruction word, without regard for error in the nominal R_{AB} resistance value. Figure 21-4 shows the equivalent circuit. We can calculate the voltage measured by the DVM by going through the complete circuit equations or we can apply some common sense simplification rules:

First, R_W and R4, the resistance of the DVM are in series. R_W is 52 ohms, which is negligible compared with the DVM's 10M ohm impedance. Hence, we'll ignore R_W.

Figure 21-4: Equivalent Circuit Voltage Divider.

Second, R1 cannot exceed about 10K ohm. R1 is in parallel with R4, 10Mohm. Since R4 is 1,000 times larger than R1, we can ignore the current through R4.

Hence, the circuit reduces to simply R1 in series with R2, with the output voltage measured at the midpoint. This is a classic voltage divider and the output voltage is:

$$V_{out} = V_{in} \frac{R1}{R1+R2}$$

In our circuit, V_{IN} is V_{DD} and we note that R1/(R1+R2) is simply the ratio of $R_B/(R_A+R_B)$. After a bit of algebra, we find that:

$$V_{out} = V_{in} \frac{D_n}{256}$$

Hence, the output voltage is D_n multiplied by a constant factor, $V_{in}/256$, or for V_{in} 5.0 volts, 19.531mV/step.

If we work through a detailed circuit analysis of the effect of these two shortcuts, it turns out the error is less than 10 microvolts, quite acceptable considering other sources of error.

```
        Main
                For i = 0 to 255
                        Low CSel
                        ShiftOut Dta,Clk,MSBPre,[Cmd_Write\8,i\8]
                        High CSel
                Next
                GoTo Main
        End
```

The code will thus generate a voltage output that goes from 0 V to just under 5 V, in steps of 19.5 mV. Figure 21-5 shows the result is as expected, with each step taking about 560 μs. Gross deviation from linearity in the MCP41010 output will show up as bends or kinks in the waveform, and none are visible. By the way, if Figure 21-5 reminds you of a digital-to-analog converter output, you are quite correct; as configured in Figure 21-3, the MCP41010 is a voltage-output DAC.

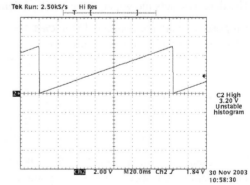

Figure 21-5: Linear voltage sweep output Ch1: N/A Ch2: wiper.

RS-232 Control of an MCP41010

Program 21-1 gives us confidence in controlling a MCP41010, but it doesn't really do much. Let's try something a bit more complex—commanding an MCP41010 by the RS-232 serial port. Our command syntax is simple:

Command	Function
Snnn (or snnn)	*Set the potentiometer to value nnn*
S? (or s?)	*Respond by sending the current setting value*
X (or x)	*Shutdown the potentiometer*

Program 21-2

```
;Program 21-2
;Control a MCP41010
;Digital potentiometer
;RS-232 Command

;Constants
;-----------
Clk             Con          A0        ;Clock
Dta             Con          A1        ;Data
CSel            Con          A2        ;Chip Select
ArraySize       Con          6         ;size of instring
Multiplier   ByteTable 100,10,1


Cmd_NOP         Con          %00000000  ;No operation
Cmd_Write       Con          %00010011  ;write value
Cmd_ShutDn      Con          %00100011  ;execute shutdown


;Variables
;-----------
i               Var          Byte      ;loop var
j               Var          Byte      ;increment counter
InStr           Var          Byte(ArraySize)
WorkStr         Var          Byte(ArraySize)
Setting         Var          Byte      ;Pot Setting
Temp            Var          Word      ;temp instring value
CmdByte         Var          Byte      ;command sent to pot

;Initialization
;--------------
EnableHSerial
SetHSerial H19200
HSerOut ["Set/Read Digital Potentiometer",13]
Setting = 0      ;initialize at 0 level
Output CSel
Low CSel

GoSub WriteSetting

Main
;------------
        ;Remind user of commands
        GoSub SendPrompt
        InStr = "        "
        ;key on 10, not 13. Assumes sending *** CR&LF ***
        ;Otherwise, input terminates on 13, and the
        ;10 appears as leading
        ;character in the next string.
        HSerIn [Str InStr\ArraySize\10]
        ;test for Setting command
        If (InStr(0) = "S") OR (InStr(0) = "s") Then
                GoSub Dispatcher            ;see what type of S command
        EndIf
```

```
                              ;Test for Shutdown command
                              If (InStr(0) = "X") OR (InStr(0) = "x") Then
                                      Setting = $00                ;dummy value
                                      CmdByte = Cmd_ShutDn   ;shut it down
                                      GoSub WriteSetting               ;send to the pot
                              EndIf

                GoTo Main

                SendPrompt               ;send prompt string
                ;--------------
                        HSerOut ["S? to Read Value Sxxx to Set Value; X for shutdown",13]
                Return

                WriteSetting    ;write to digital pot
                ;--------------
                        Low CSel                   ;enable writing
                        ShiftOut Dta,Clk,MSBPre,[CmdByte\8,Setting\8]
                        High CSel                  ;disable writing

                        ;Respond to user with message
                        ;Normal command
                        If CmdByte = Cmd_Write Then
                                HSerOut ["Set to ",DEC4 Setting,13,13]
                        EndIf
                        ;Shutdown command
                        If CmdByte = Cmd_ShutDn Then
                                HSerOut ["Shutdown Executed",13,13]
                        EndIf
                Return

                Dispatcher      ;get here only if first char is "S" or "s"
                ;-------------
                        ;Now look at the second character
                        ;If ? then report current setting
                        If InStr(1) = "?" Then
                                GoSub ReportSetting
                        EndIf

                        ;If not ? and if a numeric value, it must be a command
                        If (InStr(1) >= "0") AND (InStr(1) <= "9") Then
                                GoSub GetSetting
                        EndIf
                Return

                ReportSetting   ;output the current setting
                ;-------------
                        HSerOut ["Current Setting ",DEC4 Setting,13]
                Return

                GetSetting               ;parse the RS-232 input into numerical data
                ;-----------    ;and send it to the digital pot
                        ;right set data in WorkStr so want all 0's in unused places
                        WorkStr = "000000"
                        j = ArraySize-1
                        i = ArraySize-1

                        ;put only the numbers into WorkStr,right set
                        ;read the input string r to 1 keeping only digits 0...9
                        ;put these digits rightset into WorkStr
                        While (i >0)
                                If (InStr(i) >= "0") AND (InStr(i) <= "9") Then
                                Then
                                        WorkStr(j) = InStr(i)
                                        j = j-1
                                EndIf
                                i = i-1
                        WEnd

                        ;convert to numeric. Value of digit is its ASCII - "0"
```

```
                Temp = 0
                For i = (ArraySize-1) to (ArraySize-3) Step -1
                        Temp = Temp + Multiplier(i-3)*(WorkStr(i)-"0")
                Next

                ;If out of range, set it to 255
                If Temp > 255 Then
                        GoSub ErrorMsg
                        Temp = 255
                EndIf

                ;Write the value to the pot
                Setting = Temp
                CmdByte = Cmd_Write
                GoSub WriteSetting
        Return

        ErrorMsg
        ;-----------
                HSerOut ["Error! Set value to 255",13]
        Return
        End
```

Almost all the new code in Program 21-2 relates to acting upon the received RS-232 commands. Let's first examine the main program loop code.

```
        Main
        ;------------
                GoSub SendPrompt
                InStr = "         "
                HSerIn [Str InStr\ArraySize\10]
                If (InStr(0) = "S") OR (InStr(0) = "s") Then
                        GoSub Dispatcher            ;see what type of S command
                EndIf

                If (InStr(0) = "X") OR (InStr(0) = "x") Then
                        Setting = $00           ;dummy value
                        CmdByte = Cmd_ShutDn    ;shut it down
                        GoSub WriteSetting      ;send to the pot
                EndIf
        GoTo Main
```

After sending the prompt message, we clear the input string, **InStr**, by setting it to all blanks. (**InStr** is a byte array of length **ArraySize**, declared in the Variables section of Program 21-2.) We then read the RS-232 input into **InStr**, terminating the read when either (a) **ArraySize** number of characters has been received, or (b) the line feed character (decimal value **10**) has been received.

After the input line is finished, we then check the first letter to see what action is to be taken. If the first character of **InStr** is an "**S**" or "**s**," the action is to set the potentiometer value, so we branch execution to the subroutine **Dispatcher**. If the first character is "**X**" or "**x**," we set the command byte to the shutdown code and send it to the MCP41010 via the **WriteSetting** subroutine.

If neither the setting nor the shutdown command has been received, no command is sent to the MCP41010.

The subroutine **Dispatcher** simply determines whether the "S" command is for information (**S?**) or is a command that a changed value should be written to the MCP41010 (**Snnn** where n is a digit **0...9**).

```
        Dispatcher      ;get here only if first char is "S" or "s"
        ;-------------
                ;Now look at the second character
                ;If ? then report current setting
                If InStr(1) = "?" Then
                        GoSub ReportSetting
                EndIf

                ;If not ? and if a numeric value, it must be a command
                If (InStr(1) >= "0") AND (InStr(1) <= "9") Then
```

```
                       GoSub GetSetting
            EndIf
    Return
```

We differentiate between the two simply by looking at the second received character. If it is a "**?**" the subroutine **ReportSetting** responds with the current setting value. If it is a character in the range "**0**"..."**9**," then we convert the input string to a numerical value in the subroutine **GetSetting**.

```
GetSetting       ;parse the RS-232 input into numerical data
;-----------     ;and send it to the digital pot
            ;right set data in WorkStr so want all 0's in unused places
            WorkStr = "000000"
            j = ArraySize-1
            i = ArraySize-1

            ;put only the numbers into WorkStr,right set
            ;read the input string r to l keeping only digits 0...9
            ;put these digits rightset into WorkStr
            While (i >0)
                       If (InStr(i) >= "0") AND (InStr(i) <= "9") Then
                       Then
                               WorkStr(j) = InStr(i)
                               j = j-1
                       EndIf
                       i = i-1
            WEnd
```

Possible input strings to set the MCP41010 to the value 5 are "**S5**," "**S05**" and "**S005**." Any of these are legitimate inputs should give the correct setting. The first step in accomplishing this task is to place the digits in **InStr** following the "**S**" into a new byte array, **WorkStr**, *right-justified, with all other digits in **WorkStr** being zeros*. Thus, **S5**, **S05** and **S005** all yield a **WorkStr** value of 000005. To right justify, we simply read **InStr** one byte at a time, right-to-left, placing the read byte into **WorkStr**, starting with the rightmost position. As soon as we come to a character in **InStr** that is not "**0**"..."**9**," we stop.

Next, we convert the right justified character string in **WorkStr** into a numerical value.

```
            Temp = 0
            For i = (ArraySize-1) to (ArraySize-3) Step -1
                       Temp = Temp + Multiplier(i-3)*(WorkStr(i)-"0")
            Next
```

We remember that the numerical values of the ASCII characters "0"..."9" are in direct sequence, 48...57. Hence, to convert a numerical character to its numeric value, we simply subtract 48. To be even clearer, MBasic lets us use the character "0" in a mathematical expression and evaluates it as its numerical value, 48. Hence, to convert an individual digit character to its numerical value, we simply subtract "0" from it. Since **WorkStr** has the right justified string value, and all unused elements are set to "0" all we have to do to convert to the numerical value is multiply the rightmost digit value by 1, the next by 10 and the third by 100 and sum the total. This is efficiently accomplished with a **For i** loop. (We constructed the byte table **Multiplier** earlier as **100,10,0**.)

You may wonder why we made **WorkStr** six elements, when three would have been sufficient. The subroutine **GetSetting** is code that I use frequently, sometimes to read longer input strings, so rather than modify it for each application, I employ it intact.

```
            ;If out of range, set it to 255
            If Temp > 255 Then
                       GoSub ErrorMsg
                       Temp = 255
            EndIf

            ;Write the value to the pot
            Setting = Temp
            CmdByte = Cmd_Write
```

```
        GoSub WriteSetting

Return
```

Next **GetSetting** traps for values above the maximum setting of 255 and sets the value, in this case, to 255. (You may instead change the code to reject the setting and send an error message.) Lastly, **GetSetting** sends the setting value to the MCP41010 through a subroutine call to **WriteSetting**.

```
WriteSetting    ;write to digital pot
;--------------
        Low CSel                ;enable writing
        ShiftOut Dta,Clk,MSBPre,[CmdByte\8,Setting\8]
        High CSel               ;disable writing

        ;Respond to user with message
        ;Normal command
        If CmdByte = Cmd_Write Then
                HSerOut ["Set to ",DEC4 Setting,13,13]
        EndIf
        ;Shutdown command
        If CmdByte = Cmd_ShutDn Then
                HSerOut ["Shutdown Executed",13,13]
        EndIf
Return
```

The subroutine **WriteSetting** implements the same functionality we saw in Program 21-1, adding only the reporting information sent to the serial port.

Here's a typical command interaction with Program 21-2. Since Program 21-2 does not echo back the sent characters, to see what you are sending you must enable "local echo" in your terminal program. For clarity, I've shown in bold the commands I sent to Program 21-2, while normal characters show the response by Program 21-2. Comments I've added afterward are indicated by the ← symbol.

```
Set/Read Digital Potentiometer
S? to Read Value Sxxx to Set Value; X for shutdown
S?                                                    ← what is the current setting?
Current Setting 0
S? to Read Value Sxxx to Set Value; X for shutdown
s123                                                  ← set the pot to 123 with lower case s123
Set to 123

S? to Read Value Sxxx to Set Value; X for shutdown
s?                                                    ← what is the current setting?
Current Setting 123
S? to Read Value Sxxx to Set Value; X for shutdown
S999                                                  ← try setting to illegal value 999
Error! Set value to 255
Set to 255

S? to Read Value Sxxx to Set Value; X for shutdown
S?                                                    ← what is the current setting
Current Setting 255
S? to Read Value Sxxx to Set Value; X for shutdown
S5                                                    ← try three alternate ways of setting to 5
Set to 5                                              ← S5 is accepted correctly

S? to Read Value Sxxx to Set Value; X for shutdown
S05                                                   ← S05 is accepted correctly
Set to 5

S? to Read Value Sxxx to Set Value; X for shutdown
S005                                                  ← S005 is also accepted correctly
Set to 5
```

S? to Read Value Sxxx to Set Value; X for shutdown
X
Shutdown Executed

← **shutdown command**

S? to Read Value Sxxx to Set Value; X for shutdown
s?
Current Setting 0
S? to Read Value Sxxx to Set Value; X for shutdown
s123
Set to 123

← **shutdown reads as 0**

← **exit shutdown by setting to anything**

S? to Read Value Sxxx to Set Value; X for shutdown

The remainder of the subroutines in Program 21-2 will not be further analyzed.

Daisy Chaining Multiple MCP42010 Devices

Let's see how we extend our control to multiple dual-pot devices, such as the MCP42010, connected as shown in Figure 21-6. Our example uses two MCP42010's with the data connections in series, but the concept can be extended indefinitely.

We'll start with a very simple program to test our ability to write data to multiple MCP42010 devices. We'll write an upward stepped linear output into both pots 0 and 1 of U1 and a downward stepped linear output into both pots 0 and 1 of U2.

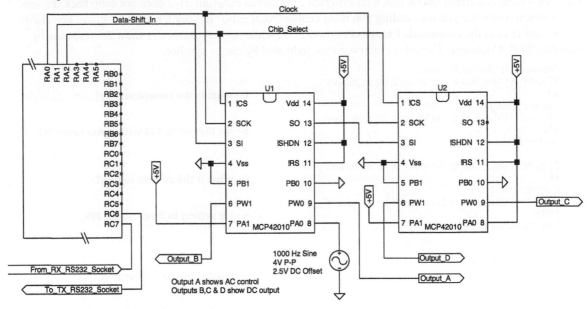

Figure 21-6: Two MCP42010 devices with daisy-chained data connection.

Program 21-3

```
;Program 21-3
;Control two MCP42010
;Digital potentiometer in tandem

;Constants
;-----------
Clk          Con      A0      ;Clock
Dta          Con      A1      ;Data
```

```
CSel            Con     A2      ;ChipSelect

Cmd_None        Con     %00000000
Cmd_WriteData   Con     %00010000
Cmd_ShutDown    Con     %00100000
Cmd_None1       Con     %00110000

Cmd_Dummy       Con     %00000000
Cmd_P0          Con     %00000001
Cmd_P1          Con     %00000010
Cmd_Both        Con     %00000011


;Variables
;-----------
i               Var     Byte    ;loop var
Temp            Var     Byte
U1_Cmd Var      Byte    ;Command for U1
U2_Cmd Var      Byte    ;Instuction for U2
U1_Inst Var     Byte    ;Command for U1
U2_Inst Var     Byte    ;Instruction for U2


;Initialization
;--------------
Output CSel
Low CSel

Main
        U1_Cmd = Cmd_WriteData + Cmd_Both
        U2_Cmd = Cmd_WriteData + Cmd_Both

        For i = 0 to 255
                U1_Inst = i
                U2_Inst = 255-i
                Low CSel
                ;Push 2nd chip data first
                ShiftOut Dta,Clk,MSBPre,[U2_Cmd\8,U2_Inst\8 |
                ,U1_Cmd\8,U1_Inst\8]
                High CSel
        Next
        GoTo Main
End
```

With the exception of how we write data to two chips, Program 12-3 requires no elaboration.

The MCP42xxx series devices (but not the MCP41xxx series) have "shift in" and "shift out" pins. Shift in (SI), as we've seen, is the data input to the device. What then is "shift out" (SO) and how might it be used? Shift out simply repeats the input data, but delayed, or shifted, by 16 bits. Let's look at this in more detail in conjunction with Figure 21-7.

Upon taking the CS line low, the internal 16-bit buffer in both U1 and U2 is cleared, that is, set to 0s. We then send the command byte and instruction byte for *U2* to *U1*. As each bit of U2's command and instruction bytes are clocked into U1's SI pin, the bits are placed into U1's buffer and the bits previously in

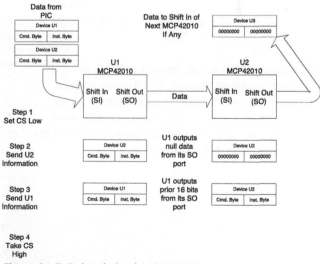

Figure 21-7: Daisy-chained MCP42010s.

U1's buffer are emitted from U1's SO pin. Since these are all 0's, U2 receives 16 0 bits. (Likewise U2 emits 16 0 bits from its SO pin as well, but since we have no U3 device, these bits remain unused. However, more MCP42xxx devices may be daisy chained from U2.)

After the first 16 bits are sent, *U1*'s buffer holds *U2*'s instructions, and *U2*'s buffer is filled with zeros. Next, still keeping the CS line low, we send *U1*'s instructions to *U1*. As the 16 bits of *U1*'s instructions arrive at *U1*'s SI pin, the bits held in its buffer (*U2*'s instructions that we just sent) are shifted out of *U1*'s SO pin into *U2*'s SI pin, and consequently are read into *U2*'s buffer.

After sending the last bit of *U1*'s instructions, we take the CS line high. At this point, *U1*'s buffer contains *U1*'s information and *U2*'s buffer contains *U2*'s information.

Taking the CS line high latches the information in their respective buffers into *U1* and *U2* and each device takes the actions corresponding to the received command byte and instruction byte.

Thus, when we daisy chain the devices together, we send the data in reverse of the device connections; information for the last device is sent first and information for the first device is sent last. A bit of reflection and study of Figure 21-7 should make the logic obvious. To verify this operation, let's look at the data going into U1's SI pin and the data out of U1's SO pin. Figure 21-8 shows four bytes of data into U1's SI pin. Figure 21-9 shows the first two bytes output from U1's SO pin are zeros, followed by two bytes of nonzero information. (You can't directly compare the data bits between Figures 21-8 and 21-9 as they were taken at different times.)

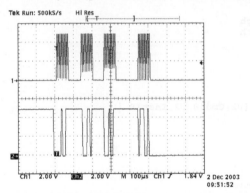

Figure 21-8: Input Data to U1's SI Pin; Ch1: Clock; Ch2: SI Pin Data.

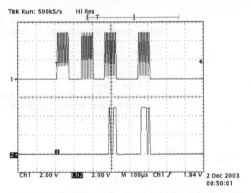

Figure 21-9: Output Data from U1's SO Pin; Ch1: Clock; Ch2: SO Pin Data.

Microchip's SI/SO design make it possible to individually address multiple SPI devices, without requiring each one to have a unique address (as we saw in Chapter 12 for 1-wire chips), using no more than three wires for communications, regardless of the number of devices being controlled.

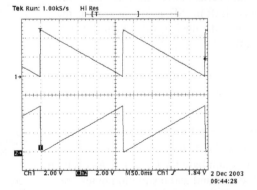

Figure 21-10: Two MCP42010 Devices in Daisy Chain; Ch1:U2, P1 Wiper; Ch2: U1 P1 Wiper.

```
For i = 0 to 255
  U1_Inst = i
  U2_Inst = 255-i
  Low CSel
  ShiftOut Dta,Clk,MSBPre,[U2_Cmd\8,U2_Inst\8,U1_Cmd\8,U1_Inst\8]
  High CSel
Next
```

As we've just seen, when we write data to U1 and U2 with the above statement, we first send *U2*'s command and instruction bytes, followed by *U1*'s command and instruction bytes. We've set U1 to linearly ramp up (**i**) and U2 (**255-i**) to linearly ramp down, and Figure 21-10 matches our expectations.

RS-232 Command of Multiple Daisy Chained MCP42010 Devices

Let's now merge our two MCP42010 daisy chain with an expanded version of Program 21-2's RS-232 control to give us independent control of all four pots.

Our command syntax is expanded from that of Program 21-4. First, we identify each of the four pots with an ID letter:

ID Letter	Pot Unit
A or a	U1, P0
B or b	U1, P1
C or c	U2, P0
D or d	U2, P1

Commands are of the form ID Letter followed by a value or "X" or a stand-alone "?" for status.

Command	Function
Qnnn (or qnnn)	Set potentiometer Q to value nnn
QX (or qx)	Shutdown potentiometer Q
?	Respond by sending the current setting value of all potentiometer values

Where Q is either A, B, C or D. For example, to set potentiometer C to value 123, the command is C123.

Program 21-4

```
;Program 21-4
;Control two MCP42010
;Digital potentiometers

;Constants
;-----------
Clk           Con    A0      ;Clock
Dta           Con    A1      ;Data
CSel          Con    A2      ;ChipSelect

ArraySize     Con    6       ;size of instring

Cmd_None             Con     %00000000
Cmd_WriteData        Con     %00010000
Cmd_ShutDown         Con     %00100000
Cmd_None1            Con     %00110000

Cmd_Dummy            Con     %00000000
Cmd_P0               Con     %00000001
Cmd_P1               Con     %00000010
Cmd_Both             Con     %00000011

Multiplier   ByteTable 100,10,1

;Variables
;-----------
```

```
i                   Var         Byte    ;loop var
j                   Var         Byte    ;loop var
Temp                Var         Word
BadData             Var         Byte    ;Is data valid?

U1_Cmd              Var         Byte    ;Command for U1
U2_Cmd              Var         Byte    ;Instuction for U2
U1_Inst             Var         Byte    ;Command for U1
U2_Inst             Var         Byte    ;Instruction for U2

A_Val               Var         Word    ;Unit A value
B_Val               Var         Word    ;Unit B value
C_Val               Var         Word    ;Unit C value
D_Val               Var         Word    ;Unit D value

InStr               Var         Byte(ArraySize)
WorkStr             Var         Byte(ArraySize)


Initialization
;--------------
        Clear   ;variables to zero
        Output CSel
        Low Csel
        EnableHSerial
        SetHSerial H19200
        HSerOut ["Set/Read Quad Digital Potentiometer",13]

        ;Initialize all four sections to 0
        U1_Cmd = Cmd_WriteData + Cmd_Both
        U1_Inst = 0
        U2_Cmd = Cmd_WriteData + Cmd_Both
        U2_Inst = 0
        GoSub WriteSetting
        GoSub ReportSetting

Main
;---------
        ;Remind user of commands
        GoSub SendPrompt
        InStr = "                "
        ;key on 10, not 13. Assumes sending *** CR&LF ***
        ;Otherwise, input terminates on 13,
        ;and the 10 appears as leading
        ;character in the next string.
        HSerIn  [Str InStr\ArraySize\10]

        GoSub Dispatcher        ;see what was received
GoTo Main

Dispatcher              ;Look at input string and
;--------------          ;act accordingly
        ;set up pseudo-Case statement

        If      (InStr(0) = "A") OR (Instr(0) = "a") Then
                GoSub GetSetting        ;get numerical value
                If BadData = 1 Then ;unknown info
                        GoSub UnknownMsg
                ELSEIF BadData = 0      ;good numerical data
                        U1_Cmd = Cmd_WriteData + Cmd_P0
                        A_Val = Temp
                        U1_Inst = A_Val
                        GoSub U2Null    ;write & set nulls

                ElseIf BadData = 2      ;shutdown
                        U1_Cmd = Cmd_ShutDown + Cmd_P0
                        ;don't care about instruction word
                        A_Val = 999
                        GoSub U2Null    ;write & set nulls
                EndIf
```

```
              EndIf

       If   (InStr(0) = "B") OR (Instr(0) = "b") Then
              GoSub GetSetting        ;get numerical value
                       If BadData = 1 Then ;unknown
                               GoSub UnknownMsg
                       ELSEIf BadData = 0    ;good numerical data
                               U1_Cmd = Cmd_WriteData + Cmd_P1
                               B_Val = Temp
                               U1_Inst = B_Val
                               GoSub U2Null
                       ElseIf BadData = 2    ;shutdown
                               U1_Cmd = Cmd_ShutDown + Cmd_P1
                               ;don't care about instruction word
                               B_Val = 999
                               GoSub U2Null
                       EndIf
       EndIf

       If   (InStr(0) = "C") OR (InStr(0) = "c") Then
       GoSub GetSetting         ;get numerical value
                       If BadData = 1 Then ;unknwon
                               GoSub UnknownMsg
                       ELSEIf BadData = 0        ;numerical data
                               C_Val = Temp
                               U2_Cmd = Cmd_WriteData + Cmd_P0
                               U2_Inst = C_Val
                               GoSub U1Null

                       ElseIf BadData = 2      ;shutdown
                               U2_Cmd = Cmd_ShutDown + Cmd_P0
                               ;don't care about instruction word
                               C_Val = 999
                               GoSub U1Null
                       EndIf
       EndIf

       If   (InStr(0) = "D") OR (InStr = "d") Then
              GoSub GetSetting        ;get numerical value
                       If BadData = 1 Then
                               GoSub UnknownMsg
                       ELSEIf BadData = 0     ;numerical data
                               D_Val = Temp
                               U2_Cmd = Cmd_WriteData + Cmd_P1
                               U2_Inst = D_Val
                               GoSub U1Null
                       ElseIf BadData = 2      ;shutdown
                               U2_Cmd = Cmd_ShutDown + Cmd_P1
                               ;don't care about instruction word
                               D_Val = 999
                               GoSub U1Null
                       EndIf
       EndIf

       If InStr(0) = "?" Then ;status querry
              GoSub ReportSetting
       EndIf
Return


U2Null  ;Shorten up dispatcher
;-----
       U2_Cmd = Cmd_None
       U2_Inst = Cmd_Dummy
       GoSub WriteSetting
       GoSub ReportSetting
Return

U1Null  ;Shorten up dispacther
;------
```

```
                U1_Cmd = Cmd_None
                U1_Inst = Cmd_Dummy
                GoSub WriteSetting
                GoSub ReportSetting
Return

WriteSetting            ;Writes to pots. Assumes
;                       U1_Cmd, U2_Cmd, U1_Inst &
;--------------         ;U2_Inst have been set before calling
                Low CSel        ;Note order - push last chip info first
                ShiftOut Dta,Clk,MSBPre,[U2_Cmd\8,U2_Inst\8, |
                U1_Cmd\8,U1_Inst\8]
                High CSel
Return

SendPrompt      ;send prompt string
;--------------
        HSerOut [13,"Annn,Bnnn,Cnnn & Dnnn to set; ?",    |
        "for values; Ax,Bx,Cx & Dx to shutdown",13]
Return

ErrorMsg            ;send error message
;-----------  where value > 255
        HSerOut ["Error! Set value to 255",13]
Return

UnknownMsg                  ;unknown command message
;-----------
        HSerOut ["Unknown command",13]
Return

ReportSetting   ;Writes all four values
;--------------
        HSerOut ["Current Settings:",13]
        HSerOut ["Unit A (U1-P0): ",Dec A_Val\3, |
        "   Unit B (U1-P1): ",Dec B_Val\3,13]
        HSerOut ["Unit C (U2-P0): ",Dec C_Val\3, |
        "   Unit D (U2-P1): ",Dec D_Val\3,13]
        HSerOut ["999 = Shutdown",13]
Return

GetSetting      ;parse the RS-232 input into numerical data
;-----------    ;and send it to the digital pot
        ;right sets data in WorkStr so want 0's in unused places
        ;Is it a shutdown command?
        If (InStr(1) = "X") OR (InStr(1) = "x") Then
                BadData = 2     ;Shutdown
                GoTo BailOut
        EndIf

        ;Is it good numerical data?
        If (InStr(1) < "0") OR (InStr(1) > "9") Then
                BadData = 1     ;not numerical data
                GoTo BailOut
        EndIf

        BadData = 0

        WorkStr = "000000"
        j = ArraySize-1
        i = ArraySize-1

        ;put only the numbers into WorkStr,right set
        ;read the input string r to 1 keeping only digits 0...9
        ;put these digits rightset into WorkStr
        While (i >0)
                If (InStr(i) >= "0") AND (InStr(i) <= "9") Then
                Then
                        WorkStr(j) = InStr(i)
                        j = j-1
```

```
              EndIf
              i = i-1
      WEnd

      ;convert to numerical value. Value ASCII - "0"
      Temp = 0
      For i = (ArraySize-1) to (ArraySize-3) Step -1
              Temp = Temp + Multiplier(i-3)*(WorkStr(i)-"0")
      Next

      ;If out of range, set it to 255
      If Temp > 255 Then
              GoSub ErrorMsg
              Temp = 255
      EndIf
      BailOut

Return

End
```

Program 21-4 tracks the earlier programs and further discussion isn't necessary. Here is a sample interaction with Program 21-4. To see the transmitted characters, local echo must be turned on in your terminal program.

As before, for clarity, I've shown in bold the commands I sent to Program 21-2, while the normal characters show the response by Program 21-2. Comments I've added afterward are indicated by the ← symbol.

Annn,Bnnn,Cnnn & Dnnn to set; ? for values; Ax,Bx,Cx & Dx to shutdown

a123   ← **Set U1,P0 to 123**
Current Settings:
Unit A (U1-P0): 123   Unit B (U1-P1): 000
Unit C (U2-P0): 000   Unit D (U2-P1): 000
999 = Shutdown

Annn,Bnnn,Cnnn & Dnnn to set; ? for values; Ax,Bx,Cx & Dx to shutdown

b123   ← **Set U1,P1 to 123**
Current Settings:
Unit A (U1-P0): 123   Unit B (U1-P1): 123
Unit C (U2-P0): 000   Unit D (U2-P1): 000
999 = Shutdown

Annn,Bnnn,Cnnn & Dnnn to set; ? for values; Ax,Bx,Cx & Dx to shutdown

c111   ← **Set U2,P0 to 123**
Current Settings:
Unit A (U1-P0): 123   Unit B (U1-P1): 123
Unit C (U2-P0): 111   Unit D (U2-P1): 000
999 = Shutdown

Annn,Bnnn,Cnnn & Dnnn to set; ? for values; Ax,Bx,Cx & Dx to shutdown

d222   ← **Set U2,P1 to 123**
Current Settings:
Unit A (U1-P0): 123   Unit B (U1-P1): 123
Unit C (U2-P0): 111   Unit D (U2-P1): 222
999 = Shutdown

Annn,Bnnn,Cnnn & Dnnn to set; ? for values; Ax,Bx,Cx & Dx to shutdown

?   ← **Query status**
Current Settings:
Unit A (U1-P0): 123   Unit B (U1-P1): 123
Unit C (U2-P0): 111   Unit D (U2-P1): 222
999 = Shutdown

Annn,Bnnn,Cnnn & Dnnn to set; ? for values; Ax,Bx,Cx & Dx to shutdown

ax   ← **Shutdown Pot A**
Current Settings:
Unit A (U1-P0): 999   Unit B (U1-P1): 123   ← **note A's value is 999**
Unit C (U2-P0): 111   Unit D (U2-P1): 222
999 = Shutdown

Annn,Bnnn,Cnnn & Dnnn to set; ? for values; Ax,Bx,Cx & Dx to shutdown

bx   ← **Shutdown Pot B**
Current Settings:
Unit A (U1-P0): 999   Unit B (U1-P1): 999   ← **note B's value is 999**
Unit C (U2-P0): 111   Unit D (U2-P1): 222
999 = Shutdown

Annn,Bnnn,Cnnn & Dnnn to set; ? for values; Ax,Bx,Cx & Dx to shutdown

Cx   ← **Shutdown Pot C**
Current Settings:
Unit A (U1-P0): 999   Unit B (U1-P1): 999
Unit C (U2-P0): 999   Unit D (U2-P1): 222   ← **note C's value is 999**
999 = Shutdown

Annn,Bnnn,Cnnn & Dnnn to set; ? for values; Ax,Bx,Cx & Dx to shutdown

Dx   ← **Shutdown Pot D**
Current Settings:
Unit A (U1-P0): 999   Unit B (U1-P1): 999
Unit C (U2-P0): 999   Unit D (U2-P1): 999   ← **note D's value is 999**
999 = Shutdown

Annn,Bnnn,Cnnn & Dnnn to set; ? for values; Ax,Bx,Cx & Dx to shutdown

Logarithmic Response for Audio Volume Control

Should we use an MCP41010 to control the volume level on an audio amplifier, we will quickly find that many step changes produce no perceptible changes in volume level but a few step changes at the extreme bottom end of the step range yield significant volume changes. This phenomenon results from human hearing's logarithmic, not linear, response to sound pressure level.

To prevent the useful volume control range from being bunched up around a few degrees shaft rotation, mechanical potentiometers are available with a nonlinear or "tapered" response. For volume controls, an "audio taper" or "log taper" pot provides perceived volume levels that change directly proportional to shaft rotation. If we do nothing special, the MCP42xxx/41xxx shows this undesired bunching, as illustrated in Figure 21-11.

Before we go further, a brief refresher on decibels may be helpful. The decibel (abbreviated dB) is a ratio relationship between two power levels, P1 and P2, such that:

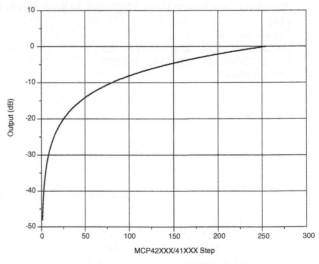

Figure 21-11: Decibel response of MCP42xxx/41xxx.

$$dB = 10\log_{10}\left(\frac{P_1}{P_2}\right)$$

Log_{10} mean the logarithm, base 10. If P1 > P2, dB is positive; if P1 < P2, dB is negative. This follows from the nature of logarithms. P1 and P2 may be in kilowatts, watts, milliwatts, microwatts, or any other power unit, as long as both are in the same units.

Instead of measuring two power levels and taking their ratio, we may measure the voltages associated with the two powers, using a formula commonly stated as:

$$dB = 20\log_{10}\left(\frac{V_1}{V_2}\right)$$

V1 and V2 are the measured voltages. Since power is proportional to the square of the voltage, the multiplier goes from 10 to 20. There is a subtlety here, though. The fundamental definition of the decibel is a ratio of two powers. Hence, the power ratio formula is always correct, but when we compute decibels from two voltage measurements, there is an implicit assumption that both voltages are across identical impedances. In practice, the identical impedance requirement is often ignored. Many times ignorance is bliss, but occasionally it will cause you trouble, so be careful when using dB calculations based on voltages.

Most authorities accept 1 dB as the minimum difference that is perceptible as a change in volume level.

Change in Sound Level (dB)	Perceived Difference
1	At the edge of perceptible change; probably imperceptible to most
3	Should be perceptible by most listeners
6	Clearly perceptible by listeners
10	Twice (or half) as loud
20	Four times (or one-fourth) as loud

We will construct, in Program 21-5, a "dB linear" arrangement, where the MCP41010's provides quantized output values that change 1 dB per level step.

One final point before we look at Program 21-5. We've only connected DC levels to the digital pots in our earlier work. Now that we are applying audio, we need to watch the applied voltage levels. The MCP42xxx/41xxx's potentiometer connections cannot be operated with a voltage below V_{SS} or above V_{DD}. Normal audio signals are AC waveforms and consequently have components that are above V_{SS} and below V_{SS}, with an average voltage of 0. Hence, we must bias the audio signal so that its average voltage is between V_{SS} and V_{DD} and so that the peak excursions do not exceed V_{DD} or drop below V_{SS}. To give maximum headroom for the audio signal, we'll use a bias voltage of $V_{DD}/2$. Figure 21-12 shows the difference between a normal audio signal (Channel 2) with approximately equal excursions above and below zero volts and a signal biased upward by approximately 2.5 V (Channel 1). Even though the audio signal has peak-to-peak amplitude of approximately 4 V, the biased signal never exceeds V_{DD}, nor drops below V_{SS}, as its maximum is approximately 4.5 V and its minimum is approximately 0.5 V. It thus meets the maximum and minimum voltage requirements of the MCP41010. The pure

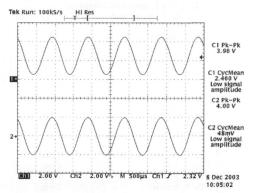

Figure 21-12: Normal and DC Biased 1000 Hz Sine Wave Signals; Ch1: Signal Biased at Approx. 2.5V; Ch2: Unbiased Signal.

AC signal, in contrast has a minimum voltage of –2 V, which violates the MCP41010's V_{SS} (0 V) minimum specification.

Figure 21-13 shows how to connect an audio signal to an MCP41010. In some cases, the signal may be properly biased from its source, in which case we may directly connect to the digital pot. If, however, the signal is AC coupled and hence requires bias, the second circuit shown in Figure 21-13 may be used. This circuit uses blocking capacitors C1 and C2 to keep the DC bias away from the audio input and output. R1 and R2, with C3 form a low impedance V_{DD}/2 bias source. R3 prevents the bias source from appreciably loading the audio source. Since no DC current flows in MCP41010, R3 may be increased in value if necessary.

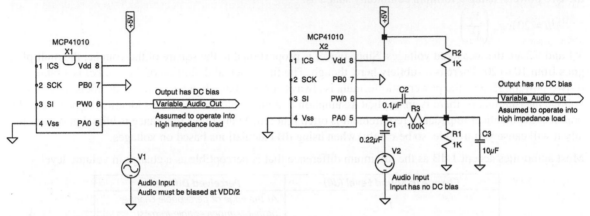

Figure 21-13: Connecting an audio signal to a digital pot.

We can't really make the linear resistance steps of an MCP41010 into logarithm-valued resistors, but we can approximate it by only permitting certain step values to be used. We'll select the values so that we have 31 steps, each step very close to 1 dB from its neighboring steps. Since the purpose of this program is to illustrate the log concept, we'll use a minimalist approach. (Program 21-5 uses the dual MCP42010 configuration of Figure 21-6, but it will work with single MCP41010 equally well, as the extra instructions are ignored.)

Program 21-5

```
;Program 21-5
;Control two MCP42010
;Digital potentiometer
;with log steps

;Constants
;-----------
Clk             Con         A0      ;Clock
Dta             Con         A1      ;Data
CSel            Con         A2      ;ChipSelect

Cmd_None        Con         %00000000
Cmd_WriteData   Con         %00010000
Cmd_ShutDown    Con         %00100000
Cmd_None1       Con         %00110000

Cmd_Dummy       Con         %00000000
Cmd_P0          Con         %00000001
Cmd_P1          Con         %00000010
Cmd_Both        Con         %00000011

;0...-30 dB steps
LogSteps ByteTable 251,224,200,178,158,141,126,112, |
100,89,79,71,63,56,50,45,40, |
```

```
        35,32,28,25,22,20,18,16,13,11,10,9,8,7

        ;Variables
        ;----------
        i               Var         Byte        ;loop var
        Temp            Var         Byte
        U1_Cmd          Var         Byte        ;Command for U1
        U2_Cmd          Var         Byte        ;Instuction for U2
        U1_Inst         Var         Byte        ;Command for U1
        U2_Inst         Var         Byte        ;Instruction for U2

        ;Initialization
        ;--------------
        Output CSel
        Low CSel

        Main
                U1_Cmd = Cmd_WriteData + Cmd_Both
                U2_Cmd = Cmd_WriteData + Cmd_Both

                For i = 0 to 30
                        U1_Inst = LogSteps(i)
                        U2_Inst = LogSteps(30-i)

                        Low CSel
                        ;Push 2nd chip data first
                        ShiftOut Dta,Clk,MSBPre,[U2_Cmd\8,U2_Inst\8, |
                        U1_Cmd\8,U1_Inst\8]
                        High CSel
                Next
                GoTo Main
        End
```

We'll concentrate on the new parts of Program 21-4. The concept is simple; we set up a byte table containing the MCP41010 step values corresponding to 1 dB increments and index into the byte table to obtain the values to load into the MCP41010. As we'll see shortly, if we extend the table to more than about 31 positions, the positions corresponding to MCP41010 steps below 8 deviate increasingly from our desired one-position-every-1dB goal.

```
        LogSteps ByteTable 251,224,200,178,158,141,126,112, |
        100,89,79,71,63,56,50,45,40, |
        35,32,28,25,22,20,18,16,13,11,10,9,8,7
```

How did we generate this byte table? Changing the MCP41010's step command changes the *voltage* at the wiper, so that tells us to use the voltage-ratio-to-dB formula. Since we want a 1 dB step, we know the ratio between the current step and the next (or prior) step is:

$$20\log_{10}\frac{V_1}{V_2} = 1 \text{ dB}$$

Divide both sides by 20

$$\log_{10}\frac{V_1}{V_2} = \frac{1}{20}$$

Now exponentiate both sides

$$10^{\log_{10}\frac{V_1}{V_2}} = 10^{0.05}$$

Since $10^{\log(x)} = x$:

$$\frac{V_1}{V_2} = 10^{0.05}$$

With a pocket calculator, or Excel, we find that $10^{0.05}$ is (to six places) 1.122018. Hence, we want the voltage output at the MCP41010's wiper to increase in steps of 1.122018:1 (or decrease in steps of 1/1.122018 when we are going down). We know that the voltage at the wiper with the input at pin A and pin B grounded is directly proportional to the step number. Hence, the step numbers in our byte table must change in the ratio 1.122018:1 as well.

When we construct a table in Excel with these ratios (each theoretical step is 1.122018 times the prior step, and the integer step is the theoretical step rounded to 0 places) we see a problem at one end of the table, where the step between 1 and 2 causes a 4dB increase. The MCP41010 only has integer steps and consequently our desired uniform step goal is not achievable.

DB Step	Theoretical Step	Integer Step
1	1.122018	1
2	1.258925	1
3	1.412538	1
4	1.584893	1
5	1.778279	2
6	1.995262	2
7	2.238721	2
8	2.511886	3

As we move up in steps, we see that we can become very close to matching the theoretical step with the MCP41010's integer steps. The complete spreadsheet calculation gives us 48 possible steps, of which we accept the 31 nonshaded as being acceptably close to 1 dB per increment. Figure 21-14 illustrates the increasing error as we try to achieve more than about 30dB range with one MCP41010.

Possible 1-12 dB			Possible 13-24 dB			Possible 25-36 dB			Possible 37-48 dB		
dB	Step	Int.	dB	Step	Int.	dB	Step	Int.	dB	Step	Int.
1	1.122018	1	13	4.466836	4	25	17.78279	18	37	70.79458	71
2	1.258925	1	14	5.011872	5	26	19.95262	20	38	79.43282	79
3	1.412538	1	15	5.623413	6	27	22.38721	22	39	89.12509	89
4	1.584893	2	16	6.309573	6	28	25.11886	25	40	100	100
5	1.778279	2	17	7.079458	7	29	28.18383	28	41	112.2018	112
6	1.995262	2	18	7.943282	8	30	31.62278	32	42	125.8925	126
7	2.238721	2	19	8.912509	9	31	35.48134	35	43	141.2538	141
8	2.511886	3	20	10	10	32	39.81072	40	44	158.4893	158
9	2.818383	3	21	11.22018	11	33	44.66836	45	45	177.8279	178
10	3.162278	3	22	12.58925	13	34	50.11872	50	46	199.5262	200
11	3.548134	4	23	14.12538	14	35	56.23413	56	47	223.8721	224
12	3.981072	4	24	15.84893	16	36	63.09573	63	48	251.1886	251

Our simple demonstration program just writes the uniform dB steps into the pot in a loop.

```
For i = 0 to 30
      U1_Inst = LogSteps(i)
      U2_Inst = LogSteps(30-i)

      Low CSel
      ;Push 2nd chip data first
      ShiftOut Dta,Clk,MSBPre,[U2_Cmd\8,U2_Inst\8, |
      U1_Cmd\8,U1_Inst\8]
      High CSel
Next
```

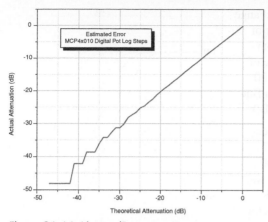

Figure 21-14: Linear dB response error.

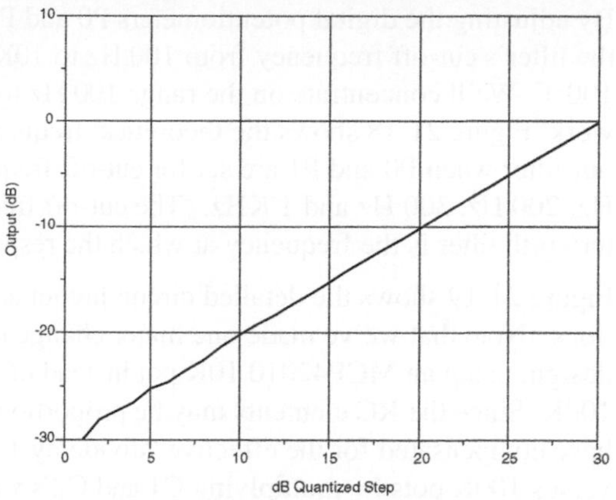

Figure 21-15: Linear dB response of Program 21-5.

A more useful program, of course, might accept RS-232 commands, or perhaps read "up" and "down" volume control buttons, or a rotary encoder, or respond to a signal from an infrared remote control.

As long as we stay in the range 0…30 dB, the response is quite close to 1dB per index step, as shown in Figure 21-15. The output steps clearly show their logarithm parentage, as seen in Figure 21-16. (See the *Ideas for Modifications to Programs and Circuits* section of this chapter for an idea on how to extend the linear dB range of a digital pot.)

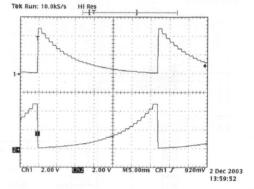

Electronically Tunable Low-Pass Filter Using MCP42010

Figure 21-16: dB Stepped Output; Ch1: dB Decreasing Steps; Ch2: dB Increasing Steps.

We shouldn't limit our thinking by envisioning digital pots strictly as computer controllable volume controls. Rather, they have many applications, and we'll look at their use to control an active low-pass filter. This design is based on Microchip's Application Note AN737 [21-2], which should be read for more detail. (Some older versions of AN737's Figure 2 schematic interchange the A and B pins in both pot sections and the circuit won't work as shown. This error was corrected with AN737 release D.)

We'll assume you have passing familiarity with active filters and operational amplifiers. If not, the reference section of this chapter identifies several excellent information sources for both subjects.

Figure 21-17 shows our conceptual filter design. It is a second order Butterworth low-pass filter, implemented with Sallen-Key topology. (If these terms mean nothing to you, pick up a copy of reference [21-6]. It's by far and away the most readable book on filter design you'll ever find.)

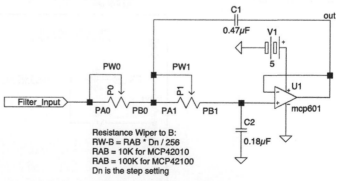

Figure 21-17: Simplified variable low-pass filter design.

By adjusting the digital potentiometers P0 and P1, we may change the filter's cut-off frequency from 100 Hz to 10KHz, a factor of 100:1. We'll concentrate on the range 100 Hz to 1 KHz in our work. Figure 21-18 shows the theoretical frequency response of our filter when P0 and P1 are set for cut-off frequencies of 100 Hz, 200 Hz, 300 Hz and 1 KHz. (The cut-off frequency for a Butterworth filter is the frequency at which the response is –3 dB.)

Figure 21-19 shows the detailed circuit layout and PIC connections. Note that we've made one major change to the AN737 design; using an MCP42010 10K pot instead of a MCP42100 100K. Since the RC elements may be proportionally scaled, we have compensated for the effective "divide by 10" of the 10K versus 100K pots by multiplying C1 and C2's values by a factor of 10, from 0.047 μF to 0.47 μF and from 0.018 μF to 0.18 μF, respectively.

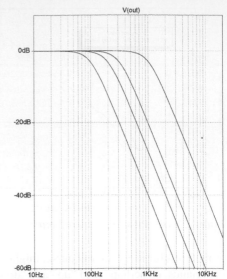

Figure 21-18: Second-order Butterworth low-pass Filter with 100 Hz, 200 Hz, 300 Hz and 1 KHz cut-offs.

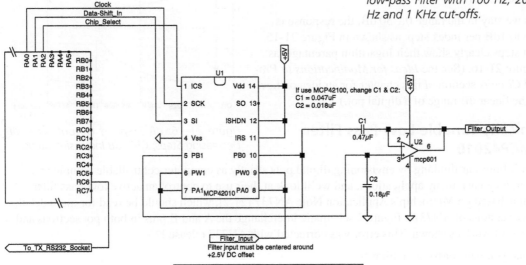

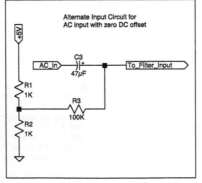

Figure 21-19: Detailed diagram— variable low-pass filter.

Application Note AN737 provides the MCP42010 settings for various cut-off frequencies:

Cut-off Frequency (Hz)	P0 Code	P1 Code
100	82	238
200	41	119
300	26	79
1000	8	24
2000	4	12
3000	3	8
10000	1	2

Since we have scaled C1 and C2 to exactly offset the difference between AN737's MCP42100 100K values, we may use these code settings without change.

Program 21-6 cycles through four cut-off frequencies, 100Hz, 200Hz, 300Hz and 1KHz, with each frequency being active for 10 seconds. The current cut-off frequency value is sent to the serial output port where it can be observed on a PC running a terminal program. During the time each cut-off frequency is active, we may measure the frequency response with signal generator and voltmeter. Since it takes longer to set up the signal generator to a new frequency than to actually take a reading, it's more efficient to measure all four response curves at a common test frequency, and then move to the next test frequency.

Program 21-6

```
;Program 21-6
;Control a MCP42010-based
;tunable low-pass filter based
;on Microchip AN737

;Constants
;-----------
Clk             Con             A0      ;Clock
Dta             Con             A1      ;Data
CSel            Con             A2      ;ChipSelect

;Define commands for the MCP42010
Cmd_None                Con             %00000000
Cmd_WriteData           Con             %00010000
Cmd_ShutDown            Con             %00100000
Cmd_None1               Con             %00110000

Cmd_Dummy               Con             %00000000
Cmd_P0                  Con             %00000001
Cmd_P1                  Con             %00000010
Cmd_Both                Con             %00000011

;For filter response measurements, set the signal
;source to desired frequency and take data as the
;filter steps through four cut-off frequencies
CutOff  WordTable       100,200,300,1000
PP0             ByteTable       82, 41, 26, 8
PP1             ByteTable       238,119,79, 24

;Variablesw
;-----------
i               Var     Byte    ;loop var
U10_Cmd Var             Byte    ;Command for U1
U10_Inst        Var     Byte    ;Command for U1
U11_Cmd Var             Byte    ;Command for U1
U11_Inst        Var     Byte    ;Command for U1

;Initialization
;--------------
Output CSel
```

```
        Low Csel

        EnableHSerial
        SetHSerial H19200        ;we will write the frequency to port
        HSerOut ["OK",13]        ;so we know to write down response data

        SetHSerial H19200
        U10_Cmd = Cmd_WriteData + Cmd_P0        ;set to Write P0
        U11_Cmd = Cmd_WriteData + Cmd_P1        ;set to Write P1

        Main
                For i = 0 to 3          ;step through the 4 frequencies
                        U10_Inst = PP0(i)       ;get the resistance step for P0
                        U11_Inst = PP1(i)       ;likewise for P1

                        Low CSel          ;Write P0
                        ShiftOut Dta,Clk,MSBPre,[U10_Cmd\8,U10_Inst\8]
                        High CSel

                        Low CSel          ;Write P1
                        ShiftOut Dta,Clk,MSBPre,[U11_Cmd\8,U11_Inst\8]
                        High CSel

                                          ;Update output
                        HSerOut["Cutoff: ",Dec CutOff(i)," Hz  ",13]
                        Pause 10000       ;10 seconds to take data
                Next
                                          ;for easier reading
                HSerOut [13,13]
                GoTo Main
        End
```

We've seen every part of this program before, in one form or another.

```
        CutOff        WordTable        100,200,300,1000
        PP0           ByteTable        82, 41, 26, 8
        PP1           ByteTable        238,119,79, 24
```

The cut-off frequency and the associated P0 and P1 settings are held in three byte tables.

```
        For i = 0 to 3          ;step through the 4 frequencies
                U10_Inst = PP0(i)       ;get the resistance step for P0
                U11_Inst = PP1(i)       ;likewise for P1

                Low CSel          ;Write P0
                ShiftOut Dta,Clk,MSBPre,[U10_Cmd\8,U10_Inst\8]
                High CSel

                Low CSel          ;Write P1
                ShiftOut Dta,Clk,MSBPre,[U11_Cmd\8,U11_Inst\8]
                High CSel

                                  ;Update output
                HSerOut["Cutoff: ",Dec CutOff(i)," Hz  ",13]
                Pause 10000       ;10 seconds to take data
        Next
```

With the **For i=0 to 3** loop, we step through the four cut-off frequencies, load the associated P0 and P1 values into the MCP41010 and make them active. The loop then pauses for 10 seconds so that we may measure the output voltage at each of several test frequencies to plot the response curve.

Figure 21-20 shows the measured data tracks the predicted response reasonably well. Since I built the circuit on a 2840 Development Board, with jumper wire connections, stray coupling and noise limits the ultimate rejection of the filter once we move past about 55–60 dB of stop band loss. The test signal had a peak-to-peak value of 4.0 V (plus a 2.5 V bias, required by the MCP42010), so with 60 dB attenuation, we must measure a 4.0 millivolt signal, not always easy in the presence of 60 Hz hum and digital noise from the PIC. Real designs on printed circuit boards require attention to layout and power supply decoupling to avoid these problems. Where we mix digital and low level analog signals on the same PCB, even greater attention to design details is necessary.

Ideas for Modifications to Programs and Circuits

* One possible way to expand the dynamic range of a dB stepping digital pot is to cascade two of them, along the lines shown in Figure 21-21. In this arrangement, P0 provides 0…30 dB attenuation, during which interval P1 is set for 0 dB attenuation. For attenuation greater than 30 dB, P0 remains at 30 db, but P1 now starts an independent 0…30dB attenuation range, thus giving a total 0…60 dB range. What happens when you try this connection with P0 and P1 both 10K devices? Since one of our underlying assumptions is that the wiper always sees a high impedance load, might P1 be loading down P0? How might this be corrected? Do you expect a different result if P1 is a 100K pot instead of a 10K? Is this 10:1 increase in impedance enough? Suppose instead P0 works into an operational amplifier buffer with essentially infinite impedance, and the output of the buffer amp drives P1. Does this provide the expected result?

* One of the problems we noted in Chapter 16 related to using a fixed frequency low-pass filter with a variable frequency DAC output. How might you adapt the digitally tunable low-pass filter and software design to track the frequency output of the DDS generator described in Chapter 16?

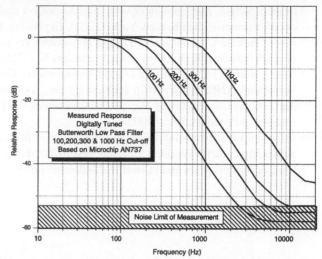

Figure 21-20: Measured response of programmable low-pass filter.

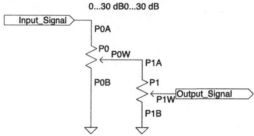

Figure 21-21: Hypothetical Extended dB range via cascaded digital pots.

References

[21-1] *MCP41xxx/42xxx Single/Dual Digital Potentiometer with SPI Interface,* Microchip Technology, Inc. Document No. DS11195C (2003).

[21-2] Baker, Bonnie C., *Using Digital Potentiometers to Design Low Pass Adjustable Filters*, AN737, Microchip Technology, Inc., Document No. DS00737 (2001).

[21-3] *MCP601/2/3/4 2.7V to 5.5V Single Supply CMOS Op Amps*, Microchip Technology, Inc., Document No. DS21314E (2003).

[21-4] Frederiksen, Thomas M., *Intuitive IC Op Amps—From Basics to Useful Applications*, National Semiconductor Technology Series, Santa Clara, CA., (1984). A classic tutorial on operational amplifiers, but unfortunately out of print.

[21-5] Jung, Walter G., *IC Op-Amp Cookbook, 3rd Ed.*, Prentice Hall PTR, (1986). A bit dated, but still a useful introduction to op amps.

[21-6] Lancaster, Don, *Active Filter Cookbook, 2nd Ed.*, Newnes, division of Elsevier Science, Woburn, MA, (1995). A classic introduction to active filters, in print since the first edition in 1975.

[21-7] Franco, Sergio, *Design with Operational Amplifiers and Analog Integrated Circuits*, McGraw-Hill Science/Engineering/Math, New York, (2001). Expensive, but an excellent book.

[21-8] Baker, Bonnie C., *Comparing Digital Potentiometers to Mechanical Potentiometers*, AN219 (Preliminary), Microchip Technology, Inc., Document No. DS000219A (2000)

[21-9] Baker, Bonnie C., *Optimizing Digital Potentiometer Circuits to Reduce Absolute and Temperature Variations*, AN691, Microchip Technology, Inc., Document No. DS00691A (2001)

[21-10] *Audio Gain Control Using Digital Potentiometers*, AN1828, Dallas Semiconductor Division of Maxim Integrated Products, Inc., (2002).

[21-11] Microchip makes available free its filter design software FilterLab, current at release 2.0, at http://www.microchip.com/1010/pline/tools/analog/software/flab/10626/index.htm. This is an excellent elementary filter design program and is well worth downloading.

Infrared Remote Controls

According to Internet lore, the average household has between five and seven remote controls. Before starting this chapter, I collected a box full to test—one Zenith, two Nakamichis, two Mitsubishis, one Samsung, one Toshiba, one Hitachi, and two Radio Shack universal remote controls—just by going through our house. Two belong to equipment long since scrapped as uneconomical to repair, but the rest are in more or less common use.

Although the first "cordless" TV remote controls used ultrasonic technology (and without electronics in the remote, no less, employing mechanically stroked tuning forks) when inexpensive LEDs became available, digital infrared transmission became the norm. (Low power radio is used for garage door openers, automobile remote locks and the like.)

With an inexpensive IR receiver module connected to a PIC, it's easy to decode most remote controls. Once decoded, we then may use the techniques we've seen in other Chapters to control a variety of devices. We might use a relay to turn a light off and on, or we might translate volume up/down buttons to adjust audio levels through a SPI-controlled MCP42010 digital potentiometer.

Finding definitive information on IR remote controls isn't easy; the consumer electronic manufacturers hold the information closely. I've pieced together this chapter from research published on the Internet by inspired hackers, supplemented with measurements of the assortment of remote controls available to me.

First, let's look at the lowest level of the transmission and reception—and at this level, all remotes work alike. When you push a button, the remote generates a particular digital control sequence and transmits it as multiple

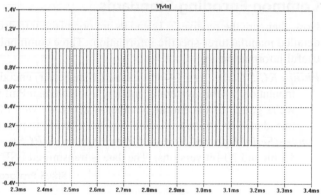

Figure 22-1: Simulated drive to IR LED for one burst—38KHz carrier.

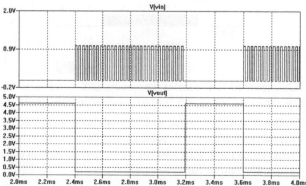

Figure 22-2: IR receiver smoothes carrier cycles into clean waveform; upper waveform: raw IR; lower waveform: IR receiver output.

bursts of IR light. Each individual burst consists of a few dozen rapid on/off cycles, as illustrated in Figure 22-1. An IR receiver detects these bursts and smooths out the individual on/off carrier cycles, thereby reconstituting the original digital sequence, as shown in Figure 22-2.

The frequency of each IR on/off cycle within a burst is the *carrier frequency*, and varies from manufacturer to manufacturer. The lowest carrier frequency commonly used is 32.75 KHz, and the highest is 56.8 KHz, but by far the most common carrier frequency is 38 KHz. IR receivers are frequency selective and to achieve reasonable range, you must match the receiver frequency to the remote's carrier frequency. (A few broadband receiver modules are available, such as Vishay Semiconductor's TSOP1100 model.) We'll talk more about IR receivers shortly.

The IR receiver also translates between light level and PIC-compatible logic level—its output is either at logic 1 or logic 0. Receivers are designed so that when *no IR light* is detected, the receiver's output is *logic 1*; and is *logic 0* when *IR light is detected*. This may seem backwards at first, but in fact we'll find it very convenient to forget about whether a light output causes a 1 or a 0 and instead concentrate on the receiver's logic output values.

Common Encoding Standards

How, then, are the receiver's 1s and 0s are organized into useful information? Here we must delve into some detailed manufacturer-specific information. Three commonly used methods of encoding the 1s and 0s into useful information are:

Philips RC-5 Code—is a 14-bit biphase (Manchester) encoded system, with a 36 KHz carrier. The documentation for this system is publicly available in a Philips Semiconductor Application Note. Because we're going to concentrate on other systems, we won't further discuss the RC-5 code. (Philips recently introduced RC-6, an expanded version of RC-5, to provide for additional data length.)

Sony—(Serial Infrared Remote Control System, SIRCS)—Originally a 12-bit code, but later expanded to 15-bit and 20-bit versions. The data is transmitted LSB first, using pulse width coding:

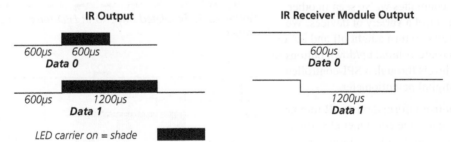

SIRCS Protocol

SIRCS uses a 40 KHz carrier frequency, and, as reflected above, 0 and 1 are distinguished by different carrier-on intervals. We won't further concern ourselves with SIRCS.

REC-80—Is a 32-bit "space width" modulated code, and is, in some sense, an inverse of SIRCS. In SIRCS, the carrier-off period is constant and the carrier-on period has two values, a narrow pulse for 0 and a wide pulse for 1. REC-80 has a constant carrier-on period and has two off-periods, a narrow pulse for 0 and a wide pulse for 1.

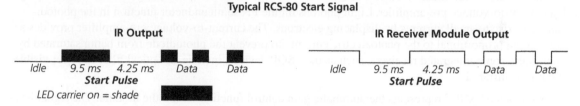

Typical REC-80 Protocol

IR Output

800 µs | 1200 µs | 800 µs
Data 0

800 µs | 1800 µs | 800 µs
Data 1

IR Receiver Module Output

800 µs | 1200 µs | 800 µs
Data 0

800 µs | 1800 µs | 800 µs
Data 1

LED carrier on = shade

The sketch of the IR receiver output suggests our decoding strategy for REC-80 signals; measure the logic high pulse duration out of the receiver and classify it as wide or narrow. We'll see, however, that in practice the narrow pulses tend to be much shorter than 1200µs and the wide pulses are somewhat shorter than 1800µs, due, in part at least, to the IR receiver's response time. (Some sources suggest the data 0 "no-carrier" period is twice the "carrier-on" time, and the data 1 "no-carrier" period is three times the "carrier-on" time. None of the remote control signals I measured came close to this 1:2:3 ratio.)

A long start signal is sent before the data bits:

Typical RCS-80 Start Signal

IR Output

Idle | 9.5 ms | 4.25 ms | Data | Data
Start Pulse

LED carrier on = shade

IR Receiver Module Output

Idle | 9.5 ms | 4.25 ms | Data | Data
Start Pulse

The normal REC-80 protocol has 32 data bits, in theory, organized as:

Typical REC-80 Data Organization Sending Order							
Byte 3		Byte 2		Byte 1		Byte 0	
Device Code		D.C. Check Byte		Function Code		F.C. Check Byte	
LSB	MSB	LSB	MSB	LSB	MSB	LSB	MSB

The check bytes are logical NOT copies of the associated data bytes. (The logical NOT operation converts 1s to 0s and 0s to 1s.) Again, however, you should be prepared for discrepancies between this typical organization and the way manufacturers actually implement their remote controls. Some, for example, might increase the possible number of device codes and function codes by dropping the check bytes, thereby yielding 16-bit device codes and function codes.

Most remote controls send the code for a key press several times, with the thought that at least one of the repeated messages will be correctly received. Repeated receptions of the same command may lead to a different error, however. For example, the power-on/power-off code is typically assigned to one "power" button. How then is the receiver to distinguish between receiving two correct copies of a code the user intends to be a single power-on command from a power-on followed by a power-off command? Some manufactures employ a "toggle bit" to distinguish between these two cases; sequential presses of a key will alternate between two codes differing in the toggle bit. Thus, two successive copies of a command received due to normal message repetition will be identical; two successive copies of a command received due to the user pressing the command twice will differ by the value of the toggle bit. None of the remote controls I worked with implemented toggle bits, however.

IR Receiver

Although it's possible to decode an IR control signal with nothing more than a photodiode or phototransistor, to work well these devices require significant ancillary circuitry. Rather than reinvent the mousetrap, we'll use an inexpensive IR receiver module. To give you an idea of the complexity of these devices, Figure 22-3 provides a high-level block diagram of a typical IR receiver, a Vishay TSOP12xx series device.

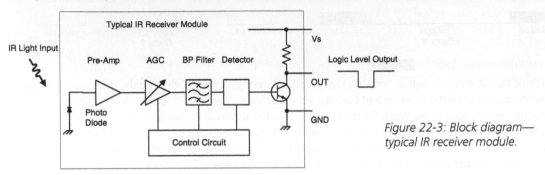

Figure 22-3: Block diagram— typical IR receiver module.

Let's see how these individual modules work, at least at a high level. The photodiode is most often employed with a current-to-voltage pre-amplifier. Light photons hit the PN semiconductor junction in the photodiode and give rise to small currents by displacing electrons. The current-to-voltage pre-amplifier provides a voltage output proportional to the photoelectric current. To prevent the photodiode from being saturated by ambient light, some integrated receivers, such as the TSOP devices, include an optical filter to reject visible light but pass infrared.

The block labeled "AGC" represents the automatic gain control function within the module. The signal level received by the photodiode changes drastically as you orient the remote at differing angles to the receiving diode, or change distance between the remote control and the receiving diode. The AGC amplifier increases its gain when the input signal is weak and decreases its gain when the input signal is strong, thereby stabilizing its output at a level roughly independent of the received signal level.

After the AGC amplifier, the incoming carrier signal is bandpass filtered. This is a very important step in reducing interference from other light sources, such as room lighting, or stray sunlight. After all, the IR receiver must differentiate a weak IR emitter from a sea of background illumination, and this is the job of the bandpass filter. (Even though the optical filter helps, many light sources, such as sunlight, have strong infrared components.) By selectively filtering on the carrier frequency, the bandpass filter rejects other light sources, such as sunlight, that are either unmodulated, or fluorescent lighting, which is modulated with a 120 Hz carrier from the AC mains power. The bandpass filter is responsible for the need to match the receiver frequency to the carrier frequency of the remote control.

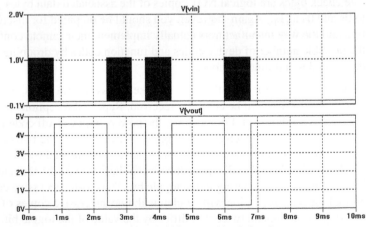

Figure 22-4: Input IR pulses yield inverted logic level output.

The output of the bandpass filter is envelope detected and drives the logic level output stage. As illustrates, the convention is that IR receivers invert the sense of the received signal; no-IR causes a logical high and IR-received causes a logical low.

> *In the remainder of this chapter, when we refer to high and low, or*
> *positive pulses and negative pulses, it is with respect to the output of the IR receiver,*
> **not the actual IR illumination of the remote control unit**.

In writing this chapter, I tested two Sharp Electronics GP1U5 receiver modules, and several Vishay TSOP12xx integrated receivers, with similar results. The GP1U5 modules have recently been replaced by GP1UD26/27/28 series devices, which are physically smaller but have similar performance characteristics. I've also seen data sheets leading me to believe that Sharp's GP1U5 modules have been cloned and still may be available if you search diligently. Vishay's TSOP series of IR receivers is available from many electronic parts suppliers, and cost around $1 in small quantities.

You should match the IR receiver's response frequency (the bandpass filter center frequency in Figure 22-3) to the carrier frequency of your remote control, within 10% or so. Operating outside this range, such as decoding a 32.75 KHz signal with a 38 KHz receiver—or vice versa—significantly reduces range. If you don't know the carrier frequency of your remote control, it's easy enough to measure it with a phototransistor using the circuit shown in Figure 22-5. The resulting waveform, when the QSE114 phototransistor is illuminated by a Mitsubishi remote control, is shown in Figure 22-6. Measuring the time interval between peaks yields a calculated carrier frequency of 37.88 KHz, so the nominal carrier frequency is 38 KHz. You may wonder why the phototransistor output isn't the nice square wave we expect from theory. The QSE114 is an IR photoreceptor, with a rise and fall time of 8µs, so we see one problem immediately; a 38 KHz square wave has an on-time of 13 µs, so we would expect to see a severely distorted output waveform to begin with. And we won't see even that speed from a QSE114 when operating with a simple resistive collector load. But, to measure the carrier frequency, all we need are consistent reference points on the waveform.

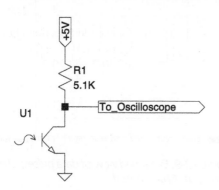

Figure 22-5: Simple circuit for measuring IR carrier frequency.

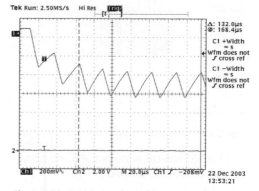

Figure 22-6: Measuring the carrier frequency of a remote control Ch1: collector of QSE114 phototransistor; Ch2: unused.

One final note and then on to some code. Although the bandpass filter reduces stray light interference, it can't solve a second major problem—background lighting saturating the photodiode and the current-to-voltage preamplifier, since the photodiode is more or less sensitive to visible as well as infrared light frequencies. You may significantly improve decoding range by adding a dark red plastic IR filter over the lens of the IR receiver. If you can't find the correct IR filter material, experiment, as some common plastics are opaque to visible light but transmit IR. A fully exposed piece of developed color negative film may also work as an IR filter. Vishay's TSOP-series integrated IR receivers have a built-in IR filter, which is another reason to prefer them to Sharp's GP-series modules.

Characterizing Wide/Narrow Pulse Intervals

Remember that we have decided to work only with REC-80 protocol, and that *high/low references are with respect to the IR receiver output.* Connect the IR receiver module to the PIC as illustrated in Figure 22-7. Let's look at the output of the IR receiver when it receives a remote control signal. Figure 22-8 shows idle, then a start pulse, followed by 32 wide or narrow data pulses, with a return to idle. We might measure the pulse width with the oscilloscope, by expanding the sweep speed, with the results illustrated in Figure 22-9. But, since our ultimate objective involves reading the remote control via a PIC, let's use the PIC to measure the received pulse widths.

The start pulse we see in Figure 22-8 consists of about 10 ms low (IR-on), followed by a 4.2 ms high (IR-off). Since we key on highs, we'll regard the start pulse as a nominal 4.2 ms high pulse preceding the actual data.

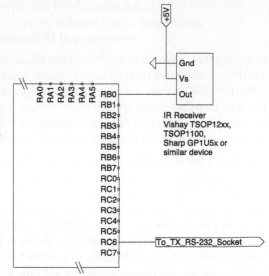

Figure 22-7: Connecting an IR receiver module to the PIC.

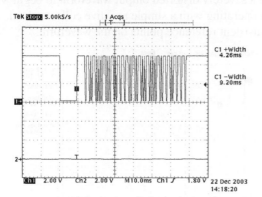

Figure 22-8: IR receiver output, REC-80 signal; Ch1: IR output Ch2: unused.

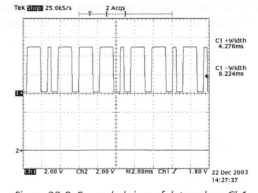

Figure 22-9: Expanded view of data pulses; Ch1: IR output; Ch2: unused.

Program 22-1

```
;Program 22-1
;IR decoder output ;to Pin B0.
;Measure positive pulse ;width and
;form histogram and dump to serial for analysis


;Variables
;----------
Width          Var     Word          ;Read pulse width
WidthArray     Var     Byte(100)     ;Data to be historgrammed
i              Var     Word          ;counter
j              Var     Byte          ;counter
```

```
;Initialization
;--------------
Clear
Input B0        ;IR Receiver to pin B0
EnableHSerial
SetHSerial H115200

Main
;---------
        ;In idle condition receiver is high
        If PortB.Bit0 = 1 Then
                GoTo Main
        EndIf
        ;press keys quickly in front of receiver
        For i = 0 to 999          ;take 1K samples

                ;measure positive pulse width; actually no-IR-output
                ;but we are inverted, so it makes sense to
                ;consider it a positive going pulse
                ;with variable width
                ;PulsIn in units of 1.4us => 18 units = 25.2us
                PulsIn B0,1,Width
                ;save in bins of 25.2uSec
                j=Width / 18
                If j > 99 Then
                        j = 99
                EndIf
                ;increase counter
                WidthArray(j) = WidthArray(j)+1
        Next

        ;dump histogram counters for later analysis
        For j = 0 to 99
          HSerOut ["Width",9,Dec(j*252)/10,9,Dec WidthArray(j),13]
        Next
        Pause 100
End
```

Our objective with Program 22-1 is to measure a selection of remote controls and see if we might identify the width threshold to distinguish a wide (logic 1) pulse from a narrow (logic 0) pulse. We do this by measuring 1000 pulses, generated when you rapidly press a random selection of remote control keys while the program is sampling data. The program then categorizes the measured widths into histogram-type width bins.

```
Main
;---------
        ;In idle condition receiver is high
        If PortB.Bit0 = 1 Then
                GoTo Main
        EndIf
```

We know that the IR receiver's idle condition is high, so we read **B0** until we detect a low, indicating data is starting to arrive. Actually, you'll see that there is sufficient random interference from florescent lighting to cause periodic low conditions, so it's a good idea to load the program into the PIC, get the remote control in position and then reset the PIC as you begin to press random buttons. (We'll see how to reduce false starts in Program 22-2.)

```
        For i = 0 to 999          ;take 1K samples
                PulsIn B0,1,Width
```

The key to reading the IR receiver is MBasic's **PulsIn** procedure. **PulsIn** measures the input pulse duration in unit steps. The unit step duration corresponds to seven instruction cycle durations, where one instruction cycle is $4/F_{osc}$.

Oscillator Frequency	Instruction Cycle Length	PulsIn Unit Step Size	Timeout Base Value
20 MHz	200 ns	1.4 μs	91.75 ms
16 MHz	250 ns	1.75 μs	114.7 ms
10 MHz	400 ns	2.8 μs	183.5 ms

PulseIn is invoked with several parameters:

```
PulseIn Pin, State {TimeoutLabel, TimeoutMultiple,} Var
```

Pin—A constant or variable that defines the pin to be measured. **PulsIn** automatically places the pin into input mode. Figure 22-7 shows we've connected the IR receiver to **B0**.

State—**PulseIn** can measure either the positive width or negative width, i.e., the width of the logic 1 signal or the width of the logic 0 signal. **State** is a variable or constant that defines which width is to be measured. If **State = 1**, we start measuring when **B0** goes from 0 to 1 and stop measuring when **B0** goes from 1 to 0, i.e., we measure the positive width. Conversely, if **State = 0**, we measure the width of the 0 level interval. Since we wish to measure the positive width, we set **State = 1**. (Don't get confused; the variable period in REC-80 is the time between LED carrier bursts; but since the IR receiver is inverting, by measuring positive intervals, we are, in fact, measuring the interval between IR light bursts.)

{Timeout Label, TimeoutMultiple}—**PulseIn** checks for the transition defined by **State**, either 0→1 or 1→0. If no such transition has been received after 65,535 unit step intervals, **PulsIn** times out, and execution passes to the next MBasic statement. Optionally it is possible to extend the time out period in steps of 65,535 unit step intervals via the **TimeoutMultiple** value. A value of 2, for example, yields a timeout period of 2 x 65,535, or 131,070 unit step intervals. Assuming our PIC has a 20 MHz clock, each unit step is 1.4μs, so the basic timeout interval is 65,535 × 1.4 μs or 91.75 ms. Upon passage of this time without an input pulse, execution will pass to **TimeoutLabel**. I found that it is not possible to set **TimeoutMultiple** without also supplying a **TimeoutLabel**, and vice versa and that any attempt to use only one of the two options produced a compiler error. We'll use these optional parameters in later programs, but not in Program 22-1.

Var—Holds the width of the measured pulse, in unit step intervals. **Var** is a 32-bit result, which calls for a long variable if you wish to measure the maximum interval. In our case, we know that the pulse length is under a few thousand microseconds, so we will use a word variable for **Var**.

```
        ;save in bins of 25.2 uSec
        j=Width / 18
        If j > 99 Then
             j = 99
        EndIf
        WidthArray(j) = WidthArray(j)+1
Next
```

Now that we have measured the positive pulse width, we wish to store its value. Our PIC doesn't have enough memory to store 1,000 word variables, and, in any event that isn't necessary, or even desirable. Instead, we categorize each measured width into one of 100 "width bins," each bin being about 25 μs wide. (With a 20 MHz clock, **Width** is measured in unit intervals of 1.4 μs duration, and 18 such units correspond to 25.2 μs.) We do this by dividing the measured width by 18 to yield the histogram "bin number" j and then incrementing **WidthArray(j)** by 1. (We trap for **j>99**, as we only have allocated 100 bins for the data.) After we have taken 1,000 samples, **WidthArray(j)** holds the number of width measurements falling within each bin value. For example, **WidthArray(20)** holds the number of pulses with measured widths between 500 and 524 μs. In statistics, this is termed a "histogram." (By using a byte array to hold bin counts, we assume, implicitly, that not more than 255 width measurements will belong to any individual bin. That's a reasonably safe assumption for 1,000 total measurements with approximately half expected to be wide and half to be narrow and with an expected spread of pulse duration.)

Execution speed is a concern in this program, and even more so in later programs in this chapter. We wish our program to not miss any positive pulses, so we must be economical with the time available before we expect the next positive pulse to start. For the REC-80 code, we know the space between pulses (the IR-on time) runs between 600 and 800 μs. That's enough time, assuming the PIC is running with a 20 MHz clock, to perform the computations we require. But, we must be careful to not burden our program with lengthy computations during this inter-pulse interval, lest the next positive pulse be missed.

```
;dump histogram counters for later analysis
For j = 0 to 99
HSerOut ["Width units ",9,Dec j*18,9,"Width us ",9, |
            Dec (j*252)/10,9,Dec WidthArray(j),13]
Next
Pause 100
```

After we have taken 1,000 samples the histogram results are sent by serial interface where we capture them with a terminal program. To convert the bin number to microseconds, we multiply by 25.2. To avoid floating point conversion, we instead multiply by 252 and divide by 10. But, for decoding it doesn't matter whether we use microseconds or unit intervals, as long as we are consistent in setting the boundary thresholds.

Here's a partial sample of the output from a Mitsubishi remote control:

Width units	576	Width µs 806	0
Width units	594	Width µs 831	10
Width units	612	Width µs 856	0
Width units	630	Width µs 882	70
Width units	648	Width µs 907	82
Width units	666	Width µs 932	104
Width units	684	Width µs 957	151
Width units	702	Width µs 982	5
Width units	720	Width µs 1008	5
Width units	738	Width µs 1033	0

```
*     *     *
```

Width units	1494	Width µs 2091	0
Width units	1512	Width µs 2116	8
Width units	1530	Width µs 2142	7
Width units	1548	Width µs 2167	80
Width units	1566	Width µs 2192	204
Width units	1584	Width µs 2217	0
Width units	1602	Width µs 2242	3
Width units	1620	Width µs 2268	4
Width units	1638	Width µs 2293	0

The displayed width value is the lower bound of the bin, *for example,* width 907 µs means between 907 and 932 µs. We can either analyze the data by inspection of the output, or we can plot it. Figure 22-10 shows the histograms for six REC-80 remote control units. The data shows that we can safely set our narrow/wide bounds as:

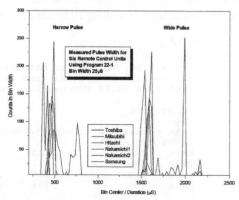

Figure 22-10: Wide and narrow pulse distributions – 6 remote control units.

	Minimum Length		Maximum Length	
Pulse Category	Units (1.4 µs)	Microseconds	Units (1.4 µs)	Microseconds
Narrow	250	350 µs	786	1100 µs
Wide	1004	1450 µs	1607	2250 µs
Start	2714	3800 µs	3429	4800 µs

One final point, oscilloscope measurements for each remote control showed a reasonably consistent start pulse of about 4500 µs.

Program 22-2

OK, enough preparatory work. Let's look at decoding a real remote control. First, though, let's work through our decoding algorithm in pseudo-code.

```
Searching:
Check input pin. If high (idle) continue checking
If input pin is low, it might be the lead-in to a start pulse
Using PulsIn check for a valid start pulse ~4.2ms
If no valid start pulse go back to Searching

Assumes we know the number of data pulses to receive = BitLength
For i = 0 to BitLength-1
        Read the pulse length
        Store the pulse length in array(bitlength) Warray
Next

Now Analyze Warray to categorize each measurement as
A Wide (W), a Narrow (N) or an Unknown (x) pulse length

populate a string array TempArray() with W,N or x
based on the pulse width windows.

Use TempArray to determine validity of data; can't have any
X values if valid.

Substitute 1 for W, 0 for N in a bit-by-bit assembly of the
Decoded value.

Write output to serial port
```

To keep things simple, we'll start with an older Mitsubishi TV/VCR remote control with a 16-bit code. Mitsubishi uses a 32.75 KHz carrier frequency, so the particular IR receiver we use is one optimized for that carrier frequency. (In reading Program 22-2, remember the | character is a line continuation.)

Program 22-2

```
;Program 22-2
;to Pin B0. Values for Mitsubishi remote control 16-bit width
;Mitsubishi is special case because it is 16-bits
;and uses 32.75 KHz carrier, not the more common 38KHz

;Constants
;--------------
;the values are in unit intervals of 1.4us which assumes
;a 20 MHz clock.
Wmin      Con          1036    ;1450us wide pulse min width
WMax      Con          1714    ;2400us wide pulse max width
NMin      Con          250     ;350us  narrow pulse min width
NMax      Con          786     ;1100us nNarrow Pulse max width


;Variables
;-----------
CodeValue       Var     Word          ;Holds the 16-bit decoded value
CodeValueR      Var     Word          ;Reversed bits within byte
TempArray       Var     Byte(16)      ;Holds W/N/x string array
i               Var     Byte          ;Counter
WArray          Var     Word(16)      ;Holds measured pulse widths
GoodFlag        Var     Byte          ;If 1, good data


;Initialization
;--------------
Clear
Input B0        ;Decoder to B0
EnableHSerial
SetHSerial H115200
```

```
HSerOut ["P 22-2",13,13]

Main
;------
        ;Unlikw REC-80 spec, Mitsubishi uses no long start pulse
        ;All we get is 16 data bits
        ;Wait for a start pulse
        i = 0
        ASM
        {
WaitForInput
                banksel PortB
                btfsc           PortB,0
                GoTo            WaitForInput
                NOP
        }

        ; pulse is followed by 16 data bits. Capture
        ;the widths to WArray
        Loop
                PulsIn B0,1,Main,1,WArray(i)
                ASM
                {
                banksel i
                incf    i&0x7F,f
                banksel PortB
                }
                If i < 16 Then Loop

;GoTo Main

        GoSub CalculateCode
        GoSub GenerateValue
        GoSub DumpRawData

        If GoodFlag = 1 Then
                HSerOut ["Good: ",Str TempArray \16,"  MSB: ", |
                        iHex CodeValue]
                HSerOut ["  LSB: ",iHex CodeValueR,13]
        ELSE
                ;Uncomment line below if want see bad data
                HSerOut ["Error: ",Str TempArray \16,"  LSB: ", |
                        iHex CodeValue,13]

        EndIf

GoTo Main

CalculateCode
;------------
        ;Scan the measured with array and classify each as
        ;N (narrow), W (wide) or x (outside W and N windows)
        For i = 0 to 15
                TempArray(i) = "x"
                If (WArray(i) > WMin) AND (WArray(i) < WMax) Then
                        TempArray(i) = "W"
                EndIf

                If (WArray(i) > NMin) AND (WArray(i) < NMax) Then
                        TempArray(i) = "N"
                EndIf
        Next
Return

GenerateValue
;------------
        ;Now calculate the 16-bit Word equal to the received data
        CodeValue = 0
        ;Set to 0 if we have bad data
        GoodFlag = 1
        For i = 0 to 15
```

```
                CodeValue.Bit0 = %0 ;overwrite if W
                ;Now convert W/N/x to numbers
                If TempArray(i) = "W" Then
                        CodeValue.Bit0 = %1
                EndIf
                If TempArray(i) = "x" Then
                        GoodFlag = 0
                EndIf
                ;No need to check N or x, pre-populated with 0s
                ;read assuming sent MSB first. This may not
                ;be the case, but for our purpose order doesn't
                ;matter as we only look for unique values
                ;Data goes in at right, move to left every entry
                ;Don't shift on last bit as no more to be added
                If i <> 15 Then
                        CodeValue = CodeValue << 1
                EndIf
        Next
        CodeValueR.Byte0 = CodeValue.Byte0 REV 8
        CodeValueR.Byte1 = CodeValue.Byte1 REV 8
Return

DumpRawData
;----------
        For i = 0 to 15
                HSerOut [Dec i,9,Dec WArray(i),13]
        Next
Return
End
```

There's something different here than in our pseudo-code; there is no code to check for a valid start pulse. That's because Mitsubishi uses a different approach, as seen in Figure 22-11. The signal goes directly from idle to data, without the prolonged start pulse seen in Figure 22-8. (It's entirely possible, of course, that the three wide pulses immediately after idle are a preamble and that the remaining 13 bits are data. That's one of the difficulties of reverse-engineering a protocol with only a limited sample of devices to test. Regardless, we will decode these as if they are valid data bits, and their status is immaterial to the technique we use in Program 22-2.)

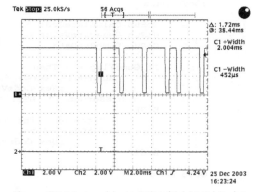

Figure 22-11: Lead-in to Mitsubishi 16-bit data; Ch1: IR receiver output; Ch2: not used.

```
        Main
        ;------

        WaitForInput
                banksel PortB
                btfsc           PortB,0
                GoTo            WaitForInput
                NOP
```

Since there is no start bit, we jump to capturing the next 16 positive pulse widths after the first time the input pin goes low. In MBasic, our wait for input routine could be written as:

```
        WaitHere
                If PortB.Bit0 = 1 Then GoTo WaitHere
                EndIf
```

The problem with a pure MBasic solution is that its execution time is marginal, given that the low time is a few hundred microseconds. Accordingly, we use a simple assembler routine that reads **PortB** and checks the status of bit 0 (**B0**) through the **btfsc** operation. If **B0** is set (**B0 = 1**) the next instruction, **GoTo WaitForInput** is executed in an endless loop. If **B0** is clear (**B0 = 0**), the **GoTo WaitForInput** instruction is skipped, a null (**NOP**) operation is executed and code execution returns to MBasic. The assembler test and branch code executes in less than 1μs with a 20 MHz clock.

We then read the duration of the next 16 high pulses using **PulsIn**. The most direct way to code this is:

```
For i = 0 to 15
        PulsIn B0,1,WArray(i)
Next
```

This construction, however, has an execution speed problem. It turns out that the combination of the **For... Next** loop overhead (162 μs) and **PulsIn**'s release time (166 μs) following the input pulse falling edge total approximately 328 μs, assuming a 20 MHz clock. Since the remote control inter-pulse dwell interval can be as short as 305 μs, we must save time someplace. We could write our own alternative **PulsIn** procedure, perhaps using timer 1, but before introducing that complexity, we first try an substitute for the **For...Next** loop. By mixing a bit of assembler with the MBasic code we shave the loop overhead from 162 μs to 73 μs, for a total execution time of 249 μs, well within our 305 μs dwell interval.

```
Loop
        PulsIn B0,1,Main,1,WArray(i)
        ASM
        {
        banksel i
        incf    i&0x7F,f
        banksel PortB
        }
        If i < 16 Then Loop
```

Our assembler routine follows that discussed in Chapter 14 and is the equivalent of **i=i+1** where **i** is a byte variable, except it executes in under 1 μs. We then capture the next pulse width in the word-length array **Warray(i)**. If we have read fewer than 16 pulses, the **if...then** test recycles to **PulsIn** to read the next input pulse. The **if...then** test in MBasic executes in about 70μs, so the combined assembler increment and **if...then** test executes in less than half the time of the **for...next** code it replaces.

After a 250 ms pause to avoid reading duplicated pulse sequences, we call the subroutine **CalculateCode** to classify each of the 16 measured pulse widths as wide, narrow or unknown.

```
CalculateCode
;-------------
        ;Scan the measured with array and classify each as
        ;N (narrow), W (wide) or x (outside W and N windows)
        For i = 0 to 15
                TempArray(i) = "x"
                If (WArray(i) > WMin) AND (WArray(i) < WMax) Then
                        TempArray(i) = "W"
                EndIf

                If (WArray(i) > NMin) AND (WArray(i) < NMax) Then
                        TempArray(i) = "N"
                EndIf
        Next
Return
```

Since we wish to display the output data for debugging, we'll use the string array **TempArray** for two purposes. First, we will use it to hold the status of each pulse, W, N or x, for wide, narrow or unknown, respectively. This array is later sent to the serial port for display. Second, we'll use it to help compute the numerical value of the received 16 bits.

Using a **For i ...Next** loop, we first assign **TempArray(i)** the value "x", or unknown. Then, we test the pulse length stored in the corresponding **Warray(i)** to see if it within our narrow or wide windows and we assign an appropriate letter—"N" or "W"—to **TempArray(i)** if it a verifiable narrow or wide pulse. If it isn't within either the wide or narrow windows, the classification stays as "x".

After completing `CalculateCode`, a typical `TempArray` looks like the following: WWWNWNWNNNNWNWNN if it has no errors, or, perhaps, WWWxWNWNNxNWNxNN if it contains pulses outside the W or N width limits. (Of course, the Ws and Ns change, depending on the key sequence pressed.)

Subroutine `GenrateValue` then calculates the word value of the 16 received bits.

```
GenerateValue
;-------------
        CodeValue = 0
        GoodFlag = 1
        For i = 0 to 15
                CodeValue.Bit0 = %0 ;overwrite if W
                If TempArray(i) = "W" Then
                        CodeValue.Bit0 = %1
                EndIf
                If TempArray(i) = "x" Then
                        GoodFlag = 0
                EndIf
                If i <> 15 Then
                        CodeValue = CodeValue << 1
                EndIf
        Next
        CodeValueR.Byte0 = CodeValue.Byte0 REV 8
        CodeValueR.Byte1 = CodeValue.Byte1 REV 8
Return
```

The returned word value, assuming the data is sent MSB-first, is in `CodeValue`. However, since we don't know if Mitsubishi sends its data in LSB or MSB order, we'll also calculate and display the LSB-first value, to be held in `CodeValueR`. We'll judge which is the correct order based on how these values change as we press different keys on the remote.

Our approach is to loop through the string array `TempArray` and, based upon the value of the character, add either a `%0` or a `%1` at the last bit in `CodeValue`. After each pass through the loop, we shift the bits in `CodeValue` one place to the left with the shift-left (`<<`) operator. Any "x" character in `TempArray`, sets the `GoodFlag` to 0, indicating the received data is corrupt. Note that since the shift-left gets us ready for the next bit, we won't execute it for the last bit, as there are no further bits to add. Hence we test for `i=15` and, when true, omit the shift-left.

Finally, we set the two bytes of `CodeValueR` equal to `CodeValue`, but with reversed bit order.

```
If GoodFlag = 1 Then
        HSerOut ["Good: ",Str TempArray \16,"   MSB: ", |
        iHex CodeValue]
        HSerOut ["   LSB: ",iHex CodeValueR,13]
ELSE
        ;Uncomment line below if want see bad data
        ;HSerOut ["Error: ",Str TempArray \16,"   MSB: ", |
        iHex CodeValue,13]
EndIf
```

The remainder of the main program loop displays the received data. If you wish to see corrupt data as well as good data, uncomment the `HserOut` statement in the `ELSE` portion of the `If` conditional. Subroutine `DumpRawData` provides the individual measured pulse durations.

Here's the output for some sample key presses. (To save space, I've suppressed the raw pulse length output data.) This particular remote control has a slide switch to select between TV and VCR, so we'll exercise both options. I've added the Mode and Key comments to Program 22-2's output and I've deleted duplicates, where each of the repeated messages automatically sent by the remote controller are correctly decoded and separately reported.

					Comments	
					Mode	Key
Good:	WWWNWNWNWNWNWNNNN	MSB:	$EA50	LSB:	$570A	VCR	Ch+
Good:	WWWNWNWNNWNWNNNNN	MSB:	$EA48	LSB:	$5712	VCR	Ch−
Good:	WWWNNNWNNWNNWNNNWN	MSB:	$E244	LSB:	$4722	VCR	Vol+
Good:	WWWNNNWNWNNWNNNWNN	MSB:	$E254	LSB:	$472A	VCR	Vol−
Good:	WWWNWNWNWNWNNNNNN	MSB:	$EA90	LSB:	$5709	VCR	0
Good:	WWWNWNWNWNNNNNNNNN	MSB:	$EA00	LSB:	$5700	VCR	1
Good:	WWWNWNWNNNNWNNNNN	MSB:	$EA10	LSB:	$5708	VCR	2
Good:	WWWNWNWNNNNNNWNNNN	MSB:	$EA08	LSB:	$5710	VCR	3
Good:	WWWNWNWNNNNNWNNNN	MSB:	$EA18	LSB:	$5718	VCR	4
Good:	WWWNWNWNNNNNNNWNNN	MSB:	$EA04	LSB:	$5720	VCR	5
Good:	WWWNWNWNNNNNWNNWNN	MSB:	$EA14	LSB:	$5728	VCR	6
Good:	WWWNWNWNNNNNNWNNWN	MSB:	$EA0C	LSB:	$5730	VCR	7
Good:	WWWNWNWNNNNNWNNWNN	MSB:	$EA1C	LSB:	$5738	VCR	8
Good:	WWWNWNWNNNNNNNNNNW	MSB:	$EA80	LSB:	$5701	VCR	9
Good:	WWWNNNWNNWNNWNWNNN	MSB:	$E250	LSB:	$470A	TV	Ch+
Good:	WWWNNNWNNWNNWNWNNN	MSB:	$E248	LSB:	$4712	TV	Ch−
Good:	WWWNNNWNNWNNWNNNWN	MSB:	$E244	LSB:	$4722	TV	Vol+
Good:	WWWNNNWNWNNWNNNWNN	MSB:	$E254	LSB:	$472A	TV	Vol−
Good:	WWWNNNWNWNWNNNNNN	MSB:	$E290	LSB:	$4709	TV	0
Good:	WWWNNNWNNNNNNNNNN	MSB:	$E200	LSB:	$4700	TV	1
Good:	WWWNNNWNNNNWNNNNN	MSB:	$E210	LSB:	$4708	TV	2
Good:	WWWNNNWNNNNNWNNNN	MSB:	$E208	LSB:	$4710	TV	3
Good:	WWWNNNWNNNNNWNNNN	MSB:	$E218	LSB:	$4718	TV	4
Good:	WWWNNNWNNNNNNNWNNN	MSB:	$E204	LSB:	$4720	TV	5
Good:	WWWNNNWNNNNNWNNWNN	MSB:	$E214	LSB:	$4728	TV	6
Good:	WWWNNNWNNNNNNWNNWN	MSB:	$E20C	LSB:	$4730	TV	7
Good:	WWWNNNWNNNNNWNNWNN	MSB:	$E21C	LSB:	$4738	TV	8
Good:	WWWNNNWNWNNNNNNNW	MSB:	$E280	LSB:	$4701	TV	9

Before deciding the most likely correct bit order, we should understand that it really doesn't matter in most cases whether we have it right or not. If we wish to use the decoded remote control to increase or decrease some value based on the TV/Channel+ or TV/Channel− key, it matters not whether we increase upon decoding **$E250** or **$470A**. All we are concerned with is matching a unique value, and that value can be the theoretically correct bit order, or its reverse; both values are unique and either serves our purpose. However, as an exercise, let's see if we can determine the correct order.

Before we decide the bit order, it appears that the high byte is a device code, where **$E2/$47** precedes control messages to the TV, and **$EA/$57** precedes messages to the VCR. If so, the low byte makes sense as a key code independent of the device address. For example, the low byte value **$90/$09** always represents the digit 0 key, regardless of whether the command is directed at the TV or the VCR. (The dual values separated by a / indicate both MSB-first and LSB-first values.)

To choose between LSB-first and MSB-first bit order, however, is more difficult. Examining the values sent when the 0…9 keys are pressed, we note that LSB-first order has fewer large value jumps than MSB-first order (the MSB-first order values associated with the 9 and 0 keys particularly stand out as unusual). On balance, it's likely the data is sent LSB first.

Finally, we also observe that, at least for the samples shown above, all key codes start with three consecutive Ws (1's) and all end with two consecutive N's (0's). It's a reasonable guess—logical, but still a guess—that these consecutive characters represent message start and message end sentinels. If so, the variable message content would be 11 bits and our view of LSB-first data order might change.

Decoding a REC-80 Controller

Let's now look at reading a true REC-80 compatible controller. Program 22-3 builds upon Program 22-2, so we'll concentrate on the differences.

Program 22-3

```
;Program 22-3
;IR decoder output
;to Pin B0. Values for Generic
;remote control 32-bit width

;Constants
;--------------
                    ;unit duration @ 20MHz clock
Smin     Con     2500    ;3500us start pulse minimum
Smax     Con     3572    ;5000us start pulse maximum
Wmin     Con     964     ;1350us Wide pulse min width
WMax     Con     1393    ;1950us Wide pulse max width
NMin     Con     214     ;300us  Narrow pulse min width
NMax     Con     571     ;800us  Narrow Pulse max width
MLen     Con     32      ;Number of data bits

;Variables
;-----------
CodeValue       Var     Long        ;holds decoded 32 bits LSB order
Temp            Var     Long        ;decoded bits MSB order
TempArray       Var     Byte(MLen)  ;Holds N/W/X strings
i               Var     Byte        ;Counter
WArray          Var     Word(MLen)  ;Holds pulse width measurements
GoodFlag        Var     Byte        ;If 1, good data

;Initialization
;--------------
Clear                   ;Zeroize
Input B0                ;Output of IR Receiver to B0
EnableHSerial
SetHSerial H115200
HSerOut ["P22-3",13,13]


Main
        i=0
        ;We first wait for start pulse
        PulsIn B0,1,Main,1,WArray(0)

        ;Is it a valid start pulse?
        If Warray(0) < Smin Then
                GoTo Main
        EndIf

        If Warray(0) > Smax Then
                GoTo Main
        EndIf

        ;Valid start pulse is followed by MLen data pulses
        Loop
                PulsIn B0,1,Main,1,WArray(i)
                ASM
                {
                banksel i
                incf    i&0x7F,f
                banksel PortB
                }
        If i < MLen Then Loop
        GoSub CalculateCode
        GoSub GenerateValue
```

```
                    If GoodFlag = 1 Then
                            GoSub ErrorCheck
                            If GoodFlag = 1 Then
                                    HSerOut ["Good: ",Str TempArray \MLen,   |
                                    "  MSB: ",iHex Temp]
                                    HSerOut ["  LSB: ",iHex CodeValue,13]
                            EndIf
                    ELSE
                            ;Uncomment line below if want bad data output
                            ;HSerOut ["Error: ",Str TempArray \MLen,         |
                            "  MSB: ",iHex Temp,13]
                    EndIf
GoTo Main

CalculateCode
;------------
            ;Scan thye measured width array and classify each
            ;as N (narrow), W (wide) or X (outside N and W windows)
            For i = 0 to (MLen-1)
                    TempArray(i) = "x" ;pre-load with bad flag
                    If (WArray(i) > WMin) AND (WArray(i) < WMax) Then
                            TempArray(i) = "W"
                    EndIf

                    If (WArray(i) > NMin) AND (WArray(i) < NMax) Then
                            TempArray(i) = "N"
                    EndIf
            Next
Return


GenerateValue
;------------
            ;Now calculate the long=to the received 32-bits
            Temp = 0
            ;Set to 0 if get bad data
            GoodFlag = 1
            ;Now convert W/N/x to values
            For i = 0 to (MLen-1)
                    Temp.Bit0 = %0
                    If TempArray(i) = "W" Then
                            Temp.Bit0 = %1
                    EndIf
                    If TempArray(i) = "x" Then
                            GoodFlag = 0
                    EndIf
                    ;No need to check N or x, since pre-populated with 0s
                    ;read assuming sent MSB first. This may not
                    ;be the case, but for our purpose order doesn't
                    ;matter as we only look for unique values
                    ;Data in at right, move to left every entry w/ new
                    ;data to follow. Skip at the end
                    If i <> (MLen-1) Then
                            Temp = Temp << 1
                    EndIf
            Next
            ;Some infor says data sent LSB first
            ;byte order is OK, but bits are reversed
            CodeValue.Byte3 = Temp.Byte3 REV 8
            CodeValue.Byte2 = Temp.Byte2 REV 8
            CodeValue.Byte1 = Temp.Byte1 REV 8
            CodeValue.Byte0 = Temp.Byte0 REV 8
Return

ErrorCheck      ;check NOT bytes
;----------
            If (CodeValue.Byte3 + CodeValue.Byte2) <> $FF Then
                    GoodFlag = 0
                    HSerOut ["Bad Check Byte 3",13]
            EndIf
```

```
If (CodeValue.Byte1 + CodeValue.Byte0) <> $FF Then
        GoodFlag = 0
        ;HSerOut ["Bad Check Byte 1",13]
EndIf
Return
End
```

First, we make the code more flexible and maintainable by defining the bit length with a constant:

```
MLen    Con         32
```

And, since the data length is 32 bits, not 16, we've changed the variables **CodeValue** and **Temp** to be type **long**, not **word**. Finally, in an attempt to cover a wide range of remote controllers, we expanded the wide and narrow window constants to match Figure 22-10.

We noted that Mitsubishi's 16-bit protocol has no start pulse, but REC-80 compatible remote controls have a start pulse. We'll accordingly modify the start conditions for Program 22-3.

```
;We first wait for start pulse
PulsIn B0,1,Main,1,WArray(0)

;Is it a valid start pulse?
If Warray(0) < Smin Then
        GoTo Main
EndIf

If Warray(0) > Smax Then
        GoTo Main
EndIf
```

We start by measuring the positive pulse width using **PulsIn**, storing the value in **Warray(0)**. Unlike in Program 22-2, however, we now define the **TimeoutMultiple** and **TimeoutLabel** . We'll set the **TimeoutMultiple** as 1, and use **Main** as the **TimeoutLabel**. Thus, if a positive pulse isn't received within 1.4 µs × 65,535 (91.7 ms) program execution jumps back to **Main**, where **PulsIn** is again executed. Consequently, we are in a continual loop, waiting for a positive pulse to arrive. When a positive pulse arrives, we save its value in **Warray(0)** and check it against the window we've defined for the start pulse, **Smin** and **Smax**. If we have received a valid start pulse, execution continues and we read the width of the next 32 data bits into **Warray**, starting with **Warray(0)**. If the possible start pulse is outside the **Smin...Smax** window, program execution returns to **Main** and we again wait for a possible start pulse.

I measured the REC-80 inter-pulse dwell time at 600 µs so we have sufficient time to perform these comparisons in MBasic. In fact, we could use a **for...next** loop to read the pulse widths. However, since our mixed assembler/MBasic code works, we'll stick with it.

After the this start related code, the remaining portion of the main program segment has only a small change related to error checking, which we'll get to shortly. The two subroutines **CalculateCode** and **GenerateValue** are identical with their namesakes in Program 22-2, save for substituting **Mlen** for the hard coded bit length. Consequently, we'll limit our analysis to the error checking subroutine, **ErrorCheck**.

As we noted in our discussion of REC-80 code format, data is sent in the form

Typical REC-80 Data Organization Sending Order							
Byte 3		Byte 2		Byte 1		Byte 0	
Device Code		D.C. Check Byte		Function Code		F.C. Check Byte	
LSB	MSB	LSB	MSB	LSB	MSB	LSB	MSB

An REC-80 check byte is formed by taking the logical NOT of the associated data byte—that is, 1's are changed to 0's and vice versa. Consequently. if we look at the data byte and its associated check byte, for every corresponding bit position, one byte has a **%0** and the other byte has a **%1**. Thus, if we add the data byte and its check byte, we always get **%11111111** or **$FF**.

We'll accordingly implement a further error check by seeing if **CodeValue.Byte3 + CodeValue.Byte2 = $FF** and if **CodeValue.Byte1 + CodeValue.Byte0 = $FF**. The new subroutine **ErrorCheck** performs this action, and upon an error the **GoodFlag** is cleared, indicating corrupted data and a brief error message is emitted over the serial port. To avoid order of precedence problems with some MBasic versions, I've used parentheses to force the **CodeVal.Byte3 + CodeValue.Byte2** and **CodeVal.Byte1 + CodeVal.Byte0** expressions to be evaluated before the **<>** comparison.

```
ErrorCheck          ;check NOT bytes
;---------
            If (CodeValue.Byte3 + CodeValue.Byte2) <> $FF Then
                    GoodFlag = 0
                    HSerOut ["Bad Check Byte 3",13]
            EndIf

            If (CodeValue.Byte1 + CodeValue.Byte0) <> $FF Then
                    GoodFlag = 0
                    ;HSerOut ["Bad Check Byte 1",13]
            EndIf
    Return
```

Let's look at the output of Program 22-3. Bold face indicates annotations I've added to the output. I've also added leading zeros to the Hitachi MSB values.

Nakamichi RM-2CDP Remote Control   **Key**
P22-3
```
Good: WWWNNWWNNNNWWNNWWNNNWWNWNWWWNN   MSB: $E619A35C   LSB: $6798C53A   Stop
Good: WWWWNNWWNNNNWWNNWWNNNWWNNWWNWNN  MSB: $E619D32C   LSB: $6798CB34   Pause
Good: WWWNNNWWNNNNWWWWNNNWWWNNNWWWNNNN MSB: $E619738C   LSB: $6798CE31   0
Good: WWWWNNWWNNNNWWWWNNNNNNNWWNWWWNN  MSB: $E619F30C   LSB: $6798CF30   1
Good: WWWNNNWWNNNNNNNWWWNNNNWWWNWWWWN  MSB: $E6190BF4   LSB: $6798D02F   2
Good: WWWNNNWWNNNNWWNNNWWNNNWWNNNWWWN  MSB: $E6198B74   LSB: $6798D12E   3
Good: WWWNNNWWNNNNWWNNWWNNNWWWNNNWWNN  MSB: $E6194BB4   LSB: $6798D22D   4
Good: WWWNNNWWNNNNNWWWNNNWWNNWWNNNNWNN MSB: $E619CB34   LSB: $6798D32C   5
Good: WWWNNNWWNNNNWWNNNWWNNWWNNNWWWWN  MSB: $E6192BD4   LSB: $6798D42B   6
Good: WWWNNWWNNNNWWNWWNNNWWNNNWNNNWWN  MSB: $E619AB54   LSB: $6798D52A   7
Good: WWWNNNWWNNNNWWNNNWWWNWNNWWNNWNN  MSB: $E6196B94   LSB: $6798D629   8
Good: WWWNNNWWNNNNNWWWNWWNNNNWWNNNWNN  MSB: $E619EB14   LSB: $6798D728   9
```

Nakamichi RM-4TA Remote Control   **Key**
P22-3
```
Good: WWWNNNWWNNNNWWNNNWWNNNWWWNWWWNN  MSB: $E619A35C   LSB: $6798C53A   Stop
Good: WWWWNNWWNNNNNWWWNNWWNNNWWWNNWNN  MSB: $E619D32C   LSB: $6798CB34   Pause
Good: WWWNNNWWNNNNNWWWNNNWNNWWWNNNWNN  MSB: $E619E31C   LSB: $6798C738   >>
Good: WWWNNNWWNNNNNWWWWWNNNWWNNWNNWNN  MSB: $E619837C   LSB: $6798C13E   <<
Good: WWWNNNWWNNNNNNNWWWWNNNWNNWWNWNN  MSB: $E61913EC   LSB: $6798C837   >|
Good: WWWNNNWWNNNNNWWWNNWNNWWNNNWWWWNN MSB: $E61943BC   LSB: $6798C23D   |<
```

Toshiba SE-R0108 Remote Control VCR/DVD   **Key**
P22-3
```
Good: WNWNNNWWNNNWNNWWWNNNNNWWNWWWWWW  MSB: $A25D48B7   LSB: $45BA12ED   Pwr
Good: WNWNNNWWNNNWNNWWWNNNNNNNWWWNNNN  MSB: $A25D07F8   LSB: $45BAE01F   VCR/DVD
Good: WNWNNNWWNNNWNNWWNNWWNNNWWWWNNNN  MSB: $A25D50AF   LSB: $45BA0AF5   0
Good: WNWNNNWWNNNWNNWWWWNNNNNNWWWWWNN  MSB: $A25D807F   LSB: $45BA01FE   1
Good: WNWNNNWWNNNWNNWWNWNNNWWWNWWWWNN  MSB: $A25D40BF   LSB: $45BA02FD   2
Good: WNWNNNWWNNNWNNWWNNNWWWWNNWWWWNN  MSB: $A25DC03F   LSB: $45BA03FC   3
Good: WNWNNNWWNNNWNNWWNNWNNWWNWWWNNWW  MSB: $A25D20DF   LSB: $45BA04FB   4
Good: WNWNNNWWNNNWNNWWNNWNNWWNNWWWWWW  MSB: $A25DA05F   LSB: $45BA05FA   5
Good: WNWNNNWWNNNWNNWWNNNWWNNWNWWWWWW  MSB: $A25D609F   LSB: $45BA06F9   6
Good: WNWNNNWWNNNWNNWWNNNWWNNWNNWWWWW  MSB: $A25DE01F   LSB: $45BA07F8   7
Good: WNWNNNWWNNNWNNWWWWNNNWWNNNWWWWW  MSB: $A25D10EF   LSB: $45BA08F7   8
```

```
Good:  WNWNNNNWNNWNWWNWWNNNNNWWNWWWW  MSB: $A25D906F  LSB: $45BA09F6   9
Good:  WNWNNNNWNNWNWWWNWWNNNWWNWWWW   MSB: $A25D28D7  LSB: $45BA14EB   Stop
Good:  WNWNNNNWNNWNWWNNNWWWNNWNWWWW   MSB: $A25DC837  LSB: $45BA13EC   >>
Good:  WNWNNNNWNNWNWWNNNWWNNNWWWWWW   MSB: $A25D9867  LSB: $45BA19E6   <<
Good:  WNWNNNNWNNWNWWNNNNNNNNWWWWWW   MSB: $A25D00FF  LSB: $45BA00FF   Pause
```

Hitachi "Remote Controller"   **Key**
P22-3

```
Good:  NNNNNWWNWWWWWWNWWWWNNNNNWWWW   MSB: $06F9E817  LSB: $609F17E8   Pwr
Good:  NNNNNWWNWWWWWWNWWNNNNNNNWWWW   MSB: $06F96897  LSB: $609F16E9   VTR/TV
Good:  NNNNNWWNWWWWWWNWWNNWWNNNWWWW   MSB: $06F99867  LSB: $609F19E6   Ch+
Good:  NNNNNWWNWWWWWWNWNNNNWWNNWWWW   MSB: $06F918E7  LSB: $609F18E7   Ch-
Good:  NNNNNWWNWWWWWWNWNNWNWNNNWWWW   MSB: $06F928D7  LSB: $609F14EB   Play
Good:  WNWNNWWNWWWWWWNWNNWNWNNNWWWW   MSB: $867928D7  LSB: $619E14EB   Rec+Play
Good:  NNNNNWWNWWWWWWNWWNNWNNNNWWWW   MSB: $06F9906F  LSB: $609F09F6   >>
Good:  NNNNNWWNWWWWWWNWNNWNNNWNWWWW   MSB: $06F950AF  LSB: $609F0AF5   <<
Good:  NNNNNWWNWWWWWWNWWNNWWNNNNWWWW  MSB: $06F9D02F  LSB: $609F0BF4   Stop
Good:  NNNNNWWNWWWWWWNWNNWNWNNWNWWWW  MSB: $06F958A7  LSB: $609F1AE5   Pause
```

None of these data reads caused a check byte error. And, that's not unexpected, as the fact that we've received 32 sequential correct width bits provides a high probability that the bits are correct. My limited test showed no 16-bit device and data codes, and no toggle bits.

If we check the MSB-first and LSB-first values against the key functions, the LSB-first values make more sense, so these manufactures have faithfully followed the REC-80 protocol. And, they seem to have made logical key assignments as well.

Numeric Key	Toshiba Key Code	Nakamichi Key Code
0	$0A	$CE
1	$01	$CF
2	$02	$D0
3	$03	$D1
4	$04	$D2
5	$05	$D3
6	$06	$D4
7	$07	$D5
8	$08	$D6
9	$09	$D7

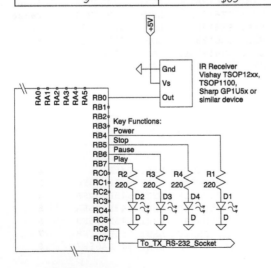

Figure 22-12: Controlling four LEDs with an IR remote control.

Toshiba encodes the key value directly (key 0 is assigned value 10, or $0A), while Nakamichi's codes track the key value as well, but with an offset of $CE.

Let's use the decoded output to actually do something. We'll only turn LEDs on and off, but the techniques we've developed in other chapters permit you to extend the concepts in Program 22-4 to control other devices. For example, you might emit SPI control signals to an MCP42010 digital potentiometer and have the output of a stereo audio signal increase or decrease in 1 dB steps in response to the Vol+ and Vol– buttons.

I've written Program 22-4 to function with a Hitachi controller, chiefly because it belongs to a no-longer-functional VCR and isn't needed elsewhere in my house. Of course, you may substitute any other codes. And, to show that bit order isn't important if—as is usually the case—we are only looking unique key codes, I've intentionally written Program 22-4 with MSB-first bit order.

Pressing the Play, Stop, Pause or Power buttons causes an LED connected to the PIC to latch on or off. Each key press reverses the status. Figure 22-12 shows the LED connection.

Program 22-4

```
;Program 22-4
;IR decoder output
;to Pin B0. Values for Generic
;remote control 32-bit width

;Constants
;--------------
                ;unit duration @ 20MHz clock
Smin      Con   2500    ;3500us start pulse minimum
Smax      Con   3572    ;5000us start pulse maximum
Wmin      Con   964     ;1350us Wide pulse min width
WMax      Con   1393    ;1950us Wide pulse max width
NMin      Con   214     ;300us  Narrow pulse min width
NMax      Con   571     ;800us  Narrow Pulse max width
MLen      Con   32      ;Number of data bits
PowerPin  Con   B4      ;LED on PowerPin
StopPin   Con   B5      ;LED on StopPin
PausePin  Con   B6      ;LED on PausePin
PlayPin   Con   B7      ;LED on PlayPin

;Decode Hitachi in MSB order
Power     Con   $06F9E817    ;Power button
StopD     Con   $06F9D02F    ;Stop button
PauseP    Con   $06F958A7    ;Pause button
Play      Con   $06F928D7    ;Play button

;Variables
;------------
CodeValue   Var   Long        ;holds decoded 32 bits
TempArray   Var   Byte(MLen)  ;holds measured widths
i           Var   Byte        ;counter
WArray      Var   Word(MLen)  ;for display
GoodFlag    Var   Byte        ;If 1, good data
FcnStr      Var   Byte(6)

;Initialization
;--------------
Clear
Input B0
;Have LEDs on B4...B7 pins
For i = B4 to B7
        Low i
Next
EnableHSerial
SetHSerial H115200
HSerOut ["P22-4",13,13]

Main
;------
```

```
                i=0
                ;We first wait for start pulse
                PulsIn B0,1,Main,1,WArray(0)

                ;Is it a valid start pulse?
                If Warray(0) < Smin Then
                        GoTo Main
                EndIf
                If Warray(0) > Smax Then
                        GoTo Main
                EndIf

                ;Valid start pulse followed by 32 data pulses
                Loop
                        PulsIn B0,1,Main,1,WArray(i)
                        ASM
                        {
                        banksel i
                        incf    i&0x7F,f
                        banksel PortB
                        }
                If i < MLen Then Loop

                GoSub CalculateCode
                GoSub GenerateValue
                If GoodFlag = 1 Then
                        GoSub ErrorCheck
                EndIf

                If GoodFlag = 1 Then
                        Pause 100       ;got valid code this time
                        FcnStr = "Unkwn "
                        ;Now we toggle pins based on received codes
                        If CodeValue = Power Then
                                Toggle PowerPin
                                FcnStr = "PWR   "
                        EndIf

                        If CodeValue = StopD Then
                                Toggle StopPin
                                FcnStr = "Stop  "
                        EndIf

                        If CodeValue = PauseP Then
                                Toggle PausePin
                                FcnStr = "Pause "
                        EndIf

                        If CodeValue = Play Then
                                FcnStr = "Play  "
                                Toggle PlayPin
                        EndIf

                        HSerOut [Str FcnStr\6, Str TempArray \MLen, |
                        "   MSB: ",iHex CodeValue,13]
                ELSE
                        ;Remove comments if you want to see bad data
                        ;HSerOut ["Error: ",Str TempArray \MLen,        |
                        "  Value: ",iHex CodeValue,13]
                EndIf

GoTo Main

CalculateCode
;------------
                ;Scan the measured width array and classify each as
                ;N (narrow), W (wide) or X (outside N or W window)
                For i = 0 to (MLen-1)
                        TempArray(i) = "x"
                        If (WArray(i) > WMin) AND (WArray(i) < WMax) Then
                                TempArray(i) = "W"
                        EndIf
```

```
              If (WArray(i) > NMin) AND (WArray(i) < NMax) Then
                      TempArray(i) = "N"
              EndIf
       Next
Return

GenerateValue
;------------
       ;Now calculate the Long=to the 32 bits
       CodeValue = 0
       ;set to 0 if we get bad data
       GoodFlag = 1
       ;Now convert W/N/X to values
       For i = 0 to (MLen-1)
              CodeValue.Bit0 = %0
              If TempArray(i) = "W" Then
                      CodeValue.Bit0 = %1
              EndIf
              If TempArray(i) = "x" Then
                      GoodFlag = 0    ;if don't want bad data
              EndIf                           ;debug could bail out here
              ;No need to check N or x, pre-populated with 0s
              ;read assuming sent MSB first. This may not
              ;be the case, but for our purpose order doesn't
              ;matter as we only look for unique values
              ;Data goes in at right, move to left
              ;every entry except last one
              If i <> (MLen-1) Then
                      CodeValue = CodeValue << 1
              EndIf
       Next
Return

ErrorCheck        ;check NOT bytes
;---------
       If (CodeValue.Byte3 + CodeValue.Byte2) <> $FF Then
              GoodFlag = 0
              HSerOut ["Bad Check Byte 3",13]
       EndIf

       If (CodeValue.Byte1 + CodeValue.Byte0) <> $FF Then
              GoodFlag = 0
              HSerOut ["Bad Check Byte 1",13]
       EndIf
Return
End
```

Since Program 22-4 duplicates much of Program 22-3, we'll concentrate on the differences.

```
PowerPin        Con       B4       ;LED on PowerPin
StopPin         Con       B5       ;LED on StopPin
PausePin        Con       B6       ;LED on PausePin
PlayPin         Con       B7       ;LED on PlayPin

;Decode Hitachi in MSB order
Power           Con       $06F9E817        ;Power button
StopD           Con       $06F9D02F        ;Stop button
PauseP          Con       $06F958A7        ;Pause button
Play            Con       $06F928D7        ;Play button
```

We've added several new constant definitions. We've aliased the pins **B4...B7** to more understandable names, and defined the four MSB-order key values we discovered in the output of Program 22-3.

```
       If GoodFlag = 1 Then
              GoSub ErrorCheck
       EndIf

       If GoodFlag = 1 Then
              Pause 100      ;got valid code this time
              FcnStr = "Unkwn "
              ;Now we toggle pins based on received codes
              If CodeValue = Power Then
```

```
                         Toggle PowerPin
                         FcnStr = "PWR    "
                  EndIf

                  If CodeValue = StopD Then
                         Toggle StopPin
                         FcnStr = "Stop   "
                  EndIf

                  If CodeValue = PauseP Then
                         Toggle PausePin
                         FcnStr = "Pause "
                  EndIf

                  If CodeValue = Play Then
                         FcnStr = "Play   "
                         Toggle PlayPin
                  EndIf

                  HSerOut [Str FcnStr\6, Str TempArray \MLen, |
                  "  MSB: ",iHex CodeValue,13]
         ELSE
                  ;Remove comments if you want to see bad data
                  ;HSerOut ["Error: ",Str TempArray \MLen,      |
                  "  Value: ",iHex CodeValue,13]
         EndIf
```

The individual pin-set/clear code is not complex. If we have received a valid 32-bit remote control signal, we test its value against the four "action" values. If an action value is received, we toggle the corresponding output pin, and define an output message string with the name of function. To avoid interpreting the built-in automatic repeat as repeated individual button presses, we've added a **Pause 100** delay statement after a successful decode.

We have slightly simplified the serial output routine to omit the LSB-first order value, but added the function identifying string. Here's a sample output from Program 22-4:

```
P22-4

PWR    NNNNNWWNWWWWWWNNWWWWWNWNNNNNWNWWW   MSB: $6F9E817
PWR    NNNNNWWNWWWWWWNNWWWWWNWNNNNNWNWWW   MSB: $6F9E817
Play   NNNNNWWNWWWWWWNWNWWNWNNNNWNWWW      MSB: $6F928D7
Play   NNNNNWWNWWWWWWNWNWWNWNNNNWNWWW      MSB: $6F928D7
Stop   NNNNNWWNWWWWWWNNWWWNWNNNNNWNWWW     MSB: $6F9D02F
Stop   NNNNNWWNWWWWWWNNWWWNWNNNNNWNWWW     MSB: $6F9D02F
Pause  NNNNNWWNWWWWWWNNWNWWNWNNNNWNWWW     MSB: $6F958A7
Pause  NNNNNWWNWWWWWWNNWNWWNWNNNNWNWWW     MSB: $6F958A7
Unkwn  NNNNNWWNWWWWWWNNWNNNNNWNNNNWNWWW    MSB: $6F9906F
Unkwn  NNNNNWWNWWWWWWNNWNWNNNNNWNWWW       MSB: $6F950AF
Unkwn  NNNNNWWNWWWWWWNNWNWNWNNNWNWWW       MSB: $6F96897
Unkwn  NNNNNWWNWWWWWWNNWWNNNWNNWNWWW       MSB: $6F99867
Unkwn  NNNNNWWNWWWWWWNNWNWWNNWNWNWWW       MSB: $6F918E7
Play   NNNNNWWNWWWWWWNWNWWNWNNNNWNWWW      MSB: $6F928D7
Play   NNNNNWWNWWWWWWNWNWWNWNNNWNWNWWW     MSB: $6F928D7
```

The four function output pins are initialized low, so the first time a function key is decoded, the associated LED illuminates. The next time, it is extinguished. Toggling continues with additional key presses. Receipt of a code other than one associated with the four defined keys triggers an "unkwn" message and the status of the four LED pins is unchanged.

I tested a breadboard version of Figure 22-11 with both a Sharp GP1U58x IR receiver module and a Vishay TSOP1238 integrated IR receiver module. Both modules are tuned to 38 KHz. I found reliable decoding at 30 feet with the GP1U58x receiver and at 36 feet with the TSOP1238 receiver. I tested the GP1U58x without an IR filter, but the TSOP1238 has an integrated IR filter.

Finally, of course the serial output data is intended to help developing, debugging and testing the program. It may be omitted if unnecessary.

Ideas for Modifications to Programs and Circuits

- Modify Program 22-4 to control an MCP42010 digital potentiometer using the circuitry and concepts developed in Chapter 21. With each press of the Vol+ button the MCP42010 should step up 1 dB; with each press of the Vol- button the MCP42010 should step down 1 dB. After each step change, store the current position in EEPROM. Upon program initiation, read the old position from EEPROM and set the MCP42010's value to the stored position.

- What else might you wish to control remotely? How would you go about it?

- Chapter 22 concentrates on receiving remote control signals. But, it's not difficult to transmit an REC-80 signal with a PIC. As a starting point, develop a simple "repeater" or range extender. After a correct signal is received—one that passes the check byte test—replay the signal to an IR diode. Here's a pseudo-code outline of how this might be accomplished:

```
Send a start pulse:
   LED-on for 9.5ms,
   LED-off for 4.5ms
Step through the array TempArray(i)
   Send LED-on for 800uS
   If TempArray(i) = W, pause for 1600uS
   If TempArray(i) = N, pause for 600uS
```

- Going beyond repeating a received remote control, use the RS-232 port to generate selected transmitted control codes. For example, a received + might generate a Vol+ IR command; a – might generate a Vol- IR command.

References

[22-1] Microchip Technologies, Inc., *Decoding Infrared Remote Controls Using a PIC16C5X Microcontroller*, AN657, Document No. DS00657A (1997).

[22-2] Sonmez, Mehmet Z., *Infrared Learner (Remote Control)*, Cypress MicroSystems Application Note AN2092, Rev. A (2002).

[22-3] Archer (Radio Shack), *Technical Data for Catalog No. 276-137*, (undated). Brief data sheet on the Sharp GP1U52X IR decoder modules.

[22-4] Vishay Semiconductors, *TSOP1100 WideBand IR Receiver Module for Remote Control Systems*, Doc. No. 82262, Rev 1.3 (2003).

[22-5] Vishay Semiconductors, *TSOP12.. IR Receiver Modules for Remote Control Systems*, Doc. No. 82013, Rev 11 (2003).

[22-6] Vishay Semiconductors, *TSOP48.. IR Receiver Modules for Remote Control Systems*, Doc. No. 82013, Rev 11 (2003).

[22-7] Sharp Electronics Components Group, *GP1UD26XK Series/ GP1UD27XK Series GP1UD28XK Series / GP1UD28YK Series Energy Saving Type Low Dissipation Current IR Detecting Unit for Remote Control*, (undated).

[22-8] An excellent starting point for IR remote control Internet research is http://www.epanorama.net/links/ir-remote.html. In addition to a brief explanation of the underlying technology, this site has dozens of useful cross-links to other sites with useful IR information.

[22-9] Innotech Systems, Inc., *A Primer on Remote Control Technology*, (undated).

[22-10] Seerden, Paul, *Using the Philips 87LPC76x microcontroller as a remote control transmitter*, AN10210, Philips Semiconductors (2003).

AC Power Control

We've learned many techniques for switching DC circuits; high side switching, low side switching, motor drivers and PWM circuits to mention a few. We even briefly looked at switching low voltage, low current AC with a bidirectional optically coupled FET in Chapter 3. This chapter explores controlling 120 V AC power line loads with a PIC. We'll use a 200 watt incandescent lamp load, but the techniques of this chapter can be extended to kilowatt level power switching.

Before we go any further, please stop and read and understand the following warning statement.

DANGER!

The circuits described in this chapter involve working with 120 V AC mains power.

SHOULD YOU COME INTO CONTACT WITH ENERGIZED CONDUCTORS, YOU MAY RECEIVE A DANGEROUS SHOCK, WITH THE POSSIBILITY OF SERIOUS INJURY OR DEATH BY ELECTROCUTION.

If you are not experienced working with 120 V power sources, please do not build the projects described in this chapter.

If you do build these projects, you must observe prudent safety precautions:

(a) Do not under any circumstances work on an energized circuit. Unplug the 120 V portion of the circuit from the power line before working on the circuit. Before touching any circuitry double check to verify that the power line is disconnected.

(b) Do not leave your breadboard circuit unattended. If you are not present, unplug the circuit from the power line.

(c) Use a 3-wire (grounded plug) connection and wiring.

(d) The design shows a 6A fuse and is intended for demonstration use with up to a 200-watt load. Do not increase the fuse size or operate with a greater power load.

(e) Do not make connections between your development board and the power control board except as shown in this chapter.

(f) Use good construction and wiring practice when you build the power control board. UNDER NO CIRCUMSTANCES INSTALL THESE PARTS ON YOUR DEVELOPMENT BOARD.

If you have any doubt about your ability to safely work on 120 V circuitry, read this chapter but don't build the associated circuitry.

CAUTION USING AN OSCILLOSCOPE WITH AC POWER CIRCUITS!

This chapter includes many oscilloscope records of both AC voltage and current. Under no circumstances should you use a PC soundcard oscilloscope to measure AC line voltages. The almost inevitable result will be a destroyed soundcard and damage to the rest of the computer circuitry.

Even if you have a laboratory oscilloscope, you must take care to ensure your oscilloscope is properly grounded and that you do not inadvertently allow the ground connector on the scope probe to contact an energized conductor.

Finally, you may be tempted to observe current waveforms using a small series resistor as a current shunt. Don't do it. Unless you know exactly what you are doing, you may damage your oscilloscope and expose yourself to dangerous risk. All of the current measurements in this chapter were taken with a Tektronix TCP202 clip-on current probe, which isolates the oscilloscope from the AC line.

Introduction to Triacs

Because they may be unfamiliar to many electronics experimenters, we'll first go through "Triacs 101."

Our principal AC power control device in this chapter is the triac. The acronym "TRIAC" is formed from the words triode (three-element) AC semiconductor switch. Although the rules of English usage say that acronyms are to be written in all capital letters, General Electric's source material and later usage favors treating triac as a normal word, written in lower case and capitalized only where required by sentence structure. And, since GE invented the triac, we'll follow its usage.

The triac is a member of the thyristor family, along with the programmable unijunction transistor (PUT), the bidirectional diode thyristor (diac) and, most notably, the silicon controlled rectifier (SCR). The common feature among all members of the thyristor family is that they have two stable states; conducting and nonconducting. Most thyristors require current injection or current removal from a gate connection to become conducting. Thyristor construction provides inherent internal positive feedback; once a thyristor starts conducting, it stays conducting. Conduction continues after gate current is removed until the current through the device reduces below some critical value (the "holding current") for a certain minimum time. Likewise, a nonconducting thyristor stays nonconducting until a trigger event occurs. (A few uncommon members of the thyristor family can be turned of by gate action under certain conditions.)

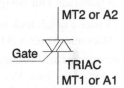

Figure 23-1: Schematic symbol for a triac.

Let's look at the particular scion of the thyristor family we'll be using, the triac. Figure 23-1 shows the schematic symbol for a triac. It has three terminals, the gate and two other terminals, identified as "main terminal 1" and "main terminal 2" or MT1 and MT2, respectively. Some European manufacturers identify these two terminals as "A1" and "A2" standing for anode 1 and anode 2. By convention, polarity for voltage reference is always with respect to MT1. Thus, if we say the gate is "positive" it means the gate is positive with respect to MT1.

The gate initiates a one-way action; a brief gate pulse turns the triac on. Once turned on, the gate cannot turn it off. Rather, the current through the triac must drop below some critical level for a specified minimum time in order for the triac to *commutate,*

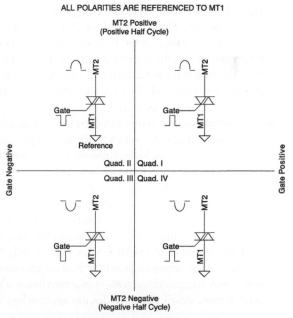

Figure 23-2: Triac triggering options.

or turn off. Since the 60 Hz AC power mains drop to zero 120 times a second, the triac is self-commutating under most circumstances. We'll look at the turn off characteristics of the triac later, as it may be impossible, without ancillary components, to turn off some types of triacs when connected to certain inductive loads.

A variant of Figure 23-2 appears in almost every triac reference, so we'll not break tradition. By convention, we map the gate and MT2 polarities on an X-Y axis. The Y-axis is the MT2 polarity while the X-axis is the gate polarity. The four possible gate/MT2 polarity combinations thus fall into four quadrants, identified by Roman numerals I, II, III and IV. Figure 23-2's quadrant assignments are summarized as follows.

Terminal MT2	Gate	
	−	+
+	*Quad. II*	*Quad. I*
−	*Quad. III*	*Quad IV*

Since both positive and negative gate drive can be used to turn on a triac, why classify the gate/MT2 relationship? First, not all triacs can be triggered in all four quadrants; some devices are not Q-IV triggerable. And, even those devices that are four-quadrant triggerable may not have identical gate drive requirements in each quadrant. Our design uses Q-I/Q-III triggering, which works with all triacs.

Let's take a quick look at the key parameters found in a triac's data sheet. We'll pay particular attention to ST Microelectronic's BTA12-600BW triac used in this chapter's circuits.

Gate current and voltage requirements—The current required to trigger conduction, I_{GT} and the voltage across the gate, V_{GT}, at this current level. The BTA12-600BW has a maximum required I_{GT} of 50 mA and V_{GT} of 1.3 V. The lower the I_{GT}, the easier it is to fire the triac. Some triacs are designed to interface directly with logic gates and can be triggered with a current of a few milliamperes. The BTA12-600TW, for example, has an I_{GT} of 5 mA. We will avoid sensitive triacs, as the tradeoff for increased gate sensitivity results in worse performance elsewhere. And, as we'll see later, we will not drive the gate directly from a PIC, so low gate current drive is not an issue in our design.

The BTA12-600BW is rated for triggering only in Quadrants I-II-III and not for Quadrant IV. The "B" version, BTA12-600B, is rated for all four quadrants. (We'll explain the difference between the BW "snubberless" triac and the B "standard" triac later in this chapter.)

Latching current—is the minimum current that the triac must carry in order to latch into conducting mode once the gate drive is removed. Its symbol is I_L. If the triac carries less than I_L current at the instant the gate drive is removed, the triac will revert to its high impedance, nonconducting state. I_L for the BTA12-600BW is specified as 70 mA.

Holding current—is the minimum current the triac must carry at all times in order to stay in conducting mode. Its symbol is I_H. When the instantaneous current drops below I_H, and no gate drive is applied, the triac reverts to its high impedance, nonconducting state. I_H for the BTA12-600BW is specified as 50 mA. The reversion to nonconducting state is, of course, what allows us to control AC current with a triac; the current drops to zero 120 times a second with 60 Hz power and if no gate drive is applied during zero crossing, the triac turns off.

dV/dt—With certain types of loads triacs can refuse to turn off. If the rate of change of voltage across the triac exceeds its maximum dV/dt rating, the triac will retrigger and will not commute. The BTA12-600BW's dV/dt rating is 1000 V/µs, significantly higher than the BTA12-600B's 400 V/µs. This is an area where the sensitive gate triac is at an enormous disadvantage; the BTA12-600TW—which requires only 5 mA triggering current—has a maximum dV/dt rating of only 20 V/µs. We'll explore this difference in more detail when we discuss snubberless versus standard triacs.

dI/dt—In addition to retriggering caused by high dV/dt, triacs are subject to retriggering due to high rate of change in current, dI/dt, a phenomena related to stored charge in the triac's junction capacitance. The BTA12-600BW's dI/dt rating is 12 A/ms without a snubber. Furthermore, there is interaction between dV/dt and dI/dt. We'll explore this in more detail when we discuss snubberless versus standard triacs.

Maximum current—The maximum RMS on-state current is $I_{T(RMS)}$ and the maximum nonrepetitive peak current is I_{TSM}. We'll see that even for an incandescent lamp load, the peak turn-on current can be many times greater than the steady state current. The BTA12-600BW is rated at 12 A steady state on current and 120 A nonrepetitive peak current.

V_{DRM}—The repetitive peak off-state voltage, i.e., the voltage that may be applied across the triac when it is in the off state before un-triggered conduction occurs. Applying more than V_{DRM} causes the triac to toggle into conduction; the triac is not necessarily damaged, but unexpected application of voltage to the load will occur.

Maximum gate current—is the maximum current that may be applied to the gate, I_{GM}. The BTA12-600BW's I_{GM} is 4 A.

V_T—is the on-state voltage across the triac. For the BTA12-600BW, the maximum V_T is 1.55 V.

Package—Triacs generate heat when switching high currents, so proper heat sinking is important. In the off state, the current through the triac is negligible—typically a few microamperes—and no thermal load is presented. When in the on state, the triac has a voltage drop of V_T. Suppose we use a BTA12-600BW to switch 12 A and the triac is used as a simple switch; off or on, with no power control implemented. When on, the triac will dissipate approximately 12 A * 1.55 V = 18.6 watts. The thermal resistance of a BTA12-600BW, junction to case, or $R_{TH(j-c)}$, is 2.3°C/watt. The maximum operating junction temperature, T_j, is 125°C. Let's see what size heat sink is required. We'll assume the ambient temperature is 35°C, and that we wish to keep the junction temperature below 100°C.

As we learned in Chapter 3, temperature and heat flow can be modeled analogously to electrical current and resistance. If the triac dissipates 18.6 watts, the temperature rise junction-to-case is 18.6 watts × 2.3°C/watt, or 43°C. In order for the junction temperature is not to exceed 100°C, the case temperature must not exceed 100° – 43°, or 57°C. If the ambient temperature is 35°C, the required heatsink thermal resistance can be calculated as (57°C – 35°C) / 18.6 watts, or 1.2°C/watt. Meeting this objective requires a serious heatsink, likely with forced air cooling. We'll keep the load down to 1.7A or less, where almost any small heatsink will suffice. (Repeating these calculations, we can determine that the total triac dissipation is 2.6 watts, corresponding to a junction temperature rise of 71°C for a small heatsink with a thermal resistance of 25°C/watt.)

You may have noticed our thermal calculation do not include a mica sheet or other electrical insulator between the heatsink and the triac. An external insulator isn't necessary because the BTA12-600BW has an insulating ceramic pad encapsulated in its TO-220 package. (The BTB series devices do not have an internal insulator.)

Snubberless versus Standard; dV/dt and dI/dt Issues

In concept, a triac is constructed from two thyristors in reverse parallel formed onto one piece of silicon. Without becoming overly involved in semiconductor physics, it is possible that at the moment of triac turn-off, unrecombined charge carriers from one thyristor cause induced gate charge in the other thyristor and thus falsely trigger conduction, a condition known as "commutation failure." Philips Semiconductor's Facts Sheet 13 [23-25] explains the commutation problem as follows:

The probability of any device failing commutation is dependent upon the rate of rise of reverse voltage (dV/dt) and the rate of decrease of conduction current (dI/dt). The higher the dI/dt the more unrecombined charge carriers are left at the instant of turn-off. The higher the dV/dt the more probable it is that some of these carriers will act as gate current. Thus the commutation capability of any device is usually specified in terms of the turn-off dI/dt and the re-applied dV/dt it can withstand at any particular junction temperature.

dI/dt

Calculating dI/dt is simple enough, assuming we are dealing with a linear load, that is, one composed of resistive, inductive and capacitive elements only. Assume that the RMS current through the triac and load is I. The instantaneous triac current i must be:

$$i = \sqrt{2}I \sin(2\pi ft + \phi)$$

where:

I is the RMS current through the triac and load (the square root of 2 multiplier is necessary since I is stated in RMS amperes.)

F is the line frequency, 60 Hz in North America.

t is the time, measured from some arbitrary starting point

ϕ is the phase shift resulting from a nonresistive load.

From introductory calculus, we know the derivative, di/dt, of this expression is:

$$\frac{di}{dt} = \sqrt{2}I \times 2\pi f \times \cos(2\pi ft)$$

We're interested in the maximum magnitude of di/dt, so we can set the $\cos(2\pi ft)$ term to 1, as $\cos(x)$ is always in the range $-1...1$. Substituting numerical values for a 60 Hz supply frequency, we find the maximum di/dt is:

$$\frac{di}{dt}_{\text{maximum}} = 533I \text{ A/sec or } 0.53I \text{ A/ms}$$

The BTA12-600BW's dI/dt rating is 12A/ms. This value corresponds to an RMS current of 22.6A, well over the devices continuous rating of 12A RMS. Hence, we need only be concerned with exceeding this device's dI/dt rating with loads that have short-term current surges, such as incandescent lamps or motors at startup.

You may be thinking "if the load is inductive, doesn't the voltage from the collapsing field induce a current in the triac that must be considered?" The answer is no, because the triac will commute only at essentially zero current and since an inductor's stored magnetic energy is proportional to the square of the current, there is no stored field when there is zero current. Hence there is no appreciable magnetic field to collapse at the instant the triac switches off.

dV/dt

The dV/dt rating relates to the rate of voltage change across the triac at the instant of commutation. If the triac is operating into a resistive load, we may use the approach we used to determine dI/dt to calculate dV/dt. The instantaneous voltage v across the triac is:

$$v = \sqrt{2}V \sin(2\pi ft)$$

where:

V is the RMS voltage across the triac (the square root of 2 multiplier is necessary since V is stated in RMS volts.)

F is the line frequency, 60 Hz in North America.

t is the time, measured from some arbitrary starting point. (No phase angle ϕ is required; we assume the phase reference is 0 with respect to the voltage.)

Likewise, dV/dt for a resistive load, 60 Hz power at 120 V RMS is:

$$\frac{dV}{dt} = \sqrt{2}V \times 2\pi f \times \cos(2\pi t)$$

The quantity dV/dt is evaluated at the time when the current is zero; for a resistive load, the current zero coincides with voltage zero—that is, at 0, 180, 360 degrees, and so on. Hence, the magnitude of $\cos(2\pi t)$ is 1 and the numerical value of dV/dt is:

$$\frac{dV}{dt} = \sqrt{2} \times 120 \times 2 \times \pi \times 60 \times 1 = 64 \times 10^3 \text{ V/s or } 0.064 \text{ V/}\mu s$$

The BTA12-600BW's dV/dt rating is 1000 V/μs, 10,000 times more than our calculated value for a resistive load. We can safely conclude that we are unlikely to have a dV/dt problem with a BTA12-600BW while switching a 120 V resistive load.

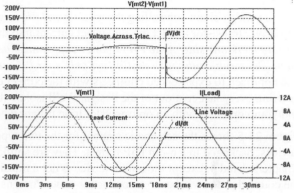

But, suppose we have an inductive load, such a motor. The current and voltage through the triac and the load are no longer in phase, as illustrated in Figure 23-3. At the time the triac snaps into non-conductive mode, the voltage across the triac is, in the worst case (for a phase angle of 90° between V and I) the peak line voltage, 170 V. And, the triac's switching speed may be on the order of a few microseconds or less. Hence, dV/dt can easily reach hundreds of volts/microsecond. The actual dV/dt depends on the load, stray inductance and capacitance and the triac's parameters.

Figure 23-3: dV/dt issues in triac turnoff of inductive load.

A common solution to dV/dt problems is to add an RC-snubber network across the triac, as shown in Figure 23-4. The RC network slows the rate of change in applied voltage and thus drops the dV/dt below the triac's rating. References [23-18] and [23-21] should be consulted for details on snubber design.

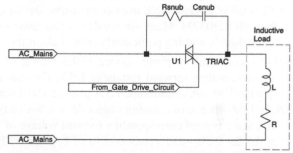

Figure 23-4: Typical snubber circuit for inductive loads.

Many new triacs, however, are built with such a high dV/dt rating that it is unlikely that an external snubber network will prove necessary even with highly inductive loads. We can see this by comparing the dV/dt ratings of the members of the BTA12-600 triac family:

Triac Variant	Description	Gate Drive (mA)	Maximum dV/dt (V/μs)
BTA12-600TW	5 mA Logic Level	5	20
BTA12-600SW	10 mA Logic Level	10	40
BTA12-600CW	35 mA Snubberless	35	500
BTA12-600BW	50 mA Snubberless	50	1000
BTA12-600C	25 mA Standard	QI-II-III 25/ Q-IV50	200
BTA12-600B	50 mA Standard	QI-II-III 50/ Q-IV: 100	400

The direct correlation between the gate sensitivity and the rated dV/dt is obvious; as the device is made more sensitive to permit lower gate current, it becomes more sensitive to false gate triggering caused by dV/dt factors. More importantly, the device we've chosen, the BTA12-600BW is 50 times less sensitive to dV/dt concerns than its logic level sister device. (All BTA12-600 devices are rated for 600 volts / 12 A continuous operation.) Thus, with almost all loads, even highly inductive loads, the snubberless device will properly commute without an external snubber circuit.

Triggering a Triac

To trigger the triac we must inject current into the gate. In the case of the BTA12-600BW, the minimum gate trigger current is 50 mA, and the peak rated gate current is 4 A, duration not to exceed 20 μs.

How do we go about controlling the triac's gate? We might connect it directly to the PIC, through an auxiliary driver transistor and let the PIC inject current into the gate. *This is a very bad idea for experimenting, as for it to work one side of the PIC's power supply must be connected to the hot power line conductor. Don't do it! The result could be to destroy not only your PIC and the ICD, but your computer as well.* We'll leave that approach to experienced engineers with the correct equipment to do it safely and where it is necessary to wring every last cent out of the design. We'll isolate the triac and the power line from the PIC, the ICD and your computer with an opto-isolator triac trigger device.

Figure 23-5 shows the schematic symbol for the two types of devices we'll use. Both are inexpensive, small devices, packaged in a 6-pin DIP format. They are similar to the optical isolators that we considered in Chapters 3 and 4, except the output device is not a transistor, but a small triac. When the LED illuminates, the coupled triac switches to a conductive state. Photons from the LED substitute for physical gate current.

By itself, the triac in the opto-isolator can't carry enough current to operate much of a load, but it's more than adequate to trigger a larger triac, such as our BTA12-600BW device. We'll use two opto-triac devices; a Vishay K3011P and a Fairchild MOC3032M. Many similar devices are available from other manufacturers and there's nothing particularly critical about these two devices.

Opto triac

K3011P—is a "random phase" device, with a typical LED current of 10 mA and a corresponding forward voltage of 1.5 V. The triac portion of the device is rated at 100 mA continuous, 1.5 A peak and will withstand 250 V.

MOC3032M—Is a zero crossing trigger device, with a typical LED current of 10 mA and a typical corresponding forward voltage of 1.25 V. The triac portion of the device is rated a peak current of 1 A, and will withstand 400 V.

Opto triac with zero crossing detector

Figure 23-5: Optically triggered triac types.

The difference between a "random phase" and a "zero crossing" device is that a random phase opto-triac may be triggered at any time during the voltage cycle. A zero crossing device has additional circuitry that inhibits the triac from being triggered, except when the AC voltage is at a zero crossing. When we examine phase and cycle control, we'll see why we need both type of devices.

Figure 23-6 shows how we use an opto-triac to trigger our main power control triac. Although the figure shows a zero crossing opto-triac, the same configuration is used for the random phase device. It's necessary to limit the opto-triac's output cur-

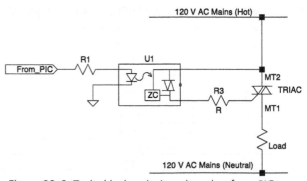

Figure 23-6: Typical isolated triac triggering from PIC.

rent with a series resistor, R1 in Figure 23-6. The maximum permitted peak current of the MOC3032M is 1.0 A, so we can calculate R3's value:

$$R3 = \frac{\sqrt{2} \times 120V}{1A} = 170\Omega$$

Since the triggering gate current is only 50 mA, this will give us more than enough gate drive. R3 should be a 1-watt resistor. The nearest 5% standard value is 180 ohms, which should be used. The worst-case instantaneous power dissipated by R3 is 160 watts (triac fired at 90 degrees) but the turn-on time of a typical triac is 5µs or less, at which time the driving voltage is removed—its source is the potential difference between MT1 and MT2, which drops to 1.5 V after the triac fires. Hence, the worst-case time averaged power dissipation of R3, assuming 120 pulses/sec, is 0.1 W. To avoid hot spots and possible resistor failure, however, R3 should be at least a 1-watt rated resistor.

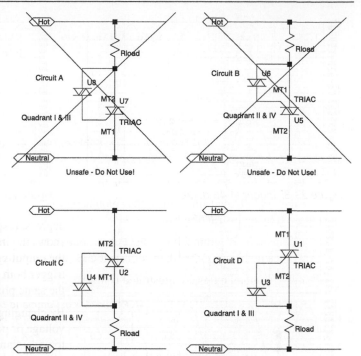

Figure 23-7: Possible triac connection arrangements.

Figure 23-7 is a simplified view of the four possible ways to connect the triac with the load and trigger drive. I prefer Figure 23-7's Circuit "D" for two reasons:

- As a matter of safety, circuits A and B are unacceptable. Contrary to good safety practice, when the triac is off, both sides of the load remain connected to the hot side of the 120 V supply mains. A switch, such as the triac, when off should isolate the load from the hot side of the mains supply and not expose the user to a source of unexpected voltage.
- Circuits C and D use the triac, when off, to isolate the load from the hot side of the mains supply and thus do not suffer the same safety problem as circuits A and B. Circuit C triggers the triac in quadrants II and IV, however, and is thus incompatible with the BTA12-600BW snubberless triac that cannot be triggered in quadrant IV. It would work with the standard member of the BTA12-600 family, however.

Phase and Cycle Control

We can use the triac in at least three control configurations:

- Off/on control, similar to a switch or relay;
- Phase control, where we turn the power on partway through an AC cycle;
- Cycle-control, where we turn the power off and on for an integer number of whole AC cycles.

Off-on control is easy to understand; the triac replaces a switch or a relay. The power to the load is either on or off for a "long" time. "Long" in this context means for more than a few cycles of AC.

Phase control is the most common control mechanism associated with triacs. To control voltage to the load, we delay firing the triac. For 60 Hz power, one half-cycle occupies 8.333 ms, so if we trigger the triac 4.167 ms after the start of a cycle, we turn the triac on halfway into the half-cycle, or at 90 degrees with respect to the full cycle. Figures 23-8, 23-9 and 23-10 illustrate triggering the triac at 45 degrees, 90 degrees and 135

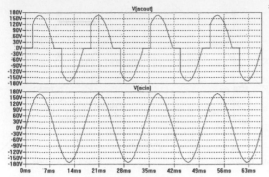

Figure 23-8: Trigger at 45 degrees.

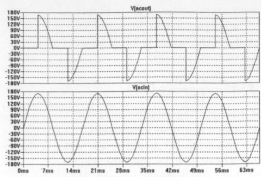

Figure 23-9: Trigger at 90 degrees.

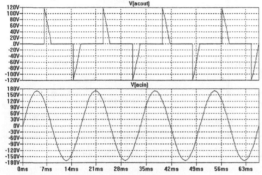

Figure 23-10: Trigger at 135 degrees.

degrees, respectively. In each figure, the bottom plot shows the input waveform and the upper plot shows the output voltage or current into a resistive load. (We trigger both the positive and negative half-cycles at the same phase delay after the zero crossings.)

By changing the triggering point, we change the voltage or power delivered to the load. It's important to understand, however, that the delivered voltage and power is not linearly related to the firing point phase. Figure 23-11 shows how the RMS and average voltages vary with respect to the firing angle for a 120 V source. The RMS value governs power-related applications, such as incandescent lamps, or resistive heaters. The average value relates to devices that function on average or peak voltage levels; such as an unregulated DC power supply, or most inexpensive analog and digital voltmeters. These meters may be calibrated in terms of RMS voltage, but actually respond to average voltage, computing RMS by multiplying the average voltage times a constant. So called "true RMS" meters are available that measure the real RMS of the waveform.

Power into a constant resistance (and a incandescent lamp is not a constant resistance) is proportional to the square of the RMS voltage, so half-power corresponds to not half RMS voltage, but rather 70.7% of RMS voltage. Figure 23-12 is a normalized version of Figure 23-11, and it adds the normalized power delivered to a constant resistance load.

Figure 23-12 shows two important points: First, to obtain almost complete load power control, we need only control the firing angle over the range of about 20…160 degrees and second, even in the range 20…160 degrees, power versus firing angle is not particularly linear. If our objective is to adjust power to the load, it requires a look-up table or other mechanism to translate the desired power level to the firing point angle.

Cycle Control involves sending groups of whole cycles of voltage to the load, as illustrated in Figure 23-13. The advantage of sending whole cycles is that all power switching is at zero crossings and hence little, if any, radio frequency interference is generated. In addition, it permits softer start of loads, such as incandescent lamps, with low cold resistance. Cycle control should not be used with unequal positive and negative half-cycles, as shown in Figure 23-14, since it introduces a DC component into the power line. A DC component can damage transformers and other devices connected to the power mains. Cycle control isn't usable with incandescent lamps, as the resulting flicker is highly objectionable.

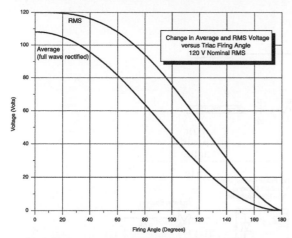

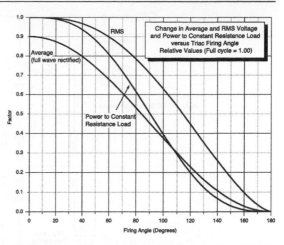

Figure 23-11: Change in average and RMS voltage versus trigger angle.

Figure 23-12: Normalized change in average and RMS voltage and power versus trigger angle.

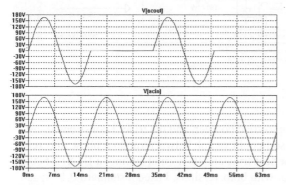

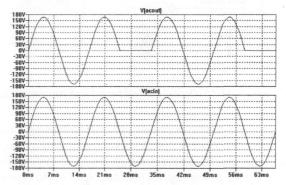

Figure 23-13: 50% power control by sending alternate cycles.

Figure 23-14: Cycle switching with DC component.

Power Control Board

Figure 23-15 shows the power control board and its interface with the Model 2840 PIC prototype board.

Under no circumstances should you build the power control board using a plug-in, solderless prototype board. Solderless prototype boards are designed to work with low current, low voltage circuits, and must not be used for high power, high voltage projects such as we are dealing with here. I'll discuss the "Manhattan-style" construction that I used to build my experimental power control board later in this section. Let's first go through the circuit design.

Power Control Board—Design

T1 is a 120 V to 10 V step-down transformer. I used a Stancor SW-210, but this is a noncritical part. Any small transformer with a secondary voltage between 5 and 15 volts RMS should work. The secondary draws less than 2 mA current, so a physically small transformer is perfectly adequate. (The SW-210 is rated at 10 V at 110 mA. A Hammond model 164D10 is a suitable substitute.) T1's secondary provides a zero-crossing reference to the PIC.

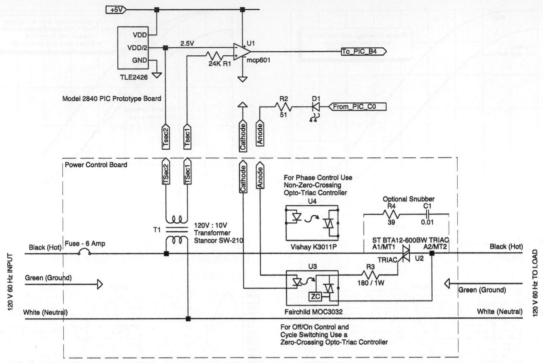

Figure 23-15: Power control board and connection to PIC.

The power control board also provides mounting for the opto-triac coupler, and the triac. I constructed my power control board with a socket for the optical coupler so I could switch between random phase and zero crossing models. I mounted both a BTA12-600BW snubberless triac and its normal sibling, a BTA12-600B triac, to the power control board so that I could experiment with both versions. I used a standard 0.1" spacing 3-pin header jack to connect the triac pins to the rest of the power control board, permitting me to unplug one triac and connect the other in a matter of seconds, after, of course, unplugging the power control board from the AC mains supply. If you are not interested in experimenting with both triacs, just install a BTA12-600BW model.

R3, the peak opto-coupler output current limiting resistor should be at least a 1-watt rated device. Although the average power dissipated by R3 is small, the peak power is 160 watts, which may cause localized heating of a physically small resistor, thereby leading to failure.

I included a 6 A fuse in my power control board as an additional safety measure. The suggested snubber is intended for experimentation and is not necessary with the recommended BTA12-600BW snubberless triac. *If you do experiment with a snubber, C1 must be rated for AC service, with a minimum of a 600 V rating as well as rated for the calculated current. A metallized polyester capacitor is often used in this application.*

This circuit is not intended for any purpose other than educating the reader. It omits many things necessary for unattended, long term operation, such as a smaller protective fuse for T1, and radio frequency interference suppression filters on both the input and output power connections.

Power Control Board—Construction

Figures 23-16 and 23-17 show my prototype board. I used "Manhattan style" construction, as described in Reference [23-24] but with one major change. Since the AC mains connections and the load connections use large wire, the super glue normally used to attach pads to the PCB substrate doesn't provide adequate mechanical strength. I used quick setting epoxy instead. I also attached the incoming power line to the substrate with a small clamp. I should have also clamped the output cable as well. Both the input and output cable are from an old computer power cord. The output cable is approximately 10 inches long and has attached to it a standard AC connection cable jack, obtainable at any good hardware store, which permits you to plug in a lamp or other load.

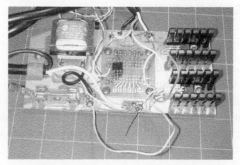

Figure 23-16: Power control board, view 1.

Figure 23-17: Power control board, view 2.

The opto-triac socket is mounted on an upside down (solder-side up) Radio Shack prototype board model 276-159. (You only need one-half of the 276-159 board.) I attached the 276-159 board with four 6-32 pan head machine screws and nuts, but you could epoxy it to the PCB substrate.

The connections between the power control board and the 2840 Development Board are made through four 18-inch long jumper wires. The jumpers are soldered to pads on the power control board with the other end stripped for insertion into the 2840 Development Board board.

The BTA12-600BW and BTA12-600B triacs are mounted with 6-32 pan head machine screws and small heatsinks. A Wakefield 231-137PAB or similar heatsink is adequate for the maximum recommended operation power level of 200 watts. Since the BTA12-600BW and BTA12-600B have internal insulators, separate mica or other insulator is not required. Most triacs do not have internal insulators, so if you substitute a different device, you may need to insulate the triac case from the heatsink and PCB substrate.

For a test load, I use a pair of 100 watt incandescent lamps, screwed into porcelain sockets mounted on a small wood base, as seen in Figure 23-18. It would be just as easy, of course, to use a standard table lamp.

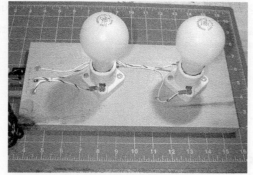

Figure 23-18: Lamp load for triac test.

Post Construction Checkout

After construction, use your ohmmeter to check your wiring in accordance with the following table. Before connecting your ohmmeter, again verify that the power control board is not connected to the AC mains power supply!

Measure Between	Should Read
Green – White Conductors on Input and Output.	> 1 Meg for both measurements. Should be an open circuit.
Green – Black Conductors on Input and Output.	> 1 Meg for both measurements. Should be an open circuit.
White Conductor to Black Conductor on Input.	Approx. 1000 ohms (depends on T1 as you are measuring T1's primary.)
White Conductor on Input to White Conductor on Output.	< 1 ohm. (Should be direction connection between these two points.)
Black Conductor on Input to Black Conductor on Output.	> 100 K (Measuring across triac; should be essentially an open circuit.)
T1 secondary to Green, Black and White conductors on both input and output side.	> 1 Meg. Should be an open circuit.
Opto LED connections to Green, Black and White conductors on both input and output side	> 1 Meg. Should be an open circuit.

Next, perform a no-conduction test. Plug the test load into the power control board's output socket. If you are using a lamp, make sure the lamp's switch is on. Position the power control board where it will not be disturbed and where you will not accidentally come in contact with it. Then plug the power control board's input power connection into the AC mains socket. Nothing should happen; the test load should remain un-powered. After this test, disconnect the power control board from the AC power mains.

Components on 2840 Prototype Board

Resistor R2 and LED D1 on the prototype board are in series with the LED contained within the opto-triac coupler. D1 provides a visual check of triac trigger pulses, while R2 limits the current through the LEDs to approximately 15 mA.

In order to know when to trigger the triac, the PIC must know when one AC half cycle ends and the next one begins. We use an MCP601 op-amp running open loop as a limiter to produce a square wave zero crossing output that follows the 60 Hz line by feeding its input with the secondary of T1. Figure 23-19 shows the input and output to the limiter. The MCP601 isn't a particularly fast op-amp and overdriving it as a limiter further slows its response. Figures 23-20 and 23-21 show a rise and fall time of around 3.6µs, but this is perfectly acceptable for our purpose, as at 60 Hz, 3.6 µs represents less than 0.1 degree.

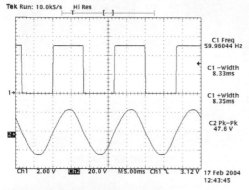

Figure 23-19: Limiter Input/Output Ch1: T1 Secondary; Ch2: MCP601 Output.

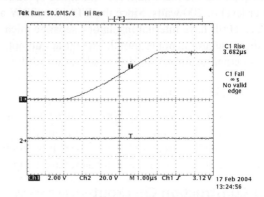

Figure 23-20: MCP601 Limiter Output Rise Time; Ch1: T1 Secondary; Ch2: MCP601 Output.

Microchip's specifications for the MCP601 state that it is permissible to drive the input pins beyond V_{DD}/V_{SS}, so long as the current drawn does not exceed 2 mA. R1, a 24K resistor, limits the peak current into U1 to 2 mA. (T1's secondary measures at approximately 48 V peak-to-peak. Small transformers with high resistance windings often have high unloaded secondary voltages.) When connecting T1's secondary, it's desirable, but not necessary, to have U1's output be high when the hot side of the AC line is positive. The circuits and software in this chapter work with either this phase, or the reverse phase, but all oscilloscope illustrations are taken with an in-phase connection. The minus input pin and the other side of T1's secondary is maintained at +2.5 V DC by a Texas Instruments TLE 2426 "rail splitter" IC. If a

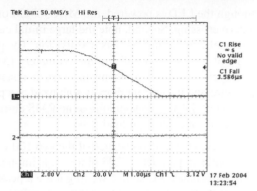

Figure 23-21: MCP601 Limiter Output Fall Time; Ch1: T1 Secondary; Ch2: MCP601 Output.

TLE2426 is not available, a resistive voltage divider could be substituted, comprised of two 470 ohm resistors in series, with 470 µF bypass capacitor from the common point of the two resistors to ground.

Programs

Program 23-1

We'll start with a simple objective; to blink the light bulb load off and on in a sequence—on for 5 seconds and off for 15 seconds. For Program 23-1, you should install the MOC3032M zero-crossing opto-triac. With the power control board *unplugged*, connect the transformer secondary jumpers and the opto-triac's LED pins as shown in Figure 23-15. Place the power control board in a safe location, where you will not accidentally touch any exposed wiring and where it not be knocked off your workbench. Connect the lamp load to the output socket. Then, connect the power board to the AC power mains and run Program 23-1.

```
;Program 23-1
;Off/On control with opto driver

;Constants
;----------
OnTime          Con     5000        ;on time (seconds)
OffTime         Con     15000       ;off Time (seconds)
OptoPin         Con     C0          ;pin with opto output

Main
        High OptoPin
        Pause OnTime
        Low OptoPin
        Pause OffTime
GoTo Main
End
```

Running Program 23-1 should flash the lamp in accordance with the **OnTime** and **OffTime** constant values. The program takes **OptoPin** (**C0**) high for 5 seconds, which triggers the opto-triac, in turn triggering the main triac, completing the circuit between the AC mains and the test load. When **OptoPin** goes low, the opto-triac is no longer triggered, the main triac commutes at the next current zero crossing and the test load is disconnected from the AC mains.

Figure 23-22 shows the relationship between the LED trigger current and the load current when we use a zero-crossing controller. Two points in Figure 23-22 deserve further discussion. First, pin **C0** goes high several milliseconds before the AC current commences. Since Program 23-1 applies power to the LED without reference to the AC zero crossing time, we see the zero crossing circuitry in the MOC3032M at work; although the LED is powered, the MOC3032M only triggers at a zero crossing. Second, note the large initial current inrush. Channel 2 in Figure 23-22 is 2A/division. The peak current draw of the 200 watt bulbs is approximately 13A. After about five cycles, the lamp filaments have reached steady state temperature and the current draw is stabilized at a peak-to-peak current of 4.4A, corresponding to an RMS current of 1.7A, as expected for a 200 watt load.

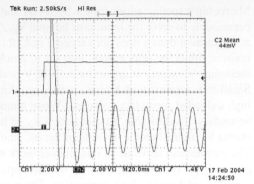

Figure 23-22: 200 Watt Incandescent Lamp Turn-On / Zero Crossing Ch1: Trigger to Triac; Ch2: Current Through Load.

I measured the cold resistance of the two 100 watt incandescent bulbs in parallel as 5.33 ohms. This implies that with a zero-impedance AC source we might see a peak current of 32A if the triac fires exactly at the peak voltage point in the AC cycle. (This can't happen, of course, so long as we use a zero-crossing trigger device.) Now, the AC mains wiring in my workshop isn't zero resistance as the circuit breaker, the AC wiring, the plug, the socket, the line cord and the triac all have some resistance. And, perhaps more importantly, the filament in a 100-watt incandescent lamp heats up quickly and its resistance increased several fold by the time the first AC waveform peak arrives. Based on a peak current of 13 A, the lamp resistance (both lamps in parallel) increased to 13 ohms by the time the peak voltage occurred, a 2.5:1 increase in 4.1 ms. (You can use this information to determine that after 4.1 ms the filament's temperature is 650°K and that its steady-state temperature is about 2500°K. The "official" color temperature of a 100 watt incandescent bulb is around 2850°K, so our measurement technique likely needs a bit more care than we casually give it here.)

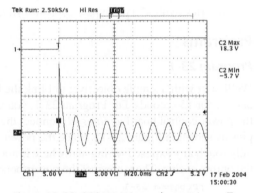

Figure 23-23: 200 Watt Incandescent Lamp Turn-On /Random Phase Trigger Ch1: Trigger to Triac; Ch2: Current Through Load.

Now, disconnect the power control board from the AC mains supply, remove the MOC3032M and install the K3011P random phase opto-triac. Reconnect the power and run Program 23-1 again. The results are identical, at least as far as the naked eye perceives the lamp illumination. However when we examine the start-up transient we see quite a difference from the zero crossing case. Figure 23-23 shows the current begins simultaneously with LED trigger, and that the peak inrush current is 18.3 A, nearly 50% above the zero-crossing case. Note that Figures 23-22 and 23 have different vertical scales.) Since the triac is triggered at a random phase point, we are almost guaranteed that the turn-on will be a nonzero voltage point. In the worst case, turn-on coincides with the AC line peak, 170 V, and the instantaneous current has a theoretical maximum value of 32A, as we calculated above, with limiting to a lesser value due to the resistances mentioned earlier. The higher inrush current with random phase starting is more stressful to both the filaments and the rest of the AC circuitry, including the triac. The fast current rise also may give rise to radio frequency interference.

Program 23-2

Next, let's try cycle controlling the load. Program 23-2 triggers the triac for a defined number of whole cycles and then idles for another defined number of cycles, and repeats this on/off ratio indefinitely.

First, disconnect the power control board from the AC mains and reinstall the MOC3032M zero-crossing opto-triac. Then, reconnect the power board to the AC power mains and run Program 23-2.

```
;Program 23-2
;Whole Cycle Control

;Constants
;----------
HTab          Con        9        ;horizontal tab
OptoPin       Con        C0       ;pin with opto output
GateDur       Con        500

;Cycle base must be a multiple of 4 to avoid
;DC in the output
CycleBase     Con        8        ;number of half-cycles in base

;Variables
;-----------------

PinStat       Var        Bit      ;Pin status
Counter       Var        Byte     ;number of half-cycles
DutyCycle     Var        Byte     ;percent cycles on
HalfCycOn      Var        Byte     ;number cycles on
HalfCycOff     Var        Byte     ;number cycles off

;Initialization
;-----------------
Low OptoPin
Output OptoPin
Counter = 0
DutyCycle = 50 ;0...100

HalfCycOn = (CycleBase * DutyCycle) / 100
HalfCycOff = CycleBase - HalfCycOn

Main
        Repeat
        Until PortB.Bit4 <> PinStat

        If Counter <=  HalfCycOn Then
                PulsOut OptoPin,GateDur
        EndIf

        If Counter >= (HalfCycOn + HalfCycOff) Then
                Counter = 0
        EndIf

        PinStat = PortB.Bit4
        Counter = Counter + 1

GoTo Main
End
```

We start by defining the "cycle base," that is, the number of half-cycles over which the off/on cycle is repeated. I've set the constant **CycleBase** to **8**, representing four full AC cycles.

```
CycleBase     Con        8        ;number of half-cycles in base

DutyCycle     Var        Byte     ;percent cycles on
HalfCycOn      Var        Byte     ;number cycles on
HalfCycOff     Var        Byte     ;number cycles off
```

We then declare three variables, **DutyCycle**, **HalfCycOn** and **HalfCycoff**. **DutyCycle** holds the percentage of "on" cycles in **CycleBase**, while **HalfCycOn** and **HalfCycOff** hold the number of half-cycles "on" and "off," respectively.

```
DutyCycle = 50 ;0...100

HalfCycOn = (CycleBase * DutyCycle) / 100
HalfCycOff = CycleBase - HalfCycOn
```

In the `Initialization` section, I've set `DutyCycle` at 50%. This means four half-cycles on, followed by four half-cycles on, then four half-cycles on, etc. `HalfCycOn` is computed as `4` and `HalfCycOff` is computed as `4`. For four cycles, the only legitimate duty cycle values are 0, 25%, 50%, 75% and 100%.

Let's now look at the main program segment.

```
Main
        Repeat
        Until PortB.Bit4 <> PinStat
```

As reflected in Figure 23-15, the output of the MCP601 limiter is connected to `B4`. The main program code monitors the status of `B4` until it changes state, which represents a voltage zero crossing in the AC mains supply. (Zero crossings occur every half cycle; which is the reason we cast the structure in terms of half-cycles, not full cycles.)

```
        If Counter <=  HalfCycOn Then
                PulsOut OptoPin,GateDur
        EndIf
```

If the number of zero crossings is less than the required number of on-cycles, Program 23-2 pulses the opto-triac to turn the main triac on for the just started half-cycle.

```
        If Counter >= (HalfCycOn + HalfCycOff) Then
                Counter = 0
        EndIf
```

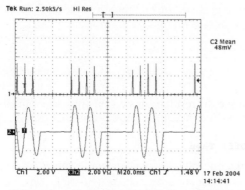

Figure 23-24: Cycle Control 50% Duty Cycle; Ch1: Trigger to Triac; Ch2: Current Through Load.

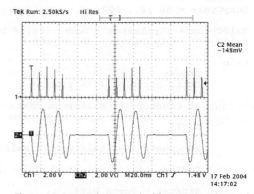

Figure 23-25: Cycle Control with Intentional DC; Ch1: Trigger to Triac; Ch2: Current Through Load.

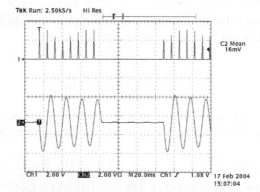

Figure 23-26: Cycle Control With Random Phase Opto DC; Ch1: Trigger to Triac; Ch2: Current Through Load.

If we have reached the end of our cycle duration, then reset **Counter**.

```
PinStat = PortB.Bit4
Counter = Counter + 1
```

Finally, at each zero crossing, refresh the status of **B4** and increment the byte variable **Counter**.

```
GoTo Main
```

Figure 23-24 shows the result. Two full cycles of current flow, followed by two cycles of no current. If the load is an incandescent lamp, by the way, this off/on cycle causes a very annoying flicker. Of course, if we were controlling a different type of load, such as a water heater, flicker is not an issue.

It's important that we keep the same number of positive and negative half-cycles in each on and off period to prevent DC current from flowing. Figure 23-25 shows the effect of setting **CycleBase = 10**, and **DutyCycle = 50**. We see two positive half-cycles and three negative half-cycles. The net average DC component of this waveform is 148 mA.

Now, disconnect the power control board from the AC mains supply, remove the MOC3032M and install the K3011P random phase opto-triac. Reconnect the power and run Program 23-2 again, but redefining **CycleBase** as **16** instead of **8**. Keep **DutyCycle** as 50. Figure 23-26 is identical with Figure 23-24; since Program 23-2 only fires the triac at a zero crossing, there is no difference in result between triggering with a random phase opto-triac or a zero-crossing opto-triac. One difference between Figures 23-24 and 23-26 is that the eight half-cycle off-period allows the filament to cool more than the four half-cycle period in Figure 23-24. Hence, we see a small inrush current increase for the first two cycles in Figure 23-26.

Program 23-3

Program 23-3 demonstrates phase control. Keep the K3011P random phase opto-triac in place for Program 23-3.

```
;Program 23-3
;Phase

;Constants
;----------
HTab        Con     9               ;horizontal tab
OptoPin     Con     C0              ;pin with opto output
ZeroPin     Con     B4              ;input pin
GateDur     Con     500             ;duration in microseconds
usPerDeg    Con     46              ;46usec per degree
OffsetCorr  Con     225             ;delay in us of program loop

;Variables
;------------------

PinStat     Var     Bit     ;Pin status
Degrees     Var     Byte    ;phase delay in degrees 0...180
TDelay      Var     Word    ;delay to fire

;Initialization
;----------------
Low OptoPin
Output OptoPin
;Degrees must >= 5 and <= 175 degrees
Degrees = 175

TDelay = Degrees * usPerDeg

If TDelay > OffsetCorr Then
        TDelay = TDelay - OffsetCorr
EndIf

Main
        Repeat
        Until PortB.Bit4 <> PinStat
```

```
          Pauseus TDelay
          PulsOut OptoPin,GateDur

          PinStat = PortB.Bit4

     GoTo Main
     End
```

We start by defining several constants:

```
     GateDur Con    500          ;duration in microseconds
     usPerDeg    Con    46           ;46usec per degree
     OffsetCorr  Con    225          ;delay in us of program loop
```

GateDur is the length of the gate drive pulse, in microseconds. For the particular opto-triac and triac that we use, 500 μs provides more than adequate gate drive duration. If **GateDur** is too long, so that you trigger late into the cycle, it's possible that gate drive will still be applied when the next cycle starts and the triac will unexpectedly latch into conduction. 500μs corresponds to approximately 11 degrees at 60 Hz, so we expect trouble for triggering delayed by 169 degrees or more. The constant **usPerDeg** is the duration of one degree of phase at 60 Hz, expressed in microseconds. (One cycle of 60 Hz is 16.667 ms, and 16.667 ms / 360 degrees yields 46.1 μs/degree.) The constant **OffsetCorr** is the length of time it takes for MBasic to execute the **Repeat...Until** statement loop in Program 23-3. This value must be considered to correctly determine the firing point and 225μs is based on a 20 MHz clock, measured as described later in this section.

```
     Degrees Var    Byte    ;phase delay in degrees 0...180
     TDelay Var     Word    ;delay to fire
```

We wish to define the firing point in degrees, but the PIC knows only time in microseconds. The variable **Degrees** holds the desired firing angle, while **TDelay** holds the time delay in microseconds between the zero crossing time and the firing time.

```
     ;Degrees must >= 5 and <= 175 degrees
     Degrees = 175

     TDelay = Degrees * usPerDeg

     If TDelay > OffsetCorr Then
          TDelay = TDelay - OffsetCorr
     EndIf
```

In the **Initialization** section, we hard code the phase angle by setting **Degree = 175**. (We'll try several values to see the difference and in Program 23-4, we enter new values through the serial port.) We then compute the time in microseconds between zero crossing and the desired firing point by the statement **TDelay = Degrees * usPerDeg**. We then must subtract the overhead delay associated with MBasic's function by the statement **TDelay = TDelay - OffsetCorr**. Since **TDelay** can't be negative, before making the overhead correction, we test its value before subtracting the overhead constant.

```
     Main
          Repeat
          Until PortB.Bit4 <> PinStat
          Pauseus TDelay
          PulsOut OptoPin,GateDur
          PinStat = PortB.Bit4
     GoTo Main
```

We use Program 23-2's approach to detecting zero crossings; continuously read and test pin **B4**'s status in the **Repeat...Until** loop. **B4** changes state only at a zero voltage crossing. Upon a zero crossing, we wait **TDelay** microseconds and then fire the triac for **GateDur** microseconds. We then repeat the wait/delay/fire process endlessly.

To determine the correct value of **OffsetCor**, we measure the time between a zero crossing and the time **OptoPin** goes high, with **Degrees** set to 0, as shown in Figure 23-27.

Figures 23-28 through 23-31 show the trigger pulses and resulting current waveforms for firing angles of 45, 90, 135 and 175 degrees respectively.

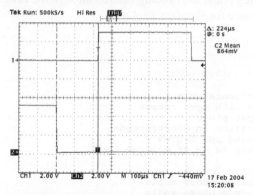

Figure 23-27: Measuring Overhead Delay in Repeat...Until Loop; Ch1: OptoPin Ch2: MCP601 Zero Crossing Detector Output.

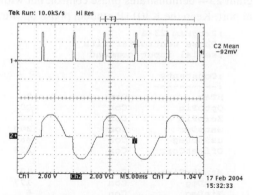

Figure 23-28: Current through Load, Program 23-3 @ 45 Degrees Ch1: Drive to Opto LED; Ch2: Load Current.

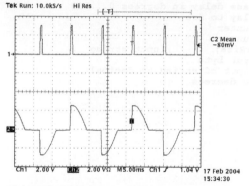

Figure 23-29: Current through Load, Program 23-3 @ 90 Degrees Ch1: Drive to Opto LED; Ch2: Load Current.

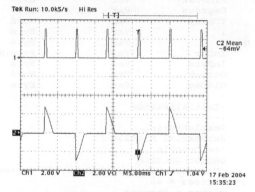

Figure 23-30: Current through Load, Program 23-3 @ 135 Degrees Ch1: Drive to Opto LED; Ch2: Load Current.

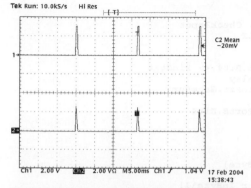

Figure 23-31: Current through Load, Program 23-3 @ 175 Degrees Ch1: Drive to Opto LED; Ch2: Load Current.

Program 23-4

Program 23-4 demonstrates phase control, but with the firing phase angle being entered through the RS-232 serial port. Keep the K3011P random phase opto-triac in place for Program 23-4.

```
;Program 23-4
;Phase controller with
;serial input command

;Constants
;----------
HTab           Con              9         ;horizontal tab
OptoPin        Con              C0        ;pin with opto output
ZeroPin        Con              B4        ;input pin
GateDur        Con              200       ;duration in microseconds
usPerDeg       Con              46        ;46usec per degree
OffsetCorr     Con              225       ;delay in us of program loop

MaxStrSize     Con              12        ;string length
MaxDeg         Con              177       ;maximum phase angle
MinDeg         Con              5         ;minimum phase angle

;Variables
;-----------------

PinStat        Var              Bit       ;Pin status
Degrees        Var              Byte      ;phase delay in degrees 0...180
TDelay         Var              Word      ;delay to fire
i              Var              Byte      ;counter
j              Var              Byte      ;counter
InStr          Var              Byte(MaxStrSize)        ;input string
InByte         Var              Byte      ;input byte
DecNum         Var              Word      ;output of string to decimal
OldDeg         Var              Byte      ;old degrees

;Initialization
;-----------------
EnableHSerial
SetHSerial H115200
Low OptoPin
Output OptoPin
;Degrees must >= 5 and <= 175 degrees
Degrees = 45
OldDeg = Degrees

GoSub SetDelay
i = 0
j = 0

Main
        HSerStat 3, CheckInput
        ReturnFromCI

        Repeat
        Until PortB.Bit4 <> PinStat
        Pauseus TDelay
        PulsOut OptoPin,GateDur

        PinStat = PortB.Bit4
GoTo Main

CheckInput
;----------
        HSerIn [InByte]
        ;Echo back
        HSerOut [Str InByte\1]

        InStr(j) = InByte
```

```
                j = j+1
                If j = (MaxStrSize-1) Then
                        j=0
                EndIf

                If InByte = 13 Then
                        If j > 0 Then
                                j = j-1 ;get rid of CR
                        EndIf
                        GoSub CRReceived
                EndIf

                HSerStat 3,CheckInput

        GoTo ReturnFromCI

        CRReceived
        ;--------------
                If j > 0 Then
                        GoSub StrToDecimal
                EndIf
                j = 0
        Return

        StrToDecimal
        ;--------------
                i = 0
                DecNum = 0
                While (InStr(i) >= "0") AND (InStr(i) <= "9")
                        DecNum = DecNum * 10
                        DecNum = DecNum + InStr(i) - "0"
                        i = i+1
                WEND
                Degrees = DecNum
                GoSub SetDelay
        Return

        SetDelay
        ;----------
                If (Degrees >= MinDeg) AND (Degrees <= MaxDeg) Then
                        TDelay = Degrees * usPerDeg

                        If TDelay > OffsetCorr Then
                                TDelay = TDelay - OffsetCorr
                        EndIf
                        HSerOut ["Was: ",Dec OldDeg,HTab,"Now ", |
                        Dec Degrees,13]
                        OldDeg = Degrees
                EndIf ;valid degrees
        Return
        End
```

The only changes between Programs 23-3 and 23-4 relate to entering and parsing the user input data.

```
        Main
                HSerStat 3, CheckInput
                ReturnFromCI
                Repeat
                Until PortB.Bit4 <> PinStat
                Pauseus TDelay
                PulsOut OptoPin,GateDur
                PinStat = PortB.Bit4
        GoTo Main
```

Program 23-4 uses code and techniques developed in Chapter 9 and used elsewhere in this book, so we'll not spend a great deal of time analyzing the code.

Note that we've added the statements:
```
        HSerStat 3, CheckInput
        ReturnFromCI
```

At the top of the main program loop, as we learned in Chapter 9, we use the undocumented function **HSerStat** to see if any characters have been received in the hardware serial input buffer. If one or more characters have been received, program execution jumps to **CheckInput**.

The code at **CheckInput**, **CRReceived** and **StrToDecimal** track the identically named routines found in Programs 9-2A, 9-2B and 9-3 and will not be further analyzed here.

After a valid number string has been received, terminated with a carriage return, subroutine **StrToDecimal** converts it to a binary number and stores the resulting value in the variable **Degrees**. Subroutine **StrToDecimal** then calls subroutine **SetDelay**, where the variable **Degrees** is further validated and used to set the triac firing angle.

```
SetDelay
;----------
        If (Degrees >= MinDeg) AND (Degrees <= MaxDeg) Then
                TDelay = Degrees * usPerDeg

                If TDelay > OffsetCorr Then
                        TDelay = TDelay - OffsetCorr
                EndIf
                HSerOut ["Was: ",Dec OldDeg,HTab,"Now ", |
                Dec Degrees,13]
                OldDeg = Degrees
        EndIf ;valid degrees
Return
```

First, we verify that **Degrees** is within the maximum and minimum range defined by the two constants **MaxDeg** and **MinDeg**, respectively. If valid, a new **TDelay** value is computed and a confirming message is sent to the user over the serial port. The new **TDelay** is implemented at the next zero crossing.

Here's a sample user interaction with Program 23-4 (I've added comments to the right):

45	← *user entered value of 45 degrees*
Was: 5 Now 45	
90	← *user entered value of 90 degrees*
Was: 45 Now 90	
177	← *user entered value of 177 degrees, greater than MaxDeg*
Was: 90 Now 177	← *no change actually took place, however*
95	← *user entered value of 95 degrees*
Was: 177 Now 95	
22	← *user entered value of 22 degrees*
Was: 95 Now 22	
111	← *user entered value of 111 degrees*
Was: 22 Now 111	
80	← *user entered value of 80 degrees*

Was: 111 Now 80

Ideas for Modifications to Programs and Circuits

- The user input in Program 23-4 is in terms of phase angle. For most applications, however, a user wishes the input value to track some controlled parameter. For example, for incandescent lamp dimming it would be more useful to enter 10 or 20 brightness steps, with each step roughly corresponding to equal changes in light output intensity. How would you do this? An tungsten filament incandescent lamp luminosity is nonlinear, with the relationship given by the following formula:

$$L_1 = L_0 \left(\frac{V_1}{V_0}\right)^{3.5}$$

where:

L_1 is the output in lumens at applied voltage V_1

L_0 is the output in lumens at the rated voltage, V_2

The reason the exponent is 3.5, not 4 as might be expected from the Stephan-Boltzman law, is that the resistance of the tungsten filament changes drastically with temperature and hence the current drawn by the lamp is a function of voltage. The power (and thus the filament temperature) thus does not vary exactly with the square of the applied voltage.

- How might you control inrush current? Might this be done by ramping up the applied voltage, such as starting at a 175 degree trigger angle and reducing it, say 5 degrees every full cycle until full voltage is reached.

- Program 23-4 has no full-on or full-off settings. Modify it to shut the down gate drive to the triac completely if the entered phase angle is 176 degrees or greater, and to apply full power if the entered phase angle is less than 5 degrees.

- For what type of loads might cycle control be suitable? For what loads is a random phase load suitable?

- How might the user interface be expanded to be more user-friendly? Which parameters does a user need to control? Would it be useful to store pre-programmed sequences, such as ramp-up and ramp down sequences, power levels and the like? These should be storable in EEPROM to preserve contents when power is removed from the PIC. How would you go about adding user-programmable sequences?

- It is possible to combine the triac circuitry of this chapter with the infrared remote control decoder of Chapter 22 and remote control lamps or other AC loads. How would you combine these concepts? You could also combine triac circuitry with temperature sensing devices, as we used in Chapter 12, to control a heater. What safety measures should be included to prevent a software or hardware problem from becoming a fire or other safety hazard?

References

[23-1] Teccor Electronics, *Fundamental Characteristics of Thyristors*, AN1001 (2004). Available for download at http://www.teccor.com/web/menuitems/downloads/appnotes.htm.

[23-2] Teccor Electronics, *Phase Control Using Thyristors*, AN1003 (2002). Available for download at http://www.teccor.com/web/menuitems/downloads/appnotes.htm.

[23-3] Teccor Electronics, *Thyristors Used as AC Static Switches and Relays*, AN1007 (2004). Available for download at http://www.teccor.com/web/menuitems/downloads/appnotes.htm.

[23-4] Teccor Electronics, *Miscellaneous Design Tips and Factss*, AN1009 (2004). Available for download at http://www.teccor.com/web/menuitems/downloads/appnotes.htm.

[23-5] Teccor Electronics, *Triggering and Gate Characteristics of Thyristors, Thyristor Design Guide* (Undated). Available for download http://www.teccor.com/web/menuitems/downloads/appnotes.htm.

[23-6] Grimm, William G., Using PIC Microcontrollers to Control Triacs, Avorex Designs (May 2003), available for download at http://www.avorex.com.

[23-7] Microchip Technology, Inc., *PICDIM Lamp Dimmer for the PIC12C508*, PICREF-4, Document No. DS40171A (1997).

[23-8] Vishay Semiconductors, *Optocoupler, Phototriac Output, 250 V V_{DRM}, K3010P / K3010PG Series*, Rev. 1.5 (Dec. 2003). Available for download at http://www.vishay.com.

[23-9] Fairchild Semiconductor Corp., *6-Pin DIP Zero-Cross Optoisolators TRIAC Driver Output (250/400 Volt Peak), MOC3031M, MOC3032M, MOC3033M, MOC3041M, MOC3042M, MOC3043M*, Document No. DS300256, (2001). Available for download at http://www.fairchildsemi.com.

[23-10] STMicroelectronics, *BTA/BTB12 and T12 Series, Snubberless, Logic Level and Standard 12A TRIACS*, Ed. 6A, (2002). Available for download at http://www.st.com/.

[23-11] Fairchild Semiconductor Corp., *Application Note AN-3003 Applications of Non Zero Crossing TRIAC Drivers Featuring the MOC3011*, Rev. 4.06,(2001). Available for download at http://www.fairchildsemi.com.

[23-12] Fairchild Semiconductor Corp., *Application Note AN-3004 Applications of Zero Voltage Crossing Optically Isolated TRIAC Driver*, Rev. 4.06,(2001). Available for download at http://www.fairchildsemi.com.

[23-13] Cox, Doug, *Interfacing to AC Power Lines, Application Note AN521*, Microchip Technology, Inc., Document No. DS00512C (1997).

[23-14] Parekh, Rakesh, *AC Induction Motor Fundamentals*, AN887, Microchip Technology, Inc., Document No. DS00887A (2003).

[23-15] Durbecq, X., *Control by a TRIAC for an Inductive Load: How to Select a Suitable Circuit*, Application Note AN308/0289, SGS-Thomson Microelectronics, (1995). Available for download at http://www.st.com/stonline/books/pdf/docs/3566.pdf.

[23-16] Ochoa, Alfredo, Lara, Alex & Gonzalez, Gabriel, *Momentary Solid State Switch for Split Phase Motors*, Application Note AND8007/D, ON Semiconductor, (1999). Available for downloading at http://www.onsemi.com.

[23-17] Zilog, Inc., *Digital Instant Water Heater, Application Note*, AN007101-0301, Document No. AP96Z8X0701. Available for download at http://www.zilog.com/docs/z8/appnotes/dig_ins_wh.pdf.

[23-18] Templeton, George, *RC Snubber Networks for Thyristor Power Control and Transient Suppression*, AN1048/D, Rev. 2, ON Semiconductor, (1999). Available for downloading at http://www.onsemi.com.

[23-19] Ochoa, Alfredo, Lara, Alex & Gonzalez, Gabriel, *High Resolution Digital Dimmer*, Application Note AND8011/D, Rev 0, ON Semiconductor, (1999). Available for downloading at http://www.onsemi.com.

[23-20] Ochoa, Alfredo, Lara, Alex & Gonzalez, Gabriel , *Solid State Control for Bi-Directional Motors*, Application Note AND8017/D, Rev 0, ON Semiconductor, (2000). Available for downloading at http://www.onsemi.com.

[23-21] Castagnet, T., New Triacs: *Is the Snubber Circuit Necessary?*, Application Note AN437/0899, STMicroelectronics (1999). Available for download at http://www.st.com/.

[23-22] Edmunds, Llew, *Heatsink Characteristics*, Application Note AN-1057, International Rectifier, Inc. (undated). Available for download at http://www.irf.com/technical-info/appnotes/an-1057.pdf.

[23-23] Philips Semiconductors, *Thyristors & Triacs — Ten Golden Rules for Success in Your Application*, Document No. 9397 750 00812 (1996). Available for download at http://www.semiconductors.philips.com/acrobat/various/SC03_TECHN_PUBS_1.pdf.

[23-24] Adams, Chuck, *Manhattan Building Techniques* (undated). Available for download at http://www.qsl.net/k7qo/manhattan.pdf. A follow up article, *Advanced Manhattan Building*, is available for downloading at http://www.qsl.net/k7qo/advmanart.pdf.

[23-25] Philips Semiconductors, *FS013 Understanding high-commutation triacs*, document order number: 9397 750 06504 (October 1999).

DC Motor Control

In earlier chapters, we learned how to control both bipolar and unipolar stepper motors. It's time now to look at a much older motor type, the permanent magnet DC motor. Simply turning a DC motor off and on with a PIC is not conceptually different than controlling any inductive load and the techniques of Chapter 3 may be used. We'll look at a more challenging aspect of DC motor control; a pulse-width modulated variable speed motor with tachometer feedback. Motors suitable for the techniques in this chapter are permanent magnet brush-type DC motors. Brushless and other specialized DC motors will not necessarily function with this chapter's circuits.

Before tackling this chapter, one small confession is in order. When tackling a new subject, whether in a formal classroom setting or at home, at the outset most of us are a bit puzzled. Then, at some magic moment, it all comes together and learning becomes a pleasure. In my undergraduate electrical engineering days, I had two classes where that gestalt moment never arrived. One was rotating machinery class (motors and generators) and the second was control systems. After the final exam in both, I said to myself, "I sure hope I never have to work with this stuff for real." As, by now you might suspect, this chapter deals with motors, generators and control theory. We'll dip our toes into the pool of learning, but we're definitely sticking to the shallow end.

Introduction to Control Theory

We're going to control a small DC permanent magnet motor with a built-in tachometer. The particular motor I used is a spindle drive motor, removed from an old 5-1/4" floppy drive. It's about 1" in diameter and 2-1/2" long and is rated at a maximum of 12 V, corresponding to about 6700 RPM. (Based on the pulley diameters, the motor ran at 2025 RPM when operating the disk spindle.) Most importantly, it has a built-in tachometer. This particular motor has an AC tachometer, which is less common in newer motors. But, along the way we'll see how the circuitry and program logic might be modified to work with other tachometer types. If you don't have an old floppy drive in your junk box, you may find a motor/tachometer in a scrap VCR, or from a surplus electronics dealer.

Controlling a DC motor is a classical application for an analog servo loop. (Indeed, DC motor speed control with electromechanical governors predates electronics.) The floppy drive used an analog feedback loop to control the spindle motor, and for many applications this remains an excellent, inexpensive solution for small motors. But, we're going to do things differently, and implement a PIC-based digital solution.

Although our experiments use a motor unattached to any load, usually we control a *system*, not an isolated element in a system such as a motor. The system may consist of something as simple as a fan blade, or as complex as a robotic arm. The feedback that our control hardware and software uses comes from the system, but not necessarily the motor. For example, to control cooling airflow over a heatsink, we might use the temperature of the semiconductors mounted on the heatsink as our feedback signal, not the RPM speed of the motor attached to the fan blades. Sometimes, of course, the motor speed is what we wish to control, such as the floppy spindle drive. (This isn't quite right is it? What we wish to control is the rotational speed of the floppy disk, not the motor. But, since the motor connects to the disk by a pulley and belt arrangement with a fixed diameter ratio, controlling the motor speed is almost as good as directly controlling the floppy disk spindle and it's much simpler mechanically.)

Since we've thrown the terms "feedback" and "control system" onto the table, let's define them. Figure 24-1 is right out of Control Systems 101; we have an input (the target pulse width), a summing point (the circle with the "X"), a control mechanism (the algorithm, drive circuitry and motor) and, finally, the feedback signal, an RPM-proportional output from the tachometer. Together, these elements constitute a control system.

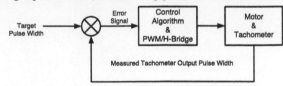

Figure 24-1: Control system for motor and tachometer.

Feedback—is a sample proportional to the output to be controlled. It doesn't have to be linearly proportional, but it's helpful if it is. In our motor control program, the feedback value is the width of the tachometer pulse, in microseconds. Since the tachometer output frequency is proportional to motor speed, the feedback value is linearly proportional to the reciprocal of motor speed.

Control input (target)—is the value that the feedback signal would be, if the output perfectly matched our objective. In our motor control program, the *target* is the width of the tachometer pulse that is output when the motor is running at exactly the desired speed. (Since the user inputs speed in RPM, we mathematically transform RPM into the target pulse width.)

Summing point—compares the feedback signal with the target and outputs the error (target minus feedback). The output of the summing point is the error signal. In our case, the summing point is a line of MBasic code where we subtract the measured pulse width from the target pulse width. This difference is the *error signal.*

Algorithm and control mechanism—consists of software and hardware that control the motor's speed in a way so as to reduce the error signal to zero. We'll try two different hardware arrangements and several software algorithms. All share the common feature, however, that the algorithm and control mechanism work such that the error signal drives the output in a direction opposite the error; if the error signal shows the speed is too high, the control mechanism changes "something" in a way that will reduce the motor's speed and vice versa. (The direction of change should be obvious; if the motor is running too fast and our control system takes that as an indication that it should run even faster, we have a runaway system, not one that controls the motor in any meaningful fashion.)

The control loop process may be clearer if we place it into pseudo-code:

```
Define target RPM
Calculate target pulse width corresponding to RPM

Main
Measure motor speed (tachometer output pulse width)
Error = Target Width - Measured Width
Motor Voltage = Motor Voltage + Function(Error)
GoTo Main
```

Function(Error) is the mechanism (algorithm and hardware combined) that takes the error signal and makes some adjustment to the voltage applied to the motor. (Don't worry about the algebraic sign of the error value; we'll make it all come out in the right direction in the adjustment function.)

With this pseudo-code in mind, let's look at the three parts of our algorithm:

Measure Motor Speed (Tachometer Output Pulse Width)

If we are to control the speed of a motor, we must have some way of measuring that speed. A wealth approaches to measuring motor speed have been devised, but four seem dominant:

- *DC tachometer*—outputs a DC voltage proportional to the motor speed. A DC tachometer is a miniature generator on the same shaft as the motor.
- *AC tachometer*—outputs an AC voltage with both its voltage and frequency proportional to the motor speed. An AC tachometer is a miniature alternator on the same shaft as the motor.
- *Optical or magnetic sensors*—produce pulses (ranging from one to several thousand) for each shaft revolution.
- *Sensorless*—Systems not employing tachometers or sensors that instead sense the motor's back electromagnetic force—which is proportional to motor speed. These are known as sensorless systems.

The oldest technology, but still widely used, is the DC tachometer, while the newest systems use optical shaft encoders, with 2048 or more positions per shaft revolution. These permit not only precise speed control, but also shaft positioning.

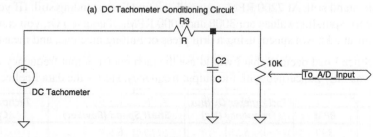

Figure 24-2 shows how we might condition a DC or an AC tachometer's output to be PIC compatible.

If our motor has a DC tachometer output, we read the voltage with the PIC's A/D converter. We must, however, filter the commutator noise with an RC low-pass filter, and use a voltage divider to scale the maximum output voltage to a value consistent with the reference voltage we use with the PIC.

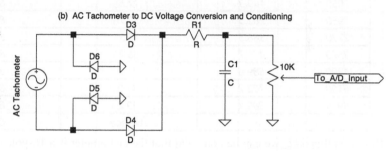

If our motor has an AC tachometer, recall that both the voltage and frequency are proportional to motor speed. Hence, we have two options; rectify and filter the tachometer output yielding a DC output proportional to motor similar to that from a DC tachometer; or measure tachometer output frequency. If we wish to measure the frequency, we must "square up" the sinusoidal tachometer output, converting it to a square wave compatible with the PIC's digital input. We'll look at the circuit of Figure 24-2(c) later in this chapter.

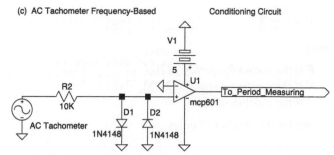

Figure 24-2: Analog tachometer conditioning circuits.

I decided to measure the tachometer's frequency. In theory, the tachometer's output frequency should be a more linear measure of shaft rotation speed than the DC rectified output. It also avoids other problems with the circuit in Figure 24-2(b), such as filtering the rectified tachometer output to remove the AC component, but still having the average DC value quickly respond to changing motor speed. As we shall see, however, measuring frequency is not without its problems. (We'll measure period, the reciprocal of frequency, but we'll generally speak about the tachometer's frequency, not period.)

Determining the Tachometer Constant

If we are going to set speed in RPM, but base our control mechanism on the tachometer output frequency, we must know the relationship between tachometer output and motor RPM. Since we don't have a manufacturer's data sheet on the motor, we'll determine it by experiment—spin the motor at a known speed and measure the tachometer output. I did this by applying a series of DC voltages between about 3.5 V and 12 V to the motor, measuring the corresponding RPM with a General Radio GR 1531AB Strobotac. (The motor speed would not stay constant, so the RPM values are "eyeball averages.") If you don't have a Strobotac handy, you can use a fluorescent lamp. Mark a line on the motor shaft and at 3600 RPM, the line will appear to stand still. At 7200 RPM, you will see two lines standing still. (If your power line frequency is 50 Hz, the corresponding values are 3000 and 6000 RPM, of course.) Or, you can spin the shaft of an un-powered motor at a known speed, using a drill press or milling machine, and measure the tachometer output.

Since I had decided that I would use the tachometer output frequency as the speed measuring mechanism, I was only concerned with the output frequency. Here's the data I collected:

RPM	Tachometer Output Frequency (Hz)	Shaft Speed (Rev/Sec)	Tachometer Output Cycles/Rev	Calculated Poles
390	52.5	6.5	8.08	16.15385
1050	149.3	17.5	8.53	17.06286
1960	262.5	32.7	8.04	16.07143
2720	368.5	45.3	8.13	16.25735
3520	471.7	58.7	8.04	16.08068
4300	589.6	71.7	8.23	16.45395
5100	702.2	85.0	8.26	16.52235
5950	809.1	99.2	8.16	16.31798
		Average	8.18	

Based on this data, we can be confident that the tachometer is a 16-pole device, and hence the tachometer constant is 8 cycles/rev. We determine this from the relationship between AC generator (or motor) poles, frequency and RPM, which is:

$$F = \frac{NP}{120}$$

where:

F is the output frequency in Hz
N is the rotational speed in RPM
P is the number of poles.

Solving for the number of poles:

$$P = \frac{120F}{N}$$

P must be an integer; you can't have a fraction of a pole in a motor or generator. Hence, P must be 16, and the tachometer constant is 8 cycles/rev.

Converting the Sinusoidal Tachometer Output to a Digital Logic Level Signal

MBasic's function **PulsIn** measures the duration of a logic-level signal (optionally starting with either the 0 to 1 transition or the 1 to 0 transition) and returns its length in unit time step periods, where one unit time step corresponds to seven machine instruction cycles. If you don't recall the details of **PulsIn**, review

Chapter 22 where we measure infrared pulse duration from remote control units. As a quick reminder, `PulsIn`'s syntax is:

```
PulseIn Pin, State {TimeoutLabel, TimeoutMultiple,} Var
```

We can't feed the sinusoidal output of the tachometer into a PIC's pin and expect `PulsIn` to work reliably. First, even at modest speed the tachometer's voltage output far exceeds the PIC's safe limit. But, suppose we can limit voltage excursions to a safe value through a series resistor and a 5.1 V Zener diode. Even so, we have another problem. In Chapter 4 we learned that the PIC reads transitions between logic 0 and logic 1 at different voltages, depending on whether the input pin is TTL compatible, or CMOS compatible or Schmitt trigger type. In any case, the transition point is at least 1.5 V or thereabouts at a minimum. Hence, the pulse width we would measure would depend upon both the amplitude of the tachometer output as well as its frequency. If you doubt this, suppose the tachometer output has a peak value of 1.4 V. The period won't be measured at all, since the gate will not change logic state. If the peak value is 1.6 V and if the input pin is a TTL compatible one with 1.5 V threshold, the logical 1 duration will be only be the length of time the instantaneous voltage exceeds 1.5 V.

The usual solution is to feed the sinusoidal signal to a "squarer," or a limiter—a circuit that converts a sine wave into a square wave. If the instantaneous output of the tachometer is positive, the squarer outputs logic 1; if it is negative, the squarer outputs logic 0. The recommended device to use as a squarer is a comparator. A comparator "compares" the voltage difference between its plus and minus inputs. If the voltage on the plus input exceeds the voltage on the minus input, the comparator outputs a logic 1; if the reverse is true, it outputs a logic 0. We're going to use an inexpensive MCP601 general purpose op-amp operating open loop instead, recognizing that its performance will be inferior to a true comparator in output slew rate and that the logic 0 to logic 1 transition will not necessarily occur as close to 0 V as it will with a true comparator.

Figure 24-3 shows the squarer circuit of Figure 24-2(c) as entered into a SPICE modeling program. The AC tachometer is modeled as a perfect sine wave source, with a series resistance of 200 ohms. I've set it for 300 Hz, with 14 V peak-to-peak output, corresponding to 2250 RPM. (The LTSpice simulator requires you to use the one-sided peak voltage as the amplitude specification for a sine source; hence Figure 24-3 shows 7 V.)

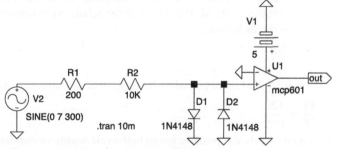

Figure 24-3: AC tachometer conditioner.

An op-amp operated open loop, i.e., without feedback, exhibits behavior similar to a comparator; if the voltage on the plus input exceeds that on the minus input (plus any offset voltage), the output is driven to the positive rail and vice versa. We've grounded MCP601's inverting input in Figure 24-3 so we expect its output to go to +5 V as soon as the tachometer's instantaneous voltage exceeds a few millivolts positive. The MCP601's input pins are limited to $V_{DD} + 0.3$ V and $V_{SS} - 0.3$ V, so we've added R2, D1 and D2 to limit maximum voltage excursion one diode drop (around 700 mV) positive or negative with respect to ground. We likely could omit D1 and D2, as the MCP601 has internal out-of-range protective circuitry so long as the input current does not exceed 2 mA, which is assured by the 10K series resistor.

Figure 24-4 shows the simulated tachometer input and MCP601 output voltages, while Figure 24-5 shows results from the real circuit operating at 1750 RPM. There is little difference between simulation and measured results.

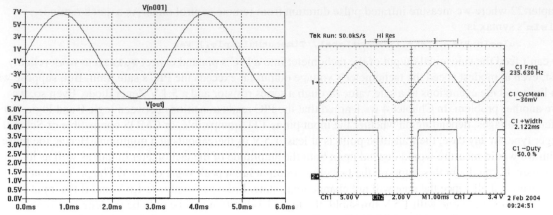

Figure 24-4: Simulation results.   Figure 24-5: Measured results.

Using MBasic's **PulsIn** function to measure the duration of the positive going output of the MCP601 gives us a value proportional to the motor RPM.

Error = Target Width – Measured Width

This part of our real code looks a lot like our pseudo-code. We calculate the error between the desired pulse width and the measured pulse width.

Let's look at how we calculate the desired pulse width. We conventionally work with motor speed in revolutions per minute, or RPM. We start with the relationship between frequency, the shaft speed and the number of poles in the tachometer:

$$F = \frac{NP}{120}$$

Since P = 16, and since period is the reciprocal of frequency, we rearrange this equation:

$$\frac{1}{F} = \frac{120}{16N}$$

We measure only the positive going half-cycle width, or one-half the period. And, beginning with version 5.3.0.0, MBasic's **PulsIn** function returns the width in unit steps, not microseconds. In order to simplify our computations, we'll work in microseconds and convert to unit steps later. Hence:

$$Period_{\mu s} = \frac{60 \times 10^6}{16N}$$

Let's see how this works. If our target RPM is 1750 we calculate the pulse width we expect **PulsIn** to measure:

$$Period_{\mu s} = \frac{60000000}{16 \times 1750} = 2143 \mu s$$

For a 20 MHz clock, **PulsIn** determines the width in units of 1.4 µs steps. 2143 µs corresponds to 1531 unit steps.

The Control Algorithm

We now have a measured pulse width, and the target pulse width, and the calculated error between the two.

Now that we have it, what do we do with the error value?

To keep things simple, we'll consider four possible control algorithms.

Bang-bang—The simplest possible control technique is to apply full voltage when the motor is running slow and cut the voltage to zero when the motor is running fast. This approach is often called "bang-bang" control, as the control parameters are always hard against the stops. Bang-bang control simplifies the electronics design—no variable voltage motor control is required. Rather, all we need is a simple on-off switch. Its corresponding drawback, of course, is that the motor's speed is always either too fast or too slow. We'll see how a bang-bang algorithm works in Program 24-4. Lest you have any doubt of the efficacy of bang-bang controls, until recently the United States Air Force's laser-guided bombs used bang-bang controlled steering vanes. (Paveway III was the first to use proportional control.)

Proportional (P)—The next step up from bang-bang control is to use the magnitude of the error signal to drive the correction amount. If the motor is very slow, then we increase the drive voltage a large amount; if the motor is just a little slow, we increase the drive voltage a smaller amount. This means, of course, that the controller must supply the motor from a variable voltage source. We'll use proportional control in Programs 24-5 and 24-6 where our source of variable voltage is the PIC's hardware pulse width modulation module. A proportional-only control can only approach, but never exactly match the set point, as the closer we get to matching it, the smaller the error and the smaller the corresponding correction. Although the residual error may be quite small, a proportional system cannot reduce it to zero.

Proportional and Integral (PI) and Proportional and Derivative (PD)—We can modify the algorithm to consider not only the error, but also the integral of the error, or the derivative of the error. We can think of a PI controller as looking at not only the instantaneous error, but also a history of recent errors. A PI controller can reduce the steady state error to zero. We could also look at the derivative of the error signal, that is, how has the error changed over time. PD systems have many implementation problems—for example, differentiation adds high frequency noise to the estimate, not a good thing at all. Consequently, PD controllers are seldom used.

Proportional, Integral and De-rivative (PID)—If we adjust the control variable based on the magnitude of the error, as well as its history and its rate of change, we have a PID controller. PID controllers are the most flexible and are very widely used. Indeed, about 90% of industrial control systems are PID.

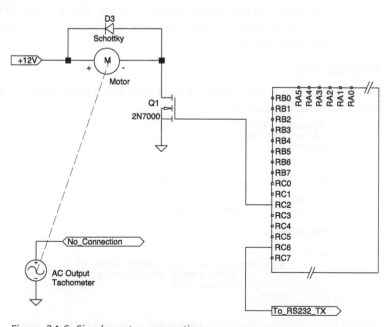

Motor Control Programs

We will not look at PI, PD or PID algorithms further as they are beyond an introductory book such as this one. Now, let's see how we implement DC motor control.

Figure 24-6: Simple motor connection.

Program 24-1

We'll start by testing our ability to turn the motor off and on. Figure 24-6 shows our circuit, which should be familiar from Chapter 3.

```
;Program 24-1
;Motor Off/On
;Constants
MtrPin Con    C2

;Initialize
Output MtrPin

Main
        High MtrPin
        Pause 1000
        Low MtrPin
        Pause 2000
GoTo Main
```

Program 24-1 turns the motor on for one second, and then turns it off for two seconds in an endless repeated cycle. When **C2** is taken high, the 2N7000 MOSFET is turned on, which makes its source-drain path look like a low value resistor thus completing the motor circuit. Since the motor windings are inductive, D3 is necessary to clamp the voltage spike when Q1 turns off. If you don't understand how this circuit works, please review Chapter 3.

We've selected pin **C2** purposefully; as we'll see in Program 24-2 the 16F876/877/A family have two hardware PWM modules, one connected to pin **C2** and the other to pin **C1**. By using **C2** for our operational test, we avoid having to rewire when using PWM techniques.

Figure 24-6 shows the motor's positive terminal connected to the positive end of the motor power supply. Many DC motors are not polarity sensitive, and cheerfully rotate in one direction with one polarity and in the reverse direction when polarity is reversed. A few DC motors, however, are mechanically constructed to function best while rotating in one direction. (Usually, this is related to the brush and commutator construction.) If your motor has polarity markings, you should follow them unless you know that it may safely be operated with reversed polarity.

Program 24-2

Now, let's vary the motor's speed through pulse width modulation. We keep the same connection arrangement as used in Program 24-1.

```
;Program 24-2
;Motor PWM Open Loop
;Constants

PWMC2        Con    0
PWMPrd       Con    2048

;Variables
DutyCyc      Var    Word

;Initialize

Main
        For DutyCyc = 0 to PWMPrd Step 25
            HPWM PWMC2,PWMPrd,DutyCyc
            Pause 1000
        Next
GoTo Main
```

We studied digital-to-analog conversion via pulse width modulation in Chapter 16, so we won't go through a ground-up review of the theory behind PWM, but instead we will recap the its fundamental concept. Let's look at three digital waveforms, A, B and C. Each waveform repeats after the eight intervals shown in the illustration.

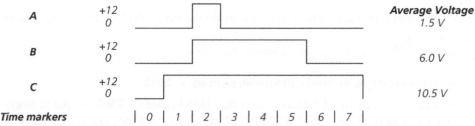

Suppose we apply these waveforms to a motor and place an average reading voltmeter across the motor windings. Waveform A will show 1.5 V, waveform B will show 6 V and waveform C will show 10.5 V. Our 12 V motor will turn slowly, if at all, when waveform A is applied, it will turn reasonably fast for waveform B and it will run nearly at maximum speed for waveform C. This assumes, of course, that each time period is relatively short, typically a few hundred microseconds or less.

MBasic supports hardware PWM through the function **HPWM**, with syntax:

```
HPWM CCPx, Period, Duty
```

CCPx—identifies the PWM generator to use. The 16F876/876A/877/877A devices, for example, have two hardware PWM generators, one hardwired to pin **C1** and the other hardwired to pin **C2**. **CCPx** may be a constant or a variable, and has two permissible values, 0 and 1:

CCPx value	PWM Generator Used	Pin Connection
0	CCP Module 1	C2
1	CCP Module 2	C1

In some printings, MBasic's User's Guide reverses the **CCPx** values required to select the CCP modules.

Period and Duty—These parameters are inter-related, so we'll look at them together. We'll use waveform B to illustrate the terminology used in **HPWM**.

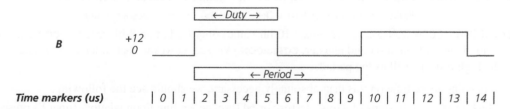

The **period** might more clearly be referred to as the "repetition period," i.e., the time between successive repetitions of the output waveform, measured from leading edge to leading edge. In the example, the period is 8 μs. **Duty** is the time that the output waveform is at logical high, 4μs in the example. A commonly used term is "duty cycle," which is the ratio of high to low periods. In the example, the duty cycle is 4 μs/8 μs or 0.50. The value is often expressed as a percentage, so our example has a duty cycle of 50%. (Occasionally duty cycle is used to express the percentage of time low, not high, so watch the context in which the term is used.)

Let's look in more detail how we go about setting **period** and **duty**.

First, the PICs fundamental time unit is *clock cycles*. I use a 20 MHz resonator in my development boards (Basic Micro ships the boards with a 10 MHz resonator.) Thus, my clock period is 50 ns (the stock 10 MHz resonator clock period is 100 ns):

$$T_{OSC} = \frac{1}{f_{osc}} = \frac{1}{20 \times 10^6} = 50 \times 10^{-9} \text{ seconds}$$

Suppose we wish the PWM period to be 100 µs. We calculate the number of clock cycles, N, required for 100 µs:

$$N = \frac{100 \times 10^{-6}}{50 \times 10^{-9}} = 2000$$

To set the PWM period to 100 µs, we invoke **HPWM** with **period = 2000**.

Second, **period** may not exceed 16383. Values greater than 16383 cause the PWM output to freeze and no compile time error message is returned. With a 20 MHz clock, this corresponds to a maximum period of 16383x50 ns, or 819.15 µs.

Duty is calculated the same way and has the same limits. Suppose we wish to generate our 100 µs period PWM waveform with a duty cycle of 25%, corresponding to an on time of 25 µs. We can calculate the required **duty** by repeating the calculation for N, but with 25 µs in the numerator, or we can simply multiply **period** by 25%, in either case yielding **duty = 500**.

Unfortunately, this summary of **period** and **duty** is not quite correct. As we expect a high level language to do, MBasic hides some of the complexity of the PWM modules in its **HPWM** function. This results in a confusing note captioned "Important Notes" in the User's Guide that attempts to explain how **duty** can be a 10-bit variable (0…1023) and yet accept 14 bit values (0…16383). Let's untangle this concept. I'll try to keep it as simple as possible, but not any simpler. (Consult Sections 8.3 of the 16F87x Data Sheet and 14.5 of the Midrange Reference Manual for more detail.)

Inside the 16F87x's PWM module, the period is determined by a combination of an 8-bit value held in register PR2, and the prescaler divide value for timer 2. The formula relating period to these two values is:

PWM Period = (PR2+1) * 4 * Tosc * (Timer 2 prescale value)

Timer 2's prescale value is either 1, 4 or 16, depending on how the prescaler control bits **T2CKPS1** and **T2CKPS0** are set in its associated control register **T2CON**.

The positive duration of the PWM output is set by a different mechanism, and its formula is:

PWM Duty (High) = (CCPRxL:CCPRxCON<5:4>) * (Timer2 prescale value)

The term (CCPRxL:CCPRxCON<5:4>) represents a 10-bit value comprised of an 8-bit register (**CCPRxL** and two bits (bits no. 5 and 4) from a second register, **CCPRxCON**.) We see, however, that timer 2's prescaler value affects the high time as well as the period.

Without spending more time on this than absolutely necessary, we should see the following:

- **Period** must move in steps of 4Tosc, 16Tosc or 64Tosc, depending upon whether timer2's prescaler is set for 1, 4 or 16.
- The period determining variable is held in an 8-bit register, which means we have only 256 possible periods for each of the three possible prescaler values, or 768 possible periods total.
- The duration of the high output is settable in steps of Tosc, 4Tosc or 16Tosc, depending upon whether timer 2's prescaler is set for 1, 4 or 16.
- The duty is always settable with four times the precision of the period.

Yet, **HPWM** lets us define both **period** and **duty** in terms of Tosc, not some multiple of Tosc. Or, does it? What actually happens is, yes, you enter values for period and duty in terms of multiples of Tosc. However, the function **HPWM** can only do that which the hardware permits. So, the values you enter for **period** and **duty** are converted to be compatible with the hardware PWM module, as shown in the following table.

Period Value				Duty Value	
HPWM period	Timer2 Prescaler	Step (Tosc)	Step (ns) for Fosc = 20 MHz	Step (Tosc)	Step (ns) for Fosc = 20 MHz
0...1023	1	4	200	1	50
1024...4095	4	16	800	4	200
4096...16383	16	64	3200	16	800

Suppose, for example, we wish to generate a PWM waveform with a period of 500 ns, and that we are using a 20 MHz oscillator. We calculate that period should be 10:

$$N = \frac{500ns}{50ns} = 10$$

Let's make the duty cycle 40%, or 200 ns. We have **period = 10** and **duty = 4**, so we invoke the hardware PWM module connected to pin C2 with **HPWM 0,10,4**. We then look at the output on pin C2 with an oscilloscope and get a surprise. We expect to see a 500 ns period, with 200 ns at logic 1 and 300 ns at logic 0. Figure 24-7 reveals, however, that the period is about 600 ns, not 500 ns, and that although the duty matches our expected 200 ns, the longer period reduces the duty cycle to 33.3%.

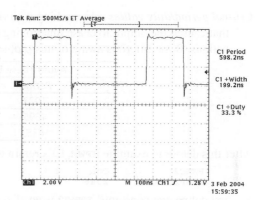

Figure 24-7: Output for **HPWM 0,10,4**.

How did this happen? For **period** between 0...1023, **HPWM** loads the 8-bit PWM register by first integer dividing **period** by 4. Hence, if we try different values of period, we see the output steps in units of four:

period	Expected Period (ns)	Actual Period (ns)
0	0	200
1	50	200
2	100	200
3	150	200
4	200	400
5	250	400
6	300	400
7	350	400
8	400	600
9	450	600
10	500	600
11	550	600
12	600	800

If we try different values of duty, say 3 instead of 4, so our function call is **HPWM 0,10,3,** we see that the duty portion of the waveform does reduce by 50 ns.

As summarized in the earlier table, we see step breaks at 1024 and 4096, where the **period** step size increases from 200 ns and becomes 800 ns and 3200 ns, respectively. And, since **duty** step size is always one-fourth the **period** step, the **duty** step increases from 50 ns to 200 ns and 800 ns at these break points.

The point to remember is that although **HPWM** lets us enter **period** and **duty** in terms of the oscillator clock period, in fact both **period** (always) and **duty** (if **period** > 1023) move in steps that are a *multiple* of the oscillator clock period.

We'll wrap up **HPWM** with two final points:

How to turn it off—Once invoked with a **HPWM** command, MBasic provides no way to turn the PWM module off. The PWM module may be turned off and the C1 or C2 pins returned to normal input/output use by clearing the associated control register with the following command.

```
CCP1Con = $0    ;turns off the PWM generator 1 (Pin C2)
CCP2Con = $0    ;turns off the PWM generator 2 (Pin C1)
```

Critical period/duty values—In working with HPWM, I found a few critical combinations of period and pulse that caused the output to cease functioning and stay frozen at logical 0. You should avoid these combinations. Basic Micro is aware of this problem and it is scheduled to be fixed in version 5.3.0.0.

Avoid These Bad Values of Period & Pulse	
Period	Pulse
1023	512 and 513
4095	2048 through 2055
16383	8192 through 8223

After this detour to examine **HPWM**, let's return to Program 24-2.

```
PWMC2        Con    0
PWMPrd  Con    2048
```

We first define two constants. **PWMC2** is an alias that we use to set the **CCPx** argument in **HPWM** to activate the PWM module attached to pin C2. **PWMPrd** is the **period** argument in **HPWM**. I've picked the value 2048, which corresponds to a period of 102 µs, or a frequency of about 10 KHz, and which is free of critical values that cause **HPWM** to fail. Why 102 µs and not some other value? As with most things in engineering, there is a trade off. To avoid acoustical noise, we must keep the frequency above the audible range. But, as we increase frequency stray capacitance and inductance start to be a problem, and switching losses increase. I selected 10 KHz as a reasonable compromise between these competing objectives.

```
Main
        For DutyCyc = 0 to PWMPrd Step 25
                HPWM PWMC2,PWMPrd,DutyCyc
                Pause 1000
        Next
GoTo Main
```

The main program loop ramps up the motor from stop to full speed by slowly varying **duty**. As we learned in Chapter 16, and summarized earlier in this chapter, by varying the on/off time of the PWM waveform, it, in effect, applies a variable voltage across the motor.

Run Program 24-2 and see what happens. Very low values of **duty** likely won't result in enough voltage to start rotating the motor shaft, so there may be a few seconds where nothing seems to be happening. Then, the motor will slowly start to rotate, gradually increasing speed to maximum. Then, it will stop and the process will begin again.

If you put an average reading meter, such as an analog voltmeter, across the motor connections, you should see the average voltage slowly ramp up. Figure 24-8 shows the voltage and current waveforms across the motor for a duty cycle of 18.1%, corresponding to an average voltage of about 2.2 V. Since the oscilloscope's channel 1 is connected to Q1's drain, a low voltage corresponds to current flow through the motor. Figure 24-9 shows the same parameters, but with a duty cycle of 92.7%, corresponding to an average voltage of 11.1 volts across the motor.

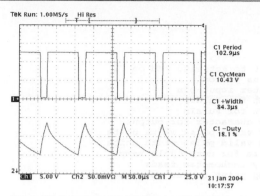

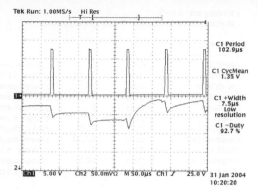

Figure 24-8: Program 24-2 Output Ch1: Q1 Drain Voltage; Ch2: Motor Current (50 mA/div).

Figure 24-9: Program 24-2 Output Ch1: Q1 Drain Voltage; Ch2: Motor Current (50 mA/div)].

Program 24-3

Now, let's wire up the tachometer circuit as shown in Figure 24-10.

Program 24-3 ramps up the motor, still in an uncontrolled, open loop mode, and reads the tachometer pulse width. We'll use Program 24-3 to verify that our tachometer circuit is functioning and see how it behaves, particularly at low speeds. We'll output several measured and computed values to the serial port for later analysis.

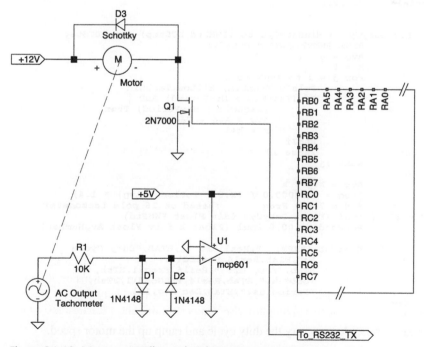

Figure 24-10: Motor controller and tachometer connection.

```
;Program 24-3
;Motor PWM versus Tach Frequency

;Constants
;-------------
PWMC2           Con     0           ;constant to activate PWM on C2
PWMPrd          Con     2048        ;for 102us period
TachPin         Con     C5          ;tachometer input pin
Low2Hi          Con     1           ;constant for PulsIn
Hi2Low          Con     0           ;constant for PulsIn
HTab            Con     9           ;the tab character for output
MinPeriod       Con     200         ;minimum valid pulse width units
MaxPeriod       Con     8000        ;maximum valid pulse width units
DCStep          Con     8           ;step size for duty cycle ramp
AvgNumber       Con     200         ;how many tachometer reads
MinDutyCyc      Con     368         ;starting point to get motor moving

;Variables
;-------------
DutyCyc Var     Word        ;used in HPWM for duty
TachWid         Var     Long        ;measured tachometer pulse width us
j               Var     Byte        ;counter
k               Var     Byte        ;counter
Avg             Var     Long        ;average pulse width
Freq            Var     Float       ;floating point tach frequency
PctGood         Var     Float       ;floating point % of good tach data
RealDC          Var     Float       ;floating point duty cycle %
RPM             Var     Float       ;computed motor RPM

;Initialize
;-------------
EnableHSerial
SetHSerial      H115200

Main
        For DutyCyc = MinDutyCyc to (PWMPrd-DCStep) Step DCStep
                HPWM PWMC2,PWMPrd,DutyCyc
                Avg = 0
                k = 0
                For j = 1 to AvgNumber
                        PulsIn TachPin, Hi2Low,TachWid
                        If (TachWid > MinPeriod) AND |
                                (TachWid < MaxPeriod) Then
                                Avg = Avg + TachWid
                                k = k+1
                        EndIf
                        Pause 10
                Next ;j

                Avg = Avg / k
                Freq = 1000000.0 / (2.0 * ToFloat (Avg) * 1.4)
                RPM = 7.5 * Freq         ;based on 16 pole tachometer
RealDC = 100.0 fmul (Float DutyCyc fdiv Float PWMPrd)
                PctGood = 100.0 fmul (Float k fdiv Float AvgNumber)

                HSerOut ["RPM: ",Real RPM \1,"Duty Cycle |
                        %:",HTab,Real4 RealDC\2, |
                        HTab,"Freq Hz: ",Real6 Freq\1,HTab, |
                        "Good %:",HTab,Real4 PctGood\1,HTab, |
                        "Period us:",HTab,Dec Avg,13]

        Next
GoTo Main
```

We start with a **For** / **Next** loop to vary the duty cycle and ramp up the motor speed.

```
Main
        For DutyCyc = MinDutyCyc to (PWMPrd-DCStep) Step DCStep
                HPWM PWMC2,PWMPrd,DutyCyc
```

There's no point stepping at duty cycles so low that the motor doesn't move. Hence, we start at **MinDutyCyc**, which for my motor is 368, corresponding to about 2.2 V average across the motor. We set the **HPWM** parameters as discussed earlier.

There's an important point about hardware-based PWM for motor control that makes it essential for motor control; it is a "set and forget" feature. This means we can set the hardware PWM to output a certain value and then execute other software functions, such as measuring the motor's speed and calculating the speed error and a correction factor while the motor continues to run. We can't do this with MBasic's software **PWM** procedure, as program execution stays with **PWM** until the statement is finished.

```
Avg = 0
k = 0
For j = 1 to AvgNumber
        PulsIn TachPin, Hi2Low,TachWid
        If (TachWid > MinPeriod) AND |
        (TachWid < MaxPeriod) Then
                Avg = Avg + TachWid
                k = k+1
        EndIf
        Pause 10
Next ;j
```

Next, we read the tachometer pulse width **AvgNumber** (200) times and add good pulse width values to **Avg**. (We'll later divide it by the number of valid counts to get the average pulse width.) We've predefined the constant **Hi2Low** to make **PulsIn** read the positive pulse width.

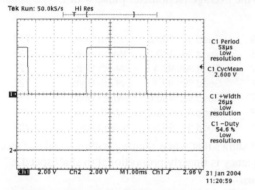

Figure 24-11: Clean Tachometer Pulse Ch1: Output of MCP601; Ch2: unused.

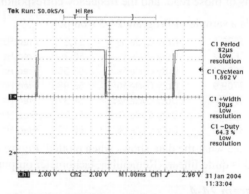

Figure 24-12: Dirty Tachometer Pulse Ch1: Output of MCP601; Ch2: unused.

When I developed Program 24-3, I found some tachometer output pulses were very clean, such as the one seen in Figure 24-11. However, particularly at lower speeds, some pulses had multiple spurious transitions at both the leading and trailing edge, as illustrated in Figure 24-12. The spurious pulses are enough to trigger **PulsIn** and give erroneous pulse width readings. Hence, we test the measured pulse width against minimum (**MinPeriod**, defined as 200 units) and maximum expected values (**MaxPeriod**, defined as 8000 units). Since I'm using a 20 MHz resonator in my 2840 Development Board, one **PulsIn** unit is 1.4 µs. If the measured pulse is within the window, we assume it is valid and add its value to **Avg**, incrementing the "good pulse" counter, **k**.

```
        Avg = Avg / k
        Freq = 1000000.0 / (2.0 * ToFloat (Avg) * 1.4)
        RPM = 7.5 * Freq        ;based on 16 pole tachometer
RealDC = 100.0 fmul (Float DutyCyc fdiv Float PWMPrd)
        PctGood = 100.0 fmul (Float k fdiv Float AvgNumber)
```

```
                    HSerOut ["RPM:   ",Real RPM \1,HTAB,"Duty Cycle |
                        %:",HTab,Real4 RealDC\2, |
                        HTab,"Freq Hz: ",Real6 Freq\1,HTab, |
                        "Good %:",HTab,Real4 PctGood\1,HTab, |
                        "Period us:",HTab,Dec Avg,13]
        Next
    GoTo Main
```

After we've read the tachometer **AvgNumber** times, we then compute several statistics.

First, we compute the average pulse width (in units) **Avg** by dividing the sum of the **PulsIn** values (held in the variable **Avg**) by the number of legitimate readings **k**. To save a variable, we store the average in **Avg**.

Freq is the frequency in Hz of the tachometer output, which we compute by taking the reciprocal of the pulse duration. Since we measure only one side of the pulse duration (the positive pulse), we must multiply the measured average duration, **Avg**, by two. To convert the duration from units to microseconds we multiply **Avg** by 1.4 and to convert from microseconds to Hz, we must compute **1000000.0 / microseconds**.

Next, we compute the motor speed **RPM**. As we discovered earlier, a 16-pole motor produces eight complete cycles per revolution. Hence, the motor speed in rev/sec is **Freq/8**. To convert to the more conventional rev/min, we multiply rev/sec by 60. Thus, **RPM = 60*Freq/8**, or, combining the constants 60 and 8, **RPM=7.5*Freq**.

Lastly, we compute the duty cycle stated in percent for the current **duty** value, the percentage of valid pulse widths of those read, and the frequency corresponding to the average tachometer pulse width.

Here's a sample of Program 24-3's output.

RPM: 3629.5	Duty Cycle %:	57.81	Freq Hz: 483.9	Good %: 100.0	Width Units:	738
RPM: 3649.2	Duty Cycle %:	58.20	Freq Hz: 486.5	Good %: 100.0	Width Units:	734
RPM: 3684.4	Duty Cycle %:	58.59	Freq Hz: 491.2	Good %: 99.4	Width Units:	727
RPM: 3709.9	Duty Cycle %:	58.98	Freq Hz: 494.6	Good %: 99.4	Width Units:	722
RPM: 3725.4	Duty Cycle %:	59.37	Freq Hz: 496.7	Good %: 100.0	Width Units:	719
RPM: 3767.3	Duty Cycle %:	59.76	Freq Hz: 502.3	Good %: 100.0	Width Units:	711
RPM: 3788.6	Duty Cycle %:	60.15	Freq Hz: 505.1	Good %: 97.9	Width Units:	707

Let's see what the data shows when plotted. Figure 24-13 shows that when running without load our motor's speed is quite linear. (Remember, tachometer output frequency is linearly proportional to shaft speed.).

Program 24-4

We've now demonstrated how to vary the motor speed through changing the PWM duty cycle and how to measure the motor's speed. It's time to put these elements together and regulate the motor's speed. Program 24-4 implements a simple bang-bang control loop.

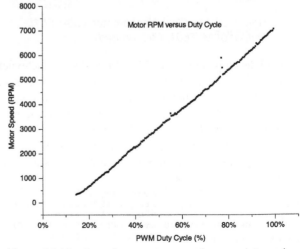

Figure 24-13: Open loop motor speed versus duty cycle.

```
;Program 24-3
;Motor PWM versus Tach Frequency
;Bang-Bang control

;Constants
;-------------
TachPin       Con    C5          ;input pin for period measurement
Low2Hi        Con    1           ;for PulsIn direction
Hi2Low        Con    0           ;for PulsIn direction
MinPeriod     Con    20          ;minimum units for valid pulse
MaxPeriod     Con    8000        ;maximum units for valid pulse
MotorPin      Con    C2          ;2N7000 driven from pin C2
RPM           Con    1750        ;Target RPM

;The AC tachometer produces 8 Hz per revolution
;To convert RPM to period, use the following
;formula

;Variables
;-------------
TachWid       Var    Word    ;Pulse width in usec
Target        Var    Word    ;target pulse width

;Initialize
;-------------
;Spin motor up for a bit to avoid
;start problems
;Assumes 1.4us/unit output from PulsIn
;The AC tachometer produces 8 Hz per revolution
;To convert RPM to unit steps, use the following
;formula
Target = 2678600/RPM

Output MotorPin
High MotorPin
Pause 100

Main
        ;We have to keep this loop as fast as possible
        ;Measure until we get a good data point
        Repeat
                PulsIn TachPin, Hi2Low,TachWid
        Until (TachWid > MinPeriod) AND (TachWid < MaxPeriod)

        ;If TachWid > Target the motor is running too slowly
        ;so we turn the drive on by taking output pin high
        ;Otherwise we are too fast, so we turn it off to slow down
        If TachWid > Target Then
                High MotorPin
        Else
                Low MotorPin

        EndIf
GoTo Main


;Initialize
;-------------
Target = 2678600/RPM
```

We start by computing the pulse width that corresponds to our target speed. (RPM is defined as a constant.) How do we arrive at this formula?

$$T_{units} = \frac{10^6}{\dfrac{RPM}{60} \times 8 \times 2 \times 1.4} \approx 2.6786 \times 10^6$$

RPMx8x2 yields the number of tachometer half-pulses per minute. The reciprocal of this quantity is the duration of each half-pulse in seconds. Multiplying by 10^6 results in microseconds and dividing by 1.4 converts microseconds to unit intervals, as returned by MBasic's **PulsIn** function with a 20MHz clock. Combining all the numerical constants, the result is **2678600/RPM**. For a target speed of 1750 RPM, the target pulse length is 2678600/1750, or 1531 unit steps.

We then spin the motor for a brief period so that the tachometer will provide accurate results. (Since the motor is driven through a 2N7000 low side switch, a high on the output pin turns the motor on.)

```
Output MotorPin
High MotorPin
Pause 100
```

The main program loop implements our simple control strategy; if the motor is running slower than the target, apply full power. If it's too fast, turn the power off.

```
Main
        Repeat
                PulsIn TachPin, Hi2Low,TachWid
        Until (TachWid > MinPeriod) AND (TachWid < MaxPeriod)
```

We don't want bad tachometer readings to cause erroneous speed commands, so we continue reading the tachometer until we get a value that's within the acceptable range.

```
        If TachWid > Target Then
                High MotorPin
        Else
                Low MotorPin
        EndIf
GoTo Main
```

Implementing our bang-bang algorithm is simple: If too slow (**TachWid > Targer**) then take **MotorPin** high, thereby applying full power to the motor. If not, the motor must be running too fast, so we remove power to it by dropping **MotorPin** low.

Pretty simple, isn't it? Let's see how well it works. Figures 24-14 and 24-15 show the voltage across the motor operating at 2000 RPM under two conditions; no load and loaded with friction created by pressing a finger lightly against the side of the spinning motor shaft. At no load, the motor maintains its speed with a voltage duty cycle of about 12.8%. When loaded with fingertip friction, the duty cycle increases to 76.3%. (Since the off/on pulses are so long—nearly 10 ms—the unloaded speed versus duty cycle graph of Figure 24-13 can't be directly applied.) I monitored the motor speed with a General Radio Strobotac while applying the load, verifying the motor stayed in speed regulation. (Remember; looking at Q1's drain, low corresponds to motor on.)

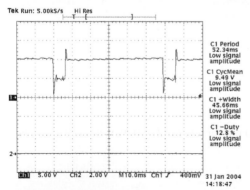 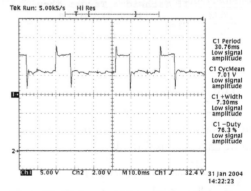

Figures 24-14: Unloaded Motor Voltage Bang-Bang Controller Ch1: 2N7000 Drain; Ch2: unused.

Figure 24-15: Loaded Motor Voltage Bang-Bang Controller Ch1: 2N7000 Drain; Ch2: unused.

Figure 24-16: *Speed Error—Bang-Bang Control*] shows motor speed error versus target motor speed. At speeds below 1500 RPM, the error climbs quite rapidly, a product of our need to wait for a valid speed measurement before making an off/on decision. At these slow speeds, we may have to make 10 or 20 or more measurements before finding a good one. We can take a measurement only once every half tachometer cycle, so even one bad measurement at slow RPM costs us several milliseconds. Thus, the control loop reaction time becomes slower than desirable and speed regulation suffers. (We suggest an alternative approach in the *Ideas for modifications to circuits and programs* section of this chapter.)

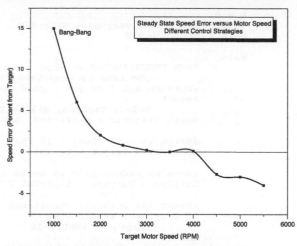

Figure 24-16: Speed error—bang-bang control.

Program 24-5

Although our simple bang-bang algorithm delivers adequate *static* speed control over the range 2000-4000 RPM, we did not test it for response to *dynamic* changes, such as quickly applied or quickly disengaged loads, or instantaneous speed change commands. We'll take a look at how proportional control reacts to speed change commands in connection with Program 24-5.

```
;Program 24-5
;Motor PWM with proportional control

;Constants
;-------------
PWMC2        Con      0         ;Set koutput for Pin C2
PWMPrd       Con      2048      ;For 20 MHz, PRF = 102.5usec
MinDutyCyc   Con      368
TachPin      Con      C5        ;PulsIn pin
Low2Hi       Con      1         ;Constant for PulsIn direction
Hi2Low       Con      0         ;Constant for PulsIn direction
MinPeriod    Con      200       ;Minimum period to be valid unit steps
MaxPeriod    Con      8000      ;Max period to be valid unit steps
                                ;above assumes 20MHz PIC clock
RPM          Con      1750      ;Target RPM
;The AC tachometer produces 8 Hz per revolution
;To convert RPM to period, use the following
;formula

;Variables
;-------------
DutyCyc Var   Word      ;Duty cycle for HPWM
TachWid       Var   Word   ;Measured tachometer width in unit steps
ErrorVal      Var   SWord  ;Error between desired and measured
K             Var   Word
Gain          Var   Byte

;Initialize
;-------------
;The AC tachometer produces 8 Hz per revolution
;To convert RPM to period, use the following
;formula
Clear
Target = 2678600/RPM
DutyCyc = MinDutyCyc              ;get motor spinning
Pause 500                        ;time to spin up
EnableHSerial
SetHSerial H115200
```

```
k = 0
;Approximation to vary gain with RPM
Gain = 100000 / RPM

Main
        HPWM PWMC2,PWMPrd,DutyCyc
                ;We have to keep this loop as fast as possible
        ;Measure until we get a good data point
        Repeat
                PulsIn TachPin, Hi2Low,TachWid
        Until (TachWid > MinPeriod) AND (TachWid < MaxPeriod)

        ;Error is + if slow, - if fast
        ErrorVal = Target - TachWid

        ;Have to reduce gain to avoid saturation
        DutyCyc = DutyCyc - ErrorVal / Gain

        ;Watch for boundary conditions
        If DutyCyc > PWMPrd Then
                DutyCyc = PWMPrd-10
        EndIf

        If DutyCyc < MinDutyCyc Then
                DutyCyc = MinDutyCyc
        EndIf

        If k < 1000 Then
                HSerOut [SDec ErrorVal,13]
                k = k+1
        EndIf
GoTo Main
```

At the outset, we see a new statement `Gain = 100000/RPM`. We'll see how this is used shortly. Otherwise, the variable and constant declarations and initialization follow the pattern set with earlier programs in this chapter.

```
Main
        HPWM PWMC2,PWMPrd,DutyCyc
```

We set `DutyCyc` at the top of our main program loop.

```
Repeat
        PulsIn TachPin, Hi2Low,TachWid
Until (TachWid > MinPeriod) AND (TachWid < MaxPeriod)
```

Next, we use the technique developed for Program 24-4 to measure the tachometer pulse width, discarding bad data.

```
;Error is + if slow, - if fast
ErrorVal = Target - TachWid
```

We now have arrived at the magic moment when we can calculate the error; it is simply the difference between `Target` and `TachWid`, that is, the target pulse width minus the measured pulse width. The sign of the error is positive if the motor is slow and negative if it is fast.

We now make some adjustment to `DutyCyc` to adjust the motor speed. We can't get a reliable corrected `DutyCyc` by simply subtracting `ErrorVal` from the current `DutyCyc` however. Why not? Our pulse period parameter, `PWMPrd`, is 2048. Hence, the maximum `DutyCyc` is 2048 and the minimum `DutyCyc` is 0. Suppose our motor is spinning at 1000 RPM, so the corresponding `DutyCyc` is around 400. For 1000 RPM, `TachWid` will be 3750 (μs). We now command it to go to 5000 RPM. For 5000 RPM, `Target` is 750, so `ErrorVal` is 750 – 3750, or –3000. Our first corrected `DutyCyc` will thus be 400 – (-3000) or 3400, which exceeds the maximum `DutyCyc` value of 2048. We could just clip out-of-range values and not permit setting `DutyCyc` to values greater than `PWMPrd` or less than 0. We'll do that as a safety measure, but it turns out that a better strategy is to divide `ErrorVal` by a value. This is the mathematical equivalent of reducing the gain in the feedback loop. Hence, we make the following adjustment in `DutyCyc`:

```
DutyCyc = DutyCyc - ErrorVal / Gain
```

And, as mentioned earlier, I've selected the divisor `Gain` to be a function of our target RPM:

```
Gain = 100000 / RPM
```

Let's finish looking at the rest of the code and then we'll examine our choice of `Gain` in more detail.

```
;Watch for boundary conditions
If DutyCyc > PWMPrd Then
        DutyCyc = PWMPrd-10
EndIf

If DutyCyc < MinDutyCyc Then
        DutyCyc = MinDutyCyc
EndIf
```

Even after we reduce the correction through dividing by `Gain`, it's still possible to generate an out-of-bounds `DutyCyc`, so we trap for any possible error and set `DutyCyc` to an appropriate value.

```
If k < 1000 Then
        HSerOut [SDec ErrorVal,13]
        k = k+1
EndIf

    GoTo Main
```

Finally, to help us understand and adjust Program 24-5, we write the first 1000 `ErrorVal` values to the serial port.

Let's see what happens if we change `Gain`. We'll first use a series of constant values, in the progression `Gain = 2,4,8,…,64`, and see what happens at start up with a target speed of 2000 RPM. We'll judge performance by looking at the first 1000 error values emitted over the serial port. A good algorithm should reduce the error to nearly zero, in a predictable fashion, reasonably quickly after startup.

Figure 24-17 shows very interesting results; if `Gain = 2`, the error never reduces to zero, but oscillates on either side of zero. Our feedback loop is under-damped, and unstable. Other `gain` values yield oscillating errors, particularly at start up, but gradually the error decreases. Although it may be difficult to judge from Figure 24-17, setting `Gain = 64` yielded quite satisfactory results.

But what happens when we set `Gain = 64` and try different target speeds? Is this a universal constant—at least for our motor and our load condition? Figure 24-18 suggests the answer is no. We see significant over-shoot for slow speeds and a slow, over-damped response for high speeds. It looks like 64 is a "magic" `gain` value only for speeds of 2000 to 4000 RPM.

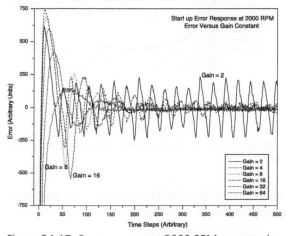

Figure 24-17: Start-up error at 2000 RPM versus gain.

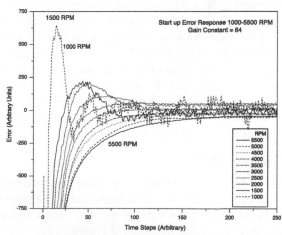

Figure 24-18: Start-up error with gain = 64 versus target motor speed.

Our conclusion is that `gain` should not be a constant, but rather should vary with RPM, being smaller with high RPM and larger for slow speeds. After some additional experimentation, I found a reasonable way to set `gain` is `Gain = 100000 / RPM`. (Since we divide the error by the value `Gain`, the effect of this equation is to increase the loop gain for higher speeds.) Figure 24-19 shows better performance over the range 1500 RPM – 5500 RPM than with any single fixed gain value.

What about static speed error? Figure 24-20 shows our simple proportional gain modification outperforms fixed gain versions when considered over the full range of speed variation.

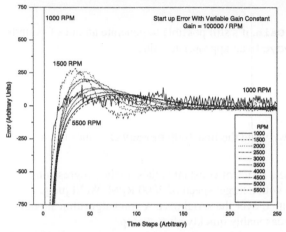

Figure 24-19: Start-up error with variable gain.

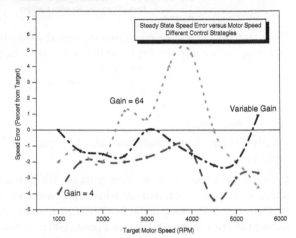

Figure 24-20: Proportional control static speed error versus changes in gain.

Finally, how does the more sophisticated proportional control compare with the bang-bang controller for static speed accuracy? Figure 24-21 shows the bang-bang approach more than holds its own if the target speed is between 2000 and 4000 RPM, but the overall winner is the proportional algorithm, with variable gain.

The performance data and algorithm structure is, of course, highly dependent upon the specific motor and the load. Don't look upon these programs as the solution to all motor control problems, but rather use these ideas as a departure point for your own work. A great deal of control system theory is devoted to calculating loop response and determining the required loop gain and time constants to produce the desired response. We'll leave these details for a future book.

Figure 24-21: Overall static speed control error comparison.

Program 24-6

We've mined about all the ore possible without a mathematical treatment of control loops, so our last program will look at proportional control using the algorithm of Program 24-5, but with the option to reverse motor direction. We'll use a SN754410 H-bridge as the motor drive, connected as shown in Figure 24-22.

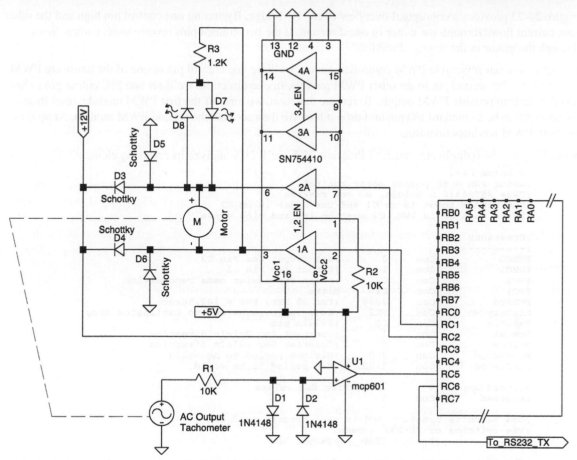

Figure 24-22: Reversing motor control circuit.

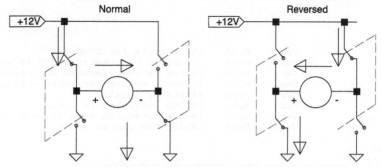

Figure 24-23: Conceptual view of H-Bridge.

Figure 24-23 provides a conceptual overview of an "H" bridge. By setting one control pin high and the other low, current flows through the motor in one direction. If the two control pins reverse state, current flows through the motor in the reverse direction.

To implement our reversible PWM controlled circuit, we'll tie one control pin to one of the hardware PWM modules and the second pin to the other PWM module. In one direction, we'll set one PIC output pin to low and the other to provide PWM output. To reverse direction, we turn off the first PWM module, reset its associated pin to be a standard I/O pin and drop it low. We then activate the second PWM module. At no time are both PWM modules operating.

Since Program 24-6 duplicates much of Program 24-5, we'll only analyze its reversing element.

```
;Program 24-6
;Motor PWM with period motor control
;Uses SN754410 H bridge, so can reverse
;One bridge input is on C1 and the other is on C2
;Hence when C1 is PWM, C2 must be low and vice versa

;Constants
;-------------
PWMC2           Con     0       ;Set output for Pin C2
PWMC1           Con     1       ;Output for Pin C1
PPC2            Con     C2      ;alias for easier name recognition
PPC1            Con     C1      ;ditto
PWMPrd          Con     2048    ;For 20 MHz, PRF = 102.5usec
MinDutyCyc      Con     512     ;Have to increase due to Darlington drop
TachPin         Con     C5      ;PulsIn pin
Low2Hi          Con     1       ;Constant for PulsIn direction
Hi2Low          Con     0       ;Constant for PulsIn direction
MinPeriod       Con     250     ;Minimum period to be valid
MaxPeriod       Con     11000   ;Max period to be valid


Forward Con     0       ;direction flag values
Reversed        Con     1

;Use hardwire constant RPM-in real program read it
;via switches or RS-232 command
RPM             Con     1750    ;Target RPM

;The AC tachometer produces 8 Hz per revolution
;To convert RPM to period, use the following
;formula. Period is in microseconds
;This also should be under user command
Target          Con     60000000 / (RPM * 2 * 8)        ;usec


;Variables
;-------------
DutyCyc                 Var     Word    ;Duty cycle for HPWM
TachWid                 Var     Word    ;Measured tachometer width in usec
ErrorVal                Var     SWord   ;Error between desired and measured
K                       Var     Word    ;Counter we use for various things
Gain                    Var     Byte    ;Actually 1/gain
Direction               Var     Byte    ;Which direction to rotate
OldDirection            Var     Byte    ;Prior direction stop if changes
PWMPin                  Var     Byte    ;Which PWM pin to use for PWM

;Initialize
;-------------
DutyCyc = MinDutyCyc
EnableHSerial
SetHSerial H115200
k = 0

;Approximation to vary gain with RPM
Gain = 100000 / RPM
```

```
                ;set direction
                Direction = Forward
                OldDirection = 99
                GoSub SetDirection
                ;May need more elaborate start up mechanism

                Main

                        HPWM PWMPin,PWMPrd,DutyCyc
                                ;We have to keep this loop as fast as possible
                        ;Measure until we get a good data point
                        Repeat
                                PulsIn TachPin, Hi2Low,TachWid
                        Until (TachWid > MinPeriod) AND (TachWid < MaxPeriod)

                        ;Error is + if slow, - if fast
                        ErrorVal = Target - TachWid

                        ;Have to reduce gain to avoid saturation
                        DutyCyc = DutyCyc - ErrorVal / Gain

                        ;Watch for boundary conditions
                        If DutyCyc > PWMPrd Then
                                DutyCyc = PWMPrd-10
                        EndIf

                        If DutyCyc < MinDutyCyc Then
                                DutyCyc = MinDutyCyc
                        EndIf

                        ;Uncomment to see start up error
                        ;If k < 1000 Then
                        ;       HSerOut [SDec ErrorVal,13]
                        ;       k = k+1
                        ;EndIf

                        ;This causes reversal every few minutes
                        ;In real program use a keyboard or switch command
                        k = k + 1
                        If k = 0 Then
                                Direction = Forward
                                GoSub SetDirection
                        EndIf

                        If k = $7FFF Then
                                Direction = Reversed
                                GoSub SetDirection
                        EndIf

                GoTo Main

                SetDirection
                        ;lf we have a change in direction, then we first
                        ;stop the motor with regenerative braking
                        ;Since tach is AC we can't determine direction from it
                        ;First flag for bad Direction value
                        If (Direction = Forward) OR (Direction = Reversed) Then
                                If OldDirection <> Direction Then
                                        ;Stop HPWM by resetting control registers
                                        ;We clear both and drop low to stop the motor
                                        ;Next application of HPWM re-initializes
                                        CCP1Con = $0
                                        CCP2Con = $0
                                        Low PPC1
                                        Low PPC2
                                        ;Give it time to stop
                                        Pause 400
                                        OldDirection = Direction
                                EndIf
```

```
                        ;Set the PWM pin to function and the other pin low
                        If Direction = Forward Then
                                PWMPin = PWMC2
                                Output PPC1
                                Low PPC1
                        EndIf
                        ;Likewise. If other, leave it the same
                        If Direction = Reversed Then
                                PWMPin = PWMC1
                                Output PPC2
                                Low PPC2
                        EndIf
                EndIf
        Return

        End
```

In order to cause a periodic reversal, I've added a simple counter routine.

```
        ;This causes reversal every few minutes

        ;In real program use a keyboard or switch command
        k = k + 1
        If k = 0 Then
                Direction = Forward
                GoSub SetDirection
        EndIf

        If k = $7FFF Then
                Direction = Reversed
                GoSub SetDirection
        EndIf
```

Every time the tachometer is successfully read, the counter **k** increments. When k = 32767, the direction reverses. Since **k** continues to increment, it eventually rolls over back to 0 after an additional 32678 steps and direction reverses again. The variable **Direction** is set to one of two predefined constants, **Forward** or **Reversed** and that value is used by subroutine **SetDirection** to configure the PWM modules and pins that implement the direction change.

```
        SetDirection
                If (Direction = Forward) OR (Direction = Reversed) Then
                        If OldDirection <> Direction Then
                                ;Stop HPWM by resetting control registers
                                ;We clear both and drop low to stop the motor
                                ;Next application of HPWM re-initializes
                                CCP1Con = $0
                                CCP2Con = $0
                                Low PPC1
                                Low PPC2
                                ;Give it time to stop
                                Pause 400
                                OldDirection = Direction
                        EndIf
```

Upon being called, subroutine **SetDirection** verifies that **Direction** contains a valid value, either **Forward** or **Reversed**. If not, control jumps to the end of subroutine and no operational code is executed.

If the direction being set is not the same as the current direction, we stop the motor and switch off the active PWM module. First, we disconnect both PWM modules by clearing their control registers. Then, we stop the motor by applying dynamic braking; by setting both control pins to the SN754410 low, both sides of the motor are connected to ground. In this condition, the motor acts as a generator into a low impedance load and rapidly comes to a stop. Lastly, we save the current direction in **OldDirection**.

```
                        ;Set the PWM pin to function and the other pin low
                        If Direction = Forward Then
                                PWMPin = PWMC2
                                Output PPC1
                                Low PPC1
```

```
                    EndIf
                    ;Likewise. If other, leave it the same
                    If Direction = Reversed Then
                              PWMPin = PWMC1
                              Output PPC2
                              Low PPC2
                    EndIf
            EndIf
      Return
```

Reversing direction requires us switch PWM modules. To accommodate this, we've changed the **HPWM** call by replacing the previous constant that identified which PWM module was to be active with a variable **PWMPin**.

```
            HPWM PWMPin,PWMPrd,DutyCyc
```

To set a direction, we assign **PWMPin** to one of two constants, **PWMC2** or **PWMC1**, which we have predefined to correspond to the values **HPWM** needs to make the PWM module attached to pin C2 or C1, respectively, active. We make the other pin an output and set it low, thus grounding the other side of the motor through the H-bridge. Figures 24-24 and 24-25 illustrate the reversal cycle.

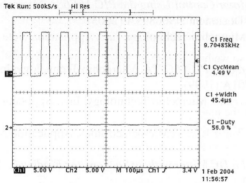

Figure 24-24: Clockwise Rotation Ch1: Input to SN754410 2A; Ch2: Input to SN754410 1A.

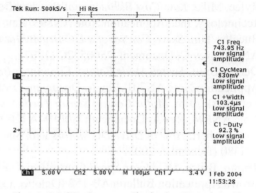

Figure 24-25: Counter-Clockwise Rotation Ch1: Input to SN754410 2A; Ch2: Input to SN754410 1A.

The SN754410 H-bridge is constructed with Darlington transistors in both the high side and low side switches. It thus has a greater voltage drop than our 2N7000 circuit, and the drop appears in both the high and low side motor connections. Hence, for a given duty cycle, less voltage appears across the motor. In our application, the increased voltage drop is not of major concern, but it might be for larger motors. Fortunately, a wide variety of H-bridge circuits are available, including those with less voltage drop and higher current ratings than the SN754410.

Ideas for Modifications to Circuits and Programs

* I've noted the problems measuring the tachometer output period at low speeds. Try implementing the full wave rectifier circuit of Figure 24-2(b). The PIC's A/D converter can then read the output of the rectifier. How should the values of R and C in Figure 24-2(b) be selected? What changes are necessary to the program logic and code to substitute an A/D reading for the period? How will the A/D and the program code handle the ripple out of the tachometer filter? How does the RC filter time constant affect speed control stability and time response?

* If we instead keep the frequency measuring approach, will a simple low-pass filter on the tachometer output improve the measurement accuracy by reducing noise? (I found that even a simple filter comprising a 0.033 μF capacitor across the tachometer terminals significantly improved low speed measurements, enabling stable results for as low as 400 RPM.)

- If you have a motor with a DC tachometer, how would you modify the program code and logic?
- If we wish to add an integral term to our error correction, we must integrate the error. One possibility is to use a simple numerical summing approach, using a circular buffer technique, using the approach we developed in Chapter 11 when we smoothed the A/D readings of the digital voltmeter project. Strictly speaking, however, numerical integration of this type assumes that all samples occur at equal time intervals. This will not be the case if the period measurements occur at differing intervals due to discarding bad readings. Is this a case where the rectification and A/D conversion approach works better? Additionally, the circular buffer discards old values at some point. This is actually good for our algorithm. How many readings should be summed in the buffer? How should the integral and proportional readings be individually weighted?

References

[24-1] Bishop, Robert H., ed., *The Mechatronics Handbook*, CRC Press LLC, (2002)

[24-2] Palmer, Mark, *Using the CCP Module(s)*, Microchip Technology, Inc., Application Note AN594, Rev. B, Document DS00594B (1997)

[24-3] Rylee, Mike, *Low-Cost Bidirectional Brushed DC Motor Control Using the PIC 16F684*, Microchip Technology, Inc., Application Note AN893, Rev. A, Document DS00893A (2003)

[24-4] Bucella, Tim, *Servo Control of a DC-Brush Motor*, Microchip Technology, Inc., Application Note AN532, Rev C, Document DS00532C (1997)

[24-5] Duane, Brett, *Switch Mode Battery Eliminator Based on a PIC16C72A*, Microchip Technology, Inc., Application Note AN701, Rev A, Document DS00701A (1999)

[24-6] Lepkowski, Jim, *Motor Control Sensor Feedback Circuits*, Microchip Technology, Inc., Application Note AN894, Rev A, Document DS00894A (2003)

[24-7] Sax, Herbert, *How to Drive DC Motors with Smart Power ICs*, STMicroelectronics, Application Note AN380 (December 2003)

[24-8] Trump, Bruce, *DC Motor Speed Controller: Control a DC Motor without Tachometer Feedback*, Burr-Brown Application Bulletin AB-152 (October 1999)

[24-9] Trump, Bruce, *Motor controller operates without tachometer feedback*, EDN, December 9, 1999, pages 170-171.

[24-10] Regan, Tim, *A DMOS 3A, 55V, H-Bridge: The LMD18200*, National Semiconductor, Application Note AN-694 (2002).

Bar Code Reader

Few things are more ubiquitous than bar codes. Indeed, they've become background noise and our eyes glide over a bar codes without actually seeing it. After you build a bar code reader, you'll seldom ignore one again.

My bar code reader is based around an inexpensive new-old-stock Hewlett Packard (now Agilent Technologies) wand, but similar wands manufactured by HP and other suppliers are widely available for a few dollars from electronics surplus dealers or on eBay. We'll learn a bit about bar codes in general and develop programs to read two bar codes, the "3 of 9" code and the "Universal Product Code, revision A," or UPC-A, the code used to identify retail merchandise.

Bar Codes "101"

The modern era of bar codes started with the grocery industry's desire to standardize product numbers in the late 1960s. After considering proposals from the leading computer manufacturers, IBM won a contract to design a coding system to implement the grocery industry's numbering system, proposing what we now know as UPC-A, or Universal Product Code, version A. Although the first UPC-A scanning system went into service in 1974, it was certainly not the first bar code. Indeed, the first United States bar code patent was granted in 1952, and the first practical use of bar codes was in the late 1960's to identify railroad boxcars. UPC-A, and its derivatives, however, are far and away the most common bar code today and have worldwide usage.

Code 3 of 9

Let's start with a simple bar code known variously as "Code 39" and "Code 3 of 9." To keep us speaking a common language, let's define a few terms.

Bars—are the dark part of the code.
Character—is a letter or digit or special symbol (such as * or $).
Element—generic reference to either a space or a bar.
Module—is the width of narrowest element (either a space or a bar) in the code
Quiet Zone—is space on either end of the bar code to isolate the bar code from its surroundings.
Spaces—are the lighter part of the code.

Figures 25-1 and 25-2 illustrate the components of Code 3 of 9 characters. Figure 25-1 shows a single character, the number 0, comprised of five bars and four spaces. Below the bar is a human readable representation of the digit. Figure 25-2

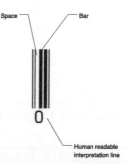

Figure 25-1: Character "0" (zero) in Code 3 of 9.

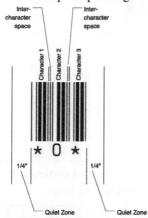

Figure 25-2: Character "0" (zero) in Code 3 of 9 with Required Start/Stop Symbols.

shows the minimum legitimate Code 3 of 9 sequence; the start character "*" followed by one or more user character(s), concluding with the stop character "*." The bar code is isolated from its surroundings by a quiet zone at both the left and right hand sides. Code 3 of 9 also requires a space between each character (we'll see that inter-character spaces are built into the UPC-A code). Finally, although the exact ratio may not be obvious, Figures 25-1 and 25-2 clearly have two element widths, which we'll call "wide" and "narrow." Nominally, wide elements are three times the width of narrow elements, although a high-density version of the code reduces this ratio to 2.2:1.

How then are characters mapped into varying bar widths and locations in Code 3 of 9? Here are the rules:

1. Wide elements are three times the width of narrow elements
2. All characters start and end with a bar and are comprised of five bars and four spaces.
3. All characters have two wide bars and one wide space. All other elements are narrow.
4. All bar codes must start and stop with the start/stop symbol, the asterisk "*."
5. A narrow space inter-character separator appears after each character (except, of course the stop symbol).
6. Wide elements are binary 1s and narrow elements are binary 0s. Whether the element is a bar or a space does not affect its binary value.
7. Any number of characters may be in a single bar code, although practical length considerations limit it to 6-10 characters in most circumstances.

We see source of the term "3 of 9," as each character has nine elements, of which three are wide. If we consider the narrowest element as having a width of one unit, the width of a complete character is 15 units, consisting of three elements of width three and six elements of width one, totaling 15 units. The *modulus* width of a character is thus 15. If we consider the inter-character space as part of the character, its modulus becomes 16.

The bar value and space value are separately evaluated according to the following table. (We won't go into why the wide and narrow elements are placed where they are; we are interested in reading the code so we will take its structure as given.)

| Bar Value | | Space Value | | | | | | | | Decimal |
| Binary | Dec Value | 1 | 2 | 4 | 7 | 8 | 11 | 13 | 14 | Binary |
		0001	0010	0100	0111	1000	1011	1101	1110	
00000	0				%		+	/	$	
10001	17	K	A	1		U				
01001	9	L	B	2		V				
11000	24	M	C	3		W				
00101	5	N	D	4		X				
10100	20	O	E	5		Y				
01100	12	P	F	6		Z				
00011	3	Q	G	7		-				
10010	18	R	H	8		.				
01010	10	S	I	9		SPACE				
00110	6	T	J	0		*				

Let's see how a particular bar value is decoded. Consider the character shown in Figure 25-1, which has the following structure:

1	2	3	4	5	6	7	8	9	10	11	12	13	14	15
1	2	3		4			5			6		7		8	9

I've added the modulus numbers (1...15) and element numbers (1...9). We decode the bar and space values (MSB first order) as:

Bars	Width	N	N	W	W	N	*Decimal Value*
	Binary	0	0	1	1	0	6
Spaces	Width		N	W	N	N	
	Binary		0	1	0	0	4

We look for a character with bar value 6 and space value 4 and see that it is the number 0.

Code 3 of 9 is widely used as an internal standard for warehousing and inventory control and it has been adopted by the United States military for its LOGMARS logistics management program.

UPC-A

Let's now turn to UPC-A, as illustrated in Figure 25-3. The figure doesn't show a quiet zone, but one is normally used with UPC-A. The UPC-A code is 12 digits long, with the digits divided into four categories:

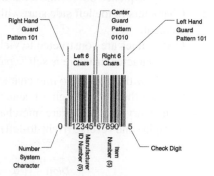

Figure 25-3: Structure of UPC-A bar code.

Number System Character—is a digit assigned according to the following scheme:

Digit	Meaning
0	Manufacturer identification numbers.
1	Reserved.
2	Random weight items marked at the store (articles where the price varies by the weight; for example meat. The code is produced in the store and attached to the article.)
3	National Drug Code (NDC) and National Health Related Items Code (HRI).
4	In-store marking without format (UPC Code which can be used without format limits).
5	UPC Coupon.
6	Manufacturer identification numbers.
7	Manufacturer identification numbers.
8	Reserved.
9	Reserved.

Manufacturer ID Number—is a five-digit number uniquely identifying the manufacturer. (The Uniform Code Council, Inc. assigns manufacturer ID numbers.)

Item Number—is a five-digit number uniquely identifying the manufacturer's product. (Once you have a manufacturer ID number, you assign your own item numbers.)

Check Digit—is a digit formed from the preceding 11 digits to help ensure data integrity. (We'll later see how the check digit is calculated.)

The element structure is divided into left and right parts, separated by the "center guard pattern." At both the left and right extremes are guard patterns. It should be obvious from Figure 25-3 that elements have varying widths. Specifically, elements have a width of 1, 2, 3 or 4 units. (We'll learn more about the detailed structure of the UPC-A code when we write programs to read it.)

The UPC-A rules are:

1. It must be exactly 12 digits, assigned as summarized above. Only digits 0...9 are permitted characters.

2. Bar codes start and finish with a guard pattern of 101.
3. The code is divided into left and right halves with a center guard pattern of 01010.
4. Element are 1, 2, 3 or 4 units wide. Each character has four elements, two bars and two spaces. The total character width is seven units.
5. The element structure maps into a 7-bit code, with spaces representing binary 0s and bars binary 1s. The element width encodes the number of 1s or 0s. A space of three units width, therefore, represents the binary sequence 000, while a bar of width two units represents the binary sequence 11.
6. Characters in the left side start with a space element; characters on the right side start with a bar element.
7. Code values are constructed to yield unique values whether read left-to-right or right-to-left.
8. The character structure is self-separating. Hence no inter-character separator need be added.

Let's see how these rules go into character decoding. As we see, digits to the left of center are coded differently than those to the right of center. In fact, the bit values in the left and right implementation are complementary—1s and 0s are interchanged. The table includes the corresponding values for both forward (left-to-right) and reverse (right-to-left) directions.

Digit	Left of Center 6	5	4	3	2	1	0	Forward Value	Reverse Value
0	0	0	0	1	1	0	1	13	88
1	0	0	1	1	0	0	1	25	76
2	0	0	1	0	0	1	1	19	100
3	0	1	1	1	1	0	1	61	94
4	0	1	0	0	0	1	1	35	98
5	0	1	1	0	0	0	1	49	70
6	0	1	0	1	1	1	1	47	122
7	0	1	1	1	0	1	1	59	110
8	0	1	1	0	1	1	1	55	118
9	0	0	0	1	0	1	1	11	104

Digit	Right of Center 6	5	4	3	2	1	0	Forward Value	Reverse Value
0	1	1	1	0	0	1	0	114	39
1	1	1	0	0	1	1	0	102	51
2	1	1	0	1	1	0	0	108	27
3	1	0	0	0	0	1	0	66	33
4	1	0	1	1	1	0	0	92	29
5	1	0	0	1	1	1	0	78	57
6	1	0	1	0	0	0	0	80	5
7	1	0	0	0	1	0	0	68	17
8	1	0	0	1	0	0	0	72	9
9	1	1	1	0	1	0	0	116	23

Let's manually decode two UPC-A digits. We'll decode two digits to illustrate UPC-A's self-spacing feature.

(Each left side character ends in a bar and starts with a space; while each right side character ends in a space and starts with a bar. Hence we always have a space between once character and its adjacent neighbor.) We'll assume the characters are from the left half and that we read them left-to-right.

Character 1							Character 2						
1	2	3	4	5	6	7	1	2	3	4	5	6	7
1		2		3	4		1		2		3	4	

I've added modulus numbers (1...7) and element numbers (1...4) for both characters.

Character	Element No.	Type	Width	Binary
1	1	Space	3	000
	2	Bar	2	11
	3	Space	1	0
	4	Bar	1	1
2	1	Space	1	0
	2	Bar	2	11
	3	Space	1	0
	4	Bar	3	111

Let's assemble the binary elements.

Character 1: 000+11+0+1 = 0001101, or decimal 13.
Character 2: 0+11+0+111 = 0110111, or decimal 55.

We check the encoding table and discover that 13 corresponds to 0 and 55 corresponds to 8. Hence the two digits are 08.

If you read the characters right-to-left, the values are:

Character 1: 1+0+11+000 = 1011000, or decimal 88.
Character 2: 111+0+11+0 = 1110110, or decimal 118.

Reading the "reverse value" entry in the table we discover that 88 decodes as 0 and 118 decodes as 8, yielding the same values as our initial left-to-right reading.

Bar Code Wand

How then do we get black and white bars on paper into a PIC? We'll use the simplest possible scanner, a bar code wand reader. To read you swipe the wand across the bar code. There's a lot of black magic that goes into a wand, so I'll assume you will use an already build commercial wand. I recently purchased two unused Hewlett Packard A000-series wands for $5.00 each from an electronics surplus house and similar wands from HP and other manufacturers are widely available in the surplus market.

My HP wand is shown in Figure 25-4. It came with a three-foot cable terminated with a 6-pin DIN plug. (The mating panel jack is a Mouser Electronics part no. 16PJ224.) Although I was able to find HP's data sheet [25-4] for the A000-series wands, my DIN plug had six pins, not five as the data sheet shows. Since you may have a similar problem, I'll go though how I identified the correct pin connection.

Figure 25-4: Bar code scanning wand.

Many wands of this era used HP's specialized LED sensor and support chip, HBCC-0500 and followed HP's recommended circuit design shown in the associated data sheet. [25-3] The wand has three connections; +5V (red), ground (black) and an open collector output (white). A fourth connection is the shield for the cable, which is normally connected to the DIN plug's shell. The wiring color code comes from HP's data sheet. Using a sharp box cutter, I carefully lengthwise slit the insulation of the cable for several inches near the plug, giving me access to the individual conductors. Sure enough, they matched HP's color code, red, black and white, and a cable shield. Using a sharp straight pin, I then pierced each wire's insulation one at a time and measured the resistance between the wire and each of the six pins of the DIN plug. After determining which pin connects to each wire, I carefully pushed the wire back into the cable sheath and repaired the slit with transparent tape. It's a good thing that I checked the connections, as HP's data sheet did not match my wand's wiring. Figure 25-5 shows how I connected the wand to my PIC, and Figure 25-6 shows the connector connection for my wand.

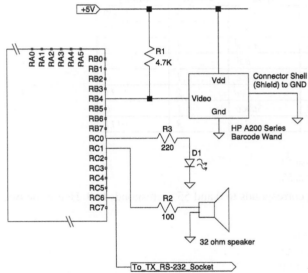

Figure 25-5: Scanning wand connection.

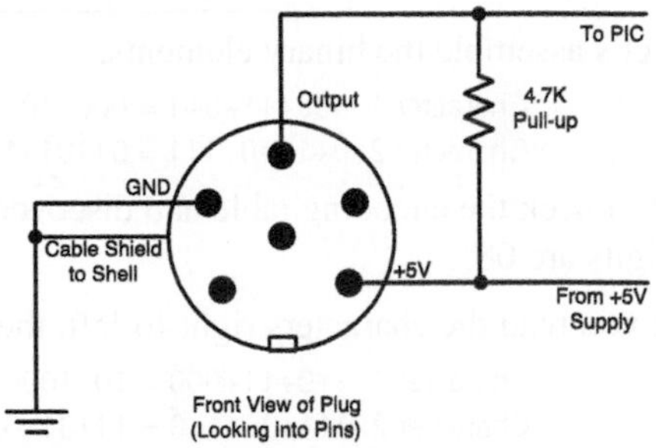

Figure 25-6: Wand Plug Connection.

Function	Wire Color	My Wand	HP A000 Data Sheet
+5V	Red	1	1
Output	White	3	2
Ground	Black	4	3
Shield	Shield Wrapping	Shell	Shell

Incidentally, I first measured the resistance between each pair of pins. I was reasonably sure that I identified the correct pin connections from the resistance measurements, but I didn't want to apply power incorrectly and destroy the wand. Hence, the process described above.

HP manufactured wands in three resolutions, and with both visible red and infrared light sources, and with and without a power-on switch. I'm pretty sure my wand is a "medium" resolution model, capable of resolving elements 0.0075 inches width, since both low and high resolution models were nonstandard products. I've found it reliable in reading Code 3 of 9 bar codes printed with 14-point type using a bar code font from IDAutomation.com.

```
;----
        For i = 0 to 255
        Next
        If OldValue <> NoEdges Then
                HSerOut [Dec NoEdges,13]
                OldValue = NoEdges
        EndIf
GoTo Main

Disable

CountEdges
;---------
        NoEdges = NoEdges+1
        Dummy = PortB
Resume
End
```

As part of the initialization code, we define the interrupt, **RbInt**, define its handler, **CountEdges** and finally enable the interrupt.

```
        OnInterrupt RbInt, CountEdges
        Enable RbInt
```

The **RbInt** is called every time pins **B4**, **B5**, **B6** or **B7** change state, whether from high-to-low or low-to-high.

```
        Input VideoIn
        SetPullups PU_on
```

We've previously aliased **VideoIn** as **B4**, the pin to which the wand output is connected. Since a change on any of **B4**, **B5**, **B6** and **B7** will cause an interrupt, we enable the PIC's internal weak pull up resistors to ensure that the unused pins **B5**, **B6** and **B7** will remain high and not cause spurious interrupts.

The main program loop executes a time killing **For i Next** loop.

```
        Main
        ;----
        For i = 0 to 255
        Next
        If OldValue <> NoEdges Then
                HSerOut [Dec NoEdges,13]
                OldValue = NoEdges
        EndIf
GoTo Main
```

At the conclusion of each loop cycle, if a different number of edge crossings is held in the variable **NoEdges** from the prior value held in **OldValue**, Program 25-1 sends the number of edges counted out on the serial port.

```
        Disable

        CountEdges
        ;---------
        NoEdges = NoEdges+1
        Dummy = PortB
        Resume
```

The interrupt procedure, when called, increments the variable **NoEdges** by one. (Don't forget to add the command **Disable** before the interrupt handler to prevent multiple calls to the interrupt procedure.) The statement **Dummy = PortB** is necessary because Microchip's PIC design does not clear the port mismatch (the mismatch causes the interrupt to be issued) unless PortB is read. Without this dummy read statement, the interrupt continues to be repeated, even though no change in PortB has occurred.

When I attempted to read a UPC-A bar code, Program 25-1 reported the following:

P 25-1

2

4

6

8

We should have seen 60 transitions (two transitions for each bar; 30 bars in a UPC-A bar code) not eight. Between the latency in interrupts and execution speed, MBasic isn't fast enough.

Program 25-2

Let's try an assembler interrupt version of Program 25-1 to count transitions.

```
;Program 25-2
;Count the number of transitions
;when scan a bar code

;Constants
;----------
VideoIn        Con     B4
Zero           Con     0

;Variables
;----------
NoEdges        Var     Byte
OldCount       Var     Byte
i              Var     Word

;Initialization
;--------------
NoEdges = 0
OldCount = 99
EnableHSerial
SetHSerial H115200
HSerOut ["P 25-2",13]
Pause 1000
Input B4
Input B5
Input B6
Input B7
SetPullups PU_on
GoSub SetupRBint
Output C0
Low C0

;Note re ASM - In MBasic ISRASM is executed upon an interrupt.
;MBasic is responsible for context savings, and RTIE.
;The "&0x7F"IS necessary to trim off
;the variable type bits added by the MBasic compiler to the ;underlying variable
address.
;=================================================================
israsm
;=================================================================
{
        ;test for source of interrupt, and only handle
        ;RBIE interrupt in this routine
        ;entry from MBasic is in Bank 0
        ;-------------------------------------------------------

        btfss   IntCon,RBIF             ;do we have a
        GoTo    EndProc         ;RBINT interrupt.
                                        ;If not jump to end

        ;--------following code executed only if valid RBINT

ClrRBIF
        movf    PortB,f
        bcf     IntCon,RBIF

        BankSel NoEdges
        Incf            NoEdges&0x7F,f ;NoEdges = NoEdges + 1

EndProc
        BankSel Zero
```

```
        }

    Main
    ;----
            For i = 0 to $200
                    Pause 10
            Next
            If OldCount <> NoEdges Then
                    HSerOut [Dec NoEdges,9, Dec (NoEdges-OldCount),13]
                    OldCount = NoEdges
            EndIf
            NoEdges = 0
            OldCount = 0
            HSerOut ["Ready",13]
    GoTo Main

    SetupRBint      ;set up the RBint interrrupt
    ;---------
            gie = 1         ;global interrupt enable (INTCON, bit 7)
            rbie = 1        ;set RBint
    Return
```

Program 25-2 tracks Program 25-1's concept, but substitutes an assembler interrupt. We've covered MBasic's **ISRASM** function in Chapters 13 and 14. If you haven't already read these Chapters, you might wish to do so before reviewing Program 25-2. Let's look at the differences between Programs 25-1 and 25-2.

```
    SetupRBint      ;set up the RBint interrrupt
    ;---------
            gie = 1         ;global interrupt enable (INTCON, bit 7)
            rbie = 1        ;set RBint
```

Since we are using an assembler-based interrupt routine, we have to enable the interrupt. Delving into Microchip's PIC16F87x data sheet, Section 12.10, we find that the interrupt on PortB change interrupt is known as the **RBIE** interrupt in assembler. **RBIE** is one of four members of the **GIE**, or general interrupt enable, family. So, to enable the interrupt on PortB change, we must enable both **RBIE** and **GIE** by setting both to **%1**. We can do this in MBasic because MBasic duplicates many assembler identifiers and lets us read and write to them in MBasic statements.

```
    israsm
    {
            btfss   IntCon,RBIF       ;do we have a
            GoTo    EndProc           ;RBINT interrupt.
                                      ;If not jump to end
            ;--------following code executed only if valid RBINT
```

Many events can trigger an interrupt, including sending or receiving information over the serial port with MBasic's **HSerOut** or **HSerIn** functions. Before executing code that responds to the **RBIE**, we accordingly verify that we have received an interrupt generated by a PortB change, not some other interrupt. This is simply accomplished by checking the **RBIF** bit in the **IntCon** register. (The "**F**" in **RBIF** indicates it's a "flag" bit associated with the PortB interrupt; the "**E**" in **RBIE** indicates it's an "enable" bit.) If **RBIF** is not set, then the interrupt is from some other cause and we immediately jump to the end of the interrupt handler via the **GoTo EndProc** statement.

```
    ClrRBIF
            movf    PortB,f
            bcf     IntCon,RBIF

            BankSel NoEdges
            Incf            NoEdges&0x7F,f ;NoEdges = NoEdges + 1
    EndProc
            BankSel Zero
    }
```

The above code fragment is executed only when there is a valid interrupt on PortB change. First, we read PortB to itself by executing **movf PortB,f**. This might seem a meaningless exercise, but as we learned in

Program 25-1, we must read PortB to reset the internal comparison storage. (The PIC compares the current value of PortB bits 4...7 against a stored copy; until we update the stored copy the mismatch that caused the interrupt continues.) Next, we reset the interrupt flag by clearing the **RBIF** bit.

Before updating **NoEdges**, we execute a **BankSel** to ensure we are in the correct RAM bank. Lastly, we update the **NoEdges** variable. Incrementing **NoEdges** by 1 through the **incf** operator is equivalent to executing the MBasic statement **NoEdges = NoEdges + 1**.

Here's the output from Program 25-2 when I scanned three Code 3 of 9 bar codes; *8*, *12* and *123*.

```
Ready
32        32
Ready
42        42
Ready
52        52
```

What should be the output from Program 25-2? The number of elements in a Code 3 of 9 bar code, as we learned earlier, is 10N-1, where N is the number of characters. Hence, the bar code *8* should have 29 elements. And, an additional transition occurs when the wand passes over the last bar and enters the quiet zone, so we've accounted for 30 transitions. What about the last two? After scanning, I lifted the wand away from the paper. The wand thus goes to black, for another transition. Then, after a brief period of black, the wand reverts to white idle, yet another transition. Hence, we should see 32 transitions, matching the output of Program 25-2. Likewise, we calculate 42 and 52 transitions for *12* and *123* bar codes. We'll have to pay attention to these extraneous transitions when we measure element widths.

Program 25-3

Before we fully decode a bar code, we must have an algorithm to measure element widths. And, it makes sense to develop and verify the algorithm in a simple stand-alone program before we go onto more elaborate programs. Hence, in Program 25-3, we'll measure the width of up to 80 bar code elements and dump the width measurements to the serial port.

As should be obvious by now, when we measure the "width" of a bar code element the result is in microseconds, not inches or millimeters. We measure the time it takes to move the wand from one edge of an element to the opposite edge. If the wand moves at a more or less constant speed across the bar code—a reasonable assumption since the distance traversed is so small—the physical width of the printed element is directly proportional to the transit time. We will trigger an element width measurement every time the wand's output status changes state, just as we did to count the number of transitions.

Let's consider how a possible algorithm might work. We'll keep track of the number of elements read in a byte variable **k** and we'll measure the width of each element in time units, storing the result in a byte array **Data(k)**. We'll set aside 80 bytes for **Data()** and define the corresponding maximum value for **k**, 79, as **kmax**. To measure element duration, we'll use timer 1, a 16-bit timer that counts up from 0000 to 65535. (We'll discuss our choice of timer 1 later.) Each transition occurs when the wand's output changes state, thereby triggering a change on PortB state interrupt.

Transition			*k*	
RBINTERRUPT				
Low-to-High (space to bar) Transition			*Initialize k=0*	
	1 1	*Reset Timer 1 to 0*	*0*	*special case at startup*
	2	*Start Timer 0*	*0*	
	3	*Clear RBIF*		
Do something else until next RBINTERRUPT				

RBINTERRUPT

High-to-Low (bar to space) Transition

2	1	Stop Timer 1	0
	2	Copy Timer 1 value to Data(k)	0
	3	Reset Timer 1 to 0	0
	4	Restart Timer 1	0
	5	k = k+1	1
	6	Clear RBIF	1

Do something else until next RBINTERRUPT

RBINTERRUPT

Low-to-High (space to bar) Transition

3	1	Stop Timer 1	1
	2	Copy Timer 1 value to Data(k)	1
	3	Reset Timer 1 to 0	1
	4	Restart Timer 1	1
	5	k = k+1	2
	6	Clear RBIF	2

Do something else until next RBINTERRUPT

RBINTERRUPT

High-to-Low (bar to space) Transition

4	1	Stop Timer 1	2
	2	Copy Timer 1 value to Data(k)	2
	3	Reset Timer 1 to 0	2
	4	Restart Timer 1	2
	5	k = k+1	3
	6	Clear RBIF	

Do something else until next RBINTERRUPT

Repeat until

A	Timer 1 overflow OR
B	k > kmax

Special Case – Timer 1 Overflow

1	Disable Change on PortB interrupt
2	Set end of collection sentinel variable

Leaving aside the special cases of the first and last elements, the algorithm isn't complex; with every transition we stop timer 1 and copy its value into **Data(k)**. We then reset timer 1 to 0, restart it and increment **k**. We know we are finished when either we run out of room in **Data(k)**, or we reach the quiet zone, as evidenced by a very wide space value.

Our choice of a timer is an important component of this approach. From Figures 25-7, 25-9 and 25-10, it looks as if the narrowest element we must measure runs around 1000 µs, and the widest element about 4000 µs. We'll allow a 2:1 safety margin to account for faster or slower wand movements or larger and smaller printed bar codes, so our target is to correctly measure a 500 μs element as well as an 8000 μs element, representing a 16:1 dynamic range. And, since we will be working in byte length integer arithmetic, we would like to have the 500 μs element correspond to a count of at least 10 or 12. This suggests a target counter "tick" interval around 50 μs. A 500 μs wide element corresponds to 10 "ticks" and an 8000 μs wide element corresponds to 160 "ticks." Both values fit into a byte variable and give reasonable resolution when it comes to deciding element lengths.

As we learned in Chapter 15, timer 1 is a 16-bit timer with an optional prescaler (1:1, 1:2, 1:4 and 1:8). My PIC uses a 20 MHz oscillator, so if timer 1 is internally clocked, and its prescaler set for 1:1, it operates at Fosc/4 and advances one count every 200 ns. Timer 1 is a 16-bit counter, so its high byte increments one count every 256 * 200 ns, or once every 51.2 μs, almost exactly our target 50 μs goal. So, we'll measure element width using **Timer1.Byte1**, or **Timer1H** in assembler.

Program 25-3 implements this algorithm.

```
;Program 25-3 -- Diagnostic and Development Program
;Read and dump bar widths for up to 80 bars/spaces

STACK = 10

;Constants
;----------
VideoIn        Con      B4        ;Pin for Wand connection
ASize          Con      80        ;size of array
AStart         Con      $A0       ;start of array Bank 1
HTab           Con      9         ;Horizontal tab char
Zero           Con      0         ;for banksel use

;Variables
;----------
Initial        Var      Byte      ;sentinel for 1st time after reset
i              Var      Byte      ;counter
k              Var      Byte      ;counter
Temp           Var      Byte      ;holds peek value

Initialization
;---------------
        EnableHSerial
        SetHSerial H115200
        HSerOut ["P 25-3",13]


        ;Port B as inputs & turn pull up on to assure states
        TRISB = $FF
        SetPullups PU_on

        GoSub SetupRBint         ;set up RBInterrupt
        GoSub SetUpTimer1        ;set up Timer 1 (no interrupts)

        GoSub ClearArray         ;fill array with zeros

        k = 0                    ;start with index = 0
        Initial = 0              ;prime for being first bar
;
;===============================================================
israsm
;===============================================================
{
        ;test for source of interrupt, and
        ;only handle RBIE interrupt in this routine
        ;-------------------------------------------
        btfss   IntCon,RBIF      ;Is this a
        GoTo    EndProc          ;RBINT interrupt.
                                 ;If not jump to end
        ;--------following code executed only if valid RBINT
```

```
ClrRBIF                          ;clear the RB interrupt flag
        movf    PortB,f          ;read to self to reset compare value
        bcf     IntCon,RBIF

        btfss   Initial&0x7F,0   ;test to see if this is 1st pass
        GoTo    FirstPass

        Call    Stop_n_Copy      ;copy Tmr1H to array
        Call    ResetTimer1
        GoTo    UpDateK

FirstPass       ;if leading edge, only start the timer
;-------
        Call    ResetTimer1      ;start the timer
        bsf     Initial&0x7F,0   ;set the sentinel
        bcf     Pir1,Tmr1IF      ;clear the (for end of scan)
        GoTo    EndProc          ;no more to do-we're out of here!

ResetTimer1     ;sets Timer 1=0 and starts it running
;-------------
        BankSel Zero             ;all in bank 0
        clrf    Tmr1H            ;clear the timer
        clrf    Tmr1L
        bsf     T1Con,Tmr1on     ;Turn timer 1 on
        Return

Stop_n_Copy     ;stop timer 1 copy high byte to mem index k
;-------------
        BankSel Zero             ;all in bank 0
        bcf     T1Con,Tmr1On     ;turn timer off
        movlw   AStart           ;get starting address into w
        BankSel k
        addwf   k&0x7F,w         ;w = i + starting address
        movwf   FSR              ;FSR holds the address to be accessed

        movf    Tmr1H,w          ;get the low byte of timer
        movwf   INDF             ;move data into memory at AStart+i
        Return

UpDateK ;k = k+1
;-------
        BankSel Zero
        incf    k&0x7F,f
        movf    k&0x7F,w         ;k=k+1. k can't exceed ASize-1
        sublw   ASize            ;test to see if k=Asize
        btfsc   STATUS,Z         ;check for equal
        clrf    k&0x7F           ;if k=ASize THEN k=0
EndProc
        BankSel Zero
}

Main
;----
        ;Time wasting loop. Initial set to 1
        ;by first scan interrupt so we use it to know
        ;scan is in progress
        Repeat
        Until Initial > 0
        ;light scan in progress LED
        High C0
        Repeat          ;interrupt works during pause
        ;we don't use interrupt per se, but we use the IF
        ;to indicate scan into overrun white space
        Until Tmr1IF = %1

        ;Time to lift the scan wand without triggering a second
        rbie = 0        ;scan. Disbles the rbie interrupt
        rbif = 0        ;clear the interrupt flag as well
        Pause 500
```

```
            rbie = 1

            Low C0                      ;scan finished

            GoSub ReadArray             ;dump scans to HSerOut
            GoSub ClearArray            ;reset to zeros

            k = 0
            Initial = 0
GoTo Main


SetupRBint       ;set up the RBint interrrupt
;---------
            gie = 1          global interrupt enable
            rbie = 1                   ;set RBint
Return

SetUpTimer1      ;set up Timer 1
;----------
            t1oscen = 0                ;disable timer 1 osc
            tmr1cs = 0                 ;set for internal clock
            t1ckps1 = 0                ;prescaler set for 1:1
            t1ckps0 = 0                ;prescaler second bit for 1:1
Return

ClearArray       ;puts zeros into array
;----------------
            For i = 0 to (ASize-1)
            ASM
            {
            movlw           AStart          ;get starting address into w
            BankSel         i
            addwf           i&0x7F,w        ;w = i + starting address
            movwf           FSR             ;FSR = address to be accessed
            BankSel         i               ;
            movlw           0x00            ;w = 0
            movwf           INDF            ;move data to address AStart+i
            BankSel         Zero            ;exit condition
            }
            Next
Return

ReadArray        ;read array to serial port
;-----------
            HSerOut [13,13]
            ;Scan memory and dump the width values out
            For i = 0 to (ASize-1)
                    PEEK (AStart+i), Temp
                    HSerOut ["i: ",Dec i,HTab,"Width: ",Dec Temp,13]
                    Pause 10
            Next
Return
End
```

As usual, we'll examine only the new aspects of Program 25-3.

Before we may use timer 1, we must first initialize it. We do this in subroutine **SetUpTimer1**, which directly follows what we learned in Chapter 15.

```
SetUpTimer1      ;set up Timer 1
;----------
            t1oscen = 0                ;disable timer 1 osc
            tmr1cs = 0                 ;set for internal clock
            t1ckps1 = 0                ;prescaler set for 1:1
            t1ckps0 = 0                ;prescaler second bit for 1:1
Return
```

Timer 1 has an optional external oscillator, which we disable, instead selecting the internal clock. Clearing both prescaler bits to 0 sets the prescale ratio at 1:1.

Before our analysis progresses further, we should consider our data structure and memory allocation. To simplify our assembler routine, we need 80 contiguous bytes of variable space. We'll use the 80 contiguous bytes of RAM starting at address **$A0** in bank 1. The default memory model used in MBasic version 5.3.0.0 reserves all RAM in bank 0 for internal use and begins allocating user variables at address **$A0** in bank 1, which would conflict with our need for 80 free bytes of RAM. To force MBasic to reserve space in bank 0 for MBasic variables, thereby freeing up bank 1 for our use, we add the compiler instruction **STACK = 10**. This frees up 28 bytes of bank 0 RAM space, which MBasic has available use for user-defined variables. Alternatively, we could install our array in bank 2, beginning at address **$120** and permit MBasic to use bank 1 memory for user variables.

We may wish to examine the ASM code generated by MBasic to verify our variable bank assignments. These techniques are discussed in Chapters 13, 14 and 15. Additionally, MBasic does not provide a way for us to declare the MBasic byte array variable **Data(80)** and have that array begin at address **$A0** (or any other address) and run for 80 contiguous bytes in a single bank. Hence, we will use **PEEK** to read the array from MBasic.

To see how we write to the fictitious array **Data(k)** in assembler, we'll examine how the subroutine **ClearArray** writes zeros to all 80 positions. (We could do this completely in MBasic via the **POKE** function, but we'll do it in assembler as a way of testing our ability to write to an array-type data structure.)

```
ClearArray       ;puts zeros into array
;---------------
        For i = 0 to (ASize-1)
            ASM
            {
            movlw          AStart              ;get starting address into w
            BankSel        i
            addwf          i&0x7F,w            ;w = i + starting address
            movwf          FSR                 ;FSR = address to be accessed
            BankSel        i                   ;
            movlw          0x00                ;w = 0
            movwf          INDF                ;move data to address AStart+i
            BankSel        Zero                ;exit condition
            }
        Next
    Return
```

We've previously defined the constants **Asize** as the array size, 80, **Astart** as the array starting address, **$A0** and **Zero** as 0. We'll mix assembler and BASIC here as well, showing the flexibility that makes MBasic so useful. This lets us wrap the assembler memory write procedure inside a MBasic **For … Next** loop.

Our clearing algorithm uses indirect memory addressing via the two special registers, **FSR** and **INDF**. The **FSR** register holds the address of the memory to be accessed. The **FSR** register is 8 bits wide and thus allows us to address memory locations **$0…$FF**, that is, all memory locations in Banks 0 and 1, including the **$A0…$EF** range we use to hold the width data. To address Banks 2 and 3, we set the **IRP** bit to 1. (We may assume the **IRP** bit is set to 0 when MBasic hands off to our ASM code.)

The actual access is made to the **INDF** register. For example, suppose we wish to clear the file at address **$A0**. First, we load **FSR** with the value **$A0**. To set the RAM memory at address **$A0** to 0, we execute the clear file instruction **clrf**, with its operand as **INDF**. Hence the instruction **clrf INDF** clears the file at the address contained in **FSR:IRP**, **$A0** in this example.

With this understanding, the first four lines of the assembler procedure should be clear:

```
            movlw          AStart              ;get starting address into w
            BankSel        i
            addwf          i&0x7F,w            ;w = i + starting address
            movwf          FSR                 ;FSR = address to be accessed
```

We load the starting address, **Astart**, into the **w** register and then add the current value of the loop counter **i** to the starting address. The sum **Astart + i** is then moved to **FSR**. As **i** goes from 0 to **Asize-1**, **FSR** indexes from **$A0** through **$EF**.

```
BankSel    I        ;
movlw      0x00     ;w = 0
movwf      INDF     ;move data to address AStart+i
BankSel    Zero     ;exit condition
```

The remainder of our assembler routine loads a 0 into **INDF**. But, as we learned, **INDF** actually accesses the file with the address previously loaded in **FSR**.

The assembler routine is the functional equivalent of:

```
For i = 0 to (ASize-1)
        Data(i) = 0
Next
```

Where **Data(i)** is the fictitious 80-element byte array starting at **$A0**.

Let's go through the main program loop.

```
Main
;----
    Repeat
    Until Initial > 0
```

The **Repeat Until** loop is a time wasting dummy activity until the start of a scan. We could be updating an LCD display or conducting other activities. Since the change on PortB interrupt is serviced by an **israsm** procedure, the latency is a few microseconds at most; unlike the case with a pure MBasic interrupt service routine called only in between MBasic statements.

The **israsm** procedure communicates that a wand scan has started by setting the sentinel variable **Initial** to **%1**. Upon the start of a scan, execution passes to a second **Repeat … Until** loop.

```
High C0
Repeat
Until Tmr1IF = %1
```

Before entering the second **Repeat … Until** loop we illuminate an LED connected to pin **C0** to inform the user that a scan has started.

The exit condition from this second **Repeat … Until** loop is a bit trickier. Timer 1 does not automatically reset and upon rollover it sets the timer 1 interrupt flag bit **Tmr1IF**. Although we do not enable the timer 1 interrupt, nonetheless, we can still use the **Tmr1IF** bit to determine when timer 1 has rolled over. And, of course, timer 1 rolls over when it attempts to measure the width of the quiet zone after measuring the last valid bar. The **Tmr1IF** bit therefore signals end-of-scan, upon which occurrence the second **Repeat … Until** loop condition fails and execution continues with the rest of the program.

```
rbie = 0
rbif = 0
Pause 500
rbie = 1
Low C0
```

The scan has ended and we wish to disable further change on PortB interrupts until we have completed analyzing the current information. Disabling the change-on-PortB interrupts immediately upon entry into the quiet zone also blocks the false wand-lift and wand-timeout transitions. My wand required about 500 ms delay to avoid false pen lift transitions. You may need to lengthen this delay slightly to match your want. HP specifies this period as "approximately one second."

```
GoSub ReadArray
GoSub ClearArray
```

The subroutine **ReadArray** dumps the width measurements to the serial port, and subroutine **ClearArray**, as we've just seen, zeros out the measurement array.

```
            k = 0
            Initial = 0
      GoTo Main
```

Finally, we reset the array index **k** and the start-of-scan sentinel, **Initial**.

Let's now delve into the interrupt handler code.

```
      israsm
      ;==============================================================
      {
              ;test for source of interrupt, and
              ;only handle RBIE interrupt in this routine
              ;------------------------------------------
              btfss   IntCon,RBIF            ;Is this a
              GoTo    EndProc               ;RBINT interrupt.
                                            ;If not jump to end
              ;--------following code executed only if valid RBINT
```

We used identical code in Program 25-2 to test for the interrupt source, and we won't repeat the analysis.

```
      ClrRBIF           ;clear the RB interrupt flag
              movf    PortB,f ;read to self to reset compare value
              bcf     IntCon,RBIF

              btfss   Initial&0x7F,0        ;test to see if this is 1st pass
              GoTo    FirstPass

              Call    Stop_n_Copy           ;copy Tmr1H to array
              Call    ResetTimer1
              GoTo    UpDateK
```

After verifying that we have a change on PortB interrupt, we conduct the PortB read and clear the interrupt flag, **RBIF**, just as we did in Program 25-2.

Next, we determine if this interrupt represents the leading edge of the first bar by examining the value of the sentinel variable **Initial**. If **Initial.Bit0 = 0**, indicating the first pass, we execute the code at **First-Pass** that resets timer 1, sets **Initial.Bit0** to **%1**, indicating initialization is finished and clears **Tmr1IF**. (We could have made the variable **k** perform double duty, but using a separate variable **Initial** clarifies our action. If we were to use **k** as a sentinel for the first pass, we test for **k=0**.)

If it is not the first pass through the interrupt service routine, then we execute two assembler subroutine calls; first to **Stop_n_Copy** and second to **ResetTimer1**. After these two subroutine calls, execution jumps to **UpDateK**.

```
      FirstPass         ;if leading edge, only start the timer
      ;-------
              Call    ResetTimer1           ;start the timer
              bsf     Initial&0x7F,0        ;set the sentinel
              bcf     Pir1,Tmr1IF           ;clear the (for end of scan)
              GoTo    EndProc               ;no more to do-we're out of here!
```

As previously stated, the code at **FirstPass** starts timer 1, sets the sentinel variable **Initial** and clears the **Tmr1IF** flag bit.

```
      ResetTimer1               ;sets Timer 1=0 and starts it running
      ;-------------
              BankSel   Zero              ;all in bank 0
              clrf      Tmr1H             ;clear the timer
              clrf      Tmr1L
              bsf       T1Con,Tmr1on      ;Turn timer 1 on
              Return
```

Subroutine **ResetTimer1** clears both the high and low bytes of timer 1 and turns the timer back on by setting the timer 1 control register bit **Tmr1on**.

```
Stop_n_Copy              ;stop timer 1 copy high byte to mem index k
;------------
        BankSel Zero                     ;all in bank 0
        bcf          T1Con,Tmr1on        ;turn timer off
        movlw        AStart              ;get starting address into w
        BankSel      k
        addwf        k&0x7F,w            ;w = i + starting address
        movwf        FSR                 ;FSR holds the address to be accessed

        movf         Tmr1H,w             ;get the low byte of timer
        movwf        INDF                ;move data into memory at AStart+i
        Return
```

Subroutine **Stop_n_Copy** first stops timer 1 by clearing the timer 1 control register bit **Tmr1on**. We then copy the high byte of timer 1, **Tmr1H**, into the memory array using the same code analyzed in subroutine **ClearArray**.

```
UpDateK ;k = k+1
;-------
        BankSel      Zero
        incf         k&0x7F,f
        movf         k&0x7F,w            ;k=k+1. k can't exceed ASize-1
        sublw        ASize               ;test to see if k=Asize
        btfsc        STATUS,Z            ;check for equal
        clrf         k&0x7F              ;if k=ASize THEN k=0
EndProc
        BankSel Zero
}
```

Lastly, we increment the array address index **k** and test its size. If **k = Asize**, we reset **k** to zero. This causes any additional scanned elements to wrap around the start of the array. Alternatively, we could end the scan by forcing a set of **Tmr1IF**. We end the **israsm** procedure by setting the bank to 0, as required before returning to MBasic.

Note that we do not preload the counter to compensate for the execution time of the setup code. Since we measure widths in 51.6 µs increments, ignoring code execution time of a microsecond or two introduces negligible error.

The last piece of code to examine is the subroutine **ReadArray**.

```
ReadArray           ;read array to serial port
;-----------
        HSerOut [13,13]
;Scan memory and dump the width values out
For i = 0 to (ASize-1)
        PEEK (AStart+i), Temp
        HSerOut ["i: ",Dec i,HTab,"Width: ",Dec Temp,13]
        Pause 10
Next
Return
```

We read the data array using MBasic's **Peek** procedure and write the widths.

Let's see the results of a scan. We'll scan the Code 3 of 9 bar code *9* and see the results. (To save space, I've reformatted the output into four columns.)

i: 0	Width: 14	i: 20	Width: 8	i: 40	Width: 0	i: 60	Width: 0
i: 1	Width: 32	i: 21	Width: 24	i: 41	Width: 0	i: 61	Width: 0
i: 2	Width: 10	i: 22	Width: 9	i: 42	Width: 0	i: 62	Width: 0
i: 3	Width: 12	i: 23	Width: 9	i: 43	Width: 0	i: 63	Width: 0
i: 4	Width: 27	i: 24	Width: 24	i: 44	Width: 0	i: 64	Width: 0
i: 5	Width: 10	i: 25	Width: 9	i: 45	Width: 0	i: 65	Width: 0
i: 6	Width: 25	i: 26	Width: 24	i: 46	Width: 0	i: 66	Width: 0
i: 7	Width: 10	i: 27	Width: 10	i: 47	Width: 0	i: 67	Width: 0
i: 8	Width: 8	i: 28	Width: 8	i: 48	Width: 0	i: 68	Width: 0
i: 9	Width: 9	i: 29	Width: 24	i: 49	Width: 0	i: 69	Width: 0
i: 10	Width: 8	i: 30	Width: 0	i: 50	Width: 0	i: 70	Width: 0
i: 11	Width: 10	i: 31	Width: 0	i: 51	Width: 0	i: 71	Width: 0
i: 12	Width: 25	i: 32	Width: 0	i: 52	Width: 0	i: 72	Width: 0
i: 13	Width: 27	i: 33	Width: 0	i: 53	Width: 0	i: 73	Width: 0
i: 14	Width: 9	i: 34	Width: 0	i: 54	Width: 0	i: 74	Width: 0
i: 15	Width: 9	i: 35	Width: 0	i: 55	Width: 0	i: 75	Width: 0
i: 16	Width: 25	i: 36	Width: 0	i: 56	Width: 0	i: 76	Width: 0
i: 17	Width: 10	i: 37	Width: 0	i: 57	Width: 0	i: 77	Width: 0
i: 18	Width: 8	i: 38	Width: 0	i: 58	Width: 0	i: 78	Width: 0
i: 19	Width: 9	i: 39	Width: 0	i: 59	Width: 0	i: 79	Width: 0

We see two things immediately. First, our timer 1 choice looks correct. The smallest bar width is 8, not unexpected since the bar code I scanned was printed with a 14-point font, but within our acceptable range. Second, we see 30 elements. Because we increment the index variable k after each scan is completed, only **k-1** widths are valid, so we should discard **Data(29)**.

Let's manually decode this data. Since we scanned a Code 3 of 9 bar code, each character has a modulus of width 15, plus one for the trailing inter-character space in the case of all characters save the last. Thus, the total modulus of this three-character bar code is 16+16+15, or 47. We sum the raw width values and find their sum is 422 "ticks" where a tick is 51.6µs. Thus, the modulus calculated in ticks is 422/47, or 8.98.

Next we classify the calculated modulus values as either narrow or wide. Then we rank each element as a bar or a space, based upon the fact that the first element is a bar. Lastly, we calculate the binary values associated with the bars and spaces.

k	Tick Width	Modulus Width	Class	Type	Binary Value Bar	Space
0	14	1.56	N	B	0	
1	32	3.56	W	S		1
2	10	1.11	N	B	0	
3	12	1.34	N	S		0
4	27	3.01	W	B	1	
5	10	1.11	N	S		0
6	25	2.78	W	B	1	
7	10	1.11	N	S		0
8	8	0.89	N	B	0	
9	9	1.00	N	S	Inter-character space	
10	8	0.89	N	B	0	
11	10	1.11	N	S		0
12	25	2.78	W	B	1	
13	27	3.01	W	S		1

(Continued)

k	Tick Width	Modulus Width	Class	Type	Binary Value Bar	Binary Value Space
14	9	1.00	N	B	0	
15	9	1.00	N	S		0
16	25	2.78	W	B	1	
17	10	1.11	N	S		0
18	8	0.89	N	B	0	
19	9	1.00	N	S	Inter-character space	
20	8	0.89	N	B	0	
21	24	2.67	W	S		1
22	9	1.00	N	B	0	
23	9	1.00	N	S		0
24	24	2.67	W	B	1	
25	9	1.00	N	S		0
26	24	2.67	W	B	1	
27	10	1.11	N	S		0
28	8	0.89	N	B	0	

Sum of Ticks	422
Modulus Units	47
Modulus Ticks	8.98

The result is:

Character Number	Bars Binary	Bars Decimal	Spaces Binary	Spaces Decimal
1	00110	6	1000	8
2	01010	10	0100	4
3	00110	6	1000	8

Checking the Code 3 of 9 value table we find that characters one and three decode as a "*" and character two decodes as the digit "9," as expected.

Program 25-4

To save space, the text of Program 25-4 is not listed but is available on the associated CD-ROM.

Now that we've established the functionality of our interrupt driven measurement algorithm, and developed a manual process for decoding the resulting measurements Program 25-4 assembles it all into a program that will read a Code 3 of 9 bar code of up to 8 characters (including the leading and trailing "*" characters).

Much of Program 25-4 is lifted without change from Program 25-3, so we'll look only at the new components.

The main program loop substitutes a call to **GoSub DecodeData** for Program 25-3's call to **ReadArray**. The interrupt service routine is unchanged from Program 25-3.

Let's look at subroutine **DecodeData**. When **DecodeData** is called, the scan is completed, width data is stored in memory starting at address **$A0** and **k** holds the last index. The subroutine **DecodeData** implements the data slicing algorithm we manually performed earlier.

```
DecodeData              ;decode the read data
;--------------
        Sum = 0         ;sum holds running total
        If k > 0 Then
                k = k-1 ;k=k+1 for last scan, so subtract 1
        ELSE
                k = 79
        EndIf
```

First, we initialize the running sum variable **sum** and correct for the last **k** increment. If we scan a length 79 bar code (*123456* for example) **k** becomes 0. We check for that condition and reset it to the proper value, 79. (Yes, it would be better to fix this problem in the interrupt service routine.)

```
        For i = 0 to k
                PEEK AStart+i,Temp
                Sum = Sum + Temp
        Next ;i
        ;HserOut ["Sum: ",Dec Sum,13]
```

We calculate the sum of the measured widths. If you wish to see the result of this calculation, uncomment the **HSerOut** statement.

```
        Modulus = 16 * (k / ElLen) -1
        Modulus = Sum / Modulus
```

The two statements above replicate our earlier manual calculation. **ElLen** is the element length, 9. We set **ElLen** at 9, not 10, because we use integer division and 9 forces the correct result, as long as the total number of elements read is 79 or less. We then calculate the modulus in terms of counter ticks, exactly as in our earlier manual example.

```
        Thsld = 2 * Modulus
```

We must determine the width threshold to distinguish wide from narrow elements. In theory, the division point should be twice the width of the narrowest element, so we set the threshold decision point **Thsld** as twice the modulus.

```
        If DataDebug = 1 Then
                HSerOut ["Total Bars/Spaces: ",Dec k,13]
                HSerOut ["Modulus: ",Dec Modulus,"    Threshold: ", |
                Dec Thsld,13]
        EndIf
```

By setting the constant **DataDebug** to 1, we can enable printing of more intermediate results.

```
        If (k < 29) Then
                GoTo NothingFound
        EndIf
```

Next, we check to see if the bar code is smaller than the smallest possible Code 3 of 9 result, which is *n* resulting in **k=29**. If **k<29**, the scan is invalid and execution jumps to the end of the subroutine.

```
        If DataDebug = 1 Then
                HSerOut [Dec i," Valid",HTab,"Thsld: ",Dec Temp,13]
                For i = 0 to (k-1)
                        PEEK (AStart+i), Temp
                        HSerOut ["i: ",Dec i,HTab,"Value: ", |
                        Dec Temp,HTab]
                        If Temp > Thsld Then
                                HSerOut ["W",13]
                        Else
                                HSerOut ["N",13]
                        EndIf
                        Pause 10
                Next
        EndIf
```

Again, as a debugging and program verification aid, we optionally can send intermediate values to the serial port.

```
        Star = Good
        ;Check for * in first position
        j = 0     ;first character
        GoSub FillWorkArray
        GoSub CalculateValues
        If (SpaceV = 8) AND (BarV = 6) Then
        Else
                Star = Bad
        EndIf

        ;Check for * in last position
        j = ((i / ElLen)-1)
        GoSub FillWorkArray
        GoSub CalculateValues
        If (SpaceV = 8) AND (BarV = 6) Then
        Else
                Star = Bad
        EndIf
```

As an error check, we verify that the first and last characters decode as "*." (We'll go through the **Calcu-lateValue** subroutine later in the analysis; for now it's enough to say it returns the bar and space numerical values in the two variables **BarV** and **SpaceV** which are then verified against the expected values of 6 and 8, respectively.)

```
        ;If good or bad send message
        If Star = Bad Then
                GoSub BadScan
        Else
                GoSub GoodScan
        EndIf
```

Depending upon the value of the first and last characters, we execute a "good scan" subroutine **GoodScan** or a "bad scan" subroutine, **BadScan**.

```
        ;if want to get more information
        If DataDebug = 1 Then
                GoSub DumpChars
        EndIf
```

Again, we have the option to dump more internal values for program development and debugging.

```
        ;loop through character at a time
        HSerOut ["Decoded as: "]
        For j = 0 to ((i / Ellen)-1)
                GoSub FillWorkArray     ;get 9 widths
                GoSub CalculateValues   ;calculate BarV and SpaceV
                ;get the char corresponding to BarV and SpaceV
                GoSub GetChar
                HSerOut [Temp]
        Next ;j
        HSerOut [13]
NothingFound
Return
```

Lastly, the main program loop outputs the decoded bar code, one character at a time. Individual characters are decoded in the three subroutines **FillWorkArray**, **CalculateValues** and **GetChar**. The decoded character is returned in the byte variable **Temp**.

Let's look at these three subroutines.

```
        ;Fill the work array. Assumes j is set before calling
        ;j is the character number
FillWorkArray
;---------------
        For k = 0 to (ElLen-1)
                PEEK (AStart+10*j+k), Temp
                If Temp > Thsld Then
                        WorkArray(k) = 1
                Else
```

```
                         WorkArray(k) = 0
                   EndIf
           Next ;k
       Return
```

Subroutine **FillWorkArray** reads the nine element widths comprising a complete character from the reserved memory block and evaluates each width as either normal or wide. The normal/wide comparison is based upon whether the modulus width is greater or less than the decision point **Tshld**. If the element evaluates as narrow, a 0 is placed into **WorkArray**, a nine-element byte array. If it evaluates as wide, a 1 is entered instead. The byte variable **j** holds the character number and is evaluated before calling **FillWork-Array**. The relationship between memory addresses, elements and characters is illustrated below.

File Address (Bank1)	Fictitious Array Equivalent	Elements	Character No.
$A0	Data(0)	1	
$A1	Data(1)	2	
$A2	Data(2)	3	
$A3	Data(3)	4	
$A4	Data(4)	5	0
$A5	Data(5)	6	
$A6	Data(6)	7	
$A7	Data(7)	8	
$A8	Data(8)	9	
$A9	Data(9)		Inter-Char Space
$AA	Data(10)	1	
$AB	Data(11)	2	
...	...	...	1
$B2	Data(18)	9	
$B3	Data(19)		Inter-Char Space
$B4	Data(20)	1	2
...	...	...	...
$EE	Data(78)	8	7
$EF	Data(79)	9	

The memory address associated with element **k** (0...8) of character **j** (0...7) is calculated as **AStart+10*j+k**.

Now that **WorkArray** is filled with the logical values for each element, subroutine **CalculateValues** numerically evaluates the bar and space values.

```
       CalculateValues              ;assumes WorkArray is filled;
                                    ;Calcultes spaceV and barV
       ;--------------
              SpaceV = WorkArray(1)*8 + WorkArray(3)*4 + WorkArray(5)*2 |
              + WorkArray(7)
              ;HSerOut ["Char: ",Dec j,HTab,"Space: ",Dec SpaceV,13]

              BarV = WorkArray(0)*16 + WorkArray(2)*8 + WorkArray(4)*4 |
              + WorkArray(6)*2 +WorkArray(8)
              ;We could add limits here, SpaceV must be <= 14
              ;and BarV <=24 but it will confuse our error data dump
       Return
```

As we learned when manually evaluating Code 3 of 9 bar codes, the bar and space values are interleaved. Subroutine **CalculateValues** de-interleaves the bar and space values and calculates the numerical value of each. There are undoubtedly more elegant ways to accomplish this, but **CalculateValues** uses a straightforward method; multiply each bar and each space element's value by its binary weighting, summing the results in separate bar and space value variables, **BarV** and **SpaceV**, respectively.

The last component of decoding the bar code scan is associating the `BarV` and `SpaceV` values with specific characters, a task subroutine `GetChar` accomplishes. If memory were not a limit, we might define a pseudo two-dimensional byte array, such as `Character(BarV,SpaceV)` and populate it with the decoded character matrix. But, memory is a concern, so we'll use a more parsimonious approach. We define four byte arrays, each holding all the characters associated with one particular space value:

```
BC4Str   ByteTable   "????7?40??29?6????18?5???3"
BC8Str   ByteTable   "???-?X*??V ?Z????U.?Y???W"
BC1Str   ByteTable   "???Q?NT??LS?P????KR?O???M"
BC2Str   ByteTable   "???G?DJ??BI?F????AH?E???C"
```

The byte table `BC4Str` holds the characters for `SpaceV = 4`, `BC8Str` for `SpaceV = 8` and so forth.

`GetChar` returns the decoded character in the byte variable `Temp`.

```
;Call with SpaceV and BarV set. Returns in Temp
GetChar ;gets the associated character
;---------
         Temp = "?"       ;overwrite if decoded
```

We set `Temp="?"` so that if we do not decode a valid Code 3 of 9 character, `Temp` indicates an error. (Code 3 of 9 does not contain the "?" symbol.)

```
         If SpaceV = 4 Then
                 Temp = BC4Str(BarV)
         EndIf

         If SpaceV = 8 Then
                 Temp = BC8Str(BarV)
         EndIf

         If SpaceV = 1 Then
                 Temp = BC1Str(BarV)
         EndIf

         If SpaceV = 2 Then
                 Temp = BC2Str(BarV)
         EndIf
```

The four `If...Then` statements slog through the four main possible values of `SpaceV` and set `Temp` based on indexing `BarV` elements into the associated byte table.

```
         ;For a few special characters, we check with nested IFs
         If BarV = 0 Then
                 If SpaceV = 7 Then
                         Temp = "%"
                 EndIf
                 If SpaceV = 11 Then
                         Temp = "+"
                 EndIf
                 If SpaceV = 13 Then
                         Temp = "/"
                 EndIf
                 If SpaceV = 14 Then
                         Temp = "$"
                 EndIf
         EndIf
Return
```

A handful of special symbols don't justify a full byte table, so we decode them by a multiple `If...Then` statements. All four of these special symbols are associated with `BarV = 0`, so we need test only the associated `SpaceV`.

The remaining new subroutines, `BadScan` and `GoodScan`, provide user feedback. A good scan produces one short beep on the speaker but a bad scan produces a double beep. In addition, a corresponding good/bad message is sent over the serial port.

```
BadScan          ;good scan - beep and message
;----------
         HSerOut ["Bad Scan",13]
         Sound SpkrPin,[BeepLen\BeepHz]
         Pause BeepLen /250
         Sound SpkrPin,[BeepLen\BeepHz]
Return

GoodScan         ;double beep and message
;-----------
         HSerOut ["Good Scan",13]
         Sound SpkrPin,[BeepLen\BeepHz]
Return
```

Let's look at some typical output of Program 25-4.

The first sample is run with **DataDebug** = **0** to suppress internal variable output. (To save space, I've reformatted the output into three columns.)

Good Scan	Bad Scan	Bad Scan
Decoded as: *0*	Decoded as: *1234?	Decoded as: *???456?
Bad Scan	Bad Scan	Good Scan
Decoded as: *1?	Decoded as: ?5678*	Decoded as: *123456*
Bad Scan	Good Scan	
Decoded as: *1?	Decoded as: *5678*	
Good Scan	Good Scan	
Decoded as: *12*	Decoded as: *12345*	
Good Scan	Bad Scan	
Decoded as: *23*	Decoded as: ?09875*	
Good Scan	Good Scan	
Decoded as: *123*	Decoded as: *09875*	
Good Scan	Bad Scan	
Decoded as: *098*	Decoded as: *???456?	

With **DataDebug** = **1**, additional information is provided. (To save space, I've reformatted the output into three columns.)

Total Bars/Spaces: 49		i: 19	Value: 18 N	i: 41	Value: 54 W	
Modulus: 18		i: 16	Value: 16 N	i: 35	Value: 18 N	
Threshold: 36		i: 17	Value: 17 N	i: 36	Value: 20 N	
i: 0	Value: 30 N	i: 18	Value: 43 W	i: 37	Value: 20 N	
i: 1	Value: 63 W	i: 19	Value: 18 N	i: 38	Value: 21 N	
i: 2	Value: 27 N	i: 20	Value: 17 N	i: 39	Value: 18 N	
i: 3	Value: 22 N	i: 21	Value: 17 N	i: 40	Value: 21 N	
i: 4	Value: 53 W	i: 22	Value: 43 W	i: 41	Value: 54 W	
i: 5	Value: 24 N	i: 23	Value: 44 W	i: 42	Value: 27 N	
i: 6	Value: 49 W	i: 24	Value: 17 N	i: 43	Value: 21 N	
i: 7	Value: 20 N	i: 25	Value: 16 N	i: 44	Value: 59 W	
i: 8	Value: 18 N	i: 26	Value: 17 N	i: 45	Value: 28 N	
i: 9	Value: 18 N	i: 27	Value: 17 N	i: 46	Value: 65 W	
i: 10	Value: 45 W	i: 28	Value: 44 W	i: 47	Value: 26 N	
i: 11	Value: 18 N	i: 29	Value: 20 N	i: 48	Value: 29 N	
i: 12	Value: 18 N	i: 30	Value: 45 W	Good Scan		
i: 13	Value: 43 W	i: 31	Value: 20 N	0	Space 8	Bar 6
i: 14	Value: 17 N	i: 32	Value: 46 W	1	Space 4	Bar 17
i: 15	Value: 15 N	i: 33	Value: 50 W	2	Space 4	Bar 9
		i: 34	Value: 20 N	3	Space 4	Bar 24
				4	Space 8	Bar 6
				Decoded as: *123*		

As you may have noted from our byte table definitions, Program 25-4 decodes the extended Code 3 of 9 values for letters A...Z as well as numbers.

Program 25-5

Now we'll turn to reading a UPC-A bar code. We'll reuse much of Program 25-4, but we'll improve the slicing and decoding algorithms, and make a few other changes to reflect the different structure of UPC-A.

To save space, the text of Program 25-5 is not listed but is available on the associated CD-ROM.

As usual, we'll concentrate on the differences between Program 25-5 and Program 25-4. Until the subroutine **Characterize**, the differences are trivial to nonexistent.

One major difference in subroutine **Characterize** is that Program 25-5 decodes and analyzes the 12 UPC-A digits one at a time throughout the entire subroutine. Program 25-4 considers individual characters only towards the end of its corresponding subroutine **Characterize**.

```
Characterize            ;read array and group and characterize
;-------------
        ;Assume scan is good; any error makes it bad
        ScanStatus = Good
        ;Scan the data based on number of characters
        k = k-1 ;increment at end is one too many
        ;A good UPC-A will have k=58 after subtraction
        ;represening 0...58 (59 total) elements
        If k <> 59 Then
                ScanStatus.Bit0 = %1
        EndIf
```

A UPC-A code is more structured than the Code 3 of 9 decoded in Program 25-4. A valid UPC-A bar code will have exactly 59 elements, so we can check for one possible error condition by verifying that we scanned exactly 59 elements. If not, we set **ScanStatus.Bit0** to **%1**. **ScanStatus.Bit0** is a bit-mapped error flag that we'll use to provide a bad scan diagnosis.

```
        Position = 0
        m = 0 ;index for digit array

        ;step through the width array, one character at a time
        ;a char is start/stop/center & the 12 digits
        For j = 0 to (NoChars-1)
                Sum = 0
                ;ChIndex holds no of elements for each character.
                ;Get the sum of the bar and space widths for all
                ;elements in the character
                For n = 0 to (ChIndex(j)-1)
                        PEEK (AStart+Position+n),Temp
                        Sum = Sum + Temp
                Next ;n
```

The **For j ... Next** loop contains the character-by-character analysis that distinguishes Program 25-5 from Program 25-4. The loop variable **j** represents the character number.

A UPC-A bar code has 12 numerical characters, plus the start, stop and center symbols. For simplicity, we'll also treat these special symbols as characters, thus giving us 15 characters to consider. The constant **NoChars** is set to 15 and the byte table **ChIndex** holds the number of bars and spaces in each symbol: **ChIndex ByteTable 3,4,4,4,4,4,4,5,4,4,4,4,4,4,3**.

When decoding Code 3 of 9 bar codes, we calculated the modulus and threshold values based upon all elements of the bar code. Program 25-5 instead calculates the modulus and threshold for each character. This improves the threshold decision, as variations in wand speed are smaller considered over a shorter span.

```
        ;ElIndex = number of elements in each character.
        ;Numbers are formed from 7 elements.
        ;Modulus is the average element width
        Modulus = Sum / ElIndex(j)
```

As we learned earlier, each UPC-A numerical character is constructed from seven elemental lengths, while the start and stop characters are length three and the center character is length five. The elemental length values are held in the byte table `ElIndex`: `ElIndex ByteTable 3,7,7,7,7,7,7,5,7,7,7,7,7,7,3`. When computing the modulus of each individual character, we divide the sum of the characters width by the elemental length: `Modulus = Sum / ElIndex(j)`.

```
;Dump data if interested
If DataDebug = 1 Then
        HSerOut [Dec Position,HTab,Dec j,HTab, |
        Dec Sum,HTab,Dec Modulus,13]
EndIf
```

As in earlier programs, we incorporate an optional debugging and development data output.

```
;If ChIndex=4, it's a digit 3 & 5 are separators
If ChIndex(j) = 4 Then
        GoSub GetDigitValue
EndIf
```

The start, stop and center characters do not have a numerical value, so they are not evaluated—just the 12 numerical characters are set to the subroutine `GetDigitValue` for evaluation. We'll study `GetDigitValue` later. For now, all we need to know is that it sets the variable `DValue` to the numerical value of character `j`. `DValue` is <u>not</u> the *character* represented by the particular bar/space combination; rather it is the *numerical value* of the bar/space combination computed using the rules we learned earlier for UPC-A valuation. `GetDigitValue` also determines the character represented by bar/space combination and stores it in the 12-element byte array `Digits()`.

```
;Determine direction by looking at first digit.
;Assume it is forward scan. If it is in the set
;RevKey, then it is a reverse scan.
If (j=1) Then
        Dir = FwdScan
        For n = 0 to 9
                If DValue = RevKey(n) Then
                        Dir = RevScan
                EndIf
        Next ;n
EndIf
```

UPC-A's inventors designed it to be scanned from either left-to-right or right-to-left. Indeed, the left half digits need not be scanned in the same direction as the right half digits. We'll implement directionless scanning in Program 25-5. To distinguish the scan direction, we test the value of the first numerical character scanned. The forward and reverse values in the decoding table, reproduced below, establish the direction of the scan.

Digit	\multicolumn{7}{c}{Left of Center}	Forward Value	Reverse Value
	6	5	4	3	2	1	0		
0	0	0	0	1	1	0	1	13	88
1	0	0	1	1	0	0	1	25	76
2	0	0	1	0	0	1	1	19	100
3	0	1	1	1	1	0	1	61	94
4	0	1	0	0	0	1	1	35	98
5	0	1	1	0	0	0	1	49	70
6	0	1	0	1	1	1	1	47	122
7	0	1	1	1	0	1	1	59	110
8	0	1	1	0	1	1	1	55	118
9	0	0	0	1	0	1	1	11	104

			Right of Center					Forward	Reverse
Digit	6	5	4	3	2	1	0	Value	Value
0	1	1	1	0	0	1	0	114	39
1	1	1	0	0	1	1	0	102	51
2	1	1	0	1	1	0	0	108	27
3	1	0	0	0	0	1	0	66	33
4	1	0	1	1	1	0	0	92	29
5	1	0	0	1	1	1	0	78	57
6	1	0	1	0	0	0	0	80	5
7	1	0	0	0	1	0	0	68	17
8	1	0	0	1	0	0	0	72	9
9	1	1	1	0	1	0	0	116	23

We test whether **DValue** is in the reverse set by comparing it to the values held in the byte table **RevKey**: **RevKey ByteTable 39,51,27,33,29,57,5,17,9,23**. We compare the first digit only against right-of-center reverse values because it's physically impossible to scan half a UPC-A bar code in one direction and half in the other in a continuous scan with a hand-held wand. Hence, the choices are left-to-right and right-to-left for the full 12 digits. Based on the result, we set the scan direction flag **Dir** to either **FwdScan** or **RevScan**.

```
;Dump data if interested
If DataDebug = 1 Then
        For n = 0 to (ChIndex(j)-1)
                PEEK (AStart+Position+n),Temp
                ;we add Modulus/2 for rounding,
                ;since integer division is smallest
                ;integer
                CWidth = (Temp + Modulus/2) / Modulus
                HSerOut [Dec (Position+n),HTab, |
                Dec Temp,HTab,Dec CWidth,13]
        Next ;n
        HSerOut [13]
EndIf
```

As in earlier programs, we incorporate an optional debugging and development data output.

```
;Position is the starting point into the width array
        Position = Position + ChIndex(j)
Next ;j
```

The last step in cycling through the bar code one character at a time is to update the variable **Position** which holds the next offset into the RAM memory block from **$A0** to **$EF** holding the scan widths.

```
;If the scan was right-to-left then reverse digits
If Dir = RevScan Then
        GoSub ReverseArray
EndIf
```

The array **Digits()**, computed in subroutine **GetDigitValue** holds the ASCII characters corresponding to the 12 UPC-A digits, held in the order of first scanned to last scanned. If we scanned left-to-right, the digits are correctly held in **Digits()**. If we scanned right-to-left, however, it's necessary to reverse the digit order, which we accomplish through a call to subroutine **ReverseArray**.

```
;write the digits
For n = 0 to (MaxDigits-1)
        HSerOut [Digits(n)]
        ;Check to see if any digits are bad
        If Digits(n) = "?" Then
                ScanStatus.Bit1 = %1
        EndIf
Next ;n
HSerOut [13]
```

We now display the digits by writing them to the serial port. The digits will be in the correct order, regardless of the direction we scanned. If any character improperly decodes, its character is set as a "?" so we use the presence "?" character to detect for a bad scan, setting the effort flag `ScanStatus.Bit1` to `%1` to reflect a digit error.

```
GoSub ComputeCheckDigit
;Use CheckDigit for error. Bad digit comes out as "?"
If CheckDigit <> (Digits(MaxDigits-1)-"0")    Then
            ScanStatus.Bit2 = %1
EndIf
```

The final error check is to compute the check digit based upon the first 11 scanned digits and compare the computed value to the scanned check digit. If the computed and scanned check digits do not match, we have another type of error and we accordingly set `ScanStatus.Bit2` to `%1`.

```
;report scan status with beeps
If ScanStatus = Good Then
            GoSub GoodScan
Else
            GoSub BadScan
EndIf
Return
```

Lastly, we execute either the `GoodScan` or `BadScan` subroutine, based upon the error flag `ScanStatus`. `ScanStatus` will only equal `Good` (a constant we defined as 0) if none of its individual error condition bits have been set.

Now, let's finish up with the subroutines called by `Characterize`.

Subroutine `GetDigitValue` is called with `j`, `AStart` and `Modulus` defined. The variable `j` is the character number, `AStart` is the memory block starting address, `$A0` in our case, and `Modulus` is the average width of the narrowest element in this particular block of four bars/spaces.

```
GetDigitValue
;-------------
      DValue = 0
      For n = 0 to ChIndex(j)-1
            PEEK (AStart+Position+n),Temp
            CWidth = (Temp + Modulus/2) / Modulus
      Next ;n
```

We read `ChIndex(j)` characters from the width memory block and calculate their width in modulus units. Since we use integer division, to obtain the equivalent of a rounding operation, we add one half of the divisor to the dividend.

An example demonstrates why this is necessary. Suppose `Modulus = 10` and the bar width, held in `Temp`, is 9. It may be obvious to us that `Temp` represents a bar of width 1, but in integer division 9/10 = 0. Biasing `Temp` upwards by one half `Modulus` causes the result to be rounded upward. A series of numerical examples illustrates the benefits of rounding upward

Temp	Modulus	If CWidth = (Temp/ Modulus)	Temp+Modulus/2	If Cwidth= (Temp+Modulus/2)/ Modulus
13	10	1	18	1
14	10	1	19	1
15	10	1	20	2
16	10	1	21	2
17	10	1	22	2

Having computed the width of the element in modules, we now convert this to a number.

```
For i = 1 to CWidth
                            DValue.Bit0 = n.Bit0
                            DValue = DValue << 1
            Next
```

Remember that left of center digits always start with a space, while right of center digits start with a bar; that spaces are binary zeros and bars are binary 1s and that the width of the bar or space is the number of 1s or 0s.

Accordingly, we set the **DValue.Bit0** to **%1** or **%0**, depending on whether the element is a space or a bar and repeat; shifting **DValue's** bits one place to the right for every unit width. Left of center characters start with a space and alternate, so when **n** is even we are working with a space; when it is odd we have a bar. (The first element is element number 0) Thus, **n.Bit0** is **%0** for spaces and **%1** for bars, which is exactly the value we wish to shift into **DValue.Bit0**.

Not so fast, you say. Using **n.Bit0** to set the decoded bit is valid for digits to the left of center but is backwards for digits to the right of center that start with spaces. That's certainly correct. However, if you carefully examine the UPC-A code chart, the only difference between left-of-center and right-of-center encoding is that the 1s and 0s are reversed, *i.e.,* right-of-center codes are the complements of left-of-center codes. Hence, ignoring bit inversion as we traverse the center still produces correct decode values.

This may be clearer if we take an example. Suppose the digit 0 appears in both left and right halves of the code. We'll call these characters 0L and 0R. If we follow the UPC-A standard strictly, we decode 0L as **%0001101**, or 13 and 0R as **%1110010**, or 114. However, by ignoring the starting element inversion, we decode 0R as **%0001101**, or 13. Whether we follow the UPC-A standard exactly, or ignore the element inversion, the character decodes identically. Should this shortcut offend your sensibilities, it would not be difficult modify subroutine **GetDigitValue** to invert the element bit value, based upon the character number.

```
            DValue = DValue >> 1    ;make up for last excess shift
```

Since we shift after each element, the final element causes an extra left shift, which we reverse with a single right shift after the element loop.

```
            If DataDebug = 1 Then
                    HSerOut [Dec DValue,HTab,iBin DValue,HTab, |
                    "Digit: ",DigitMap(DValue),13]
            EndIf
```

Again, we have an optional data dump for program debugging and development.

```
            ;store the decoded digits
            Digits(m) = DigitMap(Dvalue)
            m = m+1
            If m = MaxDigits then
                    m = 0
            EndIf
    Return
```

Finally, decode the digit into its corresponding character, using the byte table **DigitMap**.

```
DigitMap        ByteTable       "?????6???8?9?0??",    ;0
                                "?7?2???9?1?2?4??",    ;10
                                "?3?4???0???????6",    ;20
                                "?5?1???8?5?7?3??",    ;30
                                "??3?7?5?8???1?5?",    ;40
                                "6???????0???4?3?",    ;50
                                "??4?2?1?9???2???",    ;60
                                "??0?9?8???6?????"     ;70
```

We can recast this data in a more human readable format:

Code	Value	Code	Value	Code	Value	Code	Value
5	6	33	3	66	3	98	4
9	8	35	4	68	7	100	2
11	9	39	0	70	5	102	1
13	0	47	6	72	8	104	9
17	7	49	5	76	1	108	2
19	2	51	1	78	5	110	7
23	9	55	8	80	6	114	0
25	1	57	5	88	0	116	9
27	2	59	7	92	4	118	8
29	4	61	3	94	3	122	6

All other possible values in **DigitMap** are filled with the error character, "?"

An important error-checking feature in UPC-A is the check digit. The check digit is the last digit and it is calculated with the following algorithm, applied to the first 11 digits:

a. Sum the odd digits and multiply the sum by three
b. Sum the even digits and multiply the sum by one
c. Add the two values calculated in steps (a) and (b). Subtract this value from the smallest multiple of 10 that exceeds this sum. The result is the check digit.

Suppose the UPC-A is 6 39785 32176 7. The check digit is shown as 7. Let's calculate it and see if 7 is correct.

Category	Digits	Sum	Multiplier	Product
Odd	6+9+8+3+1+6	33	3	99
Even	3+7+5+2+7	24	1	24
Sum of Products				123

The smallest multiple of 10 that exceeds 123 is 130. Hence, the check digit is: 130-123 = 7. This matches the value of the check digit in the bar code.

```
;Call with the 12 values held in Digits
ComputeCheckDigit
;-----------------
        Sum = 0
        ;apply check to first 11 digits
        For i = 0 to (MaxDigits-2)
        ;odd has multiplier of 3. But odd/even is
        ;based on 1st digit being Digit No. 1, i.e., odd.
        ;Since we start from 0, must reverse odd/even
                If i.Bit0 = %0 Then
                        Sum = Sum + (Digits(i)-"0")*3
                Else
                Sum = Sum + (Digits(i)-"0")
                EndIf
                If DataDebug = 1 Then
                        HSerOut [Dec i,HTab,Dec Sum,13]
                EndIf
        Next ;i
        ;Check digit is no. to make sum = multiple of 10
        CheckDigit = 10 - (Sum // 10)
        CheckDigit = CheckDigit // 10
```

The subroutine **ComputeCheckDigit** implements this algorithm. However, instead of looking for the smallest even multiple of 10 that exceeds the sum, we take advantage of the rules of base 10 arithmetic to simplify the computation. The operation (**Sum // 10**) is the remainder after we divide the sum by 10; in our example 123//10 = 3. Subtracting this from 10 yields the check digit, 10-3 = 7. However, if **Sum//10 = 0**, the result of this subtraction is 10, not 0. Hence, we do another remainder calculation to force a checksum of 10 into the correct 0 value.

```
If DataDebug = 1 Then
        HSerOut ["Sum: ",Dec Sum,HTab,"Check Digit: ", |
        Dec CheckDigit,13]
EndIf
Return
```

Again, we can optionally see intermediate checksum calculations.

```
BadScan ;good scan - beep and message
;----------
        ;HSerOut ["Bad Scan - Error Code: ",iBin ScanStatus,13]
        Sound SpkrPin,[BeepLen\BeepHz]
        Pause BeepLen/250
        Sound SpkrPin,[BeepLen\BeepHz]

        For n = 0 to 2
                If ScanStatus.Bit0 = %1 Then
                        HSerOut [Str ErrMsg(n*16)\16,13]
                EndIf
                ScanStatus = ScanStatus >> 1
        Next ;n
Return
```

We've expanded subroutine **BadScan** to write an error message. The error messages are contained in a 48-element byte array, with each message of length 16. This lets us compactly store and write multiple error messages, based on the lower three bits of **ScanStatus**. Each bit is examined one at a time through a shift right sequence.

```
GoodScan         ;double beep and message
;-----------
        If DataDebug = 1 Then
                HSerOut ["Good Scan",13]
        EndIf
        Sound SpkrPin,[BeepLen\BeepHz]
Return
```

Subroutine **GoodScan** is the same as in Program 25-4.

```
;Reverses digits in Digits()
;Assumes i & temp are free
ReverseArray
;--------------
        ;start and left and reverse digits
        For i = 0 to ((MaxDigits/2)-1)
                Temp = Digits(i)
                Digits(i) = Digits(MaxDigits-1-i)
                Digits(MaxDigits-1-i) = Temp
        Next ;i
Return
End
```

Lastly, it's necessary to reverse the digit order where a scan is made right-to-left. We do this by swapping the first and last values in the array **Digits()**, then swapping the second and next to last values, etc. moving inward to the center of the array. We could, of course, use MBasic's **swap** function instead of the intermediate variable **temp**.

Let's see how it works in practice. Here's some sample output with **DataDebug=0** .

```
031901927370
081262224533
081262224533
639785321774
```

939785321774
Bad Bar Number
Checksum Error
914014??????
Bad Bar Number
Digit Read Error
Checksum Error
057163170074
Bad Bar Number
Checksum Error

As for Program 25-4, if we set **DataDebug=1**, we see intermediate data values. Again, I've reformatted the output into three columns to save space.

i: 0	Width: 32	i: 66	Width: 0			32	8	98	14
i: 1	Width: 18	i: 67	Width: 0			19	%10011	Digit: 2	
i: 2	Width: 25	i: 68	Width: 0			32	30	2	
i: 3	Width: 64	i: 69	Width: 0			33	14	1	
i: 4	Width: 45	i: 70	Width: 0			34	27	2	
i: 5	Width: 22	i: 71	Width: 0			35	27	2	
i: 6	Width: 20	i: 72	Width: 0						
i: 7	Width: 21	i: 73	Width: 0			36	9	99	14
i: 8	Width: 41	i: 74	Width: 0			19	%10011	Digit: 2	
i: 9	Width: 21	i: 75	Width: 0			36	30	2	
i: 10	Width: 56	i: 76	Width: 0			37	14	1	
i: 11	Width: 39	i: 77	Width: 0			38	28	2	
i: 12	Width: 35	i: 78	Width: 0			39	27	2	
i: 13	Width: 34	i: 79	Width: 0						
i: 14	Width: 18	0	0	75	25	40	10	100	14
i: 15	Width: 30	0	32	1		35	%100011	Digit: 4	
i: 16	Width: 18	1	18	1		40	16	1	
i: 17	Width: 29	2	25	1		41	13	1	
i: 18	Width: 31					42	42	3	
i: 19	Width: 16	3	1	151	21	43	29	2	
i: 20	Width: 14	13	%1101	Digit: 0					
i: 21	Width: 16	3	64	3		44	11	107	15
i: 22	Width: 58	4	45	2		49	%110001	Digit: 5	
i: 23	Width: 32	5	22	1		44	17	1	
i: 24	Width: 17	6	20	1		45	27	2	
i: 25	Width: 29					46	46	3	
i: 26	Width: 32	7	2	139	19	47	17	1	
i: 27	Width: 16	55	%110111	Digit: 8					
i: 28	Width: 14	7	21	1		48	12	110	15
i: 29	Width: 16	8	41	2		61	%111101	Digit: 3	
i: 30	Width: 15	9	21	1		48	14	1	
i: 31	Width: 15	10	56	3		49	61	4	
i: 32	Width: 30					50	20	1	
i: 33	Width: 14	11	3	126	18	51	15	1	
i: 34	Width: 27	25	%11001	Digit: 1					
i: 35	Width: 27	11	39	2		52	13	118	16
i: 36	Width: 30	12	35	2		61	%111101	Digit: 3	
i: 37	Width: 14	13	34	2		52	16	1	
i: 38	Width: 28	14	18	1		53	65	4	
i: 39	Width: 27	15	4	108	15	54	21	1	
i: 40	Width: 16					55	16	1	

(Continued)

i	Width								
i: 41	Width: 13	19	%10011	Digit: 2					
i: 42	Width: 42	15	30	2		56	14	50	16
i: 43	Width: 29	16	18	1		56	17	1	
i: 44	Width: 17	17	29	2		57	17	1	
i: 45	Width: 27	18	31	2		58	16	1	
i: 46	Width: 46								
i: 47	Width: 17	19	5	104	14	081262224533			
i: 48	Width: 14	47	%101111	Digit: 6		0	0		
i: 49	Width: 61	19	16	1		1	8		
i: 50	Width: 20	20	14	1		2	11		
i: 51	Width: 15	21	16	1		3	13		
i: 52	Width: 16	22	58	4		4	31		
i: 53	Width: 65					5	33		
i: 54	Width: 21	23	6	110	15	6	39		
i: 55	Width: 16	19	%10011	Digit: 2		7	41		
i: 56	Width: 17	23	32	2		8	53		
i: 57	Width: 17	24	17	1		9	58		
i: 58	Width: 16	25	29	2		10	67		
i: 59	Width: 25	26	32	2		Sum: 67	Check Digit: 3		
i: 60	Width: 0					Good Scan			
i: 61	Width: 0	27	7	76	15				
i: 62	Width: 0	27	16	1					
i: 63	Width: 0	28	14	1					
i: 64	Width: 0	29	16	1					
i: 65	Width: 0	30	15	1					
		31	15						
		1							

EAN-13

One final note and we'll bring the curtain down on another long chapter. Beginning in 2005, for increased international compatibility, UPC-A will be supplemented by the EAN-13 code. The good news is that EAN-13 is almost identical with UPC-A, except it adds an additional digit to the number system place, for a total of 13. Its structure is:

Number of Digits	Content
2	Number System (Country Code)
5	Manufacturer Code
5	Product Code
1	Check Digit

The not so good news is how the additional digit is added. EAN-13 maintains the same 59-element structure of UPC-A (a good thing indeed) and the same basic digit encoding (also a good thing). The extra number system digit is derived from parity computations on the five left-of-center manufacturer code digits, whereby the left-of-center digits are encoded either with odd or even parity schemes. The current UPC-A left-of-center digit encoding is the "odd parity," encoding scheme, that is, all characters are encoded with an odd number of 1s. The "even parity" scheme is the right-of-center scheme, but reversed. Depending upon which digits to the left of center are even parity encoded and which are odd parity encoded, the 13th digit may be derived. This may be though of as a superimposed five-bit binary system: OOOOO, OEOEE, etc. where OOOOO indicates all five manufacturer digits are encoded with odd parity, OEOEE indicates they are encoded in the sequence odd-even-odd-even-even parity encoding. The five-bit binary values derived from the odd/even encoding technique then are mapped into the 13th digit. For example, if the five manufacturer digits are encoded OOOOO, the 13th digit is 0. (This, by the way, yields UPC-A. And, the country code for the United States is 0, so EAN-13 reduces to UPC-A in this case.) If the five manufacturer code digits are encoded with the OEOEE pattern, the 13th digit is 1.

The check digit computation also changes, with the new 13th digit (which is the first digit sequentially, as the last digit remains the check digit) being considered an even digit for the purpose of check digit computation. Reference [25-12] provides an excellent overview of the changes between UPC-A and EAN-13 and should provide sufficient information for you to modify Program 25-5 to read both UPC-A and EAN-13 bar codes and identify which type of code has been read.

Ideas for Modifications to Programs and Circuits

- Modify Program 25-5 to read both UPC-A and EAN-13 codes. Hint: Compare the **DValue** for the five manufacturer digits against an odd/even table. Consider the digit 0 for example. If it is encoded with odd parity, it is %0001101, or 13. If it is encoded with even parity it is %0100111 or 39, assuming we scan left-to-right. If we scan the bar code right-to-left, the corresponding values would be (even parity) %1011000 (88) and (odd parity) %1110010 (114). If **DValue** = 13 or 114, the parity is odd. If **DValue** = 88 or 39, parity is even. Or, the odd/even status can be computed by simply summing the number of 1s in **DValue**. Use the parity status to set bits 0...4 of a new variable **Digit13**, and then use a byte table to determine the corresponding 13th digit value. (The 13th digit isn't the direct binary value of the odd/even superencoding.)

- How would you modify the circuitry and program to add code storage to external EEPROM? How should be data be stored and retrieved for later downloading to a PC?

- Add an LCD display to show the decoded digits, and inform the user of an error in scanning.

- Many 3 of 9 bar codes are longer than the maximum value our program accepts. How would you expand the decoding length?

- Our program structure may be categorized as "post processing" in that we scan bar width data into memory and decode its value after the scan is complete. An alternative approach decodes the bar code as the scan is in progress. Is this feasible with MBasic and appropriate assembler code where necessary for time-critical matters? How would you structure such a program? What advantages and disadvantages might it have compared with our post-processing approach?

References

[25-1] Agilent Technologies, *Elements of a Bar Code System*, Application Note 1013, (1999). Available at http://www.semiconductor.agilent.com.

[25-2] Dennon, Jack, *Linux Reads Bar Codes*, ELJonline, March 2001. Available at http://www.linuxdevices.com.

[25-3] Hewlett Packard (now Agilent Technologies), *Low Current Bar Code Digitizer IC, Technical Data HBCC-0500*, Document No. 5964-1563E, (October 1995). Contains a suggested circuit that is the basis for HP's A000 series wands.

[25-4] Agilent Technologies, *Low Current Digital Bar Code Wand Technical Data HBCS-A000 Series*, Document No. 5964-6664E (November, 1999). Covers the type of wand used in this chapter. However, the pin connection data is not necessarily accurate.

[25-5] Baker, John and Bauer, Wayne, *The Mathematics of Barcodes, A Unit for Tech Prep Mathematics Courses*, Mathematics Education Development Center, Indiana University (1999). Available at http://www.indiana.edu/~atmat/units/barcodes/barunit.htm.

[25-6] Laurer, George, Questions Pertaining to the Code and Symbol Technology, (undated). Available at http://members.aol.com/productupc/techques.html. Mr. Laurer invented the UPC bar code in the early 1970's while at IBM.

[25-7] Worth Data, Inc., *Bar Code Primer* (May 2003). Available at http://www.barcodehq.com.

[25-8] Code 39 Barcode Specification (undated). Available at http://www.barcodeman.com/info/c39_1.php3

[25-9] Thomas, Roger, *Bar Codes Demystified*, Electronics World (September 2001), reprinted at http://www.blackmarket-press.net/info/plastic/list_1.htm.

[25-10] If you wish to look up information on a product based on its UPC-A code, try http://www.upcdatabase.com/.

[25-10] By far the best of the freeware or shareware Code 3 of 9 fonts that I looked at is from IDAutomation. com, which makes available free for personal use a limited version of its Code 39 font package (restricted to numbers 0…9 and the start/stop symbol). The download is linked at http://idautomation.com/.

[25-11] A useful demo UPC-A generator program for Windows, barcodemaker.exe, is available at http://www. winbarcode.com/index.php. The bar codes generated are complete, except the word "Eval" obscures half the lead-in guard elements. The resulting bar code is perfectly readable, however.

[25-12] An excellent explanation of EAN-13 may be found is at http://www.barcodeisland.com/ean13.phtml.

Printing Your Own Bar Code

Figures 25-12 and 25-13 provide sample Code 3 of 9 and UPC-A bar codes for you to develop and test the programs in this chapter. You may easily print your own Code 3 of 9 and UPC-A bar codes, however.

3 of 9 Bar Codes

Code 3 of 9 bar codes are the easiest to print. First, you must download and install a Code 3 of 9 bar code font. One source of free 3 of 9 TrueType format bar code fonts is Barcodes, Inc., http://idautomation.com/. Download the demonstration package, which is fully functional for the digits 0…9. More comprehensive, but not free, bar code font packages are available for purchase from this supplier.

Use any Windows word processing program, such as Microsoft Word or Notepad, to type the information you wish to convert to a bar code. The first and last character must be an asterisk "" symbol. If you use Microsoft Word, the default settings cause leading and trailing asterisks to be interpreted as bold font commands and any text between the two asterisks to bolded. You may disable this automatic bolding from Word's Tools | Auto Correct menu. Select the tab "Autoformat as You Type" and uncheck the box "*bold* and _italc_ with real formatting" under "Replace as you Type." After you have typed the information to be converted to a bar code, such as *1234* and then select the text (including the leading and trailing asterisks) and change the font to the bar code font you earlier downloaded.*

Code 3 of 9

Figure 25-12: Sample Code 3-of-9 bar codes.   *Figure 25-13: Sample UPC-A bar code.*

UPC-A Bar Codes

For noncommercial experimentation, the easiest way to print a UPC-A bar code is through the Windows program barcodemaker.exe available for free download at http://www.winbarcode.com/index.php. The downloaded program is fully functional, but adds the word "eval" to the leading and trailing bars, and prints those bars half-height. The resulting bar codes properly decode.

Sending Morse Code

Although hand-sent Morse code may be archaic, it's still an interesting and important part of the hobby of Amateur Radio. We're going to build an electronic keyer to simplify the task of sending Morse code. Even if you have no interest in Morse code, don't skip this chapter, as we'll delve into decoding variable length codes using a binary weighted tree and learn a few other useful techniques as well.

A keyer translates contact closures from a *paddle* into perfectly formed and spaced dots, dashes and spaces. Spaces? Yes, spaces. Spaces—the idle time between dots and dashes and between letters and words—are as important as the dot and dash elements in sending and receiving Morse.

A *paddle*, as illustrated in Figure 26-1 is nothing more than a pair of single-pole, single throw switches, operated by pressing one lever or the other. And, since the switches of the Figure 26-1 paddle are independent, both may be closed at once. Pressing one lever sends dots, pressing the other sends dashes and pressing both sends alternating dots and dashes. (The first lever closed determines whether the first element sent is a dot or a dash.) The ability to automatically send alternating dots and dashes by squeezing both paddle levers is known in ham radio terminology as *iambic* keying and the paddle of Figure

Figure 26-1: A Morse code paddle.

26-1 is an iambic paddle. The term iambic comes from the word iamb used to describe alternating short and long syllables in poetry. From eighth grade English class you may recall that iambic pentameter means each line of the poem has five alternating short/long syllables, such as Shakespeare's Sonnet 49:

> **To leave poor me thou hast the strength of laws,**
> **Since why to love I can allege no cause.**

Based on the status of the dot and dash levers, our keyer automatically completes the correct length dot or dash, sends alternating dots/dashes, adds the correct end-of-element space and displays the sent letters (including word spaces) on an LCD display. It uses two rotary encoders to select and set important parameters, such as sending speed and the dot/dash ratio.

Morse Code 101

We first ran into Mr. Morse and his code in Chapter 17 to respond in a telephone remote control system. Let's summarize what Morse code is and how it works. Morse code is comprised not only of "dots" and "dashes" but, equally importantly, spaces. The duration of dots, dashes and spaces are defined with respect to the duration of a single dot. The relationship is:

Element	Relative Length
Dot	1
Dash	3
Inter-Element Space	1
Inter-letter Space	3
Inter-Word Space	7

Morse is a variable length code, where the most common letters are represented by the shortest combination of elements. E, for example, is the most common letter in the English alphabet and is assigned the shortest possible Morse character, a single dot, and can be transmitted in the interval of two dot lengths—one for the dot, one for the space following the dot. Q, in contrast, infrequently occurs in English and is assigned a long symbol, dash-dash-dot-dash, and requires 14 dot lengths to send—three dashes at four dot lengths each (three for the dash plus one for the following space) and one dot of two dot lengths (one for the dot, one for the space).

Here are the most common Morse characters, along with three special "prosigns" or procedure signals, $\overline{AR}$, $\overline{SK}$ and $\overline{SN}$. A prosign is sent as one letter, that is, for $\overline{AR}$ there is only one dot space between the A and the R, not three dot spaces for the separate letters "A" followed by an "R." The three prosigns mean end of message ($\overline{AR}$), end of communications ($\overline{SK}$) and ready to proceed ($\overline{SN}$) and are widely used in amateur radio Morse communications.

Char	Code		Char	Code
0	— — — — —		M	— —
1	· — — — —		N	— ·
2	· · — — —		O	— — —
3	· · · — —		P	· — — ·
4	· · · · —		Q	— — · —
5	· · · · ·		R	· — ·
6	— · · · ·		S	· · ·
7	— — · · ·		T	—
8	— — — · ·		U	· · —
9	— — — — ·		V	· · · —
A	· —		W	· — —
B	— · · ·		X	— · · —
C	— · — ·		Y	— · — —
D	— · ·		Z	— — · ·
E	·		?	· · — — · ·
F	· · — ·		"/"	— · · — ·
G	— — ·		,	— — · · — —
H	· · · ·		=	— · · · —
I	· ·		$\overline{SK}$	· · · — · —
J	· — — —		$\overline{AR}$	· — · — ·
K	— · —		$\overline{SN}$	· · · — ·
L	· — · ·			

Let's get a few more concepts out of the way. Amateur radio operators transmit Morse by off/on keying of a transmitter, a technique usually called "CW" or continuous wave transmission. Dots and dashes represent "key down" output; spaces are transmitted as no signal or "key up." This technique is better described as ICW, or interrupted continuous wave transmission, but that's not what hams call it. Modern transmitters or transceivers have a control voltage of 12 V or less that, by convention, is pulled to ground to put the transmitter in the "key down" or transmit state. For operator convenience, keyers usually provide an audio signal, or *sidetone*, that tracks the transmitter keying, as it's almost impossible to accurately send even moderate speed Morse without acoustical feedback. Finally, although the theoretically perfect dot/dash ratio is 1:3, some operators prefer lighter (shorter dots) or heavier (longer dots) characteristics. (It may also be necessary to adjust the weight to compensate for the turn on/turn off time of the particular transmitter with which the keyer is to be used.) The dot/dash ratio is called *weighting*, or *weight*, and we'll make it adjustable in our final designs. (The term weight itself comes from the days of mechanical semiautomatic keys that used a vibrating lever to create dots; the speed of the dots was changed by moving a weight closer or further from the spring pivot; while the dot duration was changed by increasing or decreasing the weight.)

Programs

Figure 26-2 shows the circuit arrangement for Program 26-1. The pull-up resistors, R1 and R2 are 1K, so that 5 mA current flows through the paddle contacts. We wish to keep the current through the paddle contacts sufficiently high to ensure reliable contact, and 5 mA is a reasonable value, but may not be enough for some contact materials. Additionally, a paddle is usually connected to the keyer through a connecting cable several feet long and hence is subject to stray 60 Hz AC pickup. By keeping the

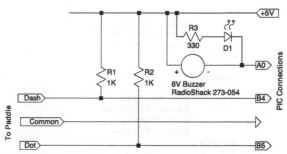

Figure 26-2: Circuit configuration for Program 26-1.

pull-up resistor values in the 1K and less range, we reduce stray voltages that otherwise might be read as false paddle closures. (In a useable version, we must also be concerned about transmitted radio frequency energy being coupled into the keyer, and we may find it necessary to add radio frequency chokes and bypass capacitors to all I/O leads.)

The buzzer is a piezo-electric sounder, Radio Shack part no. 273-054 that operates on 6 V. It puts out more than adequate volume for our brief test. (We only use the buzzer in Program 26-1; if you don't want to spend the money to buy one, skip to Program 26-2.) A key-down condition drops **A0** low, thereby sounding the buzzer and illuminating LED D1.

Program 26-1

```
;Program 26-01
;minimalist keyer

;Constants
;----------------
WPM             Con     25       ;speed in WPM
TonePin         Con     A0       ;Pin for beep/Morse output
DashPin         Con     B4       ;from paddle
DotPin          Con     B5       ;from paddle

;Variables
;----------
ELtime          Var     Word     ;Element length in milliseconds
DashCon         Var     PortB.Bit4
DotCon          Var     PortB.Bit5

;Initialization
;--------------
Input DotPin
Input DashPin
Output TonePin
ELtime = 1200 / WPM   ;From PARIS = 50 Elements = 1 Word
```

```
        Low TonePin
        Pause 250
        High TonePin

Main
                If DashCon=%0 Then
                        GoSub DashOut
                EndIf
                If DotCon = %0 Then
                        GoSub DotOut
                EndIf
        GoTo Main

DotOut
;------
                Low TonePin
                Pause ElTime
                High TonePin
                Pause ElTime
        Return

DashOut
;------
                Low TonePin
                Pause 3*ElTime
                High TonePin
                Pause ElTime
        Return
        End
```

To keep the focus on concepts, we define the code speed as a constant.

```
        WPM             Con     25      ;speed in WPM
```

Morse code speed is universally defined in "words per minute" and the "standard word" "PARIS" is comprised of 50 elements, including inter-character, inter-letter and inter-word spaces, where an element is equal to one dot length. I've set the constant **WPM** at 25, for 25 words per minute. Next we calculate the dot length, our base length of time, in milliseconds.

```
        ELtime = 1200 / WPM  ;From PARIS = 50 Elements = 1 Word
```

The divisor, 1200, is derived from the standard 50 elements = 1 word reference. Thus, the elements per second (EPS) is related to the speed in words per minute (WPM) by:

$$EPS = \frac{WPM \times 50 el/word}{60 \sec/min}$$

Or, restating in milliseconds and inverting the result to yield milliseconds per element:

$$t = \frac{60 \times 1000}{WPM \times 50} = \frac{1200}{WPM}$$

where:

t is the duration of an element, in milliseconds

WPM is the code speed in words per minute.

ElTime is, therefore, the duration in milliseconds, of the basic element of Morse, one **ElTime** is a dot (and inter-element space) length and **3*ElTime** is a dash length.

```
        Main
                If DashCon=%0 Then
                        GoSub DashOut
                EndIf
                If DotCon = %0 Then
                        GoSub DotOut
                EndIf
        GoTo Main
```

The main program loop scans the two paddle contacts and if either is closed, branches to a "send a dot" or a "send a dash" subroutine. Suppose both the dash and dot contacts are closed. How does the iambic action function? It's simple; the main program loop cycles through dash/dot sequences. Whichever paddle lever is closed first determines the first element sent; and the loop structure above automatically adds iambic functionality.

```
DotOut
;------
        Low TonePin
        Pause ElTime
        High TonePin
        Pause ElTime
Return

DashOut
;------
        Low TonePin
        Pause 3*ElTime
        High TonePin
        Pause ElTime
Return
```

The two output subroutines take the output pin low, thereby sounding the buzzer, keeping it low for one element length (dot) or three element lengths (dash). The **Pause ELTime** statement adds one inter-element space at the end of each dot or dash. It's up to the operator, of course, to insert an appropriate inter-word or inter-letter space.

Program 26-2

Let's improve Program 26-1 in two ways. First, we replace the buzzer with a small speaker and let the PIC generate the sidetone. Second, we improve the paddle sensing to better fit with human physiology. Figure 26-3 shows the revised schematic.

Save, perhaps for diode D1, Figure 26-3 is self-explanatory. Pin C1 is one of the 16F87x/A family's PWM hardware module outputs and toggles between high and low at 1600 Hz. We let the PWM run con-

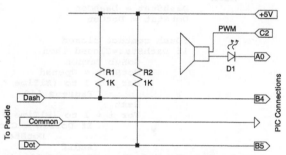

Figure 26-3: Circuit configuration for Program 26-1.

tinuously, and take pin **A0** low to gate the speaker on. Since a speaker produces sound for either polarity of current flow, this scheme will not work without D1. D1 serves a second purpose of a visual indicator that **A0** is taken low.

```
;Program 26-02
;second stage keyer
;Add memory and sidetone generator

;Constants
;----------------
WPM             Con     30          ;speed in WPM
OutPin          Con     A0          ;Pin for sidetone
DashPin         Con     B4          ;from paddle
DotPin          Con     B5          ;from paddle
SideTonePer     Con     12500       ;for 1600 Hz based on 20 MHz clock
DutyCycle       Con     SideTonePer / 2
Opened          Con     %1          ;for keying pin status
Closed          Con     %0

;Variables
;----------
ELtime          Var     Word        ;Element length in milliseconds
DashCon         Var     PortB.Bit4
```

```
DotCon          Var     PortB.Bit5
Mode            Var     Byte
LastEle         Var     Byte    ;holds last element dot or dash
HalfDot         Var     Word    ;duration of one half dot in ms
HalfDash        Var     Word    ;duration of half dash in ms
DotStat         Var     Bit     ;status of dot lever
DashStat        Var     Bit     ;status of dash lever
i               Var     Word    ;loop var

Initialization
;--------------
        ;Setup IO. Dot & Dash have pull ups
        Input DotPin
        Input DashPin
        Output OutPin

        ;set up element timings.
        ELtime = 1200 / WPM  ;From PARIS = 50 Elements = 1 Word
        HalfDot = ElTime / 2
        HalfDash = 3 * HalfDot

        High OutPin
        ;set sidetone frequency
        HPWM 0,SideTonePer,DutyCycle

        DotStat  = Opened
        DashStat = Opened

Main
        DashStat = DashCon
        DotStat = DotCon

        ;Dash contact closed
        If DashStat=Closed Then
                GoSub TurnOn
                DotStat = Opened
                For i = 1 to (ElTime * 3) ;HalfDash
                        Pauseus 140
                Next
                For i = 1 to (ElTime * 3) ;HalfDash
                        If DotCon = Closed Then
                                DotStat = Closed
                        EndIf
                        Pauseus 60
                Next

                ;Pause HalfDash
                GoSub TurnOff
        EndIf

        ;Dot contact closed
        If DotStat = Closed Then
                GoSub TurnOn
                DashStat = Opened
                For i = 1 to ElTime
                        ;If DashCon = Closed Then
                                ;DashStat = Closed
                        ;EndIf
                        Pauseus 300
                Next
                        For i = 1 to ElTime
                        If DashCon = Closed Then
                                DashStat = Closed
                        EndIf
                        Pauseus 200
                Next
                GoSub TurnOff
        EndIf
GoTo Main
```

```
TurnOn
;------
        Low OutPin
Return

TurnOff
;--------
        High OutPin
        Pause ElTime
Return
End
```

Let's start with something simple; the sidetone generator. We'll use one of the 16F87x/A PWM hardware modules to output a square wave (duty cycle 50%). If you need a refresher on MBasic's **HPWM** function, please review Chapter 24. Remember that once we start the hardware PWM running, it continues to function independently of whatever software function we may have the PIC perform. This "set and forget" feature is an important part of our strategy in the sidetone function. We establish two constants, **SideTonePer** for the PWM period and **DutyCycle** for the duty cycle period.

```
SideTonePer     Con     12500   ;for 1600 Hz based on 20 MHz clock DutyCycle Con
SideTonePer / 2
```

The maximum possible period and duty cycle values for MBasic's **HPWM** function are 16838, with both the period and duty cycle variables defined in terms of clock periods. I use a 20 MHz resonator in my development boards, so the clock period is 50 ns. This means the maximum period the PWM module can output is 16383 x 50ns, or 819µs. Since frequency is 1/period, the corresponding frequency is 1.22 KHz. I used a 2" diameter speaker in my layout, and I found it yielded satisfactory volume when the frequency was between 1.5 and 2 KHz. I settled upon 1600 Hz as the sidetone frequency. The duty cycle determines the duration of the positive PWM output; for a square wave it is one-half the period.

```
HPWM 0,SideTonePer,DutyCycle
```

Once the period and duty cycle values are determined, we turn on the hardware PWM module connected to pin C2 with the **HPWM** function, with the module selection byte set to 0. The speaker is gated off and on by pin A0, as discussed earlier.

We're building an iambic keyer, that alternates dots and dashes automatically when both paddle contacts are closed. But, you can't squeeze both paddle levers forever—at some point the alternating sequence must end and either repetitive dots or dashes are sent, or an end-of-character space is sent. A very good Morse operator can work at 50 WPM or faster, where an element length is 24 ms or less. To correctly send Morse at this speed, good hand coordination is necessary and it's critical that the keyer read the paddle contacts at the correct time. The second improvement in Program 26-2 is when the paddle contact status is read. Reference [26-2] contains a detailed explanation of the importance of reading the status of the paddle contacts at precisely the correct time. As reference [26-2] reads:

> *If at the mid point of an element the opposite paddle is still depressed, then the alternating element will be sent after the space. If you can let go of the opposite paddle before this critical time (the midpoint), then you won't get anything from that paddle, unless you re-close it before the finish of the space.*

Let's see how to implement this midpoint sampling.

```
Main
        DashStat = DashCon
        DotStat = DotCon
```

At the top of the main program loop we read both paddle contacts and store their status in the variables **DashStat** and **DotStat**. (We've already aliased **PortB.Bit4** and **PortB.Bit5** to **DashCon** and **DotCon**, respectively.)

We could have read both contacts simultaneously by reading all PortB bits in one statement, along the lines of the following pseudo-code:

```
PortStatus = PortB
DashStat = PortStatus.Bit4
DotStat = PortStatus.Bit5
```

Since a read only takes 50 µs, however, reading the two contacts in sequence in practice works just as well as reading them simultaneously.

To implement midpoint sampling, we break the dot and dash elements into halves. In pseudo-code our algorithm is:

```
If Dash contact is closed, take the sidetone control low
Wait for one half a dash length
        During the second half of the dash length, read the dot
        Contact. If the dot contact is closed at any time
        during the second half of the dash, store the status as dot closed
After the second half of the dash, take the sidetone high
Wait one element length for inter-element spacing
```

The following code implements this pseudo-code algorithm.

```
        ;Dash contact closed
        If DashStat=Closed Then
            GoSub TurnOn
            DotStat = Opened
            For i = 1 to (ElTime * 3) ;HalfDash
                Pauseus 140
            Next
            For i = 1 to (ElTime * 3) ;HalfDash
                If DotCon = Closed Then
                        DotStat = Closed
                EndIf
                Pauseus 60
            Next

            ;Pause HalfDash
            GoSub TurnOff
        EndIf
```

To break the dash into two halves, we've chosen to use two **For...Next** loops. Since **ElTime** is calculated in milliseconds, it makes sense to set the duration of each half-period loop at 0.5 ms or 500 µs. I've adjusted the two **Pauseus** arguments to make each **For...Next** loop execute in exactly 500 µs. Since the second half **For...Next** loop has extra statements reading **DotCon** and performing the **If...Then** test, it requires less idle time in its **Pauseus** statement.

To calibrate the loop time, and determine the proper arguments for the **Pauseus** statements, add timing statements to the each loop half, along the following lines:

```
        If DashStat=Closed Then
            GoSub TurnOn
            DotStat = Opened
            bsf      PortB,0
            For i = 1 to (ElTime * 3) ;HalfDash
                Pauseus 140
            Next
            Bcf      PortB,0
```

The two added assembler statements set **B0** high for the duration of the **For...Next** loop, thus permitting the length of time it takes to execute to be measured with an oscilloscope. We've discussed this technique elsewhere and also described an alternative measuring technique using the PIC's internal counter to profile code duration. Either method can be used to adjust the **For...Next** loop to execute in exactly 500 µs.

During the second half **For...Next** loop, we read the dot contact every loop cycle. If we find the dot contact closed at any time during the second **For...Next** loop, we set the status variable **DotStat** to closed. Contact closures during the first half **For...Next** loop are not read, in keeping with our desired algorithm.

We implement the same strategy for dashes, sampling, of course, the status of the dash contact after one half dot duration has passed.

```
If DotStat = Closed Then
        GoSub TurnOn
        DashStat = Opened
        For i = 1 to ElTime
                ;If DashCon = Closed Then
                        ;DashStat = Closed
                ;EndIf
                Pauseus 300
        Next
                For i = 1 to ElTime
                If DashCon = Closed Then
                        DashStat = Closed
                EndIf
                Pauseus 200
        Next
        GoSub TurnOff
EndIf
```

It's interesting to see the difference in time delays required. The only difference between the dot and dash **For...Next** statements is that the dash **For** statement includes the computation **3*ElTime**. This single computation consumes about 160 µs overhead.

The remainder of Program 26-2 holds the speaker gating subroutines.

```
TurnOn
;------
        Low OutPin
Return


TurnOff
;--------
        High OutPin
        Pause ElTime
Return
```

Note that we add the inter-element off period of one element length to subroutine **TurnOff**.

Program 26-3

Program 26-2 works well, and allows sending high quality Morse. But, with both the speed and weight set by constants, it's usable only for experimentation. Let's now add two rotary encoders that allow us to change speed and weight.

We'll use the interrupt-driven encoder reading scheme introduced in Program 6-3 and 6-4. If you haven't read Chapter 6 yet, and in particular the analysis of these two programs, you may wish to do so now. Since the rotary encoders must be connected to interrupt-enabled pins, we must rearrange our circuitry a bit, including moving the two paddle connections and the gating pin for the speaker. Figure 26-4 shows the revised schematic.

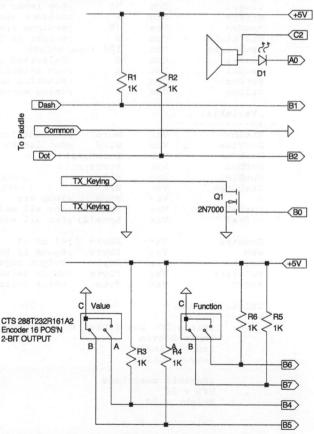

Figure 26-4: Circuit configuration for Program 26-3.

In addition to the two rotary encoders, we've added a 2N7000 MOSFET to key a transmitter. When in key-down condition, the 2N7000's gate is taken high by pin **B0**, thereby driving the 2N7000 into conduction and turning on the transmitter.

The rotary encoder connection is identical with that developed in Chapter 6, Figure 6-13 and will not be further discussed in this chapter.

Although we only adjust two parameters—speed and weight—rather than dedicate one encoder to each variable, we'll set them up as a *function selection* and *value selection* pair. We'll find this helpful when we consider making more parameters user-adjustable in Program 26-4. Rotating the function selector toggles between speed and weight, while rotating the value encoder changes the value of the selected parameter, speed or weight.

```
;Program 26-03
;second stage keyer with variable speed and weight

;Constants
;----------------
OutPin          Con     A0          ;Pin for sidetone
DashPin         Con     B1          ;from paddle
DotPin          Con     B2          ;from paddle
SideTonePer     Con     12500       ;for 1600 Hz based on 20 MHz clock
DutyCycle       Con     SideTonePer / 2
Opened          Con     %1          ;for keying pin status
Closed          Con     %0          ;key lever closed
MinWPM          Con     5           ;minimum speed
MaxWPM          Con     75          ;maximum speed
MinWeight       Con     0           ;weight is % change
MaxWeight       Con     100 ;max weight
MinSel          Con     0           ;Selection range
MaxSel          Con     1           ;max selection range
FcnEnc          Con     1           ;Function encodere
ValEnc          Con     0           ;Value encoder

;Variables
;----------
ELtime          Var     Word        ;Element length in milliseconds
DotTime         Var     Word        ;dot length (allow for weight)
DashCon         Var     PortB.Bit1
DotCon          Var     PortB.Bit2
DotStat         Var     Bit             ;status of dot lever
DashStat        Var     Bit             ;status of dash lever
i               Var     Word        ;loop var
CurVal          Var     Byte(2)     ;for all selections
OldVal          Var     Byte(2)     ;for all selections

Counter         Var     SByte       ;+1 or -1
WPM             Var     SByte       ;Speed in WPM
Weight          Var     SByte       ;Weight adjustment
Selector        Var     SByte       ;which selection
EncID           Var     Byte        ;which rotary encoder

Initialization
;--------------
        ;Setup IO. Dot & Dash have EXTERNAL pull ups
        Input DotPin
        Input DashPin
        Output OutPin

        ;Default positions
        WPM = 20
        Weight = 50

        Selector = 0

        ;set up element timings.
```

```
        ELtime = 1200 / WPM   ;From PARIS = 50 Elements = 1 Word
        DotTime = (ElTime * Weight) / 50

        High OutPin
        ;set sidetone frequency
        HPWM 0,SideTonePer,DutyCycle

        DotStat  = Opened
        DashStat = Opened


        ;Since we have two encoders, need to keep track of both
        CurVal(ValEnc) = (PortB & %00110000) / 16
        OldVal(ValEnc) = CurVal(ValEnc)

        CurVal(FcnEnc) = (PortB & %11000000) / 64
        OldVal(FcnEnc) = CurVal(FcnEnc)

        ;Interrupts on change of either encoder
        OnInterrupt RBINT, ReadEncoder
        Enable RBINT

        ;2N7000 is on B0. Initialize in key up
        Output B0
        bcf       PortB,0

Main
        ;Read the paddle
        DashStat = DashCon
        DotStat = DotCon

        ;Dash contact closed
        ;Break key down into first and second half. In first half
        ;ignore opposite lever. Sample opposite only in last half
        ;Remember ElTime is in milliseconds.
        If DashStat=Closed Then
                GoSub TurnOn     ;key down
                DotStat = Opened         ;assume opposite not activated
                ;The i loop takes 500us to execute

                For i = 1 to (ElTime * 3) ;HalfDash
                        Pauseus 115      ;adjusted for clock speed
                        ;and execution time
                Next                     ;so total time = 500us
                ;start the second half. 500us to execute

                For i = 1 to (ElTime * 3) ;HalfDash
                        If DotCon = Closed Then
                                DotStat = Closed
                        EndIf
                        Pauseus 45       ;adjusted for clock
                        ;and execution time
                Next                     ;so total time = 500us
                GoSub TurnOff
        EndIf


        ;Dot contact closed
        ;Dash contact closed
        ;Break key down into first and second half. In first half
        ;ignore opposite lever. Sample opposite only in last half
        ;Remember ElTime is in milliseconds.
        ;Difference in timing is due to 3* factor in dash loop
        If DotStat = Closed Then
                GoSub TurnOn             ;key down
                DashStat = Opened               ;assume not activated
                ;Likewise, each pass through the i loop takes 500us
                ;Dot time is adjusted by weight
                For i = 1 to DotTime
                        Pauseus 280                 ;needed to make 500us
```

```
                              Next
                              ;Start second half and look at opposite lever
                              For i = 1 to DotTime
                                      If DashCon = Closed Then
                                              DashStat = Closed
                                      EndIf
                                      Pauseus 190      ;needed to make 500us
                              Next
                              GoSub TurnOff
              EndIf
      GoTo Main

      ;start sidetone and key down
      TurnOn
      ;------
              Low OutPin               ;sidetone
              Bsf     PortB,0 ;take gate of 2N7000 high; key down
      Return

      ;turn sidetone off and key up
      TurnOff
      ;--------
              High OutPin              ;sidetone off
              bcf     PortB,0 ;2N7000 gate low; key up
              Pause ElTime    ;wait inter-element space
      Return

      Disable

      ;ReadEncoder is called whenever B4...B7 changes as one of the two
      ;rotary encoders is rotated
      ReadEncoder
      ;----------
              Pause 10        ;debounce
              CurVal(ValEnc) = (PortB & %00110000) / 16     ;value encoder
              CurVal(FcnEnc) = (PortB & %11000000) / 64     ;function encoder

              ;Which encoder has been changed? Encoders are identifed as
              ;0 and 1. 0 is value encoder, 1 is function selection
              If CurVal(ValEnc) <> OldVal(ValEnc) Then
                      EncID = ValEnc
              EndIf
              If CurVal(FcnEnc) <> OldVal(FcnEnc) Then
                      EncID = FcnEnc
              EndIf

              ;Based on last value of encoder, determine direction
              If CurVal(EncID) <> OldVal(EncID) Then
              Branch CurVal(EncID), [S0,S1,S2,S3]
              AfterBranch
              ;If function select encoder, update the selector value
              If EncID = FcnEnc Then
                      GoSub UpDateSelector
              EndIf
              ;If the value encoder, update speed or weight
                      If EncID = ValEnc Then
                              Branch Selector, [UpDateWPM,UpDateWeight]
                              AfterSelector
                      EndIf
          EndIf
      Resume

      ;The following four routines decode the encoder value and
      ;based on the old and new value determine the direction of
      ;shaft rotation.
      S0
      ;----------
              If OldVal(EncID) = 2 Then
                  GoSub ClockWise
              ELSE
```

```
            GoSub CounterClockWise
        EndIf
GoTo AfterBranch

S1
;-----------
        If OldVal(EncID) = 0 Then
            GoSub ClockWise
        ELSE
            GoSub CounterClockWise
        EndIf
GoTo AfterBranch

S2
;-----------
        If OldVal(EncID) = 3 Then
            GoSub ClockWise
        ELSE
            GoSub CounterClockWise
        EndIf
GoTo AfterBranch


S3
;-----------
        If OldVal(EncID) = 1 Then
            GoSub ClockWise
        ELSE
            GoSub CounterClockWise
        EndIf
GoTo AfterBranch

;If clockwise rotation, increase value
;will use Counter to update values later
ClockWise
;--------------
        OldVal(EncID) = CurVal(EncID)
        Counter = 1
Return

;If counterclockwise decrease
;Use Counter to update values later
CounterClockWise
;--------------
        OldVal(EncID) = CurVal(EncID)
        Counter = -1
Return

;If function = WPM, we can update WPM by adding the current
;counter value to the current WPM. Also check for bad values
;of WPM
UpDateWPM
;----------
        WPM = WPM + Counter
        If WPM < MinWPM Then
                WPM = MinWPM
        EndIf
        If WPM > MaxWPM Then
                WPM = MaxWPM
        EndIf
        ELtime = 1200 / WPM   ;From PARIS = 50 Elements = 1 Word
        DotTime = (ElTime * Weight) / 50
GoTo AfterSelector

;Selector determines which variable is updated by
;the value encoder.
UpDateSelector
;--------------
        Selector = Selector + Counter
        If Selector < MinSel Then
```

```
                        Selector = MinSel
             EndIf
             If Selector > MaxSel Then
                        Selector = MaxSel
             EndIf
     Return

     ;Weight determines the dot length. Keep dash length determined
     ;by WPM, but allow dot length to vary if the operator wants to
     ;change it.
     UpDateWeight
     ;-------------
             Weight = Weight + Counter
             If Weight < MinWeight Then
                        Weight = MinWeight
             EndIf
             If Weight > MaxWeight Then
                        Weight = MaxWeight
             EndIf
             DotTime = (ElTime * Weight) / 50
     GoTo AfterSelector

     End
```

Since we allow user settable parameters, we must set maximum and minimum permitted values. We define a series of appropriate constants. We also associate a maximum and minimum value for the selection encoder.

```
     MinWPM          Con     5        ;minimum speed
     MaxWPM          Con     75       ;maximum speed
     MinWeight       Con     0        ;weight is % change
     MaxWeight       Con     100 ;max weight
     MinSel          Con     0        ;Selection range
     MaxSel          Con     1        ;max selection range
```

For convenience, we also define a few other constants. (I've omitted discussing constants we've seen in Program 26-2.)

```
     ;Constants
     ;----------------
     Opened          Con     %1       ;for keying pin status
     Closed          Con     %0       ;key lever closed
     FcnEnc          Con     1        ;Function encodere
     ValEnc          Con     0        ;Value encoder
```

The function and value encoders are identified by numbers, 0 and 1, so we create more meaningful aliases for these numbers. Likewise, we define **Opened** and **Closed** to represent the status of the paddle levers.

In a small change from Program 6-4, we'll keep the current and prior encoder values in a two two-element arrays, such that, for example, the last value of the function encoder is held in **OldVal(FcnEnc)**.

```
     CurVal          Var     Byte(2) ;for all selections
     OldVal          Var     Byte(2) ;for all selections

     Counter         Var     SByte    ;+1 or -1
     WPM             Var     SByte    ;Speed in WPM
     Weight          Var     SByte    ;Weight adjustment
     Selector        Var     SByte    ;which selection
     EncID           Var     Byte     ;which rotary encoder
```

And, since we can both increment and decrement the speed, weight and selector variables, we'll make these signed byte type. The variable **Counter** holds the result of a change in encoder shaft angle, +1 for clockwise rotation, and –1 for counterclockwise rotation.

```
     Initialization
     ;--------------
             ;Setup IO. Dot & Dash have EXTERNAL pull ups
             Input DotPin
             Input DashPin
             Output OutPin
```

```
;Default positions
WPM = 20
Weight = 50
Selector = 0
```

We establish the paddle connections as inputs, the sidetone gate as an output and define default values for speed and weight.

```
;set up element timings.
ELtime = 1200 / WPM   ;From PARIS = 50 Elements = 1 Word
DotTime = (ElTime * Weight) / 50
```

We have added a new timing variable, **DotTime**, and a new parameter, **Weight**. **Weight** determines the duration of a dot, with respect to a theoretically ideal dot length. **Weight** ranges from 0...100, with 50 corresponding to the ideal dot length. If **Weight=25**, the dot length is one-half normal; if **Weight=100**, the dot length is twice normal.

```
High OutPin
;set sidetone frequency
HPWM 0,SideTonePer,DutyCycle

DotStat  = Opened
DashStat = Opened
```

The previous initialization follows Program 26-2.

```
;Since we have two encoders, need to keep track of both
CurVal(ValEnc) = (PortB & %00110000) / 16
OldVal(ValEnc) = CurVal(ValEnc)

CurVal(FcnEnc) = (PortB & %11000000) / 64
OldVal(FcnEnc) = CurVal(FcnEnc)

;Interrupts on change of either encoder
OnInterrupt RBINT, ReadEncoder
Enable RBINT
```

The above code tracks Program 6-4 and reads both encoders to establish their position at start-up. By setting both the old and current values equal as part of **Initialization**, we avoid a false update at start-up. After reading the current encoder position we enable the change on PortB interrupt, **RBINT**, so that should either rotary encoder be rotated, program execution will go to the interrupt service routine **ReadEncoder**.

```
;2N7000 is on B0. Initialize in key up
Output B0
bcf      PortB,0
```

Finally, we initialize the pin used for transmitter keying, **B0**. Note that we've used an assembler statement to set this pin to 0. We could have accomplished this in MBasic by via a **Low B0** statement, but as a variation, we'll use the much faster assembler call.

We'll skip all of the **Main...GoTo Main** portion of Program 26-3, except for the **DotTime** related portion, as we've already analyzed this code for Program 26-2.

```
For i = 1 to DotTime
      Pauseus 280              ;needed to make 500us
Next
```

Since we wish to set the dot duration independently from the dash duration, we've modified the dot related **For...Next** loops by substituting **DotTime** for **ElTime**.

The **ReadEncoder** interrupt routine is executed when either rotary encoder is moved to a new position. We've analyzed this routine in Chapter 6 and won't repeat it here, save for the parts specific to our application.

```
;rotary encoders is rotated
ReadEncoder
;----------
      Pause 10      ;debounce
      CurVal(ValEnc) = (PortB & %00110000) / 16      ;value encoder
```

```
          CurVal(FcnEnc) = (PortB & %11000000) / 64      ;function encoder
```

We don't know which encoder is changed, so we read both.

```
          ;Which encoder has been changed? Encoders are identifed as
          ;0 and 1. 0 is value encoder, 1 is function selection
          If CurVal(ValEnc) <> OldVal(ValEnc) Then
                    EncID = ValEnc
          EndIf
          If CurVal(FcnEnc) <> OldVal(FcnEnc) Then
                    EncID = FcnEnc
          EndIf
```

To identify which has changed, we compare the current value with the last value. We set the encoder ID variable **EncId** to **ValEnc** or **FcnEnc**, depending on which encoder has changed value.

```
          ;Based on last value of encoder, determine direction
          If CurVal(EncID) <> OldVal(EncID) Then
          Branch CurVal(EncID), [S0,S1,S2,S3]
```

The four possible values of an encoder are 0,1,2 and 3. We branch to routines **S0**…**S3** based upon the current value of the encoder to determine the direction the shaft was rotated to arrive at the current value. **S0** corresponds to a current value of 0, and so on.

```
          AfterBranch
          ;If function select encoder, update the selector value
          If EncID = FcnEnc Then
                    GoSub UpDateSelector
          EndIf
```

If the changed encoder is the function encoder, we update **Selector**. **Selector** determines whether the value encoder changes speed (**Selector =0**) or weight (**Selector=1**)

```
          ;If the value encoder, update speed or weight
                    If EncID = ValEnc Then
                              Branch Selector, [UpDateWPM,UpDateWeight]
                              AfterSelector
                    EndIf
          EndIf
     Resume
```

Based on the value of **Selector**, we branch to either **UpDateRPM** or **UpDateWeight**.

The four routines **S0**…**S3** are similar, so we'll only look at one, **S0**.

```
     S0
     ;----------
          If OldVal(EncID) = 2 Then
               GoSub ClockWise
          ELSE
               GoSub CounterClockWise
          EndIf
     GoTo AfterBranch
```

If the current encoder value is 0, and its last value was 2, then the shaft rotation must have been clockwise. If not, the rotation was counterclockwise. We go to an appropriate subroutine.

```
     ClockWise
     ;--------------
          OldVal(EncID) = CurVal(EncID)
          Counter = 1
     Return

     ;If counterclockwise decrease
     ;Use Counter to update values later
     CounterClockWise
     ;--------------
          OldVal(EncID) = CurVal(EncID)
          Counter = -1
     Return
```

Clockwise rotation sets **Counter** to 1; counterclockwise sets it to –1.

When a changed position is found on either encoder, execution goes to an update routine, **UpDateSelec-tor**, **UpDateWPM** and **UpDateWeight**. **UpDateSelector** is a true subroutine, while **UpdateWPM** and **UpDateWeight** are pseudo-subroutines, called with a **Branch** statement and returning to the line after the **Branch** via a **GoTo**. Since all three update functions are similar, we'll examine only one, **UpdateWPM**.

```
;If function = WPM, we can update WPM by adding the current
;counter value to the current WPM. Also check for bad values
;of WPM
UpDateWPM
;----------
        WPM = WPM + Counter
        If WPM < MinWPM Then
                WPM = MinWPM
        EndIf
        If WPM > MaxWPM Then
                WPM = MaxWPM
        EndIf
        ELtime = 1200 / WPM  ;From PARIS = 50 Elements = 1 Word
        DotTime = (ElTime * Weight) / 50
GoTo AfterSelector
```

To update the variable **WPM**, we add **Counter** to it. We then check to see if the updated **WPM** exceeds the maximum permitted **WPM**, or is below the minimum **WPM**. If these limits are breached, we set **WPM** to either the maximum or minimum values, as appropriate. If we change **WPM**, we also must calculate a new **ElTime**, and a new **DotTime**, using the same approach we did at initialization.

Finally, we have modified the subroutines **TurnOn** and **TurnOff** to include setting and clearing the transmitter keying pin **B0** connected to the 2N7000 gate. We use an assembler op code to set and clear B0 to execute the command as quickly as possible. In reality, the extra 50µs or so that it would take if we replaced **bsf PortB,0** by its equivalent MBasic statement **High B0** will not produce an appreciable delay.

```
TurnOn
;------
        Low OutPin              ;sidetone
        Bsf     PortB,0 ;take gate of 2N7000 high; key down
Return

;turn sidetone off and key up
TurnOff
;--------
        High OutPin             ;sidetone off
        bcf     PortB,0 ;2N7000 gate low; key up
        Pause ElTime   ;wait inter-element space
Return
```

Program 26-4

Although Program 26-3 is functional enough to use on the air, we can do much better. What improvements will we make?

- Add an LCD display showing the speed and weight.
- Add a user-selectable left/right hand option and display.
- Decode the sent Morse and display it on the LCD.
- Save the speed, weight and hand selection when powered down.

Figure 26-5 shows the configuration for our improved functionality keyer. It follows Figure 26-3, but with new pin assignments necessitated by the LCD. The LCD connection is similar to the one we examined in Chapter 5 so we won't repeat the analysis. I used a Displaytech 204A LCD, a 20 × 4 display. Although Program 26-4 uses only lines 1 and 4, we'll see some possible uses for the extra two lines in this chapter's Ideas for Modifications to Programs and Circuits section.

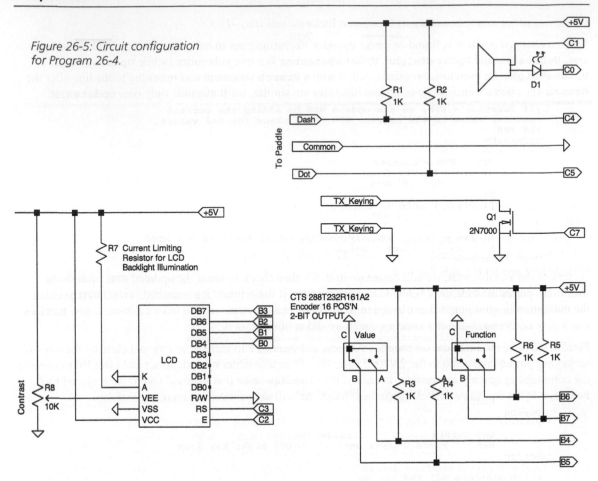

Figure 26-5: Circuit configuration for Program 26-4.

Program 26-4 is very long. Rather than follow our normal structure of listing the program in full, followed by a section-by-section analysis, we'll skip the full listing. Program 26-4, like all other programs in this book, is available in an electronic copy in the associated CD-ROM. If you need a listing to follow while reading the analysis, you may print one from the program file.

We've seen almost every line of code in Program 26-4 before, either in Program 26-3, or in other chapters. Hence, we'll concentrate on new concepts.

Data structure—In Program 26-3, we kept the speed and weight values in separate variables, **WPM** and **Weight**. To make Program 26-4 more extensible and allow easy introduction of more user-settable parameters, we'll define an array, **Param**, to hold all user parameters, and the maximum and minimum permissible values for each parameter. **Param** is structured as:

```
Param(x) holds the current value, such as speed, weight, etc.
Param(x+1) holds the maximum allowed value for Param(x)
Param(x+2) holds the minimum allowed value for Param(x)
```

The index **x** has values 0, 3 and 6 in our application. Thus, the speed in WPM is at **Param(0)**, the maximum allowed speed setting is at **Param(1)** and the minimum allowed speed is at **Param(2)**. To make it easier to write and understand the code, we also define several associated constants:

```
WPM            Con      0       ;WPM selector in Param()
MinWPM         Con      5       ;minimum speed
MaxWPM         Con      75      ;maximum speed

Wt             Con      3       ;Weight selector in Param()
MinWeight      Con      0       ;weight is % change
MaxWeight      Con      100     ;max weight

Hand           Con      6       ;Hand selector in Param()
Normal         Con      0       ;left or right hand paddle config
Reversed       Con      1       ;reverse left/right

MinOff         Con      2       ;offset of min value in Param()
MaxOff         Con      1       ;offset of max value in Param()
```

To see how defining these constants make the resulting code easier to follow, let's look at initializing the speed, weight and hand. (**Hand** determines whether the pressing the left paddle lever sends dots or dashes. Left-handed operators, such as the author, generally prefer the left lever to send dashes and the right dots, the reverse of settings used by right-handed operators.)

```
Param(WPM)=0   : Param(WPM+MaxOff)=MaxWPM    : Param(wpm+MinOff)=MinWPM
Param(Wt)=0    : Param(Wt+MaxOff)=MaxWeight  : Param(Wt+MinOff)=MinWeight
Param(Hand)=0  : Param(Hand+MaxOff)=Reversed : Param(Hand+MinOff)=Normal
```

We've set the three parameters to **0** because the next step is to retrieve their stored values from EEPROM. You should see, however, that the meaning of **Param(WPM)** is easier to understand than **Param(0)**.

Implementing Left/Right Reversal and Identifying Spaces

Let's look at a simplified version of our main program loop.

```
Main
NormalSense
;==========
          DashStat = DashCon            ← dash to dash
          DotStat = DotCon              ← dot to dot

          If DashStat=Closed Then
          …
          EndIf

          If DotStat = Closed Then
          …
          EndIf
          GoTo EndMain

ReversedSense
;=============
          DashStat = DotCon             ← dash to dot reversal
          DotStat = DashCon             ← dot to dash reversal

          If DashStat=Closed Then
          …
          EndIf

          If DotStat = Closed Then
          …
          EndIf

EndMain
          IdleCount = IdleCount + 1
          If IdleCount = LetterSpace Then
                  GoSub Spacing
          EndIf
Branch Param(Hand), [NormalSense,ReversedSense]
```

How we handle left/right reversal is straightforward. We have identical key down timing loops, except that one set reverses the dot/dash sense. Which loop we branch to depends upon the value of **Param(Hand)**. In addition to reversing the dot/dash sense at the top of each timing loop, we also reverse the variables when reading the opposite lever inside the second half loops, as can be seen in comparing two corresponding sample code sections:

Normal	Reversed

<pre>If DashStat=Closed Then
    GoSub TurnOn
 	
    Morse(j) = Dash
    DotStat = Opened
    For i = 1 to (ElTime * 3)
 	
    Pauseus 115
    Next
 	
    For i = 1 to (ElTime * 3)
        If DotCon = Closed Then ←
 	
    DotStat = Closed
        EndIf
        Pauseus 45
    Next
 	
    GoSub TurnOff
EndIf</pre>
	<pre>If DashStat=Closed Then
	    Morse(j) = Dash
	    GoSub TurnOn
	    DotStat = Opened
	
	    For i = 1 to (ElTime * 3)
	        Pauseus 115
	
	    Next
	
	    For i = 1 to (ElTime * 3)
	        If DashCon = Closed Then ←
	
	    DotStat = Closed
	        EndIf
	        Pauseus 45
	
	    Next
	
	    GoSub TurnOff
	EndIf</pre>

A similar reversal is in the **DotStat = Closed** loops of both the normal and reversed sections.

Close examination of the sample code shows a new statement, **Morse(j)=Dash**, as well as some code added at the bottom of the Main program loop. All are related to displaying the sent Morse on the LCD. Let's see how we accomplish this.

First, we define an array, **Morse**, and four associated constants:

```
Dot        Con    -1     ;for Array Morse
Dash       Con    1      ;in array of send elements
Nil        Con    0      ;is classed as dot/dash or nothing
MorseLen   Con    8      ;length of array holding dots/dash/nil

Morse      Var    SByte(MorseLen) ;holds elements
```

Assume, for the moment, that we start with **Morse** filled with **Nil** (0) values. We have added **Morse(j)=Dash** and **Morse(j)=Dot** statements to each dash and dot loop, so that **Morse(j)** will be filled with the dot/dash values as the operator sends them. (We increment the element counter **j** as part of subroutine **TurnOff**.)

This leaves two important questions open:

- Since Morse is a variable length code, how do we know when one character has been sent and the next starts?
- How do we translate the array **Morse()** into characters to display on the LCD?

Let's treat these questions in order. The first question is a bit misleading. While Morse is a variable length code, it does have a unique end-of-character signal—an interval equal to three element lengths where neither a dot nor a dash has been sent. Remember, to think of Morse as only comprising dots and dashes isn't correct; as we said earlier, spaces are an integral component.

So then, how do we know when neither a dot nor a dash has been sent for three element lengths, and how do we distinguish that from the word space, which is no dot or dash sent for seven element lengths? Let's take a look at the bottom of **Main** program loop again.

```
        IdleCount = IdleCount + 1
        If IdleCount = LetterSpace Then
                GoSub Spacing
        EndIf
```

The variable **IdleCount** increments every pass through the main program loop. If we reset **IdleCount** to 0 every time we send a dot or a dash, and if we know how long it takes to pass through a complete loop when neither the dot or dash levers are pressed, i.e., the loop is idling, we can calculate how many **IdleCounts** correspond to an end-of-character space. I've measured the time it takes for execute Program 26-4's loop when neither paddle lever is pressed, based on a 20 MHz clock, as 586 microseconds. We've defined a new constant **IdleTime** to hold this value.

```
        IdleTime        Con     586     ;microseconds for idle loop execution
```

We now calculate how many idle passes it takes to identify an end-of-character space and store this value in a word variable **LetterSpace**. In order to give the operator a bit of margin for sending non-perfect code (remember—this spacing is not automatically determined), we'll set the end-of-character threshold at 2.5 element lengths, not 3.

```
        LetterSpace = (ELTime * 1500)/IdleTime
```

The factor 1500 comes from (a) **ElTime** is in milliseconds, while **IdleTime** is in microseconds, so we must multiply **ElTime** by 1000 to convert to microseconds and (b) one inter-element space, equal to **ElTime**, is automatically inserted as part of the subroutine **TurnOff** at the end of each dot or dash. The main loop is not idle while **TurnOff** is executing, so we must subtract one element length from our calculation, thus making our threshold 1.5 element lengths, not 2.5.

Let's turn our attention to the subroutine **Spacing**, which is called when **IdleCount = LetterSpace**, that is, when 2.5 element lengths have passed without a dot or dash being sent.

```
        Spacing
        ;---------
                ;Following avoids junk at start of LCD readout
                If First = FirstYes Then
                        Return
                EndIf
```

In order to prevent stray characters from appearing when the program initializes, we avoid calling subroutine **Spacing** until the first dot or dash has been sent. The variable **First** accomplishes this trap, as we initialize it as **FirstYes** in the initialization section and set it to **FirstNo** every time a dot or dash is sent. Hence, before the first dot or dash is sent, a call to **Spacing** results in an immediate **Return**.

```
                IdleCount = 0
```

We must reset **IdleCount** to 0 to be ready for the next space measurement.

```
                WordSpace = WordSpace + 1
                ;Is pause long enough to be a word space?
                ;If so, trigger a word space out and reset counter
                If WordSpace = WordRatio Then
                        WordSpace = 0
                        If EndWord = EndYes Then
                                EndWord = EndNo
                                TempStr = " "
                                GoSub ShowChar ;put it on LCD
                        EndIf
                EndIf
```

Recall that a word space is seven elements long, nominally. Hence, we can determine whether we have a word space or simply the end-of-character space by using a supplemental counter, **WordSpace** to see how many consecutive end-of-character spaces have occurred. How many times is **Spacing** called before we decide a letter space has been sent? Here we have to deal with another boundary condition:

0	1	2	3	4	WordSpace=WordSpace+1
	2.5 El Len 1 Char Space	4.0	5.5	7.0	Total elapsed time since end of dot or dash in element lengths
Dot or Dash	1.0	1.5	1.5	1.5	1.5 Element/space length

Spacing is called first 2.5 element lengths after the end of the preceding dot or dash. The next time it is called is 4.0 element lengths after the end of the preceding dot or dash, and the time after that 5.5 element lengths. Since we wish to give ourselves a bit of a margin for operator error, we'll say that any idle time over 5.5 element lengths is a word space and accordingly define the constant **WordRatio** at 3. Since we increment **WordSpace** before the equality test, setting **WordRatio** at 3 means that the test succeeds on the third call to **Spacing**.

Suppose **Spacing** has been called, but insufficient idle time has occurred to constitute a word space. Hence, we have an end-of-character space and thus are ready to convert the values in **Morse()** to a character and display it on the LCD. At this point, we have the character's dot/dash/nil sequence loaded into the array **Morse**, and we have determined that a valid end-of-character space has been received.

Let's suppose the letter is "C," or dash/dot/dash/dot. **Morse()** holds:

0	1	2	3	4	5	6	7	Element
Dash	Dot	Dash	Dot	Nil	Nil	Nil	Nil	Character
1	−1	1	−1	0	0	0	0	Morse() Value

If we diagram the letters A…Z in Morse, as shown in Figure 26-6 we see an interesting relationship. Figure 26-6's structure is known as a "binary tree" in computer science terms. (The tree is upside down, with the stem in the air and the branches downward.) At each node (element), we make a decision based upon the next element; go along the dot branch or the dash branch until the next element has a nil value. We have then decoded the letter.

We might implement a binary tree through a series of **If…Then** statements. Let's see how that might work to decode "C."

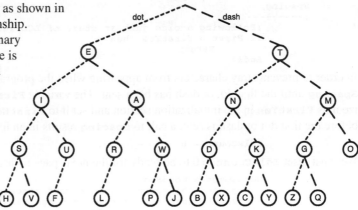

Figure 26-6: *Morse code binary tree.*

```
If Morse(0)=Dash Then
        If Morse(1) = Dot Then
                If Morse(2) = Dash Then
                        If Morse(3)= Dot Then
                                If Morse(4) = Nil Then
                                        Character = "C"
                                EndIf
                        EndIf
                Endif
        EndIf
EndIf
```

For simplification, I've omitted all the other conditional branches in this nested **If...Then** statement that allow us to decode T,N,M,D,K,G,O,B,X,Y,Z and Q as well as C starting with a dash at **Morse(0)**, but it should be clear that nested **If...Then** statements are, at best, a singularly inelegant method to decode Morse code, not to mention being slow in execution.

Some computer languages facilitate tree structures, such as support for linked lists, but MBasic doesn't. That's not surprising, considering PICs are not intended for heavy-duty data processing. And, we can devise a quick and efficient way to decode Morse without recourse to esoteric data structures. We start by noting the similarity of a binary tree to a binary number. If we say the starting point (root) value is 128; that every layer down represents one binary shift to the right and that a dot represents subtraction and a dash represents addition, we can map each Morse character into a unique number between 0...255.

Let's see how this works for "C."

	0	1	2	3	4	5	6	7	Element
	Dash	Dot	Dash	Dot	Nil	Nil	Nil	Nil	Character
	1	−1	1	−1	0	0	0	0	Morse() Value
	64	32	16	8	4	2	1	0	Binary Weight
+128	+64	−32	+16	−8	0	0	0	0	Sum

The letter "C" thus corresponds to 128 + 64 − 32 + 16 − 8, or 168. We then can index into a byte table that has the letter "C" at its 168th position and quickly retrieve the correct decoded character.

This weighted binary tree conversion for A...Z, 0...9 and common punctuation marks and prosigns is as follows:

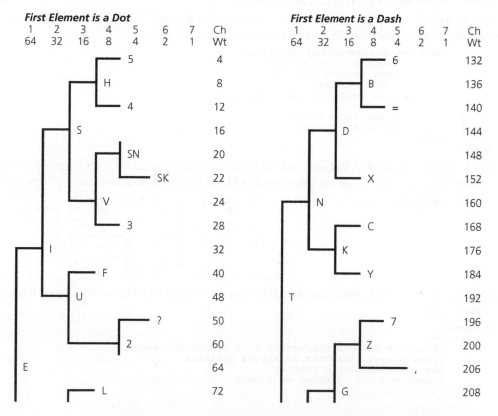

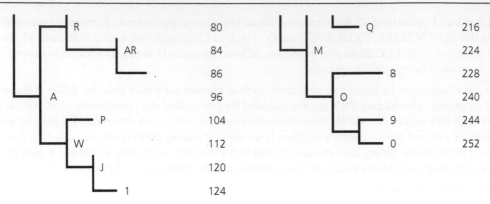

R	80	Q 216
AR	84	M 224
	86	8 228
A	96	O 240
P	104	9 244
W	112	0 252
J	120	
1	124	

If the next element is a dot, move upward one level; if it is a dash, move downward one level

Based on this approach, calculating the binary tree value for **Morse()** is simple; we multiply each element value by that level's weight and sum the total:

```
If EndLetter = EndYes Then      ;we have an end of the letter
    Temp = 128
    For j = 0 to (MorseLen-1)
            Temp = Temp + Morse(j)* TreeW(j)
                    Morse(j) = Nil ;clear for next letter
    Next
```

Note that we earlier defined the level weightings in the byte array **TreeW**:

```
TreeW  ByteTable $40,$20,$10,$8,$4,$2,$1
```

The values in **TreeW** are stated in hex, where **$40** corresponds to decimal 64, etc. You should also note our earlier choice of 1,-1 and 0 for dash, dot and nil values was not accidental.

```
;now get corresponding character
        TempStr = DeCode(Temp)
        ;Put it onto the LCD
        GoSub ShowChar
        ;Get ready for next letter
        j = 0
        EndLetter = EndNo
        EndWord = EndYes
    EndIf
Return
```

We have defined another byte table, **DeCode**, to hold the corresponding letter for each binary weighted Morse character value. Characters that are unassigned in Morse are given the underscore "_" character.

```
DeCode ByteTable        "___5__H___4__S__^_%_V___3___",
                        "I_____F_____U_?_____2___",
                        "E_____L_____R___+_._____*___",
                        "A_____P_____W_____J___1___",
                        "___6__B___=__D___/__X_____",
                        "N_____C_____K_____Y_____",
                        "T___7__Z_____,_G_____Q_____",
                        "M___8_____O___9_____0___"
```

The final step is to display the decoded character at the top line of the LCD display. We do this through subroutine **ShowChar**.

```
ShowChar
;---------
        ;Hold the decoded characters in a length 20 array
        ;Insert next character at Mindex position
        BarrelAry(Mindex) = TempStr
        LineNo = 1      ;display on line 1
```

Our LCD has 20 characters/line and four lines. We'll use line 1, the top line, to display the sent character. And, for a more elegant display, we'll write the text left to right, but as soon as the line is filled with the first 20 characters, we'll insert new characters in the rightmost position and scroll the other characters one position to the left. To accomplish the scrolling, we'll hold the incoming characters in the 20-element byte array **BarrelAry**. (This structure is not quite a classical barrel array, since we introduce a new character at the end. However, its structure and use is similar to a barrel array; hence its variable name.) The variable **Mindex** holds the next open position in **BarrelAry** and the new character is inserted at **BarrelAry(Mindex)**.

```
;Position LCD cursor & write the character array
LCDWrite RegSel\Clk, LCDNib,[ScrRAM+LineOff(LineNo)]
For k = 0 to (BarrelSize-1)
LCDWrite RegSel\Clk, LCDNib,[Str BarrelAry(k)\1]
Next
```

Now that the new character has been stored at the desired position, we move the cursor to the start of Line 1 and we write the contents of **BarrelAry** to the LCD, one character at a time. We have previously defined the byte array **LineOff** to hold the line offset memory index. We've extensively covered the relationship between LCD memory address and display position in Chapter 5 and we won't repeat it here.

```
;We start at left side of screen, write chars to right.
;When reach the last position, we keep adding chars at
;right and shift all prior
;chars one position to the left.
;Hence we only increase Mindex if it is less than 19.
;Once it hits 19 we keep it there.
If MIndex < (BarrelSize-1) Then
    MIndex = MIndex + 1
EndIf
```

We've initialized **Mindex** to 0 in Program 26-4's **Initialization** section and filled **BarrelAry** with the space character. Accordingly we increment **Mindex** until it reaches 19. (As our usual practice, we've defined the length of **BarrelAry** in terms of a constant **BarrelSize**, which is set to 20. Hence, the last valid array index is **BarrelSize-1**.) When **Mindex=BarrelSize-1**, we have filled all 20 positions of the Line 1 with information and we then add new characters at the right hand end shift the older characters left by one place to accommodate the new letter. Hence the above code increments **Mindex** only until it reaches 19, at which point it freezes at 19.

```
;Shift all chars to the left to make room for new ones
;that come in at the end.
If MIndex >= (BarrelSize-1) Then
    For k = 0 to (BarrelSize-2)
            BarrelAry(k) = BarrelAry(k+1)
    Next
EndIf
```

The above code moves every character one position to the left, overwriting **BarrelAry(0)**.

```
LineNo = 4
LCDWrite RegSel\Clk,LCDNib, |
    [ScrRAM+LineOFf(LineNo)+CursorPsn(Selector)]
Return
```

Finally, we return the LCD cursor to the status line, Line 4, and position it over the currently selected parameter.

Let's see how we display the speed, weight and hand variables in Line 4. Subroutine **LCD_UpDateStatus** is called whenever any of the three user-adjustable parameters are altered. Since the process for reading the rotary encoders is largely unchanged from Program 26-3, we will concentrate only on the LCD display aspect.

```
LCD_UpDateStatus
;------------------
        LineNo = 4
        LCDWrite RegSel\Clk,LCDNib,[ScrRAM+LineOFf(LineNo)]
        LCDWrite RegSel\Clk,LCDNib,[DEC Param(WPM)\2," WPM ", |
        DEC Param(Wt)\3," Wt "]
```

```
If Param(Hand) = Normal Then
        LCDWrite RegSel\Clk, LCDNib,["Normal"]
EndIf
If Param(Hand) = Reversed Then
        LCDWrite RegSel\Clk, LCDNib,["Revs'd"]
EndIf
LCDWrite RegSel\Clk,LCDNib,[ScrRAM+LineOff(LineNo)+CursorPsn(Selector)]
Return
```

Subroutine **LCD_UpDateStatus** applies the principles we learned in Chapter 5; it moves the cursor to Line 4 and writes the current value of speed and weight. Depending upon the hand setting, the last text message written is the word "normal" or "reversed."

Figure 26-7 shows the LCD display. The square cursor shows the currently selected parameter that may be changed through the value encoder. Figure 26-8 shows the 2N7000's gate voltage while sending the letter "K" at 20 WPM.

One last addition to Program 26-4 is that the current values of speed, weight and hand are saved to EEPROM and are read at program launch, thereby preserving changes during power off.

Figure 26-7: LCD display showing decoded text and status display,

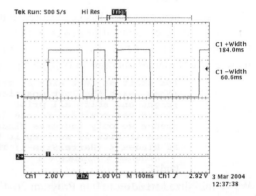

Figure 26-8: Output while sending letter "K".

Three subroutines, **GetParams**, **SaveParams** and **FirstPass** provide EEPROM functionality.

```
;Save parameters to EEPROM
SaveParams
;-----------
        For n = 0 to 2
                Write n,Param(n*3)
        Next
Return
```

SaveParams writes the three values, one byte at a time, to the first three EEPROM positions. Note that since the parameter values are held in **Param(0)**, **Param(3)** and **Param(6)**, we index with **n*3**. Since we define the associated maximum and minimum values as constants, we need not save any other elements of **Param()**.

```
;Read params from EEPROM
GetParams
;----------
        ;Get parameters from EEPROM
        For n = 0 to 2
                Read n,Param(n*3)
        Next
```

We read the saved parameters in a process that tracks the save routine.

```
;We now take advantage of fact that at program time
;MBasic zeros out the EEPROM. Hence we test for this
;condition as WPM will be < MinWPM. We only need to
;initialize to defaults once when programming; in
;operation, changed values are written to eeprom
;automatically and FirstPass is not necessary
```

```
        If Param(WPM) < Param(WPM+MinOff) Then
                GoSub FirstPass
        EndIf
    Return
```

Reading parameters from EEPROM is complicated since we must distinguish between developing the program, where the start-up sequence is invoked by MBasic's programmer and where the start-up sequence results from normal power on/off sequence. At program time, MBasic fills all 256 EEPROM cells with the value **$FF**, overwriting their prior content. Hence, if we simply read EEPROM to get stored parameters, our code will fail at programming time, even though it might work perfectly once programmed. We detect for launch-after-programming by testing the value of **Param(WPM)**. Since **Param** is an array of type **SByte**, the **$FF** value will be interpreted as –1. The test for **Param(WPM) < Param(WPM+MinOff)** will therefore be true only for the first run after programming. In this case, Program 26-4 executes subroutine **FirstPass**.

```
    ;First pass is called only when EEPROM is zeroized during Programming
    ;It provides default parameters
    FirstPass
    ;-------
            Param(WPM) = 20
            Param(Wt) = 50
            Param(Hand) = Normal

            GoSub SaveParams
    Return
```

Subroutine **FirstPass** substitutes default values and saves those values to EEPROM. At this point, powering the PIC off and back on, or pressing the 2840 Development Board's reset button will correctly read the EEPROM stored values.

Ideas for Changes to Programs and Circuits

- The algorithm we've used in Programs 26-2 through 26-4 looks at the status of the opposite paddle lever beginning halfway through the current element. If the opposite contact is closed at any time during the second half, the opposite element will be automatically sent when the current element is completed. This is known as "Mode B" keying in ham radio terminology. "Mode A" keying works differently. If the paddle contact is released, sending stops after completion of the current character. Selection between Modes A and B is a matter of personal preference. How would you modify Program 26-4 to add a Mode A/B selector? Where should be mode status be displayed?
- Another mode is "bug emulation" where the keyer emulates a mechanical semi-automatic key. Dots are automatically sent as long as the dot lever is pressed, but the operator must make dashes manually; pressing and holding the dash lever gives a continuous output. How would you add this to Program 26-4?
- A valuable keyer feature is message recording and playback. Since Program 26-4 decodes the letter or prosign corresponding to the operator's transmission, how would you go about saving the message in EEPROM and allowing the operator to erase or send the message? The 256 bytes of EEPROM, less whatever is reserved for parameter saving, might be divided into 4 to 8 messages, of perhaps 40 to 80 characters length. (Not all memories need identical length.) It's also desirable to be able to "chain" memory messages, so that message 1 can link to, for example, message 3. How would you do this? What changes to the user interface would be appropriate?
- With a 20 MHz clock, the lowest sidetone frequency that can be generated by one PWM module is around 1.2 KHz. Sidetone frequency is another matter of personal preference, but I like a sidetone of 400 Hz to 500 Hz. Since a 16F876A has two PWM generators, how might we produce a 400 Hz sidetone, when neither generator will go below 1200 Hz? Suppose we connect one side of the speaker to one PWM generator and the other side to the second PWM generator. And, suppose one PWM outputs 10.000 KHz and the second outputs 10.400 KHz. The speaker will reproduce the sum (20.400 KHz)

and difference (400 Hz) between these two frequencies. The sum frequency will be above the normal hearing range, but the difference produces our desired sidetone frequency. This will, of course, require us to gate one or both of the PWM generators to give off/on control, but we've seen how that is accomplished. Since sidetone frequency is a personal preference item, it should be user selectable. Ideally, the user should be able to select sidetone frequency and volume. (Review Chapters 21 and 17 for an idea on how a small audio amplifier with PIC-controlled volume level might be added to the Program 26-4 and its associated circuit.)

References

[26-1] American Radio Relay League, *The ARRL Handbook for Radio Communications*, 81st Ed., Newington, CT. (2004) Now in its 81st edition, the ARRL handbook is the starting point reference for all things amateur radio.

[26-2] Adams, Chuck, *Iambic Keying 101—Part 1*, http://www.n9vv.com/k7qo-a-b-keying.html. (April 2002).

CHAPTER 27

A Morse Code Reader

We've seen in Chapter 26 how a PIC can help send Morse code. Now, it's time to put a PIC to work reading Morse. And, even if you're not interested in receiving Morse with a PIC, read this chapter anyway to learn a bit about phase locked loops and an adaptive algorithm to account for timing variations in hand sending.

Sending and Receiving Morse

We won't repeat the Morse code introductory material presented in Chapters 17 and 26. But, let's consider briefly how Morse code is sent and received. In Chapter 26, we learned that amateur radio operators use off/on transmitter keying to send Morse; a dot or a dash represents transmitter on; spaces are transmitter off time. In the amateur radio fraternity, this on/off modulation scheme is known as CW, or "continuous wave" transmission. In the computer and data communications world, on/off modulation is called OOK (on-off keying) or ASK (amplitude shift keying). We've seen OOK before—in Chapter 22 to be precise, as it's part of the infrared remote control communications protocol. And, of course, there's nothing limiting us to sending Morse by OOK; we could use any of many modulation protocols, such as frequency shift keying (dots and dashes are represented by one frequency; spaces by a second frequency) or phase shift keying or many other modulation techniques. As a practical matter, however, we'll concern ourselves with OOK, as that's what is used in practice.

Since OOK implies turning a *transmitter* on and off, to detect the signal we require a *receiver*. And, since Morse is used most widely in the amateur radio short wave frequencies (amateur operators have access to a number of frequency bands between 1.8 and 30 MHz; see reference [27-2] for more details), we'll need a short wave receiver. We'll assume you already have a receiver, and if not reference [27-3] may help you select one. Since Morse is traditionally received as an on-off audio tone, the receiver must convert the on-off signal to an on-off audio tone. This is done through a "beat frequency oscillator" or BFO built into the receiver. Older receivers had a switch labeled BFO on/off, while newer ones simplify things by automatically selecting the proper circuitry when either the CW or single sideband mode is engaged. (Single sideband, or SSB as it's usually called, is a method of voice transmission that also requires a BFO; a receiver in either CW or SSB mode will receive Morse.)

We'll assume that you have a suitable receiver and that you now have the sounds of Morse code echoing out of the speaker. How then to convert the audio tone to a logic level signal that our PIC will recognize? We'll build a circuit called, logically enough, a "tone detector" that converts the presence or absence of an audio tone to a logic level signal. There are many ways to build a tone detector; our design employs an Exar XR-2211 phase locked loop.

Figure 27-1 is an overview of our XR-2211 circuit. The XR-2211 is the receiving half (demodulator) of a modem device; it receives frequency shift keyed data signals and translates the two tones into logic level output signals. (Modem stands for modulator-demodulator; the XR-2211 performs the demodulation half of the modem function, its companion chip, the XR-2206 performs the modulation function.) An auxiliary function built into the XR-2211, however, is tone detection to determine whether the modem is currently receiving a signal, and that's the part we'll use in our Morse decoder circuit.

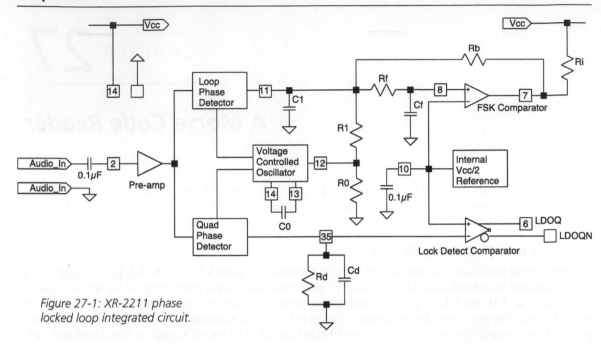

Figure 27-1: XR-2211 phase locked loop integrated circuit.

Don't let the apparent complexity of the XR-2211 scare you off; we've seen its underlying principle before. Before we dig into the details of the XR-2211, let's start at the beginning—what is a phase locked loop, or PLL? Figure 27-2 shows a conceptual PLL arrangement; an input signal (at frequency F_{INPUT}), a phase detector, a phase detector output conditioning circuit and a voltage-controlled oscillator or VCO with output frequency F_{VCO}. The VCO's frequency range includes F_{INPUT}.

Let's start with the phase detector, which we may consider as a balanced frequency mixer; its output is the sum and difference of the two input signals, F_{INPUT} and F_{VCO} in Figure 27-2, i.e., $(F_{INPUT}+F_{VCO})$ and $(F_{INPUT}\text{-}F_{VCO})$. Let's concentrate on the difference signal, $(F_{INPUT}\text{-}F_{VCO})$. *This difference signal represents the error between the input signal and the VCO.* This difference signal then is conditioned—and possibly level shifted—and fed into the VCO. The control signal output the conditioning system then drives the VCO frequency F_{VCO} to match the input frequency F_{INPUT}. If this sounds familiar—it's because it's precisely the same control loop arrangement we saw in Chapter 24 where we measured the motor speed, compared it with the target frequency (the *error*) and increased or decreased the motor voltage to drive the error to zero. Here we measure the difference (the *error*) between the input signal (our target) and the VCO and we either increase or decrease the control voltage to drive the error to zero.

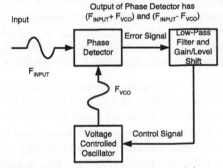

Figure 27-2: Conceptual phase lock loop.

How then do we use a PLL to determine when an audio tone is present or not? Let's consider what happens when no signal is present, which means the input is noise. F_{INPUT} can be though of as a collection of random frequency components that quickly come and go. The VCO tries to synchronize to the input noise, but its ability to track quickly varying noise is limited by the conditioning stage that includes a low-pass filter element. (Remember the output of the phase detector also includes an $(F_{INPUT}+F_{VCO})$ component? At a minimum, we must attenuate this component with a low-pass filter, but we'll see that other criteria dictate design of the low-pass filter.) Hence, the output of the phase detector is a random time varying signal, which, if the input is true random noise, will have an average value of zero.

Now, suppose a tone is present and the VCO locks to it. The feedback mechanism works, F_{VCO} matches F_{INPUT} in both frequency and phase and the output of the phase detector is ... zero.

Wait a minute, you say, both the locked and unlocked conditions are zero? How do we distinguish a locked zero from an unlocked zero? It turns out that with real phase detectors, we don't have quite such a difficult task. Suppose our phase detector response is that shown in Figure 27-3, which may implemented from an XOR logic gate. If the input signal is noise and if the loop filter keeps the VCO from tracking the noise, then the phase relationship between the input signal and the VCO will be random, uniformly varying from −180 to +180 degrees. Inputting uniform random phases from −180 to +180 degrees to the phase detector of Figure 27-3 results in a random output, but one with an *average value of Vcc/2.* (The low pass loop filter can be thought of as an electronic averaging circuit. You can

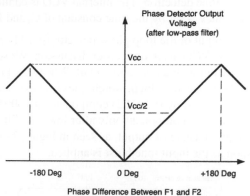

Figure 27-3: XOR-type phase detector response.

compute the output mathematically or graphically from Figure 27-3, or you can simply note that its output is positive for any phase difference except 0.) Now, suppose we have a tone input and the VCO is locked in frequency and phase to the input signal. In this case, the phase error is constant at 0 degrees, and the output of the phase detector is 0 V. Thus, the phase detector of Figure 27-3 allows us to easily distinguish between noise (average output Vcc/2) and locked to a signal (average output 0). A voltage comparator with one input on the low pass filtered phase detector output voltage and the other input biased at Vcc/4 will thus toggle its output between lock and no-lock conditions. In keeping with the introductory level of this book, we won't consider other types of phase detectors, but the concept we've illustrated in Figure 27-3 applies to many of them; the average output voltage is different for random phase noise compared with the locked condition.

Tone Detector Circuit

Figure 27-4 shows the finished tone detector circuit. Let's take a quick pass through how it works before we zero in on setting the component values.

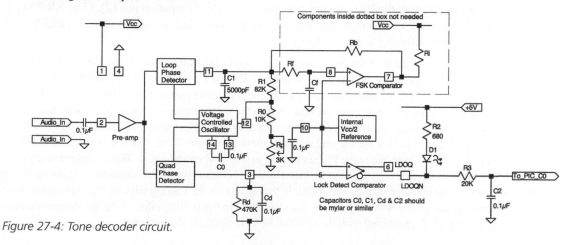

Figure 27-4: Tone decoder circuit.

The audio input signal is coupled to the XR-2211's internal preamplifier through a 0.1µF blocking capacitor (pin 2). The preamplifier is used as a limiter so the incoming sine wave signal is converted to a square wave, with the 10 mV being the minimum signal level for the onset of limiting. The preamp's output feeds two phase detectors, one for loop control (output on pin 11) and a "quadrature" phase detector (output on pin 3) for tone detection. The internal VCO is connected to both phase detectors. The VCO's center (unlocked) frequency is set by the time constant of C_0 and R_0.

The output of the loop phase detector (pin 11) contains both the sum and difference frequencies of the input and the VCO, as in our earlier discussion. We see this in Figure 27-5 when we look at the loop phase detector output with a tone burst input. When no signal is present, the phase detector output is dominated by the VCO frequency which, which idles is at the center frequency, 800 Hz; when the tone is present, the output is dominated by (F1 + F2) component, at 1600 Hz in this case, as soon as the PLL locks up. (We can see the $F_{INPUT} - F_{VCO}$ component only after low-pass filtering out the $F_{INPUT} + F_{VCO}$ component.) An expanded look at the phase detector output, as seen in Figure 27-6, shows immediate change in the phase detector output as soon as the input tone burst is applied.

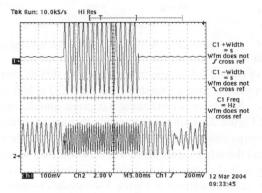

Figure 27-5: Ch1: 800 Hz tone burst input; Ch2: phase detector output (pin 11).

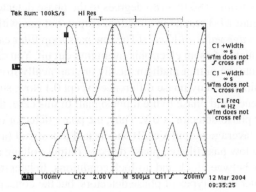

Figure 27-6: Expanded view of phase detector output; Ch1: 800 Hz tone burst input; Ch2: phase detector output (pin 11).

The sum component is attenuated by C_1 and the difference component F_{INPUT}-F_{VCO} (the error signal) is applied as a correcting current to the VCO control port through R_1 (VCO input on pin 12). (The XR-2211's variable frequency oscillator is current controlled, not voltage controlled, so we should call it a CCO, not a VCO, but we'll continue referring to it as a VCO.)

When the VCO is synchronized with the input signal, the output of the quadrature phase detector is driven to Vcc; when the VCO is unsynchronized it is close to ground. R_D and C_D provide filtering for the quadrature phase detector output. Figure 27-7 shows the quadrature phase detector output responding to a tone burst input. The output of the quadrature phase detector is connected to the minus input of a comparator with the plus input connected to an internal voltage reference that is approximately at Vcc/2 (internal reference voltage output at pin 10). The comparator has both a normal and a complementary output. Thus, when an input tone is present, the normal output (pin 6) is high when the PLL is in lock and low when unlocked. The complementary output (pin 5) is low when in lock and high when unlocked. Figure 27-8 show how the comparator cleans up the ripple on the quadrature phase detector output. The multiple transitions on the trailing edge of the comparator output are a direct consequence of the relatively gentle transition time in the trailing edge of the quadrature detector output seen in Figure 27-7, combined with the residual 800 Hz ripple in the output.

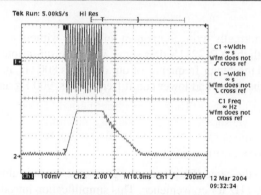

Figure 27-7: Quadrature phase detector output; Ch1: 800 Hz tone burst input; Ch2: quadrature phase detector output (pin 3).

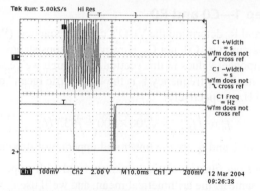

Figure 27-8: Output of quadrature comparator; Ch1: tone burst input; Ch2: quadrature comparator inverted output (pin 5).

The lock detector comparator is of open collector design and hence requires a pull-up resistor. R2/D1 pulls up the comparator output and provides, through LED D1, a visual indication of tone detection. The RC network comprised of R3 and C2 provides a small amount of low pass filtering to reduce edge glitches from the comparator output. Figure 27-9 shows the effect of the RC filter when decoding off-the-air signals mixed with noise. Although not nearly as effective as a multipole low-pass filter, the RC network knocks the worst edges off the spurious transitions.

The design parameters I used are:

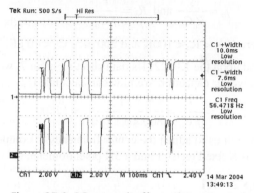

Figure 27-9: RC network effect; Ch1: output of quadrature comparator (pin 5); Ch2: PIC input following RC filter.

Parameter	Value
Center Frequency	800 Hz
Tracking Bandwidth	±100 Hz

The center frequency is somewhat a matter of personal preference. Most Morse operators like to listen to a tone in the 800–1000 Hz range. Although my personal preference is for a 400 Hz beat note, our design goes with the traditional 800 Hz frequency. My choice of a ±100 Hz lock bandwidth is based upon experience; the narrower the bandwidth the longer it takes for the tone detector to lock up and release. For 60 WPM Morse, the dot length is 20 ms, using the formula we derived in Chapter 26. To reduce distortion, we will require the tone decoder to lock up in less than 25% of the shortest element duration, or 5 ms. Since the lock time is approximately 1/total bandwidth, to meet this objective the total bandwidth must not be less than 1/.005, or 200 Hz total, or ±100 Hz.

Designing the circuit requires us to calculate values for C0, C1, Cd, R0, R1 and Rd, as well as some ancillary components. Although the XR-2211's data sheet covers these calculations, I found an inconsistency or two that initially lead me to incorrect values for several components. Let's go through the calculations. You may wish to have the XR-2211 data sheet in front of you.

Step 1—C0 and R0

C_0 and R_0 determine the VCO's center frequency. (We normally write component references without a subscript, i.e., C0, not C_0. However, for consistency with the XR-2211's data sheet, we'll use subscripted component values in this section.) Although we've said the center frequency of our design is 800 Hz, in reality the center frequency f_0 is the "geometric mean" of the upper and lower cutoff frequencies, 900 Hz and 700 Hz, respectively:

$$f_0 = \sqrt{f_L f_U} = \sqrt{700 \times 900} = 793.7 Hz$$

where f_L and f_U are the lower and upper cutoff frequencies.

Since our filter is relatively narrow band with respect to the center frequency, the geometric mean is almost the same as the arithmetical mean, and we'll use $f_0 = 800$ Hz for convenience. This simplification introduces less than 1% error in component values.

The XR-2211 data sheet provides the following relationship among f_0, R_0 and C_0:

$$f_0 = \frac{1}{R_0 C_0} \text{ and } R_0 = \frac{1}{f_0 C_0}$$

Where $10K < R_0 < 100K$ and $200 \text{ pF} < C_0 < 10 \text{ μF}$

Based upon the XR-2211's data sheet's Figure 7, we'll pick C_0 as 0.1 μF. C_0 should be a high quality Mylar or polyester capacitor. We now can calculate R_0:

$$R_0 = \frac{1}{f_0 C_0} = \frac{1}{800 \times 0.1 \times 10^{-6}} = 12500$$

To permit us to adjust the center frequency to compensate for C_0's tolerance, we'll use a series combination of a 10K fixed resistor and a 3K potentiometer.

Step 2—R1

R_2 defines the bandwidth. The correct equation relating R_2 and bandwidth is:

$$R_2 = \frac{R_1 f_0}{\Delta f}$$

Where Δf is the *one-sided* bandwidth—that is, 100 Hz. (The design example at page 16 of the XR2211's data sheet confuses single-sided and double-sided bandwidth values and introduces an extra factor of 2 into the numerator of this equation.)

$$R_2 = \frac{R_1 f_0}{\Delta f} = \frac{12500 \times 800}{100} = 100K$$

Based on measured data, I found that the calculated R_2 was too high and that 82K provided exactly the desired ±100 Hz bandwidth. We'll discuss how to verify performance of your completed circuit later.

Step 3—C1

C_1 is part of the control loop feedback circuit's low-pass filter. It has three purposes: first to attenuate the $(F_{INPUT} + F_{VCO})$ component from the phase detector; second to average the $(F_{INPUT} - F_{VCO})$ component out of the phase detector and third to govern the step response of the control loop. Chapter 24, Figures 24-17, 18 and 19 show what may happen to a motor control loop in response to an abrupt change in controlled or reference values; the loop may undershoot, overshoot or oscillate. The same phenomena occur in our PLL loop. In control system language, the reaction of the loop to a change is called "damping." A loop that undershoots is over-damped, while one that oscillates may be under-damped. We'll skip the math behind it, simply following Exar's recommendation to size C_1 for a damping factor of 0.5, where the relationship between C_1 and the damping factor is:

$$C_1 = \frac{1250 \cdot C_0}{R_1 \cdot \zeta^2}$$

Where ζ is the damping factor.

$$C_1 = \frac{1250 \cdot C_0}{R_1 \cdot \zeta^2} = \frac{1250 \times 0.1 \times 10^{-6}}{100000 \times 0.5^2} = 5 \times 10^{-9} = 5nF$$

I used a 5 nF (5000 pF) Mylar capacitor for C_1.

Step 4—R_D and C_D

Exar recommends R_D be set at 470K, and C_D sized using the following equation:

$$C_D > \frac{16}{\Delta f}$$

C_D is in μF.

$$C_D > \frac{16}{\Delta f} = \frac{16}{200} = 0.075 \mu F$$

(There's another inconsistency in the XR-2211 data sheet in this equation; it appears that Δf here refers to the double-sided bandwidth.)

In addition to the lock up time of the PLL, which we set through R_1 and C_1, C_D sets the lock up time of the tone detector comparator. Thus, for the tone detector output to reflect tone capture two events must occur: (a) the main PLL loop (comprising the phase detector with Pin 12 as its output) must lock up and (b) the tone detect phase detector output tracks the lockup of the main PLL loop and C_D charges sufficiently so that the voltage at Pin 3 exceeds the internal reference voltage and the lock detector comparator changes state. Hence, we wish to make C_D as small as feasible, consistent with the minimum value we've calculated. I set C_D at 0.1 μF. C_0 should be a Mylar or other temperature stable capacitor.

The final components we must determine are R_2, R_3 and C_2. R_2 is the pull-up resistor for the open collector lock detection comparator. We also use R_2 to limit the current through the lock detection LED, D1. Exar rates the comparator output to sink a maximum of 5 mA, so we use the technique discussed in Chapter 3 to determine that R2 should be approximately 700 ohms. We'll use the nearest standard 5% value, 680 ohms. As we'll see later in this chapter, under some conditions, particularly if the input tone is at one edge of the lock range, the comparator output has some false transitions. R_3/C_2 partially de-glitch the comparator output. Since we've designed the PLL for a response time of 5 ms, we'll likewise set the time constant of R_3/C_2 for 5 ms. Picking a convenient value of 0.1 μF for C_2, we determine R_3 should be 20K.

Adjusting the Circuit

After construction, it's necessary to adjust R_P to center the VCO. You may also have to change R_1 to obtain the design bandwidth. Since the VCO is not brought out to a pin, we must use an indirect method to adjust R_P. While monitoring the output of the tone detector comparator (pin 5) vary the input frequency and note the upper and lower drop out frequencies or the upper and lower pull-in frequencies. (The drop-out frequency is measured by starting with a frequency in the passband and then observing when tone detector drops out. The pull-in frequency is measured by starting at a frequency out-of-lock and then adjusting the frequency until the tone detector shows lock.)

Adjust R_p until the upper or lower frequencies (either pull-in or drop-out) are centered on 800 Hz. The difference between the upper and lower pull-in frequencies should be approximately 200 Hz. If you can't adjust R_p to center the upper and lower frequencies to 800 Hz, change R_0; reduce it to increase the center frequency or increase it to decrease the center frequency.

If the bandwidth (upper pull-in frequency minus the lower pull-in frequency) is significantly less than 200 Hz, then you will need to adjust R_1. I wouldn't adjust R_1 unless the bandwidth was more than 50 Hz wider or narrower than the 200 Hz target. To increase the bandwidth, reduce R_1; to decrease the bandwidth, increase R_1. You may see interaction between bandwidth and center frequency adjustments, so keep adjusting the center frequency and bandwidth determining components until the center frequency and bandwidth meet your requirements.

Programs

Program 27-0

We'll start out cautiously; collecting some statistics on Morse code as sent on the ham bands and as received by our tone decoder circuit. Figure 27-10 shows how the XR-2211 tone decoder connects to the PIC for Programs 27-1 and 27-2. The statistics we're concerned with are the dot and dash lengths for typical amateur operator sent Morse and we'll use these to develop an appropriate decoding algorithm. We can't, of course, use exactly the same approach we did in Chapter 26, where we distinguish unambiguously between dots and dashes by whether the dot or dash paddle lever is pressed. Instead, we have to measure the length of the re-

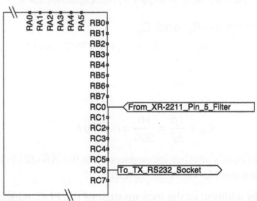

Figure 27-10: Connections for programs 27-0, −1 and −2.

ceived code elements and decide which are dots and which are dashes, based solely upon their relative lengths. Recall from Chapter 26 the recommended element and spacing lengths for perfect Morse:

Element	Relative Length
Dot	1
Dash	3
Inter-Element Space	1
Inter-letter Space	3
Inter-Word Space	7

Why collect these statistics, you might ask? After all, we know the dot/dash ratio is 1:3 and the space ratios are also well defined. If everyone sent computer-generated code, with perfect element lengths we could skip collecting statistics and proceed directly to coding our program. However, hams send Morse with a variety of technologies, ranging from up/down telegraph keys that would have been at home in an Civil War era telegrapher's office to computer programs that generate perfect code. The majority of hams likely use an electronic keyer, similar in concept to the one we developed in Chapter 26, which ensures consistent dot/dash ratios, but even so, the operator is responsible for manually inserting letter and word spaces. And, not all operators are equally proficient in their ability to operate their chosen instruments; otherwise it would not have been necessary for the amateur radio community to coin a special code signal QLF, or "now try sending with your left foot" to describe those attempting Morse with unacceptable results.

Program 27-0 measures tone pulse lengths and outputs the measurements over the serial interface for post-collection analysis.

```
;Program 27-0
;collect stats on typical Morse sending

;Constants
;----------
;Set max and min width in milliseconds
MinWidth        Con     15      ;less is noise
MaxWidth        Con     360     ;more is interference
CPinIn          Con     C0      ;constant for input pin

;Variables
;---------
PWidth          Var     Word            ;width in milliseconds
VPinIn          Var     PortC.Bit0      ;variable for input pin

;Initialization
;--------------
EnableHSerial
SetHSerial H115200
HSerOut ["P27-0",13]
Input CPinIn
PWidth = 0

Main
;====
        If VPinIn = %0 Then
                While VPinIn = %0
                        PWidth = PWidth + 1
                        PauseuS 790
                WEND
        EndIf
        If (PWidth >= MinWidth) AND (PWidth <= MaxWidth) Then
                HSerOut [Dec PWidth,13]
        EndIf
        PWidth = 0
GoTo Main
End
```

Unless we need a special feature of a particular pin, such as B0's interrupt-on-change feature or we must have our pins in sequence, such as for data transfer to or from an LCD, our choice of an input pin is, to a large extent, arbitrary. Here, however, we pick pin C0 to connect to the XR-2211 tone decoder because the C-series input pins have CMOS level Schmitt-trigger inputs and thus offer more immunity to noise or glitches in the tone decoder's output than the TTL-level B-series input pins. If you can't recall the difference between a CMOS level Schmitt-trigger input pin and a TTL-level input pin, you may wish to reread Chapter 4.

As usual, we define constants and variables, in that order.

```
;Constants
;----------
;Set max and min width in milliseconds
MinWidth        Con     15      ;less is noise
MaxWidth        Con     360     ;more is interference
CPinIn Con      C0      ;constant for input pin
```

If you've ever tuned around a short wave band, particularly in the summer, you've noticed static crashes, clicks and pops. These noise bursts are usually quite short—a few milliseconds long—but the tone decoder can momentarily lock onto these bursts and read a false character. Hence, we'll discard any tone duration measured less than **MinWidth**, which I've set to 15 ms. Likewise, we don't want a steady signal to be measured, so we'll discard durations exceeding **MaxWidth**, which I've set to 360 ms. I've based both **MinWidth** and **MaxWidth** on expected values for reasonable code speeds. Recall from Chapter 26 that a dot length is given by:

$$W_{DOT} = \frac{1200}{WPM}$$

669

Where WPM is the code speed in words per minute and W_{DOT} is the dot duration in milliseconds. For a maximum expected code speed of 60 WPM, W_{DOT} is 20 ms, so I've set **MinWidth** a bit below this value. If we set the slowest speed we are interested in measuring at 10 WPM, the dot length is 120 ms and a properly proportioned dash will be 360 ms. Hence, I've set **MaxWidth** at 360 ms.

I've also defined the constant **CPinIn** as alias for **C0**.

```
;Variables
;----------
PWidth   Var      Word           ;width in milliseconds
VPinIn   Var      PortC.Bit0     ;variable for input pin
```

Program 27-0 only has two variables; one to hold the measured pulse width, **PWidth**, and one to check the status of the input pin, **VPinIn**. My naming convention added "**C**" for constant to **PinIn** to yield the constant **CPinIn** and "**V**" for variable to yield the variable **VpinIn**.

The objective in Program 27-0 is to read the width of the tone pulse, to verify that it is within the acceptable range and if so write its width to the serial port.

```
Main
;====
        If VPinIn = %0 Then
                While VPinIn = %0
                        PWidth = PWidth + 1
                        PauseuS 790
                WEND
        EndIf
```

We use a **If…Then** test to monitor the status of the pin connected to the XR-2211 tone detector output. If the input pin goes low, the XR-2211 has detected an input tone and we measure the duration of the low by timing how long the input pin stays low.

When I developed Program 27-0 under MBasic 5.2.1.1, I could not use its **PulsIn** timing function to measure Morse width because its maximum returned value, 65535 μs, wasn't long enough to measure slow Morse. (Version 5.3.0.0 remedies that problem, but to show an alternative, we'll keep the timing approach I developed to use with version 5.2.1.1.) How then to measure a pulse value out to several hundred milliseconds? We could write an assembler interrupt routine that uses the 16F87x's internal 16-bit timer 1 to generate a tick every millisecond that updates a word variable "tick holder." The pulse width measurement loop would then zero the tick holder upon entrance and read its value upon exit. We'll use a simpler technique; adding a "do nothing" **Pauseus** statement to the measurement loop so that every pass through the loop executes in exactly 1ms. By incrementing **PWidth** every pass through the **While…Wend** loop, **PWidth** automatically measures the duration in milliseconds. This is the same approach we used in Chapter 26 to add a specific delay when generating Morse code, and the technique we discussed there for calibrating the **Pauseus** statement can be used for Program 27-0.

Of course, there's nothing requiring us to measure the tone pulse width in milliseconds as part of decode the incoming characters; we could delete the **Pauseus** statement and measure the duration in "loop cycles," where one loop cycle is however long it takes the **While…Wend** loop to complete one pass. After all, we distinguish dots from dashes by their relative lengths, not by some absolute duration. If a dot is 345 loop cycles in duration and the next element measures 1050 loop cycles, it's most likely a dash, regardless of how long a loop cycle is in terms of milliseconds or microseconds. However, in Program 27-3, we will display the decoded Morse signal's estimated code speed in WPM. For this we require the pulse width to be determined in milliseconds. Hence, we'll use Program 27-0 as an opportunity to write and debug a millisecond measuring routine.

```
If (PWidth >= MinWidth) AND (PWidth <= MaxWidth) Then
        HSerOut [Dec PWidth,13]
EndIf
PWidth = 0
GoTo Main
```

After the incoming code element ends, we then test the measured with to ensure it is valid and, if valid, write the measured duration to the serial port.

Let's tune around the ham bands, find a station sending Morse and adjust the receiver dial until the detection LED pulses in synchronization with the incoming audio. You may find that noise triggers the LED equally well with the received signal. *If so, you should reduce the audio level to the XR-2211 until with no Morse code signal present (but with background noise), the LED remains off and does not flash.* Our simplified circuit is quite sensitive to noise and audio overload, so it's important that you follow this step in adjusting the input signal level. Then, you should see an output similar to the following:

```
118
309
22
15
31
61
196
62
206
68
60
137
169
176
40
64
156
```

It appears that dots are around 60-70 ms and dashes are around 150-200 ms but we see quite a few values far removed from these ranges. Let's look at some sample data I recorded with Program 27-0. Figure 27-11 is a sample of good sounding Morse; it was pleasant to listen to with good element length and spacing. The data shows a dash/dot ratio quite close to the ideal 3:1 value. Figure 27-12 is from the other end of the quality spectrum; poor quality sending with dash lengths all over the place. I could copy it by ear, but it was fatiguing to listen to. The operator's dots were quite consistent in length, but his dash duration varied more that 2:1, with the average dash/dot ratio being 4.3:1. This operator used a mechanical semi-automatic key (colloquially known as a "bug") with a vibrating spring to automatically create dots but with manually sent dashes. Figure 27-13 shows similar data from five stations, expressed in terms of average element length in order to place stations sending at different speeds into a comparable format. This data, although limited, confirms what many years of ham radio operation had taught me; there's a huge variation in quality of Morse transmission. We therefore prefer an algorithm that robustly differentiates dots from dashes and deals with varying space lengths. We shall see, however, that even a relatively simple averaging technique works reasonably well with a wide variety of Morse reception.

What we desire is a threshold value; elements longer than the threshold are dashes and shorter elements are dots. Of course, the threshold must be adaptive; changing between operators and also tracking during any single transmission to adjust for changed speeds. Figure 27-11 and 12 show the results of two of many possible threshold methodologies:

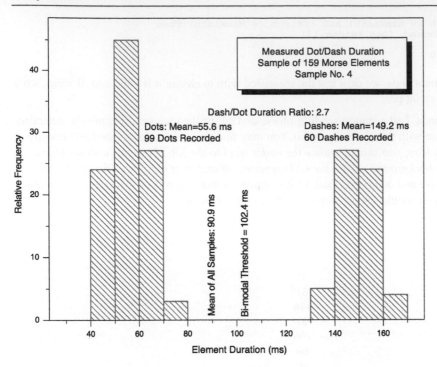

Figure 27-11: Element length distribution; high quality Morse.

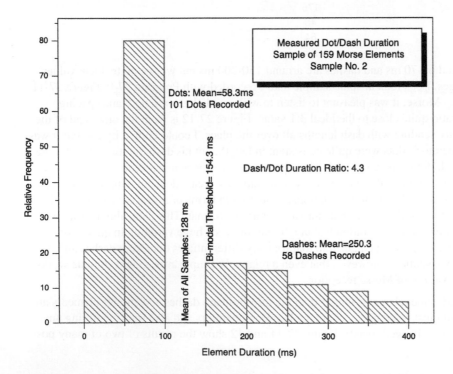

Figure 27-12: Element length distribution; poor quality Morse.

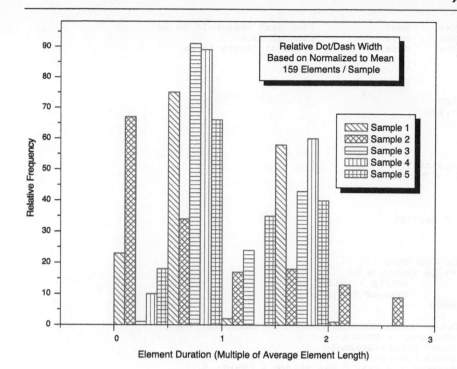

Figure 27-13: Normalized element length distribution.

- Average all elements and set the threshold at the average. If the number of dots equals the number of dashes and if the dash/dot ratio is 3:1, this establishes the threshold exactly where desired, halfway between the dot and dash lengths.
- Categorize the elements as "long" and "short" and find the average of the long elements and the average of the short elements and set the threshold halfway between the two averages.

Program 27-1

Looking at more data than shown in Figures 27-11 and 12, I determined that a simple arithmetical average of the element lengths provides a reasonable threshold value and can be executed in MBasic significantly faster than computing individual dot/dash means. Program 27-1 tests this threshold approach.

```
;Program 27-1
;Average Threshold

;Constants
;----------
;Set max and min width in milliseconds
MinWidth        Con     15              ;less is noise
MaxWidth        Con     360             ;more is interference
TimeMult        Con     3               ;timeout multiple
CPinIn          Con     C0              ;constant for input pin
ArySize         Con     32              ;average array size
DefaultWPM      Con     20              ;starting WPM

;Variables
;---------
PWidth          Var     Byte            ;width in milliseconds
VPinIn          Var     PortC.Bit0      ;variable for input pin
wArray          Var     Word(ArySize)   ;holds widths
i               Var     Byte            ;counter
k               Var     Byte            ;counter for wArray
WPM             Var     Byte            ;code speed WPM
```

```
Threshold        Var     Word            ;dot/dash threshold in ms
Sum              Var     Word            ;sum of data
ElLen            Var     Byte            ;element length in ms

;Initialization
;--------------
EnableHSerial
SetHSerial H115200
HSerOut ["P27-1",13]
Input C0
PWidth = 0

Threshold = 100
For i = 0 to (ArySize-1)
        wArray(i) = Threshold
Next

k = 0
Sum = Threshold * ArySize

Main
;====
        If VPinIn = %0 Then
                While VPinIn = %0
                        PWidth = PWidth + 1
                        PauseuS 790
                WEND
        EndIf
        If (PWidth >= MinWidth) AND (PWidth <= MaxWidth) Then
                wArray(k) = PWidth
                GoSub UpDateThreshold
                GoSub UpDateWPM
                HSerOut ["Width: ",Dec PWidth,9,"Thld: ", |
Dec Threshold,9,"WPM: ",Dec WPM,13]
        EndIf
        PWidth = 0
GoTo Main

;Recalulate the WPM and ElLength values
UpDateWPM
;------------
        ;Threshold is avg of dot & dash 32 samples
        ;so it is = 2 dot lengths
        ElLen = Threshold /2
        WPM = 1200 / ElLen
Return

;update threshold based on current value of dot/dash averages
UpDateThreshold
;--------------
        ;wArray is a cicular buffer and we use the
        ;technique of Chapter 11 to simplify averages
        Sum = Sum + wArray(k)
        k = k+1
        If k = ArySize Then
                k = 0
        EndIf
        Threshold = Sum /ArySize
        Sum = Sum - wArray(k)
        HSerOut ["k: ",Dec k,9,"Sum: ",Dec Sum,13]
Return
End
```

We've defined several new constants and variables to hold the information we need to calculate the tone duration:

```
ArySize Con     32              ;average array size

wArray Var      Word(ArySize)   ;holds average widths
Threshold       Var     Word    ;dot/dash threshold in ms
```

```
Sum              Var     Word        ;sum of data
ElLen            Var     Byte        ;element length in ms
```

We'll collect tone duration measurements in a 32-element word length array, **wArray**. We'll then compute the mean value of **wArray** and use that as the threshold, holding the result in the variable **Threshold**. The variable **Sum** holds the current sum of all 32 measured values.

```
Threshold = 100
For i = 0 to (ArySize-1)
        wArray(i) = Threshold
Next

k = 0
Sum = Threshold * ArySize
```

We preload **Warray** and **Sum** with default values to secure better results when receiving the first 32 elements.

```
Main
;====
        If VPinIn = %0 Then
                While VPinIn = %0
                        PWidth = PWidth + 1
                        PauseuS 790
                WEND
        EndIf
```

Programs 27-1 shares a common timing structure with Program 27-0 and no further discussion is required.

```
        If (PWidth >= MinWidth) AND (PWidth <= MaxWidth) Then
                wArray(k) = PWidth
                GoSub UpDateThreshold
                GoSub UpDateWPM
                HSerOut ["Width: ",Dec PWidth,9,"Thld: ",  |
Dec Threshold,9,"WPM: ",Dec WPM,13]
        EndIf
        PWidth = 0
GoTo Main
```

If the measured tone duration is a valid value, we store it in **wArray(k)** and call subroutines **UpDateThreshold** to calculate the current threshold variable **Threshold** and **UpDateWPM** to compute the corresponding code speed variable **WPM**. These values are then written to the serial port.

```
;update threshold based on current value of dot/dash averages
UpDateThreshold
;--------------
        ;wArray is a cicular buffer and we use the
        ;technique of Chapter 11 to simplify averages
        Sum = Sum + wArray(k)
        k = k+1
        If k = ArySize Then
                k = 0
        EndIf
        Threshold = Sum /ArySize
        Sum = Sum - wArray(k)
        HSerOut ["k: ",Dec k,9,"Sum: ",Dec Sum,13]
Return
```

In Chapter 11 we learned an efficient method of computing the arithmetical average. The brute force method of computing an arithmetical average of **wArray** is to sum all 32 values and divide the sum result by 32 each time we take a new measurement. Thus each computation requires 32 additions and one division. Using the technique described in Chapter11, we structure **wArray** as a circular buffer, in other words, it holds the last 32 values with the most recent value overwriting the oldest value. We also save the sum of the readings in the variable **Sum**. To compute the current average of all 32 values, we update the sum, re-compute the average and then subtract the oldest value from the sum. The number of required add/subtract operations is reduced from 32 additions to one addition and one subtraction. Both require one division. If this brief explanation isn't adequate, check Chapter 11 for a thorough analysis of the algorithm.

```
;Recalulate the WPM and ElLength values
UpDateWPM
;------------
        ;Threshold is avg of dot & dash 32 samples
        ;so it is = 2 dot lengths
        ElLen = Threshold /2
        WPM = 1200 / ElLen
Return
```

The remaining subroutine computes the estimated code speed in words per minute. This computation assumes that **Threshold** is computed as two dot lengths, which, as we saw earlier, is itself based upon the assumption that the last 32 elements comprise equal numbers of dots and dashes and that the dash/dot duration is 3:1. Neither of these assumptions is likely to be completely valid, so we should use the computed code speed as only an estimated value.

Here's a sample of Program 27-1's output. This data was taken after the program had been running for a few moments with my receiver tuned to an acceptable quality Morse signal.

Width: 164	*Thld: 121 WPM: 20*
k: 27	*Sum: 3869*
Width: 193	*Thld: 124 WPM: 19*
k: 28	*Sum: 3832*
Width: 63 Thld: 122 WPM: 19	
k: 29	*Sum: 3932*
Width: 200	*Thld: 126 WPM: 19*
k: 30	*Sum: 3900*
Width: 68 Thld: 125 WPM: 19	
k: 31	*Sum: 3949*
Width: 149	*Thld: 126 WPM: 19*
k: 0	*Sum: 3937*
Width: 71 Thld: 125 WPM: 19	
k: 1	*Sum: 3837*
Width: 71 Thld: 125 WPM: 19	
k: 2	*Sum: 3722*

Based on this limited sample, it looks like our threshold calculation does what is required. None of the measurements shorter than the threshold value seem to be real dashes and none that are longer than the threshold seem to be real dots.

Program 27-2

Now that we have confidence in our threshold algorithm, let's merge Program 27-1 with the Morse code decoding elements of Program 26-4. We'll output the decoded Morse to the serial port and use the same connections are for Program 27-1.

```
;Program 27-2
;Decode Morse and send to serial port
;Assumes PLL analog-to-digital decoder
;is tied to pin CpinIn

;Constants
;----------
;Set max and min width in milliseconds
MinWidth      Con      15      ;less is noise
MaxWidth      Con      500     ;more is interference
TimeMult      Con      3       ;timeout multiple
CPinIn        Con      C0      ;constant for input pin
ArySize       Con      32      ;average array size

Dot           Con      -1      ;for Array Morse
Dash          Con      1       ;in array of send elements
Nil           Con      0       ;is classed as dot/dash or nothing
MorseLen      Con      8       ;len array holding dots/dash/nil
```

```
EndYes          Con     0         ;was a character sent? Yes
EndNo           Con     1         ;no. Used to terminate spaces
WordRatio       Con     3         ;letter spaces = word space
IdleTime        Con     510       ;uss for idle loop execution
LSpace          Con     850       ;letter space multipler
;LetterSpace = (Threshold * ;LSpace)/IdleTime

;Decode holds the tree form of the decoded character
DeCode          ByteTable        "____5___H___4___S___^_%_V___3___",
                                 "I_____F_____U_?_____2___",
                                 "_E_____L_____R__+_._____*_____",
                                 "_A_____P_____W_____J__1____",
                                 "____6___B___=___D___/___X_____",
                                 "_N_____C_____K_____Y_____",
                                 "_T__7___Z_____,_G_____Q_____",
                                 "M___8_____O_9_____0____"

;TreeW holds the tree weightings
TreeW   ByteTable $40,$20,$10,$8,$4,$2,$1

;Variables
;---------
PWidth          Var     Word              ;width in milliseconds
VPinIn          Var     PortC.Bit0        ;variable for input pin
wArray          Var     Word(ArySize)     ;holds widths
i               Var     Byte              ;counter
j               Var     Byte              ;Counter for weight
k               Var     Byte              ;counter for wArray
WPM             Var     Byte              ;code speed WPM
Threshold       Var     Word              ;dot/dash threshold in ms
Sum             Var     Word              ;sum of data
Morse           Var     SByte(MorseLen)   ;holds elements
IdleCount       Var     Word              ;counter -- idle period
LetterSpace     Var     Word              ;passes through idle
                                          ;to equal a letter space
m               Var     Byte              ;counter for Morse(m)
EndLetter       Var     Byte              ;is letter finished?
EndWord         Var     Byte              ;is word finished?
Temp            Var     Byte              ;value of Morse char
TempStr         Var     Byte              ;Character
WordSpace       Var     Byte              ;number of word spaces

Initialization
;--------------
        ;Change as desired but not too slow
        EnableHSerial
        SetHSerial H115200
        HSerOut ["P27-2",13]
        Input CPinIn   ;for PLL input
        PWidth = 0                 ;pulse width

        ;Threshold = 2 dot lengths for perfect code
        Threshold = 120         ;corresponds to 20 wpm
        ;Initialized value
        LetterSpace = (Threshold * LSpace)/IdleTime

        ;Holds measured lengths of dot/dash pulses
        For i = 0 to (ArySize-1)
                wArray(i) = Threshold
        Next

        ;holds dot/dash decisioned values
        For i = 0 to (MorseLen-1)
                Morse(i) = Nil
        Next
        ;zero loop/timer counters
        IdleCount = 0
        WordSpace = 0
        k = 0
```

```
              Sum = Threshold * ArySize
              m = 0
              ;Output C1 ;use for idle time measurement
              ;uncomment if need to calibrate loop

Main
;====
;Wait for a dot/dash on input pin
;If it occurs, measure the length. Set loop for
;1 ms and count number of times through the loop
              If VPinIn = %0 Then
                      While VPinIn = %0          ;1ms for 20 MHz PIC
                              PWidth = PWidth + 1
                                                  ;Don't modify unless change
                              PauseuS 790              ;the PauseuS constant!
                      WEND
              EndIf

              ;Check for bad pulse widths
              If (PWidth <= MaxWidth) AND (PWidth > MinWidth) Then
                      wArray(k) = PWidth
                      ;Classify as dot or dash and save in PWidth
                      ;Same as for Programs in Chapter 26.
                      If PWidth > Threshold Then
                              Morse(m) = Dash
                      Else
                              Morse(m) = Dot
                      EndIf
                      m = m+1 ;ready for next dot/dash
                      If m = MorseLen Then
                              m = 0
                      EndIf
                      GoSub UpDateThreshold ;update threshold
                      ;Zero space counters since we have a
                      ;valid dot/dash
                      IdleCount = 0
                      WordSpace = 0
                      EndLetter = EndYes      ;maybe end of letter?
              EndIf
              PWidth = 0        ;clear for next pulse

              ;How many passes through in idle do we have?
              ;LetterSpace is based on time to execute this loop.
              ;If idle time = Letter space we have an end of
              ;letter and possibly an end of word
              IdleCount = IdleCount + 1
              If (IdleCount >= LetterSpace) Then
                      GoSub Spacing
              EndIf

              ;Following to measure idle time with 'scope on C1
              ;Uncomment if need to calibrate loop time
              ;bsf    PortC,1
              ;NOP
              ;NOP
              ;NOP
              ;bcf PortC,1
GoTo Main

;Recalulate the WPM and ElTimegth values
UpDateWPM
;------------
              ;Threshold is avg of dot & dash 32 samples
              ;so it is = 2 dot lengths
              WPM = 600 / Threshold
              LetterSpace = (Threshold * LSpace)/IdleTime
Return

;update based on current value of dot/dash averages
;This assumes #dots ~ #dashes. Usually true
```

678

```
      ;Called every letter space
      UpDateThreshold
      ;---------------
              ;wArray is a cicular buffer and we use the
              ;technique of Chapter 11 to simplify averages
              Sum = Sum + wArray(k)
              k = k+1
              If k = ArySize Then
                      k = 0   ;wrap around for k
              EndIf
              ;new average
              Threshold = Sum /ArySize
              ;must subtract oldest average from sum, and
              ;oldest average is k+1.
              Sum = Sum - wArray(k)
      Return


      ;Subroutine Spacing is called when we have a letter space
      ;Spacing checks to see if this is a letter space or a
      ;word space.
      Spacing
      ;--------
              IdleCount = 0   ;reset when this trips
              WordSpace = WordSpace + 1 ;times Spacing called
              ;we update at end of full letter

              GoSub UpDateWPM                 ;get new WPM as well

              ;If Spacing is called enough, must be a word space
              ;If it is a word space, we output a space char
              ;Following executed only if end of word
              ;--not end of letter
              If WordSpace = WordRatio Then
                      WordSpace = 0
                      If EndWord = EndYes Then
                              ;Use EndWord to prevent multiple spaces
                              EndWord = EndNo  ;sent one space no more
                              TempStr = " "    ;have a Morse character
                              HSerOut [" "]

                      EndIf
              EndIf

              ;Following executed if end of letter
              ;and not end of word
              ;The conversion routine follows program in Chapter 26
              If EndLetter = EndYes Then
                      Temp = 128                      ;binary tree mid value
                              ;go up/down tree based on dot/dash
                      For j = 0 to (MorseLen-1)
                              Temp = Temp + Morse(j)* TreeW(j)
                                      Morse(j) = Nil
                                      ;clear for next letter
                      Next
                      ;have index--get corresponding character
                      TempStr = DeCode(Temp)
                      ;Send it out
                      HSerOut [TempStr]
                      ;Ready for next dot or dash
                      m = 0
                      ;Letter done? Word Done?
                      EndLetter = EndNo
                      EndWord = EndYes
              EndIf

      Return

      End
```

Since a great deal of the nuts and bolts in Program 27-2 duplicates Program 26-4, we'll concentrate on the differences. Let's look at the main program loop:

```
Main
;====
If VPinIn = %0 Then
        While VPinIn = %0        ;takes 1ms for 20 MHz PIC
                PWidth = PWidth + 1;Don't modify unless change
                PauseuS 790               ;the PauseuS constant!
        WEND
EndIf
```

The measurement routine is identical with our earlier code and needs no further analysis.

```
;Check for bad pulse widths
If (PWidth <= MaxWidth) AND (PWidth > MinWidth) Then
        wArray(k) = PWidth
        ;Classify as dot or dash and save in PWidth
        ;Same as for Programs in Chapter 26.
        If PWidth > Threshold Then
                Morse(m) = Dash
        Else
                Morse(m) = Dot
        EndIf
```

To determine if a tone pulse is a dot or a dash, we compare it with the current **Threshold**. If longer, it's a dash; if shorter, it's a dot. We've defined the constants **Dash** and **Dot** as 1 and –1, respectively so we may use the same numerical binary tree evaluation algorithm used in Program 26-4. (And, of course, we've also defined the constant **Nil** as 0 as representing neither a dot nor a dash.)

This program structure ignores the inter-element space. Immediately after we have classified an element as a dot or a dash, the main program loop is ready to classify the next received element. We do not attempt to measure the space between elements and determine if the space is approximately one dot duration. Ignoring inter-element space helps properly decode ill-sent Morse and simplifies our programming task. It does, however, introduce a potential problem with false characters caused by glitches on the XR-2211 output pin, or by noise bursts.

```
m = m+1 ;ready for next dot/dash
If m = MorseLen Then
        m = 0
EndIf
```

After classifying the tone pulse, we update the array counter **m** that we use to index into the array **Morse()** that holds the numerical **Dash**, **Dot** and **Nil** values.

```
GoSub UpDateThreshold ;update threshold
;Zero space counters -- valid dot/dash
IdleCount = 0
WordSpace = 0
EndLetter = EndYes       ;maybe letter end?
EndIf
```

We recalculate **Threshold** based on the most recent element length through a subroutine call to **Update-Threshold**. We use **IdleCount** in the same way it's used in Program 26-4; it measures how many times the main loop is executed when no tone input is detected. Accordingly, we zero **IdleCount** whenever we have a valid tone input. Likewise, **WordSpace** works the same way in Program 27-2 as it does in Program 26-4; it determines whether a word space has occurred by measuring how many inte-element spaces have occurred. **WordSpace** accordingly must be set to zero after each tone input measurement. **EndLetter** also tracks its Program 26-4 functionality; it lets us emit only one letter space regardless of how long the main loop remains idle.

```
PWidth = 0       ;clear for next pulse

;How many passes through in idle do we have?
;LetterSpace is based on
```

```
;time to execute this loop. If idle time = Letter space
; we have an end of letter and possibly an end of word
        IdleCount = IdleCount + 1
        If (IdleCount >= LetterSpace) Then
                GoSub Spacing
        EndIf
```

To time the "key up" space intervals, we use the technique developed in Chapter 26; we measure the time it takes to execute the main program loop and compute how many times the program loop must execute to equal one inter-letter space. This value is held in the variable **LetterSpace**. (We'll see how **LetterSpace** is calculated when we look at subroutine **UpDateWPM**.) When the main loop has idled without either a dot or dash having been received long enough for one letter space to pass, subroutine **Spacing** is called.

```
        ;Following for measuring idle time with 'scope on C1
        ;Uncomment if need to calibrate loop time
        ;bsf    PortC,1
        ;NOP
        ;NOP
        ;NOP
        ;bcf PortC,1
GoTo Main
```

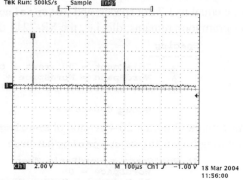

As in Program 26-4, we measure the time it takes for the main program loop to execute in idle once to obtain a calibration factor that we then define as the constant **IdleTime**. To make this measurement, we uncomment the five lines of code shown above, along with the line in **Initialization** that makes C1 an output pin:

```
        ;Output C1 ;use for idle time measurement
        ;uncomment if need to calibrate loop
```

We then connect an oscilloscope to Pin C1 and measure the time between the brief positive pulses, as seen in Figure 27-14. Using the calibrated cursor feature of my oscilloscope, I determined that the idle time was 510 µs. Hence, I added the following declaration to the **Constants** section:

Figure 27-14: Measuring the idle execution time for main program loop, program 27-2.

```
        IdleTime     Con     510      ;us for idle loop execution
```

If you don't have an oscilloscope, you may use the execution timing monitor developed in Chapter 14.

Subroutine **UpDateThreshold** is identical with the Program 27-1 version, so we'll not discuss it further.

```
        ;Recalulate the WPM and ElTimegth values
UpDateWPM
;-------------
        ;Threshold is avg of dot & dash 32 samples
        ;so it is = 2 dot lengths
        WPM = 600 / Threshold
        LetterSpace = (Threshold * LSpace)/IdleTime
Return
```

Subroutine **UpDateWPM** is called after **Threshold** is updated. Since **Threshold** is equal—given the caveats we mentioned earlier—to two dot lengths, we calculate the corresponding code speed by **WPM = 600 / Threshold**.

Finally, we calculate a new value for **LetterSpace**, based on the new **Threshold**. When we have made **LetterSpace** passes through the main loop while in idle, subroutine **Spacing** is called.

The value of **Threshold** represents two dot lengths, measured in milliseconds. **IdleTime** is the number of microseconds it takes to execute one pass through the main program loop when idling. If we wished to call subroutine **Spacing** exactly at 3 dot lengths of idle time (not a good idea, by the way), we would set

LetterSpace = 1.5 * 1000 * Threshold / IdleTime. The factor 1000 converts **Threshold** in milliseconds to **Threshold** in microseconds.

As usual, we'll wish to add a bit of a safety margin and consider that a letter space occurs at about 1.7 dot lengths. This could be done by: **LetterSpace = (1.7/2) * 1000 * Threshold / IdleTime**. Or, we can simplify it into:

$$\texttt{LetterSpace = (Threshold * LSpace)/IdleTime}$$

Where in the **Constants** section, we defined **LSpace** as 850:

```
LSpace          Con              850
```

The value 1.7 is based on subjective factors and experimenting with different values. Setting the inter-letter time to two dot lengths caused a surprisingly large error rate on many transmissions. Likewise, values less than 1.5 caused too many cases of not properly distinguishing letter spaces from inter-element spaces.

Subroutine **Spacing** is quite similar—but not identical—to subroutine **Spacing** in Program 26-4. Let's go through it, concentrating on the differences. First, we must remember that **Spacing** is called after an idle time equal to 1.7 dot spaces, and that **Spacing** is called thereafter every 1.7 dot spaces so long as the main program loop remains idle, i.e., the XR-2211 does not detect a tone input.

```
Spacing
;--------
        IdleCount = 0  ;rest - called when this trips
        WordSpace = WordSpace + 1 ;times Spacing called
        ;we update at end of full letter

        GoSub UpDateWPM              ;get new WPM as well
```

As with the Program 26-4 version of **Spacing**, we determine whether a letter space interval has occurred by counting how many times **Spacing** has been called, using the variable **WordSpace**.

```
If WordSpace = WordRatio Then
        WordSpace = 0
        If EndWord = EndYes Then
                ;Use EndWord to prevent multiple spaces
                EndWord = EndNo  ;sent one space no more
                TempStr = " "    ;have a Morse character
                HSerOut [" "]

        EndIf
EndIf
```

If **Spacing** has been called enough times that **WordSpace** = **WordRatio**, we know an end of word has occurred and we should emit a blank space. We've defined **WordRatio** as 3 in the **Constants** section:

```
WordRatio       Con              3
```

Since we increment **WordSpace** before making the comparison, and since **WordSpace** is reset to zero after each dot or dash, we deem a word space to have occurred after **Spacing** is called three times, corresponding to 5.1 dot spaces. (We don't have the same nonuniform problem in Program 27-2 seen in Program 26-4 resulting from the extra dot space the keyer automatically inserts after each dot or dash.) This value seems to work well in practice.

In order to prevent word spaces from being endlessly emitted while the main program loop idles, we use variable **EndWord** as a sentinel to detect multiple passes. After a letter space is sent, **EndWord** is set to **EndNo**. No further letter space characters will be emitted until **EndWord** is reset to **EndYes**, which occurs when either a dot or a dash is received.

```
;Following code executed only if end of letter
;and not end of word
;The conversion routine follows program in Chapter 26
If EndLetter = EndYes Then
```

```
      Temp = 128                    ;binary tree mid value
      ;go up/down tree based on dot/dash
      For j = 0 to (MorseLen-1)
            Temp = Temp + Morse(j)* TreeW(j)
            Morse(j) = Nil ;clear for next letter
      Next
      ;Have numerical index--get corresponding character
      TempStr = DeCode(Temp)
      ;Send it out
      HSerOut [TempStr]
      ;Ready for next dot or dash
      m = 0
      ;Letter done? Word Done?
      EndLetter = EndNo
      EndWord = EndYes
EndIf

Return
```

The remainder of **Spacing** follows closely the version we examined in Chapter 26. To understand our algorithm, start with the Morse alphabet.

0	— — — — —	M	— —
1	· — — — —	N	— ·
2	· · — — —	O	— — —
3	· · · — —	P	· — — ·
4	· · · · —	Q	— — · —
5	· · · · ·	R	· — ·
6	— · · · ·	S	· · ·
7	— — · · ·	T	—
8	— — — · ·	U	· · —
9	— — — — ·	V	· · · —
A	· —	W	· — —
B	— · · ·	X	— · · —
C	— · — ·	Y	— · — —
D	— · ·	Z	— — · ·
E	·	?	· · — — · ·
F	· · — ·	"/"	— · · — ·
G	— — ·	.	· — · — · —
H	· · · ·	,	— — · · — —
I	· ·	=	— · · · —
J	· — — —	$\overline{SK}$	· · · — · —
K	— · —	$\overline{AR}$	· — · — ·
L	· — · ·	$\overline{SN}$	· · · — ·

Let's suppose for the moment that **Spacing** has been called, but insufficient idle time has occurred to constitute a word space. Hence, we have an end-of-character space and thus are ready to convert the values in **Morse()** to a character and display it on the LCD. At this point, we have the character's dot/dash/nil sequence loaded into the array **Morse**, and we have determined that a valid end-of-character space has been received.

Let's suppose the letter is "C," or dash/dot/dash/dot. `Morse()` holds:

0	1	2	3	4	5	6	7	Element
Dash	Dot	Dash	Dot	Nil	Nil	Nil	Nil	Character
1	-1	1	-1	0	0	0	0	Morse() Value

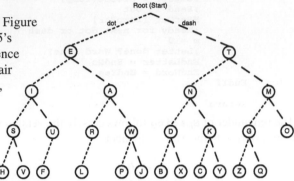

If we diagram the letters A…Z in Morse, as shown in Figure 27-15 we see an interesting relationship. Figure 27-15's structure is known as a "binary tree" in computer science terms. (The tree is upside down, with the stem in the air and the branches downward.) At each node (element), we make a decision based upon the next element; go along the dot branch or the dash branch until the next element has a nil value. We have then decoded the letter.

We might implement a binary tree through a series of **If…Then** statements. Let's see how that might work to decode "C."

Figure 27-15: Morse code binary tree.

```
If Morse(0)=Dash Then
    If Morse(1) = Dot Then
        If Morse(2) = Dash Then
            If Morse(3)= Dot Then
                If Morse(4) = Nil Then
                    Character = "C"
                EndIf
            EndIf
        Endif
    EndIf
EndIf
```

For simplification, I've omitted all the other conditional branches in this nested **If…Then** statement that allow us to decode T,N,M,D,K,G,O,B,X,Y,Z and Q as well as C starting with a dash at **Morse(0)**, but it should be clear that nested **If…Then** statements are, at best, a singularly inelegant method to decode Morse code, not to mention being slow in execution.

Some computer languages facilitate tree structures, such as support for linked lists, but MBasic doesn't. That's not surprising, considering PICs are not intended for heavy-duty data processing. And, we can devise a quick and efficient way to decode Morse without recourse to esoteric data structures.

We start by noting the similarity of a binary tree to a binary number. If we say the starting point (root) value is 128; that every layer down represents one binary shift to the right and that a dot represents subtraction and a dash represents addition, we can map each Morse character into a unique number between 0…255.

Let's see how this works for "C."

	0	1	2	3	4	5	6	7	Element
	Dash	Dot	Dash	Dot	Nil	Nil	Nil	Nil	Character
	1	-1	1	-1	0	0	0	0	Morse() Value
	64	32	16	8	4	2	1	0	Binary Weight
+128	+64	-32	+16	-8	0	0	0	0	Sum

The letter "C" thus corresponds to 128+64-32+16-8, or 168. We then can index into a byte table that has the letter "C" at its 168th position and quickly retrieve the correct decoded character.

This weighted binary tree conversion for A...Z, 0...9 and common punctuation marks and prosigns is as follows.

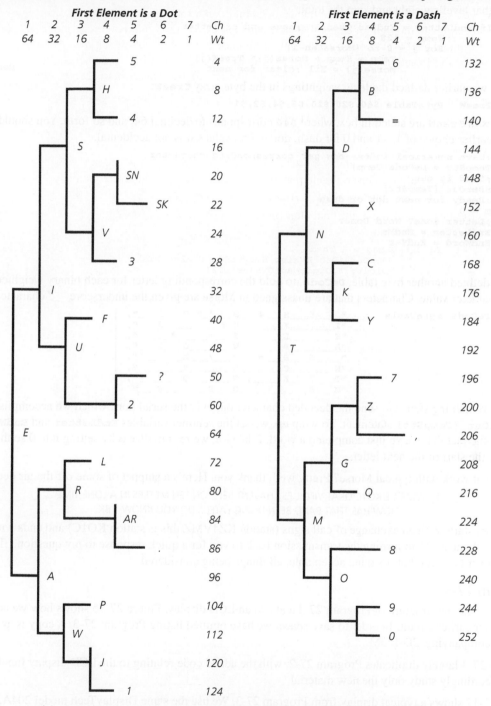

First Element is a Dot								**First Element is a Dash**							
1	2	3	4	5	6	7	Ch	1	2	3	4	5	6	7	Ch
64	32	16	8	4	2	1	Wt	64	32	16	8	4	2	1	Wt

First Element is a Dot:

- 5 — 4
- H — 8
- 4 — 12
- S — 16
- SN — 20
- SK — 22
- V — 24
- 3 — 28
- I — 32
- F — 40
- U — 48
- ? — 50
- 2 — 60
- E — 64
- L — 72
- R — 80
- AR — 84
- . — 86
- A — 96
- P — 104
- W — 112
- J — 120
- 1 — 124

First Element is a Dash:

- 6 — 132
- B — 136
- = — 140
- D — 144
- — 148
- X — 152
- N — 160
- C — 168
- K — 176
- Y — 184
- T — 192
- 7 — 196
- Z — 200
- , — 206
- G — 208
- Q — 216
- M — 224
- 8 — 228
- O — 240
- 9 — 244
- 0 — 252

If the next element is a dot, move upward one level; if it is a dash, move downward one level

Based on this approach, calculating the binary tree value for **Morse()** is simple; we multiply each element value by that level's weight and sum the total:

```
If EndLetter = EndYes Then     ;have end of letter
        Temp = 128
        For j = 0 to (MorseLen-1)
                Temp = Temp + Morse(j)* TreeW(j)
                Morse(j) = Nil ;clear for next          Next
```

Note that we earlier defined the level weightings in the byte array **TreeW**:

```
TreeW  ByteTable $40,$20,$10,$8,$4,$2,$1
```

The values in **TreeW** are stated in hex, where **$40** corresponds to decimal 64, and so forth. You should also note our earlier choice of 1, –1 and 0 for dash, dot and nil values was not accidental.

```
;Have numerical index--now get corresponding character
TempStr = DeCode(Temp)
;Send it out
HSerOut [TempStr]
;Ready for next dot or dash
m = 0
;Letter done? Word Done?
EndLetter = EndNo
EndWord = EndYes
EndIf
Return
```

We have defined another byte table, **DeCode**, to hold the corresponding letter for each binary weighted Morse character value. Characters that are unassigned in Morse are given the underscore "_" character.

```
DeCode ByteTable      "____5___H___4___S___^_%_V___3____",
                      "I_____F_____U_?_____2___",
                      "E_____L_____R___+_._____*____",
                      "A_____P_____W_____J___1____",
                      "____6___B___=___D___/___X_____",
                      "N_____C_____K_____Y_____",
                      "T___7___Z_____,_G_____Q_____",
                      "M___8_____O___9_____0__"
```

The only remaining step is to send the decoded character out over the serial port, which we accomplish with the **HSerOut [TempStr]** statement. To wrap up, we set the sentinel variables **EndLetter** and **EndWord** to reflect the fact that we've just completed a word. Likewise, we re-initialize **m** by setting it to 0 so that it's ready for the start of the next letter.

How does it work with typical Morse? Pretty well, thank you. Here's a snippet of some off-the-air decoding:

KZ4VMB DE KO1C OKAT VIFGIL ES I HAVENT BEEN ON TEN METERS IN A LONP TIME

HOW HAS THAT BAND BEEN DOING LATELY, DO YOU KNOW ? BK

The extract starts with an exchange of call signs (station KZ4VMZ this is station KO1C) and ends with a "BK" which means "I'm turning the transmission back to you for a quick response to my question." The decoding isn't perfect, but it's quite acceptable, all things being considered.

Program 27-3

Let's now take the structure of Program 27-2 and add an LCD display. Figure 27-16 shows how we add the LCD to our earlier circuit. In order to save space, we have omitted listing Program 27-3. A copy is available in the accompanying CD-ROM.

Program 27-3 largely duplicates Program 27-2, with the added code relating to the LCD display functions. We'll accordingly study only the new material.

Figure 27-17 shows a typical display from Program 27-3. We use the same DisplayTech model 204A, 20x4 LCD discussed in Chapter 26, and a display arrangement similar to Program 26-4. The top line displays

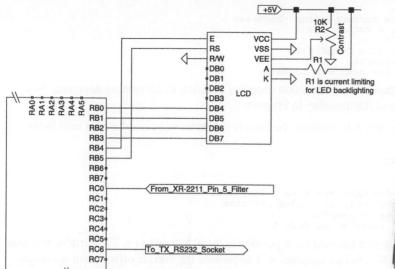

Figure 27-16: Configuration for Program 27-3.

Figure 27-17: Program 27-3 display.

the received text, which is added left-to-right until the 20th character is received, at which point it reverts to scrolling mode. The bottom line shows the estimated code speed in words per minute.

We've added several constants and variables related to the display:

```
RegSel          Con     B5      ;LCD control
Clk             Con     B4      ;LCD control
BarrelSize      Con     20      ;length of barrel array for display

;Index into LCD llines for a 4x20 display
LineOff ByteTable $0,$0,$40,$14,$54

LCDNib          Var     PortB.Nib0      ;LCD data transfer port
LineNo          Var     Byte            ;LCD Line number
;(assumes 20x4 display)
BarrelAry       Var     Byte(BarrelSize)    ;for LCD
Mindex          Var     Byte                ;for barrel counter
```

RegSel, **Clk** and **LCDNib** define the pins used to communicate with the LCD, while the byte table **LineOff** provides the starting memory addresses for lines 1 through 4. Decoded Morse characters are held in the byte array **BarrelAry()**. In each case, we used identical constant and variable identifiers as we did in Program 26-4.

Let's look at fragments of the subroutine **Spacing** to see the LCD display changes:

```
If WordSpace = WordRatio Then
        WordSpace = 0
        If EndWord = EndYes Then
                ;Use EndWord to prevent multiple spaces
                EndWord = EndNo         ;sent one space--
                TempStr = " "   ;have Morse character
                HSerOut [" "]
                GoSub ShowChar
```

The only addition to the above code segment is the subroutine call **GoSub ShowChar**.

```
If EndLetter = EndYes Then
        Temp = 128              ;binary tree mid value
        For j = 0 to (MorseLen-1)        ;go up/down tree
                Temp = Temp + Morse(j)* TreeW(j)
                        Morse(j) = Nil ;clr next letter
```

```
            Next
            ;now get corresponding character
            TempStr = DeCode(Temp)
            ;Send it out
            HSerOut [Str TempStr\1]
            GoSub ShowChar
            ;Ready for next dot or dash
```

Likewise, we have added **GoSub ShowChar** to the main decoding function in subroutine **Spacing**. Note that we have retained the serial output functionality in Program 27-3.

Let's look at subroutine **ShowChar**. When it is called, the current character to be displayed is held in the byte variable **TempStr**.

```
        ;Put sent char onto LCD
        ShowChar
        ;---------
                ;Hold the decoded characters in a length 20 array
                ;Insert next character at Mindex position
                BarrelAry(Mindex) = TempStr
                LineNo = 1       ;display on line 1
```

We insert the character to be displayed in the next open position in array **BarrelAry**. The variable **MIndex** holds the next open position index. We also set **LineNo = 1** to provide the correct offset when we index into **LineOff**.

```
                ;Code takes about 7.5 ms
                ;Position LCD cursor & write the character array
            LCDWrite RegSel\Clk, LCDNib,[ScrRAM+LineOff(LineNo), |
        Str BarrelAry\BarrelSize]
```

We write all 20 characters in **BarrelAry** at once to the top line of the LCD, unlike in Program 26-4 where we write the letters one at a time. It turns out that Program 27-3 struggles to keep the LCD updated when the incoming code speed is faster than about 30 WPM or so. Writing all 20 characters of **BarrelAry** at once saves several milliseconds and improves the LCD update performance.

```
            ;We start at left side of screen, write chars to right.
            ;When reach last position, add chars at right
            ;and shift all prior chars one position to the left.
            ;Hence we only increase Mindex if it is less than 19.
            ;Once it hits 19 we keep it there.
            If MIndex < (BarrelSize-1) Then
                MIndex = MIndex + 1
            EndIf
```

We've initialized **Mindex** to 0 in Program 27-3's **Initialization** section and filled **BarrelAry** with the space character. Accordingly we increment **Mindex** until it reaches 19. (As our usual practice, we've defined the length of **BarrelAry** in terms of a constant **BarrelSize**, which is set to 20. Hence, the last valid array index is **BarrelSize-1**.) When **Mindex=BarrelSize-1**, we have filled all 20 positions of the Line 1 with information and we then add new characters at the right hand end shift the older characters left by one place to accommodate the new letter. Hence the above code increments **Mindex** only until it reaches 19, at which point it freezes at 19.

```
                ;Shift all chars 1 position to the left to make room
                ;for new ones that come in at the end.
                    ;Following code takes 8 ms
            If MIndex >= (BarrelSize-1) Then
                For kk = 0 to (BarrelSize-2)
                        BarrelAry(kk) = BarrelAry(kk+1)
                Next
            EndIf
```

The above code moves every character one position to the left, overwriting **BarrelAry(0)**.

```
                ;LineNo = 4
                ;LCDWrite RegSel\Clk,LCDNib,[ScrRAM+LineOFf(LineNo)]
        Return
```

Finally, we position the cursor back to the start of line 4, ready for an updated code speed display.

We've also added a subroutine call to **UpDateStatus** at the end of subroutine **Spacing**:

```
            GoSub UpDateWPM
            GoSub UpDateStatus
```

Subroutine **UpDateStatus** displays the newly calculated code speed at line 4 of the LCD.

```
;Updates the status line
UpDateStatus
;-----------
            LineNo = 4
            LCDWrite RegSel\Clk,LCDNib,[ScrRAM+LineOFf(LineNo)]
            LCDWrite RegSel\Clk,LCDNib,[DEC WPM\2," WPM   "]
      Return
```

Since the remainder of Program 27-3 duplicates Program 27-2, no further analysis is necessary.

Ideas for Changes to Programs and Circuits

* In addition to displaying the current estimated code speed, it would be interesting to display the dash/dot ratio as a measure of the quality of code being sent. How might this be done? Since we have distinguished between dots and dashes in our main loop, we could establish two new word arrays; say **DotAry** and **DashAry**, each of length 16. Each dot or dash length would be stored in **DotAry** or **DashAry** in addition to **WArray**, and the arithmetical average of both arrays (say **DotAvg** and **DashAvg**) calculated using the same running sum technique employed to determine **Threshold**. Every time the speed display is updated, compute the dash/dot ratio as **DashAvg / DotAvg**. Of course, it will either be necessary to perform this computation in floating point, or add keep it an integer calculation, but multiply **DashAvg** by 10 to obtain one decimal point of accuracy, i.e., **(DashAvg * 10) / DotAvg**. Before the value is displayed, a decimal point can be added at an appropriate point.

* How might the dot/dash decision threshold algorithm be improved upon? Is there a algorithm that **will** accurately compute the separate arithmetical average of each part of a bi-modal data set?

* I originally planned to use the top three lines of the 20x4 display for received code, with a 60-character scrolling display. However, the time required to update three lines didn't permit accurate decoding high speed Morse. How might we expand the memory and preserve scrolling? Suppose we update the display only at word breaks, where additional time is available before the next code element is received. This produces a somewhat jerky display, of course. Would it make sense to implement both a smooth scrolling 60-character display, switching to a word-based display when the received code speed exceeds a certain value? Or, should certain time sensitive routines be recoded in mixed assembler and MBasic?

* To reduce the effect of tone decoder output glitches, our circuit adds an RC filter, comprised of R3 and C2. Would an active low-pass filter, implemented around an MCP601 op-amp yield a better result? Would a variable cut-off low-pass filter offer more improvement, where the cutoff frequency could be adjusted by the PIC, based upon the speed of the received signal? Can the programmable low-pass filter discussed in Chapter 21 be used for this purpose? Should any of the filter component values in the XR-2211 be adjusted for better performance at low to moderate code speeds? If so, how might these be placed under PIC control?

References

[27-1] EXAR Corporation, *XR-2211 FSK Demodulator/Tone Decoder*, rev. 3.01 (June 1997). Available for downloading at http://www.exar.com (requires free registration to access data sheets.)

[27-2] American Radio Relay League, *The ARRL Handbook for Radio Communications*, 81st Ed., Newington, CT. (2004) Now in its 81st edition, the ARRL handbook is the starting point reference for all things amateur radio.

[27-3] Osterman, Fred, *Shortwave Receivers Past & Present Communications Receivers 1942-1997, 3rd ed.*, Universal Radio Research, Columbus OH (April 1998). This compendium covers almost every model short wave receiver manufactured between 1942 and 1997, and is a valuable reference for a new entrant to short wave radio that wishes to purchase a new or used receiver.

[27-4] Helms, Harry L., Shortwave Listening Guidebook: *The Complete Guide to Hearing the World, 2nd ed.*, Universal Radio Research, Columbus OH, (August 1993). Good introduction to overall short wave listening, although a bit dated.

CHAPTER **28**

Weather Station and Data Logger

Now that we've learned how to read analog voltages, to store data in an external EEPROM, to read digital temperature sensors and to write an interactive serial program, let's assemble these disparate concepts into a data logging mini-weather station.

We'll measure temperature, relative humidity and barometric pressure and save the data to external EEPROM for later extraction. In order to keep the program to manageable length, we'll not implement the full interactive serial functions, but what you've learned in Chapter 9 will let you add it, should you so desire. And, of course, you might decide to add an LCD display and additional temperature sensors and have a self-contained indoor/outdoor weather station.

Sensor Selection

We've limited our weather station sensor suite to the three commonest elements: relative humidity, barometric pressure and temperature. For each element, we have a wide selection of sensor types, but to keep things simple, we'll limit our sensor selection to integrated modules. We'll first review our choice of sensors, next see how they connect to the PIC and finally write the code to make it work.

Humidity

As we learned in introductory science classes, air is a mixture of gases including, most importantly oxygen and nitrogen, but also water vapor. Relative humidity is defined as the ratio of actual vapor pressure of water to the saturation vapor pressure, i.e., the maximum amount of water vapor the air can hold before it precipitates out as liquid water droplets or mist. Relative humidity is customarily expressed as a percentage, from 0 to 100%. Since the saturation carrying capacity of air is proportional to temperature, the same absolute quantity of water vapor results in differing relative humidity as the temperature changes.

Almost all inexpensive relative humidity sensors rely upon absorption of water by a material that changes properties in response to the absorption. A few sensors are constructed from material that changes resistance when exposed to water vapor, but the most popular relative humidity sensor today relies upon changes in dielectric constant. When the correct material is between the two conducting plates, the result is a sensor with a capacitance proportional to relative humidity. By measuring the sensor's capacitance, we may determine the relative humidity with surprisingly good accuracy, typically within ±2%.

Humirel's HS-1100 sensor, for example, has a nominal capacitance of 180 pF at 55% relative humidity, and varies approximately 0.34 pF per each 1% change in relative humidity. (The variation is not quite linear, and Humirel provides a four-term polynomial equation to fit capacitance to relative humidity.) It's possible to measure capacitance with a PIC by several techniques, such as measuring the charging time of an RC network with a known resistance; or by measuring the frequency of an oscillator where the sensor's capacitance is one of the frequency determining elements. From the measured capacitance we may calculate the relative humidity after going through a few mathematical manipulations.

To simplify our project, however, we'll use a pre-built relative humidity module from Humirel, the HM-1500. This module combines a capacitive relative humidity sensor with other circuitry to yield a DC voltage output proportional to relative humidity. Without correction, our humidity reading will be within ±3% over the range 30%…80% and within ±5% over the full operating range of 10%…95% relative humidity. If increased accuracy is desired, Humirel provides an equation to correct for nonlinearity and temperature. We'll keep things simple, and use the uncorrected linear equation:

$$V_{out} = 25.68RH + 1079$$

V_{out} is the output voltage in millivolts from the HM-1500

RH is the relative humidity in percent

Since we wish to determine RH, knowing V_{out}, we solve for RH and (restating in volts, not millivolts) find:

$$RH = \frac{V_{out} - 1.087}{0.02568}$$

The HM-1500's output voltage ranges from 1.3 V at 10% RH to 3.55V at 95% RH.

The HM-1500 is a bit expensive at $30 each in single lots, compared with the bare HS-1101 sensor at $12.50 so this is a case where we pay for convenience. An alternative to the HM-1500 would be one of Honeywell's HIH-3610 series integrated sensor modules, which are slightly less expensive at $22 each in single lots. The HIH-3610 also has a DC voltage output proportional to relative humidity and should be substitutable for the HM-1500 with only a minor change to the voltage-to-relative humidity equation.

We'll read the HM-1500's output with a 16F87x's A/D converter. A quick back-of-the envelope error budget shows that the 16F87x's 10-bit A/D resolution is more than adequate for this task; the sensor accuracy is ±2%, and, as we learned in Chapter 11, a good reference voltage yields an expected error of ±8mV or so over the range 0…4.096 V. The sensor output range is approximately 2.2 V, so we estimate the measurement error over the sensor output range is 8mV/2.2V, or approximately 0.4%. This is about five times better than the sensor error, so we'll consider it acceptable.

Barometric Pressure

Barometric pressure is the pressure exerted by the atmosphere above the sensor. The measured pressure is a function of both the weather conditions and the elevation of the sensor above (or below) sea level. For historical reasons, barometric pressure has long been expressed in inches of mercury, as the original barometers measured air pressure against a column of mercury. At sea level, the standard barometric pressure is 29.92 inches. The most commonly used metric unit of air pressure is the hectopascal (hPa), but its older name, the millibar, is still in widespread use. We'll use the term millibar. (The preferred metric term for pressure measurement is the kilopascal, or KPa, as "hecto" (×100) multipliers are nonstandard.) The relationship among inches of mercury, millibars, kilopascals and pounds per square inch is:

	PSI	KPa	Inches Hg	mbars
PSI	1	6.8948	2.0360	68.9476
KPa	0.1450	1	0.2953	10.000
Inches Hg	0.4912	3.3864	1	33.8639
mbars	0.0145	0.1000	0.0295	1

To convert, for example, 30.00 inches Hg to mbar, multiply by 33.8639, yielding 1015.9 mbar. (If we were concerned with high precision measurements, we would learn that manometer-type units, such as inches Hg, or inches of water are not well defined; for example, they depend on the density of the mercury, the temperature, the local gravity, etc. But at our level of precision, don't worry.)

The automotive industry is the largest consumer of pressure sensors, used in electronic engine control, among other things, and we'll take advantage of the cost benefits of the associated volume production. Freescale Semiconductor, Inc., a recent spin-off of Motorola sensor business, is a leading producer of pressure sensors, so we'll look at their products in detail. Freescale's pressure sensors use piezoelectric technology; a thin flexible silicon diaphragm is ion implanted with a material that changes resistance when pressure is applied, thus forming a strain gauge. When current is passed through the strain gauge, a voltage proportional to diaphragm deflection (and hence the pressure differential between the two sides of the diaphragm) results. If one side of the diaphragm is in a vacuum, the deflection is proportional to the absolute pressure on the exposed side.

As might be expected, the raw output of the strain gauge requires calibration and signal conditioning to compensate for temperature effects and to amplify the relatively small voltage versus pressure change. Fortunately, Freescale offers many pressure sensors with integrated signal conditioning and calibration. The particular sensor we'll use is a MPX6115A6U, which costs about $17.50 in single lots. This device is an 8-pin surface mount, small outline package, but can easily be mounted to a Surfboard 6103 3-connection surface mount adapter, as only pins 2, 3 and 4 of the MPX6115A6U are used. The small outline package pin spacing is 0.1" so it's quite manageable without special surface mount tools or procedures. The Surfboard 6103 adapter plugs directly into a standard 0.1" breadboard, such as Basic Micro's 2840 development board.

The MPX6115A6U operates over a range of 15 kPa to 115 kPa, or 150 mbar to 1150 mbar. Freescale states the relationship between output voltage and pressure as:

$$V_{out} = V_S (0.009P - 0.095) \pm Error$$

V_{out} is the output in volts

V_S is the source supply voltage in volts

P is the pressure in kPa

Solving for P in millibars, and ignoring the error component for now, we find:

$$P_{mbar} = \frac{V_{out} + 0.095V_S}{0.090V_S}$$

Over the calibrated range of 150 mbar to 1150 mbar, the voltage output, assuming V_S is 5.000 V, ranges from 4.700 V (1150 mbar) to 0.200 V (150 mbar). The question then becomes "over what range might we expect normal barometric pressure to vary?"

It appears that the lowest barometric pressure ever recorded (in a typhoon) was 25.68 inches Hg (869.63 mbar), and the highest pressure (sea level equivalent) was 32.01 inches Hg (1083.8 mbar), recorded in Siberia, corrected for a station elevation of 263 meters.

Before going further, we must take a brief detour. By convention, barometric pressure is adjusted or "reduced" to sea level equivalent, so data may be compared without knowledge of the station elevation above mean sea level. However, we'll be measuring absolute pressure, which includes elevation effects. The absolute pressure in Denver, Colorado, at an elevation of approximately 5300 feet AMSL, will be lower than in the basement workshop in my house in suburban Washington DC with an elevation of 320 feet AMSL. The barometric pressure adjustment factor for elevation can be determined the relationship between elevation and air pressure change, which is approximately:

$$Z = 62900 \log_{10}\left(\frac{P_0}{P}\right)$$

Where:

> Z is the elevation in feet AMSL
> P_0 is the sea level pressure
> P is the pressure at elevation Z

Since we know Z and P_0 we solve for P:

$$P = \frac{P_0}{10^{\frac{Z}{62900}}}$$

To develop an elevation correction, we'll use the standard atmosphere sea level pressure, 29.92 inches Hg. We then calculate P for the altitude and subtract from 29.92 inches Hg to obtain the elevation correction. (This is a simplified correction methodology, but is acceptable for our purposes.) I've tabulated a sample of elevation corrections below.

Elevation (feet AMSL)	in Hg offset
0	0.00
100	0.11
200	0.22
320	0.35
500	0.54
1000	1.08
2000	2.11
5000	5.00
5300	5.28
10000	9.17

The elevation correction is added to the absolute pressure. For example, if my MPX6115A6U has an output voltage of 4.025 V and the power supply voltage is 5.000 V, what's the sea level corrected barometric pressure?

1. Calculate pressure in mbar using equation $P_{mbar} = \dfrac{V_{out} + 0.095 V_S}{0.090 V_S}$. The result is 1000.0 mbar.
2. Calculate the pressure in inches of mercury by multiplying mbar by 0.02953. The result is 29.53 inches Hg.
3. Add the elevation correction. My ground elevation is 320 feet, so the correction is 0.35 inches Hg, resulting in a corrected value of 29.53 + 0.35, or 29.87 inches Hg.

With this understanding, we can now estimate the voltage output range of our pressure sensor. Based on the weather records mentioned earlier, the maximum absolute pressure expected is very unlikely to exceed 32 inches Hg. (Weather records are kept based on elevation-corrected readings, so we can't be 100% sure of this statement.) The minimum absolute pressure is dependent upon the observing altitude. We'll arbitrarily set this at 5000 feet, and assume the corrected pressure is 26 inches Hg, corresponding to an absolute pressure of 21 inches Hg. Converting these values to mbar and calculating the expected sensor output, we find the maximum expected voltage is 4.401 V and the minimum expected voltage is 2.725 V, a range of 1.649 V from maximum to minimum.

Let's see how these values fit with our 10-bit A/D converter. First, we notice that the precision 4.096 V reference used in Chapter 11's design no longer works, as we need to measure up to 4.401V. (Assuming we don't scale the MPX6115A's output, of course.) Let's assume, for our analysis, that we will use a 4.6 V reference. And, let's assume for simplicity of design we set the A/D's V_{ref} to ground. (We could improve resolution about 2:1 if we set V_{ref} to, say, 2.5 V, but at the cost of additional circuitry.) Each of the 1024 voltage steps is thus 4.5 mV and the 1.649 V output range encodes into 367 steps. Since this voltage range corresponds to 11 inches Hg, each A/D step is 0.03 inches Hg. The A/D error, including errors in the voltage reference and analog circuitry to scale the voltage reference to 4.6V, corresponds to about 0.06 inches Hg.

How does this error budget relate to the overall sensor accuracy? Freescale rates its MPX6115A sensor as ±1.5% of V_{FSS} over the full pressure and temperature range. V_{FSS} is the full-scale (maximum rated pressure) voltage output, nominally 4.75V, so the error voltage is ±71.25 mV. The error is not "1.5% of the pressure reading." Rather, every voltage reading is subject to an error of ±71.25 mV. If the actual absolute pressure is 32 inches Hg, corresponding to a nominal 4.401 V, the actual voltage output could be as low as 4.3298 V (corresponding to 31.53 inches Hg) or as high as 4.4723 V (32.46 inches Hg). This constant error voltage corresponds to an error of ±0.47 inches Hg, regardless of the pressure. Considering the inherent sensor error, our A/D error budget of ±0.06 inches Hg looks acceptable.

Temperature

We used Maxim's DS18B20 one-wire temperature sensors in Chapter 12, but we'll reconsider that decision for this chapter. Since both the humidity and barometric pressure sensors are analog, we could stay with an "all analog" sensor suite and use an analog temperature sensor, such as National Semiconductor's LM34C, with an output of 10.0 mV/°F. With an A/D resolution of about 4.5 mV/step, we could read temperature with a resolution of 0.45°F/0.25°C. However, additional circuitry is required to sense temperatures below 32°F. Instead, we'll use Maxim's I²C-protocol DS1624 temperature sensor / EEPROM. The DS1624 combines a temperature sensor with a resolution of 0.03125°C with 256 bytes of serial EEPROM. (No, that's not 256 *Kbytes* of memory; it's 256 *bytes*.) We won't use the EEPROM.

We explored I²C communications with external EEPROM serial memory chips in Chapter 18, so we'll concentrate on the differences between the DS1624 and the 24-series memory chips.

First, we note the pin connections differ significantly from those of generic 24-series devices:

Pin Name	DS1624 8-pin DIP Pin Number	Function and Comments
SDA	1	Serial Data. **This is an open drain pin and must have a pull-up resistor**. 10K is a suitable value for Standard Speed data transfer.
SCL	2	Serial Clock.
NC	3	No connection. The DS1624 does not have a Write Protect pin.
V_{SS}	4	Ground.
A2	5	Chip address bit A2.
A1	6	Chip address bit A1.
A0	7	Chip address bit A0.
V_{DD}	8	Plus power supply; +5V in our breadboard designs.

The DS1624 does not have a write protect pin.

Let's look at a DS1624 control byte *as used with MBasic's* **I2Cout** *and* **I2Cin** *procedures*. (If you don't recall the nuances of MBasic's **I2Cout** and **I2Cin** procedures, now's a good time to review Chapter 18.) It's comprised of three elements:

DS1624 Control Byte							
Device Family Code				Chip Addressing Information			8/16 Bit
Bit 7	Bit 6	Bit 5	Bit 4	Bit 3	Bit 2	Bit 1	Bit 0
1	0	0	1	A2	A1	A0	0

Recall that the family code for 24-series EEPROMs is **$A** or **%1010**; the DS1624's family code is **$9** or **%1001**. The DS1624 uses all three chip address bits for chip addressing; there are no block bits or unused bits.

The control byte's bit 0 determines whether the EEPROM address information sent immediately following the control byte is a single 8-bit byte (256 possible addresses) or a 16-bit word (65536 possible addresses). If bit 0 is 0, the address is sent as an 8-bit byte; if bit 0 is 1, the address is sent as a 16-bit word. The DS1624 requires an 8-bit value to immediately follow the control byte, so we must set Bit 0 to **%0**.

As with 24-series EEPROM devices, the control byte definition in the DS1624's data sheet refers to bit 0 as the "read/write" or R/W bit, and says that that it determines the direction of data transfer; if 0, the master sends and if 1, the slave sends. *However, MBasic, for reasons of compatibility with other versions of PIC BASIC, automatically inserts the correct R/W bit and instead uses bit 0 of the control byte for address length information.* It's easy to get confused, so attention to detail is critical.

Perhaps the most confusing difference between the DS1624 and 24-series EEPROM is that when sending data from the master PIC to the slave I²C device, a DS1624 requires a command byte that immediately follows the control byte and precedes any data byte:

Device	1st Byte	2nd Byte	3rd Byte	4th Byte
24-Series EEPROM	Control Byte	Address Byte	Address Byte if 16-bit	Data Byte
DS1624	Control Byte	Command Byte	Data Byte	

The command byte has five legitimate values:

Command Byte	Action
$17	Access the 256-byte EEPROM memory. After this command byte, the next byte is the memory address to be accessed.
$AC	Access the configuration register, either to read stored values or set values.
$AA	Read the last temperature result. Two bytes are returned by the DS1624.
$EE	Start a temperature conversion. If in continuous mode, $EE initiates continuous operation. No data value follows a $EE.
$22	Stop temperature conversion. If in continuous mode, $22 halts continuous operation until a $EE is received. No data value follows a $22

The configuration register is a single byte, with only two options:

Bit 7	Bit 6	Bit 5	Bit 4	Bit 3	Bit 2	Bit 1	Bit 0
Done	1	0	0	1	0	1	1-shot

Done—If Done = **%1**, the commanded temperature conversion is complete; if Done = **%0**, the conversion is still in progress.

1-Shot—The DS1624 operates in two modes; continuous and one-shot. In continuous mode (1-Shot = **%0**) the DS1624 continuously makes temperature conversions; as soon as one is completed, the next one starts. In one-shot mode (1-Shot = **%1**), temperature conversions are made only when commanded by the $EE command byte. If the DS1624 is in continuous mode, however, a command byte $22 will terminate continuous mode until a $EE command is received.

The configuration register is nonvolatile, so its value is retained when power is removed. The factory configuration is continuous mode.

Using the command byte may be clearer with sample code to start a conversion, read the temperature result and stop conversions:

```
TCtrl       Con     %10011110       ;control word & address combined
ReadT       Con     $AA             ;read temperature
StartC      Con     $EE             ;start temperature conversion
```

```
StopC          Con    $22          ;stop temp conversion

RawTemp        Var    Word         ;holds binary temperature output
;Start Conversion
I2COut MDta,MClk,ErrorWMsg,MCtrlB,StartC,[StartC]
;wait for conversion to complete
Pause 1000
;Read the temperature
I2Cin MDta,MClk,ErrorRMsg,MCtrlB,ReadT, |
   [RawTemp.Byte1,RawTemp.Byte0]
;stop conversion to save power
I2COut MDta,MClk,ErrorWMsg,MCtrlB,StopC,[StopC]
```

For the purpose of this code fragment, assume we've defined the data and clock pins as **MData** and **MDlock**, respectively, and that two error routines have been defined, **ErrorWMsg** for write errors and **ErrorRMsg** for read errors.

In Chapter 18, when writing to a serial EEPROM, our **I2Cout** command was

```
I2COut Dta,Clk,ErrorMsg,RCtrl,i+iStart,[Ti]
```

Where **i+iStart** is the memory address and **Ti** is a byte value being written to the address.

To send a command byte to the DS1624, we simply substitute the command byte for the address:

```
I2COut MDta,MClk,ErrorWMsg,MCtrlB,StartC,[StartC]
```

And, since **I2Cout** requires a value be sent, we resend the command byte as a dummy value.

Rather than repeatedly read the configuration register to determine when the temperature conversion is complete, we wait one second, the maximum conversion time, for it to be completed.

```
Pause 1000
;Read the temperature
I2Cin MDta,MClk,ErrorRMsg,MCtrlB,ReadT, |
   [RawTemp.Byte1,RawTemp.Byte0]
```

The **I2Cin** procedure works the same was as the **I2Cout**; we substitute the command byte for the address byte and then read the two-byte temperature data.

```
I2COut MDta,MClk,ErrorWMsg,MCtrlB,StopC,[StopC]
```

To stop conversions, we send the $22 command byte as a substitute for the address byte in an **I2Cout** procedure, again using the stop conversion constant as a dummy data value to satisfy **I2Cout**'s requirement of at least one data element.

The last difference we'll concern ourselves with is the returned temperature data. The DS1624's distant cousin, the DS18B20, returns a 12-bit temperature value, right justified in two successive bytes that, in Chapter 12, we read as a word length result. The DS1624 returns a *13-bit* temperature value, *left* justified (shown for +25.0625°C):

Byte 1								Byte 0							
128	64	32	16	8	4	2	1	.5	.25	.125	.0625	.03125	X	X	X
B15	B14	B13	B12	B11	B10	B9	B8	B7	B6	B5	B4	B3	B2	B1	B0
0	0	0	1	1	0	0	1	0	0	0	1	0	0	0	0

The value is calculated as:

```
16 + 8 + 1 + 0.0625 = 25.0625
```

The reason the data is left justified is that it permits us—by reading byte 1 only—to get the temperature directly in degrees Celsius, without any manipulation. For many purposes, a one-degree step is adequate. However, we wish to use the full 13-bit resolution of the DS1624, so we must deal with the left justified data. One of many ways to simplify the result is to shift the data in Byte 0 three places to the right and treat the resulting Byte 0 value as representing 0.03125°C steps.

```
RawTemp.Byte0 = RawTemp.Byte0 >> 3
```

After this shift, our sample value looks like:

Byte 1 = 25								Byte 0 = 2							
B15	**B14**	**B13**	**B12**	**B11**	**B10**	**B9**	**B8**	**B7**	**B6**	**B5**	**B4**	**B3**	**B2**	**B1**	**B0**
0	0	0	1	1	0	0	1	0	0	0	0	0	0	1	0

The value of byte 0 is 2, so the fractional part of the temperature is 2 * 0.03125°C, or 0.0625°C. Byte 1, holding the integer value, remains 25, so the resulting temperature is 25.0625°C, exactly as it should be. We can compute the temperature as **RawTemp.Byte1 + 0.03125 * RawTemp.Byte0**. We'll see how this works with MBasic's floating-point library when we examine the code later in this chapter.

As with the DS18B20, the DS1624 uses "a 16-bit sign-extended two's complement number" when reporting temperatures below 0°C. We'll use the routines developed in Chapter 12 to convert negative readings and to convert Celsius to Fahrenheit.

Connecting the Sensors and Memory

Our weather station is the most complex circuit we've yet built, so we'll approach our circuit analysis in modules.

Voltage Reference

Our analysis of the MPX6115 pressure sensor shows that we need to measure up to 4.4 V to read the designed maximum air pressure. We'll use the same precision 4.096 V reference, a Linear Technologies LT1634-CCZ-4.096 device as our master voltage reference. We have two options: to scale the MPX6115's output down to a maximum value below 4.096 V, or to scale

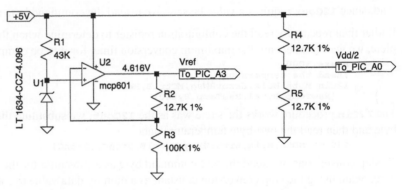

Figure 28-1: Voltage reference and VDD scaling section.

the reference voltage, V_{REF+}, up to about 4.6 V. We'll choose to scale the reference voltage up, as it's the simpler option. In Chapter 11, we used an MCP601 op-amp unity gain buffer to provide a low impedance 4.096 V reference to the PIC; here we will add two resistors to change the MCP601 from unity gain to a gain of approximately 1.127, giving us a reference voltage of 4.616V. Figure 28-1 shows the resulting circuit.

We analyze the voltage scaling circuit by employing a basic op-amp rule; in a closed loop negative feedback amplifier, such as that in Figure 28-1, the voltages at the inverting (−) and noninverting (+) inputs to the op amp are identical. Since the noninverting voltage is +4.096 V, the inverting input must also be +4.096 V. But, the inverting input is connected to a voltage divider comprising R2 and R3. Hence, the voltage at the output of the op-amp must be greater than 4.096 V. The gain of the op-amp circuit is simply the voltage division ratio of the resistive voltage divider network:

$$A_V = \frac{R2 + R3}{R3}$$

Where A_V is the resulting DC gain.

To be able to measure 4.401 V from the MPX6115, we need something greater at the V_{REF+} input to the PIC's A/D converter. The reference voltage can't get too close to the +5 V supply, or else we run into problems with the MCP601. (The MCP601 is a rail-to-rail output device, but even so, it's a good idea to maintain some margin between output and supply voltage.) If we split the difference between the +5 V supply and the maximum voltage to be measured, we arrive at about 4.7 V, so we need (R2 + R3)/R3 to be about 1.147.

Any error in R2 and R3 directly translates into a voltage reference error. The LTC1634 is ±0.05%, so ideally R2 and R3 have at least this good a tolerance. Precision resistors of 0.05% or better tolerance are available, but they are not inexpensive and are not off-the-shelf items at most supply houses. Even 0.1% tolerance resistors are not commonly stocked. (Mouser Electronics is one supplier; 0.1% resistors are priced about $1.00 each in single lot quantities. 0.01% resistors run about $6.00 each in single lot quantities.) So, what to do? For experimentation, we'll use 1% tolerance resistors and accept the results. (I cheated a bit here; I used 1% resistors, but measured the voltage reference output with a 5.5 digit digital voltmeter. The program uses the measured value, not the calculated reference value.) And, I didn't have the 1% resistor combinations necessary for a voltage gain of 1.147, so I used what I had, 100K and 12.7K 1% resistors, yielding a theoretical gain of 1.127, for an output of 4.616 V. I measured the actual output voltage at 4.6235 V, about 0.16% in error from the theoretical value.

It's not a good idea to substitute 5% carbon film resistors, with the thought that you can either hand-select a pair that match the desired value, or simply measure the output voltage and use it in the calculations. Metal film precision resistors have a much better temperature coefficient than carbon film resistors and are also more stable over time.

The remaining part of Figure 28-1 is a 1:2 voltage divider for measuring the +5 V supply. We must measure the supply voltage because the MPX6115 output-voltage-versus-pressure equation has in it a term for the supply voltage. A look at the data sheet for the 7805 regulator used in the 2840 development board will convince you that the 7805 was designed for cost and reliability, not precision. Again, we should use at least 1% resistors for the voltage divider. I had several 12.7K ohm 1% resistors, so I used a pair for R4 and R5. These values are not particularly critical and almost any two equal value resistors up to 15K ohm or so could be used. In fact, R4 and R5 need not be equal in value if you alter the scaling constant in the program.

Sensor Suite

The three sensors connect as shown in Figure 28-2.

Don't forget to configure the address pins on the DS1624. I've arbitrarily set them all high (because it's easy to run jumpers from pin 8 to 7 to 6 to 5) so the DS1624's address is %111 or $7. If you change the address jumpers, you'll have to make a corresponding change in the program constant definition section.

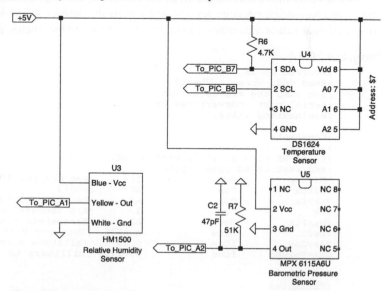

Figure 28-2: Sensor suite.

Real-Time Clock and Serial EEPROM

Figure 28-3 show the connection for the DS1302 real-time clock and the AT24C512 512kbit serial EEPROM.

For a refresher on these devices, review Chapters 12 (real-time clock) and 18 (AT24C512). Both device connections directly duplicate the designs in those chapters, save for changes to pin connections to the PIC.

I've set the AT24C512's address pins to %00, or $0. Although it's not directly obvious from the separate drawings, the clock and data lines of the AT24C512 are paralleled with those of the DS1624 temperature sensor. The two devices are selectively addressed through differences in their control byte device family bits, as well as their addresses.

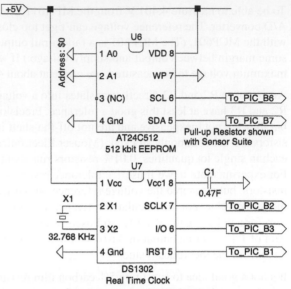

Figure 28-3: Real-time clock and serial EEPROM.

Combined Schematic

The combined schematic, Figure 28-4 integrates the three individual elements we separately reviewed. (Don't forget to connect the RS-232 port.) I've used a 16F877A in this design, but a 16F877 would work just as well, as would a 16F876 or 16F876A.

Initial Tests

There's a lot to wire up on a small plug board, and it's easy to make a mistake, so we'll first write several simple programs to verify that the individual elements of our circuit are properly connected and functioning.

First we'll verify the A/D components of the circuit are working. Program 28-1A reads $V_{DD}/2$ and the humidity and barometric pressure sensors, and reports the outputs in raw A/D units, in computed volts and in computed final values, i.e., relative humidity in percent and barometric pressure in inches Hg and mbar.

Program 28-1A

```
;Program 28-1A
;Read Vdd, Vpres & Vhumid
;and dump the values out on
;serial port. Convert raw to
;engineering units.

;Constants
;------------------
;Keeps RE pins digital, Vref on AN3
;and rest of AN pins inputs
ADSetup         Con         $83       ;see Chapter 11
VddPin          Con         A0        ;Vdd/2
VrhPin          Con         A1        ;Relative humidity pin
Clk             Con         2         ;A/D conversion clock
VbarPin         Con         A2        ;Barometric pin
Vref            fCon        4.6235    ;Measured Vref
ElevCor         fCon        0.35      ;altitude correction for 320 ft
mBarToInHg      fCon        0.02953   ;millibars to inches Hg


;Variables
;------------------
```

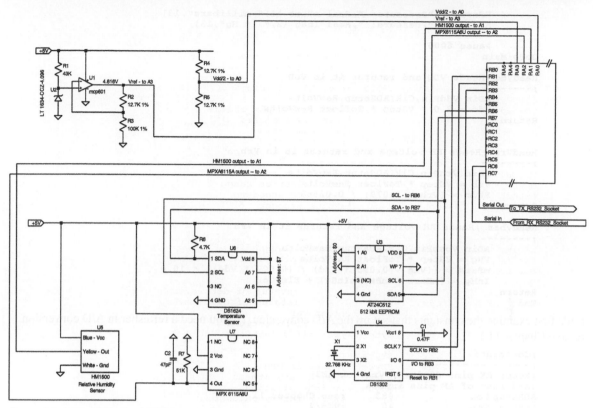

Figure 28-4: Combined schematic.

```
RawVolts        Var     Word    ;Read voltage
Vstep           Var     Float   ;volts per A/D step
Vdd             Var     Float   ;actual Vdd
Vrh             Var     Float   ;Output of RH sensor
VHg             Var     Float   ;Output of Barometric sensor
InHg            Var     Float   ;Inches of Mercury pressure
mBars           Var     Float   ;Output in Bars (kPa/10)
Rhum            Var     Float   ;Relative humidity
i               Var     Byte    ;delay counter

;Initialization
;-------------------
        EnableHSerial
        SetHSerial H38400        ;serial output
        HSerOut ["P 28-1A",13]
        HSerOut ["V/Step: ",Real VStep,13]

        Vstep = Vref / 1024.0 ;can't compute as constant

Main
        GoSub ReadVdd
        HSerOut ["Vdd: ",Real Vdd \3," Raw: ",Dec RawVolts,13]

        GoSub ReadVrh
        HSerOut ["Vrh: ",Real Vrh \3," Raw: ",Dec RawVolts,13]
        HSerOut ["RH: ",Real Rhum \1,"%",13]

        GoSub ReadVbar
        HSerOut ["Vbar: ",Real VHg \3," Raw: ",Dec RawVolts,13]
```

```
            HSerOut ["Pressure: ",Real mBars \1," millibars",13]
            HSerOut ["Pressure: ",Real InHg \2," in Hg",13]

        Pause 6000
GoTo Main

ReadVdd ;Reads VDD and returns it in VDD
;--------
        ADin VddPin,Clk,ADSetup,RawVolts
        Vdd = 2.0 * VStep * ToFloat RawVolts
Return


ReadVrh ;Reads RH voltage and returns it in Vrh
;--------
        ADin VrhPin,Clk,ADSetup,RawVolts
        Vrh = VStep * ToFloat RawVolts
        Rhum = (Vrh - 1.079) / 0.02568
Return

ReadVbar ;Reads RH voltage and returns it in Vrh
;--------
        ADin VbarPin,Clk,ADSetup,RawVolts
        VHg = VStep * ToFloat RawVolts
        mBars = ((VHg + 0.095 * Vdd) / (0.009 * Vdd)) * 10.0
        InHg = (mBars * mBarToInHg) + ElevCor
Return
End
```

Let's first examine the constants that set up the A/D conversion. (If you need a refresher in A/D conversion, review Chapter 11.)

```
;Constants
;------------------
;Keeps RE pins digital, Vref on AN3
;and rest of AN pins inputs
ADSetup Con            $83     ;see Chapter 11
VddPin  Con            A0      ;Vdd/2
VrhPin  Con            A1      ;Relative humidity pin
Clk          Con       2          ;A/D conversion clock
VbarPin Con            A2      ;Barometric pin
```

We have three analog voltages to read and we wish to use an external V_{REF+} with V_{REF-} at ground. And, we wish our 10-bit output to be right justified, since we intend to use all 10 bits. These requirements, as reflected in the table below, extracted from Chapter 11, are met with an A/D setup value of $83. This constant when used with MBasic's **ADin** procedure sets pins A0, A1 and A2 to analog input, pin A3 to V_{REF+} and sets V_{REF-} to ground and right-justifies the returned 10-bit value. We also alias the three analog input pins to more useful names.

Hex for	Hex for	Ports (AN = Analog -- RA/RE = Register A/E (Digital)								Volt Refs		
Left Just.	Right Just.	AN7 RE2	AN6 RE1	AN5 RE0	AN4 RA5	AN3 RA3	AN2 RA2	AN1 RA1	AN0 RA0	Vref+	Vref–	Chan/ Refs
03	83	D	D	D	A	Vref+	A	A	A	RA3	Vss	4/1
		Shaded cells apply only to 16F877 and 16F877A.										

We set the conversion clock value to 2, representing Tosc/32, since our 16F877A uses a 20 MHz resonator and Tosc/32 is the only permitted conversion clock value for that clock speed. If this material is unfamiliar, please review Chapter 11.

```
Vref        fCon    4.6235   ;Measured Vref
ElevCor fCon    0.35            ;altitude correction for 320 ft
mBarToInHg  fCon    0.02953  ;millibars to inches Hg
```

Next, we define three floating point constants:

Vref is the measured reference voltage at pin A3, 4.616 V nominal. If you can't measure the reference voltage with a digital voltmeter with 0.25% accuracy or better, just set **Vref** to the nominal value 4.616, or whatever value it should have, based on your selection of R2 and R3 in Figure 28-1.

ElevCor is the barometer adjustment to correct to sea level readings made at an elevation of 320 feet, based upon the elevation of my workshop. **ElevCor** is in inches Hg and is derived as discussed earlier in this chapter.

MBartoInHg is the constant that translates pressure in millibars (hPa) to inches Hg. It is from the table presented earlier in this chapter.

(We'll deal with the variable assignments as we use them in each routine.)

```
;Initialization
;------------------
        SetHSerial H38400        ;serial output
        HSerOut ["P 28-1A",13]
        HSerOut ["V/Step: ",Real VStep,13]

        Vstep = Vref / 1024.0 ;can't compute as constant
```

In addition to setting up the serial output port in the initialization section, we also calculate one term that is unchanged throughout our calculations; how many volts each of the 1024 A/D output steps represents. We define this parameter as a variable **Vstep** and calculate its value during initialization. **Vstep** is 4.5151 mV. (Of course, we could have instead used a pocket calculator to determine that **Vstep** is 0.0045151 and then defined it as a constant **Vstep fCon 0.0045151**. It's better programming practice, though, to use constants to define fundamental values and let the compiler calculate derivative values. This makes program maintenance easier.)

```
Main
        GoSub ReadVdd
        HSerOut ["Vdd: ",Real Vdd \3,"  Raw: ",Dec RawVolts,13]

        GoSub ReadVrh
        HSerOut ["Vrh: ",Real Vrh \3,"  Raw: ",Dec RawVolts,13]
        HSerOut ["RH: ",Real Rhum \1,"%",13]

        GoSub ReadVbar
        HSerOut ["Vbar: ",Real VHg \3,"  Raw: ",Dec RawVolts,13]
        HSerOut ["Pressure: ",Real mBars \1," millibars",13]
        HSerOut ["Pressure: ",Real InHg \2," in Hg",13]

        Pause 6000
GoTo Main
```

The main program loop consists of a series of calls to subroutines where the heavy lifting is performed. After each subroutine call, we write the values of the corresponding variables to the serial port where we use the data to debug the program and to verify that each sensor is properly functioning.

```
ReadVdd ;Reads VDD and returns it in VDD
;--------
        ADin VddPin,Clk,ADSetup,RawVolts
        Vdd = 2.0 * VStep * ToFloat RawVolts
Return
```

We use MBasic's **ADin** procedure to read the voltage on pin **VddPin** (**A0**). The resulting word value is returned in the variable **RawVolts**. We then convert **RawVolts** (which is in A/D units, 0…1023) to an actual voltage by first converting **RawVolts** to a floating point number with the **ToFloat** operator and then multiplying it by the voltage per A/D step value, **VStep**, we calculated in the initialization routine. Finally, since we sense V_{DD} through a 2:1 voltage divider comprised of R4 and R5, we multiply by 2.0 to yield the true V_{DD}.

```
ReadVrh ;Reads RH voltage and returns it in Vrh
;--------
          ADin VrhPin,Clk,ADSetup,RawVolts
          Vrh = VStep * ToFloat RawVolts
          Rhum = (Vrh - 1.079) / 0.02568
      Return
```

We use a similar approach to read the HM-1500 humidity sensor output voltage. After reading the A/D count in **RawVolts**, we convert it to floating point voltage by the same process employed in **ReadVdd**. We then use the equation developed earlier in this chapter to convert the measured output voltage to a relative humidity percentage.

$$RH = \frac{V_{out} - 1.087}{0.02568}$$

The MBasic statement **Rhum = (Vrh - 1.079) / 0.02568** implements the conversion equation, with the resultant relative humidity percentage stored as a floating point value in **Rhum**.

```
ReadVbar ;Reads RH voltage and returns it in Vrh
;--------
          ADin VbarPin,Clk,ADSetup,RawVolts
          VHg = VStep * ToFloat RawVolts
          mBars = ((VHg + 0.095 * Vdd) / |
          (0.009 * Vdd)) * 10.0
          InHg = (mBars * mBarToInHg) + ElevCor
      Return
```

Likewise, we read the voltage output from the MPX6115 pressure sensor and convert it to a floating point value, stored in **VHg**. We then convert the measured output voltage to an absolute pressure using the equation developed earlier in this chapter.

$$P_{mbar} = \frac{V_{out} + 0.095V_S}{0.090V_S}$$

The MBasic statement **mBars = ((VHg + 0.095 * Vdd) / (0.009 * Vdd)) * 10.0** implements this equation and stores the resulting pressure in mbar in the variable **mbars**. (Note that the MBasic statement actually calculates the pressure in kPa and then multiplies it by 10.0 to yield a result in hPa or mbar. This is the functional equivalent of the equation, of course.)

The remaining steps to achieve a sea level corrected pressure in inches Hg require us to convert from mbar to inches Hg and add the elevation correction. We accomplish this with the MBasic statement **InHg = (mBars * mBarToInHg) + ElevCor**.

I've decided to keep the pressure in mbars uncorrected. You might wish to correct it as well.

Here's a sample output from Program 28-1A:

```
Vdd: 4.966  Raw: 550
Vrh: 2.027  Raw: 449
RH: 36.9%
Vbar: 3.901  Raw: 864
Pressure: 978.2 millibars
Pressure: 29.26 in Hg

Vdd: 4.966  Raw: 550
Vrh: 2.027  Raw: 449
RH: 36.9%
Vbar: 3.887  Raw: 861
Pressure: 975.2 millibars
Pressure: 29.17 in Hg
```

```
Vdd: 4.957  Raw: 549
Vrh: 2.018  Raw: 447
RH: 36.5%
Vbar: 3.910  Raw: 866
Pressure: 981.8 millibars
Pressure: 29.37 in Hg
```

As expected, we see the ±1 count jitter in the raw output when reading **Vdd** and **Vrh**. The barometric sensor output bounces around more than ±1 count, however. This is likely a combination of stray signals being coupled in the breadboard and some real variation.

Program 28-1B

Now that we have the sensor suite installed and functioning, we can verify that the AT24C512 serial EE-PROM memory chip is working.

```
;Program 18-1B to read/write
;selectively to 512Kbit 24C512 Serial EEPROM
;I2C communications ;Demonstrate random read/write

;Constants
;---------
Clk     Con     B6       ;clock is on A0
Dta     Con     B7       ;data on A1
                ;have to add the page select to Ctrl
                ;16-bit address word, so ends in 1
RCtrl   Con     %10100001        ;control word starting point
Display Con     0 ;set to 0 to block writing values to serial port
Adr0    Con     0 ;Address of 24C512 A0=A1=0
                ;has full 64K write address space

;Variables
;----------
CtrlB           Var     Byte    ;control byte
i               Var     Word    ;index
j               Var     Byte    ;byte value to/from 25FC512
ShouldBe        Var     Byte    ;calculated read value

;Initialization
;--------------
;Communicate with outside world via serial
EnableHSerial
SetHSerial H38400

;Demonstrate random access byte at a time
;24C512 is organized as two 32 byte pages.
;Up to 4 chips are addressed in the control byte
HSerOut [13,"Into Write",13]

Pause 50

For i = $0000 to $FFFF
        CtrlB = RCtrl + Adr0
        ;To write something distinctive to RAM
        ;we write the index plus the offset
        j = i.LowByte + i.HighByte
        ;Write to addressed memory
        I2COut Dta,Clk,ErrorMsg,CtrlB,i,[j]
        Pause 5 ;needed for write time

        If Display > 0 Then
                HSerOut ["Write  i: ",Dec i,"  Data: ",Dec j,13]
                Pause 25        ;needed if write output
        EndIf

        If (i//256) = 0 Then
                HSerOut ["Write  i: ",Dec i,"  Data: ",Dec j,13]
                Pause 25        ;needed if write output
```

```
                EndIf
        Next

        j= 0
        Pause 50
        HSerOut ["End of Write",13]
        Pause 50

        ;Now read it back
        For i = $0000 to $FFFE
                ;Set up the control byte in the same way
                CtrlB = RCtrl + Adr0
                ShouldBe = i.LowByte + i.HighByte
                I2CIn Dta,Clk,ErrorMsg,CtrlB,i,[j]

                If Display > 0 Then
                        HSerOut ["Read  i: ",Dec i," Data: ",Dec j,13]
                        Pause 25        ;needed if write output
                EndIf

                If (i//256) = 0 Then
                        HSerOut ["Read  i: ",Dec i," Data: ",Dec j,13]
                        Pause 25        ;needed if write output
                EndIf

                If j <> ShouldBe Then
                        HSerOut ["Read  i: ",Dec i," Data: ",Dec j]
                        HSerOut [" <-- Error!",13]
                        Pause 25
                EndIf
        Next
        HSerOut ["End of Read",13]
        Pause 50

    ;GoTo Main
    End

    ErrorMsg
        HSerOut ["Error!"]
    End
```

Program 28-1B is a direct clone of Program 18-2B, changed only to reflect different data and clock pins, so we will not reanalyze it. If you don't understand the logic or code of Program 28-1B, please review Program 18-2B and the related text in Chapter 18.

The output of Program 28-1B is shown below, with repetitive pieces suppressed.

```
Into Write
Write i: 0  Data: 0
Write i: 256  Data: 1
Write i: 512  Data: 2
Write i: 768  Data: 3
Write i: 1024  Data: 4
...
Write i: 65024  Data: 254
Write i: 65280  Data: 255
End of Write
Read i: 0  Data: 0
Read i: 256  Data: 1
Read i: 512  Data: 2
...
Read i: 64768  Data: 253
Read i: 65024  Data: 254
Read i: 65280  Data: 255
End of Read
```

There should be no error messages generated when running Program 28-1B.

Program 28-1C

```
;Program 28-1C to test reading a DS1624 EEPROM/Thermometer
;I2C communications. In parallel with an AT24C512 512kbit EEPROM

;Constants
;---------
Clk      Con      B6       ;clock is on A0
Dta      Con      B7       ;data on A1
                           ;have to add the page select to Ctrl
                           ;8-bit address word, so ends in 0
TCtrl             Con      %10011110      ;control word & address combined
AccMem Con        $17      ;access memory
AccConf Con       $AC      ;access configuration
ReadT             Con      $AA ;read temperature
StartC Con        $EE ;start temperature conversion
StopC             Con      $22      ;stop temp conversion

;Variables
;----------
CtrlB             Var      Byte     ;control byte
RawTemp Var       Word     ;index
PlusFlag          Var      Byte     ;plus/minus flag
Centigrade        Var      Float    ;Floating point C temp
Fahrenheit        Var      Float    ;Floating point F temp

;Initialization
;--------------
;Communicate with outside world via serial
SetHSerial H38400
RawTemp = 0

Main
        CtrlB = TCtrl
        ;start temperature conversion
        I2COut Dta,Clk,ErrorMsg,CtrlB,StartC,[StartC]
        Pause 1000
        I2CIn Dta,Clk,ErrorMsg,CtrlB,ReadT, |
        [RawTemp.Byte1,RawTemp.Byte0]
        ;data is left justified
        RawTemp.Byte0 = RawTemp.Byte0 >> 3
        ;stop conversion to save power
        I2COut Dta,Clk,ErrorMsg,CtrlB,StopC,[StopC]

        PlusFlag = 1
        If RawTemp.HighBit = 1 Then    ;check for < 0C
           ;Subtract from 0 to read below 0 value
           RawTemp = $0000-RawTemp
           PlusFlag = 0
        EndIf

        ;Returned value is in Centigrade steps of 0.03125 Deg C.
        ;This differs from the DS18B20 we used in Chapter 12!
        Centigrade = (ToFloat RawTemp.Byte0) * 0.03125
        Centigrade = Centigrade + ToFloat RawTemp.Byte1

        If PlusFlag = 0 Then            ;below zero, multiply by -1
                Centigrade = -Centigrade
        EndIf

        ;Standard conversion formula to get to F from C.
        Fahrenheit = (Centigrade * 1.8) + 32.0          ;convert F
        HSerOut ["Temp: ",Real Centigrade \2,"C  ", |
                Real Fahrenheit \2,"F",13]
        Pause 5000
GoTo Main

ErrorMsg
```

```
        HSerOut ["Error!"]
End
```

As usual, we start by defining constants that we'll use to communicate with the DS1624

```
;Constants
;---------
Clk        Con    B6           ;clock is on A0
Dta        Con    B7           ;data on A1
TCtrl      Con    %10011110        ;control word & address combined
AccMem     Con    $17          ;access memory
AccConf    Con    $AC          ;access configuration
ReadT      Con    $AA          ;read temperature
StartC     Con    $EE          ;start temperature conversion
StopC      Con    $22          ;stop temp conversion
```

Although we won't use them in this chapter, we've defined all five possible command bytes. We also construct the **I2Cin/I2Cout** control byte, following the methodology developed earlier in this chapter:

Element	Value
Family Code	%1001
Address (Pins A0/A1/A2 at +5V)	%111
8/16 Bit Selection (8-bit)	%0

The control byte is thus **%10011110**.

Let's look at reading the DS1624.

```
Main
        CtrlB = TCtrl
        I2COut Dta,Clk,ErrorMsg,CtrlB,StartC,[StartC]
```

As we learned earlier, the DS1624 protocol is control byte: command byte: optional data byte. **I2Cout** transmits this sequence if we substitute the command byte for the address byte. **I2Cout** requires a data byte, so we use the command byte as a dummy data element. The **StartC** (**$EE**) command byte starts a temperature conversion.

```
        Pause 1000
```

Rather than poll the DS1624 to determine when the temperature conversion is completed, we'll just wait one second. If time is critical, it's possible to poll the DS1624 and determine when the conversion is completed. (The typical conversion time is 400 ms, but the maximum time is 1000 ms.)

```
        I2CIn Dta,Clk,ErrorMsg,CtrlB,ReadT, |
        [RawTemp.Byte1,RawTemp.Byte0]
```

We use the same approach when reading the returned temperature value; substitute the command byte for the address byte in the **I2Cin** procedure. To read the temperature data, the command byte is **ReadT** (**$AA**). Since the returned temperature is in two bytes, we read each separately. The order is MSB first, so we read **RawTemp.Byte1**, followed by **RawTemp.Byte0**.

```
        RawTemp.Byte0 = RawTemp.Byte0 >> 3
```

As we learned when studying the DS1624, the 13-bit data is returned left justified. To make it easier to handle, we right justify **RawTemp.Byte0** by shifting its contents three bits to the right using MBasic's shift right operator.

```
        I2COut Dta,Clk,ErrorMsg,CtrlB,StopC,[StopC]
```

After reading the data, we stop further conversions with the **StopC** (**$22**) command byte.

```
        PlusFlag = 1
        If RawTemp.HighBit = 1 Then    ;check for < 0C
            ;Subtract from 0 to read below 0 value
            RawTemp = $0000-RawTemp
            PlusFlag = 0
        EndIf
```

After reading the 13-bit raw temperature information and right-justifying, we now check to see if it represents a below zero Celsius value. Where the temperature is below zero, the data is "two's complemented." We check for this by looking at the high order bit of the returned data, which will be a 1 only if the data is below zero. If the temperature is below zero, we remove the complement by subtracting the returned value from $0000 and set the plus flag to 0.

This code fragment is identical with the technique we used in Chapter 12 to read the DS18B20 one-wire temperature sensor, where it is explained in more detail.

```
;Returned value is in Centigrade steps of 0.03125 Deg C.
;This differs from the DS18B20 we used in Chapter 12!
Centigrade = (ToFloat RawTemp.Byte0) * 0.03125
Centigrade = Centigrade + ToFloat RawTemp.Byte1

If PlusFlag = 0 Then          ;below zero, multiply by -1
        Centigrade = -Centigrade
EndIf
```

We next convert the returned value to an actual Celsius temperature. `RawTemp.Byte1` holds the temperature in integer degrees, while `RawTemp.Byte0` holds the fractional part of the temperature, in increments of 0.03125°C. We convert to a floating point temperature in degrees Celsius by multiplying `RawTemp.Byte1` by 1.000 and multiplying `RawTemp.Byte0` by 0.03125 and summing the two products. The resulting temperature value is held in the floating point variable `Centigrade`.

```
;Standard conversion formula to get to F from C.
Fahrenheit = (Centigrade * 1.8) + 32.0        ;convert F
HSerOut ["Temp: ",Real Centigrade \2,"C  ", |
        Real Fahrenheit \2,"F",13]
Pause 5000
GoTo Main
```

Lastly, we convert the Celsius temperature to Fahrenheit and store the result in the floating point variable **Fahrenheit**.

Sample output from Program 28-1C appears below. (To test the response of the DS1624, I briefly put my finger on it—in a few seconds the reading increased by about 4°F.)

```
Temp: 22.59C 72.66F
Temp: 22.65C 72.78F
Temp: 24.28C 75.70F
Temp: 24.90C 76.83F
Temp: 24.75C 76.54F
Temp: 24.09C 75.36F
Temp: 23.68C 74.63F
Temp: 23.46C 74.24F
Temp: 23.31C 73.96F
Temp: 23.21C 73.79F
Temp: 23.15C 73.68F
```

Program 28-1D

The remaining element to test is the DS1302 real time clock. Program 28-1D exercises the DS1302 and outputs the date and time to the serial port.

```
;Program 28-1D. Reads a Dallas/Maxim DS1302
;Real time clock.
;Also writes the current time to the clock

;Varibles
;---------------
RTCCmd          Var     Byte            ;Command byte
Temp            Var     Byte            ;Temporary variable
i               Var     Byte            ;counter
```

```
TimeData        Var     Byte(7)         ;ss:mm:hh:dd:mm:yy
OldSeconds      Var     Byte            ;last read seconds

;For convienence we alias the time array
Seconds         Var     TimeData(0)     ;Seconds
Minutes         Var     TimeData(1)     ;Minutes
Hours           Var     TimeData(2)     ;Hours
Date            Var     TimeData(3)     ;Day of Month
Month           Var     TimeData(4)     ;Month
Day             Var     TimeData(5)     ;Day of Week
Year            Var     TimeData(6)     ;Year

;Constants
;--------------                         () are DIP8 numbers
Clk             Con     B2              ;Clock pin (7)
Dta             Con     B3              ;Data pin  (6)
Reset           Con     B1              ;Reset pin (5)

CtrlReg Con     %00111

;Initial load of Sec  Min  Hrs  Date MO   Day   YR
PreSet ByteTable $00, $05, $10, $08, $01, $05, $04
;These are in BCD, hence November (Month 11)
;is entered as $11, not 11.

;Command to select charger settings
Charger Con     %01000
;1 diode, 2K charging resistance. See
;Figure 5 of DS1320 data sheet
Battery Con     %10100101

;For Day and Month names, use the following. Since
;Sun=1 and Jan=1 we use nul at the beginning instead of
;subtracting 1.
DayName ByteTable "NulSunMonTueWedThuFriSatSun"
MoName  ByteTable "NulJanFebMarAprMayJunJulAugSepOctNovDec"

Initialization
;--------------
        EnableHSerial
        SetHSerial H38400
        HSerOut ["Pgm 28-01d",13]

        ;To set the clock we first disable write protection.
        ;Then send the clock set information, and then restore
        ;write protection.

        ;Disable DS1302 WriteProtection by 0 in Bit7
        Temp = $00
        RTCCmd = CtrlReg
        GoSub PreSetData

        Goto NoSet      ;skip if data already set

        ;Set the date and time from the Preset byte table
        For i = 0 to 6
                Temp = Preset(i)
                RTCCmd = i
                GoSub PreSetData
        Next
NoSet

        ;SInce we have a Super Cap backup, we want charger ON
        Temp = Battery
        RTCCmd = Charger
        GoSub PreSetData

        ;DS1320 Write protection back on
        Temp = $80
        RTCCmd = CtrlReg
        GoSub PreSetData
```

```
Loop
;----------
        ;Read the clock output. Check the seconds.
        ;If changed, write time
        GoSub ReadData
        ;Write in form:
        ;13:10:45 11/24/03  Mon 24-Nov-2003 to the serial port
        ;New seconds, so write output
        If OldSeconds <> Seconds Then
        HSerOut [Dec2 (BCD2BIN Hours)\2,":",  |
        Dec2 (BCD2BIN Minutes)\2,":",Dec2 (BCD2BIN Seconds)\2]
        HSerOut ["  ",Dec2 (BCD2BIN Month)\2,"/",  |
        Dec2 (BCD2BIN Date)\2,"/",Dec2 (BCD2BIN Year)\2]
        HSerOut ["    ",Str DayName(Day*3)\3,"  "]
        HSerOut [Dec2 (BCD2BIN Date)\2,"-",  |
        Str MoName(3*BCD2BIN Month)\3,"-20",  |
        Dec2 (BCD2BIN Year)\2,13]
    EndIf
GoTo Loop

PreSetData               ;Routine to upload data to the DS1302
;-----------
        High Reset
        ;The %0\1,RTCCmd\5,%10\2 results in the equivelent of
        ;sending 10 RTCCMD 0. Each element is sent in element order
        ;but bit reversed, as required by the DS1320.
        ShiftOut Dta,Clk,LSBFIRST,[%0\1,RTCCmd\5,%10\2,Temp\8]
        Low Reset
Return

ReadData
;----------
        ;Reset pin must be high for  read
        High Reset
        ;%10111111 is BURST SEND REGISTER command,
        ;so time is 8 bytes
        ;the last byte is the control register,
        ;which we don't read or use
        ShiftOut Dta,Clk,LSBFirst,[%10111111\8]
        OldSeconds = Seconds
        ;Now actually read the burst send data.
        ShiftIn Dta,Clk,LSBPRE, [Seconds\8,Minutes\8,  |
        Hours\8,Date\8,Month\8,Day\8,Year\8]
        Low Reset
Return
```

Except for changes to pin assignments, Program 28-1D duplicates Program 12-4. Since Chapter 12 presents a detailed analysis of the DS1302 real time clock and Program 12-4, we won't further discuss it.

Program 28-1D writes the time and date, every second, to the serial port. Sample output appears below.

```
08:43:57 01/19/04  Mon  19-Jan-2004
08:43:58 01/19/04  Mon  19-Jan-2004
08:43:59 01/19/04  Mon  19-Jan-2004
08:44:00 01/19/04  Mon  19-Jan-2004
08:44:01 01/19/04  Mon  19-Jan-2004
08:44:02 01/19/04  Mon  19-Jan-2004
08:44:03 01/19/04  Mon  19-Jan-2004
```

Program 28-2

Programs 28-1A through 28-1D verify the temperature, humidity and barometric pressure sensors are correctly functioning, that we may write to and read from the external EEPROM serial memory and that the real time clock can be set and read. We are now ready to assemble these four elements into a weather station and data logger.

Before we delve into the code, let's recap what we expect this program to accomplish:

- Read the real time clock and the three weather sensors. Output the time, date, temperature, relative humidity and barometric pressure to the serial port.
- Periodically save the date, time, temperature, relative humidity and barometric pressure to the serial EEPROM.
- Upon command from the user, read the saved data from EEPROM and send it over the serial port. (The command is the letter "o" or "O" followed by a carriage return.)

Since Program 28-2 is based largely upon the subroutines of Programs 28-1A through 28-1D, let's use pseudo-code to see how these subroutines fit into our program structure.

```
Main
        GoSub ReadTime - ReadTime constructed from Pgm 28-1D

        Is it time to output data? If not goto Main

        GoSub ReadTemp - ReadTemp constructed from Pgm 28-1C

        GoSub ReadVbar - From Program 28-1A
        GoSub ReadVrh - From Program 28-1A

        GoSub WriteTime - Writes time/date to RS-232

        GoSub WriteWxData       - Write weather data to RS-232

        GoSub PackTimeInfo - Compress date/time to use less memory
        GoSub SavePackTime - Save to EEPROM from Program 28-1B

        GoSub PackReadings - Compress weather data
        GoSub SavePackData - Save to EEPROM from Program 28-1B

        Read serial port for input command to write saved data
        (Based on menu program from Chapter 9)
        If found unpack and send data to serial port
        (Read routine based on Program 28-1B)

GoTo Main
```

Program 28-2 is very long and largely duplicates this chapter's earlier program. Rather than follow our normal structure of listing the program in full, followed by a section-by-section analysis, we'll skip the full listing. Program 28-2, like all other programs in this book, is available in an electronic copy in the associated CD-ROM. If you need a listing to follow while reading the analysis, you may print one from the program file.

To avoid duplication, we will not analyze the subroutines that are derived from Programs 28-1A through 28-1D. We'll start with the main program loop and overall program flow.

Program 28-2	Comment
Main	
HSerStat 3, CheckInput	Check the serial input to see if the user has sent a command to the program. If data is in the serial input buffer, execute the code at label **CheckInput**. **HserStat** is an undocumented MBasic procedure discussed in Chapter 9 and summarized later in this chapter.

`ReturnFromCI`	Label `ReturnFromCI`. Execution returns here after the `CheckInput` code executes. This is a way of faking a GoSub/Return where MBasic doesn't allow a GoSub/Return construction.

`GoSub ReadTime`	Read the DS1302 real-time clock.

```
If OldMins <> Minutes Then
        MinCount = MinCount + 1
        OldMins = Minutes
EndIf
```

Keep track of how many minutes have passed since the last write to EEPROM in `MinCount`. We do this by checking the DS1302's minute value. If it is different than the minute value in the last DS1302 read, a new minute has started. We then increment `MinCount`.

```
If MinCount = Mins Then
        MinCount = 0
ELSE
        GoTo Main
EndIf
```

`Mins` is a constant that defines the number of minutes that should elapse between EEPROM saves. If the elapsed minutes, held in `MinCount`, does not match `Mins`, then we go back to `Main` and wait for either a user input to the serial port, or for `MinCount` to equal `Mins`. If MinCount = Mins then we reset MinCount to 0 and go on to execute the remainder of the code.

The current program structure writes to the serial port only when data is saved, i.e., the weather data is sent to the serial port only as frequently as the data is saved to EEPROM.

`GoSub ReadTemp`	Read the DS1624 temperature sensor.

`GoSub ReadVbar`	Read the MPX6115 barometric pressure sensor.

`GoSub ReadVrh`	Read the HM1500 relative humidity sensor.

`GoSub WriteTime`	Write the date and time to the serial port.

`;GoSub WriteEngData`	`WriteEngData` is a debugging and experimentation subroutine that writes raw A/D readings and voltage values to the serial port. I've commented it out, but if you have problems with one or more sensors, remove the comment symbol and rerun the program.

`GoSub WriteWxData`	Write the weather information to the serial report.

`;GoSub WxToOscope`	Another debugging and analysis aid. MBasic's integrated development environment includes a graphing analysis tool, under Tools | Oscilloscope. If the comment symbol is removed so the subroutine **WxToOscope** is executed, the relative humidity, temperature (F) and barometric pressure (in Hg) are written to the serial port in a format compatible with the Oscilloscope utility.
`GoSub PackTimeInfo`	If we save the date/time information in the form it is emitted by the DS1302, it would require 6 bytes. We reformat the data and bit-pack it into 4 bytes in the subroutine **PackTimeInfo**.
`GoSub SavePackTime`	Subroutine **SavePackTime** saves the bit-packed date/time to EEPROM.
`GoSub PackReadings`	Subroutine **PackReadings** reformats and bit-packs the temperature, barometric pressure and relative humidity readings into 4 bytes.
`GoSub SavePackData`	Subroutine **SavePackData** saves the bit-packed weather data to EEPROM.
`RecordNo = RecordNo + 1`	**RecordNo** is the index that keeps track of how many packed date/time and weather data records have been written to EEPROM.

`If RecordNo = $1FFE Then`
`        Recordno = $0`
`EndIf`
	The EEPROM's capacity is 512kbits, or 65,536 bytes. This corresponds to 8192 date/time/ weather records of 8 bytes length (0...8191). We found problems writing to the last byte of EEPROM in Chapter 18, so we limit ourselves to 8191 records (0...8190. Since we increase after writing, we just test for 8191 and reset when it is reached. For convenience, I've written this in hex as $1FFE.

`GoTo Main`	Go back to **Main** and do it all over again.

Let's look at the subroutines we have not seen elsewhere.

We'll start with the routines to read user information received over the serial port. Program 28-2 uses a stripped down version of the full menu/input program we developed in Chapter 9.

First, consider the first statement in the main program loop: **HSerStat 3, CheckInput**

HSerStat is an undocumented MBasic command that lets you read and modify the status of both the receive and transmit hardware serial buffers. **HserStat** works with **HSerIn** and **HSerOut** and assumes you have initialized the hardware serial routines with the **EnableHSerial** compiler directive and **SetHSerial** command. Its command syntax is:

```
HserStat exp,{label}
```

Exp is a value from 0...6 that determines the function to be executed.

Command	Function
0	Clear the input buffer
1	Clear the output buffer
2	Clear both the input and output buffers
3	If the input buffer has data, execute a GoTo label
4	If the input buffer is empty, execute a GoTo label
5	If the output buffer has data, execute a GoTo label
6	If the output buffer is empty, execute a GoTo label

The statement **HSerStat 3, CheckInput** is executed every millisecond or so, as we read the current time from the DS1302 and loop back to **Main**, unless it is time to read the sensors and save data to the EEPROM. Since we invoke **HSerStat** with an argument of 3, if there is data in the receive buffer, execution will jump to the label **CheckInput**.

CheckInput is not a true subroutine, as it's not possible to use a subroutine call with **HSerStat**, but the **GoTo ReturnFromCI** statement lets it acts similar to a true subroutine, as the label **ReturnFromCI** immediately follows the **HSerStat** statement.

```
CheckInput
;--------------
        HSerIn [InByte]
        ;Echo back
        HSerOut [Str InByte\1]

        InStr(jj) = InByte

        jj = jj+1
        If jj = (MaxStrSize-1) Then
                jj=0
        EndIf

        If InByte = 13 Then
                GoSub CRReceived
        EndIf

        HSerStat 3,CheckInput

    GoTo ReturnFromCI
```

Once we know a character is waiting in the serial input buffer, we then read it into the byte array **InStr**. We check for two possible conditions (a) that the number of bytes received exceeds the size of **InStr** or (b) that we have received an end-of-line indicator (carriage return, with a decimal value of 13). If the number of bytes received reaches **InStr**'s size, we reset the index to 0. If a carriage return is received, we then execute the subroutine **CRReceived** to see if the received characters are a valid command. We use the **HSerStat** procedure to loop back through **CheckInput** until all characters in the buffer have been read. At that point, **HSerStat 3, CheckInput** no longer executes the **GoTo CheckInput**.

Subroutine **CRReceived** is far more complex than necessary if our goal is simply to determine if the character "o" or "O" has been received. **CRReceived** is, in fact, part of the menu/data serial entry program we developed in Chapter 9. Rather than modify it, we'll use the complete subroutine, since we know it works.

```
CRReceived
;---------
```

```
              For jj = 0 to (MaxStrSize-1)
                      If (InStr(jj) < " ") OR (InStr(jj) > "z") Then
                              InStr(jj) = $0 ; non-printing chars
                      EndIf
                      ;Convert lower case to upper case
                      If (InStr(jj) >= "a") and (InStr(jj) <= "z") Then
                              InStr(jj) = InStr(jj) - $20
                      EndIf
              Next ;jj
```

Here we convert any lower case letters to upper case, simplifying our command syntax.

```
              ;how long is the valid part of the string?
              k = 0
              While InStr(k) >= " "
                      k = k+1
              WEND
```

The byte variable **k** holds the length of the received string, excluding nonprinting characters that are below the space character in the ASCII collating sequence.

```
              If k > 0 Then
                      k = InStr(0) - "?"
                      If k > 27 Then
                              k = 1
                      EndIf
                      Branch IndexArray(k), [BR_Unk,BR_Out]
              EndIf
```

As we learned in Chapter 9, our approach to a general-purpose dispatcher is based on a menu system that looks at the first character of the user's response and indexes that character into a table to determine which function it should execute. Permissible first character values are "?," "@," and the letters "A" through "Z." We convert the first character into a numerical value, ranging from 0 ("?") to 27 ("Z") by subtracting "?" from the received character. If the first character is outside that range, we force it to be 1, corresponding to "@" which we then may treat as "unknown command" and issue an appropriate response to the user.

IndexArray() is a 28 element byte table that relates the first letter of possible commands to an index value. We defined **IndexArray** in the constants section of Program 28-2, but we'll repeat it here. (I've slightly modified **IndexArray**'s format to better fit the printed page.)

```
      ;              ? @ A B C D E F G H I J K L
      IndexArray  ByteTable 0,0,0,0,0,0,0,0,0,0,0,0,0,0,  |
        0,0,1,0,0,0,0,0,0,0,0,0,0,0
      ;M N O P Q R S T U V W X Y Z
```

Suppose the first letter received is "O." We convert this letter to its ordinal value via **"O" - "?"**. The result is 16, which is stored in **k**. We've given **IndexArray(16)** the value 1. Hence the statement **Branch IndexArray(k), [BR_Unk,BR_Out]** evaluates to **Branch 1, [BR_Unk,BR_Out]** and program execution jumps to the label **BR_Out**. This may seem overly complex, and it is if all we need do is detect the letter "o" or "O." But, as a part of a complex user interaction system, as we developed in Chapter 9, it is an efficient way of adding, deleting or rearranging the command syntax without being forced to modify code. Instead all we need do is change a table entry or two and add or delete a short corresponding routing function.

```
              BR_Unk
                      GoSub Sub_Unk
                      GoTo EndCmd
```

If the first letter received does not correspond to a legitimate character, its index value in **IndexArray** is 0, which causes a branch to **BR_Unk**, which then informs the user that he has entered an unknown command. Note also that if only a carriage return has been received (**k=0**), **BR_Out** is still executed.

```
              BR_Out
                      GoSub Sub_Out
```

```
                    GoTo EndCmd
```

If the received letter is "o" or "O," execution will jump to **BR_Out** and subroutine **Sub_Out** is called.

```
        EndCmd
                HSerStat 0
                j = 0
        Return
```

Before returning, we clear the received buffer with a **HSerStat 0** command.

The subroutine **Sub_Out** reads all stored values from the EEPROM unpacks the data and writes the time/date and weather information to the serial output port. All of these functions are performed by calls to functional subroutines, and we'll defer their analysis until after we go through the data packing process.

```
        Sub_Out ;Dump EEPROM to serial
        ;------------
                For AdrConst = 0 to (RecordNo-1)
                        HSerOut [Dec AdrConst,9]
                        ReadAdr = AdrConst * 2 * WriteInc
                        GoSub ReadPackTime
                        GoSub UnpackTimeInfo
                        GoSub WriteUnpackedTime

                        ReadAdr = ReadAdr + WriteInc
                        GoSub ReadPackData
                        GoSub UnPackReadings
                        GoSub WriteUnpackedWxData
                Next
        Return
```

Data is stored in records, with each record consisting of two four-byte packed variables; the first holding the time/date and the second holding weather data. **AdrConst** is an index for the number of saved records and has a maximum value of 8190. **ReadAdr** is the byte address for the EEPROM read, i.e., **ReadAdr** ranges from 0...65527 for the AT24C512.

WriteInc (standing for Write Increment) is a constant equal to four, and is the number of bytes in each half of a record. **ReadAdr** thus increments in steps of four, with two increments for each **AdrConst** step, for example, **AdrConst = 0, ReadAdr = 0** and **4; AdrConst = 1, ReadAdr = 4** and 8, and so forth. When **AdrConst = 0,8,12**... we write the date/time information; when it is **4,12,16**... we write the weather information. This arrangement is reflected in the following data map:

24C512 Address	Contents	Record No.	AdrConst	ReadAdr
65535 65528	Unused	Unused	Unused	Unused
65527 65524	Packed Wx Data	Record 8190	8190	65524
65523 65520	Packed Time			65520
...	...	...	...	...
15 12	Packed Wx Data	Record 1	1	12
11 8	Packed Time			8
7 4	Packed Wx Data	Record 0	0	4
3 0	Packed Time			0

Let's now turn to how we pack and unpack the time/date and weather information.

The DS1302 data structure is a seven-byte record, with each byte containing a two-digit BCD value. We don't need the day-of-the-week information, but we are still left with six BCD bytes, 1 byte per field; hour: minute: second and day / month / year. Looking at the minimum number of bits required for each field, we see that a full byte is wasteful:

Field	Minimum/Maximum Value	Minimum Bits Required
Seconds	0...59	6
Minutes	0...59	6
Hours	0...23	5
Day of Month	1...31	5
Months	1...12	4
Years	01...00 [2001...2100]	6*
Total		32

The DS1302 correctly calculates leap year through 2100, so we might like the year field to run through 2100. Unfortunately, this requires 7 bits, and pushes the total up to 33 bits. To keep to a 4-byte record length, we'll shave 1 bit off the year field and thus limit our potential year range to 00...63. The resulting packed data map is shown below.

3 1	3 0	2 9	2 8	2 7	2 6	2 5	2 4	2 3	2 2	2 1	2 0	1 9	1 8	1 7	1 6	1 5	1 4	1 3	1 2	1 1	1 0	0 9	0 8	0 7	0 6	0 5	0 4	0 3	0 2	0 1	0 0
6 bits						4 bits				5 bits					5 bits					6 bits						6 bits					
Year						Month				Day of Month					Hours					Minutes						Seconds					
Byte 3										Byte 2								Byte 1							Byte 0						

Packing time/date information into the four-byte record is a two-step process; first we convert the BCD value of each of six field values to straight binary values. Then, we insert the relevant bits of the binary values into the time/date record. Since a type long variable in MBasic is four bytes, it's convenient to make the record a type long variable, **PackedTime**. Subroutine **PackTimeInfo** bit packs the six time/date fields into **PackedTime**.

```
PackLen ByteTable $5,$5,$4,$4,$3,$0,$5

PackTimeInfo    ;Put date/time info in bit-packed long
;-----------
        PackedTime = 0
        k = 0
        For i = 0 to 6
                If i <> 5 Then
                        TempByte = BCD2BIN TimeData(i)
                        For j = 0 to PackLen(i)
                                k = k+1
                                PackedTime.Bit31 = TempByte.Bit0
                                If k <> 31 Then
                                        PackedTime = PackedTime >> 1
                                EndIf
                                TempByte = TempByte >> 1
                        Next ;j
                EndIf
        Next ;i
    Return
```

We start by clearing **PackedTime** by setting it to 0. Remember that the time and date information is stored in the seven-element byte array **TimeData**, in order seconds to years and that **TimeData(5)** is the day of the week, a field we do not wish to save.

For each **TimeData** element, except for **TimeData(5)**, we first convert from its native BCD format to a binary and hold the binary value in **TempByte**. We then copy **TempByte.Bit0** to **PackedTime.Bit31**. We next shift both `TempByte` and `PackedTime` 1 bit to the right and repeat the copy until we have copied all the required bits of `TempByte` into `PackedTime`. After one element of `TimeData` is shifted into `Packed-Time`, we get the next element, convert it from BCD to binary and repeat the copy/shift process.

To see how this works in more detail, let's look at an example. Suppose the time is 33 seconds. Since `TimeData(0)` holds the seconds value in BCD, `TimeData(0).Nib1 = 3` and `TimeData(0).Nib0 = 3..`Let's see how this is converted to decimal and shifted into `PackedTime`.

As we learned in Chapter 7, BCD data is organized as one digit 0...9 in each nibble of a byte. Thus, `Time-Data(0)`, with a value of 33 seconds holds the following binary pattern:

TimeData(0)							
7	6	5	4	3	2	1	0
Nibble 1				Nibble 0			
Value = $3				Value = $3			
0	1	1	0	0	1	1	0

After `TempByte = BCD2BIN TimeData(0)`, `TempByte` contains:

TempByte							
7	6	5	4	3	2	1	0
0	0	1	0	0	0	0	1

We now start shifting `TempByte` into `PackedTime`.

After the first bit copy, with the copied bit shown shaded:

TempByte							
7	6	5	4	3	2	1	0
0	0	1	0	0	0	0	1

PackedTime													
31	30	29	28	27	26	25	24	23	22	21	...	1	0
1	0	0	0	0	0	0	0	0	0	0		0	0

Now we shift both `TempByte` and `PackedTime` one bit to the right and copy `TempByte.Bit0` to `Packed-Time.Bit31` again:

TempByte							
7	6	5	4	3	2	1	0
0	0	0	1	0	0	0	0

PackedTime													
31	30	29	28	27	26	25	24	23	22	21	...	1	0
0	1	0	0	0	0	0	0	0	0	0		0	0

And again:

TempByte							
7	6	5	4	3	2	1	0
0	0	0	0	1	0	0	0

PackedTime													
31	*30*	*29*	*28*	*27*	*26*	*25*	*24*	*23*	*22*	*21*	*...*	*1*	*0*
0	0	1	0	0	0	0	0	0	0	0		0	0

After the sixth copy and shift:

TempByte							
7	*6*	*5*	*4*	*3*	*2*	*1*	*0*
0	0	0	0	0	0	0	1

PackedTime													
31	*30*	*29*	*28*	*27*	*26*	*25*	*24*	*23*	*22*	*21*	*...*	*1*	*0*
1	0	0	0	0	1	0	0	0	0	0		0	0

Unpacking the data is just the reverse of packing it.

```
UnPackTimeInfo             ;Unpack date/time info
;----------------
        For i = 6 to 0 Step -1
            If i <> 5 Then
                TempByte = 0
                For j = 0 to PackLen(i)
                        TempByte.Bit0 = PackedTime.Bit31
                        PackedTime = PackedTime << 1
                        If j <> PackLen(i) Then
                                TempByte = TempByte << 1
                        EndIf
                Next ;j
                TimeData(i) = BIN2BCD TempByte
            EndIf
        Next ; i
    Return
```

Note that we work in the reverse order; starting with years—**TimeData(6)**—and work down to seconds, **TimeData(0)**. And, instead of shifting to the right, we shift to the left. Lastly, since we save the data in the **TimeData** array, we restore the binary unpacked value to BCD with MBasic's **BIN2BCD** operator.

We take a slightly different approach with the weather sensor data, however. We could save the raw A/D readings for the humidity and pressure sensors, for a total of 20 bits, and save the most significant 12 bits of the 13-bit temperature value (resulting in a resolution of 0.0625°C). These three values could then be bit-packed into 4 bytes just as we did with the time data. There is a slight problem with this approach, since the barometric pressure reading must be calculated using knowledge of V_{DD}, which we have no room to save if we stick with a 4-byte record length. We could assume V_{DD} is 5.00 V, and in fact the typical excursion of V_{DD} from 5.00 V results in a relatively small error in the calculated barometric pressure.

As an exercise in trying a different approach, we'll quantize the calculated temperature into 0.25°F steps, the barometer into 0.01 inches Hg steps and the relative humidity into 0.1% steps.

Field	Steps	Minimum/Maximum Value	Minimum Bits Required
Temperature (F)	*0.25°F*	*-40...216*	*10*
Barometric Pressure (in Hg)	*0.01 inches Hg*	*0...4096*	*12*
Relative Humidity %	*0.1%*	*0...1000*	*10*
Total			*32*

Our bit map is therefore:

31	30	29	28	27	26	25	24	23	22	21	20	19	18	17	16	15	14	13	12	11	10	09	08	07	06	05	04	03	02	01	00
10 bits										10 bits										12 bits											
Temperature °F										Relative Humidity %										Barometric Pressure inches Hg											
Byte 3										Byte 2									Byte 1								Byte 0					

And, while we are at it, you'll see a different approach to bit packing. The subroutine **PackReadings** quantizes the temperature, humidity and barometric pressure and bit packs their values into **PackedData** in just four lines of code.

```
PackReadings
;----------------
        IntTemp = ToInt ((Fahrenheit + 40.0) * 4.0)
        IntHum = ToInt (Rhum * 10.0)
        IntBaro = ToInt (InHg * 100.0)

        PackedData = 4194304 * IntTemp + 4096 * IntHum + IntBaro
Return
```

We declare three word variables, **IntTemp**, **IntHum** and **IntBaro** to hold the quantized values of Fahrenheit temperature, relative humidity percentage and barometric pressure in inches Hg, respectively. We quantize humidity by multiplying by 10 and converting to an integer, and we quantize barometric pressure by multiplying by 100 and converting to an integer. Thus, for example, 35.1% relative humidity is converted to 351, and 30.31 inches Hg is converted to 3031. Converting back simply requires us to divide by 10 or 100. 351 Is converted to 35.1 after dividing by 10, and 3031 is converted to 30.31 by dividing by 100.

Because the temperature may go below zero—and our simplistic system can't deal with negative numbers—we first bias the temperature up by 40 degrees, then multiply by 4 and finally convert it to an integer. The bias offset permits us to properly encode any temperature between –40°F and +216°F. For example, suppose the temperature is 72.34 degrees. This is quantized into 4 * (72.34 + 40), or 449. We convert back by first subtracting 160 and dividing the remainder by 4. The result is 72.25°F.

We could use the bit shifting approach in **PackTimeInfo** to insert the desired bits from these three integer values into the long variable **PackedData**. Instead, because there are only three variables to add, we multiply each by its base value and add to **PackedData**.

Multiplying by 4096 is the functional equivalent of shifting left by 12 bits. Likewise, multiplying by 4194304 is equivalent to shifting left by 22 bits. Let's take an example. We start with **PackedData = 0**.

31	30	29	28	27	26	25	24	23	22	21	20	19	18	17	16	15	14	13	12	11	10	09	08	07	06	05	04	03	02	01	00
0	0	0	0	0	0	0	0	0	0	0	0	0	0	0	0	0	0	0	0	0	0	0	0	0	0	0	0	0	0	0	0
10 bits										10 bits										12 bits											
Temperature °F										Relative Humidity %										Barometric Pressure inches Hg											
Byte 3										Byte 2									Byte 1								Byte 0					

We'll assume the relative humidity is 35.1%, or 351 when quantized.

IntHum = 351															
15	14	13	12	11	10	9	8	7	6	5	4	3	2	1	0
0	0	0	0	0	0	0	1	0	1	0	1	1	1	1	1

Computing 351 * 4096 yields 1437696, or **%101011111000000000000**.

In MBasic's internal 32-bit numerical calculation stack, 1437696 is:

3	3	2	2	2	2	2	2	2	2	2	2	1	1	1	1	1	1	1	1	1	1	0	0	0	0	0	0	0	0	0	0
1	0	9	8	7	6	5	4	3	2	1	0	9	8	7	6	5	4	3	2	1	0	9	8	7	6	5	4	3	2	1	0
0	0	0	0	0	0	0	0	0	0	0	1	0	1	0	1	1	1	1	1	1	1	0	0	0	0	0	0	0	0	0	0

When we add this result to **PackedData**, we see that the relative humidity bits have been placed exactly where we want them.

3	3	2	2	2	2	2	2	2	2	2	2	1	1	1	1	1	1	1	1	1	1	0	0	0	0	0	0	0	0	0	0
1	0	9	8	7	6	5	4	3	2	1	0	9	8	7	6	5	4	3	2	1	0	9	8	7	6	5	4	3	2	1	0
0	0	0	0	0	0	0	0	0	0	0	1	0	1	0	1	1	1	1	1	1	1	0	0	0	0	0	0	0	0	0	0

10 bits										10 bits										12 bits											
Temperature °F										Relative Humidity %										Barometric Pressure inches Hg											
Byte 3										Byte 2									Byte 1									Byte 0				

Unpacking the weather data is the reverse process; but with a twist or two.

```
UnPackReadings
;---------------
        IntBaro = PackedData // 4096
        InHg = (ToFloat IntBaro) / 100.0
        mbars = (InHg - ElevCor) / mBarToInHg

        IntTemp = PackedData / 4194304
        Fahrenheit = (ToFloat (IntTemp - 160)) / 4.0

        IntHum = (PackedData / 4096) // 1024
        Rhum = (ToFloat IntHum) / 10.0
Return
```

Remember that the **/** operator returns the integer quotient and **//** returns the integer remainder. If the divisor is of the form 2^N, the remainder is the value resulting from masking off all the bits to the left of the $N-1^{th}$ bit. Hence the statement **IntBaro = PackedData // 4096** returns the rightmost 12 bits (bits 0…11) of **PackedData**, since $4096 = 2^{12}$.

Likewise, if the divisor is of the form 2^N, the quotient is the value you get by masking off all bits to the right of the $N-1^{th}$ bit and right justifying the result. Hence, the statement **IntTemp = PackedData / 4194304** returns the leftmost 10 bits (bits 22...31) of **PackedData** and right justifies the result since $4194304 = 2^{22}$. Only 10 bits are returned because **PackedData** is a 32-bit variable.

To extract the middle 10 bits, we just combine the division and remainder operators. First, we extract the leftmost 22 bits with **(PackedData / 4096)**, therefore leaving us with the temperature and humidity right justified in a 32-bit buffer in MBasic's internal mathematic stack. We then extract the rightmost 10 bits of this residue with the remainder operator **//1024**. We use 1024, or 2^{10}, because we wish to extract 10 bits.

The remainder of the subroutine converts the extracted integer values to floating point numbers by dividing by the quantization constant we used in subroutine **PackReadings**. In the case of the Fahrenheit temperature, we also subtract the bias value.

We now return to saving and reading the packed data. Before calling **SavePackTime**, we have updated **RecordNo** to the next value. Using the techniques we learned in Chapter 18, we write the four bits of **PackedTime** to the serial EEPROM.

```
SavePackTime
;-----------
        MCtrlB = MRCtrl + MAdr0
        ;Write to addressed memory
        MemAdr = RecordNo * 2 * WriteInc
        I2COut MDta,MClk,ErrorWMsg,MCtrlB,MemAdr, |
        [PackedTime.Byte3, PackedTime.Byte2, |
        PackedTime.Byte1, PackedTime.Byte0]
        Pause 5
Return
```

Likewise, we simply write the four bytes of **PackedData** using the same techniques. We advance the byte address variable **MemAdr** as discussed earlier.

```
SavePackData
;-----------
        MCtrlB = MRCtrl + MAdr0
        MemAdr = RecordNo * 2 * WriteInc + WriteInc
        I2COut MDta,MClk,ErrorWMsg,MCtrlB,MemAdr, |
        [PackedData.Byte3, PackedData.Byte2, |
        PackedData.Byte1, PackedData.Byte0]
        Pause 5

    Return
```

The remaining new parts of the code involve writing the unpacked date/time and weather information to the serial port.

```
WriteUnpackedWxData
;------------------
        HSerOut [Real Fahrenheit \2," F",9, |
        Real Rhum \1," %RH",9,Real inHg \2," in Hg",13,10]
    Return

WriteUnpackedTime
;------------------
        HSerOut [Dec2 (BCD2BIN Hours)\2,":", |
        Dec2 (BCD2BIN Minutes)\2,":",Dec2 (BCD2BIN Seconds)\2,9]
        HSerOut [" ",Dec2 (BCD2BIN Month)\2,"/", |
        Dec2 (BCD2BIN Date)\2,"/",Dec2 (BCD2BIN Year)\2,9]
    Return
```

Both these subroutines mimic the ones we used to write from the freshly read data in Programs 28-2 and 12-4 and need no further explanation.

Program 28-2 also provides an optional special format data output, through an optional call to **WxToOscope** subroutine.

```
WxToOscope       ;output to IDE's Oscilloscope utility
;-------------
        TempLong = ToInt (Fahrenheit * 10.0)
        j = 0
        GoSub OutputToScope
        TempLong = ToInt (inHg * 10.0)
        j = 1
        GoSub OutputToScope
        TempLong = ToInt (RHum * 10.0)
        j = 2
        GoSub OutputToScope
    Return

OutputToScope
;--------------
        HSerOut [Chan1 + j,TempLong.Byte0,TempLong.Byte1, |
        TempLong.Byte2,TempLong.Byte3]
        HSerOut [13]
    Return
```

MBasic's integrated development environment includes a utility program, Oscilloscope, accessed from the Tools menu. Although Oscilloscope is undocumented in MBasic User's Guide, it is documented in a downloadable Application Note at Basic Micro's website. Data written to the Oscilloscope utility must be an integer type **long**. Data is written from the serial port, and starts with a "channel number," which ranges from 15 to 30 and is followed by the 4 bytes of the long variable, and ends with a carriage return (13). Received lines not starting with a valid channel number are assumed to be text messages and are displayed in Oscilloscope's message window. Oscilloscope plots each channel with a different color.

Since Oscilloscope plots only integer values, we'll need to scale the floating point data, as well as converting it to integer—otherwise all we would see are integer steps in temperature, relative humidity and barometric pressure. To provide better scales, therefore, we'll multiply the temperature (in Fahrenheit) by 10; multiply the relative humidity percentage by 10 and the barometric pressure by 10. This allows us to plot a scale of 0...1000 and see all three sensor outputs on one screen. Figure 28-5 shows the result. Figures 28-6 and 28-7 show how adjusting the vertical scale and offset allow you to expand otherwise small changes in the output values.

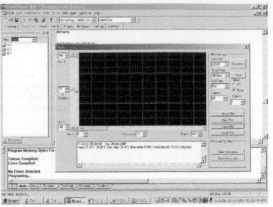

Figure 28-5: Temperature, relative humidity and barometric pressure displayed on oscilloscope.

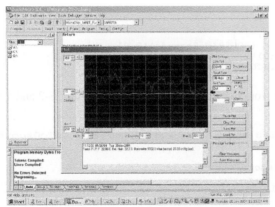

Figures 28-6: Expanded view of barometric pressure and relative humidity.

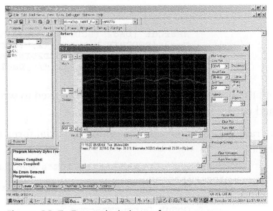

Figure 28-7: Expanded view of temperature.

Before we wrap up this very long Chapter, let's look at the output format. First, we let the program run and output the time/date and weather data once every minute.

```
15:49:05  01/18/04  Sun  18-Jan-2004
Temp 73.17 F  22.87 C  Rel. Hum. 36.5 %  Barometer 979.2 mbar (uncor)  29.26 in Hg (cor)
15:50:00  01/18/04  Sun  18-Jan-2004
Temp 73.06 F  22.81 C  Rel. Hum. 36.3 %  Barometer 980.3 mbar (uncor)  29.29 in Hg (cor)
15:51:00  01/18/04  Sun  18-Jan-2004
Temp 72.94 F  22.75 C  Rel. Hum. 36.2 %  Barometer 981.8 mbar (uncor)  29.34 in Hg (cor)
15:52:00  01/18/04  Sun  18-Jan-2004
Temp 72.89 F  22.71 C  Rel. Hum. 35.8 %  Barometer 981.8 mbar (uncor)  29.34 in Hg (cor)
15:53:00  01/18/04  Sun  18-Jan-2004
Temp 72.83 F  22.68 C  Rel. Hum. 35.8 %  Barometer 981.8 mbar (uncor)  29.34 in Hg (cor)
```

Now, we send the command to output the saved data, an "o" or "O" followed by a carriage return (enter).

```
o
0       15:49:05    01/18/04 73.00 F    36.5 %RH 29.25 in Hg
1       15:50:00    01/18/04 73.00 F    36.2 %RH 29.28 in Hg
2       15:51:00    01/18/04 72.75 F    36.1 %RH 29.33 in Hg
3       15:52:00    01/18/04 72.75 F    35.7 %RH 29.33 in Hg
4       15:53:00    01/18/04 72.75 F    35.7 %RH 29.33 in Hg
```

We see the data matches the real time data, except for variations due to rounding.

Another real time report arrives.

15:54:00 01/18/04 Sun 18-Jan-2004
Temp 72.78 F 22.65 C Rel. Hum. 36.0 % Barometer 983.9 mbar (uncor) 29.40 in Hg (cor)

Again, we send a "o" and dump the saved data to the serial port.

o
0	15:49:05	01/18/0473.00 F	36.5 %RH29.25 in Hg
1	15:50:00	01/18/0473.00 F	36.2 %RH29.28 in Hg
2	15:51:00	01/18/0472.75 F	36.1 %RH29.33 in Hg
3	15:52:00	01/18/0472.75 F	35.7 %RH29.33 in Hg
4	15:53:00	01/18/0472.75 F	35.7 %RH29.33 in Hg
5	15:54:00	01/18/0472.75 F	36.0 %RH29.39 in Hg

If we keep the data save interval at one minute, we have capacity for 8191 minutes, or 5.68 days. That time, of course, can be extended proportionally by lengthening the time between data saves. (We'll see an approach suggested in the *Ideas for Modifications to Programs and Circuits* that might double the data save capacity.)

Let's see how the saved data looks when examined over several days. Figure 28-8 plots about two days worth of compressed/decompressed data. The data was taken in my basement workshop and shows the drop in temperature at night when the cutback thermostat drops the first floor temperature. The rise in indoor relative humidity on the 18th is a result of freezing rain and high outdoor humidity.

Figure 28-9 compares two days of the barometric pressure I recorded with the National Weather Service's data for Dulles, VA, about 10 miles from my house. As reflected in Figure 28-10, my MPX6115 appears to be biased low by about 0.17 inches Hg. It's a simple matter to add an additional 0.17 inches Hg offset to the reported values. The bias is not uniform, of course, so an even better correction would be one that includes both intercept and slope factors, as shown in Figure 28-11. The corrected pressure for my sensor is:

$$P_{CORRECTED} = 1.02959 P_{RAW} - 0.60568$$

Where both $P_{CORRECTED}$ and P_{RAW} are in inches Mercury, and P_{RAW} is corrected for altitude. This correction equation is valid, of course, only for my MPX6115, and, even with this understanding, it would be wise to collect more than two day's data before accepting this equation.

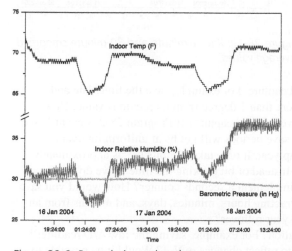

Figure 28-8: Recorded weather data.

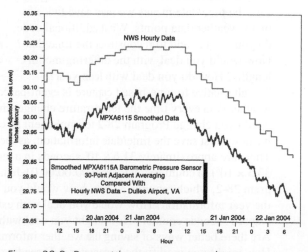

Figure 28-9: Barometric pressure comparison.

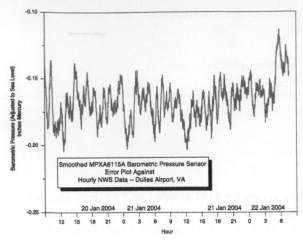

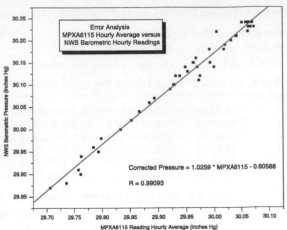

Figure 28-10: Error plot; MPX6115 versus NWS data.

Figure 28-11: Linear regression fit to correct MPX6115 reading.

Note that the corrections are applied to averaged data. Figure 28-12 shows the reading-to-reading variation makes any individual reading suspect.

Ideas for Modifications to Programs and Circuits

- Program 28-2 uses time-based capture. Since we know the start time and since each capture occurs at constant intervals, is it necessary to save the time/date? Instead, all we need save is the start date/time and the save interval. From that information, we can reconstruct the each sample date/time, without error. Not having to save the time/date means we may save twice as many weather data points. What additional code do you need to add to reconstruct the time/date? How would you deal with the differing month lengths? How do you deal with leap year?

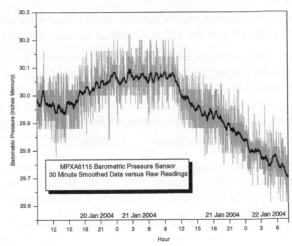

Figure 28-12: Raw 1-minute and 30-minute smoothed average values.

- The alternative to time-based capture is event-based capture. For example, save the time/date and weather data every time the temperature changes more than 1 degree from its previous value. How would you change Program 28-2 to implement an event-based capture? If Program 28-2 is event-based, we now must save the time/date information, as the save periods will not be at uniform intervals.

- There are approximately 31.6×10^6 seconds in a leap year. It only takes 25 bits (2^{25} is approximately 33.5×10^6) to count seconds from 00:00:00 01 Jan. Instead of bit packing the time as we do in Program 28-2, which requires 32 bits, how would you implement a seconds counter? How would you save the year information? How would you deal with extracting hours, minutes, days, and months from an elapsed seconds counter? How might you deal with year rollover?

- Are there better ways of saving the weather information? For example, saving the raw A/D readings? How might you modify the barometer A/D reading to reflect changes in V_{DD}? Would it be possible to create a fictitious 10-bit A/D value that assumes V_{DD} is 5.000 V and yields the same computed barometric pressure as the actual A/D reading with a correction for V_{DD}?

- Bit packing is a first order compression technique and provides reasonable returns, in our example, for modest programming difficulty. Would more computationally intensive data compression work? For example, suppose the time/date information was delta encoded; instead of repeating the unchanged parts of the date and time with every record, we might just record changes. The first record contains the full time/date, while the next record contains only the changed values from the prior record, and so forth. Would the same approach work for the weather data? Delta encoding can be quite efficient, but it likely forces us to variable length records, either variable byte length, or even variable bit length. How would you structure a variable length record structure within the confines of a AT24C512 serial EEPROM?

- To provide an interactive menu driven system, merge the ideas of Program 9-3 with those of this chapter.

- An LCD display would, of course, make a user-friendlier weather station. Which parameters should be reported? To reduce the reading-to-reading variation in the barometric pressure and relative humidity, should these values be averaged? If so, over what period?

- How would you add external temperature and humidity sensors? How would you expand the memory?

- How would you go about reading the MPX6115 sensor with improved resolution without a higher resolution A/D converter? It is possible to derive a second voltage from the 4.096 V precision reference through a second op-amp running with a gain of less than 1.00, and use that as V_{REF-}. Figure 28-13 is one possible circuit. All resistors should be 1% tolerance or better. If V_{REF+} is 4.6085 V and V_{REF-} is 3.5853 V, to what resolution may the MPX6115 be read? What is the maximum and minimum absolute pressure range?

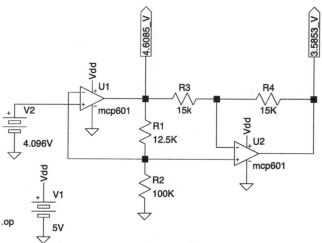

Figure 28-13: Possible dual-voltage reference source.

References

[28-1] Honeywell International, Inc., *Thermoset Polymer-based Capacitive Sensors, Application Sheet*, (undated) available at http://www.honeywell.com/sensing.

[28-2] Humirel , Inc., *Technical Data Relative Humidity Sensor HS 1100 / HS 1101,* Document HPC001, Rev. 7, June 2002, available at http://www.humirel.com/Humidity_temperature_sensors.php

[28-3] Humirel, Inc., *Technical Data Relative Humidity ModuleHM 1500,* Document HPC009, Rev H., October 2001, available at http://www.humirel.com/Humidity_temperature_sensors.php.

[28-4] Honeywell International, Inc., *Humidity Sensor HIH-3610 Series,* (undated), available at http://www. honeywell.com/sensing.

[28-5] Honeywell International, Inc., *Humidity Sensor HIH Series,* (undated), available at http://www.honeywell.com/sensing.

[28-6] Motorola Semiconductor, Sensor Device Data Book DL200/D, Rev. 5 (Jan 003), available at http://e-www.motorola.com/files/sensors/doc/data_lib/DL200.pdf. This document contains the data sheet for the MPX6115A series pressure sensors, as well as other useful information.

[28-7] Dallas Semiconductor Corp. division of Maxim Integrated Products, Inc., *DS1624 Digital Thermometer and Memory*, (July 20, 1999) available at http://pdfserv.maxim-ic.com/en/ds/DS1624.pdf.

[28-8] National Semiconductor Corp., *National Semiconductor's Temperature Sensor Handbook, 2nd Ed.* (undated).

[28-9] Basic Micro, Inc., *Application Note: Oscilloscope*, (undated), available at http://www.basicmicro.com/downloads/docs/oscope%20app%20note.pdf.

Migrating from v5.2.1.x to 5.3.0.0 and the Undocumented MBasic

Migrating from v5.2.1.x to 5.3.0.0

In early 2005, Basic Micro plans to release MBasic version 5.3.0.0. I've been given access to a prerelease version of 5.3.0.0 and the printed program listings in this book are compatible with 5.3.0.0. The CD-ROM contains both 5.2.1.1 and 5.3.0.0 program versions.

Version 5.3.0.0 fixes a number of small bugs in earlier editions, but also adds new functions and significantly modifies the way many existing functions operate. I've identified many of these changes throughout this book and this chapter summarizes those discussions in one location.

ADIN16—**ADIN16** is a new hardware command. **ADIN16** reads a 10-bit A/D value enough times to create an averaged 16-bit result, thereby approximating a 16-bit AD converter under certain conditions (though not perfectly). **ADIN16**'s syntax is identical with **ADIN**. Program 11-1A provides an example of **ADIN16** usage.

MiddleMult renamed FractionalMultiply—The function **MiddleMult** has been renamed **Fractional-Multiply**.

Changes in floating point operators—The special operators for floating point arithmetic (**FMUL**, **FDIV**, and so on.) are deleted and replaced with normal +, –, / and * operators.

- **FADD** is now **+**
- **FSUB** is now **–**
- **FMUL** is now *****
- **FDIV** is now **/**
- **FNEG** is replaced by prefixing the variable with a – sign

In addition, **TOINT**, **TOFLOAT** and **FLOATTABLE** are new operators:

- **FLOAT** is renamed **ToFloat**
- **INT** is renamed **ToInt**
- **FloatTable** is used to define a floating point data table just like **ByteTable** is used for byte-length data tables.

It's no longer necessary to include the **FLOAT** conversion operator when defining constants; instead the new operator **FCon** is used to define floating point constants:

v. 5.2.1.1:   `FloatingConstant    Con    Float 1.00`
v. 5.3.0.0:   `FloatingConstant    FCon 1.00`

User-defined function—User-defined functions are now support with both parameters and a returned variable.

```
MyFunction [var1,var2,var3...]{,returnval}
                    ;Function code.
            return {returnval}
```

To call a user-defined function use **GoSub**:

```
Gosub MyFunction[var1,var2,var3...]{,returnval}
```

Here's a simple example showing how to define and use a user-defined function:

```
EX        Var             Word
FX        Var             Word
GX        Var             Word
HX        Var             Word
QX        Var             Word

Clear

Main
          EX = 4
          FX = 5
          HX = EX * FX                      ;HX=20

          GX = 0                            ;GX =0
          GoSub MyFunction [3,4],GX         ;GX=20
          GoSub MyFunction [30,40],GX       ;GX = 1200
          GoSub MyFunction [128,256],GX     ;GX = 32768

          GoSub MyFunction [HX,16],QX       ;QX=320 & GX=320
Goto Main

;Define Function Prototype
;--------------------------
MyFunction      [EX,FX]
          GX=EX*FX
Return GX
```

Several points should be noted:

* Either place the function definition at the end of the code, or "jump around" the function definition with a **GoTo** operation, so that the function code is not executed except as a consequence of the **GoSub** statement.
* Variables used in the function definition are still global variables. Consequently any changes made to any variable in the function still have global visibility to the rest of your code.
* As with built-in functions, arguments may be variables or constants.
* The limit on the number of command arguments is the size of the stack command. The default stack size is 17, which also happens to be the largest permitted stack size.
* There is no error checking on the number of arguments in the **GoSub MyFunction** operation. If you use fewer arguments than the number in the function prototype definition, the unused arguments are undefined (they could be any value); if you use more arguments than in the function prototype definition, the extra values are junked.

PIC parameter header—The standard header is no longer accepted and must be set in the configuration box.

```
;Versions earlier than 5.3.0.0
CPU = 16F876
MHZ = 20
CONFIG 16250
```

If you load a BAS file created in an earlier version, MBasic 5.3.0.0 will alert you and ask if it should automatically convert the file to the new version. Answer the dialog box "Yes" and you will be presented with the program configuration dialog box so that you can confirm the setup constants. The principal change in the dialog box from earlier versions is that the PIC's clock frequency must be entered in *Hz*, not *MHz*. A series of buttons predefine the common clock speeds, for example, pressing the "20 MHz" button enters 20000000 into the clock speed box. Or, you may enter a custom speed directly.

The new header information is entered as a comment line at the top of the BAS file, but is not displayed in MBasic's code editor. The header may be seen if you open a BAS file with a text editor:

```
;%CONFIG% 16F877A $1312d00 $3f3a $0 $0  $0  $0  $0  $0  $0  $0
$0  $0  $0  $0  $0  $0  $0  $0
```

32-bit function parameters—Earlier versions limited some function arguments, to 16-bit values. With version 5.3.0.0, all built-in function arguments may now be up to 32 bits.

Timing parameter changes—Most timing parameters are now in microseconds, not milliseconds. This change is related to the 32 bit argument extension, and permits a maximum time interval of 2^{32} μs, or about 4,295 seconds. Specifically:

- **Sound**—duration is in μs not ms.
- **Sound2**—duration is in μs not ms.
- **SPMotor**—delay time is in μs not ms.
- **DTMFOut**—on-time and off-time in 0.5 μs intervals not ms.
- **DTMFOut2**—on-time and off-time in μs not ms.
- **FreqOut**—duration in 0.5 μs intervals not ms.
- **Pause**—not changed – duration remains in ms.
- **Pauseus**—accepts 32-bit arguments
- **PulsIn**—Resolution is *unit steps*, with a minimum of 16 steps. Each *unit step* is 7 machine cycles, or 1.4 μs for a 20-MHz clock. (This interval corresponds to minimum measurable duration of about 22 μs.) The returned value is the duration in unit steps, as a 32-bit value. The default timeout is still 65535. However, the timeout value is in terms of *unit step* intervals, so the maximum timeout is 65535 * unit step duration.
- **PulsOut**—duration in μs; maximum value is 2^{32} μs.
- **PWM**—cycles can be up to 2^{32} (duty remains 0…255).
- **RCTime**—Resolution is *unit steps*, with each step 17 machine cycles, or 3.4 μs for a 20 MHz clock. (The minimum measurable duration is 1 unit step.) The returned value is duration in unit steps, as a 32-bit value. The default timeout is still 65535. However, the timeout value is in terms of *unit step* intervals steps, so the maximum timeout is 65535 * unit step duration.
- **Sleep**—maximum 2^{32} seconds.
- **HPWM**—remains limited by hardware—no change from 5.2.1.1.

Optional half-step motor operation—A new compiler directive optionally switches **SPMotor** to half-step mode. Since it is a compile-time option, you can't switch back and forth during program execution.

```
;To switch to half step mode add this to your program.

SPMOTOR_HALF con 0

;can make this constant any value since it simply has to b
;defined.
;When using half step mode one cycle equals 8 half steps.
```

New LCDInit command—A shortcut to LCD initialization is added to version 5.3.0.0:

```
LCDINIT RegSel\Clk{\RdWrPin},LCDNib
```

LCDInit runs the initialization specified by Hitachi for its LCD controllers, which has been adopted as the *de facto* LCD controller standard. You no longer need the **LCDINIT1**, **LCDINIT2** arguments in the first **LCDWRITE** command if you use **LCDInit** first. The older initialization technique continues to work in version 5.3.0.0.

EnableHSerial compiler directive—A new **EnableHSerial** compiler directive is required before using **HSerIn** or **HSerOut**. Otherwise **HserOut** and **HSerIn** are unchanged.

New SerIn and SerOut constants—The function constants associated with **SerIn** and **SerOut** have been replaced. (The old constants are no longer recognized.)

Sense		Mode		Data Bits		Parity		Stop Bits		Speed	
Ltr	Fcn	Ltr	Fcn	Ltr	Fcn	Ltr	Fcn	Ltr	Fcn	(with leading _)	
N	Normal	Blank	Normal	7	7 data bits	N	None	1	1 stop bit	_300	_19200
										_600	_21600
I	Inverted	O	Open (Bus mode)	8	8 data bits	E	Even	2	2 stop bits	_1200	_24000
										_2400	_26400
						O	Odd			_4800	_28800
										_7200	_31200
										_9600	_33600
										_12000	_36000
										_14400	_38400
										_16800	_57600

You construct the appropriate constant by concatenating the six elements in the table. For example, the baud-mode constant for inverted, 8 data bits, no parity, 1 stop bit, standard mode is **I8N1_9600**. (Normally, you will use the standard output configuration, that is, not open collector, where the configuration character is a blank, so in most cases your constant will have five elements, not six.)

Stack size and location—Version 5.3.0.0 handles the stack differently, and by default assigns all Bank 0 RAM to the stack. The main implication of this is that by default no user variable will be in Bank 0 and hence any assembler reference to variables must deal with bank selection. (Although not its purpose, this enforces good programming practice.) Alternatively, you may manually define the stack to a smaller size to free up Bank 0 RAM. The second change is that stack size is stated in type **Long** (4 bytes), whereas in 5.2.1.1 stack is stated in type **Word** (2 bytes).

To change the stack allocation, use the stack compiler directive. For example, to allocate 40 bytes of stack space, use the following compiler directive:

```
STACK = 10
```

New Function GetTimer1—the function **GetTimer1** reads the timer 1 high (**TMR1H**) and low bytes (**TMR1L**) and returns the values in a single word variable. (This function is in version 5.2.1.1 but is not documented.)

New Modes for ShiftIn & ShiftOut—The **ShiftIn** and **ShiftOut** functions have eight new mode constants:

	Mode Constant	Bit Order	Clock / Data Relationship	Speed (Machine Cyles)	Speed (With 20 MHz clock)
High Speed (all new in 5.3.0.0)	FASTMSBPRE	MSB first	Data valid on leading edge	25	200 KHz (5 µs/bit)
	FASTLSBPRE	LSB first			
	FASTMSBPOST	MSB first	Data valid on falling edge		
	FASTLSPOST	LSB first			
Slow Speed (all new in 5.3.0.0)	SLOWMSBPRE	MSB first	Data valid on leading edge	100	50 KHz (20µs/bit)
	SLOWLSBPRE	LSB first			
	SLOWMSBPOST	MSB first	Data valid on leading edge		
	SLOWLSBPOST	LSB first			
Normal Speed { } indicates backwards compatible for older versions	MSBPRE {MSBFIRST}	MSB first	Data valid on leading edge	50	100 KHz (10 µs/bit)
	LSBPRE {LSBFIRST}	LSB first			
	MSBPOST	MSB first	Data valid on leading edge		
	LSBPOST	LSB first			

These mode constants have the same bit order and bit timing sequence as the non-"FAST" or non-"SLOW" prefixed constants, but with different output speeds. The output speed is approximately 50,100 and 200 KHz with Slow, Normal and Fast modes on a 20 MHz PIC but proportionally slower with slower clocks.

Two new alias constants associated with **ShiftIn** and **ShiftOut** are defined in version 5.3.0.0, **MSBFIRST** and **LSBFIRST**, which are the same as **MSBPRE** and **LSBPRE**, respectively. Note that there is no equivalent speed-up for one-wire support.

Pre-defined macros in ISRASM—Version 5.3.0.0 removes the restriction preventing MBasic-defined assembler macros inside an **ISRASM** block.

New definition of _MHZ—The assembler *constant* _MHZ was changed from a byte-length representation of the processor clock in MHz to a type **Long** defining the clock frequency in Hz. A 20 MHz clock assembles as **_MHZ EQU 20** in version. 5.2.1.1. but as **_MHZ EQU 20000000** in v. 5.3.0.0.

Redefined order of precedence—Version 5.3.0.0's order of precedence is revised to more closely align with other languages.

Order	MBasic 5.3.0.0	MBasic 5.2.1.1
1st (Highest)	DEC2BCD, BCD2DEC, TOINT, TOFLOAT, NOT, ~, ABS, SIN, COS,-, DCD, NCD, SQR, RANDOM	NOT, ABS, SIN, COS, - (NEG), DCD, NCD, SQR, Random, FNEG, INT, FLOAT, DEC2BCD, BCD2DEC
2	REV, DIG	<, <=, =, >, >=, <>
3	MAX, MIN	AND, OR, XOR
4	*, **, */, /, //	REV, DIG
5	+, -	<<, >>
6	<<, >>	MAX, MIN
7	<, <=, =, >, >=, <>	&, ⌐ ^, ‾&, ⌐ ‾^
8	&, ⌐ ^	*, **, */, /, //, FMUL, FDIV
9 (Lowest)	AND, OR, XOR	+, −, FSUB, FADD

New assembler—Beginning with version 5.3.0.0, MBasic no longer uses Microchip's MPASM assembler and linker, instead using GPASM. As of the time this chapter is written, the GPASM assembler does not accept computations in **banksel**. Hence, the construction **banksel variable&0x1FF** fails; but **banksel variable** works and provides correct banking. Also, **banksel variable+1** to reach the upper byte (where variable is type word) also fails. Likewise **banksel .0** fails—the argument can't be a direct numerical value—but we can define new named constant **zero con 0** and then use **banksel zero**.

New reserved words—new reserved words with version 5.3.0.0 include:

P0...P31	[ATOM pin names]
ADIN16	[new function]
ToInt	[renamed conversion operator]
ToFloat	[renamed conversion operator]
GetTimer1	[undocumented function]

Undocumented MBasic

While writing this book, I found a number of features that are not covered in the current User's Guide. Although I've discussed most of these undocumented features in earlier chapters, it's useful to have a central repository of the "Undocumented MBasic." Basic Micro is rewriting the MBasic User's Guide to include this material as well as to reflect version 5.3.0.0 changes.

The page references in this chapter are to the User's Guide, revision 5.3, released 2003.

Function	Page	Summary
System Setup— Compile Optimize Size/Speed	18	*Feature—Compile Optimize Size/Speed—Added after Manual Written*

The System Setup dialog box discussion does not reflect the current compiler release, as it omits the Compiler Optimize Size/Speed selection. Figure 29-1 shows the new optimization selection box at the bottom. You may select speed or size optimization.

What's the difference between optimizing for speed or size? Program compiled with "Optimize for Speed" run about 15–25% faster than the same program compiled with "Optimize for Size." For large programs, "Optimize for Size" consumes roughly 40% to 50% less *token memory*. The exact numbers depend on the mix of program statements used. Figures 29-2 and 29-3 show speed versus memory requirements for two sample programs compiled with version 5.2.1.1. Figure 29-2 is based upon a program with repeated copies of one simple mathematical computation—multiplying two bytes and storing the product in a third byte number. Figure 29-3 is based upon a program with repeated instances of 10 mathematical operations, including floating point, square root and trigonometric computations. The data is from a 16F877 with a 20 MHz resonator.

In reading Figures 29-2 and 29-3, don't forget that the required program memory is the sum of the *token memory* and the *library memory*. Broadly speaking, the required library memory is a function of how many different types of MBasic functions and operations your program uses, while the required token memory is proportional to the number of program statements. "Optimizing for Size" trades off speed and a *larger* library size for *smaller* token memory consumption. Thus, a very short program may actually consume *more* memory as well as run slower under "Compile for Size" compared with "Compile for Speed." Larger programs usually don't show this anomaly.

Figure 29-1: Compiler optimization option in Tools | System Setup Dialog Box.

Function	Page	Summary
Line Continuation Symbol |	42	*An undocumented feature is the "line continuation symbol," |*

The vertical bar | character permits long lines to be continued if it appears before the paragraph break when entering programs. Although used in some sample programs, it is never formally documented in the User's Guide. An example of the line continuation symbol is:

```
HSerOut ["This is going to be a very long line ",Dec i,   |
" continued into two lines",13]
```

Function	Page	Summary
Array Variable Usage	48	*Reminder that array variables are not checked for bounding errors.*

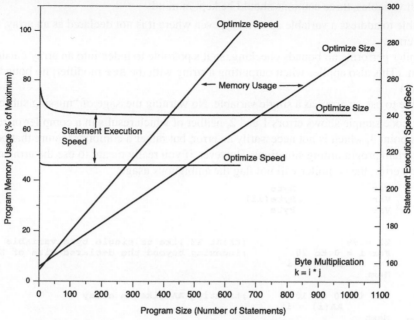

Figure 29-2: Memory versus speed for compiler optimization selection; simple program.

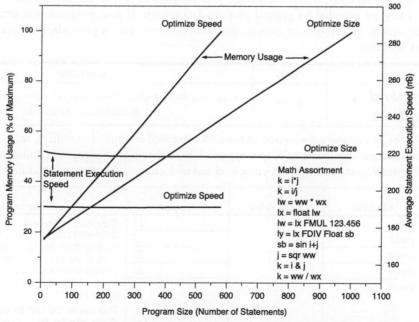

Figure 29-3: Memory versus speed for compiler optimization selection; complex program.

In dealing with variables, three cautions should be kept in mind:

1. It is possible to address a variable as an array, even where it is not declared as an array, without any compiler error;

2. The compiler performs no bounds checking, so it's possible to index into an array outside its declared dimension. (This also applies when outputting a string with the **str** modifier; no bounds checking is performed.)

3. It's possible to use an array as a single variable. No warning message of "missing subscript" is issued.

The following code sample shows errors 1 and 2, neither of which results in a compiler or run-time error. It also illustrates point 3, which is not necessarily an error, but rather a compiler feature that permits greater flexibility in certain programming situations. However, if you really meant to use the array as an array but simply made an error, the compiler will not flag the ambiguous usage.

```
AA          Var             Byte
ZZ          Var             Byte(11)
i           Var             Byte

Main
            ZZ = 99                 ;treat ZZ like as single byte variable
            For i = 0 to 20         ;indexing beyond the declared size of ZZ
                ZZ(i) = i
            Next

            For i = 0 to 10         ;treating AA like an array
                AA(i) = i
            Next
        GoTo Main
        End
```

As you should understand from Chapter 14, over running the array boundaries results in accessing register files not reserved for that array. It may even cause your program to access special purpose register files, which are assuredly not intended for general-purpose data storage. If your program uses array variables and you experience strange data results, or even program freezes or restarts, a good place to look for trouble is with your array indexing.

Function	Pages	Summary
Command Modifiers	53–55	*Omits many command modifiers; also command modifiers may be used as stand-alone functions.*

Chapter 9 provides an extensive discussion of the undocumented command modifiers, including a sample program to demonstrate how the same output appears with different formatting. The following table, reproduced from Chapter 9, summarizes the documented and undocumented command modifiers.

Modifier Family	Modifier	Base Value	Output	Comments
Decimal (Dec) (and others)	**Dec**		123	The modifier **Decn** where n is a number determines the number of digits to display, starting from the right. Note that specifying more digits than required to display the number does not cause blanks or leading zeros. **Dec** without an "n" displays all digits of the number.
	Dec1		3	
	Dec2		23	
	Dec3		123	
	Dec4	123		**This same "add n" to define displayed digits applies to:** **Dec, sDec** **Hex, iHex, sHex, isHex** **Bin, iBin, sBin, isBin**
			123	

Modifier Family	Modifier	Base Value	Output	Comments
Decimal (Dec) and Others	`Dec x\2`		123	Leading zeros can be added by appending a \n modifier to pad with leading zeros to n places.
	`Dec x\3`		123	
	`Dec x\4`		0123	
	`Dec x\5`	x = 123	00123	**This same "\n" to define leading digits applies to:**
	`Dec x\6`		000123	`Dec, sDec` `Hex, iHex, sHex, isHex` `Bin, iBin, sBin, isBin`
Real (floating point numbers)	`Real`		123.4559936523	Extra digits are resolution error. Without any modifier, we get the full 14 characters.
	`Real1`		3.4559936523	Used without anything else, Realn works the same way as for Dec. However, **n** applies only to digits to the left of the decimal point.
	`Real2`		23.4559936523	
	`Real3`		123.4559936523	
	`Real4`	123.456	123.4559936523	
	`Real4 Var\1`		123.4	To set the digits displayed on the right of the decimal point, use the /n syntax. (**Var** is a type Long)
	`Real4 Var\2`		123.45	
	`Real4 Var\3`		123.455	
	`Real4 Var\4`		123.4559	
Signed Decimal	`sDec`	−123	−123	If value is <0, − sign appended; no + sign if >0.
		+123	123	
Signed Hex	`sHex`	7B	7B	If value is <0, − sign appended; no + sign if >0.
		−7B	−7B	
Indicated Signed Hex	`isHex`	7B	$7B	If value is <0, − sign appended; no + sign if >0. Minus sign appears after the $ hex indicator.
		−7B	$-7B	
Signed Binary	`sBin`	1111011	1111011	If value is <0, − sign appended; no + sign if >0.
		−1111011	−1111011	
Indicated Signed Binary	`isBin`	1111011	%1111011	If value is <0, − sign appended; no + sign if >0. Minus sign appears after the % binary indicator.

It's also possible to use these modifiers as functions. For example:

```
InStr Var Byte(4)
InVal Var Word

InVal = 1234
InStr = Dec InVal
```

After the operation, **InStr** contains the string "1234" with the remainder of its length (if any) padded with 0's.

When using the repeat option, for example, **HSEROUT [Str i\30]**, no length checking is performed. If **i** is only two characters long, nonetheless, all 30 will be returned. The returned characters are from register files starting with the base address of the variable and continues for the desired repetition number of bytes.

Function	Page	Summary
Pin Variables	57	Reversed sense from Basic Stamp usage; extra reserved words.

The Basic Stamp compatibility command for input/output setting **DIRxx** is reversed from that found in the Basic Stamp. In the Basic Stamp to set for output, for example, **DIR0 = 1**, but in MBasic to set for output **DIR0 = 0**. (This reversal is mentioned in a note in the Basic Stamp compatibility section at page 237)

The User's Guide list of reserved words (pages 241-252) implies that Basic Stamp compatibility has been extended, as reserved words go from **INA/OUTA/DIRA/...INF/OUTF/DIRF** and pin functions stop at **IN31/OUT31/DIR31**. However, these additional pins are not implemented in the **IN/OUT/DIR** functions.

Function	Page	Summary
#Include	63	*Undocumented feature*

#Include is an undocumented MBasic feature that allows you to break up a large program into more manageable sections and then read them into the master program, just as if they were typed in directly.

To incorporate include files into the master program, the following structure is used:

```
#Include "P-XX-Constants.inc" ;Holds constant definitions
#Include "P-XX-Vars.inc"      ;Holds variable declarations
```

You must be careful when breaking code into **#Include** files—MBasic depends, to some extent, on order of entry, so the order and contents of the **#Include** files is important. Consider the following sequence:

```
X       Var Byte
X =     ZZ
ZZ      Con     5
```

This code fragment produces an error, as when the compiler reaches the statement **X=ZZ**, **ZZ** has not yet been defined. (The actual error message is "unexpected token type.")

The proper order requires the variables and constants to be declared before they may be used.

```
ZZ      Con     5
X       Var Byte

X =     ZZ
```

Further, if the **#Include** file has subroutines, it should not be executed until the individual subroutines are properly called.

Function	Page	Summary
Sin and Cos Functions	73	*Returned value not clear*

MBasic's **sin** and **cos** function return a signed byte value in the range $-127...127$.

Function	Page	Summary
BIN2BCD and BCD2DEC	73	*Renamed functions*

The functions identified in the User's Guide as DEC2BCD and BCD2DEC have been renamed:

DEC2BCD is **BIN2BCD**
BCD2DEC is **BCD2BIN**

Function	Page	Summary
ToInt	75	*Returns floor, not rounded value*

The `ToInt` function, which converts from floating point to integer is a floor function, and does not round. For example, `ToInt (1.99) = 1` and `ToInt (2.001) = 2`. To force rounding add 0.5 before `ToInt` conversion, e.g., `ToInt (x + 0.5)` yields the integer value of **x**, rounded to the nearest integer.

Function	Pages	Summary
I2Cin and I2Cout	*109-112*	*Missing information in User's Guide;* *problems with I2Cout under certain conditions*

There is an implicit requirement that the data being read or written is in byte form. Hence it is possible to correctly write and read individual bytes, or array of bytes. However, type word, long, etc. fail to correctly read and write, but no error message is provided.

Caution—the Control Word, bit 0 is correctly discussed in the User's Guide as an 8/16 bit address flag. Don't be confused by documentation for memory chips that say Bit 0 of the control word is R/W. MBasic automatically sets the R/W flag bit.

Caution—If the design has multiple serial chips in parallel, multiple byte writes or reads that straddle a chip boundary will fail without error message.

The discussion in the User's Guide, page 113, states that the 24C04 is "word (16-bit) addressing." The 24C04 is a 512 byte (4K bit) device. It uses 8-bit addressing, with the 9th bit (A8) being set as bit 1 in the command word. The sample code correctly shows the command byte ending with 0, thus indicating an 8-bit address device.

The code in the User's Guide, pages 114–115, puts two address bits into the command word (A8 at bit 1 and A9 at bit2). But, the 24C04 does not use A9 addressing bit.

Function	Page	Summary
Random	*150*	*Error in sample program*

The first User's Guide, page 150, sample program always returns the same value; 51.

```
;Following sample program from User's Guide fails
;don't use it
Temp var word

Main
        Temp = Random 25
        HSerOut [DEC Temp,13]
        Pause 1000
Goto Main
```

The second sample program in the User's Guide, page 150, is correct:

```
;This sample program works correctly
Temp var word

temp = 25

Main
        Temp = Random temp      ;25 is the seed value
        HSerOut [DEC Temp,13]
        Pause 1000
Goto Main
```

However, the output of the sample program has, at places, a progressive pattern that may not be satisfactory for all random number requirements. Following are the first eleven results—the numbers approximately double with each output.

```
51
102
204
409
819
1638
3276
6553
13107
26214
52428
```

It's important that the result of **Random** be a word length variable. If you declare **Temp var byte**, the output repeats after only four cycles:

```
51
102
204
153
51
102
204
153
51
102
204
153
```

If you need a byte length random variable, keep the result of **Random** in a word length variable and use the lower byte as the byte length random, for example, **ByteRandomVar = Temp.LowByte**.

Function	Page	Summary
SerOut *also HSerOut*	175	*String output does not stop automatically for 0*

The User's Guide states that when outputting a string-formatted variable, such as **[STR TempStr]**, the characters of **TempStr** will be outputted until a $00 byte is found. Neither **SerOut** nor **HSerOut** behave this way. Rather, when outputting a string, the output goes on for 255 characters, and $00 bytes are sent as null characters. To force the output to stop at $00, you must add an explicit \$0 end-of-line command as in the following example:

```
StrVar Var   Byte(20)
i      Var   Byte

Clear   ;fill with zeros
SetHSerial H115200
StrVar = "Hello!"

Main

        HSerOut ["No EOL: ",9,Str StrVar,13]
        HSerOut ["W/ EOL: ",9,Str StrVar\20\0,13]
        Pause 1000
Goto Main
```

The first serial output line results in the following output:

> No EOL: Hello!<00><00><00><00><00><00><00><00> *[continued for several lines]*

The second serial output line results in the following output:

> W/ EOL: Hello!

The construction `Str StrVar\20\0` causes output to stop at the earlier of: (a) 20 characters; or (b) the first 0 character.

Function	Page	Summary
Sleep	*186*	*Must enable watchdog timer at compile time*

In order for `Sleep` to work, you must enable the Watchdog Timer at compile time via the IDE's Configuration dialog box. When in sleep mode, an MBasic interrupt will not end the sleep period, a product of the way MBasic checks for interrupts. An `ISRASM` procedure will be executed, however.

Function	Pages	Summary
Xin and Xout	*203–210*	*Correct certain constants and add explanatory material*

Chapter 20 provides an extensive discussion of X10 interfacing including examples of `Xin` and `Xout`.

The code `X_Units_On %10000` is misidentified in the User's Guide. There is no function "All Units On" in X10; the function is actually "All Units Off" and corresponds to %10000. This library constant should be renamed "X_Units_Off."

In the User's Guide, page 203, the description of parameters for `Xin` says `VAR` "is a variable/constant used to store the results…" Since `Xin` reads from the interface unit, of course, the data may only be stored in a variable, and the reference to `VAR` being a constant should be removed.

The table for Key Codes (User's Guide, page 208) is mislabeled "unit" instead of Key Code.

There is no table for House Codes, so I've reproduced the one from Chapter 20:

B3	B2	B1	B0	Decimal Value	House Code
0	1	1	0	6	X_A
0	1	1	1	14	X_B
0	1	0	0	2	X_C
0	1	0	1	10	X_D
1	0	0	0	1	X_E
1	0	0	1	9	X_F
1	0	1	0	5	X_G
1	0	1	1	13	X_H
1	1	1	0	7	X_I
1	1	1	1	15	X_J
1	1	0	0	3	X_K
1	1	0	1	11	X_L
0	0	0	0	0	X_M
0	0	0	1	8	X_N
0	0	1	0	4	X_O
0	0	1	1	12	X_P

As developed in Chapter 20, the binary values shown for both Unit and Key Codes are reversed in the User's Guide. For example, **x_1** has a value of Decimal 12, Hex $C. This corresponds to %01100. The User's Guide shows **x_1** = %00110, which is decimal 6 or hex $6.

This confusion arises from bit order. Official X10 literature assumes codes are sent LSB first. Since MBasic's **Xout** and **Xin** operators appear to send or receive the bits left-to-right (MSB first) starting with the rightmost 4 (or 5 for units and key codes) bits, it is necessary to reverse the bit order from that shown in the User's Guide.

For example, to send the code for Unit 1, the bits must be sent in the following order 0-1-1-0-0. If these bits are held in a MBasic byte variable and are sent left-to-right, the MBasic byte value must be decimal 12 ($C). In fact, the value of the constant **x_1** is $C. However, the bit value shown in the User's Guide is %00110. This corresponds to decimal 6 ($6), not the actual value of **x_1**.

The triplet bit repetition interval in **Xout** appears to be hard coded for 60 Hz. This might cause problems in areas with 50 Hz power, where three phase power is used.

The User's Guide data pin description is in error. The data pin is set as an output, not input. No pull-up resistor is required.

Function	Page	Summary
HSerIn and HSerOut	214	*Errors in sample code and added warnings*

Chapter 9 provides an extensive discussion of serial input and output, including examples of **HSerIn** usage.

The User's Guide sample code is:

```
HSERIN 1000,main, [temp]
```

The command arguments are reversed; the correct code should read:

```
HSERIN main, 1000, [temp]
```

HserOut works by transferring the bytes to be sent to a 47-byte output buffer (assuming 16F87x/87xA devices) as quickly as possible. The hardware USART then sends the buffer contents with minimal support from MBasic thereafter. Execution is passed back to the next statement after **HserOut** as soon as the buffer is loaded, not when the buffer is emptied, as what MBasic support is required is supplied via an interrupt handler. It is possible, therefore, to get into trouble with the following:

```
…other code …
HserOut ["This is a very long string message"]
END      ;end of program
```

The **End** statement is executed before the "This is a very long string message" has completed sending. The result is program termination, and a truncated string transmission. The fix is to add a **Pause** statement between **HserOut** and the **End** statement.

MBasic's **Clear** operator interacts with **SetHSerial**.

The following construction fails:

```
SetHSerial H38400
Clear
HSerOut ["P 28-50",13,13]
```

The desired string is not emitted over the serial port. To prevent this problem, the **Clear** command must precede **SetHSerial**:

```
Clear
SetHSerial H38400
HSerOut ["P 28-50",13,13]
```

The above construction works.

Function	Page	Summary
HSerStat	214	Undocumented function for HSerOut and HSerIn

Chapter 9 provides sample code using the undocumented function **HSerStat**. **HSerStat** is invoked with the following syntax:

HSERSTAT cmd{,label}

CMD value	Function
0	Clear Input buffer(label not used)
1	Clear Output buffer(label not used)
2	Clear both buffers(label not used)
3	If Data available in input buffer goto label
4	If no Data available in input buffer goto label
5	If Data is waiting to be sent goto label
6	If no Data is waiting to be sent goto label

Function	Page	Summary
HPwm	215	Pins reversed in User's Guide; how to turn HPWm off; critical values for period and duty may cause erroneous output

The **CCPx** pin identifications are reversed in the User's Guide with the correct assignments for 16F87x/87xA devices being:

CCPx Value	Pin
0	C2
1	C1

The User's Guide omits how to turn off the hardware PWM module and restore the pins C1 or C2 to normal I/O use. This may be accomplished by clearing the associated control registers:

```
CCP1Con = $0          ;turns off the PWM generator 1 (Pin C2)
CCP2Con = $0          ;turns off the PWM generator 2 (Pin C1)
```

When using the **HPwm** function, it is not necessary to first set the desired pin to **Output**.

Function	Page	Summary
Interrupts	224	Latency in MBasic interrupts

Chapter 10 provides detailed coverage of interrupts in MBasic code. The User's Guide does not explicitly state that MBasic interrupts only are called at the following execution of each MBasic statement. Thus, the interrupt polling is not true "real time" due to latency in the MBasic statement execution. This is seen at its worst if a PAUSE statement is executed; the interrupt will not be executed until after the pause completes.

Function	Page	Summary
RBINT	225	*Note concerning reading PortB to clear the mismatch*

The port mismatch that causes the **RBINT** to be issued is not cleared unless PortB is read. Usually the interrupt handler will read PortB as part of its normal code and the interrupt causing condition will be reset. If the interrupt handler does not read PortB—it might just count interrupt calls, for example—the **RBINT** will continue to be serviced and spurious interrupt calls will occur. In this case, add a dummy PortB read, such as:

```
Dummy          Var          Byte

RBINTServiceRoutine
    ...other code
    Dummy = PortB
Resume
```

Function	Page	Summary
Macros	None	*Undocumented feature*

Chapter 14 covers MBasic's undocumented in-line assembler macro features, as well as certain faster operating MBASC macros. These macros are discussed, and examples provided in Chapter 14.

Standard Name	Macro Name
Pin Assignment Functions	
High	@High
Low	@Low
Input	@Input
Output	@Output
Timing Functions	
Pause	@MSDelay
Pauseus	@USDelay
Assembler Only	
GoTo	@GoTo
Return	@Return
BankSel	@Bank
Call	@Call
	@RatedDelay

The following table shows the speed increase possible by using these macros. The data is based on a 16F876A with a 20 MHz resonator.

Standard MBasic		MBasic Library Macros	
Name	*Execution Speed (μs)*	*Name*	*Execution Speed (μs)*
High	28.6	@High	0.8
Low	28.4	@Low	0.8
Input	28.0	@Input	0.6
Output	28.4	@Output	0.6

Function	Page	Summary
ISRASM	None	*Undocumented feature*

MBasic supports assembler handling of interrupt procedures. Chapter 15 is devoted to ISRASM procedures, so for this chapter's purposes, we'll just note that ISRASM is an undocumented feature of MBasic.

Parts List

Suppliers

Parts for the experiments in this book are available from many suppliers. In this appendix's detailed chapter-by-chapter component section, I'll identify sources for components that you might have trouble finding; otherwise any of the larger new component dealers should have everything you need. Incidentally, DigiKey and Mouser, at least, crosslink their on-line catalogs with manufacturer datasheets.

Suppliers that I've had success with are as follows:

New Components

Short Name	URL	Company Name/Address/ Telephone	Comments
DigiKey	http://www.digikey.com	DigiKey Corporation 701 Brooks Ave South Thief River Falls, MN 56701 800-344-4539	Very large stock; new components only. In most cases, same day shipment; Service charge for orders under $25.00.
Mouser	http://www.mouser.com	Mouser Electronics, Inc. 1000 North Main Street Mansfield, Texas 76063 800 346-6873	Large stock; new components only; no minimum order amount.; no service charge.
Allied	https://www.alliedelec.com	Allied Electronics, Inc. 7410 Pebble Drive Fort Worth, Texas 76118 1-800-433-5700	Old line distributor, very large stock; local branches in many cities; $50 minimum order.
Newark	http://www.newark.com	Newark InOne 4801 N. Ravenswood Chicago, IL 60640-4496 800-263-9275	Old line distributor; very large stock; local branches in many cities; Service charge for orders under $25.00; Common ownership with Farnell in UK and can supply any item in Farnell catalog to US purchaers.
Arrow	http://www.arrow.com/	Arrow Electronics, Inc. 50 Marcus Drive Melville, New York 11747 631-847-2000.	Large stock, including some products hard to find from other distributors; Local branches in many cities; $25 minimum order.
JameCo	http://www.jameco.com	Jameco Electronics 1355 Shoreway Road Belmont, CA 94002 800-831-4242.	Mail order only; smaller stock than above; has some new closeout or discontinued parts; Processing fee for under $20.

Short Name	URL	Company Name/Address/ Telephone	Comments
SmartHome	http://www.smarthomeusa.com	Smart Home Systems 127 East Main Street, Smithtown, NY 888-843-9103	X-10 modules only.

Overstock, Discontinued and Surplus Components

As with any source of overstocked, discontinued and surplus components, stock varies from moment to moment. Hence, what's available today may not be tomorrow and vice versa.

Short Name	URL	Company Name/Address/ Telephone	Comments
BG Micro	http://www.bgmicro.com/	B.G. Micro 555 N. 5th St. Suite #125 Garland, Texas 75040 800-276-2206	Useful source of LCD displays in addition to general parts and modules.
All Elect	http://www.allcorp.com/	ALL ELECTRONICS CORP. P.O. Box 567 Van Nuys, CA 91408-0567 888-826-5432	Electromechanical as well as electronic parts.
Debco	http://www.debcoelectronics. com/	Debco Electronics Inc. 4025 Edwards Road Cincinnati, Ohio, USA, 45209 800-423-4499	Carries some hard to find linear and digital semiconductors.
Goldmine	http://www.goldmine-elec.com/	The Electronic Goldmine PO Box 5408 Scottsdale, AZ 85261 800-445-0697	Carries LCDs along with other components. $10 minimum order and $6 minimum shipping.
MP Jones	http://www.mpja.com/	Marlin P. Jones & Assoc., Inc. P.O. Box 530400 Lake Park, Florida 33403 800-652-6733	Another source for LCD displays; good for power supplies as well; minimum order $15.

Generic Components Required

The chapter-by-chapter parts list concentrates on major or special components. In addition, you will require what can be called "garden variety" components. I recommend you have on hand:

Resistors—10 each, ¼ watt, carbon film resistors of value: 100, 220, 470, 680, 1K, 2.2K 4.7K, 6.8K and 10K ohm. Resistor values used in many schematics in this book may vary slightly from these values, but one of these values can be substituted in almost all instances. For example, I frequently use 5.1K pull-up resistors but a 4.7K works just as well in this application. You may wish to purchase a resistor assortment, such as Jameco's 81832CL ($27.95 including cabinet) or 103165CL (resistors only $12.95).

Capacitors—10 each .01 µF and 0.1 µF disc capacitors, 50 volt rating. You may wish to purchase a disc capacitor assortment, such as Jameco's 81859CL ($35.95 including cabinet) or 130232CL (capacitors only $17.95).

Semiconductors—you will need a small quantity (at least 10 each) of a few inexpensive general purpose semiconductors, such as:

NPN low power switching transistors—I use 2N4401 devices, in TO-92 package. You can purchase 25 2N4401s from DigiKey, PN 2N4401-ND, for $2.97.

PNP low power switching transistors—I use 2N4403 devices in TO-92 package. You can purchase 25 2N4403s from DigiKey, PN 2N4403D26ZCT-ND, for $2.97.

N-Channel Low Power MOSFET—I use 2N7000 devices in TO-92 package. You can purchase 10 2N7000s from DigiKey, PN 2N7000FS-ND, for $2.16.

Low power silicon signal diodes—I use 1N4148 devices in DO-35 package. You can purchase 10 1N4148s from DigiKey, PN 1N4148MSCT-ND, for $0.53, or 100 for $4.20.

Power Diodes—A classic 1N4001, 50V, 1A power diode is used in a few circuits. You can purchase 10 1N4001s from DigiKey, PN 1N4001MSCT-ND, for $0.43.

Schottky Diodes—an inexpensive Schottky diode, rated at 40 V, 1A is the 1N5891. You can purchase 10 1N5891s from DigiKey, PN 1N5819IR-ND, for $2.31.

LEDs—it's useful to have a few different color LEDs on hand, green, red and yellow, for example. The LEDs should be rated for 5 to 15 mA. You may prefer the larger T-1¾ size LEDs, or smaller T1 devices. (The number following the "T" is the diameter of the LED in units of 1/8 inch. The LEDs on the 2840 development board and the ICD programmer board are T1¾ devices.) I usually purchase LEDs from one of the surplus dealers, but all new component dealers carry them as well.

OpAmp—The operational amplifiers used in this book are type MCP601 from Microchip Technologies. These amplifiers are optimized to operate with a single +5V supply. MCP601s are available from DigiKey, PN MCP601-I/P-ND, for $0.68 each.

PICs—PICs are carried by DigiKey, Mouser, Jameco and others. Some PICs are also available from Basic Micro. I suggest you acquire and have on hand two 16F877A and two 16F876A devices. When ordering these devices, carefully watch the package type—the 2840 Development Boards work only with DIP packaged devices, DigiKey part numbers PIC16F876A-I/SP-ND ($7.05 each) and PIC16F877A-I/P-ND ($7.70). Both are 20 MHz parts.

Resonators—I prefer to run my PICs with 20 MHz resonators, type ZTT. These are available from DigiKey, PN X909-ND, for $0.54 each, or 10 for $4.50.

Extra Breadboard—Although the 2840 Development Board includes a 3" × 1" (300 contacts) solderless breadboard, it's not big enough to hold many of the circuits in this book, requiring you to split the circuit between the 2840 Development Board and an auxiliary solderless board. I recommend at least a 740 contact auxiliary solderless board, and a 1500 contact or larger board will make prototyping work easier.

LCD Display—several of the projects in this book require an LCD. I recommend a 4-line, 20 characters per line display, often called a 20x4 or a 204 model. These can be expensive to purchase new, so you may wish to check the surplus outlets for suitable displays. Connecting the LCD to a solderless development board requires at least 9 pins, and in some cases 16 pins. The most convenient way I've found to connect the LCD to a solderless board is with a 16-pin ribbon cable, one end terminated in a 16-pin DIP plug, the other end free for you to solder to the pads on the LCD, Digikey part numbers C2PXG-1618-ND (18" long for $2.10 each) or C2PXG-1606-ND (6" long for $1.84 each) are suitable. (These cables are "value added" parts from DigiKey and are assembled only when ordered, so the stock count will show 0 when you order one.)

Unless otherwise noted, all components are "through hole" type. All prices are as of the writing of this text, Spring 2004, and are based upon small quantity purchases.

Specific Components

In addition to the generic items described above, this listing identifies other specific components required for each chapter.

Chapter 1

None.

Chapter 2

None.

Chapter 3

Qty	Mfg No	Description	Catalog No	Supplier	Unit Price	Comments
1	MV5491A	Two color LED	MV5491A-ND	DigiKey	$0.51	
1	G2RL-24DC-12	Omron Relay 12V coil	Z147-ND	DigiKey	$4.03	To demonstrate switching of an inductive load; needs external 12V supply.
1	TIP31	NPN power transistor	497-2602-5-ND	DigiKey	$0.69	
1	TIP120	NPN Darlington power transistor	497-2539-5-ND	DigiKey	$0.74	
1	IRF510	Power N-channel MOSFET	IRF510-ND	DigiKey	$1.08	
1	IPS021	Intelligent MOSFET low side power switch	IPS021-ND	DigiKey	$2.00	
1	IRF9510	Power P-channel MOSFET	IRF9510-ND	DigiKey	$1.16	
1	IPS511	Intelligent MOSFET high side power switch	IPS511-ND	DigiKey	$3.00	
1	4N25	Opto-isolator NPN	4N25FS-ND	DigiKey	$0.48	
1	PS710A-1A	Optically isolated bilateral power MOSFET switch	PS710A-1A-ND	DigiKey	$6.50	
1	GF0401M	8-ohm speaker, 40mm diameter	GF0401M-ND		$2.49	Almost any speaker may be substituted.
1	2222 021 33471	470uf 6.3V electrolytic capacitor radial leads	4000PHCT-ND	DigiKey	$0.47	

Chapter 4

Qty	Mfg No	Description	Catalog No	Supplier	Unit Price	Comments
4	EVQ-21409K	Panasonic SPST normally open push button switch	P12223SCT-ND	DigiKey	$0.33	
1	VEL16KEY	Velleman 16 key (4x4) keypad	SWT1067	BG Micro	$4.95	Can be connected via 8-pin ribbon connector with DIP8 plug.

Chapter 5

Qty	Mfg No	Description	Catalog No	Supplier	Unit Price	Comments
1	DMC-16202NYJ-LY-AKE-BG	Optrix LED LCD Display 2x16 w/ LED backlight	73-1032-ND	DigiKey	$16.90	Shop around for surplus LCD modules.
1	DMC-20481NY-LY-ABE	Optrix LED LCD Display 4x20 w/ LED backlight	73-1094-ND	DigiKey	$30.30	Use 4x20 as alternate to 2x16; will use 4x20 display for other chapters.
1	3386F-103	Bournes 10K trimpot	3386F-103-ND	DigiKey	$1.29	

These LCD modules are expensive. It isn't necessary to purchase both a 2 × 16 and a 4 × 20 display; if you purchase only one, buy a 4 × 20 display. LCD displays are often available from surplus dealers at a much lower price than quoted above. For experimentation, it isn't necessary to use a backlighted display, so it's possible to save some money with a non-backlighted display.

Chapter 6

Qty	Mfg No	Description	Catalog No	Supplier	Unit Price	Comments
1	CD10RL1CK	Hex rotary DIP switch	CKN3045-ND	DigiKey	$4.89	Alternative is 4 single pole switches.
1		LCD				As in Chapter 5.
2	288T232R161A2	CTS rotary encoder 16 position w/ detent	CT3002-ND	DigiKey	$2.58	

Chapter 7

Qty	Mfg No	Description	Catalog No	Supplier	Unit Price	Comments
4	MAN4410A	Green common anode 7-segment display 0.4" digit	MAN4410A-ND	DigiKey	$0.98	Almost any common anode display may be substituted.
1	ULN2003A	NPN Darlington transistor array	296-1979-5-ND	DigiKey	$0.64	
4		SPST switches				As in Chapter 4.

Chapter 8

Qty	Mfg No	Description	Catalog No	Supplier	Unit Price	Comments
1	AIRPAXLB82773-M1	Airpax unipolar stepper motor 5V, 48 steps/rev		Jameco	$4.29	Almost any small 5V unipolar stepper may be substituted
1	ULN2003A					As in Chapter 7.
4	1N5242DO35	12V Zener diode	1N5242DO35MSCT-ND	DigiKey	$0.21	

Chapter 9

None.

Chapter 10

Qty	Mfg No	Description	Catalog No	Supplier	Unit Price	Comments
1		SPST switches				As in Chapter 4.
1		Rotary encoder				As in Chapter 6.

Chapter 11

Qty	Mfg No	Description	Catalog No	Supplier	Unit Price	Comments
1	LT1634CCZ-4.096	LT1634CCZ-4.096 voltage reference	LT1634CCZ-4.096-ND	DigiKey	$3.88	
3	LDS-M514RI-RA	7-Segment Display Lumex	67-1512-ND	DigiKey	$1.98	
3	LDS-M541RI-RA	Capital Advanced Technologies 33117 7-lead Surface Mount to 0.1" SIP (Use for breadboarding LED displays)	33117CA-ND	DigiKey	$3.22	
1	3386F-203	Bournes 20K trimpot	3386F-203-ND	DigiKey	$1.29	

In addition, several resistors and capacitors other than the generic stock values are necessary. Please review the associated schematics.

Chapter 12

Qty	Mfg No	Description	Catalog No	Supplier	Unit Price	Comments
2	18B20	One-wire digital temperature sensor	DS18B20-ND	DigiKey	$5.04	
1	DS1302	Digital real time clock	DS1302-ND	DigiKey	$3.31	
1	ECS-.327-8-14	32.786 KHz crystal	X803-ND	DigiKey	$2.63	Should be a 6pf crystal; the specified unit is 8pf. DigiKey does not stock the 6pf version.
1	A0820-2R5474	0.47F backup power storage capacitor	283-2497-ND	DigiKey	$2.40	

Chapter 13

None.

Chapter 14

None.

Chapter 15

None.

Chapter 16

Qty	Mfg No	Description	Catalog No	Supplier	Unit Price	Comments
1	TLC7528CN	Dual 8-bit multiplying DAC	296-1871-5-ND	DigiKey	$3.31	
1	LTC1062	Switched capacitor lowpass filter	LTC1062CN8-ND	DigiKey	$6.13	
1	LC411CN	OpAmp JFET input dual supply	LF411CN-ND	DigiKey	$1.35	

In addition, several resistors and capacitors other than the generic stock values are necessary. Please review the associated schematics.

Chapter 17

Qty	Mfg No	Description	Catalog No	Supplier	Unit Price	Comments
1		Touchtone Pad				See note 1.
1	MC145436A	Motorola DTMF decoder		Debco	$4.29	See note 2.
	CM8870P1	Mitel DTMF Decoder		Debco	$1.79	See note 2.
1		3.58 MHz crystal	3.579545 Mhz Crystal Colorburst (HC49)	Debco	$0.99	
1	CH1837A	Cermetek Microelectronics Telephone coupler module	N/A	Arrow	$18.67	

1. The test circuit uses a touchtone pad from an old Western Electric 500 series desk telephone. If you are unable to find a similar touchtone pad, you may use a second PIC and the **DTMFout** function to generate test signals. Alternatively, it is possible to generate the touchtone test signals by temporarily connect a standard touchtone telephone with a voltage source and an audio transformer.
2. At the time Chapter 17 was written, both the Motorola MC145436A and Mitel CM8870P1 DTMF decoder integrated circuits were available from Debco. At the time this parts list is compiled, Debco shows show both as out-of-stock. The MC145436A chip is officially discontinued, but may still be available from other suppliers. The CM8870P1 DTMF decoder IC is an alternative to the MC145436A but is not pin-for-pin interchangeable.

Chapter 18

Qty	Mfg No	Description	Catalog No	Supplier	Unit Price	Comments
1	24LC16B	16K serial EEPROM	24LC16B/P-ND	DigiKey	$0.63	
2	24LC256	256K serial EEPROM	24LC256-I/P-ND	DigiKey	$2.25	
1	24FC515	512K serial EEPROM	24FC515-I/P-ND		$4.68	
1	AT24C512	512K serial EEPROM	AT24C512-10PI-2.7-ND	DigiKey	$6.19	
1	74ACT843	9-bit latch	512-74ACT843SP	Mouser	$2.55	
1	CY7C128A-35PC	2K parallel RAM 35ns	428-1045-ND	DigiKey	$4.95	

Chapter 19

Qty	Mfg No	Description	Catalog No	Supplier	Unit Price	Comments
1	SN754410	Quad half bridge	296-9911-5-ND	DigiKey	$1.88	
1		10 ohm 5w power resistor	588-25J-10	Mouser	$1.37	See note 1
2		series current limiting resistor				See note 2
1		Bipolar stepper motor				See note 3

1. Used only to demonstrate clean microstep current.
2. Resistor value and dissipation rating depends upon motor winding resistance, voltage rating and supply voltage.
3. The unipolar motor used in Chapter 8 may be used in bipolar mode for Chapter 19.

Chapter 20

Qty	Mfg No	Description	Catalog No	Supplier	Unit Price	Comments
2	TW523	X-10 powerline transceiver	PSC-05	SmartHome	$19.95	See note 1.
1	PAM-02	Appliance control module	PAM-02	SmartHome	$13.95	
1		4x20 LCD				As in Chapter 5.
1		18B20 temperature sensor				As in Chapter 12.

1. Two transceivers are required to send data from one location and receive at a second location.

Chapter 21

Qty	Mfg No	Description	Catalog No	Supplier	Unit Price	Comments
1	MCP41010	10K digital potentiometer	MCP41010-I/P-ND	DigiKey	$1.68	
2	MCP42010	Dual 10K digital potentiometer	MCP42010-I/P-ND	DigiKey	$2.53	
1	0.18 µF cap	Metalized polyester capacitor	E1184-ND	DigiKey	$0.39	
1	0.47 µF cap	Metalized polyester capacitor	495-1112-ND	DigiKey	$0.41	

Chapter 22

Qty	Mfg No	Description	Catalog No	Supplier	Unit Price	Comments
1	QSE114	Photo transistor	QSE114-ND	DigiKey	$0.55	See note 1.
1	TSOP1100	Vishay IR receiver - broadband	782-TSOP1100	Mouser	$1.50	See note 2.
1	TSOP1238	Vishay IR receiver - 38 KHz version	782-TSOP1238	Mouser	$1.08	See note 2.

1. Only necessary for demonstration.
2. You may use either a broadband IR receiver, such as Vishay's TSOP1100, or, for better performance, a narrow band IR receiver matched to your remote control. The remote control I used has a 38 KHz carrier, and the narrow band receiver is a TSOP1238.

Chapter 23

Qty	Mfg No	Description	Catalog No	Supplier	Unit Price	Comments
1	K3011P	Random phase opto-triac	78-K3011P	Mouser	$0.53	
1	MOC3032	Zero crossing opto-triac	512-MOC3032M	Mouser	$1.24	
1	BTA12-600BW	Snubberless triac- 12A 600V	511-BTA12-600BW	Mouser	$0.96	
1	SW210	Transformer - Hammond Transformer substitute for SW-210.	546-164D10	Mouser	$8.44	See note 1.
1	TLE2426	Vcc/2 rail splitter	296-1994-ND	DigiKey	$1.30	
1	H312006	6A 3AG fuse	5760-12006	Mouser	$0.39	
1	441-R345A	3AG fuse holder	441-R345A	Mouser	$0.93	

1. Noncritical item. We use the transformer to provide an isolated 60 Hz zero crossing reference, so it draws no appreciable secondary current. The transformer secondary should be 10V RMS to 30V RMS.
2. Mechanical parts needed include a piece of unused single sided printed circuit board approximately 6" × 4", and AC power cord, plug and receptacle. The latter items may be obtained at a local hardware store.

Chapter 24

Qty	Mfg No	Description	Catalog No	Supplier	Unit Price	Comments
		Motor with AC tachometer				*See note 1.*
1	*SN754410*					*As in Chapter 19.*

1. The motor used in Chapter 24 has an AC tachometer and was removed from an old full height 5¼ inch floppy drive. Similar drives are available from scrapped computers for the asking.

Chapter 25

Qty	Mfg No	Description	Catalog No	Supplier	Unit Price	Comments
1		*Bar code wand*				*See note 1.*

1. I purchased the HP A200 bar code wand reader used in Chapter 25 from Goldmine in January 2004, but their stock is currently exhausted. Similar bar code wands are available from other surplus houses and frequently appear on eBay.

Chapter 26

Qty	Mfg No	Description	Catalog No	Supplier	Unit Price	Comments
1		*6V buzzer*	*273-054*	*Radio Shack*	*$2.99*	*See note 1.*
1		*Speaker*				*As in Chapter 3.*
2		*Rotary encoder*				*As in Chapter 6.*
1		*LCD*				*As in Chapter 5.*
1	*16F628*	*16F628*	*PIC16F628-20/P-ND*	*DigiKey*	*$3.88*	*See note 2.*

1. The 6 V buzzer is only used for preliminary tests.
2. Only preliminary programs are developed with the 16F628. Programming the 16F628 requires an 0818 Development Board.

Chapter 27

Qty	Mfg No	Description	Catalog No	Supplier	Unit Price	Comments
1	*XR2211*	*PLL decoder IC*	*34999*	*Jameco*	*$1.79*	*Exar - See note 1.*
or	*NJM2211*	*PLL decoder IC*	*NJM2211D-ND*	*DigiKey*	*$0.93*	*Second source of XR2211.*
3	*0.01 µF*	*0.1 µF metalized polyester*	*B32529C103J*	*DigiKey*	*$0.19*	
1	*4700 pf*	*4700 pf metalized polyester*	*B32529C472J*	*DigiKey*	*$0.19*	*Substitute for 5000 pf in Fig 27-4. See note 2.*
1	*3386F-1-502*	*5K trimpot*	*3386F-502-ND*		*$1.29*	*Substitute for 3K in Fig 27-4. See note 2.*

1. Chapter 27 was developed with an XR2211 device; a second, lower price, replacement is the NJM2211 manufactured by JRC, but it has not been tested with Chapter 27's circuitry.
2. A 4700 pF (0.0047 µF) metalized polyester capacitor may be used in place of the 5000 pF shown in Figure 27-4. Likewise, a 5K trimpot may be used in place of the 3K trimpot shown in Figure 27-4.

Chapter 28

Qty	Mfg No	Description	Catalog No	Supplier	Unit Price	Comments
1	HM1500	Humidity sensor	HM1500-ND	DigiKey	$30.50	
1	MPX6115A6U	Pressure sensor	MPXA6115A6U-ND	DigiKey	$17.57	
1	6103	Surfboard adapter	6103CA-ND	DigiKey	$1.48	
1	LT1634CCZ-4.096					As in Chapter 11.
3		12.7K 1% 1/4w resistor	12.7KXBK-ND		$0.54	See note 1.
1		100K 1% 1/4w resistor	100KXBK-ND		$0.54	See note 1.
1	DS1624	Temperature sensor	DS1624-ND		$9.04	
1	DS1302	real time clock				As in Chapter 12.
1		32 KHz crystal				As in Chapter 12.
1		0.47F aerogel backup capacitor				As in Chapter 12.
1	AT24C512	512k serial EEPROM				As in Chapter 18.

1. Price is for minimum purchase quantity, 5 units.

Function Index

The following index identifies programs in which the indicated MBasic function is used at least once. I've omitted common functions, such as **If...Then**, **For...Next**, as they are used in almost every program.

BIN2BCD & BCD2BIN
Program 07-3
Program 12-5; Program 12-6
Program 28-2

Floating Point Math
Program 06-03; Interrupt Version; Program 06-04;
Program 08-03;
Program 10-1; Program 10-1A; Program 10-1B; Program 10-2; Program 10-5;
Program 10-6;
Program 11-1; Program 11-2;
Program 12-2; Program 12-3; Program 12-5; Program 12-6;
Program 14-01-B; Program 14-01-C; Program 14-01; Program 14-01A; Program 14-100; Program 14-101; Program 14-102;
Program 14-103; Program 14-104; Program 14-105; Program 14-106; Program 14-107; Program 14-108; Program 14-109;
Program 14-110; Program 14-111; Program 14-112; Program 14-113; Program 14-114; Program 14-115; Program 14-116;
Program 14-118; Program 14-119-A;
Program 14-119-B; Program 14-4;
Program 15-01-A; Program 15-01; Program 15-02;
Program 16-1; Program 16-2;
Program 18-3;
Program 20-05;
Program 24-2A; Program 24-3; Program 25-2;
Program 25-3; Program 25-4; Program 25-5;
Program 26-03; Program 26-04;
Program 28-1A; Program 28-1C; Program 28-2;

ADin

Program 4-01; Program 4-02; Program 4-05

Program 06-01; Program 06-02; Program 06-04

Program 07-3

Program 09-1f

Program 11-1; Program 11-2

Program 12-5

Program 19-03

Program 20-05; Program 20-06

Program 21-2 Program 21-4 Program 21-6

Program 25-3; Program 25-4; Program 25-5

Program 28-1A; Program 28-1C; Program 28-2

Branch

Program 06-01; Program 06-02; Program 06-03 - Interrupt Version; Program 06-03

Program 06-04

Program 09-3

Program 10-2

Program 16-3

Program 17-2; Program 17-3z

Program 26-03; Program 26-04

Program 28-2

Button

Program 04-04

DTMFOut

Program 16-3

FreqOut

Program 16-3

I2Cin / I2Cout

Program 18-1; Program 18-2; Program 18-2A; Program 18-2B; Program 18-3

Program 28-1B; Program 28-1C; Program 28-2

ISRASM

Program 15-01-A; Program 15-01; Program 15-02

Program 16-2

Program 19-03

Program 25-2; Program 25-3; Program 25-4; Program 25-5

LCDRead / LCDWrite
Program 5-01; Program 5-02; Program 5-03; Program 5-04; Program 5-05
Program 06-04
Program 27-3
Program 26-04

OWin / OWout
Program 12-1; Program 12-2; Program 12-3; Program 12-5; Program 12-6
Program 20-05

PulseIn
Program 22-1; Program 22-2; Program 22-3; Program 22-4;
Program 24-3; Program 24-4; Program 24-5; Program 24-6

PulsOut
Program 16-3
Program 18-4; Program 18-5
Program 23-2; Program 23-3; Program 23-4

PWM
Program 16-3

ReadDM
Program 06-04

Serin / SerOut
Program 09-1A; Program 09-1B; Program 09-1c; Program 09-1d; Program 09-1E; Program 09-1f;
Program 09-1g

ShiftIn / ShiftOut
Program 12-0; Program 12-4; Program 12-5; Program 12-6
Program 21-1; Program 21-2; Program 21-3; Program 21-4; Program 21-5; Program 21-6
Program 28-1D; Program 28-2

Sound
Program 16-3
Program 25-4; Program 25-5

SPMotor
Program 08-01

Xin / Xout

Program 20-01 Program 20-02; Program 20-03; Program 20-04; Program 20-05; Program 20-06

SetHSerial / HSerin / HSerOut

Program 06-01; Program 06-02; Program 06-03 - Interrupt Version; Program 06-03

Program 09-2A; Program 09-2B; Program 09-3

Program 10-2; Program 10-3; Program 10-5

Program 11-1

Program 12-1; Program 12-2; Program 12-3; Program 12-4; Program 12-5; Program 12-6

Program 14-01-B; Program 14-01-C; Program 14-01; Program 14-01A; Program 14-100; Program 14-101;
Program 14-102; Program 14-103; Program 14-104; Program 14-105; Program 14-106; Program 14-107;
Program 14-108; Program 14-109; Program 14-110; Program 14-111; Program 14-112; Program 14-113;
Program 14-114; Program 14-115; Program 14-116; Program 14-118; Program 14-119-A;
Program 14-119-B; Program 14-4

Program 17-1; Program 17-2; Program 17-3z

Program 18-1; Program 18-2; Program 18-2A; Program 18-2B; Program 18-3; Program 18-4; Program 18-5

Program 19-03

Program 20-02; Program 20-03; Program 20-05; Program 20-06; Program 20-07

Program 21-2; Program 21-4; Program 21-6

Program 22-1; Program 22-2; Program 22-3; Program 22-4

Program 23-2; Program 23-3; Program 23-4

Program 24-1; Program 24-2; Program 24-2A; Program 24-3; Program 24-5; Program 24-6

Program 25-1; Program 25-2; Program 25-3; Program 25-4; Program 25-5

Program 27-0; Program 27-1; Program 27-2; Program 27-3

Program 28-1A; Program 28-1B; Program 28-1C; Program 28-1D; Program 28-2

HSerStat

Program 09-2A; Program 09-3

Program 23-4

Program 28-2

HPWM

Program 19-01; Program 19-02

Program 24-2; Program 24-2A; Program 24-3; Program 24-5; Program 24-6

Program 26-02; Program 26-03; Program 26-04

SetPullups

Program 25-1; Program 25-2; Program 25-3; Program 25-4; Program 25-5

Interrupts Handled in MBasic

Program 06-03 - Interrupt Version; Program 06-04

Program 10-1; Program 10-1A; Program 10-1B; Program 10-2; Program 10-4; Program 10-5

Program 10-6

Program 17-3z

Program 25-1

Program 26-03; Program 26-04

SetHSerial / HSerOut / HSerIn

Program 06-01; Program 06-02; Program 06-03 - Interrupt Version; Program 06-03;

Program 09-2A; Program 09-2B; Program 09-3;

Program 10-2; Program 10-3; Program 10-5;

Program 11-1;

Program 12-1; Program 12-2; Program 12-3; Program 12-4; Program 12-5; Program 12-6;

Program 14-01-B; Program 14-01-C; Program 14-01; Program 14-01A; Program 14-100;

Program 14-101; Program 14-102; Program 14-103; Program 14-104; Program 14-105; Program 14-106; Program 14-107;

Program 14-108; Program 14-109; Program 14-110; Program 14-111; Program 14-112; Program 14-113; Program 14-114;

Program 14-115; Program 14-116; Program 14-118; Program 14-119-A; Program 14-119-B; Program 14-4;

Program 17-1; Program 17-2; Program 17-3z;

Program 18-1; Program 18-2; Program 18-2A; Program 18-2B; Program 18-3; Program 18-4;

Program 18-5;

Program 19-03;

Program 20-02; Program 20-03; Program 20-05; Program 20-06; Program 20-07;

Program 21-2; Program 21-4; Program 21-6;

Program 22-1; Program 22-2; Program 22-3; Program 22-4;

Program 23-2; Program 23-3; Program 23-4;

Program 24-1; Program 24-2; Program 24-2A; Program 24-3; Program 24-5; Program 24-6;

Program 25-1; Program 25-2; Program 25-3; Program 25-4; Program 25-5;

Program 27-0; Program 27-1; Program 27-2; Program 27-3;

Program 28-1A; Program 28-1B; Program 28-1C; Program 28-1D; Program 28-2;

About the Author

Jack R. Smith

For nearly 30 years, Mr. Smith worked in the area of mixed legal-technical telecommunications; bringing both electrical engineering and legal background to bear on problems. Involved with first experimental cellular radio systems in the United States, he subsequently worked with start-up cellular and personal communications system radio operators in the United States, Europe and Asia. He is a co-founder of the telecommunications consulting company TeleworX and has shaped its radio propagation and network design software.

He received a BSEE Degree from Wayne State University, Detroit MI in 1968, and a Juris Doctor degree, also from Wayne State University, in 1976. He is a long time amateur radio operator, holding the callsign K8ZOA and is active on digital modes.

Index

Numbers and Symbols

#Include 738
$00 bytes 740
// returns the integer remainder 722
/ operator returns the integer quotient 722
1-wire 231, 389
  bytes are transmitted and received from lowest address to highest address 235
  CRC (cyclic redundancy code) 235
  family code 234
  family number 235
  global address 238
  global addressing 237
  MBasic currently supports only the initial 14 kb/s speed 1-wire protocol 234
  permissible mode values 234
  presence pulse 232
  reset 238
  serial number 234, 235
1-wire bus 240
  selectively read 240
24LC16B 392, 393
24LC256 389
2840 development board 8, 151, 154, 202, 212, 217
  how the PC is to be connected to the 159
2840 prototype board 554
2N7000 142
7-segment LED driver circuit 218
7805 699

A

A/D reading 720
A/D step 695
A/D units 703
absolute pressure 693, 704
accuracy 212
active filters 511
active low-pass filter 511
AC power
  optically coupled FET 542
  warning statement 542
addlw 434
address byte 708

address pins 699
ADIN 218, 729
ADin 703, 756
ADIN16 218, 729, 733
All Units Off 741
analog-to-digital 337
analog-to-digital conversion 211, 335, 702
  8-bit A/D conversion 214
analog-to-digital converter (ADC) 6, 211, 336, 408
  accuracy 212
  ADIN 216
  clock constant selection 215
    internal RC oscillator 216
  errors 212
    differential linearity error 212
    errors in the reference voltages 213
    gain error 213
    integral linearity error 212
    internal 213
    noise 213
    offset error 213
    quantization error 213
    settling time 213
    source impedance 213
  input function
    ADIN 216, 218
    ADIN16 216, 218
  input pin constant selection 216
  internal selection switches 214
    select internal or external negative reference voltage 214
    select internal or external positive reference voltage 214
    which of several possible pins connects to the A/D converter module 214
  LT1634CCZ-4.096 voltage reference 213
  MCP601 op-amp 213
  minimum time to convert 215
    acquisition time, TACQ 215, 216
    analog-to-digital conversion time, TAD 215, 216
  offset voltage 213
  resolution 212
    quantization error 212

result 214
  left justified 214
  right-justified 214
  signal to noise ratio 337
  two voltage references 212
    negative reference 212
analog voltages 702
Array Variable Usage 734
ASK (amplitude shift keying) 661
assembler code
  _@RateDelay Macro 294
    decfsz 295
    delay function 294
    GoTo 295
    NOP 295
  bank selection 291
    ASM 293
      _@bank 293
    BankSel 293
      illegal label 293
      MPASM assembler 293
      usage changed slightly beginning with MBasic version 5.3.0.0 293
      usage in MBasic before version 5.3.0.0 293
      version 5.3.0.0 GPASM assembler 293
    determine which bank MBasic has placed a variable 291
    STACK command 292
    stack size is specified in increments of 4 bytes 292
    variable assignment procedure 292
  page selection 294
    long calls 294
    long jumps 294
  word variable 303
    high byte 303
    low byte 303
assembler instructions
  ADDLW 267
  ADDWF 267
  field 266
    b 266
    d 266
    f 266
    k 266

PC 266
PD 266
TO 266
W 266
x 266
movlw 260
movwf 261, 264
rotate
    RRF 267
    RRL 267
SUBLW 267
SUBWF 267
assembler interrupt service routine 317
    advantages over MBasic 317
    interrupts 317
        accommodating multiple interrupts 327
        clear the interrupt flag
            ASM {...} construction 321
            in-line assembler macros 321
            INTCON 321
            INTE 321
            INTF 321, 325
        context restoration 322
        context saving 322
        context switching 322
        control bit
            "enable" bit 318
            disabled 318
            enabled 318
            higher level interrupt control bits 319
            one dedicated control bit 318
        control register
            INTCON 319
        device control 318
        enable 325
        enable and disable interrupts 320
            _@bcf 320
            _@bsf 320
            ASM {...} assembler code to set or clear enable bits 320
            Banksel 326
            banksel 327
            bcf 320
            bsf 320, 326
            cannot enable or disable the interrupt using MBasic's Enable and Disable functions 320
            decfsz 326
            Ext_H2L 320
            GoTo 326
            in-line assembler to set or clear enable and setup bits 320
            Intcon.bit4 320
            Inte 320
            IntEdg 320
            ISRASM 320, 331
            OnInterrupt 320
            Option_Reg 320
            SetExtInt 320
            SetTmr0 320
        set or clear the setup and enable bits from MBasic 320
        using MBasic commands 320
    enable global and peripheral interrupts 332
    external (RB0) 325, 332
        btfsc 332
    global interrupts (GIE) 318, 334
    global interrupt enable bit 318, 319, 325, 326
    Intcon 321
    Inte 321
    interrupt event 321
        flag bit 322
        FSR 322
        HSerIn 322
        HSerOut 322
        interrupt vector 322
        latency 321
        PCLath 322
        RCIE 322
        Status 322
        TXIE 322
        W 322
    interrupt flag 332
        intf 333
    interrupt flag is cleared at processor reset 321
    INTF 323
    Intf bit 326
    ISRASM 318, 321, 322
        automatically performs context switching 322
        latency 331
        reentrant 322
    OnInterrupt 317
    Option_Reg 321
    peripheral family interrupt enable bit 319
    peripheral interrupts 325
    peripheral interrupts (PEIE) 318
    peripheral interrupt bit (PEIE) 325
    peripheral interrupt enable (PIE) registers 319
        PIE1 319
        PIE1 bit 0 319
        PIE2 319
    peripheral interrupt family 318
    peripheral interrupt registers (PIR) 319
        PIR1 319
        PIR1 bit 0 319
        PIR2 319
    PIE and PIR registers
        register addresses align 319
    status bit
        status flag 318
    timers 1 and 2 332
        fire their interrupts simultaneously 334
        in the order established by the test/jump statements 334
    members of the peripheral interrupt family 332
    timer 1 319, 327, 331, 332, 333
        banksel 334
        btfsc 332
        enable bit 319
        interrupt flag 319
        make timer 1 reset 334
        Tmr1IF 333
    timer 2 328, 331, 332, 333
        banksel 333
        btfsc 332
        interrupt flag 333
        PIR1 333
        postscaler 333
        prescaler 333
        shut timer 2 down by clearing the Tmr2On 333
        T2Con 333
        Tmr2 333
        Tmr2If 333
        Tmr2On 333
        turn timer 2 on 333
    ISRASM 317, 323, 325, 326, 327, 332, 333, 334
        @bank 322
        banksel 322, 323
        disable interrupts 331, 332
        End 326
        may place the ISRASM code anywhere 326
    ISRASM code 322
assembler language programming 260
    address 261
    banking 264
        2-bit bank address 264
        7-bit opcode relative address 264
        assembler macro, BankSel 264
        Banksel 264
        bank address bits 264
        destination code 264
        file addresses 264
        macro 264
            advantages 264
            BankSel 264
        opcode 264, 265
            mov 264
            movwf 264
        special function register bits RB1 and RP0 264
        status register 264
    comparing MBasic, assembler and machine instructions 260
    definition 260
    file location 261
    opcodes 267
        ADDLW 267
        ADDWF 268
        ANDWF 269
        AND function 268, 269
            "mask off" bits 268
            AND function truth table 268

logical AND 268
    test multiple bits to see if all are zero 269
BCF 269
BSF 269
BTFSC 270
BTFSS 270
CALL 271
    label 271
CLRF 271
CLRW 271
CLRWDT 271
COMF 272
DECF 272
DECFSZ 272
GOTO 273
INCF 273
INCFSZ 273
IORLW 274
IORWF 274
MOVF 275
MOVLW 275
MOVWF 276
NOP 276
OR function truth table 274, 275
RETFIE 276
RETLW 276
    assembler equivalent of MBasic's ByteTable 276
    look-up table 276
RETURN 277
RLF 277
RRF 277
SLEEP 278
SUBLW 278
SUBWF 278
SWAPF 279
XORLW 279
XORWF 279
PIC's instruction set 260
special purpose register
    Status register 266
        !Borrow 267
        !PD 266
        !TO 266
        C Carry/!Borrow 267
        DC Digit carry/!Borrow 267
        IRP 266
        Z 266
stepwise substitution 413
storage register (file) 261
value 261
assembler terminology
    address 261
    file 261
    literal 261
    move 260, 261
    opcode 261
        movlw 261
        movwf 261
    register 261
    working register is called w 261

AT24C512 700
AT24C512 serial EEPROM 705
automatically convert the file to the new version 730
automatic scaling 3-digit digital voltmeter 218
    7-segment LED driver circuit 218

B
banksel 733
Bank 0 RAM 732
barometric pressure 691, 692
barometric pressure adjustment factor 693
bar codes 595
    bars 595, 601, 618, 619, 626
        value 596
    bar value 596, 597
    byte table 620
    character 595
        complementary 598
        left side character 599
        right side character 599
    check digit 597, 627, 628
        calculated with the following algorithm 627
    code 39 595
    code 3 of 9 595, 596, 597, 600, 601, 606, 614, 615, 616, 617, 619, 622
    compute the check digit 625
    EAN-13 630, 631
        check digit computation changes 631
        left-of-center digits 630
        odd parity 630
        parity computations 630
    element 595, 596, 597, 601, 606, 607
        center guard 601
        center guard pattern 597, 598
        elemental lengths 623
            center character 623
            start and stop characters 623
        element length 617
        guard patterns 597
        left and right parts 597
    element duration
        bank 0 611
        bank 1 611
        bank 2 611
        clrf 611
        compiler instruction 611
            STACK = 10 611
        do not preload the counter 614
        FSR 611, 612
        INDF 611, 612
        indirect memory addressing 611
        interrupts
            disable 612
        IRP 611
        israsm 612
        PEEK 611

POKE 611
    timer 1 606, 607, 608, 610, 612
        prescaler 608, 610
        Timer1.Byte1 608
        Timer1H 608
    timer 1 interrupt flag 612
        Tmr1IF 612
element widths 606, 619
    measure the width of a bar code element 606
    three are wide 596
    width of 1, 2, 3 or 4 units 597
integer division 625
inter-character space 596
    modulus 596
interrupt 617
interrupts
    BankSel 606
    GIE 605
    HSerIn 605
    HSerOut 605
    incf 606
    interrupt on PortB 605
    RBIE 605
    RBIF 605, 606
interrupt handler code 613
    interrupt flag 613
        isram 614
        RBIF 613
        Tmr1H 614
        Tmr1IF 613, 614
        Tmr1on 614
    timer 1 613, 614
item number 597
left-of-center 626
manufacturer ID number 597
    Uniform Code Council 597
MBasic version 5.3.0.0
    memory model 611
module 595
modulus 597, 615, 617, 622, 623, 625
        modulus width 596, 619
number system character 597
quiet zone 595, 596, 597, 612
RAM memory 624
right-of-center 626
rounding 625
rules 596
scan direction 623
    directionless scanning 623
    left-to-right 623, 624
    right-to-left 623, 624, 628
spaces 595, 601, 618, 619, 626
    value 596
space value 596, 597
start character 596
stop character 596
timer 1 615
transition
    interrupt 602, 603
        Disable 603

enable 603
port mismatch 603
RBint 602
RbInt 603
Universal Product Code, revision A
(UPC-A) 595, 596, 597, 598,
601, 603, 622, 623, 624, 626,
627, 630, 631
EAN-13 630
start, stop and center symbols 622
wand 595, 599, 601, 602, 606, 607,
612, 624
HBCC-0500 600
how I identified the correct pin connection 599
wand-lift 612
wand-timeout 612
wand reverts to white idle 606
weak pull up resistors 603
bar graph display 80
Basic Stamp 18
backwards compatibility with the 18
Basic Stamp compatibility 738
baudmode constant 732
BCD2BIN 719
BCD2DEC 738
BCD2DEC is BCD2BIN 738
BIN2BCD 720
Bin2BCD 116, 117
BIN2BCD & BCD2BIN 755
BIN2BCD and BCD2DEC 738
binary coded decimal (BCD) 116, 117,
718, 719, 720
packed BCD 117
binary coded decimal (BCD) to binary
719
binary weighted tree 633
bit order 742
bolt-in assembler functions 295
ASM 297
routines 296
Stack = 10 compiler directive 297
using an in-line macro or _@macro
297
stepwise refinement 295
bolt-in replacements 281
Branch 756
BSR 453
BSR System X10 453
buffers 176
Butterworth low-pass filter 511
Button 756
byte length random 740

C

capacitance 691
Capture16L2H 207
Capture4L2H 207
capture mode 203
capture function is executed independently 207
capture interrupt flag 207

timer 1 203
using the capture interrupt 203
carriage return 445, 715, 716
CCP1 203, 208
CCP1Con 578, 743
CCP1INT 203
CCP1Int 208, 210
CCP2 203, 208
CCP2Con 578, 743
CCP2INT 203
CCP2Int 208
CCPR1 203
CCPR2 203
CCPRxCON 576
CCPRxL 576
CCPRxL:CCPRxCON 576
CCP module 208
central processing unit (CPU) 1
clamp 574
clamping diode 35
clamping diode should be a fast
switching device, such as a
Schottky diode 35
clamp diode 110
Clear 742
clear function
must precede SetHSerial 158
clock (SCL) 390
clock frequency must be entered in Hz,
not MHz 730
CMOS compatible 571
command byte 708
command bytes 708
common cathode displays 108
CompareSpecial 210
compare mode 203, 208
CCP1Int 210
CompareSpecial 210
SetCompare 208, 210
timer 1 203
timer 2 203
compiler directive 731, 732
Compiler Optimize Size/Speed 734
compiler vs. interpreter 25
complementary metal oxide semiconductor (CMOS) configuration
28
ComTest 152
Configuration dialog box 741
constants 16
continuation symbol 15
control byte 708
conversion clock 702
cosine function 421, 425
cosine transition current waveform 424
counters 194
CRC (cyclic redundancy code) 234
CTS 163
CTS and received data 163
CY7C128A 415

D

Darlington transistor 39, 110, 135
turn off time 40
data (SDA) 390
data logger 691
data memory 1, 2, 261
banks 261
files 261
opcode's destination bit 262
registers 261
register files
general-purpose registers 261, 262,
263
accesses 263
special function register 261, 262
CCP1CON 263
INTCON 263
SSPSTAT 263
unimplemented address 261
working or W register 262
data RAM 6
DCE (data communications equipment)
159
DC circuits 542
high side switching 542
low side switching 542
PWM 542
DC motor 567
comparator 571
MCP601 572
op-amp 571
tachometer 571
control algorithm 572
bang-bang 573, 582, 584, 585
dynamic changes 585
speed change commands 585
static speed control 585
proportional, integral and derivative
(PID) 573
PID controller 573
proportional (P) 573
bang-bang 573
hardware pulse width modulation 573
residual error 573
proportional and integral (PI) and
proportional and derivative
(PD) 573
derivative of the error 573
error 573
integral of the error 573
PD controllers 573
PD systems 573
PI controller 573
control system 568
algorithm and control mechanism
568
error signal 568
control input (target) 568
control loop 568
feedback 568
tachometer 568
summing point 568

compares feedback signal with
target 568
error (target minus feedback) 568
error signal 568
converting the sinusoidal tachometer
output to a digital logic level
signal 570
MBasic
PulsIn 570, 571
PulsIn's syntax 571
unit time step 570
duty cycle 575
feedback 567, 568
MBasic
PulsIn 572
sinusoidal signal
limiter 571
squarer 571
comparator 571
squarer circuit 571
tachometer 571
tachometer output frequency
period 569
DC motor control programs 573
bang-bang control loop 582
commutator 574
hardware pulse width modulation
(PWM) 575, 576, 581
duty cycle 575, 576, 582
HPWM 575, 576
MBasic's software PWM procedure
581
PWM output
duration of the high output 576
formula 576
PWM period 576
repetition period 575
set and forget 581
hardware PWM modules 574, 576,
577, 578
associated control register 578
CCP1Con 578
CCP2Con 578
duty cycle 577, 580
how to turn it off 578
HPWM 577, 578, 581
oscillator clock period 578
period 577
PR2 576
formula relating period to these
two values 576
prescaler divide value for timer
2 576
formula relating period to these
two values 576
prescale control bits
T2CKPS0 576
T2CKPS1 576
T2CON 576
timer 2 576
prescale 576
tachometer 579, 581, 582

PulsIn 581, 582
spurious pulses
PulsIn 581, 582
spurious transition 581
DC motor speed
AC tachometer 569
condition a DC or an AC tachometer's
output to be PIC compatible
569
DC tachometer 569
commutator noise 569
low pass filter 569
duty cycle 584
dynamic braking 592
error 572
oscillating errors 587
static speed error 588
gain in the feedback loop 586
oscillating errors 587
optical or magnetic sensors 569
over-damped 587
overshoot 587
proportional control 588
hardware PWM modules 590
reversible PWM controlled circuit
590
SN754410 588, 592, 593
PWM module 593
sensorless systems 569
systems not employing tachometers
569
tachometer output frequency 569,
570
determining the tachometer constant 570
GR 1531AB Strobotac 570
relationship between AC generator
(or motor) poles, frequency and
RPM 570
Strobotac 584
unit steps 572
DC permanent magnet motor 567
AC tachometer 567
brush-type 567
built-in tachometer 567
pulse-width modulated (PWM) variable speed 567
tachometer feedback 567
Dec 736
DEC2BCD 738
DEC2BCD is BIN2BCD 738
decfsz 434
decoding variable length codes 633
default stack size is 17 730
detent torque 121
development boards 13
caution in connecting devices to the
+5 V regulated supply in Basic
Micro's development boards
71
in-circuit programming 14
models 0818 and 2840 23

model 0818 for 8- and 18-pin DIP 13
model 2840 for 28- and 40-pin DIP
13
dial tone 371
dielectric constant 691
digital-to-analog 337
digital-to-analog conversion (DAC)
335, 336, 338, 339, 340, 341,
342, 348, 349, 352, 353, 574
alternative analog output solutions
Pauseus 357
pulse width modulation (PWM)
352
PulsOut 355, 357
Sound 355, 356
invokes 356
Branch 357
digitize a time variant signal 337
direct digital synthesis (DDS) 335,
338
jump table 348, 352
arbitrary waveform 352
literal 348
retlw 348, 352
output frequency 342
phase accumulator 342, 345, 347,
348
phase advance 349
phase step 342, 348
phase step interval 347
relationship amongst the various
parameters 342
samples per second 342
sample rate 348
error analysis 337
error sources
AC errors in the low-pass filter 337
errors in the low-pass filter 337
linearity error 337
output buffer amplifier error 337
quantization error 337
reference voltage error 337
formula developed in Chapter 11 for
calculating the ADC step size
applies to a DAC 335
GoTo 357
integer conversion 344
floating point 344
phase accumulator 344
interrupts 348
GIE 348
ISRASM 348, 349
bit clear 348
bit set 348
PEIE 348
timer 2 347, 348
constants 347
postscaler 347
timer 2 is enabled 348
low-pass filter 338, 356
10 KHz low-pass 340
multiple feedback design 340

1 KHz low-pass 339
Bessel 338
Butterworth 338
active RC low-pass filter 340
Cauer (elliptical) 338
Chebyshev 338
linear phase 338
LT1634CCZ-4.096 voltage reference 336
LTC1062 339, 340, 349, 356
Butterworth active RC low-pass filter 340
outputs 340
ratio pin 339
LTC 1062 switched capacitor low-pass filter 338
multiple feedback active RC filter 338
Nyquist's sampling rate 338
Nyquist's sampling theorem 337
highest frequency that may be generated without error is one-half the sampling rate 337
programmable attenuator 336
pulse width modulation (PWM) 352, 353, 356
quantization error 336
relationship between PWM and FreqOut 355
resolution 336
signal-to-noise ratio (SNR) 337
Nyquist sampling range 337
relationship between the maximum signal to noise ratio and the number of bits in the sample 337
SNRdb 337
spectrum analysis plots 349
TLC7528 339
chip select pin 339
current mode 339
current-to-voltage converter 339
DAC A/B pin 339
voltage mode 339
write enable pin 339
TLC7528 parallel loading dual 8-bit 336, 337
quantization noise 337
digital input 53
current that flows in or out of an input pin is due to leakage 57
hardware debouncing 58
integrated circuit debouncers 58
simple circuit 58
input pins represent a high input impedance 57
internal pull-up resistors 57
RB0...RB7 57
pull-up resistors 57, 58
reading a keypad 63
matrix switches 63
program to test keypad reading 64

pull-down 63
sealing current 58
software debouncing 60
switch 58
built-in switch debounce procedure Button 60
illustrates how Button may be used 61
debounce the switch 58
mechanical switch operates, the same contact bounce 58
software debouncing 60
switch bounce 58
wetting current 58
digital potentiometers 487
active low-pass filter 511
audio taper 506
Butterworth filter 512
Butterworth low-pass filter 511
Sallen-Key 511
commanding an MCP41010 by the RS-232 serial port 493
convert the input string to a numerical value 496
line feed 495
command bytes 491
dummy command 491
set the tap setting 491
shutdown mode 491
command selection bits 490
control interfaces 487
1-wire 487
three-wire (SPI) 487, 489, 491
MCP41xxx/42xxx family 487
Microchip's SI/SO design 500
cut-off frequency 512
decibel 506, 507
change in sound level (dB) 507
perceived difference 507
power ratio formula 507
minimum difference that is perceptible as a change in volume 507
DS1302 489, 490, 491
LSB first 491
LSBFIRST 491
LSBPRE 491
good and bad points of digital versus mechanical 488
tapered response 506
log taper 506
MCP41010 489, 490, 491, 492, 493, 495, 496, 497, 506, 507, 508, 509, 510
1 dB step 509
accuracy of the relative steps 491
blocking capacitors 508
byte table 509, 510
connect an audio signal 508
convert the input string to a numerical value 496
digital-to-analog converter 492

linearity 492
line feed 495
maximum and minimum voltage requirements 507
multiple dual-pot devices 498
relative accuracy error 491
right justify 496
to convert a numerical character to its numeric value 496
voltage-output digital-to-analog converter 492
volume level 506
MCP41xxx 488, 491, 499
MCP41xxx/42xxx devices 487, 488
residual resistance 487
resistance between terminals A, B and the wiper 488
resistance of the "wiper" connection 487
MCP42010 498, 501, 512, 513
SPI 517
MCP42100 513
MCP42xxx 488, 499, 500
command byte 499
CS 499, 500
data in reverse of the device connections 500
instruction byte 499
shift in (SI) pins 499
shift out (SO) pins 499
MCP42xxx/41xxx 490, 491, 506, 507
cannot be operated with a voltage below VSS or above VDD 507
chip select pin 491
data
before the clock pulse 491
MSB first 491
Microchip's SI/SO 500
multiple dual-pot devices 498
output voltage is Dn multiplied by a constant factor 492
potentiometer 487
potentiometer selection bits 491
rheostat 487
statement
ShiftOut 491
MSBPRE 491
tapered response 506
variable resistance 487
variable resistor 487
variable voltage divider 487
voltage divider 492
volume control 487
digital signal levels 53
input logic level specifications 54
input voltage against the output value 54
logical high 53
logical low 53
RC3/RC4 Schmitt 54
Schmitt trigger inputs 54

correctly read an input signal in the presence of large noise voltages 57
hysteresis 56
  noise rejection 56
special Schmitt trigger inputs 54
threshold voltage 53
TTL 54
TTL level 53
undefined region 53
VIH -- The minimum voltage on an input pin that will be read as a logical high 53
VIL -- The maximum voltage on an input pin that will be read as a logical low 53
diode thyristor (diac) 543
direct digital synthesis (DDS) 338, 340, 341, 342, 345, 348, 349
  clock jitter 345
  definition 340
  external digital-to-analog conversion (DAC) 338
  FREQOUT 338
  FreqOut 356
  hardware PWM 339
  minimum frequency step 348
  numerically controlled oscillator 340
  phase advance 341
  pulse width modulated (PWM) 338
  PULSOUT 338
  PWM procedure 338
    does not use the PIC's hardware PWM 339
  SOUND 338
Disable 100, 210
DS12B20 235
DS1302 709, 718
DS1302 real-time clock 243, 244, 246, 247, 248, 249, 250, 251, 252, 256, 257
  accuracy 244
  aliased 247
  binary coded decimal (BCD) 247, 252
    BCD2BIN 252
  charger control register 250
  command byte 251
    clock/calendar burst read 251
    ShiftIn 251
    ShiftOut 251
  command byte structure 247
  exchange data with 247
    bytes of data 247
    command byte 247
  HserOut 252
  initialize 247
  internal charger 247
  LSB-first bit order 249
  LSBFIRST 249
    ShiftIn 249
    ShiftOut 249

LSBPRE 257
  ShiftIn 257
  ShiftOut 257
master write protect bit 248
register structure 246
  clock burst register 246
  control register 246
  date (day of month) 246
  day of week 246
  hours 246
  minutes 246
  month 246
  seconds 246
  trickle charger control 246
  year 246
reset line 244
reset pin 251
reverse current isolation diodes 250
rising edge 244
serial clocked data 258
  falling edge of the clock pulse 255
  least significant bit first 258
  MCP41010 digital potentiometer 258
  most significant bit first 258
  rising edge of the clock pulse 255
  rising or falling clock and bit order 258
  ShiftIn 258
  ShiftOut 258
  ShiftOut/ShiftIn 258
  valid on the falling edge 258
  valid on the rising edge 258
ShiftIn 251
ShiftOut 249, 251, 257
  modes 249
    FASTLSBPRE 249
    FASTLSPOST 249
    FASTMSBPOST 249
    FASTMSBPRE 249
    LSBPOST 249
    LSBPRE{LSBFIRST} 249
    MSBPOST 249
    MSBPRE{MSBFIRST} 249
    SLOWLSBPOST 249
    SLOWLSBPRE 249
    SLOWMSBPOST 249
    SLOWMSBPRE 249
supercapacitors 251
  leakage current 251
three-wire serial connection 243
  ShiftIn 243
  ShiftOut 243
trickle charger control bits (TCS) 250
writing data 257
  ShiftIn 257
DS1624 700, 708, 709
DS18B20 709
DS18B20 temperature sensor 231, 232, 235, 236, 237, 238, 239, 240, 242, 252, 470
  18B20 234

accuracy 236
command 237
  convert temperature 237, 238
  copy scratchpad 237
  read power supply 237
  read scratchpad 237
  recall EEPROM 237
  write scratchpad 237
command values 237
configuration register 236
  high (TH) temperature alarm 237
  low (TL) temperature alarm 237
convert temperature command 238
convert the floating point Celsius temperature to Fahrenheit 239
global address 238
increased resolution means longer conversion time 236
one-wire protocol 231
  "presence" pulse 232
  64-bit serial number 233
    addressed either individually or globally 233
  master writes to the slave 233
    "issuing a time slot" 233
    "time slot" coding 233
  permissible mode values 234
  reset pulse 232
  serial number 234
scratchpad memory 238
scratchpad memory contents 237
selectively address 242
  selective address version 242
  universal address version 242
sign bit 238
variable resolution 236
DS18S20 231
DTE (data terminal equipment) 159
DTMFOut 731, 756
DTMFOut2 731
dual-tone multi-frequency (DTMF) 360
  ASCII collating sequence 383
  inter-office 360
duty cycle 423, 424
dynamic braking 592

E
EEPROM memory 3, 101, 102, 103, 389, 393, 394, 398, 403, 407, 408, 695, 715, 717
  24AA16 390
  24FC515 394, 401
  24LC16B 390, 393, 394, 395, 396, 399
    block select bits do not change from 000 during the multi-byte I2Cout 395
    cannot span a block boundary 395
    chip address bits 394
    chip address pins 390
  24LC515 399
    address bank selection 401

bank selection 399
  control byte 401
  I2Cout 401
  nonconfigurable chip select 399
AT24C256 393
AT24C512 393, 401, 403, 406, 407, 408
  I2Cout 407
  linear 64Kbyte address space 401
clock (SCL) 390
control byte 392
  family type 392
    10-bit control sequence 393
  read/write (R/W) bit 393
    Bit 0 of the control byte for address length information 393
    MBASIC automatically inserts the correct R/W bit 393
data (SDA) 390
extend PIC's internal RAM by swapping data between RAM and EEPROM 403
external 691
Group A 392
Inter IC bus (I2C) 392, 393, 403
  I2C support functions 392
    I2Cin 392, 395, 408
    I2Cout 392, 395, 408
MBasic
  Random function 407
multiple serial EEPROM
  24LC256 396, 398
    chip address 396, 397
    chip boundaries 398
    control byte 398
    I2Cin 398
    I2Cout 398
    memory is organized 397
  no delay is required after reading the EEPROM 395
  paralleling multiple serial memory chips 395
  serial EEPROM 389, 408
    chip address bits 393
    standard identification system for generic serial EEPROM chips 390
  SRAM 408
  static RAM 409
  time to write to memory 394
  write the strings to EEPROM 101
  writing to EEPROM 101
    WriteDM 101, 102
EEPROM read 717
EIA/TIA RS-232E 151
EnableHSerial 87, 157, 176, 731
end-of-line character 166
end-of-line command 740
EPROM 3
EPROM and EEPROM 3
  one time programmable (OTP) 3
Epson/Seiko's SED1278 67

Equ 267
explicit end-of-line definition 167
external EEPROM 691, 695, 711
external memory
  24LC16B 392
  24LC256 389
  Inter IC bus (I2C) 389
    24LC256 389
  parallel access memory 408

F
FADD 729
family code 234
  CRC (cyclic redundancy code) 234
FASTLSBPRE 732
FASTLSPOST 732
FASTMSBPOST 732
FASTMSBPRE 732
fast switching--sound from a PIC 50
  driving the speaker directly from a PIC 50
  emitter follower to drive the 3.2 ohm speaker 51
  self-contained sounder, such as the Sonalert 50
FDIV 729
flash memory 3
FLOAT 729
Floating Point Math 755
FLOATTABLE 729
floor function 739
flow control 160, 164
FMUL 729
FNEG 729
Fourier transform (FFT) 408
FractionalMultiply 729
Freescale Semiconductor, Inc. 693
FreqOut 731, 756
FSUB 729
function arguments 731
function arguments may now be up to 32 bits 731

G
GetTimer1 733
global addressing 237
global variables 730
GoSub 730
GoTo 434, 730
GPASM 413, 432, 733
Gray code 94

H
half-step mode 731
hardware pulse width modulation (PWM) 420, 423
hardware PWM module's period 423
hardware serial input/output code module 87
  EnableHSerial 87
  HserOut 87
  SetHSerial 87

Harvard architecture model 2, 261
high-side digit switches 109
high power bipolar low side switching 39
  why VCE(SAT) is so poor 39
high power MOSFET low side switching 40
  high side switching 42
    small PNP switch 42
  intelligent power MOSFET switch 41
  IPS021 41
    turn-on and turn-off speeds 41
high side digit selection switch 108
high side segment drivers 108
high side switching 31, 42
  adding an 2N4401 NPN transistor to control the base of Q1 42
  central problem in high side switching with a PNP transistor or a positive MOSFET (PMOS) 42
high power high side switching 43
  auxiliary n-channel gate control device 43
  integrated high side MOSFET switch 43
  IPS511 43
    determine the status of the IPS511 we would implement the following pseudo-code 45
    status pin 43
  small PNP switch 42
    calculate the base resistor 42
high voltage programming (HVP) 23
Hitachi HD44780 controller/driver chip 67
HM-1500 692
holding torque 121
hookswitch 371
House Codes 741
how to turn off the hardware PWM module 743
HPWM 424, 576, 581, 731, 758
HPwm 743
HSerIn 151, 155, 158, 159, 175, 176, 210, 714, 731
  buffers 176
HSerIn/HserOut 157
HSerIn and HSerOut 742
HSerOut 158, 175, 176, 210, 617, 714, 731
  buffers 176
HserOut 87, 151, 252, 742
HSerStat 158, 159, 175, 176, 714, 715, 717, 743, 758
  undocumented command 158, 714
humidity sensors 691
HWPM 352
hybrid 371
HyperTerminal 87
Hyperterminal 152

I

I2Cin 389, 697, 708
I2Cin/I2Cout control byte 708
I2Cin / I2Cout 756
I2Cout 389, 697, 708
I2C device 696
in-circuit programming 14
in-line assembler 281
incremental encoders 92
inductive load 567
inductive time constant 128
InitLCD 90
InitLCD1 74
InitLCD2 74
input signal 194
input switches 86
  position encoders 91
  relative encoder 95
    reading a relative encoder 95
      change from earlier versions of
        MBasic 97
  rotary encoders 91
    absolute encoders 91
    code output 94
      difference between hexadecimal
        and Gray code 94
      Gray code 94
      straight hexadecimal binary code
        94
    construction 92
      mechanical encoders 92
      optical encoders 92
      quadrature encoder 93
      quadrature sensor 93
      relative encoders 91, 92
        incremental encoders 92
    shaft encoders 91
InStr 168
instruction word length 3
INT 729
INTASM 733
integer division 625
integer quotient 722
integer remainder 722
intelligent power MOSFET switch 41
internal pull-up resistors 57
internal RC oscillator 216
interrupt 99, 317
  MBasic will not break execution of a
    statement to service an interrupt
    100
  RBINT type interrupt 100
  turn off additional interrupts 100
Interrupts 743
Interrupts Handled in MBasic 759
interrupt handler 99, 104
  "enable" the RBINT interrupt 100
  Disable 100
  Enable RBINT 100
  OnInterrupt 100
  Resume 100
  turn off additional interrupts 100
interrupt service routine (ISR) 207,

210, 317
Inter IC bus (I2C) 389
  fast mode 389
  high-speed mode 389
  I2C support functions 392
    I2Cin 392
    I2Cout 392
      uses its internal address counter
        392
  standard mode 389
    I2Cin 389
    I2Cout 389
IPS021 41
IPS511 43
  determine the status of the IPS511 we
    would implement the following
    pseudo-code 45
  status pin 43
IR filter 521, 522
IR receiver 517, 518, 519, 520, 521,
    522, 523, 524, 526
  burst 524
    carrier frequency 518
  common encoding standards 518
    Philips RC-5 Code 518
      biphase (Manchester) 518
    REC-80 518, 522, 524, 525, 532,
      534
      check byte 534
      data byte 534
      data organization sending order
        534
      error checking 534
      histogram 525
      inter-pulse dwell time 534
      protocol 519
      PulsIn 534
      space width modulated 518
      start pulse 534
    Sony (Serial Infrared Remote Con-
      trol System, SIRCS) 518
      LSB first 518
      pulse width coding 518
    typical REC-80 data organization
      sending order 519
      check bytes 519
      device codes 519
      function codes 519
  data pulse 522
  detects these bursts 518
  GP-series 522
  GP1U5 521
  GP1U58x 540
  GP1UD26/27/28 521
  idle condition 523
  IR filter 521
  MBasic
    order of precedence problems 535
    PulseIn 524
    PulsIn 523, 524
      unit steps 523
  Mitsubishi 534

start pulse 534
photodiode 520
phototransistor 520
refer to high and low, or positive
    pulses and negative pulses, it is
    with respect to the output of the
    IR receiver 521
start pulse 522
start signal 519
  TSOP-series 522
  TSOP1238 540
  TSOP12xx 520, 521
    automatic gain control (AGC) 520
    bandpass filter 520, 521
    optical filter 520
    photodiode 520
    unit intervals 525
    unit step intervals 524
      histogram 524, 525
IR remote control 517
  bursts of IR light 518
  carrier frequency 521
    if you don't know, easy to measure
      it 521
  MCP42010
    SPI 537
  photoreceptor 521
  phototransistor 521
  power-on/power-off code 519
  QSE114 521
  toggle bit 519
  width threshold 523
    histogram 523
isolated switching 31, 45, 62
  optically isolated MOSFET 49
    PS710A-1A is a high power MOS-
      FET optoisolator 49
    switching high currents in micro-
      second times 49
    undesired transients and oscilla-
      tions 49
  optical coupler 62
  optical isolated NPN switch 48
    low power optoisolator 48
    optical couplers 48
    optoisolators 48
  read a switch closure that cannot have
    a common ground with your
    PIC circuit 62
  relay switching 46
    contact bounce 48
    good things about relays 46
    minimum contact load 48
    not so good things about relays 46
    operate time 47
    pull-in time 47
    Reed relays 47
      for contact bounce 47
      turns on 47
      turn off time 47
    release time 47, 48
    snubbing diode 46

test the contact closure and oper-
ate/release time 46
voltage spike even a small inductive
load generates 46
switched through a relay 62
ISP-PRO 8, 10, 11, 14, 23
universal adapter 11
ISP-PRO programmer 10
ISRASM 741, 744, 756
interrupt 345
israsm 614

K

Key Codes 741

L

latency 208, 743
Latency in MBasic interrupts 743
LCD/configuration switch 88
continuation character 90
duplexing the switch and LCD 89
InitLCD 90
same pins to read four switch contacts
and output data to an LCD
module 88
LCDINIT 73
new in version 5.3.0.0 73
LCDInit 74, 731
LCDINIT1 731
LCDINIT2 731
LCDRead / LCDWrite 757
LCDWRITE 731
LCDWrite 73, 75
LCD initialization 481, 731
LCD modules 67
40 × 4 display 74
backlighting 68
electroluminescent 68
LED 68
much longer lived than electroluminescent devices, with typical
expected lifetimes well over
50,000 hours 69
two major disadvantages 68
reflective 68
connection to PIC 69
4-bit (nibble) transfer mode 71
calculate the LED current limiting
resistor's value 71
display contrast 70
enable (E) pin 71
read/write (R/W) pin 71
register select pin 70
crystal technology 68
film super twisted nematic (FSTN)
68
super twisted nematic (STN) 68
twisted nematic (TN) 68
custom characters 80
bar graph display 80
character generator memory
(CGRAM) 80, 81, 84

create custom characters 81
write these custom characters into
CGRAM 82
flashing the text off and on 78
font selection 79
extended font set 79
write the extended character 79
initialization constants 74
CLEAR 74
HOME 74
SCR 74
TWOLINE 74
initialization message with LCDWrite
73
initialize an LCD using the older
method 73
InitLCD1 74
InitLCD2 74
LCDINIT 73
new in version 5.3.0.0 73
LCDWrite 73, 75
LCD display sizes 67
character size 68
dot matrix size 68
numbers of characters 67
LCD environmental considerations
69
extended temperature LCDs 69, 70
memory layout for 16 × 2 and 20 × 2
displays 74
Off 79
physical connections between the PIC
and the LCD module 73
SCR 79
SCRBLK 79
SCRCUR 79
SCRCURBLK 79
ScrLeft 75, 77
ScrRAM 77
ScrRight 75, 77
selecting an LCD module 67
vacuum fluorescent display (VFD) 69
writing at a defined location 77
LED 32
calculate the series current limiting
resistor 32
connect two LEDs to one pin 33
constant voltage drop 32
MV5491A two-pin dual LED 33
produce four states in a 2-pin dual
LED 33
the series current limiting resistor
where the LED is on when the
PIC driving pin is high 33
LED displays 107
alphanumeric multisegment 107
dot matrix 107
seven-segment LED displays 107
Bin2BCD 116, 117
binary coded decimal (BCD) 116,
117
packed BCD 117

common anode 108
high side digit selection switch
108
individual low side digit selectors
108
common cathode 108
high side segment drivers 108
low side digit selector 108
high-side digit switches 109
high-side drivers 108
low-side drivers 108
map the desired digits to the segments 112
mounting style 107
one reason to prefer common anode
displays 108
TRISA 112
TRISB 112
left justified 708
length checking 737
library memory 734
Linear Technology LT1634CCZ-4.096
precision source 220
Liquid Crystal Displays 67
literal 434
logical high 53
logical low 53
low-pass filter 511
low-side drivers 108
2N7000 142
low side digit selector 108
low side digit selectors 108
low side switching 31, 36
base to emitter junction voltage, VBE
37
determine case temperature 36
hFE or DC current gain 37
high power bipolar low side switching 39
simple low side switch 36
switching speed 37
stored charge effect 37
turn-off delay 37
low voltage device 7
low voltage programming (LVP) 22, 23
LSB-first 540
LSBFIRST 733
LSBPOST 732
LSBPRE 733
LSBPRE{LSBFIRST} 732
LSB first 742
LT1634-CCZ-4.096 698
LT1634CCZ-4.096 voltage reference
213, 336
LVP programming pin selection 22

M

"modulus" operator uses the // symbol
84
machine
comparing MBasic, assembler and
machine instructions 260

macro 264
  advantages 264
Macros 744
macros 281
  assembler macros versus ASM compiler directive 289
  "trim off" these excess compiler bits 290
    logical AND 290
  7-bit register address 289
    extract the 7-bit register address from the 16-bit compiler-generated version 290
  warning messages 289
    Invalid RAM location 289
    register in operand not in bank 0 289
  definition 281
  ISRASM 294
    before version 5.3.0.0 294
  library macros
    address trimming 290
    _@clrf 290
    _@opcode 290
    ASM 290
MBasic 729
  "modulus" operator uses the // symbol 84
  // or remainder function 435
  // or remainder operator 342
    MOD or "modulus" function 342
    remainder division 345
  1-wire protocol
    MBasic currently supports only the initial 14 kb/s speed 1-wire protocol 234
    OWIn 234
    OWOut 234
  adding assembler code 281
    ASM directive 282, 287
      ASM 282
    assembler opcode macro 282
      opcode 282
    definition 281
    interrupt handler 282
      ISRASM 282
    library macro 281, 285
      @Bank 285
      @Call 285
      @GoTo 285
      @High 285, 286
      @Input 285, 286
      @Low 285, 286
      @MSDelay 285
      @Output 285, 286
      @RatedDelay 285
      @Return 285
      @TblJmp 285
      @USDelay 285
    address trimming 290
    undocumented feature 281

macro
  @Call 287
  @MSDelay 287
  @TblJmp 287
  @USDelay 287
  macro name is usually formed by adding a "@" prefix to the MBasic normal name 285
  measure statement execution speed 282
ADin 702, 703
assembler language 281
  ASM 284
  Call 284
  GoSub 284
  latency 283
  prescaler 283
  timer 1 283, 284
    StartTime 284
    StopTime 284
assembler module
  StartTime 284
    bcf 284
    clrf 284
    return 284
    t1con 284
    timer 1 284
    tmr1h 284
    tmr1l 284
    turn timer 1 on 284
  StopTime 284
    bcf 284
    stop timer 1 284
    t1con 284
    timer 1 284
assembler opcodes 281
BankSel 413
bank selection bits 287
Basic Stamp 18
  backwards compatibility with the 18
BIN2BCD 720
Branch instruction 184
Branch label 184
built-in waveform output functions 352
  FreqOut 352
  PulsOut 352
  PWM 352
  Sound 352
CCP 423
clear function
  must precede SetHSerial 158
clear to send (CTS) 161
comparing MBasic, assembler and machine instructions 260
compiled language or an interpreted language? 25
compiler vs. interpreter 25
counters 194
counter section
  input signal 194

three possible sources 194
default is to reserve 100% of bank 0 memory for its internal stack 292
default stack setting 334
different ways we can assign a pin to an output and to make its value 0 19
Disable 210
DTMFOut 387
DTMFOut2 387
enable the interrupts 332
flag clearing 332
FREQOUT procedure 338, 352, 353
from version 5.3.0.0. onwards 156
  SerIn 156
  SerOut 156
from version 5.3.0.0 onward, MBasic's SPMotor can be commanded to operate in either full step or half step modes 138
from version 5.3.0.0 onward, we may define a floating point constant directly 205
hardware flow control 161
  SerIn 161
  SerOut 161
hardware handshaking 160
hardware PWM 575
how MBasic handles pins and ports 17
HPWM 423, 639
HSerIn 151, 210, 714
HSerOut 210, 714
HserOut 151
HSerStat 714, 715, 717
  undocumented MBasic command 714
HWPM 352
I2Cin 389, 695
I2Cout 389, 695, 708
I2C read procedure 395
  I2Cin 395
  I2Cout 395
in-line assembler macros 281
  assembler opcodes 281
in-line macros for the assembler mathematical operations 299
  high 299
  low 299
inputs/outputs modifiers 17
  Bin 169
  Dec 169
  Decn 169
  Hex 169
  iBin 169
  iHex 169
  isBin 169, 170
  isHex 169, 170
  Real 170
  Realn 170
  sBin 169, 170

sDec 169, 170
sHex 169, 170
undocumented portions of input/
    output modifiers 169
input and read its value 20
InStr
    Branch 184
    convert this letter to an ordinal digit
        183
interrupts 187, 188
    capture 203
        Capture16L2H 207
        Capture4L2H 207
        capture function is executed
            independently 207
        capture interrupt flag 207
        SetCapture 205, 207
    CCP1Int 208
    CCP2Int 208
    compare 203, 208
        CCP1Int 210
        CompareSpecial 210
        SetCompare 208, 210
    definition 187
    Disable 193, 202
    Enable 191, 193
    enable 188, 190
    Enable ExtInt 190
    EXTINT 189
        SetExtInt 190
    ExtInt 190, 192
        Disable 191
    interrupt requests collide 202
    interrupt service routine (ISR) 187,
        190, 191, 193, 194, 207, 210
        internal stack to overflow 191
        lag time between the interrupt ac-
            tion and start-of-execution 191
        latency 191, 208
        Resume 190, 191
        SerIn 187
        significant restriction 187
        timer 1 208
        written purely in MBasic 187
    MBasic Identifier 188
    Microchip acronym 188
        interrupt flag 188
    OnInterrupt 190, 206, 210, 317
    options for the interrupt 190
    program with an interrupt 188
        disable 188
        enable the interrupt 188
        link the interrupt source with the
            name of the associated interrupt
            service routine 188
    RBINT 192, 193, 194
    Resume 193
    rotary encoder 192
    switch bounce 192
    using the capture interrupt 203
INxx, OUTxx and DIRxx 18
ISRASM 605

library constants 208
    CompareInt 208
    CompareOff 208
    CompareSetHigh 208
    CompareSetLow 208
    CompareSpecial 208
library mode constants 206
    Capture16L2H 206
    Capture1H2L 206
    Capture1L2H 206
    Capture4L2H 206
    CaptureOff 206
LVP checkbox in MBasic's configura-
    tion setup dialog box 23
LVP programming pin selection 22
MBasic is an optimizing compiler
    and will not allocate storage for
    variables that are unused within
    MBasic code 325
    Clear 325
modifiers
    undocumented feature 172
        may be used as functions to
            convert a numerical value to its
            string representation 172
numerical input variables 173
    automatic buffer 173
    indicated input 173
        case conversion 173
    nonnumerical characters 173
    Real 173
    signed value 173
    termination of input 173
ONBOR brownout reset command
    387
order of precedence problems 535
Oscilloscope 723
    utility program 723
        undocumented 723
output modifiers 169
OWin 470
Owout 470
parsing input strings 178
    convert a numerical string of the
        form nnnn to a binary number
        178
    convert a string of the form we
        might use to represent a date or
        time, hh:mm:ss or dd/mm/yy to
        three separate numbers. 178
PEEK 611
pins 17
    designator AN0 21
    pins through a sequential number-
        ing system 18
    pin addresses 20
        input 20
        low 20
        output 20
    pin architectures 28
    pin assignments we have discussed
        to this point are run time alter-

able 21
    pin variables 20
    program time 21
        12F629 example of pins config-
            ured at program time 21
    testing for a pin value condition 21
POKE 611
ports 17
    file 17
    GPIO (general purpose input/out-
        put) 17
    letters from A...E identify ports 17
    register 17
    treated internally by the PIC's CPU,
        and by MBasic, as byte (8-bit)
        variables 17
    variable 17
port or a pin as an address or as a
    variable 19
predefined baudmode constants 156
preselector 207
procedures for serial communications
    156
    hardware (USART)-based 157
        Clear 158
        EnableHSerial 157
        good and bad points of 158
        HSerIn 157, 158, 159, 175
        HSerIn/HserOut 157
        HSerOut 157, 158, 175
        HSerStat 158, 159, 175
        interaction between Clear and
            SetHSerial 158
        receive input 157
        SerIn 158
        SetHSerial 157
        syntax for HSerIn 158
        syntax for SetHSerial 157
        universal synchronous/asyn-
            chronous receiver/transmitter
            (USART) 157
        USART transmit output 157
    software-based
        flow control 157, 160
        good and bad points of 158
        parity error 157
        SerDetect 157
        SerIn 156, 157, 159, 160
        SerOut 156, 157, 159, 160
        syntax for SerIn 156
        syntax for SerOut 157
        Timeout 157
program time 21
    12F629 example of pins configured
        at program time 21
    configuration dialog for 12F629
        – oscillator configuration 22
Pu_Off 23
Pu_On 23
PulsIn 572, 581
    unit steps 572
PULSOUT 338

PulsOut 355, 412, 415
PWM 581
PWM procedure 338, 339, 352, 353
  HWPM 339
RBint 602
Read Function 214
  ADIN 214
  ADIN16 214
register address 287
remainder 84
request to send (RTS) 161
Reverse Polish Notation 25
revised baudmode constants 156
SerIn 151, 156
SerOut 151, 156
SetPullUps 23, 57
SetTmr1 206
SIN 344
SOUND 338
Sound 355, 356
standard modifier/expression rules
  392
string output
  does not check the length 170
  undocumented 170
Terminal 162
timers 187, 194
  comparator 195
  counter 195
  definition 187, 194
  how to get "short count" 195
  input signal 194
  interrupt condition 195
  interrupt source 195
  post-scalers 187
  postscaler 195, 196
  prescaler 187, 194, 196
  timer 0 187, 195
    TMR0 195
  timer 1 187, 195, 196, 197, 201,
    202, 209, 210
    clock source 202
    measure the execution time of
      various MBasic statements 196
    preload 202
    preload is lost with every rollover
      and must be re-entered after
      every interrupt 202
    preselector division ratio 197
    required preload value 202
    setting timer 1's counter division
      ratio 202
    setup timer 1 197
    Tmr1H 195, 202
    Tmr1Int 202
    Tmr1L 195, 202
  timer 2 187, 195, 198, 209
    16 possible postscale ratios 199
    comparator 200
    errors in mode library constants
      199
    interrupt, Tmr2Int 200

interrupt interval 200
interrupt service routine (ISR)
  200
  maximum interrupt period 200
  mode library constants 199
  setup function 199
  TMR2 195
turn counter on/off 195
timers and interrupts work together
  198
ToFloat 703
undocumented numerical modifiers
  173
USART transmit output 157
variables 287
  arrays 289
    may occupy more than one bank,
      but within the array no variable
      will be split across banks 289
  bits/nib index 288
  compiler 287
    base address 288
    compiler assigns the variable a
      16-bit identifier 288
    will not split a multi-byte type
      variable across different banks
      289
  compiler control bits 288
  multi-byte type 288
  the way MBasic passes variable
    information onto the assembler
    287
    GPASM 287
    MPASM 287
variable address
  options in using MBasic variable
    addresses in assembler code
    290
versions prior to 5.3.0.0 323, 398
  I2Cin and I2Cout processes freeze
    when addressing memory
    65535 398
  ISRASM procedures 322
  MPASM assembler 323
    necessary to obtain a 9-bit regis-
      ter address for banksel 323
version 05.2.1.1 208
version 5.2.1.1 349
version 5.3.0.0 445, 572
  ADIN16 216
  GPASM assembler 323
  memory model 611
version 5.3.0.0 friendly 349
version 5.3.0.0 onward 432
  GPASM 432
will not break execution of a statement to service an interrupt
  100
Xin 453, 454
Xout 453, 454
MBasic876 8, 10, 11
2840 development board 8

ISP-PRO 8
MBasic code
  AByte = 0 298
  AByte = Abyte + 1 300
  AByte = AByte - 1 301
    decf 301
    Status <Z> bit 302
  AByte = AByte - Constant 304
    subwf 304
  AByte = BByte + CByte 302
    addwf 302
  AByte = Constant 299
  AWord = 0
    298
  AWord = AWord + 1 300
    btfsc 300
    Status 300
    Status <Z> 300
  AWord = AWord - 1 301
    btfsc 302
    movf 302
  AWord = AWord - Constant 305
    subwf 305
  AWord = BWord + CWord 303
    @bank 303
    addwf 303
    btfsc 303
    incf 303
    Status <C> 303
  AWord = Constant 299
  Byte Table 311
    alternative to MBasic's ByteTable
      311
    Call 312
    call 311
    computed GoTo 311
    GoTo 312
    retlw 311
  Byte Variable Array Read and Write
    313
    file selection register 313
    FSR 313, 314
    FSR/indf 313
    INDF 314
    indf 313
    indirect file register 313
    indirect memory addressing 313
    IRP 313, 314
    keyword conflict between the as-
      sembler operator HIGH and
      MBasic's High function 314
    movlw 314
    RP1:RP0 313
    Status <7> 313
  Division (Special) 307
    @GoTo 308
    decfsz 308
    RRF 307
    Status <C> 307
  For ... Next Loop 310
    @GoTo 310
    btfsc 310

movf 310
movlw 310
Status <Z> 310
Remainder (Special) 306
remainder operation, // 306
Set / Clear Port Bit 308
@bank 309
bfc 308
bfs 308
MBasic command
High 308
low 308
Output 309
Set / Clear Port Bit and Set to Output
309
bcf 309
bsf 309
TRIS 309
TRISA 309
TRISB 309
Swap AByte and BByte 305
Swap 305
MBasic compiler 10
certain things are omitted in the schematics 17
choice of PIC
MBasic876 17
substitute a 16F876 or 876A
17
constants 16
continuation symbol 15
development board 11
development boards 13
in-circuit programming 14
model 0818 for 8- and 18-pin DIP
PICs 13
model 2840 for 28- and 40-pin
DIP 13
high voltage programming (HVP) 23
ISP-PRO programmer 10
low voltage programming (LVP) 22,
23
MBasic Professional 10
note on serial ports 11
professional version 10
programming style 15
prototype boards 13
in-circuit programming 14
model 08/18 prototype board 13
model 28/40 prototype board 13
diagram of 14
software 10
standard assumptions 16
standard program Layout 15
standard version compiler 10
subroutine names 16
verb-noun 16
taxonomy of MBasic functions and
procedures 11, 12, 13
updated MBasic compiler (version
5.3.0.0) 10
variables 16

adjective-noun 16
index or counting variables 16
MBasic compiler software 10
MBasic function 755
MBasic interrupt 741
MBasic Professional 10
MBasic version 5.3.0.0 729
MCP41010 digital potentiometer 258
MCP601 698, 699
MCP601 op-amp 213
MCP601 op-amp buffers 220
measure capacitance with a PIC 691
memory and voltage designators 4
MF signaling 361
Microchip's General Purpose PIC 2
24LC16B 390
instruction word length 3
line name 2
base-line (12-bit) 2
high-end (16-bit) 2
mid-range (14-bit) 2
program counter 436
Microchip Technology, Inc. 1
literal 267, 434
microstep 425
microstepping
bit shifting
rotate left through carry flag (RLF)
435
MiddleMult 729
milli-Newton-meters 129
modem 159
modifiers as functions 737
Morse code 375, 377, 633, 641, 652,
654, 655, 656, 661, 668, 671,
684, 687
amateur radio 661
ASK (amplitude shift keying) 661
binary tree 654, 656
binary weighted 656
byte table 656
continuous wave transmission (CW)
634, 661
dash/dot 637
dashes 633, 634, 637, 639, 640, 641,
651, 652, 653, 654, 655, 656
dash branch 654
dash duration 647
decoding variable length codes 633
binary weighted tree 633
dot/dash 652
dot/dash ratio 633, 635, 668
theoretically perfect dot/dash ratio
635
dots 633, 634, 635, 637, 639, 640,
647, 651, 652, 653, 654, 655,
656
dot branch 654
dot duration 647
EEPROM 651, 658, 659
At program time, MBASIC fills all
256 EEPROM cells with the

value $FF, overwriting their
prior content 659
current values of speed, weight and
hand saved to EEPROM 658
read at program launch 658
elements per second (EPS) is related
to the speed in words per minute (WPM) 636
encoder 646, 648, 649
end-of-character signal 652
end-of-character space 653, 654
function and value encoders 646
function encoder 648
function selection 642
how Morse code is sent and received
661
HPWM 639
iambic 637, 639
iambic paddle 633
inter-element space 653
interrupt 641
keyer 633, 635, 639, 649, 668, 682
keyers 635
key down 635
level weighting 656
map each Morse character into a
unique number 655
milliseconds per element 636
Morse characters 634
Morse code
control loop
over-damped 667
nil 656
OOK (on-off keying) 661
paddle 633, 635, 639, 646, 647, 651,
653
Pauseus 640
pro-signs 634
procedure signals, AR, SK and SN
634
prosigns 655
pull-up resistors 635
PWM hardware module 637, 639
RBINT 647
reading Morse code 661
adaptive algorithm 661
amateur radio 661
arithmetical average 673
arithmetical mean 666
ASK (amplitude shift keying) 661
bandwidth 668
interaction with center frequency
668
bandwidth and center frequency
interaction 668
beat frequency oscillator (BFO)
661
binary number 684
binary tree 684
byte table 684
center frequency 665, 666, 668
interaction with bandwidth 668

circular buffer 675
complementary output 664
computes the estimated code speed 676
control loop 666
  damping 666
  damping factor 667
  oscillate 666, 667
  overshoot 666
  under-damped 667
  undershoot 666
control signal 662
dash 682
dash/dot ratio 671
dashes 668, 670, 671, 680, 681, 683, 686
dash branch 684
demodulation 661
demodulator 661
design parameters
  center frequency 665
  tracking bandwidth 665
dots 668, 670, 671, 680, 681, 682, 683, 686
dot branch 684
dot length 665, 681
double-sided bandwidth 667
double sided bandwidth 666
drop-out frequency 668
easure the time it takes for the main program loop to execute in idle 681
element length distribution 672
end-of-character space 683
end of word 682
error 662
error signal 664
frequency shift keyed data signals 661
geometric mean 666
HSerOut 686
inter-element space 680
inter-letter time 682
letter space interval 682
lock 667
lock bandwidth 665
lock detector 665
lock range 667
lock up 667
lock up time 667
lock up time of the tone detector comparator 667
loop control 664
loop phase detector 664
MBASIC 5.2.1.1 670
  PulsIn 670
MBasic 5.3.0.0 remedies problem 670
normal output 664
numerical binary tree 680
one-sided bandwidth 666
OOK (on-off keying) 661

out-of-lock 667
Pauseus 670
phase detector 662, 663, 664
phase locked loop (PLL) 661, 662, 666, 667
  in lock 664
  locks 664
  unlocked 664
pull-in frequency 667, 668
pull-up resistor 667
quadrature phase detector 664
receiver 661
Schmitt-trigger 669
scrolling 687
single sideband mode (SSB) 661
single sided bandwidth 666
step response 666
threshold value; elements longer than the threshold are dashes and shorter elements are dots 671
timer 1 670
tone decoder 665, 669
tone detection 661, 664
tone detector 661, 663, 667
  XR-2211 661, 662, 664, 665, 666, 667, 668, 669, 670, 671, 680, 682
tone detect phase detector 667
tracking bandwidth 665
TTL-level 669
upper and lower drop out frequencies 667
upper and lower pull-in frequencies 667
voltage-controlled oscillator (VCO) 662, 663, 664, 666
  adjust RP to center 667
weighted binary tree 685
word space 682, 683
XOR logic gate 663
recommended element and spacing lengths 668
rotary encoder 641, 642, 647, 657
  interrupt-driven 641
scrolling 657
sidetone 635, 639, 647
spaces 633, 634, 652
speed 636, 642, 650, 658
speed, weight and selector 646
square cursor 658
standard word "PARIS" 636
value encoder 648
value selection 642
variable length code 634, 652
weight 635, 641, 642, 648, 650, 658
weighted binary tree conversion 655
weighting 635
words per minute 636
word space 652, 653, 654
Motorola 693
movf 434

movlw 434
movwf 434
MPASM assembler and linker 733
MPX6115 699, 725
MPX6115A6U 693
MSBFIRST 733
MSBPOST 732
MSBPRE 733
MSBPRE{MSBFIRST} 732
MSB first 742

N
National Electrical Manufacturers Association (NEMA) 124
New definition of _MHZ 733
New reserved words 733
nibble 368
nominal capacitance 691
no error checking 730
Nyquist sampling rate 211, 345

O
offset voltage 213
one-wire protocol 231
  64-bit serial number 233
    addressed either individually or globally 233
  bytes are transmitted and received from lowest address to highest address 235
  CRC (cyclic redundancy code) 235
  family code 234
  family number 235
  global addressing 237
  OWIn 231, 234, 238
  OWOut 231, 234
  parasitic powering 233
  permissible mode values 234
  reset 238
  reset pulse 232
  serial number 234, 235
OnInterrupt 100, 210
OOK (on-off keying) 661
open loop 120
open systems interconnection (OSI) reference model 151
operational amplifiers 511
optically coupled FET 542
optically isolated MOSFET 49
  PS710A-1A is a high power MOSFET optoisolator 49
  switching high currents in microsecond times 49
    undesired transients and oscillations 49
optical coupler 62
optical couplers 48
optical tachometers 92
Optimize for Size 734
Optimize for Speed 734
Optimizing for Size 734
optional data byte 708

optoisolators 48
  low power optoisolator 48
order of entry 738
order of precedence 733
order of precedence rules for arithmetic
    433
oscillating errors 587
OSI stack 151
output buffer 742
OWIn 234, 238
OWin / OWout 757
OWOut 234

P
P0...P31 733
packed variables 717
parallel access memory 408
  CY7C128A 409, 410, 415
    '843 latch 415
    74ACT574 409
    74ACT843
      9-bit latch 409
      latch enable (LE) pin 409
    address inputs 409
    bidirectional I/O pins 409
      high impedance state 409
    control pins 409
      chip enable (CE) 409
      output enable (OE) 409
      write enable (WE) 409
    Fourier transform (FFT) 408
    SRAM 408, 409, 410, 413, 416
      tri-state 416
    static RAM 409
parameters and a returned variable 729
parameter Mdelay 136
parasitic powering 233
parsing input strings 178
Pause 731
Pauseus 357, 731
peripherals 2
phase accumulator 348
phase advance 349
phase and cycle control 549
phase locked loops 661
photo-diode 48
photodiode 520
phototransistor 48, 520, 521
PIC's instruction set 260
PICmicro microcontrollers 1
  "CCP" hardware module 203
  16C781 338
  16C782 338
  absolute maximum ratings for
      16F87X PICs 29, 30
  current that flows in or out of an input
      pin is due to leakage 57
  data memory 261
  fast switching--sound from a PIC 50
  high side switching 31
  how do I pick one? 7
  input pin

CMOS compatible 571
Schmitt trigger type 571
TTL compatible 571
input pins represent a high input
    impedance 57
interrupt 317
isolated switching 31, 45
  optical isolated NPN switch 48
  relay switching 46
    good things about relays 46
    not so good things about relays
        46
    Reed relays 47
    test the contact closure and oper-
        ate/release time 46
ISP-PRO 23
LED 32
  calculate the series current limiting
      resistor 32
  connect two LEDs to one pin 33
  constant voltage drop 32
  MV5491A two-pin dual LED 33
  produce four states in a 2-pin dual
      LED 33
  the series current limiting resistor
      where the LED is on when the
      PIC driving pin is high 33
  when the PIC driving pin is low 32
low side switching 31, 36
  base to emitter junction voltage,
      VBE 37
  determine case temperature 36
  hFE or DC current gain 37
  simple low side switch 36
  switching speed 37
    stored charge effect 37
    turn-off delay 37
low voltage device 7
main elements inside a PIC 1
  central processing unit 1
  data memory 1, 2
  peripherals 2
  program memory 1, 2, 3
    EPROM and EEPROM 3
    extended voltage 3
    flash memory 3
    memory and voltage designa-
        tors 4
    partial list of package designa-
        tors 4
    read-only 3
    standard voltage 3
    temperature range designators 4
maximum safe parameters apply to
    the 16F87x series 29
physical connections between the PIC
    and the LCD module 73
PICS supported by MBasic 4
PICs with a multipart identifier 2
  case style 2
  family number 2
  maximum clock frequency 2

program memory type 2
silicon die layout revision suffix 2
temperature range 2
program memory 261
pull-up resistors 23
  internal pull-up resistors 23
small N-Channel MOSFET switch
    37
  difference between the MOSFET
      and NPN bipolar transistor 38
  high power MOSFET low Side
      switching 40
  high power MOSFET low side
      switching 40
  MOSFETs, gate charge 39
  relationship between RDS(ON) and
      the gate voltag 38
  speed 7
  state of input pins RB4...RB7 change
      99
  switching inductive loads 34
    calculate the inductive spike level
        and decay time analytically, but
        it's much easier to use a SPICE
        simulation program 35
    clamping diode should be a fast
        switching device, such as a
        Schottky diode 35
    peak voltage spike and current
        decay times interact for the
        circuit 35
    switching speed 37
    Zener diode 35
  terminal software 152
  timers 7
  USART 7
  weak pull-up resistors for Port B 23
Pico Electronics 453
PICS supported by MBasic 4, 5, 6
piezoelectric 693
pin architectures 28
  complementary metal oxide semicon-
      ductor (CMOS) configuration
      28
  RA4 is different; it is configured as an
      open drain MOSFET 29
pin saving techniques 86
  read a 16-state switch 86
  read a configuration switch 86
Pin Variables 737
port mismatch 744
position encoders 91
potentiometer 487
power-on/power-off code 519
power control board 551, 552
  2840 prototype board 554
    limiter 554
    MCP601 554, 555
    opto-triac 554
    TLE2426 555
    TLE 2426 rail splitter 555
    zero crossing 554

BTA12-600B 552
BTA12-600BW snubberless triac 552
construction 553
 BTA12-600B 553
 BTA12-600BW 553
 heatsinks 553
 Wakefield 231-137PAB 553
 post construction checkout 553
 no-conduction test 554
design 551
K3011P 559
 gate drive duration 560
 opto-triac 559, 560
 random phase 559
 zero-crossing 559
 triac 560
 trigger pulses 561
 zero crossing 559, 560
 zero voltage 560
Manhattan-style construction 551
Manhattan style construction 553
MOC3032M 559
opto-coupler 552
opto-triac 553
opto-triac coupler 552
Radio Shack prototype board model
 276-159 553
snubber 552
triac 552
Pre-defined macros in INTASM 733
prescaler 608
preselector 207
pressure sensor 704
pressure sensors 693
programmable unijunction transistor
 (PUT) 543
programming style 15
program configuration dialog box 730
program counter 265, 436
program instruction 265
program memory 1, 2, 3, 6, 261, 265,
 734
 call
 GoSub 265
 EEPROM 3
 EPROM 3
 EPROM and EEPROM 3
 one time programmable (OTP) 3
 extended voltage 3
 flash memory 3
 interrupt vectors 266
 jump execution 265
memory and voltage designators 4
opcodes
 jump opcodes
 call 265
 GoTo 266
 goto 265
pages 265
partial list of package designators 4
PCLATH register 265
PCL register 265

program counter 265
program instruction 265
read-only 3
reset vectors 266
stack memory 266
 call operations 266
 interrupt operations 266
standard voltage 3
temperature range designators 4
program words 6
pseudo-code 23, 24
 define the problem 24
 document the problem and the solution 24
 program modularly, proceeding from
 the top down 24
 think first, code later 24
 writing good code into one rule 24
Pu_Off 23
Pu_On 23
pull-down 63
pull-up resistors 23
 internal pull-up resistors 23
PulseIn 757
pulse width modulation (PWM) 352,
 354, 355, 357, 420, 424, 574
 1-bit DAC 352
 duty cycle 424
 FREQOUT 352
 FreqOut 355
 linearity problems 353
 RC low-pass filter 353
PulsIn 581, 731
PulsOut 731, 757
PWM 731, 757
PWM module 593

Q
quadrature encoder 98
quantization error 212, 213, 336
quantized value 721

R
RAM memory 389
 parallel static RAM (SRAM) 389
Random 739, 740
random number 739
RBINT 744
RBINT type interrupt 100
RCTime 731
RC filter 354
 calculating values 355
RC network
 -3 dB frequency 355
read/write (R/W) bit 393
ReadDM 757
Real 737
real-time clock 700
real-time clock and serial EEPROM
 700
real time clock 709, 711, 712
received data 163

records 717
Redefined order of precedence 733
registered data access arrangement
 (DAA) 372
relative encoder 95
 quadrature encoder 98
 reading a relative encoder 95
 change from earlier versions of
 MBasic 97
relative humidity 691, 692, 704
relative humidity module 692
remainder 84
remainder operator 722
remote controls 517
 Hitachi 536, 537
 check byte 536
 device and data codes 536
 LSB-first 536
 MSB-first 536, 537
 REC-80 536
 toggle bits 536
 infrared transmission 517
 IR receiver module 517
 LSB-first 540
 Mitsubishi 526, 528, 534
 bit order 531
 LSB-first 530, 531, 536
 LSB or MSB order 530
 MSB-first 530, 531, 536
 btfsc 528
 carrier frequency 526
 device and data codes 536
 device code 531
 execution speed problem 529
 HserOut 530
 key code 531
 message end 531
 message start 531
 PulsIn 529
 timer 1 529
 REC-80 536
 shift-left (<<) 530
 start pulse 528
 toggle bits 536
 Nakamichi 535, 537
 check byte 536
 device and data codes 536
 key code 536
 LSB-first 536
 MSB-first 536
 REC-80 536
 toggle bits 536
 QSE114 521
 Toshiba 535, 537
 check byte 536
 device and data codes 536
 keycode 536
 LSB-first 536
 MSB-first 536
 REC-80 536
 toggle bits 536
repeat option 737

reset pulse 232
resolution 212
Resume 100
Reverse Polish Notation 25
rheostat 487
right-justifies 702
right justified 702
right justifying 709
rising edge 244
rotary encoders 91, 192
  absolute encoders 91
  code output 94
    difference between hexadecimal
      and Gray code 94
    Gray code 94
    straight hexadecimal binary code
      94
  construction 92
    mechanical encoders 92
    optical encoders 92
  incremental encoders 92
  quadrature encoders 93
  quadrature sensor 93
  relative encoders 91
rotor 121, 124
RS-232 serial interface 151, 154, 159
  2840 development board 154
    carrier detect (DCD) 154
    clear to send 154
    data set ready 154
    data terminal ready 154
    receive data (RX) 154
    request to send 154
    ring indicator 154
    signal ground 154
    transmit data (TX) 154
  bus mode 156
  carriage return 152
    code value 152
  clock skew 155
  connect to your PC 151
    ComTest 152
    terminal software 152
      com port 152
      data bits 152
      end of line character(s) 152
      handshaking 152
      Hyperterminal 152
      parity 152
      stop bits 152
      Terminal 152
  control signals
    Assert/Negate 153
    Asserted/Unasserted 153
  convert the RS-232 levels to logic
    levels compatible with our PIC
    153
  data bits
    even parity 155
    odd parity 155
    parity 155
    parity bit 155

data signals
  Marking 153
  Spacing 153
EIA/TIA RS-232E 151
flow control 160
HSerIn 151, 155
HserOut 151
inversion between RS-232 and logic
  153
level converters 153
  relationship between the ±15 V RS-
    232 levels and the 0/+5 V logic
    employed by PICs 153
  SP232ACN 153
line feed
  code value 152
maximum permitted error between
  the sending and receiving data
  rates 155
physical layer 151
serial transmission 154
  asynchronous 154
  synchronization 155
  synchronous 154
SerIn 151, 155
  syntax for SerIn 156
SerOut 151
SP232ACN data sheet 154
SP232ACN level converters 153
standard pin connections 154
start (space) bit 154
stop (mark) bit 154
Str modifier 166
  parameters 166
Teletype 153
USB-to-serial adapter 151
RTS/CTS handshaking 163

S

salient poles 124
Sallen-Key 511
sample rate 428
Samsung's KS0066 67
Schmitt trigger inputs 54
  correctly read an input signal in the
    presence of large noise voltages
    57
  special Schmitt trigger inputs 54
Schmitt trigger type 571
scratchpad memory 238
ScrLeft 75, 77
ScrRAM 77
ScrRight 75
sensor suite 699
serial 151
serial EEPROM 389, 408, 697, 700,
  712, 722
serial input buffer 715
serial peripheral interface (SPI) 231
  1-wire 231
    "presence" pulse 232
    reset pulse 232

DS1302 real time clock 231
DS18B20 temperature sensor 231
one-wire protocol 231
  OWIn 231
  OWOut 231
  three-wire 231
serial ports, note on 11
serial terminal program 87
  HyperTerminal 87
  Terminal 87
SerIn 151, 155, 156, 158, 159, 163,
  164, 165, 176, 731
  doesn't mean that we can send a
    string of characters at 57600
    bits/sec and expect them to be
    correctly received 165
  flow control 164, 165
  I/O modifier 165
  read a string of multiple characters
    164
  syntax for SerIn 156
Serin / SerOut 757
SerIn and HSerIn
  differences between 176
SerIn and SerOut constants 731
SerOut 151, 156, 159, 163, 164, 731
SerOut also HSerOut 740
SetCapture 207
SetCompare 208, 210
SetHSerial 87, 157, 176
  set the desired baud rate 176
SetHSerial / HSerin / HSerOut 758
SetHSerial / HSerOut / HSerIn 759
SetPullUps 23, 57
SetPullups 758
SetTmr1 206
seven-segment LED displays 107
shaft encoders 91
Sherline Products, Inc. 93
ShiftIn 733
ShiftIn / ShiftOut 757
ShiftOut 733
ShiftOut/ShiftIn 258
  rising or falling clock and bit order
    258
ShiftOut/ShiftIn -- rising or falling
  clock and bit order 258
Signal System No. 7 361
silicon controlled rectifier (SCR) 543
simple circuit 58
Sin and Cos 738
Sleep 731, 741
SLOWLSBPOST 732
SLOWLSBPRE 732
SLOWMSBPOST 732
SLOWMSBPRE 732
small N-Channel MOSFET switch 37
  difference between the MOSFET and
    NPN bipolar transistor 38
  high power MOSFET low side
    switching 40
  MOSFETs, gate charge 39

relationship between RDS(ON) and the gate voltage 38
SMBus 54
SN754410 147, 148, 588, 592, 593
superior ratings 147
SN754410 H-bridge 132, 147
snubber 35
snubbing diode 35, 46, 110
software debouncing 60
Sound 731, 757
Sound2 731
SP232ACN level converters 153
special purpose switching 50
spectrum analysis plots 349
speed 7
speed or size optimization 734
speed versus memory requirements 734
SPI (3-wire) 389
SPMotor 135, 138, 141, 147, 731, 757
when stopped 138
SpMotor 133
SRAM 409
STACK 732
stack 732
stack command 730
stack compiler directive 732
Stack size and location 732
stack size is stated in type Long 732
standard barometric pressure 692
standard header 730
standard program layout 15
start (space) bit 154
static RAM 409
static torque 121
stator 121, 124
stepper motors 120, 418, 428, 567
"bolt-in" MBasic-to-assembler 429
advantages 120
Assembler Interrupt Service Routine (ISR) 447
assembler routines 429
compiler control bits 433
force the compiler to allocate space 433
GPASM 447
stepwise refinement 429
banksel 432, 433
benefits and drawbacks of unipolar and bipolar connections 123
bipolar 122, 133, 147, 418, 440, 567
advantage of bipolar motors 133
conceptual view of controlling a bipolar motor 131
H-bridge 418
rotor 418
SN754410 H-bridge 132, 147
stator 418
two-phase 122
bit set file
bcf 435, 437
bsf 437
btfsc 437

bit shifting 434
carry flag 438
mask off bits by the bitwise AND operator 435
modulus 435
rotate left through carry flag (RLF) 434
rotate right through carry flag (RRF) 434
rotate through carry 435, 438
more properly named "re-circulation" 435
Status 435
carriage return 445
carriage return / line feed combination 445
CCP 423
clock speed value is converted into a constant 445
cosine 438
cosine function 421, 425
cosine transition current waveform 424
Darlington transistor 135
detent torque 121
digital-to-analog converter 419
disadvantages 120
duty cycle 423, 424, 425, 437
four phase 123
full step mode 418, 421
GPASM 432, 447
banksel 432
half step mode 418
hardware PWM 439
hardware PWM module 438
hardware PWM module's period 423
high voltage spike 138
holding torque 121
HPWM module 424
HserOut 446
HWPM module 428
identifying stepper motors 124
improve the current release time by adding Zener diodes in reverse series 142
inductive current rise 135
initialization string 445
interrupt 440
ISRASM 440
interrupt service routine (ISR) 447, 448
16-bit comparison 449
addwf 448
bit pattern
full step 448
half step 448
decf 448
incf 448, 449
ISRASM 445
page boundaries 448
retlw 448
ISRASM 446, 447

L/R drive 418
L/R time constant 419, 439
mask literal 438
MBasic
// operator is the modulus or remainder function 435
hardware pulse width modulation (PWM)
HPWM 423
HPWM 423
duty cycle 423
period 423
Pauseus 438
MBasic variables
access a word variable 447
BankSel 447
bank selection 447
base address 447
Clear 444
compiler control 447
high order byte is at the base address plus one 447
instructions
BankSel 447
incf 447
microstepping 418, 419, 420
full step 419
microsteps 418, 425, 426, 434
pauseus 432
MPASM 432
National Electrical Manufacturers Association (NEMA) 124
open loop 120
operation modes 129
compile-time option 138
full step 130
full wave 131
half-step 130, 131
nonuniform torque 131
half step 131
routines 138
wave 130, 131, 133
parameter Mdelay 136
period 425
PF35-48 unipolar 422
programs 133
current release improvement a Zener diode makes over using the ULN2003A's internal clamping diode 134
improve the current rise time 136
SPMotor 134, 135
SpMotor 133
program counter 436
PWM modules 423, 428
duty cycle 437
reading a stepper specification sheet 126
ambient temperature range -- operating and temperature rise 129
drive mode 126
excitation mode 126

holding torque 129
  milli-Newton-meters 129
  ounce-inches 129
mass 129
rotor inertia 129
starting pulse rate, max and slewing
    pulse rate max 129
steps per revolution 126
step angle 126
step angle tolerance 126
voltage and winding resistance 126
winding inductance 127
  inductive time constant 128
registers
  addlw 434
  AND 438
  btfsc 436
  call 436
  CCP1CON 437
  CCPR1L 437, 438
  decfsz 434
  GoTo 434, 436
  HPWM 437
  incf 436
  jump table 436, 437
  movf 434
  movlw 434
  movwf 434, 436
  PCL 436
  PCLATH 436
  retlw 436
rotation direction 135
rotor 121, 124, 418
salient poles 124
sample rate 428
Schottky voltage clamp diodes 148
select a series resistance and voltage
    136
SN754410 147, 148
  superior ratings 147
SN754410 bipolar driver circuit 419,
    420, 421, 422, 428, 440
  enable pins 419, 420
  enable voltage 428
  pulse width modulation (PWM)
    modules 420, 422
  duty cycle 420
SN754410 quad half-H bridge device
    147
SPMotor 138, 141, 147
  SPMOTOR_Half CON 0 138
SpMotor 133
  when stopped 138
STACK 433
static torque 121
stator 121, 124, 418
stepper pattern in a subroutine 143
the relationship between R1, V1 and
    Imotor 136
timer 1 440, 445, 446, 449
  HSerIn
    interrupt 447

interrupt 445, 446, 447
  PIR1 447
  tmr1if 447, 448
  preload 446, 448
two-phase 122
types of stepper motors 123
  can stack 123
  hybrid 123, 124
    industry-standard case sizes 124
    rotor 124
    stator 124
  permanent magnet 123
  tin can 123
  variable reluctance 124
ULN2003A 135, 138
ULN2003A Darlington transistor
    driver 133
unipolar 122, 123, 129, 133, 418,
    440, 567
  center taps 123
  four phase 123
  L/R time constant of the bipolar
    connected winding is twice that
    of the unipolar connection 148
  unipolar stepper 133
  winding inductance increases 148
variable constant current 419
wave mode 418
stepwise refinement 295
stepwise substitution 413
stop (mark) bit 154
straight-through cable 159
straight hexadecimal binary code 94
Str modifier 166
  explicit end-of-line definition 167
  length to the Str modifier 167
  parameters 166
subroutine names 16
  verb-noun 16
supercapacitors 251
  leakage current 251
Surfboard 6103 693
switching speed 37
  turn-on and turn-off speeds 41
switch bounce 58, 192
  built-in switch debounce procedure
    Button 60
System Setup dialog box 734

T

tachometer 582
  constant 570
  determining the tachometer constant
    570
  feedback 568
taxonomy of MBasic functions and
    procedures 11, 12
Teletype 152, 153
temperature 691, 695
temperature, humidity and barometric
    pressure 711
temperature conversion 697

Terminal 152, 162
threshold voltage 53
thyristor family 543
  diode thyristor (diac) 543
  holding current 543
  programmable unijunction transistor
    (PUT) 543
  silicon controlled rectifier (SCR) 543
  triac 543
time/date 717
timers 7, 187, 194
  definition 194
  timer 0 187
  timer 1 187, 208, 209, 210
  timer 2 187, 209
    prescale 331
timing parameters 731
TOFLOAT 729
ToFloat 217, 703, 733
toggle bit 519
TOINT 729
ToInt 729, 733, 738, 739
token memory 734
Touch-Tone 360, 363, 364, 367, 371,
    372, 373, 387
  ASCII collating sequence 383
  ASCII collation order 385
  CH1837A 374, 375, 376, 377, 387
    latency 376
    off-hook 377
    OffHook pins 375
    RI pins 375, 376
  connecting a common Western
    Electric / Lucent Touch-Tone
    dial to function as a stand-alone
    Touch-Tone generator 361
  digit receiver 371
  DS18B20 temperature sensor 387
  dual-tone multi-frequency (DTMF)
    360
    inter-office 360
  DV (digit valid) 362
  four-wire 371
  generating Touch-Tone signals 361
    DTMFout 361
    DTMFOut2 361
  hybrid 371
  LM386 374
  MBasic
    DTMFOut 387
    DTMFOut2 387
    ONBOR brownout reset command
      387
  MC145436A 374
  MF signaling 361
  Morse code 375, 377, 382, 385, 386
    "pro-sign" Morse characters 383
    di-bit encoding 383, 385
    duration of dots, dashes and spaces
      are defined 382
    Morse code sending 382
    Morse code speed 376

standard word 376
words per minute 376
variable length code 382
X10 receiving modules 381
X10 transceiver 381
overview of the traditional single-party telephone connection 370
answering a call 371
dial tone 371
FCC established set of rules 372
hookswitch 371
hybrid 372
off-hook 372
on-hook 371, 372
originating a call 371
ringing voltage 371, 372
subscriber loop 371
characteristic impedance 372
nominal characteristic impedance 372
protected from over-voltage and over-current 372
transmit limit 373
registered data access arrangement (DAA) 372
CH1837A 372
CH1840 372
MC145436A 372
MC154436A 372
MH88422 372
XE0068 372
ringing voltage 376
Signal System No. 7 361
timer 1 375
interrupt service routine 375
prescaler 375
Touch-Tone decoder IC 361
MC145436 361
byte table 364
MC145436A 361, 363, 367
DV pin 364
maximum permitted signal level 363
read the MC145436A 363
MCP145436A 362
output 362
MT8870 361
nibble 368
two-wire to four-wire converter 371
triac 544, 545, 547, 549, 558, 560
anode 1 (A1) 543
anode 2 (A2) 543
BTA12-600B 544
dV/dt 544
standard triac 544
BTA12-600BW 544
dV/dt 544
quadrants I-II-III 544
quadrant IV 544
snubberless triac 544
BTA12-600TW
dV/dt 544

commutate 543
commutation 546
commutation failure 545
commute 546
cycle controlling 556
DC component 559
dI/dt 545, 546
BTA12-600BW 545, 546
commute 546
linear load 546
load is inductive 546
snubberless 545
standard 545
stored charge 545
unrecombined charge carriers 545
di/dt 546
dV/dt 544, 545, 546, 547
BTA12-600 548
BTA12-600BW 547, 548
resistive load 547
commute 544
false gate triggering 548
gate current 546
gate sensitivity 548
inductive load 547
inductive loads 548
resistive load 546
snubber 548
snubberless 544, 548
snubber network 547
standard 544
dv/dt 547
numerical value of dv/dt 547
resistive load 547
dV/dt and dI/dt issues 545
four-quadrant triggerable 544
gate 543, 544
gate current 544
IGT 544
gate drive requirements 544
positive and negative gate drive 544
voltage across the gate 544
VGT 544
gate/MT2 relationship 544
heat sinking 545
BTA12-600BW 545
junction temperature 545
junction to case RTH(j-c) 545
thermal resistance 545
VT 545
holding current IH 544
introduction 543
K3011P 556, 562
inrush current 556
opto-triac 562
random phase 556
start-up transient 556
zero-crossing 556
latching current IL 544
main terminal 1 (MT1) 543
main terminal 2 (MT2) 543, 544

maximum current 545
BTA12-600BW 545
maximum nonrepetitive peak current ITSM 545
maximum RMS-on state current IT(RMS) 545
maximum gate current IGM 545
BTA12-600BW 545
MCP601 558
opto-triac 558
zero crossing 558
MOC3032M 555, 556
cold resistance 556
color temperature 556
initial current inrush 556
opto-triac 555, 556
relationship between the LED trigger current and the load current when we use a zero-crossing controller 555
zero-crossing 556
on-state voltage VT 545
BTA12-600BW 545
phase and cycle control 549
cycle-control 549
cycle control 550
DC component 550
radio frequency interference 550
softer start 550
off-on control 549
off/on control 549
phase control 549
average value 550
firing angle 550
power versus firing angle 550
RMS 550
RMS and average voltages vary with respect to the firing angle 550
triggering point 550
true RMS meters 550
phase control 562
power control board 552
Q-I/Q-III 544
Q-IV triggerable 544
quadrants 544
quadrants II and IV 549
BTA12-600BW 549
self-commutating 544
snubberless versus standard 545
thyristors 545
trigger 554
triggering a triac 548
BTA12-600 549
BTA12-600BW 548
snubberless 549
difference between zero crossing and random phase 548
gate 548
gate current 549
MT1 549
MT2 549

K3011P 548
    random phase device 548
MOC3032M 548
    zero crossing trigger device 548
optical isolators 548
opto-isolator 548
opto-triac 548
quadrants II and IV 549
quadrant IV 549
random phase opto-triac 548
zero crossing device 548
VDRM 545
ways to connect the triac with the
    load and trigger drive 549
TRISA 112
TRISB 112
TTL compatible 571
TTL level 53
turn-off delay 37
two's complemented 709

U

ULN2003A 135, 138
ULN2003A Darlington transistor driver
    133
ULN2003A NPN Darlington array
    109, 110
undefined region 53
undocumented command modifiers 736
Undocumented MBasic 733
Unit and Key Codes 742
universal adapter 11
Universal Product Code, revision A
    (UPC-A) 595
universal synchronous/asynchronous
    receiver/transmitter (USART)
    157
updated MBasic compiler (version
    5.3.0.0) 10
USART 7, 742
USART transmit output 157
user-defined function 730
User-defined functions 729

V

vapor pressure 691
variables 16
    adjective-noun 16
    index or counting variables 16
variable resistance 487
variable resistor 487
variable voltage divider 487
version 05.2.1.1
    SetCompare
        error that reverses the byte order
            208
vertical bar | character 734
VIH -- The minimum voltage on an
    input pin that will be read as a
    logical high 53
VIL -- The maximum voltage on an
    input pin that will be read as a

logical low 53
voltage-controlled oscillator (VCO)
    voltage-controlled oscillator (VCO)
        lock and no-lock conditions 663
voltage divider 699
voltage scaling 698
volume control 487
von Neumann's architecture 2

W

WaitStr modifier 168, 169
    nonmatching characters 168
warning statement 542
water vapor 691
weather station 691
    7805 699
    address pins 699
    bit-packed 720
    external EEPROM 711
    MPX6115 698
    packing time/date information into
        the four-byte record 718
    quantized value 721
    real time clock 711, 712
    sensor suite 691, 699
        AT24C512 serial EEPROM 705
        barometric pressure 691, 692, 694
            A/D converter 694
            A/D error 695
            A/D step 695
            absolute pressure 693, 694, 704
            barometric pressure adjustment
                factor 693
            corrected pressure 694
            EEPROM 695
            elevation correction 694
            elevation of the sensor 692
            error budget 695
            error voltage 695
            estimate the voltage output range
                of our pressure sensor 694
            Freescale Semiconductor, Inc.
                693
            hectopascal (hPa) 692
            I2Cout 695, 697
            inches of mercury 692
            LT1634-CCZ-4.096 698
            Maxim's DS18B20 695, 697,
                698
            Maxim's I2C-protocol DS1624
                695, 696, 697, 698, 708
            MCP601 698, 699
            millibar 692
            Motorola 693
            MPX6115 698, 699, 704
            MPX6115A6U 693, 694
            piezoelectric 693
            pressure sensor 704
            pressure sensors 693
            quantize barometric pressure 721
            reduced to sea level equivalent
                693

relationship amongst inches of
    mercury, millibars, kilopascals
    and pounds per square inch
    692
relationship between elevation
    and air pressure change 693
sensor error 695
standard barometric pressure 692
strain gauge 693
surface mount adapter 693
Surfboard 6103 693
command byte 708
control byte 708
DS1302 709
optional data byte 708
relative humidity 691, 692, 704
    A/D converter 691
    capacitance 691
    dielectric constant 691
    estimate the measurement error
        692
    HM-1500 692, 704
    Honeywell's HIH-3610 692
    humidity sensors 691
    Humirel's HS-1100 sensor 691,
        692
    measure capacitance with a PIC
        691
    nominal capacitance 691
    quantize humidity 721
    relative humidity module 692
    uncorrected linear equation 692
    water vapor 691
relative humidity sensor 692
temperature 691, 695
    24-series EEPROM 696
    A/D resolution 695
    address byte 708
    analog temperature sensor 695
    command byte 696, 697, 708
    configuration register 696, 697
    continuous mode 696
    control byte 696
    convert the returned value to an
        actual Celsius temperature 709
    data byte 696
    DS1624 708, 709
    DS18B20 709
    EEPROM 695, 696
    external EEPROM 695
    family code 696
    fractional part 698
    I2C 696
    I2C-protocol 695
    I2Cin 695, 697, 708
    left justified 697, 708
    National Semiconductor's
        LM34C 695
    one-shot mode 696
    pin connections 695
    real time clock 709
    right justifying 709

temperature conversion 697
two's complemented 709
serial EEPROM 712
temperature, humidity and barometric
pressure 711
verify the A/D components 700
voltage divider 699
voltage scaling 698
write 723
WriteDM 101, 102

X

X-10 messages 453, 473
function code 453, 456
MBasic provides library constants
453
Xin 453
Xout 453
house code 453, 456
MBasic provides library constants
453
Xin 453
Xout 453
unit code 453, 456
MBasic provides library constants
453
Xin 453
Xout 453
X-10 technology 453, 454, 456, 457,
458, 459, 460, 465, 469, 473,
475, 480, 486
appliance module 460
bit order 457
byte table 465, 466, 481
complements 484
LSB-first 458, 465, 481
LSB first 457, 484
most significant bit 457
MSB-first 465
MSB first 457, 484
serial interface 457
LSB first 457
start bits 483
data 456
data block 456
house block 456

LSB-first order 456
start block 456
data transmission via the X-10 interface 469
DS18B20 470, 471, 474
X-10 link 471
how X-10 commands are sent 456
LCD initialization 481
MBasic functions
binary equivalents 458
MSB first order 459
library constants 457, 461, 465,
469
cross-references with the names
and bit sequences followed in
X-10 documents 457
unit code 465
OWin 470
Owout 470
Real modifier as a function 474
undocumented 474
Xin 454, 455, 457, 458, 459, 468,
469, 473
software version 480
Xout 454, 455, 457, 458, 459, 460,
461, 465, 473, 476
PAM02 appliance control module
454
PL513 468
protocol 453, 461, 473, 483
16-channel telemetry system 473
bit level data flow in the X-10 455
start signal 461
PSC05 454
Real modifier as a function 474
receiving X-10 signals 468
signal bridge 486
three-phase power 455
zero crossing 455, 461, 468, 482
burst structure 461
transmit-only PL513 device 454
TW523 transceiver 454, 455, 456,
457, 459, 460, 468, 481, 482,
486
Function Code 456
House Code 456

Idle Time 456
opto-isolator 460
power strips 486
receiving modules 486
RJ11 four-pin modular connector
454
Start Code 456
Unit Code 456
XOR function 484
X-10 code sniffer 480
X-10 control module 453
addressed by both house and unit
codes 453
all call 453, 459, 461, 466, 485
appliance control module 454
channel code 473
data code 484
end code 473
false code rejection 462
function code 456, 458, 464, 465,
466, 469, 484
X_Off 469
function name 485
house code 453, 456, 461, 464,
466, 469, 470, 473, 484, 485
key code 461
starting code 483
start code 473, 483
unit code 453, 456, 458, 461, 464,
465, 466, 484, 485
X-10 Pro model PAM02 appliance
modules 459
X-10 Pro PSC05 454
X-10 transceiver module 454
X-10 transmitter 454
X_Units_Off 741
X10 742
X10 interfacing 741
X10 USA 453
Xin 741, 742
Xin / Xout 758
Xin and Xout 741
Xout 741, 742

Z

Zener diode 35

ELSEVIER SCIENCE DVD-ROM LICENSE AGREEMENT

PLEASE READ THE FOLLOWING AGREEMENT CAREFULLY BEFORE USING THIS DVD-ROM PRODUCT. THIS DVD-ROM PROD-UCT IS LICENSED UNDER THE TERMS CONTAINED IN THIS DVD-ROM LICENSE AGREEMENT ("Agreement"). BY USING THIS DVD-ROM PRODUCT, YOU, AN INDIVIDUAL OR ENTITY INCLUDING EMPLOYEES, AGENTS AND REPRESENTATIVES ("You" or "Your"), ACKNOWLEDGE THAT YOU HAVE READ THIS AGREEMENT, THAT YOU UNDERSTAND IT, AND THAT YOU AGREE TO BE BOUND BY THE TERMS AND CONDITIONS OF THIS AGREEMENT. ELSEVIER SCIENCE INC. ("Elsevier Science") EXPRESSLY DOES NOT AGREE TO LICENSE THIS DVD-ROM PRODUCT TO YOU UNLESS YOU ASSENT TO THIS AGREEMENT. IF YOU DO NOT AGREE WITH ANY OF THE FOLLOWING TERMS, YOU MAY, WITHIN THIRTY (30) DAYS AFTER YOUR RECEIPT OF THIS DVD-ROM PRODUCT RETURN THE UNUSED DVD-ROM PRODUCT AND ALL ACCOMPANYING DOCUMENTATION TO ELSEVIER SCIENCE FOR A FULL REFUND.

DEFINITIONS

As used in this Agreement, these terms shall have the following meanings:

"Proprietary Material" means the valuable and proprietary information content of this DVD-ROM Product including all indexes and graphic materials and software used to access, index, search and retrieve the information content from this DVD-ROM Product developed or licensed by Elsevier Science and/or its affiliates, suppliers and licensors.

"DVD-ROM Product" means the copy of the Proprietary Material and any other material delivered on DVD-ROM and any other human-readable or machine-readable materials enclosed with this Agreement, including without limitation documentation relating to the same.

OWNERSHIP

This DVD-ROM Product has been supplied by and is proprietary to Elsevier Science and/or its affiliates, suppliers and licen-sors. The copyright in the DVD-ROM Product belongs to Elsevier Science and/or its affiliates, suppliers and licensors and is protected by the national and state copyright, trademark, trade secret and other intellectual property laws of the United States and international treaty provisions, including without limitation the Universal Copyright Convention and the Berne Copyright Convention. You have no ownership rights in this DVD-ROM Product. Except as expressly set forth herein, no part of this DVD-ROM Product, including without limitation the Proprietary Material, may be modified, copied or distributed in hardcopy or machine-readable form without prior written consent from Elsevier Science. All rights not expressly granted to You herein are expressly reserved. Any other use of this DVD-ROM Product by any person or entity is strictly prohibited and a violation of this Agreement.

SCOPE OF RIGHTS LICENSED (PERMITTED USES)

Elsevier Science is granting to You a limited, non-exclusive, non-transferable license to use this DVD-ROM Product in ac-cordance with the terms of this Agreement. You may use or provide access to this DVD-ROM Product on a single computer or terminal physically located at Your premises and in a secure network or move this DVD-ROM Product to and use it on another single computer or terminal at the same location for personal use only, but under no circumstances may You use or provide access to any part or parts of this DVD-ROM Product on more than one computer or terminal simultaneously.

You shall not (a) copy, download, or otherwise reproduce the DVD-ROM Product in any medium, including, without limitation, online transmissions, local area networks, wide area networks, intranets, extranets and the Internet, or in any way, in whole or in part, except that You may print or download limited portions of the Proprietary Material that are the results of discrete searches; (b) alter, modify, or adapt the DVD-ROM Product, including but not limited to decompiling, disassembling, reverse engineering, or creating derivative works, without the prior written approval of Elsevier Science; (c) sell, license or otherwise distribute to third parties the DVD-ROM Product or any part or parts thereof; or (d) alter, remove, obscure or obstruct the display of any copyright, trademark or other proprietary notice on or in the DVD-ROM Product or on any printout or download of portions of the Proprietary Materials.

RESTRICTIONS ON TRANSFER

This License is personal to You, and neither Your rights hereunder nor the tangible embodiments of this DVD-ROM Product, including without limitation the Proprietary Material, may be sold, assigned, transferred or sub-licensed to any other person, including without limitation by operation of law, without the prior written consent of Elsevier Science. Any purported sale, assignment, transfer or sublicense without the prior written consent of Elsevier Science will be void and will automatically terminate the License granted hereunder.

TERMS

This Agreement will remain in effect until terminated pursuant to the terms of this Agreement. You may terminate this Agreement at any time by removing from Your system and destroying the DVD-ROM Product. Unauthorized copying of the DVD-ROM Product, including without limitation, the Proprietary Material and documentation, or otherwise failing to comply with the terms and conditions of this Agreement shall result in automatic termination of this license and will make available to Elsevier Science legal remedies. Upon termination of this Agreement, the license granted herein will terminate and You must immediately destroy the DVD-ROM Product and accompanying documentation. All provisions relating to proprietary rights shall survive termination of this Agreement.

LIMITED WARRANTY AND LIMITATION OF LIABILITY

NEITHER ELSEVIER SCIENCE NOR ITS LICENSORS REPRESENT OR WARRANT THAT THE INFORMATION CONTAINED IN THE PROPRIETARY MATERIALS IS COMPLETE OR FREE FROM ERROR, AND NEITHER ASSUMES, AND BOTH EXPRESSLY DISCLAIM, ANY LIABILITY TO ANY PERSON FOR ANY LOSS OR DAMAGE CAUSED BY ERRORS OR OMISSIONS IN THE PROPRIETARY MATERIAL, WHETHER SUCH ERRORS OR OMISSIONS RESULT FROM NEGLIGENCE, ACCIDENT, OR ANY OTHER CAUSE. IN ADDITION, NEITHER ELSEVIER SCIENCE NOR ITS LICENSORS MAKE ANY REPRESENTATIONS OR WARRANTIES, EITHER EXPRESS OR IMPLIED, REGARDING THE PERFORMANCE OF YOUR NETWORK OR COMPUTER SYSTEM WHEN USED IN CONJUNCTION WITH THE DVD-ROM PRODUCT.

If this DVD-ROM Product is defective, Elsevier Science will replace it at no charge if the defective DVD-ROM Product is returned to Elsevier Science within sixty (60) days (or the greatest period allowable by applicable law) from the date of shipment.

Elsevier Science warrants that the software embodied in this DVD-ROM Product will perform in substantial compliance with the documentation supplied in this DVD-ROM Product. If You report significant defect in performance in writing to Elsevier Science, and Elsevier Science is not able to correct same within sixty (60) days after its receipt of Your notification, You may return this DVD-ROM Product, including all copies and documentation, to Elsevier Science and Elsevier Science will refund Your money.

YOU UNDERSTAND THAT, EXCEPT FOR THE 60-DAY LIMITED WARRANTY RECITED ABOVE, ELSEVIER SCIENCE, ITS AFFILIATES, LICENSORS, SUPPLIERS AND AGENTS, MAKE NO WARRANTIES, EXPRESSED OR IMPLIED, WITH RESPECT TO THE DVD-ROM PRODUCT, INCLUDING, WITHOUT LIMITATION THE PROPRIETARY MATERIAL, AN SPECIFICALLY DISCLAIM ANY WARRANTY OF MERCHANTABILITY OR FITNESS FOR A PARTICULAR PURPOSE.

If the information provided on this DVD-ROM contains medical or health sciences information, it is intended for professional use within the medical field. Information about medical treatment or drug dosages is intended strictly for professional use, and because of rapid advances in the medical sciences, independent verification f diagnosis and drug dosages should be made.

IN NO EVENT WILL ELSEVIER SCIENCE, ITS AFFILIATES, LICENSORS, SUPPLIERS OR AGENTS, BE LIABLE TO YOU FOR ANY DAMAGES, INCLUDING, WITHOUT LIMITATION, ANY LOST PROFITS, LOST SAVINGS OR OTHER INCIDENTAL OR CON-SEQUENTIAL DAMAGES, ARISING OUT OF YOUR USE OR INABILITY TO USE THE DVD-ROM PRODUCT REGARDLESS OF WHETHER SUCH DAMAGES ARE FORESEEABLE OR WHETHER SUCH DAMAGES ARE DEEMED TO RESULT FROM THE FAILURE OR INADEQUACY OF ANY EXCLUSIVE OR OTHER REMEDY.

U.S. GOVERNMENT RESTRICTED RIGHTS

The DVD-ROM Product and documentation are provided with restricted rights. Use, duplication or disclosure by the U.S. Government is subject to restrictions as set forth in subparagraphs (a) through (d) of the Commercial Computer Restricted Rights clause at FAR 52.22719 or in subparagraph (c)(1)(ii) of the Rights in Technical Data and Computer Software clause at DFARS 252.2277013, or at 252.2117015, as applicable. Contractor/Manufacturer is Elsevier Science Inc., 655 Avenue of the Americas, New York, NY 10010-5107 USA.

GOVERNING LAW

This Agreement shall be governed by the laws of the State of New York, USA. In any dispute arising out of this Agreement, you and Elsevier Science each consent to the exclusive personal jurisdiction and venue in the state and federal courts within New York County, New York, USA.